# Fuzzy Topology

## ADVANCES IN FUZZY SYSTEMS — APPLICATIONS AND THEORY

**Honorary Editor:** Lotfi A. Zadeh (*Univ. of California, Berkeley*)
**Series Editors:** Kaoru Hirota (*Tokyo Inst. of Tech.*),
George J. Klir (*Binghamton Univ.–SUNY*),
Elie Sanchez (*Neurinfo*),
Pei-Zhuang Wang (*West Texas A&M Univ.*),
Ronald R. Yager (*Iona College*)

---

Vol. 1: Between Mind and Computer: Fuzzy Science and Engineering
(*Eds. P.-Z. Wang and K.-F. Loe*)

Vol. 2: Industrial Applications of Fuzzy Technology in the World
(*Eds. K. Hirota and M. Sugeno*)

Vol. 3: Comparative Approaches to Medical Reasoning
(*Eds. M. E. Cohen and D. L. Hudson*)

Vol. 4: Fuzzy Logic and Soft Computing
(*Eds. B. Bouchon-Meunier, R. R. Yager and L. A. Zadeh*)

Vol. 5: Fuzzy Sets, Fuzzy Logic, Applications
(*G. Bojadziev and M. Bojadziev*)

Vol. 6: Fuzzy Sets, Fuzzy Logic, and Fuzzy Systems: Selected Papers by Lotfi A. Zadeh
(*Eds. G. J. Klir and B. Yuan*)

Vol. 7: Genetic Algorithms and Fuzzy Logic Systems: Soft Computing Perspectives
(*Eds. E. Sanchez, T. Shibata and L. A. Zadeh*)

Vol. 8: Foundations and Applications of Possibility Theory
(*Eds. G. de Cooman, D. Ruan and E. E. Kerre*)

Vol. 9: Fuzzy Topology
(*Y. M. Liu and M. K. Luo*)

Vol. 10: Fuzzy Algorithms: With Applications to Image Processing and Pattern Recognition
(*Z. Chi, H. Yan and T. D. Pham*)

Vol. 11: Hybrid Intelligent Engineering Systems
(*Eds. L. C. Jain and R. K. Jain*)

Vol. 12: Fuzzy Logic for Business, Finance, and Management
(*G. Bojadziev and M. Bojadziev*)

Vol. 15: Fuzzy-Logic-Based Programming
(*Chin-Liang Chang*)

*Forthcoming volumes:*

Vol. 13: Fuzzy and Uncertain Object-Oriented Databases: Concepts and Models
(*Ed. R. de Caluwe*)

Vol. 14: Automatic Generation of Neural Network Architecture Using Evolutionary Computing
(*Eds. E. Vonk, L. C. Jain and R. P. Johnson*)

Advances in Fuzzy Systems – Applications and Theory Vol. 9

# Fuzzy Topology

**Liu Ying-Ming**
**Luo Mao-Kang**

*Sichuan Union University, China*

*Project supported by*
*the National Natural Science Foundation of China,*
*the Science Foundation of the State Education Commission of China and*
*the Mathematical Center of the State Education Commission of China*

World Scientific
Singapore • New Jersey • London • Hong Kong

*Published by*

World Scientific Publishing Co. Pte. Ltd.

P O Box 128, Farrer Road, Singapore 912805

*USA office:* Suite 1B, 1060 Main Street, River Edge, NJ 07661

*UK office:* 57 Shelton Street, Covent Garden, London WC2H 9HE

**Library of Congress Cataloging-in-Publication Data**

Liu, Ying-ming.

Fuzzy topology / Liu Ying-ming, Luo Mao-kang.

p. cm. -- (Advances in fuzzy systems -- applications and theory; vol. 9)

Includes bibliographical references and index.

ISBN 9810228627

1. Fuzzy topology. I. Luo, Mao-kang. II. Title. III. Series.

Advances in fuzzy systems ; vol. 9

QA611.2.L58 1997

514'.322--dc21 97-17915

CIP

**British Library Cataloguing-in-Publication Data**

A catalogue record for this book is available from the British Library.



This book is printed on acid-free paper.

Printed in Singapore by Uto-Print

# Contents

# Preface

In order to study the control problems of complicate systems and dealing with fuzzy information, American cyberneticist L. A. Zadeh introduced fuzzy set theory in 1965, describing fuzziness mathematically for the first time. Following the study on certainty and on randomness, the study of mathematics began to explore the previously restricted zone — fuzziness. Fuzziness is a kind of uncertainty. Since the 16th century, probability theory has been studying a kind of uncertainty — randomness, i.e. the uncertainty of the occur of an event; but in this case, the event itself is completely certain, the only uncertain thing is whether the event will occur or not, the causality is not completely clear now. However, there exists another kind of uncertainty — fuzziness, i.e. for some events, it cannot be completely determined that which cases these events should be subordinated to (e.g., they have already occurred or have not occurred yet), they are in a nonblack and nonwhite state; that is to say, the law of excluded middle in logic cannot be applied any more. Which case an event should be subordinated to, in mathematical view, is just that which set the "element" standing for the event should belong to. However, in mathematics, a set $A$ can be equivalently represented by its characteristic function — a mapping $\chi_A$ from the universe $X$ of discourse (region of consideration, i.e. a larger set) containing $A$ to the 2-value set $\{0,1\}$; that is to say, $x$ belongs to $A$ if and only if $\chi_A(x) = 1$. But in "fuzzy" case the "belonging to" relation $\chi_A(x)$ between $x$ and $A$ is no longer "0 or otherwise 1," it has a degree of "belonging to," i.e. membership degree, such as 0.7. Therefore, the range has to be extended from $\{0,1\}$ to $[0,1]$; or more generally, a lattice $L$, because all the membership degrees, in mathematical view, form a ordered structure, a lattice. A mapping from $X$ to a lattice $L$ called a generalized characteristic function describes the fuzziness of "set" in general. A fuzzy set on a universe $X$ is simply just a mapping from $X$ to a lattice $L$.

Thus, fuzzy set extended the basic mathematical concept — set. In view of the fact that set theory is the cornerstone of modern mathematics, a new and more general framework of mathematics was established. Fuzzy mathematics is just a kind of mathematics developed in this framework, and fuzzy topology is just a kind of topology developed on fuzzy sets. Hence, in a certain sense, fuzzy mathematics is a kind of mathematical theory which contains wider content than the classical theory.

Denote the family of all the fuzzy sets on the universe $X$, which takes lattice $L$ as the range, by $L^X$. From the partial order in $L$, it is easy to equip $L^X$ with a partial order pointwisely, then $L^X$ is also a lattice. Substituting inclusion relation by the order relation in $L^X$, we introduce a topological structure naturally into $L^X$. So that fuzzy topology is a common carrier of ordered structure and topological structure. According to the point of view of Bourbakian school, there are mainly three large kinds of structures in mathematics — topological structure, algebraic structure, and ordered structure. Fuzzy topology fuses just two large structures — ordered structure and topological structure, or more specifically, it is a kind of topology on lattice. Therefore, even if we consider only its pure mathematical significance but not its practical background, fuzzy topology also has important value to research.

With regard to topology on lattice, we can trace the work back to that of Ehresmann in the later part of 1950s. He claimed that lattices possessing certain kinds of distributivities can be studied as a kind of generalized topological structure, but it does not matter if they are indeed lattices of open sets or not. Combining this point of view with intuitionistic logic, the so-called locale theory is formed. It is a kind of topology on complete Heyting algebra (a lattice!). In intuitionistic logic, the axiom of choice is excluded; correspondently, in locale theory, the study related to "points" is avoided and its method emphasizes constructibility. This is an important characteristic of this kind of topology on lattice, and P. T. Johnstone asserted that locale theory is a kind of "pointless topology."

Corresponding to the "pointless feature" of locale theory, fuzzy topology, another branch of topology on lattice, naturally possesses "pointlike" structure. This structure is a basic characteristic in fuzzy topology. To illustrate this point, we can take the problem of "membership relation" between a point and a set in fuzzy topology as an example. In classical topology, this relation is simple and clear: "An open set is a neighborhood of a point if and only if this point belongs to this open set." In the early period of fuzzy topology, "membership relation" was similarly defined. But serious differences soon surfaced from this basic definition in fuzzy topology. For example, under this definition, one point can belong to a union of some sets but does not belong to any one of them. This caused many other relative problems. In other words, theso-called "multiple choice principle" no longer holds. This obstacle was not overcome until 1977. In this year, theory of quasi-coincident neighborhood system produced by Liu Ying-ming made a breakthrough in this respect. As is well known, neighborhood structure can be decomposed as:

"Structure of Open Sets + Membership Relation between Point and Set."

The "membership relation" corresponding to traditional neighborhood systems is just "relation of belonging to." For making the membership relation between point and set in fuzzy topology satisfy the very basic "multiple choice principle" mentioned above and some other obvious requirements, Liu proved that in the framework of fuzzy sets this membership relation can only be the so-called "quasi-coincidence relation," but in general not "relation of belonging to." The neighborhood structure corresponding to the quasi-coincidence relation is just the quasi-coincident neighborhood system. In this theory, a "point" can be in the outside of its neighborhood structure — quasi-coincident neighborhood. This kind of topology construction, in which points do not "belong to" their neighborhood structure, was investigated early in 1916 by French Frechét (see his monograph *Espaces Abstract*). The research was summarized later as a chapter: "V-space Theory" in Sierpinski's monograph *General Topology*. However, in V-space theory, the intersection of a set and its complement is always empty. Since the law of excluded middle is no longer valid in fuzzy sets, this property no longer holds either. Therefore, fuzzy topology and V-space theory are two kinds of completely different theories. This example also shows us that the study of fuzzy topology can deepen our understanding of some most basic structure (e.g., neighborhood structure) in classical mathematics.

On the other hand, both fuzzy topology and locale theory are topologies on lattice, and there naturally exists relatively close connection between them. A fuzzy topology is naturally a locale; but for a definite range $L$, when will a locale be isomorphic to a fuzzy topology on $L^X$? The characterizations of this sufficient and necessary condition are quite interesting. In addition, when a fuzzy topology is considered as a locale, the relation between fuzzy topological properties and the correspondent localic ones is another significant topic.

In fact, pointlike structure of fuzzy sets is a kind of behavior of their level structures, or in the other word, stratifications. For every fuzzy set $A : X \longrightarrow L$ and every element $a$ of $L$, the set $A_{[a]} = \{x \in X : A(x) \geq a\}$ is a $a$-level of $A$. Possessing level structures is the most essential characteristic of fuzzy sets. Obviously, $A$ can be formed from its stratifications. Thus, the relation between fuzzy sets and ordinary sets (stratifications) is established. Maybe the converse problem is more interesting: For a certain family of ordinary sets, how to construct a fuzzy set (a mapping!) from the family with some desired properties? To show the effects of stratification and the associate method, let us observe the following classical Hahn–Dieudonné–Tong insertion theorem. Let $X$ be a topological space, $f, g : X \longrightarrow [0,1]$ upper semicontinuous function and lower semicontinuous function from $X$ to unit interval respectively, $f \leq g$, then $X$ is normal if and only if there exists a continuous function $h : X \longrightarrow [0,1]$ such that $f \leq h \leq g$. The proof of this theorem, in other words, the determination of the inserting function $h$, is pointwisely obtained, fulls of analytic techniques, its argument considerably complicated. Now we substitute the range [0,1] with a certain kind of lattice $L$ and investigate this problem in lattice-valued case. We need to construct the desired lattice-valued inserting mapping (a fuzzy set!). At this time, since it is difficult to use analytic techniques in lattice $L$, the method of pointwisely constructing functions is invalid. Based on the understanding of the set theoretical relations and the topological relations among a mapping and its stratifications, we used the stratification method to construct mappings level by level, successfully solved this inserting mapping problem and generalized this noted theorem in classical mathematics. Compared with the original analytic techniques, this proof based on the stratification method is more effective, and is natural, simple and conceptual. This example also shows us that the study of fuzzy topology can offer us new methods and stronger conclusions. Although the peculiar level structure of fuzzy topological space makes some problems complicated, however, it is just level structure itself which makes fuzzy topological spaces possess more abundant properties, making the relation between fuzzy topology and other branches of classical mathematics closer.

Since topology on lattice is a kind of theory developed on lattice, it involves many problems on ordered structure. For example, complete distributivity of lattices is a pure algebraic problem, but applying the results of fuzzy topology, we gave out analytic and topological characterizations of complete distributive lattices and hence established a connection between algebra and analysis (topology).

From the previous exposition, we know that fuzzy topology is a generalization of topology in classical mathematics, but it also has its own marked characteristics. These examples mentioned above show that it can deepen the understanding of basic structure of classical mathematics (such as neighborhood structure), offer new methods and results (such as stratification method and lattice-valued Hahn–Dieudonné–Tong insertion theorem), and obtain significant results of classical mathematics (such as the analytical and topological characterizations of complete distributive lattices, etc.). Moreover, it also has applications in some important respects of science and technology. In Zadeh's fundamental paper "Fuzzy Sets," the investigation on convex subsets occupies one third of the whole of the paper and was applied to pattern recognition. We used results of fuzzy topology to consummate one of Zadeh's theorem on fuzzy convex sets.

In the early period of the study of fuzzy topology, it was based on the similarity of it with another branch of topology on lattice — locale theory. Some scholars used ideas and methods of locale theory to study some problems not involving "points" in fuzzy topology. Many interesting and beautiful results were obtained, such as normality, Uryshon lemma, uniformity, etc. The marked characteristic of these work is certainly "pointless." But many problems on fuzzy topological spaces unavoidably involve "point," such as separation, embedding theory, etc. Fuzzy topological spaces naturally possess pointlike structure. Hence "method of pointed disposition" is indispensable and powerful in the study of fuzzy topology, and has been rapidly and widely spread. In fact, many beautiful work in fuzzy topology is "pointed." Most of these pointed work were done by Chinese scholars, who published more than 200 related papers in international publications and proceedings.

A large part of the contents of this monograph are based on the results of Chinese scholars, some of whose works have been reported in international congresses or symposiums. These work are summarized here for the first time. Meanwile, compared with general topology, almost all the contents of Kelley's noted book *General Topology* are generalized in fuzzy set framework and synthesized in our monograph. The manuscript of this book (in Chinese) was based on a series of lectures given in Sichuan University in the 1980s. We hope the monograph will be useful for our colleagues involved in the teaching and research in the new branch.

# Chapter 1

# Preliminaries

Stratification is the most essential character of fuzzy sets distinguishing themselves from ordinary sets and reflecting their practical background. Usually, they form a partially ordered structure, a lattice. Therefore, as the necessary preliminaries, concepts and operations in lattices relative to fuzzy topology are introduced in this chapter. In Section **1.3**, a special method to characterize the local property of a completely distributive lattice, called *minimum set method*, is introduced to convert the global property of a completely distributive lattice into a local one.

## 1.1 Sets and Lattices

It is supposed that the reader has known basic concepts and results of general topology. But for unifying terminologies and symbols, we still introduce some of basic concepts in this section:

**1.1.1 Definition** Let $X$ be a set.

Denote the cardinal number of $X$ by $|X|$, the family of all the subsets of $X$ by $\mathcal{P}(X)$, the family of all the finite subsets of $X$ by $[X]^{<\omega}$, the identical mapping on $X$ by $id_X : X \to X$ or $id$ for short if it will not cause any confusion.

For every $A \subset X$, denote the complementary set $\{x \in X : x \notin A\}$ of $A$ by $X\backslash A$ or $A^c$; denote the characteristic function of $A$ on $X$ by $\chi_A : X \to \{0,1\}$ or $Chr(A) : X \to \{0,1\}$, i.e. $\chi_A(x) = Chr(A) = 1$ for $x \in A$ and $\chi_A(x) = Chr(A) = 0$ for $x \in X\backslash A$; symble $i : A \hookrightarrow X$ means $i$ is a *inclusion* mapping, i.e. $i(a) = a$ for every $a \in A$.

Let $X$, $Y$ be sets, $A \subset X$, $B \subset Y$, $f : X \to Y$ a mapping.

Denote

$$dom(f) = X, \quad ran(f) = Y, \quad f[A] = \{f(a) : a \in A\}, \quad img(f) = f[X],$$

and the family of all the mappings from $X$ to $Y$ by $Y^X$, called a *power set* of $Y$.

A *reverse mapping* of $f$ is a mapping $g : Y \to X$ such that $gf = id_X$, $fg = id_Y$. Since a reverse mapping, if exists, is obviously unique, so denote the reverse mapping of $f$, if exists, by $f^{-1}$.

Sometimes, for an arbitrary mapping $f : X \to Y$, we also use the symbol $f^{-1}$ to denote a mapping from $\mathcal{P}(Y)$ to $\mathcal{P}(X)$ as follows:

$$f^{-1}(B) = \{x \in X : f(x) \in B\}, \qquad \forall B \subset Y.$$

Define *restrictions* of $f$ as follows:

Define $f|_A : A \to Y$ as $f|_A(a) = f(a)$, $\forall a \in A$;

If $img(f) \subset B$, define $f|^B : X \to B$ as $f|^B(x) = f(x)$, $\forall x \in X$;

If $f[A] \subset B$, define $f|_A^B : A \to B$ as $f|_A^B(a) = f(a), \forall a \in A$.

Let $X$ be a set, $A \subset X$. An *operation* $P$ on $X$ is a mapping

$$P: \bigcup_{T \in \mathcal{T}} X^T \to X,$$

where $\mathcal{T}$ is a set of sets. Define the *restriction of an operation* $P: \bigcup_{T \in \mathcal{T}} X^T \to X$ on $A$ as

$$P|_{\bigcup_{T \in \mathcal{T}} A^T}: \bigcup_{T \in \mathcal{T}} A^T \to X,$$

where $A^T$'s are isomorphically regarded as subsets of $X^T$.

Let $X$ be a set, $A \subset X$, $P$ an operation on $X$. $A$ is called *closed under* $P$, if for the restriction $R$ of $P$ on $A$, $img(R) \subset A$.

Let $\mathcal{A} = \{X_t : t \in T\}$ be a family of sets. Denote their *product* (or some times called *cartesian product*), i.e. the set of all the mappings $f: T \to \bigcup_{t \in T} X_t$ such that $\forall t \in T$, $f(t) \in X_t$, by $\prod \mathcal{A}$ or $\prod_{t \in T} X_t$, and for every $s \in T$, define a mapping $p_s: \prod_{t \in T} X_t \to X_s$ by $p_s(f) = f(s)$. An element $f \in \prod_{t \in T} X_t$ is also often denoted as $(x_t)_{t \in T}$, where $x_t = f(t) \in X_t$, $\forall t \in T$. Each set $X_t$ is called a *coordinate set* and each mapping $p_s$ is called a *projection* or a *coordinate projection*. Particularly, if $\mathcal{A}$ is finite, $\mathcal{A} = \{X_0, X_1, \cdots, X_n\}$, $\prod \mathcal{A}$ is often denoted by $X_0 \times X_1 \times \cdots \times X_n$, and its elements are also denoted as $(x_0, x_1, \cdots, x_n)$, where $x_i \in X_i$, $\forall i < n$. For every $s \in T$, a subset $\tilde{X}_s$ of $\prod_{t \in T} X_t$ is called a *slice* in the product $\prod_{t \in T} X_t$ *parallel* to $X_s$, if $p_s[\tilde{X}] = X_s$ and $p_t(x) = p_t(y)$ for every $t \in T \setminus \{s\}$ and every pair $x, y \in \tilde{X}_s$; more exactly, for every point $x \in \tilde{X}_s$, also call $\tilde{X}_s$ the *slice* in $\prod_{t \in T} X_t$ *at point* $x$ *parallel to* $X_s$, denote it by $sl(x, s)$.

Let $X$ be a set. A *relation* $R$ on $X$ is a subset of $X \times X$. For every two $x, y \in X$, define $xRy \iff (x, y) \in R$. The *reverse* of $R$, denoted by $R^{op}$, is also a relation on $X$, defined as $xR^{op}y \iff yRx$, $\forall x, y \in X$.

**1.1.2 Definition** Let $X$ be a set. A family $\mathcal{T} \subset \mathcal{P}(X)$ is called a *topology* on $X$, if it satisfies following three conditions:

(TP1) $\emptyset, X \in \mathcal{T}$.

(TP2) $\forall \mathcal{A} \subset \mathcal{T}$, $\bigcup \mathcal{A} \in \mathcal{T}$.

(TP3) $\forall \mathcal{B} \in [\mathcal{T}]^{<\omega}$, $\bigcap \mathcal{B} \in \mathcal{T}$.

Dually define *co-topology* on $X$.

A set $X$ equipped with a topology $\mathcal{T}$ is called a *topological space*, usually denoted by $(X, \mathcal{T})$. The elements of $\mathcal{T}$ are called *open subsets*, and the complementary subsets of open subsets are called *closed subsets*.

Also call a set $X$ equipped with a co-topology $\mathcal{C}$ a *topological space*.

For a topological space $(X, \mathcal{T})$ and $\mathcal{B} \subset \mathcal{T}$, call $\mathcal{B}$ a *base* of $\mathcal{T}$, if

$$\mathcal{T} = \{\bigcup \mathcal{A} : \mathcal{A} \subset \mathcal{B}\};$$

call $\mathcal{B}$ a *subbase* of $\mathcal{T}$, if the family

$$\{\bigcap \mathcal{F} : \mathcal{F} \in [\mathcal{B}]^{<\omega} \setminus \{\emptyset\}\}$$

is a base of $\mathcal{T}$.

For a set $X$ and a family $\mathcal{B} \subset \mathcal{P}(X)$, denote by $Grt(\mathcal{B})$ the topology on $X$ generated by $\mathcal{B}$, i.e.

$$Grt(\mathcal{B}) = \{\bigcup\{\bigcap \mathcal{F} : \mathcal{F} \in \mathcal{G}\} : \mathcal{G} \subset [\mathcal{B}]^{<\omega} \setminus \{\emptyset\}\}.$$

Dually define *base* and *subbase* of a co-topology.

For a topology (co-topology, resp.) $\mathcal{T}$ on a set $X$ and a subfamily $\mathcal{B} \subset \mathcal{T}$, we say $\mathcal{T}$ is *generated* by $\mathcal{B}$, if $\mathcal{B}$ is a base or a subbase of the topology (co-topology, resp.) $\mathcal{T}$.

Let $(X, \mathcal{T})$ be a topological space, $A \subset X$. By condition (TP2), define the *interior of* $A$ as the largest open subset contained in $A$, i.e. the union of all the open subsets contained in $A$, denote it by $int(A)$ or $A^\circ$. By the dual of condition (TP2), define the *closure of* $A$ as the smallest closed subset containing $A$, i.e. the intersection of all the closed subsets containing $A$, denote it by $cl(A)$ or $A^-$ .

For every $x \in X$. An open subset $U$ of $X$ is called a *neighborhood* of $x$, if $x \in U$. The family of all the neighborhoods of $x$ is called the *neighborhood system* of $x$, denoted by $\mathcal{N}_\mathcal{T}(x)$ or $\mathcal{N}(x)$ for short.

Denote the set of all the real numbers (equipped with ordinary topology on real line) by **R**, the set of all the integral numbers by **Z**, the set of all the natural numbers by **N** , the set of all the rational numbers (equipped with the relative topology as a subspace of **R**) by **Q**.

Denote infinite ordinal numbers in their original order by $\omega_0, \omega_1, \cdots$, infinite cardinal numbers in their original order by $\aleph_0, \aleph_1, \cdots$ respectively. Particularly, denote $\omega_0$, i.e. the set of all the nonnegative integral numbers, by $\omega$ for short.

**1.1.3 Definition** Let $P$ be a set. $\leq$ is a relation on $P$.

$\leq$ is called *reflexive*, if $a \leq a$ for every $a \in P$; called *transitive*, if $a \leq b$, $b \leq c$ in $P$ imply $a \leq c$; called *antisymmetric*, if $a \leq b$ and $b \leq a$ in $P$ imply $a = b$.

$\leq$ is called a *preorder* on $P$, if $\leq$ is reflexive and transitive. A set $P$ equipped with a preorder $\leq$ is called a *preordered set.*

In a preordered set $P$, if $a \leq b$ but $a \neq b$, denote this case by $a < b$.

Two elements $a$, $b$ in a preordered set $P$ are called *comparable*, if $a \leq b$ or $b \leq a$; are called *incomparable*, denoted by $a \perp b$, if $a \not\leq b$, $b \not\leq a$.

For a preordered set $P$ with preorder $\leq$, define its *dual preordered set* as the set $P$ equipped with preorder $\leq^{op}$ called the *dual preorder* of $\leq$:

$$\forall a, b \in P, \quad a \leq^{op} b \Longleftrightarrow b \leq a.$$

Denote the dual preordered set of $P$ by $P^{op}$.

For a subset $P_0$ of a preordered set $P$, define its preorder as the restriction of the preorder of $P$ on $P_0$.

A preorder $\leq$ is called a *partial order* on $P$, if it is antisymmetric. A set $P$ equipped with a partial order $\leq$ is called a *partially ordered set*, and usually called a *poset* for short.

A poset $C$ is called a *totally ordered set* or a *chain*, if all its elements are comparable to each other. Then the partial order is called a *total order*

A poset $P$ is called a *well-ordered set*, if for every nonempty subset $A$ of $P$, there exists $a_0 \in A$ such that $a_0 \leq a$ for every $a \in A$. Then the partial order is called a *well order*

**1.1.4 Definition** Let $P$ be a preordered set, $A \subset P$, $a \in A$. $a$ is called a *minimal element* of $A$, if there not exists $b \in A$ such that $b \leq a$ and $b \neq a$; called the *smallest element* of $A$, denoted by $min\ A$, if $b \geq a$ for every $b \in A$.

Dually define *maximal element* of $A$, the *largest element max* $A$ of $A$.

**1.1.5 Proposition** *Let A be a set. Then*

$$\begin{aligned} A \text{ is a well-ordered set} &\Longrightarrow A \text{ is a chain} \\ &\Longrightarrow A \text{ is a poset} \\ &\Longrightarrow A \text{ is a preordered set.} \end{aligned}$$ □

**1.1.6 Definition** Let $P$ be a preordered set, $A_0 \subset A \subset P$. $b \in P$ is called an *upper bound* of $A$, if $a \leq b$ for every $a \in A$. $b \in P$ is called a *lower bound* of $A$, if $a \geq b$ for every $a \in A$. $A_0$ is called *up-cofinal* in $A$, or *cofinal* in $A$ for short, if for every $a \in A$, there exists $a_0 \in A_0$ such that $a_0 \geq a$. $A_0$ is called *down-cofinal* in $A$, if for every $a \in A$, there exists $a_0 \in A_0$ such that $a_0 \leq a$.

**1.1.7 Definition** A preordered set $P$ is called *order dense*, if for every pair $a, b \in P$ such that $a < b$, there exists $c \in P$ such that $a < c < b$.

Concerning partial ordered set, there are three lemmas in the following equivalent to the Axiom of Choice, which will be used in this book and the related proofs are omitted:

**1.1.8 Lemma (Zorn's Lemma)** *If every chain in a poset P has an upper bound, then P has a maximal element.* □

**1.1.9 Lemma (Kuratowski's Lemma)** *Every chain in a poset P is contained in a maximal chain in P.* □

**1.1.10 Theorem (Zermelo's Theorem on well-ordering)** *For every set A there exists a relation < which well-orders the set A.* □

**1.1.11 Definition** Let $P$ be a poset, $A \subset P$. Define

$$\uparrow A = \{b \in P : \exists a \in A,\ b \geq a\}, \quad \downarrow A = \{b \in P : \exists a \in A,\ b \leq a\}.$$

$A$ is called an *upper set*, if $\uparrow A = A$; a *lower set*, if $\downarrow A = A$. For a singleton $\{a\}$, denote $\uparrow a = \uparrow\{a\}$, $\downarrow a = \downarrow\{a\}$.

Also respectively denote $\uparrow A$, $\downarrow A$, $\uparrow a$, $\downarrow a$ by $\uparrow_P A$, $\downarrow_P A$, $\uparrow_P a$, $\downarrow_P a$ to emphasize the poset $P$.

**1.1.12 Proposition** *Let P be a poset, $A \subset C \subset P$. Then*

(i) $\uparrow_C A = (\uparrow_P A) \cap C$.

(ii) $\downarrow_C A = (\downarrow_P A) \cap C$.

**Proof** Need only prove (i):

$$c \in \uparrow_C A \Longleftrightarrow c \in C \subset P,\ \exists a \in A,\ c \geq a \Longleftrightarrow c \in (\uparrow_P A) \cap C.$$ □

**1.1.13 Definition** Let $L$ be a poset, $A \subset L$. An element $x \in L$ is called the *join* (or the *least upper bound*, or the *supremum*) of $A$, denoted by $\bigvee A$ (or $sup\ A$, or $sup_P\ A$ to emphasize the poset $P$), if

(i) $x$ is an upper bound of $A$,

(ii) if $y$ is an upper bound for $A$, then $x \leq y$.

If $A$ is finite, call $\bigvee A$ (if it exists) a *finite join* . If $A$ consists of two elements $a$ and $b$, write $a \vee b$ for $\bigvee\{a, b\}$ for convenience.

The dual notions *meet, greatest lower bound, infimum, finite meet* and dual symbols $\bigwedge$, $inf\ A$, $inf_P\ A$, $\wedge$ are dually defined.

**1.1.14 Remark**

(1) By the antisymmetry of partial order, a join or a meet in a poset, if it exists, must be unique. That is just the reason of our writing "the join", "the meet" but not

"a join", "a meet" in Definition **1.1.13**.

(2) If a poset $P$ has the smallest element $p_0$, then for the empty set $S = \emptyset \subset P$, by the definition, clearly $\bigvee S = p_0$. Dually, if $P$ has the largest element $p_1$, then for $S = \emptyset$, strictly follow Definition **1.1.13**, we obtain $\bigwedge S = p_1$.

**1.1.15 Example** Simple posets are often represented as figures consisting of dots and line segments, the following figure shows us an example in this way. In this figure (usually, in this kind of figures), dots represent elements, small circle means there is no dot (then there is no element) at that position. Two elements $a$, $b$ are comparable if and only if they are connected with a line segment, and, the higher dot represents the larger element:

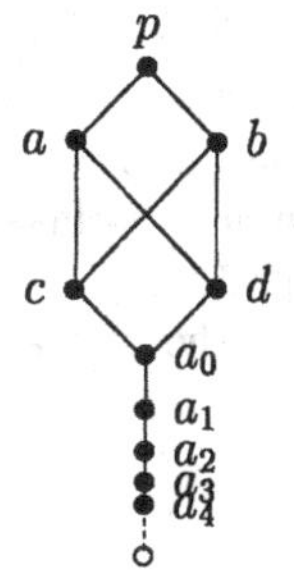


Denote this poset by $P$. From the preceding figure, we can find that $P$ has the largest element $p$ and $a \vee b = p$, but $a \wedge b$ does not exist. Oppositely, $c \vee d$ does not exist but $c \wedge d = a_0$. $P$ has an ordered sequence $a_0 \geq a_1 \geq a_2 \geq a_3 \geq \cdots$, but $P$ has not the smallest element and any meet (greatest lower bound) for any infinite subset.

**1.1.16 Definition** Let $L$ be a poset.

$L$ is called a *join-semilattice* if every join for a finite subset of $L$ exists; particularly, the *smallest element* exists as the join of empty subset.

$L$ is called a *meet-semilattice* if every meet for a finite subset of $L$ exists; particularly, the *largest element* exists as the meet of empty subset.

$L$ is called a *lattice*, if it is both a join-semilattice and a meet-semilattice.

Denote the smallest element and the largest element of $L$ by $0_L$ and $1_L$ respectively. If it will not cause any confusion, denote them by 0 and 1 for short respectively.

**1.1.17 Remark** Certainly, according to the definition, a lattice will always be nonempty; but on the other hand, it can consist of only one element — the smallest element coincides with the largest one. This kind of lattice, called *trivial lattice*, will not be considered in our investigation. Therefore, in the sequel, every lattice is always supposed to possess at least two elements: the distinguished smallest element 0 and largest element 1.

**1.1.18 Remark**

(1) Some authors[11, 25, 168] defined semilattices and lattices by joins of and meets of two elements (equivalently, nonempty finite subsets) but without the smallest

element and the largest element. Since smallest elements and largest elements are always needed in our investigation, we adopt the prededing definition.

(2) Let $L$ be a set, $0 \in L$, $\vee: L \times L \to L$ a symmetric binary operation on $L$, then it is not difficult to prove the following conclusion:

*Define a relation $\leq$ on $L$ as $a \leq b \Longleftrightarrow a \vee b = b$, then $(L, \vee, 0)$ is a commutative monoid in which every element is idempotent if and only if $\leq$ is a partial order on $L$, $L$ forms a join-semilattice under the partial order $\leq$ such that $a \vee b$ is just the join of $a$ and $b$ for all $a, b \in L$ and $0$ is the smallest element.*

Because of this reason, some authors defined join-semilattice from the angle of commutative monoid and meet-semilattice dually. We want to emphasize the effect of ordered structures in lattices, so defined them from the angle of order.

**1.1.19 Example** (i) Real unit interval $[0,1]$ with its own paritial order (total order, in fact) is a typical example of lattice. Fuzzy set theory is just establised with it as the value domain. Because of its basic importance, especially in possibility theory, and its abundant and nice property, many auothers still investigate various problems in fuzzy topology with value domain $[0,1]$.

(ii) Take $L = \{0, a, b, 1\}$ and define the partial order on it as follows:

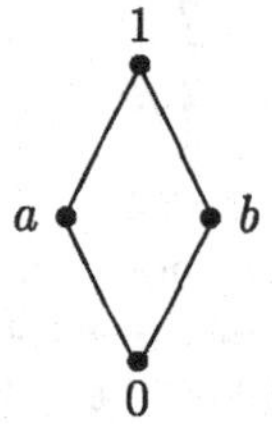


then $L$ is a lattice, called the *diamond-type lattice*. Since it is the simplest lattice not totally ordered, it is frequently used in $L$-fuzzy topology.

With preceding most basic definition of lattice, one can straight deduce the following most basic properties of lattice:

**1.1.20 Proposition** *Let $L$ be a lattice. Then the following conclusions hold:*

(i) $\forall a \in L,\ a \vee a = a,\ a \wedge a = a$.

(ii) $\forall a, b \in L,\ a \wedge b \leq a, b \leq a \vee b$.

(iii) $\forall a, b \in L,\ a \leq b \Longleftrightarrow a \vee b = b \Longleftrightarrow a \wedge b = a$.

(iv) $\forall a, b \in L,\ a \wedge b < a, b < a \vee b \Longleftrightarrow a \perp b$.

(v) $\forall a, b, c \in L,\ (a \vee b) \vee c = a \vee (b \vee c),\ (a \wedge b) \wedge c = a \wedge (b \wedge c)$.

(vi) *For each family $\mathcal{A}$ of subsets of $L$, $\bigvee\{\bigvee A: A \in \mathcal{A}\} = \bigvee \bigcup\{A: A \in \mathcal{A}\}$ if all the concerned joins exist.*

(vii) *For each family $\mathcal{A}$ of subsets of $L$, $\bigwedge\{\bigwedge A: A \in \mathcal{A}\} = \bigwedge \bigcup\{A: A \in \mathcal{A}\}$ if all the concerned meets exist.* □

Just as the similarity between join-semilattices and monoids discussed above in Remark **1.1.18** (2), latties also behave in some degree similarly to rings. Then some properties or problems are naturally aroused by the relation between two kinds of basic operations — joins and meets. The most basic and natural one is the following

**1.1.21 Definition** Let $L$ be a lattice. $L$ is called *distributive* , if $L$ satisfies following two conditions (FD1) and (FD2), called *finitely distributive law*:

(FD1) $\forall a, b, c \in L, \quad a \wedge (b \vee c) = (a \wedge b) \vee (a \wedge c)$,

(FD2) $\forall a, b, c \in L, \quad a \vee (b \wedge c) = (a \vee b) \wedge (a \vee c)$.

A distributive lattice $L$ is also called *finitely distributive* , or possessing *distributivity* or *finite distributivity* .

One can find that the conditions (FD1) and (FD2) of finite distributive law are dual, but in fact any one of them has been enough to the definition:

**1.1.22 Proposition** *A lattice $L$ satisfies* (FD1) *if and only if it satisfies* (FD2).

**Proof** Suppose $L$ be a lattice satisfying (FD1), $a, b, c \in L$, then

$$\begin{aligned}(a \vee b) \wedge (a \vee c) &= ((a \vee b) \wedge a) \vee ((a \vee b) \wedge c) && \text{(by the finitely distributive law)}\\ &= a \vee ((a \vee b) \wedge c) && \text{(by \textbf{1.1.20} (ii), (iii))}\\ &= a \vee ((a \wedge c) \vee (b \wedge c)) && \text{(by the finitely distributive law)}\\ &= (a \vee (a \wedge c)) \vee (b \wedge c) && \text{(by \textbf{1.1.20} (v))}\\ &= a \vee (b \wedge c).\end{aligned}$$

Dually prove the sufficiency. □

An other important property of lattices is completeness, as is defined in the following:

**1.1.23 Definition** Let $L$ be a poset.

$L$ is called a *complete join-semilattice* if every join for an arbitrary subset of $L$ exists; particularly, the smallest element exists as the join of empty subset.

$L$ is called a *complete meet-semilattice* if every meet for an arbitrary subset of $L$ exists; particularly, the largest element exists as the meet of empty subset.

$L$ is called a *complete lattice*, if it is both a complete join-semilattice and a complete meet-semilattice.

Similar to the case of finitely distributive law, there still exists some kind of symmetry here:

**1.1.24 Proposition** *Let $L$ be a poset, then the following conditions are equivalent:*

(i) *$L$ is a complete lattice.*

(ii) *$L$ has the smallest element and $\forall A \subset L,\ A \neq \emptyset,\ \bigvee A$ exists in $L$.*

(iii) *$L$ has the largest element and $\forall A \subset L,\ A \neq \emptyset,\ \bigwedge A$ exists in $L$.*

(iv) *$L$ is a lattice and $\forall A \subset L,\ A \neq \emptyset,\ \bigvee A$ exists in $L$.*

(v) *$L$ is a lattice and $\forall A \subset L,\ A \neq \emptyset,\ \bigwedge A$ exists in $L$.*

**Proof** Need only note that $\bigvee \emptyset = 0$, $\bigwedge \emptyset = 1$ and for every nonempty subset $A \subset L$,

$$\bigvee A = \bigwedge \{x \in L : \ \forall a \in A,\ x \geq a\},$$

$$\bigwedge A = \bigvee \{x \in L : \ \forall a \in A,\ x \leq a\}.$$ □

**1.1.25 Definition** Let $L$ be a complete lattice, $C \subset L$. $C$ is called a *join-generating set* of $L$ or a *generating set* of $L$ for short, if $\forall a \in L$, $\exists C_a \subset C$ such that $\bigvee C_a = a$; called a *strictly join-generating set* of $L$ or a *strictly generating set* for short, if $\forall a \in L$, $\exists C_a \subset C \backslash \{a\}$ such that $\bigvee C_a = a$.

Dually define the dual notions *meet-generating set* of $L$, *strictly meet-generating set* of $L$.

**1.1.26 Proposition** *Let $L$ be a complete lattice, $C \subset L$. Then*

(i) *$C$ is a join-generating set of $L \iff \forall a, b \in L,\ a \not\leq b,\ \exists c \in C,\ c \leq a,\ c \not\leq b$.*

(ii) *$C$ is a strictly join-generating set of $L \iff \forall a, b \in L,\ a \not\leq b,\ \exists c \in C,\ c < a,\ c \not\leq b$.*

(iii) *$C$ is a meet-generating set of $L \iff \forall a, b \in L,\ a \not\leq b,\ \exists c \in C,\ c \not\geq a,\ c \geq b$.*

(iv) *$C$ is a strictly meet-generating set of $L \iff \forall a, b \in L,\ a \not\leq b,\ \exists c \in C,\ c \not\geq a,\ c > b$.*

**Proof** Only prove (iv), the others are similar.

($\Longrightarrow$) $\forall a, b \in L$ such that $a \not\leq b$. Since $C$ is a strictly meet-generating set of $L$, for $b \in L$, $\exists C_b \subset C \backslash \{b\}$ such that $\bigwedge C_b = b \not\geq a$. So $\exists c \in C_b$ such that $c \not\geq a$. Since $C_b \subset C\backslash\{b\}$, $\bigwedge C_b = b$, so $c > b$.

($\Longleftarrow$) $\forall b \in L$, take $C_b = C \cap ((\uparrow b)\backslash\{b\})$. If $\bigwedge C_b \neq b$, then denote $a = \bigwedge C_b$, $a \not\leq b$, $\exists c \in C$ such that $c \not\geq a$, $c > b$. Then $c \in C_b$, $\bigwedge C_b = c \wedge \bigwedge C_b = c \wedge a < a = \bigwedge C_b$, a contradiction. □

**1.1.27 Proposition** *Let $L$ be a complete lattice, $L_0 \subset L$. Then*

(i) *If $L_0$ is a join-generating set of $L$, then for every $A \subset L$, there exists $A_0 \subset A$ such that $|A_0| \leq |L_0|$ and $\bigvee A_0 = \bigvee A$.*

(ii) *If $L_0$ is a meet-generating set of $L$, then for every $A \subset L$, there exists $A_0 \subset A$ such that $|A_0| \leq |L_0|$ and $\bigwedge A_0 = \bigwedge A$.*

**Proof** Need only prove (i). (ii) is dual.

Since $L_0$ is a join-generating set of $L$, $\forall a \in A$, $\exists C(a) \subset L_0$ such that $\bigvee C(a) = a$. Take $C = \bigcup_{a \in A} C(a) \subset L_0$, then $\bigvee C = \bigvee A$. $\forall c \in C$, $\exists D(c) \in A$ such that $c \in C(D(c))$ and hence

$$c \leq \bigvee C(D(c)) = D(c) \leq \bigvee A. \tag{1.1}$$

Then $A_0 = \{D(c) :\ c \in C\} \subset A$, $|A_0| \leq |C| \leq |L_0|$. By $\bigvee C = \bigvee A$ and inequalities (1.1), $\bigvee A_0 = \bigvee A$. □

**1.1.28 Theorem** *Let $L$ be a complete lattice, $L_0, L_1 \subset L$. Then*

(i) *$L_0$, $L_1$ are join-generating sets of $L \implies \exists L_2 \subset L_1$ such that $L_2$ is a join-generating set of $L$ and $|L_2| \leq |L_0|^2$.*

(ii) *$L_0$, $L_1$ are strictly join-generating sets of $L \implies \exists L_2 \subset L_1$ such that $L_2$ is a strictly join-generating set of $L$ and $|L_2| \leq |L_0|^2$.*

(iii) *$L_0$, $L_1$ are meet-generating sets of $L \implies \exists L_2 \subset L_1$ such that $L_2$ is a meet-generating set of $L$ and $|L_2| \leq |L_0|^2$.*

(iv) *$L_0$, $L_1$ are strictly meet-generating sets of $L \implies \exists L_2 \subset L_1$ such that $L_2$ is a strictly meet-generating set of $L$ and $|L_2| \leq |L_0|^2$.*

**Proof** Only prove (i), the others are similar.

Since $L_1$ is a join-generating set of $L$, $\exists C :\ L_0 \to \mathcal{P}(L_1)$ such that $\bigvee C(a) = a$. Since $L_0$ is a join-generating set of $L$, $\exists D :\ L_1 \to \mathcal{P}(L_0)$ such that $\bigvee D(c) = c$. Take $E :\ L_0 \to \mathcal{P}(L_0)$ as

$$E(a) = \bigcup\{D(c) :\ c \in C(a)\}, \quad \forall a \in L_0,$$

then $\forall a \in L_0$, $\forall d \in E(a) \subset L_0$, $\exists f(a, d) \in C(a) \subset L_1$ such that $d \in D(f(a, d))$ and

$$d \le \bigvee D(f(a,d)) = f(a,d) \le \bigvee C(a) = a. \tag{1.2}$$

Take $L_2 = \{f(a,d) : a \in L_0, d \in E(a)\}$, then $L_2 \subset L_1$ and since every $E(a)$ is contained in $L_0$, $|L_2| \le |L_0| \times |L_0| = |L_0|^2$.

Denote $F = \bigcup\{E(a) : a \in L_0\}$, then $\forall a \in L_0$,

$$a = \bigvee C(a) = \bigvee\{\bigvee D(c) : c \in C(a)\} = \bigvee(\bigcup\{D(c) : c \in C(a)\}) = \bigvee E(a).$$

Since $L_0$ is a join-generating set of $L$, so is $F$. By the definition of $L_2$ and inequality (1.2), $L_2$ is a join-generating set of $L$. □

**1.1.29 Definition** Let $L$ be a complete lattice. $L$ is called *infinitely distributive*, if $L$ satisfies both following two conditions (IFD1) and (IFD2), called *the 1st infinitely distributive law* and *the 2nd infinitely distributive law* respectively:

(IFD1) $\forall a \in L$, $\forall B \subset L$, $a \wedge \bigvee B = \bigvee_{b \in B} (a \wedge b)$,

(IFD2) $\forall a \in L$, $\forall B \subset L$, $a \vee \bigwedge B = \bigwedge_{b \in B} (a \vee b)$,

An infinitely distributive lattice $L$ is also called possessing *infinite distributivity*.

Distinguished from finitely distributive law, the 1st infinitely distributive law (IFD1) is not equivalent to its dual form (IFD2), this point can be shown by the following example:

**1.1.30 Example** (i) Let $L$ be the set of all the open subsets of real line $\mathbf{R}$, equip it with the *inclusion order*, i.e. $\forall U, V \in L$, $U \le V \iff U \subset V$, then $L$ is a complete lattice with $0_L = \emptyset$ and $1_L = \mathbf{R}$. Clearly, $\forall U, V \in L$, $\forall \mathcal{U} \subset L$, $\bigvee \mathcal{U} = \bigcup \mathcal{U}$, $U \wedge V = U \cap V$, and $\bigwedge \mathcal{U} = int(\bigcap \mathcal{U})$, i.e. the largest open set contained in $\bigcap \mathcal{U}$. Then

$$V \wedge \bigvee \mathcal{U} = V \cap \bigcup \mathcal{U} = \bigcup\{V \cap U : U \in \mathcal{U}\} = \bigvee\{V \wedge U : U \in \mathcal{U}\},$$

$L$ satisfies the 1st infinitely distributive law (IFD1). But take $V = (-1,0) \cup (0,1)$, $U_i = (-\frac{1}{i}, \frac{1}{i})$ for $i = 1,2,3,\cdots$, then

$$\bigcap\{U_i : i = 1,2,3,\cdots\} = \{0\},$$

$$\bigwedge\{U_i : i = 1,2,3,\cdots\} = int(\bigcap\{U_i : i = 1,2,3,\cdots\}) = int(\{0\}) = \emptyset,$$

$$V \vee \bigwedge\{U_i : i = 1,2,3,\cdots\} = V \cup \emptyset = V = (-1,0) \cup (0,1).$$

On the other hand,

$$\forall i = 1,2,3,\cdots, \quad V \vee U_i = (-1,1),$$

$$\bigwedge\{V \vee U_i : i = 1,2,3,\cdots\} = int((-1,1)) = (-1,1),$$

so

$$V \vee \bigwedge\{U_i : i = 1,2,3,\cdots\} \ne \bigwedge\{V \vee U_i : i = 1,2,3,\cdots\},$$

the 2nd infinitely distributive law (IFD2) not hold in $L$.

(ii) To avoid possible confusion, for $a, b \in [0,1]$, in this example, we denote the point in $[0,1]^2$ with the first coordinate $a$ and the second coordinate $b$ by $\langle a, b\rangle$.

Take $L \subset [0,1]^2$ as follows:

$$L = [0,1) \times [0,1) \vee \{\langle 1,1\rangle\},$$

and equip it with the restriction of the partial order of the product lattice $[0,1]^2$, i.e.

$$\forall \langle a,b\rangle, \langle c,d\rangle \in L, \quad \langle a,b\rangle \le \langle c,d\rangle \iff a \le c,\ b \le d.$$

Denote the two projections from $L$ to $[0,1]$ by $p_1$, $p_2$ respectively. Then for every $A \subset L$, $\bigvee A$ and $\bigwedge A$ exist and

$$\bigvee A = \begin{cases} \langle \bigvee_{a\in A} p_1(a), \bigvee_{a\in A} p_2(a)\rangle, & \bigvee_{a\in A} p_1(a), \bigvee_{a\in A} p_2(a) < 1, \\ \langle 1,1\rangle, & \text{otherwise,} \end{cases}$$

$$\bigwedge A = \langle \bigwedge_{a\in A} p_1(a), \bigwedge_{a\in A} p_2(a)\rangle.$$

So $L$ is a complete lattice. Since the finite joins and finite meets in $L$ are uniform with the correspondent ones in $[0,1]^2$, $L$ is finitely distributive.

But $L$ does satisfies (IFD1). In fact, for $a = \langle \frac{2}{3}, \frac{2}{3}\rangle \in L$, $B = \{\langle \frac{1}{3}, \frac{n}{n+1}\rangle : n < \omega\} \subset L$,

$$a \wedge \bigvee B = \langle \tfrac{2}{3}, \tfrac{2}{3}\rangle \wedge \langle 1,1\rangle = \langle \tfrac{2}{3}, \tfrac{2}{3}\rangle,$$

$$\bigvee_{b\in B} (a \wedge b) = \bigvee_{n<\omega} \langle \tfrac{1}{3}, \tfrac{2}{3} \wedge \tfrac{n}{n+1}\rangle = \langle \tfrac{1}{3}, \tfrac{2}{3}\rangle,$$

$$a \wedge \bigvee B \neq \bigvee_{b\in B} (a \wedge b).$$

Using (IFD1) and (IFD2) twice, one can easily show following

**1.1.31 Proposition** *Let $L$ be a complete lattice. Then*

(i) *$L$ satisfies the 1st infinitely distributive law* (IFD1) *if and only if*

$$\forall A, B \subset L, \quad \bigvee A \wedge \bigvee B = \bigvee_{a\in A,\ b\in B} (a \wedge b).$$

(ii) *$L$ satisfies the 2nd infinitely distributive law* (IFD2) *if and only if*

$$\forall A, B \subset L, \quad \bigwedge A \vee \bigwedge B = \bigwedge_{a\in A,\ b\in B} (a \vee b). \qquad \square$$

With this distinction between (IFD1) and (IFD2), one can certainly define two different kinds of complete lattices possessing property (IFD1) or (IFD2) respectively. In fact, complete lattice with property (IFD1) will just be an important object in this book. But the lattices here now we will concentrate on are defined in

**1.1.32 Definition** Let $L$ be a complete lattice. $L$ is called *completely distributive*, if $L$ satisfies following two conditions (CD1) and (CD2), called *completely distributive law*:

$$\forall \{\{a_{i,j} : j \in J_i\} : i \in I\} \subset \mathcal{P}(L)\backslash\{\emptyset\},\ I \neq \emptyset,$$

$$\text{(CD1)} \quad \bigwedge_{i\in I}(\bigvee_{j\in J_i} a_{i,j}) = \bigvee_{\varphi\in\prod_{i\in I} J_i} (\bigwedge_{i\in I} a_{i,\varphi(i)}),$$

$$\text{(CD2)} \quad \bigvee_{i\in I}(\bigwedge_{j\in J_i} a_{i,j}) = \bigwedge_{\varphi\in\prod_{i\in I} J_i} (\bigvee_{i\in I} a_{i,\varphi(i)}).$$

The conditions (CD1) or (CD2) is also called *complete distributivity*.

**1.1.33 Remark**

(1) By Proposition **1.1.22**, a lattice satisfying any one of (IFD1) and (IFD2) is finitely distributive, so an infinitely distributive lattice is finitely distributive. But Example **1.1.30** shows that the converse is false.

(2) Taking $I = \{0,1\}$, $J_0 = \{0\}$ in (CD1) and (CD2), one can easily verify (CD1)$\Longrightarrow$(IFD1), (CD2)$\Longrightarrow$(IFD2). So a completely distributive lattice must be infinitely distributive. The converse is not true, a relative counterexample **1.2.19** (ii) is constructed in next section.

Although two infinitely distributive laws are not equivalent to each other, but completely distributive laws (CD1) and (CD2), just as (FD1) and (FD2), each one implies the other:

**1.1.34 Theorem** *A complete lattice L satisfies* (CD1) if and only if it satisfies (CD2).

**Proof** By the symmetry between (CD1) and (CD2), need only prove (CD1) $\Longrightarrow$ (CD2).

$\forall i \in I$, $\forall \varphi \in \prod_{i\in I} J_i$, we clearly have $\bigwedge_{j\in J_i} a_{i,j} \le a_{i,\varphi(i)} \le \bigvee\limits_{i'\in I} a_{i',\varphi(i')}$. So that

$$\bigvee_{i\in I}\Big(\bigwedge_{j\in J_i} a_{i,j}\Big) \le \bigwedge_{\varphi\in\prod\limits_{i\in I} J_i}\Big(\bigvee_{i\in I} a_{i,\varphi(i)}\Big).$$

Now we prove the converse of this inequality by applying (CD1) to its right side. Denote $D = \prod_{i\in I} J_i$; $\forall \varphi \in D$, denote $E_\varphi = I$; $\forall i \in E_\varphi$, denote $x_{\varphi,i} = a_{i,\varphi(i)}$. Then by (CD1),

$$\begin{aligned}\bigwedge_{\varphi\in\prod\limits_{i\in I} J_i}\Big(\bigvee_{i\in I} a_{i,\varphi(i)}\Big) &= \bigwedge_{\varphi\in D}\Big(\bigvee_{i\in E_\varphi} x_{\varphi,i}\Big)\\ &= \bigvee_{\psi\in\prod\limits_{\varphi\in D} E_\varphi}\Big(\bigwedge_{\varphi\in D} x_{\varphi,\psi(\varphi)}\Big)\\ &= \bigvee_{\psi\in\prod\limits_{\varphi\in D} E_\varphi}\Big(\bigwedge_{\varphi\in\prod\limits_{i\in I} J_i} a_{\psi(\varphi),\varphi(\psi(\varphi))}\Big).\end{aligned}$$

To prove that the last part is smaller than or equals to the left side of (CD2), we need only prove that $\forall \psi \in \prod_{\varphi\in D} E_\varphi$, $\exists i \in I$ such that $\{a_{\psi(\varphi),\varphi(\psi(\varphi))} : \varphi \in \prod_{i\in I} J_i\} \supset \{a_{i,j} : j \in J_i\}$. But it must be true. Otherwise, $\forall i \in I$, $\exists \varphi_0(i) \in J_i$ such that $a_{i,\varphi_0(i)} \notin \{a_{\psi(\varphi),\varphi(\psi(\varphi))} : \varphi \in \prod_{i\in I} J_i\}$. Then we obtain a mapping $\varphi_0 \in \prod_{i\in I} J_i$. But for $i = \psi(\varphi_0) \in I$, $a_{i,\varphi_0(i)} \in \{a_{\psi(\varphi),\varphi(\psi(\varphi))} : \varphi \in \prod_{i\in I} J_i\}$, this is a contradiction! So what we should do for the proof have been done. □

**1.1.35 Example**

(i) Take $L$ as the unit interval $[0,1]$, then $L$ is completely distributive. To show this fact, we verify (CD1) for $L = [0,1]$. $\forall i \in I$, $\forall \varphi \in \prod_{i\in I} J_i$, we clearly have $\bigvee_{j\in J_i} a_{i,j} \ge a_{i,\varphi(i)} \ge \bigwedge\limits_{i'\in I} a_{i',\varphi(i')}$. So that

$$\textstyle\bigwedge_{i\in I}(\bigvee_{j\in J_i} a_{i,j}) \ge \bigvee_{\varphi\in\prod_{i\in I} J_i}(\bigwedge_{i\in I} a_{i,\varphi(i)}),$$

what we need is to verify its converse. $\forall i \in I$, denote $b_i = \bigvee_{j\in J_i} a_{i,j}$. For an arbitrary $\varepsilon > 0$, $\forall i \in I$, by the definition of $b_i$'s and $\{b_i : i \in I\} \subset \mathbf{R}$, we can take $\varphi_\varepsilon(i) \in J_i$ such that $b_i - \varepsilon < a_{i,\varphi_\varepsilon(i)} \le b_i$. Then

$$\Big(\bigwedge_{i\in I} b_i\Big) - \varepsilon = \bigwedge_{i\in I}(b_i - \varepsilon) \le \bigwedge_{i\in I} a_{i,\varphi_\varepsilon(i)} \le \bigwedge_{i\in I} b_i,$$

$$(\bigwedge_{i\in I} b_i) - \varepsilon \le \bigvee_{\varphi\in\prod_{i\in I} J_i} (\bigwedge_{i\in I} a_{i,\varphi(i)}) \le \bigwedge_{i\in I} b_i.$$

Since this inequality holds for every $\varepsilon > 0$, it can only be

$$\bigvee_{\varphi\in\prod_{i\in I} J_i} (\bigwedge_{i\in I} a_{i,\varphi(i)}) = \bigwedge_{i\in I} b_i.$$

(ii) For any set $X$, one can verify that $\mathcal{P}(X)$ equipped with inclusion order is a completely distributive lattice.

Although each kind of distributivity is strictly stronger than the previous one in the sequence: finite distributivity, infinite distributivity, complete distributivity, but they coincide to each other in a special case:

**1.1.36 Proposition** *Let $L$ be a finite lattice, then the following conditions are equivalent:*

(i) *$L$ is finitely distributive.*
(ii) *$L$ is infinitely distributive.*
(iii) *$L$ is completely distributive.* □

## 1.2 Operations on Lattices

**1.2.1 Definition** Let $P_0$, $P_1$ be posets, $f:\ P_0 \to P_1$ a mapping.

$f$ is called *order preserving*, or a *poset homomorphism*, if $\forall a, b \in P_0,\ a \le b \Longrightarrow f(a) \le f(b)$.

$f$ is called a *poset isomorphism*, if $f$ is bijective and, both $f:\ P_0 \to P_1$ and $f^{-1}:\ P_1 \to P_0$ are poset homomorphisms.

$f$ is called a *poset embedding*, if $f$ is injective and, both $f:\ P_0 \to P_1$ and $f^{-1}|_{img(f)}:\ img(f) \to P_0$ are poset homomorphisms.

Say $P_0$ and $P_1$ are *isomorphic as posets* to each other, if there exists a poset isomorphism from one of them to the other.

Say $P_0$ can be *embedded as a poset* into $P_1$, if there exists a poset embedding from $P_0$ to $P_1$.

Use notion "order preserving," first of all, we can prove the following very useful Total Ordering Theorem:

**1.2.2 Theorem**(Total Ordering Theorem) *Let $P$ be a poset. Then there exists a chain $C$ and a bijection $f:\ P \to C$ which is order preserving.*

**Proof** We shall use transfinite induction to define $C$, and then define $f$. Suppose the partial order on $P$ is $\le$. By the Zermelo's Theorem on well-ordering, there exists an ordinal number $\alpha$ and a bijection $g:\ \alpha \to P$. Take $C_0 = \{g(0)\}$ and define a total order $\le_0$ on $C_0$ by $g(0) \le_0 g(0)$, then $a, b \in C_0 \Longrightarrow a \le_0 b$. Let $\gamma < \alpha$. Suppose for every $\sigma < \gamma$, we have defined $C_\sigma = g[\sigma + 1] \subset P$ and a total order $\le_\sigma$ on $C_\sigma$ such that $\forall \sigma, \sigma_1, \sigma_2 < \gamma$, the following two conditions are fulfilled:

(a) $a, b \in C_\sigma,\ a \le b \Longrightarrow a \le_\sigma b$,

(b) $\sigma_1 \le \sigma_2 \Longrightarrow \le_{\sigma_1} \subset \le_{\sigma_2}$ .

Then $\leq_* = \bigcup_{\sigma<\gamma} \leq_\sigma$ is a total order on $C_* = \bigcup_{\sigma<\gamma} C_\sigma = g[\gamma]$ and

$$a, b \in C_*, \ a \leq b \Longrightarrow a \leq_* b. \tag{1.3}$$

Denote

$$A_* = \{a \in C_* : \ a \leq g(\gamma)\}, \quad A^* = \{a \in C_* : \ a \geq g(\gamma)\}, \quad A = C_* \backslash (A_* \cup A^*),$$

then since $g(\gamma) \notin C_*$, $C_*$ is exactly the disjoint union of $A_*$, $A^*$ and $A$. Take $C_\gamma = C_* \cup \{g(\gamma)\} = g[\gamma + 1]$, define a relation $\leq_\gamma$ on $C_\gamma$ as follows:

$$\begin{aligned} \forall a, b \in C_\gamma, \quad a \leq_\gamma b \iff & (1)\ a, b \in C_*, \ a \leq_* b && \text{or} \\ & (2)\ a = b = g(\gamma) && \text{or} \\ & (3)\ a = g(\gamma), \ b \in A^* && \text{or} \\ & (4)\ a = g(\gamma), \ b \in A, \ \exists c \in A^*, \ c \leq_* b && \text{or} \\ & (5)\ a \in A_*, \ b = g(\gamma) && \text{or} \\ & (6)\ a \in A, \ b = g(\gamma), \ \forall c \in A^*, \ c \not\leq_* a. \end{aligned}$$

Then it is not hard to verify that no matter $A_*$, $A^*$ and $A$ are empty or not, $\leq_\gamma$ is a total order on $C_\gamma$. Suppose $a, b \in C_\gamma$, $a \leq b$. If $a, b \in C_*$, by condition (1.3), $a \leq_* b$, it is just the previous case (1). If $a = b = g(\gamma)$, it is just the previous case (2). If $a = g(\gamma) \neq b$, then $b \in A^*$, it is just the previous case (3). If $a \neq g(\gamma) = b$, then $a \in A_*$, it is just the previous case (5). So in a word, we always have $a \leq_\gamma b$. Hence condition (a) is fulfilled for $C_\gamma$ and $\leq_\gamma$. As for condition (b), it obviously holds for every two $\sigma_1, \sigma_2 \leq \gamma$. Therefore, by the transfinite induction, for every $\sigma < \alpha$, there exists a total order $\leq_\sigma$ on $C_\sigma = g[\sigma + 1]$ such that conditions (a) and (b) hold for arbitrary $\sigma, \sigma_1, \sigma_2 < \alpha$. Define $C = g[\alpha] = P$, $\leq_w = \bigcup_{\sigma<\alpha} \leq_\sigma$, then $\leq_w$ is clearly a total order on $C$ and

$$a, b \in P, \ a \leq b \Longrightarrow a \leq_w b.$$

Take $f: \ P \to C$ as $f(a) = a$ for every $a \in P$, then $C$ is a chain, $f$ is bijective and order preserving. □

**1.2.3 Definition** Let $L_0$, $L_1$ be lattices, $f: \ L_0 \to L_1$ a mapping.

$f$ is called *finite join preserving* (*finite meet preserving*, a *lattice homomorphism* respectively), if $\forall A \in [L_0]^{<\omega}$, $f(\bigvee A) = \bigvee_{a \in A} f(a)$ $\ (f(\bigwedge A) = \bigwedge_{a \in A} f(a)$, $f$ is both finite join preserving and finite meet preserving respectively).

$f$ is called a *join-semilattice isomorphism* (*meet-semilattice isomorphism*, *lattice isomorphism* respectively), if $f$ is bijective and, both $f: \ L_0 \to L_1$ and $f^{-1}: \ L_1 \to L_0$ are finite join preserving (finite meet preserving, lattice homomorphisms respectively).

$f$ is called a *join-semilattice embedding* (*meet-semilattice embedding*, *lattice embedding* respectively), if $f$ is injective and, both $f: \ L_0 \to L_1$ and $f^{-1}|_{img(f)}: img(f) \to L_0$ are finite join preserving (finite meet preserving, lattice homomorphisms respectively).

Say $L_0$ and $L_1$ are *isomorphic as join-semilattices* (*as meet-semilattices*, *as lattices* respectively) to each other, if there exists a join-semilattice isomorphism (meet-semilattice isomorphism, lattice isomorphism respectively) from one of them to the other.

Say $L_0$ can be *embedded as a join-semilattice* (*meet-semilattice*, *lattice* respectively) into $L_1$, if there exists a join-semilattice embedding (meet-semilattice embedding, lattice embedding respectively) from $L_0$ to $L_1$.

**1.2.4 Definition** Let $L_0$, $L_1$ be complete lattices, $f: L_0 \to L_1$ a mapping.

$f$ is called *complete join preserving* or *arbitrary join preserving* (*complete meet preserving* or *arbitrary meet preserving*, a *complete lattice homomorphism* respectively), if $\forall A \subset L_0$, $f(\bigvee A) = \bigvee_{a\in A} f(a)$ $(f(\bigwedge A) = \bigwedge_{a\in A} f(a)$, $f$ is both complete join preserving and complete meet preserving respectively).

$f$ is called a *complete join-semilattice isomorphism* (*complete meet-semilattice isomorphism*, *complete lattice isomorphism* respectively), if $f$ is bijective and, both $f: L_0 \to L_1$ and $f^{-1}: L_1 \to L_0$ are complete join preserving (complete meet preserving, complete lattice homomorphisms respectively).

$f$ is called a *complete join-semilattice embedding* (*complete meet-semilattice embedding*, *complete lattice embedding* respectively), if $f$ is injective and, both $f: L_0 \to L_1$ and $f^{-1}|_{img(f)}: img(f) \to L_0$ are complete join preserving (complete meet preserving, complete lattice homomorphisms respectively).

Say $L_0$ and $L_1$ are *isomorphic as complete join-semilattices* (*as complete meet-semilattices*, *as complete lattices* respectively) to each other, if there exists a complete join-semilattice isomorphism (complete meet-semilattice isomorphism, complete lattice isomorphism respectively) from one of them to the other.

Say $L_0$ can be *embedded as a complete join-semilattice* (*complete meet-semilattice*, *complete lattice* respectively) into $L_1$, if there exists a complete join-semilattice embedding (complete meet-semilattice embedding, complete lattice embedding respectively) from $L_0$ to $L_1$.

**1.2.5 Note** Note that in the preceding definitions, every finite join preserving mapping preserves 0, every finite meet preserving preserves 1, and hence every lattice homomorphism preserves both 0 and 1.

The reasonableness of kinds of embeddings defined above are guaranteed by the following two theorems (**1.2.6** and **1.2.7**):

**1.2.6 Theorem** *Let $P_1$ be a poset, $f: P_0 \to P_1$ be a mapping. Then*

(i) *$P_0$ is a poset, $f$ is order preserving $\Longrightarrow$ $img(f)$ is a poset.*

(ii) *$P_0$ is a join-semilattice, $f$ is finite join preserving $\Longrightarrow$ $img(f)$ is a join-semilattice.*

(iii) *$P_0$ is a meet-semilattice, $f$ is finite meet preserving $\Longrightarrow$ $img(f)$ is a meet-semilattice.*

(iv) *$P_0$ is a lattice, $f$ is a lattice homomorphism $\Longrightarrow$ $img(f)$ is a lattice.*

(v) *$P_0$ is a complete join-semilattice, $f$ is arbitrary join preserving $\Longrightarrow$ $img(f)$ is a complete join-semilattice.*

(vi) *$P_0$ is a complete meet-semilattice, $f$ is arbitrary join preserving $\Longrightarrow$ $img(f)$ is a complete meet-semilattice.*

(vii) *$P_0$ is a complete lattice, $f$ is a complete lattice homomorphism $\Longrightarrow$ $img(f)$ is a complete lattice.* □

**1.2.7 Theorem** *Let $L_0$, $L_1$ be posets, $f: L_0 \to L_1$ a mapping.*

(i) *If $L_0$, $L_1$ are join-semilattices, then $f$ is a join-semilattice embedding if and only if $f$ is injective and finite join preserving.*

(ii) *If $L_0, L_1$ are join-semilattices, $f$ is a join-semilattice embedding, then $img(f)$ is closed under finite joins in $L_1$.*

(iii) *If $L_0$, $L_1$ are join-semilattices, then $f$ is a join-semilattice isomorphism if and only if $f$ is bijective and finite join preserving.*

(iv) *If $L_0$, $L_1$ are meet-semilattices, then $f$ is a meet-semilattice embedding if and only if $f$ is injective and finite meet preserving.*

(v) *If $L_0$, $L_1$ are meet-semilattices, $f$ is a meet-semilattice embedding, then $img(f)$ is closed under finite meets in $L_1$.*

(vi) *If $L_0$, $L_1$ are meet-semilattices, then $f$ is a meet-semilattice isomorphism if and only if $f$ is bijective and finite meet preserving.*

(vii) *If $L_0$, $L_1$ are lattices, then $f$ is a lattice embedding if and only if $f$ is a injective lattice homomorphism.*

(viii) *If $L_0$, $L_1$ are lattices, $f$ is a lattice embedding, then $img(f)$ is closed under finite joins and finite meets in $L_1$.*

(ix) *If $L_0$, $L_1$ are lattices, then $f$ is a lattice isomorphism if and only if $f$ is a bijective lattice homomorphism.*

(x) *If $L_0$, $L_1$ are complete join-semilattices, then $f$ is a complete join-semilattice embedding if and only if $f$ is injective and arbitrary join preserving.*

(xi) *If $L_0$, $L_1$ are complete join-semilattices, $f$ is a complete join-semilattice embedding, then $img(f)$ is closed under arbitrary joins in $L_1$.*

(xii) *If $L_0$, $L_1$ are complete join-semilattices, then $f$ is a complete join-semilattice isomorphism if and only if $f$ is bijective and arbitrary join preserving.*

(xiii) *If $L_0, L_1$ are complete meet-semilattices, then $f$ is a complete meet-semilattice embedding if and only if $f$ is injective and arbitrary meet preserving.*

(xiv) *If $L_0$, $L_1$ are complete meet-semilattices, $f$ is a complete meet-semilattice embedding, then $img(f)$ is closed under arbitrary meets in $L_1$.*

(xv) *If $L_0, L_1$ are complete meet-semilattices, then $f$ is a complete meet-semilattice isomorphism if and only if $f$ is bijective and arbitrary meet preserving.*

(xvi) *If $L_0$, $L_1$ are complete lattices, then $f$ is a complete lattice embedding if and only if $f$ is a injective complete lattice homomorphism.*

(xvii) *If $L_0$, $L_1$ are complete lattices, $f$ is a complete lattice embedding, then $img(f)$ is closed under arbitrary joins and arbitrary meets in $L_1$.*

(xviii) *If $L_0$, $L_1$ are complete lattices, then $f$ is a complete lattice isomorphism if and only if $f$ is a bijective complete lattice homomorphism.*

**Proof** Only prove (xi) and (x), the others are similar.

In the following proof, denote the operation $\bigvee$ in a lattice $L$ by $\bigvee_L$ to emphasize it be carried out in $L$.

(xi) $\forall B \subset img(f)$, since $f$ is injective, $\exists A \subset L_0$ such that $B = f[A]$. Then

$$\bigvee_{img(f)} B = \bigvee_{img(f)} f[A] \geq \bigvee_{L_1} f[A] = f(\bigvee_{L_0} A).$$

But $f(\bigvee_{L_0} A) \in img(f)$ and $\forall a \in A,\ f(\bigvee_{L_0} A) \geq f(a)$, so

$$f(\bigvee_{L_0} A) \geq \bigvee_{img(f)} f[A] = \bigvee_{img(f)} B.$$

Combining the previous inequalities, we get

$$\bigvee_{img(f)} B = \bigvee_{L_1} f[A] = \bigvee_{L_1} B.$$

(x) The condition is obviously necessary. Suppose $B$ and $A$ as they are in the proof of (xi), then by (xi) proved above and $f$ being injective,

$$f^{-1}|_{img(f)}(\bigvee_{img(f)} B) = f^{-1}(\bigvee_{L_1} f[A]) = f^{-1}(f(\bigvee_{L_0} A)) = \bigvee_{L_0} A$$
$$= \bigvee_{L_0} f^{-1}|_{img(f)}[f[A]] = \bigvee_{L_0} f^{-1}|_{img(f)}[B]. \qquad \square$$

**1.2.8 Definition** Let $L$ be a lattice, $L_0 \subset L$. $L_0$ is called a *sublattice* of $L$, if the inclusion $i: L_0 \hookrightarrow L$ is a lattice embedding; called a *sub-complete-lattice*, if $i: L_0 \hookrightarrow L$ is a complete lattice embedding; called a *complete sublattice*, if the inclusion $i: L_0 \hookrightarrow L$ is a complete lattice homomorphism.

Clearly, joins and meets in lattices are uniquely determined by partial orders. On the other hand, by Remark **1.1.18** (2) and the naturally true dual, joins and meets also uniquely determine partial orders. Therefore, following conclusions are obvious:

**1.2.9 Theorem**

(i) *Arbitrary lattices $L_0$ and $L_1$ are isomorphic as lattices if and only if they are isomorphic as posets.*

(ii) *Arbitrary complete lattices $L_0$ and $L_1$ are isomorphic as complete lattices if and only if they are isomorphic as lattices if and only if they are isomorphic as posets.* $\square$

Regard posets as small categories, apply the categoical concept "adjunction" on mappings between posets as follows:

**1.2.10 Definition** Let $A$, $B$ be posets, $f: A \to B$, $g: B \to A$ be order preserving mappings. $f$ is called a *left adjoint* of $g$ and $g$ is called a *right adjoint* of $f$, denoted by $f \dashv g$, if

$$\forall a \in A, \forall b \in B, \quad f(a) \le b \iff a \le g(b).$$

We also call $f \dashv g$ a *adjunction.*

**1.2.11 Proposition** *Let $A$, $B$ be posets, $f: A \to B$, $g: B \to A$ be order preserving mappings. Then*

(i) $f \dashv g \iff \forall a \in A, \forall b \in B, \quad a \le gf(a), \ fg(b) \le b.$

(ii) $f \dashv g \Longrightarrow fgf = f, \ gfg = g.$

(iii) $f \dashv g \Longrightarrow f$ *and $g$ restrict to a bijection between* $\{a \in A: \ a = gf(a)\}$ *and* $\{b \in B: \ b = fg(b)\}$.

**Proof** (i) ($\Longrightarrow$) $\forall a \in A, \ f(a) \le f(a) \Longrightarrow a \le gf(a)$. $\forall b \in B, \ g(b) \le g(b) \Longrightarrow fg(b) \le b$.

($\Longleftarrow$) $f(a) \le b \Longrightarrow a \le gf(a) \le g(b) \Longrightarrow a \le g(b) \Longrightarrow f(a) \le fg(b) \le b \Longrightarrow f(a) \le b$.

(ii) By (i), $\forall a \in A, \ f(a) \le fgf(a)$, so $f \le fgf$. But for $b = f(a)$, we have $fgf(a) \le f(a)$, then $fgf \le f$. Hence $f = fgf$.

Similarly prove $g = gfg$.

(iii) Denote $A^* = \{a \in A: \ a = gf(a)\}$, $B^* = \{b \in B: \ b = fg(b)\}$. $\forall a \in A^*$, by (ii), $fg(f(a)) = f(a)$, hence $f(a) \in B^*$, $f[A^*] \subset B^*$. Similarly, $g[B^*] \subset A^*$. So we can define $f^* = f|_{A^*}^{B^*}$, $g^* = g|_{B^*}^{A^*}$. $\forall b \in B^*$, by (ii), $g(b) = gf(g(b))$, $g(b) \in A^*$, so $b = f(g(b)) = f^*(g(b))$, $f^*$ is surjective. $\forall a_1, a_2 \in A^*$. If $f^*(a_1) = f^*(a_2)$, then $a_1 = gf(a_1) = g(f^*(a_1)) = g(f^*(a_2)) = gf(a_2) = a_2$, so $f^*$ is injective.

$g^*$ is similar to $f^*$. □

**1.2.12 Definition** Let $L$ be a lattice, $\alpha \in L$. $\alpha$ is called *prime*, if $\alpha < 1$ and $\forall a, b \in L,\ \alpha \geq a \wedge b \Longrightarrow \alpha \geq a$ or $\alpha \geq b$; called *irreducible* or *meet-irreducible*, if $\alpha < 1$ and $\forall a, b \in L,\ \alpha = a \wedge b \Longrightarrow \alpha = a$ or $\alpha = b$.

The dual notions *co-prime*, *join-irreducible* are dually defined.

A join-irreducible element of $L$ is also called an *molecule* in $L$, and an *atom* in $L$ is a minimal element in $L \backslash \{0\}$.

For every $A \subset L$, denote the set of all the prime elements of $L$ in $A$ by $pr(A)$, the set of all the irreducible elements of $L$ in $A$ by $ir(A)$, the set of all the molecules of $L$ in $A$ by $M(A)$; especially, the sets of all the prime elements, all the irreducible elements and all the molecules in $L$ are denoted by $pr(L)$, $ir(L)$, $M(L)$ respectively.

**1.2.13 Proposition** *Let L be a lattice. Then*

(i) *Every atom of L is a molecule.*

(ii) *Every prime element of L is irreducible.*

(iii) *Every co-prime element of L is join-irreducible.*

(iv) *If L is distributive, then for each element $a \in L$, a is prime if and only if a is irreducible.*

(v) *If L is distributive, then for each element $a \in L$, a is co-prime if and only if a is join-irreducible.*

**Proof** Only prove (i), (ii) and (iv).

(i) Let $a$ is an atom of $L$, $a = b \vee c$. If $a \neq b, c$, then $a > b, c$. Certainly $b, c \neq 0$, then this contradicts the minimality of $a$ in $L \backslash \{0\}$.

(ii) Suppose $\alpha \in pr(L)$, $\alpha = a \wedge b$, then $\alpha \geq a$ or $\alpha \geq b$. But $\alpha = a \wedge b \leq a, b$, so $\alpha = a$ or $\alpha = b$. Since $\alpha \in pr(L)$, $\alpha \neq 1$, so it is irreducible.

(iv) Suppose $L$ is distributive, $\alpha \in L$ is irreducible, $\alpha \geq a \wedge b$. Then by Proposition **1.1.20** (iii) and the distributivity, $\alpha = \alpha \vee (a \wedge b) = (\alpha \vee a) \wedge (\alpha \vee b)$. Since $\alpha$ is irreducible, $\alpha = \alpha \vee a$ or $\alpha = \alpha \vee b$. So by Proposition **1.1.20** (iii), $\alpha \geq a$ or $\alpha \geq b$. By $\alpha \neq 1$, $\alpha \in pr(L)$.

As for the necessity, it has been proved in (ii). □

**1.2.14 Definition** Let $L$ be a lattice. An element $x \in L$ is called a *complement* of an element $a \in L$, if $x \wedge a = 0$, $x \vee a = 1$.

A mapping $': L \to L$ is called *order-reversing*, if $\forall a, b \in L,\ a \leq b \Longrightarrow a' \geq b'$; called an *involution* on $L$, if $'' = id_L : L \to L$; called a *complementary operation*, if $\forall a \in L$, $a'$ is a complement of $a$.

For an order-reversing involution $'$ on $L$ and every $A \subset L$, denote $A' = \{a' : a \in A\}$.

Let $L_0$, $L_1$ be lattices equipped with involutions (complementary operations respectively), $f : L_0 \to L_1$ a mapping. $f$ is called *involution preserving* (*complementary operation preserving* or *complement preserving* respectively), if $\forall a \in L,\ f(a') = f(a)'$.

**1.2.15 Proposition** *Let L be a lattice, $'$ an order-reversing involution on L, then*

(i) $M(L)' = pr(L)$.

(ii) $pr(L)' = M(L)$. □

**1.2.16 Proposition** *Let $L$ be a lattice, $'$ an order-reversing involution on $L$, then following equalities hold if all the joins and meets involved exist:*

(DM1) $(\bigvee_i a_i)' = \bigwedge_i a_i'$,

(DM2) $(\bigwedge_i a_i)' = \bigvee_i a_i'$. □

Equalities (DM1) and (DM2) are called *De Morgan's Law.*

**1.2.17 Theorem** *Let $L$ be a distributive lattice.*

(i) *For each element $a \in L$, if there exists a complement $x$ of $a$, then $x$ is unique.*

(ii) *If there exists a complementary operation $'$ on $L$, then $'$ is unique.*

(iii) *Every complementary operation on $L$ is an order-reversing involution.*

**Proof** (i) Suppose $x, y$ are complements of $a \in L$, then by the distributivity,

$$x = x \wedge 1 = x \wedge (y \vee a) = (x \wedge y) \vee (x \wedge a) = (x \wedge y) \vee 0 = x \wedge y,$$
$$x = x \vee 0 = x \vee (y \wedge a) = (x \vee y) \wedge (x \vee a) = (x \vee y) \wedge 1 = x \vee y.$$

By Proposition **1.1.20** (iii), $x \leq y \leq x$. Since $\leq$ is a partial order, $x = y$.

(ii) By (i), the conclusion is clear.

(iii) Suppose $' : L \to L$ is a complementary operation. $\forall a, b \in L,\ a \leq b$, then

$$b' = b' \wedge 1 = b' \wedge (a' \vee a) = (b' \wedge a') \vee (b' \wedge a) \leq (b' \wedge a') \vee (b' \wedge b) =$$
$$= (b' \wedge a') \vee 0 = b' \wedge a'.$$

By Proposition **1.1.20** (ii) and (iii), we have $b' \leq a'$, $'$ is order-reversing. Prove it is an involution:

$$a'' = a'' \wedge 1 = a'' \wedge (a' \vee a) = (a'' \wedge a') \vee (a'' \wedge a) = 0 \vee (a'' \wedge a) = a'' \wedge a,$$
$$a'' = a'' \vee 0 = a'' \vee (a' \wedge a) = (a'' \vee a') \wedge (a'' \vee a) = 0 \wedge (a'' \vee a) = a'' \vee a.$$

Similar to (i), it proves $a'' = a$. □

Because of the reason shown above, for an element of a lattice, we say "the complement" of it but not "a complement" of it in the sequel.

**1.2.18 Definition** Let $L$ be a distributive lattice. $L$ is called a *Boolean algebra*, if there exists a complementary operation $'$ on $L$.

Theorem **1.2.17** (i) and (ii) show that in a Boolean algebra the complementary operation $'$ is uniquely determined by the partial order, it follows that any lattice homomorphism between two Boolean algebras must be a *Boolean algebra homomorphism*, i.e. it preserves complementary operation.

**1.2.19 Example**

(i) For any set $X$, $\mathcal{P}(X)$ with inclusion order is a Boolean algebra, the complement of an element $A$ of $\mathcal{P}(X)$ is just the complementary set of $A$ in $X$ in the sense of set theory.

(ii) Now we'll construct a very useful Boolean algebra $Reg(\mathbf{R})$ possessing following properties:

(1) $Reg(\mathbf{R})$ is a complete Boolean algebra;

(2) $Reg(\mathbf{R})$ is infinitely distributive;

(3) $Reg(\mathbf{R})$ has no irreducible element, and hence neither prime element;

(4) $Reg(\mathbf{R})$ has not join-irreducible element, and hence neither co-prime element;

(5) $Reg(\mathbf{R})$ is atomless;

(6) $Reg(\mathbf{R})$ is not completely distributive.

Denote the topology of real line $\mathbf{R}$ by $\Omega(\mathbf{R})$. $U \in \Omega(\mathbf{R})$ satisfying $U^{-\circ} = U$ is called an *regular open subset.* Define $Reg(\mathbf{R})$ as the family of all the regular open subsets of $\Omega(\mathbf{R})$, i.e.

$$Reg(\mathbf{R}) = \{U : U \in \Omega(\mathbf{R}),\ U^{-\circ} = U\},$$

equip it with inclusion order.

Now we prove preceding conclusions step by step:

(A) $\forall A \subset \mathbf{R},\ A^{-\circ} \in Reg(\mathbf{R})$.

Since $A^{-\circ-} \geq A^{-\circ}$, so $A^{-\circ-\circ} \geq A^{-\circ}$. For another direction, we have $A^{-\circ} \leq A^{-}$, so $A^{-\circ-} \leq A^{--} = A^{-}$, and hence $A^{-\circ-\circ} \leq A^{-\circ}$, it follows $A^{-\circ-\circ} = A^{-\circ}$, $A^{-\circ} \in Reg(\mathbf{R})$.

(B) $Reg(\mathbf{R})$ is a complete lattice.

$\forall \mathcal{A} \subset Reg(\mathbf{R})$, define

$$\bigvee \mathcal{A} = (\bigcup \mathcal{A})^{-\circ}, \quad \bigwedge \mathcal{A} = (\bigcap \mathcal{A})^{-\circ},$$

then by (A), these two operations are well-defined, $Reg(\mathbf{R})$ with these two operations is indeed a complete lattice.

(C) $\forall a, b \in Reg(\mathbf{R}),\ a \wedge b = a \cap b$.

Since $(a\cap b)^{-} \geq a\cap b$, so $a\wedge b = (a\cap b)^{-\circ} \geq (a\cap b)^{\circ} = a\cap b$. For the other direction, $(a\cap b)^{-} \leq a^{-} \cap b^{-}$, so $a\wedge b = (a\cap b)^{-\circ} \leq (a^{-}\cap b^{-})^{\circ} = a^{-\circ}\cap b^{-\circ} = a\cap b,\ a\wedge b = a\cap b$.

(D) $Reg(\mathbf{R})$ satisfies (IFD1).

$\forall a \in Reg(\mathbf{R})$, $\forall B \subset Reg(\mathbf{R})$, clearly $a \wedge \bigvee B \geq \bigvee_{b\in B} (a \wedge b)$. Prove the converse of this inequality:

$$\begin{aligned}
& a \wedge \bigvee B = a \cap (\bigcup B)^{-\circ} && \text{(by (C) and (B) proved above)}\\
=\ & a \cap \bigcup\{U : U \in \Omega(\mathbf{R}),\ U \subset (\bigcup B)^{-}\} && \text{(by definition of interior)}\\
=\ & a \cap \bigcup\{U : U \in \Omega(\mathbf{R}),\ U \cap (\bigcup B)^{-\,c} = \emptyset\}\\
=\ & a \cap \bigcup\{U : U \in \Omega(\mathbf{R}),\ U \cap (\bigcup B)^{c\,\circ} = \emptyset\}\\
=\ & a \cap \bigcup\{U : U \in \Omega(\mathbf{R}),\ U \cap \bigcup\{V : V \in \Omega(\mathbf{R}),\ V \subset (\bigcup B)^{c}\} = \emptyset\}\\
& && \text{(by definition of interior)}\\
=\ & a \cap \bigcup\{U : U \in \Omega(\mathbf{R}),\ U \cap \bigcup\{V : V \in \Omega(\mathbf{R}),\ V \cap \bigcup B = \emptyset\} = \emptyset\}\\
\leq\ & a \cap \bigcup\{U : U \in \Omega(\mathbf{R}),\ U \cap \bigcup\{V \cap a : V \in \Omega(\mathbf{R}),\ (V \cap a) \cap \bigcup B = \emptyset\} = \emptyset\}\\
=\ & \bigcup\{U \cap a : U \in \Omega(\mathbf{R}),\ U \cap a \cap \bigcup\{V : V \in \Omega(\mathbf{R}),\ V \cap (a \cap \bigcup B) = \emptyset\} = \emptyset\}\\
\leq\ & \bigcup\{U : U \in \Omega(\mathbf{R}),\ U \cap \bigcup\{V : V \in \Omega(\mathbf{R}),\ V \cap (a \cap \bigcup B) = \emptyset\} = \emptyset\}\\
=\ & \bigcup\{U : U \in \Omega(\mathbf{R}),\ U \cap (a \cap \bigcup B)^{c\,\circ} = \emptyset\}\\
& && \text{(by definition of interior)}\\
=\ & \bigcup\{U : U \in \Omega(\mathbf{R}),\ U \cap (a \cap \bigcup B)^{-\,c} = \emptyset\}\\
=\ & \bigcup\{U : U \in \Omega(\mathbf{R}),\ U \subset (a \cap \bigcup B)^{-}\}\\
=\ & (a \cap \bigcup B)^{-\,\circ} && \text{(by definition of interior)}\\
=\ & (\bigcup_{b\in B} (a \cap b))^{-\,\circ}\\
=\ & \bigvee_{b\in B} (a \wedge b). && \text{(by (B) and (C) proved above)}
\end{aligned}$$

So $a \wedge \bigvee B = \bigvee_{b \in B} (a \wedge b)$, *Reg*(**R**) satisfies (IFD1).

(E) *Reg*(**R**) is a Boolean algebra.

$\forall a \in Reg(\mathbf{R})$, define $a' = a^{c\,\circ}$, then

$$a' \wedge a = a^{c\,\circ} \cap a = a^{-\,c} \cap a \leq a^c \cap a = \emptyset,$$

$$a' \vee a = (a^{c\,\circ} \cup a)^{-\,\circ} = (a^{-\,c\,-} \cup a^-)^{\circ} = (a^{-\,\circ\,c} \cup a^-)^{\circ} = (a^c \cup a^-)^{\circ} \geq = \mathbf{R}.$$

So $'$ is indeed a complementary operation on *Reg*(**R**). In (D) it has been proved that *Reg*(**R**) satisfies (IFD1), so by Remark **1.1.33** (1), *Reg*(**R**) is distributive, thus a Boolean algebra.

(F) *Reg*(**R**) is infinitely distributive.

Since *Reg*(**R**) is a Boolean algebra and satisfies (IFD1), by Theorem **1.2.17** (iii) and Proposition **1.2.16**, it satisfies (IFD2) as well.

(G) *Reg*(**R**) has no irreducible element.

$\forall U \in Reg(\mathbf{R}) \backslash \{1\}$, then $U \neq \mathbf{R}$, $U' \neq \emptyset$, $U' = U^{-\circ\prime} = U^{\prime\circ-}$, $U'$ contains a nonempty open subset $U'^{\circ}$ of **R**. So $\exists u \in U'$ and $\varepsilon > 0$ such that $[u-3\varepsilon, u+3\varepsilon] \subset U'$. Let $V = U \cup (u-2\varepsilon, u-\varepsilon)$, $W = U \cup (u+\varepsilon, u+2\varepsilon)$, then $\forall x \in U$, $\forall y \in (u-2\varepsilon, u-\varepsilon) \cup (u+\varepsilon, u+2\varepsilon)$, $|x-y| \geq \varepsilon$, $U^- \cap ([u-2\varepsilon, u-\varepsilon] \cup [u+\varepsilon, u+2\varepsilon]) = \emptyset$,

$$\begin{aligned} V^{-\,\circ} &= (U \cup (u-2\varepsilon, u-\varepsilon))^{-\,\circ} = (U^- \cup [u-2\varepsilon, u+\varepsilon])^{\circ} \\ &= U^{-\,\circ} \cup (u-2\varepsilon, u-\varepsilon) = U \cup (u-2\varepsilon, u-\varepsilon) = V. \end{aligned}$$

Similarly, $W^{-\,\circ} = W$. Hence $V, W \in Reg(\mathbf{R})$. Since $V, W > U$ but $V \cap W = U$, $U$ is not irreducible.

(H) Conclusions (4) and (5) hold.

By (E) and (G) proved above.

(I) *Reg*(**R**) is not completely distributive.

We prove (CD2) is not satisfied by *Reg*(**R**). Since the set **Q** of all the rational numbers is countable, so it can be denoted by $\mathbf{Q} = \{r_i : i < \omega\}$. Take $I = \omega$, $\forall i \in I$, $J_i = \omega$. $\forall i \in I$, $j \in J_i$, take $a_{i,j} = (r_i - \frac{1}{j+1}, r_i + \frac{1}{j+1})$, then clearly $a_{i,j} \in Reg(\mathbf{R})$, and $\bigvee_{i \in I}(\bigwedge_{j \in J_i} a_{i,j}) = \emptyset$. But since **Q** is dense in **R**, $\bigwedge_{\varphi \in \prod_{i \in I} J_i}(\bigvee_{i \in I} a_{i,\varphi(i)}) = \mathbf{R}$, (CD2) not hold.

**1.2.20 Definition** A relation $\leq$ on a set $D$ is called *directed*, if for every finite $D_0 \subset D$, there exists $d_0 \in D$ such that $d \leq d_0$ for every $d \in D_0$. A set $D$ equipped with a directed preorder $\leq$ is called a *directed set* or *up-directed set* in $D$.

A subset $I$ of a poset $P$ is called an *ideal* in $P$, if $I$ is both a lower set and an up-directed set; called a *proper ideal* in $P$, if $I$ is an ideal and $I \neq P$; called an *ultra-ideal*, if $I$ is a maximal proper ideal; called an *principal ideal* in $P$, if there exists $a \in P$ such that $I = \downarrow a$.

The dual notions *down-directed set*, *filter* in a poset $P$, *proper filter* in a poset $P$, *ultra-filter* in a poset $P$ and *principal filter* in a poset $P$ are dually defined.

Denote the set of all the ideals in $P$ by $Idl(P)$, the set of all the filters in $P$ by $Flt(P)$, equip them with inclusion order.

Let $L$ be a lattice, $I \subset L$. $I$ is called a *prime ideal* in $L$, if $I$ is a proper ideal in $L$ and $\forall a, b \in L$, $a \wedge b \in I \Longrightarrow a \in I$ or $b \in I$.

The dual notion *prime filter* is dually defined.

Denote the set of all the prime ideals in $L$ by $PrIdl(P)$, the set of all the prime filters in $P$ by $PrFlt(P)$.

**1.2.21 Theorem**

(i) *Every up-directed set in a poset is not empty.*

(ii) *Every down-directed set in a poset is not empty.*

(iii) *Every ideal in a lattice is not empty.*

(iv) *Every filter in a lattice is not empty.*

(v) *A subset $I$ of a lattice $L$ is an ideal if and only if $I$ is a lower set and $a, b \in I \Longrightarrow a \vee b \in I$ for all $a, b \in I$.*

(vi) *A subset $F$ of a lattice $L$ is a filter if and only if $F$ is an upper set and $a, b \in F \Longrightarrow a \wedge b \in F$ for all $a, b \in F$.*

(vii) *For every lattice $L$, $Idl(L)$ is a complete lattice, $pr(Idl(L)) = PrIdl(L)$ and $\forall \{I_t : t \in T\} \subset Idl(L)$, $\forall I_0, I_1 \in Idl(L)$,*

$$\bigvee_{t\in T} I_t = \{a \wedge \bigvee S : \ a \in L, \ S \in [\bigcup_{t\in T} I_t]^{<\omega}\},$$

$$\bigwedge_{t\in T} I_t = \bigcap_{t\in T} I_t,$$

$$I_0 \wedge I_1 = \{a \wedge b : \ a \in I_0, \ b \in I_1\}.$$

(viii) *For every lattice $L$, $Flt(L)$ is a complete lattice, $pr(Flt(L)) = PrFlt(L)$ and $\forall \{F_t : t \in T\} \subset Flt(L)$, $\forall F_0, F_1 \in Flt(L)$,*

$$\bigwedge_{t\in T} F_t = \{a \vee \bigwedge S : \ a \in L, \ S \in [\bigcup_{t\in T} F_t]^{<\omega}\},$$

$$\bigvee_{t\in T} F_t = \bigcup_{t\in T} F_t,$$

$$F_0 \vee F_1 = \{a \vee b : \ a \in F_0, \ b \in F_1\}.$$

(ix) *For every complete lattice $L$, $\forall \{I_t : t \in T\} \subset Idl(L)$,*

$$\bigwedge_{t\in T} I_t = \{\bigwedge_{t\in T} \varphi(t) : \ \varphi \in \prod_{t\in T} I_t\}.$$

(x) *For every complete lattice $L$, $\forall \{F_t : t \in T\} \subset Flt(L)$,*

$$\bigvee_{t\in T} F_t = \{\bigvee_{t\in T} \varphi(t) : \ \varphi \in \prod_{t\in T} F_t\}.$$

(xi) *For every distributive lattice $L$, $\forall \{I_t : \ t \in T\} \subset Idl(L)$,*

$$\bigvee_{t\in T} I_t = \{\bigvee S : \ S \in [\bigcup_{t\in T} I_t]^{<\omega}\}.$$

(xii) *For every distributive lattice $L$, $\forall \{F_t : \ t \in T\} \subset Flt(L)$,*

$$\bigwedge_{t\in T} F_t = \{\bigwedge S : \ S \in [\bigcup_{t\in T} F_t]^{<\omega}\}.$$

**Proof** It is sufficient to prove only the conclusions about up-directed sets and ideals, the others are dual.

(i): Suppose $D$ is a directed set, then for the finite subset $\emptyset \subset D$, $\exists d_0 \in D$ such that $d \leq d_0$ for every $d \in \emptyset$. So $D$ is not empty.

(iii): By (i).

(v): Sufficiency is clear. Suppose $I$ is an ideal in lattice $L$, $a, b \in I$, then since $I$ is an up-directed set, $\exists c \in I$ such that $c \geq a, b$. So $c \geq a \vee b$. By $\downarrow I = I$, $a \vee b \in I$.

(vii): Certainly, $Idl(L)$ has the smallest element $\{0\}$ and the largest element $L$. $\forall \{I_t : \ t \in T\} \subset Idl(L)$, by (v) proved above, $\bigcap_{t\in T} I_t$ is an ideal. Since each lower bound of $\{I_t:\ t\in T\}$ in $Idl(L)$ must be contained in $\bigcap_{t\in T} I_t$, so $\bigwedge_{t\in T} I_t=\bigcap_{t\in T} I_t$. By Proposition **1.1.24** (iii), $Idl(L)$ is a complete lattice.

$\forall\{I_t:\ t\in T\}\subset Idl(L)$, denote $I=\{a\wedge\bigvee S:\ a\in L,\ S\in[\bigcup_{t\in T} I_t]^{<\omega}\}$, then $\forall b\in\downarrow I$, $\exists a\in L$, $\exists S\in[\bigcup_{t\in T} I_t]^{<\omega}$ such that $b\le a\wedge\bigvee S$. So $b=b\wedge(a\wedge\bigvee S)=(b\wedge a)\wedge\bigvee S\in I$, $I$ is a lower set. $\forall a_1,a_2\in L$, $\forall S_1,S_2\in[\bigcup_{t\in T} I_t]^{<\omega}$, by Proposition **1.1.24** (iii),

$$(a_1\wedge\bigvee S_1)\vee(a_2\wedge\bigvee S_2)\le(\bigvee S_1)\vee(\bigvee S_2)=1\wedge\bigvee(S_1\cup S_2)\in I.$$

Since $I$ is a lower set, so $(a_1\wedge\bigvee S_1)\vee(a_2\wedge\bigvee S_2)\in I$. By Proposition **1.1.24** (iii), $I$ is an ideal. Clearly $I\supset\bigcup_{t\in T} I_t$ and every ideal in $L$ containing $\bigcup_{t\in T} I_t$ must contain $I$, we obtain $\bigvee_{t\in T} I_t=I$.

$\forall I_0,I_1\in Idl(L)$, denote $I=\{a\wedge b:\ a\in I_0,\ b\in I_1\}$, then $\forall c\in\downarrow I$, $\exists a\in I_0$, $b\in I_1$ such that $c\le a\wedge b$. So $c\le a,b$, $c\in I_0$, $c\in I_1$, $c=c\wedge c\in I$, $I$ is a lower set. $\forall a_1,a_2\in I_0$, $\forall b_1,b_2\in I_1$, $(a_1\wedge b_1)\vee(a_2\wedge b_2)\le(a_1\vee a_2)\wedge(b_1\vee b_2)\in I$. Since $I$ is a lower set, it follows $(a_1\wedge b_1)\vee(a_2\wedge b_2)\in I$, $I$ is an ideal. Certainly $I\subset I_0\cap I_1$. On the other hand, $\forall a\in I_0\cap I_1$, $a=a\wedge a\in I$, so $I=I_0\cap I_1=I_0\wedge I_1$.

$\forall I\in pr(Idl(L))$. Suppose $I\notin PrIdl(L)$, then $\exists a,b\in L\backslash I$ such that $a\wedge b\in I$, then by the conclusion just proved, $I\ge\downarrow(a\wedge b)=(\downarrow a)\wedge(\downarrow b)$, $I\ge(\downarrow a)$ or $I\ge(\downarrow b)$. That is to say $a\in I$ or $b\in I$, a contradiction. On the other hand, $\forall I\in PrIdl(L)$, if $I\notin pr(Idl(L))$, then $\exists I_0,I_1\in Idl(L)$ such that $I\ge I_0\wedge I_1$ but $I\not\ge I_0,I_1$. Then $\exists a\in I_0$, $b\in I_1$ such that $a,b\notin I$. By what just prove above, we have $a\wedge b\in I$, contradicts to $I\in PrIdl(L)$.

(ix): Similar to its finite case in the proof of (vii).

(xi): Directly deduced from (vii). □

**1.2.22 Theorem** *Let L be a lattice, I an ideal in L, then the following conditions are equivalent:*

(i) *$I$ is a prime ideal.*
(ii) *$L\backslash I$ is a filter.*
(iii) *$L\backslash I$ is a prime filter.*
(iv) *The mapping $f:\ L\to\{0,1\}$ with kernel $f^{-1}(0)=I$ is a lattice homomorphism.* □

**1.2.23 Proposition** *Let L be a lattice, $a\in L$, then*

(i) *$a$ is a prime element in $L$ if and only if $\downarrow a$ is a prime ideal in $L$.*
(ii) *$a$ is a co-prime element in $L$ if and only if $\uparrow a$ is a prime filter in $L$.* □

Concerning ideals in lattices there are following important *Maximal Ideal Theorem* and *Prime Ideal Theorem* (their original forms are Corollary **1.2.25** and Corollary **1.2.29** respectively). Since Maximal Ideal Theorem is equivalent to the Axiom of Choice[155, 150, 120, 121, 52] and Prime Ideal Theorem is strictly weaker than the Axiom of Choice but is still not provable in ZF set theory,[29] they play essential roles in set theory and lattice theory.

**1.2.24 Theorem** (Maximal Ideal Theorem)

*Let L be a lattice, $I\in Idl(L)$, $F\in Flt(L)$, $I\cap F=\emptyset$, then $\{J\in Idl(L):\ J\supset I,\ J\cap F=\emptyset\}$ has a maximal element, i.e. there exists an ideal of L which is maximal amongst those containing I and disjoint F.*

**Proof** Straightforwardly apply Zorn's Lemma to the family $\{J \in Idl(L) : J \supset I,\ J \cap F = \emptyset\}$. □

**1.2.25 Corollary** *Every lattice has a ultra-ideal.*

**Proof** Take $I = \{0\}$, $F = \{1\}$ in Maximal Ideal Theorem (**1.2.24**). □

**1.2.26 Lemma** *Let $L$ be a distributive lattice, $I \in Idl(L)$, $F \in Flt(L)$, $I_0$ is a maximal element of $\{J \in Idl(L) : J \supset I,\ J \cap F = \emptyset\}$, then $I_0$ is prime.*

**Proof** Since $I_0 \cap F = \emptyset$ and $1 \in F$, $I_0$ is a proper ideal. Suppose $a_1, a_2 \in L$, $a_1 \wedge a_2 \in I_0$, denote $J_1 = I_0 \vee \downarrow a_1$, $J_2 = I_0 \vee \downarrow a_2$. Since both $J_1$ and $J_2$ have contained $I$, to prove $a_1 \in I_0$ or $a_2 \in I_0$ we need only prove that one of $J_1$ and $J_2$ must be disjoint from $F$, because then by the maximality of $I_0$, $I_0$ must be one of $J_1$ and $J_2$. If both $J_1$ and $J_2$ meet $F$, then by Theorem **1.2.21** (xi), $\exists x_1, x_2 \in I_0$, $y_1 \leq a_1$, $y_2 \leq a_2$ such that $x_1 \vee y_1 \in (I_0 \vee \downarrow a_1) \cap F$, $x_2 \vee y_2 \in (I_0 \vee \downarrow a_2) \cap F$. Since $F \in Flt(L)$, it must contain

$$\begin{aligned}(x_1 \vee y_1) \wedge (x_2 \vee y_2) &= (x_1 \wedge x_2) \vee (x_1 \wedge y_1) \vee (x_2 \vee y_1) \vee (y_1 \wedge y_2) \\ &\leq (x_1 \wedge x_2) \vee x_1 \vee x_2 \vee (a_1 \wedge a_2) \in I_0.\end{aligned}$$

But $I_0$ is a lower set, so by this inequality $I_0$ also contains $(x_1 \vee y_1) \wedge (x_2 \vee y_2)$. Then $I_0$ is not disjoint from $F$, a contradiction. □

**1.2.27 Corollary** *Every ultra-ideal in a distributive lattice is prime.* □

Then Prime Ideal Theorem is deduced as a direct corollary of Lemma **1.2.26**:

**1.2.28 Theorem** (Prime Ideal Theorem)
*Let $L$ be a distributive lattice, $I \in Idl(L)$, $F \in Flt(L)$, $I \cap F = \emptyset$, then $\{J \in Idl(L) : J \supset I,\ J \cap F = \emptyset\}$ contains a prime element, i.e. there exists a prime ideal of $L$ containing $I$ and disjoint $F$.* □

Just as prove Corollary **1.2.25**, we have

**1.2.29 Corollary** *Every distributive lattice has a prime ideal.* □

The verification of the following result is straightforward:

**1.2.30 Proposition** *Let $\{P_t : t \in T\}$ be a family of preordered sets. Then the relation $\leq$ on $\prod_{t\in T} P_t$ defined by*

$$\alpha, \beta \in \prod_{t\in T} P_t,\quad \alpha \leq \beta \Longleftrightarrow \forall t \in T,\ \alpha(t) \leq \beta(t)$$

*is a preorder on $\prod_{t\in T} P_t$.* □

By the preceding proposition, we can reasonablly give out the following

**1.2.31 Definition** Let $\{P_t : t \in T\}$ be a family of preordered sets. Define the *product preorder* on $\prod_{t\in T} P_t$ by:

$$\forall \alpha, \beta \in \prod_{t\in T} P_t,\quad \alpha \leq \beta \Longleftrightarrow \forall t \in T,\ \alpha(t) \leq \beta(t).$$

Define the *product preordered set* of $\{P_t : t \in T\}$ as $\prod_{t\in T} P_t$ equipped with the product preorder on it.

Particularly, for a family $\mathcal{A}$ of directed sets, or of posets, or of lattices, call $\prod \mathcal{A}$ with the product preorder on it the *product directed set* of $\mathcal{A}$, the *product poset* of $\mathcal{A}$, the *product lattice* of $\mathcal{A}$ respectively.

**1.2.32 Theorem**

(i) *Let $\mathcal{D} = \{D_t : t \in T\}$ be a family of directed sets, then the product directed set $\prod \mathcal{D}$ is a directed set.*

(ii) *Let $\mathcal{P} = \{P_t : t \in T\}$ be a family of posets, then the product poset $\prod \mathcal{P}$ is a poset.*

(iii) *Let $\mathcal{L} = \{L_t : t \in T\}$ be a family of lattices, then the product lattice $\prod \mathcal{L}$ is a lattice, and for arbitrary $\alpha, \beta \in \prod_{t\in T} L_t$, the join $\alpha \vee \beta$ and the meet $\alpha \wedge \beta$ of $\alpha$, $\beta$ fulfill the following relations:*

$$\forall t \in T, \quad (\alpha \vee \beta)(t) = \alpha(t) \vee \beta(t), \quad (\alpha \wedge \beta)(t) = \alpha(t) \wedge \beta(t).$$

(iv) *Let $\mathcal{L} = \{L_t : t \in T\}$ be a family of complete lattices, then the product lattice $\prod \mathcal{L}$ is a complete lattice, and for arbitrary $A \subset \prod_{t\in T} L_t$, the join $\bigvee A$ and the meet $\bigwedge A$ of $A$ fulfill the following relations:*

$$\forall t \in T, \quad (\bigvee A)(t) = \bigvee_{p\in A} p(t), \quad (\bigwedge A)(t) = \bigwedge_{p\in A} p(t). \qquad \square$$

Moreover, we can establish the following stronger and useful results:

**1.2.33 Theorem** *Let $\{L_t : t \in T\}$ be a family of preordered sets. Then*

(i) *$\prod_{t\in T} L_t$ is a directed set $\Longleftrightarrow \forall t \in T$, $L_t$ is a directed set.*

(ii) *$\prod_{t\in T} L_t$ is a poset $\Longleftrightarrow \forall t \in T$, $L_t$ is a poset.*

(iii) *$\prod_{t\in T} L_t$ is a lattice $\Longleftrightarrow \forall t \in T$, $L_t$ is a lattice.*

(iv) *$\prod_{t\in T} L_t$ is a complete lattice $\Longleftrightarrow \forall t \in T$, $L_t$ is a complete lattice.*

(v) *$\prod_{t\in T} L_t$ is a distributive lattice $\Longleftrightarrow \forall t \in T$, $L_t$ is a distributive lattice.*

(vi) *$\prod_{t\in T} L_t$ is a complete lattice satisfying* (IFD1) *$\Longleftrightarrow \forall t \in T$, $L_t$ is a complete lattice satisfying* (IFD1).

(vii) *$\prod_{t\in T} L_t$ is a complete lattice satisfying* (IFD2) *$\Longleftrightarrow \forall t \in T$, $L_t$ is a complete lattice satisfying* (IFD2).

(viii) *$\prod_{t\in T} L_t$ is a completely distributive lattice $\Longleftrightarrow \forall t \in T$, $L_t$ is a completely distributive lattice.* $\square$

**1.2.34 Definition** Let $\alpha$ be an ordinal number, $\mathcal{L} = \{L_\beta : \beta < \alpha\}$ a family of lattices, $\forall \beta < \alpha$, $\leq_\beta$ the partial order of $L_\beta$. Because if elements of $\mathcal{L}$ are not pairwise disjoint we can always isomorphically replace them with pairwise disjoint ones $L_\beta \times \{\beta\}$'s, without loss of generality, suppose $L_\beta$'s are pairwise disjoint. Take an element $1 \notin \bigcup_{\beta<\alpha} L_\beta$, define

$$\lhd\mathcal{L} = \mathop{\lhd}_{\beta<\alpha} L_\beta = (\bigcup_{\beta<\alpha} L_\beta \backslash \{1_{L_\beta}\}) \cup \{1\},$$

$$\mathop{\lhd}_{\beta<\alpha} \leq_\beta = (\bigcup_{\beta<\alpha} (\leq_\beta \backslash (L_\beta \times \{1_{L_\beta}\}))) \cup (\bigcup_{\beta<\gamma<\alpha} (L_\beta \backslash \{1_{L_\beta}\}) \times (L_\gamma \backslash \{1_{L_\gamma}\})) \cup ((\lhd\mathcal{L}) \times \{1\}).$$

If $\alpha$ is finite, $\alpha = n$, $\lhd\mathcal{L}$ is also denoted by $L_0 \lhd L_1 \lhd \cdots \lhd L_n$.

The proof of following proposition needs some carefulness, but is straightforward:

**1.2.35 Proposition** *Let $\mathcal{L} = \{L_\beta : \beta < \alpha\}$ be a family of lattices, $\forall \beta < \alpha$, $\leq_\beta$ the partial order of $L_\beta$. Then*

(i) *$\mathop{\lhd}_{\beta<\alpha} \leq_\beta$ is a partial order on $\lhd\mathcal{L}$.*

(ii) *With partial order $\mathop{\lhd}_{\beta<\alpha} \leq_\beta$, $\lhd\mathcal{L}$ is a lattice.*

(iii) *With partial order $\mathop{\lhd}_{\beta<\alpha} \leq_\beta$, $\lhd\mathcal{L}$ is a complete lattice if and only if $\forall \beta < \alpha$, $L_\beta$ is a complete lattice.*

(iv) *With partial order* $\underset{\beta<\alpha}{\lhd}\leq_\beta$, $\lhd\mathcal{L}$ *is a distributive lattice if and only if* $\forall\beta<\alpha$, $L_\beta$ *is a distributive lattice.*

(v) *With partial order* $\underset{\beta<\alpha}{\lhd}\leq_\beta$, $\lhd\mathcal{L}$ *is a complete lattice satisfying* (IFD1) *if and only if* $\forall\beta<\alpha$, $L_\beta$ *is a complete lattice satisfying* (IFD1).

(vi) *With partial order* $\underset{\beta<\alpha}{\lhd}\leq_\beta$, $\lhd\mathcal{L}$ *is a complete lattice satisfying* (IFD2) *if and only if* $\forall\beta<\alpha$, $L_\beta$ *is a complete lattice satisfying* (IFD2). □

Therefore, we can reasonably introduce the following

**1.2.36 Definition** Let $\mathcal{L}=\{L_\beta:\ \beta<\alpha\}$ be a family of lattices, $\forall\beta<\alpha$, $\leq_\beta$ the partial order of $L_\beta$. Define the *well-ordered sum* of $\mathcal{L}=\{L_\beta:\ \beta<\alpha\}$ as $\lhd\mathcal{L}$ equipped with the partial order $\underset{\beta<\alpha}{\lhd}\leq_\beta$.

## 1.3 Characterizations of Complete Distributivity

As an important ordered structure in topology on lattice, completely distributive lattice has had various characterizations and representations. Some of them, e.g. topological characterization and analysis characterization, as our results, will be deduced in Section **4.5**. Those which closely relate to fuzzy topology are introduced in this section. As for others, e.g. characterizations and representations in theory of continuous posets, will be omitted.

**1.3.1 Definition** Let $L$ be a complete lattice. Define a relation $\preceq$ in $L$ as follows: $\forall a,b\in L$, $a\preceq b$ if and only if $\forall S\subset L$, $\bigvee S\geq b\Longrightarrow\exists s\in S$ such that $s\geq a$. $\forall a\in L$, denote $\beta_L(a)=\{b\in L:\ b\preceq a\}$, $\beta^*{}_L(a)=M(\beta_L(a))$; or denote them respectively by $\beta(a)$ and $\beta^*(a)$ for short.

$\forall a\in L$. $D\subset\beta(a)$ is called a *minimal set* of $a$, if $\bigvee D=a$.

**1.3.2 Proposition** *Let $L$ be a complete lattice, $a\in L$, then* $\beta^*(a)=\beta(a)\cap M(L)$.□

**1.3.3 Definition** Let $L$ be a complete lattice, $a\in L$, $A,B\subset L$. $A$ is called a *cover* of $a$, if $\bigvee A\geq a$. $A$ is called a *proper cover* of $a$, if $\bigvee A=a$. $B$ is called a *refinement* of $A$, or say $B$ *refines* $A$, if for every $b\in B$, there exists $a\in A$ such that $b\leq a$.

**1.3.4 Theorem** *Let $L$ be a complete lattice, $a\in L$, $D\subset L$. Then $D$ is a minimal set of $a$ if and only if $D$ is a proper cover of $a$ and $D$ is a refinement of every cover of $a$.* □

Dually we can introduce the following notions[168] and, obtain the dual conclusions related to minimal sets directly in the sequel.

**1.3.5 Definition** Let $L$ be a complete lattice. Define a relation $\succeq$ in $L$ as follows: $\forall a,b\in L$, $a\succeq b$ if and only if $\forall S\subset L$, $\bigwedge S\leq b\Longrightarrow\exists s\in S$ such that $s\leq a$. $\forall a\in L$, denote $\underline{\beta}(a)=\{b\in L:\ b\succeq a\}$, $\underline{\beta}^*(a)=pr(a)\cap\underline{\beta}(a)$.

$\forall a\in L$. $D\subset\underline{\beta}(a)$ is called a *maximal set* of $a$, if $\bigwedge D=a$.

**1.3.6 Theorem** *Let $L$ be a complete lattice, $a,b,c,d\in L$, $\{a_i:\ i\in I\}\subset L$. Then*

(i) $a\preceq b\Longrightarrow a\leq b$.

(ii) $a\leq b\preceq c\leq d\Longrightarrow a\preceq d$.

(iii) $\beta(a)$ *is a lower set.*

(iv) *$a$ has a minimal set* $\Longleftrightarrow$ *$\beta(a)$ is the largest minimal set of $a$.* □

**1.3.7 Theorem** *Let $L$ be a complete lattice, $a,b,c,d\in L$, $\{a_i:\ i\in I\}\subset L$. Then*

(i) $a \succsim b \Longrightarrow a \geq b$.
(ii) $a \geq b \succsim c \geq d \Longrightarrow a \succsim d$.
(iii) $\underline{\beta}(a)$ *is an upper set.*
(iv) *$a$ has a maximal set* $\Longleftrightarrow$ $\underline{\beta}(a)$ *is the largest maximal set of $a$.* □

**1.3.8 Theorem** *Let $L$ be a complete lattice, $\alpha \in L$, then the following conditions are equivalent:*

(i) *Every minimal set of $\alpha$ is an up-directed set.*
(ii) *One of the minimal sets of $\alpha$ is an up-directed set.*
(iii) $\alpha \in M(L)$.

**Proof** (i)$\Longrightarrow$(ii): Clear.

(ii)$\Longrightarrow$(iii): Suppose $D$ is a minimal set of $\alpha$ which is up-directed. If $\alpha \notin M(L)$, then $\exists a, b \in L$ such that $\alpha = a \vee b$ but $\alpha \neq a, b$, i.e. $\alpha \nleq a, b$. Since $D$ is a minimal set of $\alpha$, $\alpha = \bigvee D$, so $\exists x, y \in D$ such that $x \nleq a$, $y \nleq b$. Since $D$ is up-directed, $\exists z \in D$ such that $z \geq x, y$, and hence $z \nleq a, b$. But $D$ is a minimal set of $\alpha = a \vee b$, for $z \in D$, $\exists c \in \{a, b\}$ such that $c \geq z$, i.e. $z \leq a$ or $z \leq b$, this is a contradiction.

(iii)$\Longrightarrow$(i): Suppose $D$ is a minimal set of $\alpha$. $\forall x, y \in D$, denote $D_1 = \{z \in D: \ z \geq x\}$, $D_2 = \{z \in D: \ z \ngeq x\}$, then $\bigvee D_1 \vee \bigvee D_2 = \bigvee D = \alpha$. Since $\alpha \in M(L)$, $\alpha = \bigvee D_1$ or $\alpha = \bigvee D_2$. But if $\alpha = \bigvee D_2$, by $x \in D$ and $D$ being a minimal set of $\alpha$, $\exists z \in D_2$ such that $z \geq x$, this contradicts the definition of $D_2$. So $\alpha = \bigvee D_1$, and by $y \in D$ and $D$ being a minimal set of $\alpha$, $\exists z \in D_1$ such that $z \geq y$. That is to say, $z \geq x, y$. Therefore, $D \neq \emptyset$, $D$ is an up-directed set. □

**1.3.9 Theorem** *Let $L$ be a complete lattice, $\alpha \in L$, then the following conditions are equivalent:*

(i) *Every maximal set of $\alpha$ is a down-directed set.*
(ii) *One of the maximal sets of $\alpha$ is a down-directed set.*
(iii) $\alpha \in pr(L)$. □

With the concept of minimal set, we can establish the following important and useful characterization theorem of completely distributive lattices:

**1.3.10 Theorem** *Let $L$ be a complete lattice, then following conditions are equivalent:*

(i) *$L$ is completely distributive.*
(ii) *$\forall a \in L$, there exists a minimal set of $a$.*
(iii) *$\forall a \in L$, $\beta(a)$ is a minimal set of $a$.*
(iv) *$\forall a \in L$, $\beta^*(a)$ is a minimal set of $a$.*

**Proof** First of all, we prove that for every complete lattice $L$, every $a \in L$ and $\mathcal{A}_a = \{A \subset L: \ \bigvee A \geq a\}$,

$$\mathcal{A}_a = \{\{a_{i,j}: \ j \in J_i\}: \ i \in I\} \Longrightarrow \beta(a) = \{\bigwedge_{i \in I} a_{i,\varphi(i)}: \ \varphi \in \prod_{i \in I} J_i\}, \tag{1.4}$$

where different indexes indicate different sets or elements.

$\forall i \in I$, denote $A_i = \{a_{i,j}: \ j \in J_i\}$, then $\mathcal{A}_a = \{A_i: \ i \in I\}$. Denote $B = \{\bigwedge_{i \in I} a_{i,\varphi(i)}: \ \varphi \in \prod_{i \in I} J_i\}$. $\forall b \in \beta(a)$, $\forall i \in I$, since $A_i \in \mathcal{A}_a$, by $b \preceq a$, $\exists \varphi(i) \in J_i$ such that $b \leq a_{i,\varphi(i)} \in A_i$. By Theorem **1.3.6** (i), $b \in \downarrow a \in \mathcal{A}_a$, so $\exists i_0 \in I$ such that $A_{i_0} = \downarrow a$, we can take $\varphi(i_0) \in J_{i_0}$ such that $a_{i_0,\varphi(i_0)} = b$. Hence $\varphi \in$

$\prod_{i\in I} J_i$, $b = \bigwedge_{i\in I} a_{i,\varphi(i)} \in B$. On the other hand, $\forall b \in B$, suppose $\varphi_0 \in \prod_{i\in I} J_i$ such that $b = \bigwedge_{i\in I} a_{i,\varphi_0(i)}$. $\forall A \in \mathcal{A}_a$, $\exists i_0 \in I$ such that $A = A_{i_0} = \{a_{i_0,j} : j \in J_{i_0}\}$. So $b \leq a_{i_0,\varphi_0(i_0)} \in A_{i_0} = A$. It follows $b \preceq a$, $b \in \beta(a)$. Hence $\beta(a) = B$.

(i)$\Longrightarrow$(iii): Need only to prove $\forall a \in L$, $\bigvee \beta(a) = a$. Suppose $L$ is completely distributive, then by condition (1.4) proved above, (CD1) and $\downarrow a \in \mathcal{A}_a$,

$$\bigvee \beta(a) = \bigvee_{\varphi\in\prod\limits_{i\in I} J_i} \bigwedge_{i\in I} a_{i,\varphi(i)} = \bigwedge_{i\in I} \bigvee_{j\in J_i} a_{i,j} = \bigwedge_{i\in I} \bigvee A_i = \bigwedge_{A\in\mathcal{A}_a} \bigvee A = a.$$

(iii)$\Longrightarrow$(i): Suppose $\forall i \in I$, $A_i = \{a_{i,j} \colon j \in J_i\} \subset L$. Take $a = \bigwedge_{i\in I} \bigvee_{j\in J_i} a_{i,j}$, $\mathcal{A}_a = \{A \subset L : \bigvee A \geq a\}$, we have $\{A_i : i \in I\} \subset \mathcal{A}_a$. So we can suppose $\mathcal{A}_a = \{A_i : i \in I_0\}$, $I_0 \supset I$. $\forall i \in I_0 \backslash I$, still denote $A_i = \{a_{i,j} : j \in J_i\}$. $\forall \psi \in \prod_{i\in I_0} J_i$, take $\varphi \in \prod_{i\in I} J_i$ such that $\forall i \in I$, $\varphi(i) = \psi(i)$, we have $\bigwedge_{i\in I_0} a_{i,\psi(i)} \leq \bigwedge_{i\in I} a_{i,\varphi(i)}$. Since $\beta(a)$ is a minimal set of $a$, by $\downarrow a \in \mathcal{A}_a$ and condition (1.4) proved above,

$$\bigwedge_{i\in I} \bigvee_{j\in J_i} a_{i,j} = a = \bigvee \beta(a) = \bigvee_{\psi\in\prod\limits_{i\in I_0}} \bigwedge_{i\in I_0} a_{i,\psi(i)} \leq \bigvee_{\varphi\in\prod\limits_{i\in I}} \bigwedge_{i\in I} a_{i,\varphi(i)}.$$

Since the other direction of the inequality is obvious, (CD1) has been proved. By Theorem **1.1.34**, $L$ is completely distributive.

(iii)$\Longleftrightarrow$(ii): By Theorem **1.3.6** (iv).

(iii)$\Longrightarrow$(iv): Suppose (iii) holds, then it can be verified that

$$\forall a, b \in L,\ a \preceq b \Longrightarrow \exists c \in L,\ a \preceq c \preceq b. \tag{1.5}$$

In fact, denote $S = \{d \in L :\ \exists c \in L,\ d \preceq c \preceq b\}$, then by (iii), $\bigvee S = \bigvee\{\bigvee \beta(c) : c \in \beta(b)\} = \bigvee \beta(b) = b$. By $a \preceq b$, $\exists d \in S$ such that $d \geq a$. Conclusion follows from Theorem **1.3.6** (ii).

By the definition of minimal set, need only prove $\forall a \in L$, $\bigvee \beta^*(a) = a$. Suppose it not hold, denote $a_0 = \bigvee \beta^*(a)$, then by Theorem **1.3.6** (i), $a_0 < a$. Since (iii)$\Longrightarrow$(i) has been proved above, by Theorem **1.1.34**, $L^{op}$ is also a completely distributive lattice. For the relations $\leq$, $<$, $\preceq$ and operations $\uparrow$, $\bigvee$ in $L$, denote their duals in $L^{op}$ by $\leq^{op}$, $<^{op}$, $\preceq^{op}$ and $\uparrow^{op}$, $\bigvee^{op}$ respectively. Apply (i)$\Longrightarrow$(ii) proved above on $L^{op}$, for $a <^{op} a_0$ in $L^{op}$, $\exists b \preceq^{op} a_0$ such that $b \not\leq^{op} a$. Inductively apply implication (1.5) on $b \preceq^{op} a_0$, we obtain a series $\{b_i : i < \omega\} \subset L^{op}$ such that

$$b \preceq^{op} \cdots \preceq^{op} b_1 \preceq^{op} b_0 \preceq^{op} a_0. \tag{1.6}$$

Take $F = \bigcup_{i<\omega} \uparrow^{op} b_i$. Since $b \not\leq^{op} a$, clearly $a \notin F$, so by Kuratowski's Lemma, $L^{op}\backslash F$ contains a maximal chain $C$ containing $\{a\}$. Denote $m = \bigvee^{op} C$, then if $m \in F$, $\exists i < \omega$ such that $b_i \leq^{op} m$. By condition (1.6) and Theorem **1.3.6** (ii), $b_{i+1} \preceq^{op} m = \bigvee^{op} C$, $\exists c \in C$ such that $b_{i+1} \leq^{op} c$, $c \in F$. But $c \in C \subset L^{op}\backslash F$, this is a contradiction. So $m \notin F$, $m \in L^{op}\backslash F = L\backslash F$.

Now we prove this element $m$ is a molecule of $L$. In $L$, if $m = x \vee y$ but $m \neq x, y$, then $m > x, y$, $m <^{op} x, y$. Since $C$ is a maximal chain in $L^{op}\backslash F$ and $m = \bigvee^{op} C$, both $x$ and $y$ cannot be in $L^{op}\backslash F = L\backslash F$, and hence $x, y \in F$. But $F$ is clearly a filter in $L^{op}$, i.e. an ideal in $L$, so $m = x \vee y \in F$, this contradicts the conclusion "$m \in L\backslash F$" proved above.

Therefore, the original supposition "$\bigvee \beta^*(a) < a$" is false.

(iv)$\Longrightarrow$(iii): By the definition of minimal set. □

**1.3.11 Theorem** *Let $L$ be a complete lattice, then following conditions are equivalent:*

(i) *$L$ is completely distributive.*

(ii) *$\forall a \in L$, there exists a maximal set of $a$.*

(iii) *$\forall a \in L$, $\beta(a)$ is a maximal set of $a$.*

(iv) *$\forall a \in L$, $\beta^*(a)$ is a maximal set of $a$.* □

**1.3.12 Corollary** *Let $L$ be a completely distributive lattice. Then*

(i) *$M(L)$ is a join-generating set of $L$, i.e. $\bigvee M(\downarrow a) = a$ for every $a \in L$.*

(ii) *$\forall a, b \in L$, $a \leq b$ if and only if $M(\downarrow a) \subset M(\downarrow b)$.* □

**1.3.13 Corollary** Let $L$ be a completely distributive lattice. Then $pr(L)$ is a meet-generating set of $L$. □

By the previous characterizations of completely distributive lattices, the following useful insertion theorem can be proved:

**1.3.14 Theorem** (Insertion Property of Completely Distributive Lattice)

*Let $L$ be a completely distributive lattice. Then*

(i) $\forall a, b \in L$, $a \preceq b \Longrightarrow \exists c \in M(L)$, $a \preceq c \preceq b$.

(ii) $\forall a \in L$, $\forall b \in M(L)$, $a \neq b$, $a \preceq b \Longrightarrow \exists c \in M(L)$, $c \neq a$, $a \preceq c \preceq b$.

**Proof** (i) Take $S = \bigcup\{\beta(c) : c \in \beta^*(b)\}$, then by the virtue of Theorem **1.3.10**,

$$\bigvee S = \bigvee\{\bigvee \beta(c) : c \in \beta^*(b)\} = \bigvee \beta^*(b) = b.$$

By $a \preceq b$, $\exists d \in S$ such that $d \geq a$, i.e. $\exists c \in \beta^*(b)$, $\exists d \in \beta(c)$ such that $a \leq d$. Then by $d \preceq c$, $c \in \beta^*(b)$ and Theorem **1.3.6** (ii), we have $a \preceq c \preceq b$ and $c \in M(L)$.

(ii) By (i), $\exists c_0 \in M(L)$ such that $a \preceq c_0 \preceq b$. If $c_0 = a \neq b$, then by Theorem **1.3.6** (i), $c_0 < b$, by Theorem **1.3.10** (iii), $\bigvee \beta(b) = b > c_0$, $\exists c_1 \in \beta(b)$, $c_1 \not\leq c_0$. By Theorem **1.3.10** (iv) and (iii)$\Longleftrightarrow$(i) in Theorem **1.3.8**, $\exists c \in \beta^*(b)$, $c \geq c_0, c_1$. So $a = c_0 \preceq c_0 < c$. Apply Theorem **1.3.6** (ii), we get $c \neq a$, $c \in M(L)$, $a \preceq c \preceq b$. □

**1.3.15 Theorem** *Let $L$ be a completely distributive lattice. Then*

(i) $\forall a, b \in L$, $a \succeq b \Longrightarrow \exists c \in L$, $a \succeq c \succeq b$.

(ii) $\forall a \in L$, $\forall b \in pr(L)$, $a \neq b$, $a \succeq b \Longrightarrow \exists c \in L$, $c \neq a$, $a \succeq c \succeq b$. □

**1.3.16 Remark** People can find that Theorem **1.3.10** gives us some locally characterizations for the global property of completely distributive lattices, this point makes it very useful in the pointed approach of fuzzy topology and become the base of minimal set theory.

The embryonic form of minimal set appeared in Hutton's work [38] with "$\forall S \subset L$, $\bigvee S = b \Longrightarrow \exists s \in S$ such that $s \geq a$" replacing "$\forall S \subset L$, $\bigvee S \geq b \Longrightarrow \exists s \in S$ such that $s \geq a$" in Definition **1.3.1** and called "minimal family" there. Clearly it is weaker than minimal set, and was proved that could not be used to characterize completely distributivity, neither even finite distributivity of a complete lattice. In later years, Chinese scholars developed this motivation and made deep study on this topic,[67, 165] formed the theory of minimal set. Theorem **1.3.10**, its main part was completed by Guo-jun Wang,[168] and Theorem **1.3.8**, obtained for completely distributive lattices with an order-reversing involution by Yu-wei Peng and Dong-sheng Zhao independently,[125, 198] are two typical results in this theory.

Completely distributive lattices can also be characterized via continuous lattices.

For the notions and results on continuous lattice referred here, see [25].

**1.3.17 Definition** Let $L$ be a complete lattice, $a, b \in L$. We say $a$ is *way below* $b$, denoted by $a \ll b$, if for every up-directed set $S$ in $L$, $\bigvee S \geq b \Longrightarrow \exists s \in S$ such that $s \geq a$. Denote $\Downarrow a = \{b \in L: b \ll a\}$, $\Uparrow a = \{b \in L: a \ll b\}$.

**1.3.18 Theorem** *Let $L$ be a complete lattice, $a, b, c, d \in L$. Then*

(i) $a \preceq b \Longrightarrow a \ll b \Longrightarrow a \leq b$.

(ii) $a \leq b \ll c \leq d \Longrightarrow a \ll d$.

(iii) $\forall a \in L$, $\Downarrow a \in Idl(L)$.

(iv) $\forall a \in L$, $\beta(a) \subset \Downarrow a \subset \downarrow a$.

**Proof** (i) Apply the definition of $\ll$ to $S = \downarrow b$.

(ii) Straightforwardly from the definition.

(iii) By (ii), $\Downarrow a$ is a lower set. Clearly $0 \in \Downarrow a$, $\Downarrow a$ is not empty. $\forall b, c \in \Downarrow a$, then for every up-directed set $S \subset L$ such that $\bigvee S \geq a$, $\exists x, y \in S$ such that $x \geq b$, $y \geq c$. Since $S$ is up-directed, $\exists s \in S$ such that $s \geq x, y$, i.e. $s \geq x \vee y \geq b \vee c$. So $b \vee c \in \Downarrow a$, $\Downarrow a \in Idl(L)$.

(iv) By (i). □

**1.3.19 Definition** A *continuous lattice* is a complete lattice $L$ satisfying $\forall a \in L$, $\bigvee \Downarrow a = a$.

**1.3.20 Theorem** (Insertion Property of Continuous Lattice)

*Let $L$ be a continuous lattice, then $\forall a, b \in L$, $a \neq b$, $a \ll b \Longrightarrow \exists c \in L$, $c \neq a$, $a \ll c \ll b$.*

**Proof** Take $S = \{c \in L: \exists d \in L, c \ll d \ll b\}$. We intend to prove $S$ is an ideal. Clearly, $0 \in S$, $S$ is not empty. By Theorem **1.3.18** (ii), $S$ is a lower set. $\forall c_1, c_2 \in S$, $\exists d_1, d_2 \in L$ such that $c_1 \ll d_1 \ll b$, $c_2 \ll d_2 \ll b$. By Theorem **1.3.18** (ii) and (iii), $c_1, c_2 \ll d_1 \vee d_2 \in \Downarrow b$, $c_1 \vee c_2 \in \Downarrow (d_1 \vee d_2)$. Hence $c_1 \vee c_2 \in S$, $S$ is up-directed, $S \in Idl(L)$. Therefore, since $L$ is a continuous lattice,

$$\bigvee S = \bigvee\{\bigvee \Downarrow d: d \ll b\} = \bigvee \Downarrow b = b.$$

By $a \ll b$, $a \in S$, $\exists c_0 \in L$, $a \ll c_0 \ll b$.

If $c_0 \neq a$, then the proof is completed by taking $c = c_0$; otherwise, $c_0 < b$, by $\bigvee \Downarrow b = b$, $\exists c_1 \in \Downarrow b$, $c_1 \not\leq c_0$. By Theorem **1.3.18** (iii), $\exists c \in \Downarrow b$, $c \geq c_0, c_1$, so $a \ll c_0 < c$. By Theorem **1.3.18** (ii), $a \ll c \ll b$ and $a \neq c$. □

**1.3.21 Lemma** *Let $L$ be a continuous lattice, $a, b \in L$, $a \not\leq b$, then $\exists F \in Flt(L)$ such that*

(i) $F \supset \Uparrow a$, $F \cap \downarrow b = \emptyset$;

(ii) $\forall x \in L \backslash F$, *$L \backslash F$ has an irreducible and maximal element $\hat{x} \geq x$.*

**Proof** (i) Since $L$ is a continuous lattice, $\bigvee \Downarrow a = a$, $\exists c \in \Downarrow a$ such that $c \not\leq b$. Inductively apply insertion theorem for continuous lattice (Theorem **1.3.20**) on $c \ll a$, get a series $\{c_i : i < \omega\} \subset L$ such that

$$c \ll \cdots \ll c_1 \ll c_0 \ll a. \qquad \square$$

Take $F = \bigcup_{i<\omega} \uparrow c_i$, then clearly $F \in Flt(L)$, $F \supset \Uparrow a$. If $F \cap \downarrow b \neq \emptyset$, then $\exists i < \omega, \exists d \in \uparrow c_i \cap \downarrow b$. By Theorem **1.3.18** (i), $a \leq c_i \leq d \leq b$, $a \leq b$, a contradiction.

(ii) $\forall x \in L \backslash F$, $\{x\}$ is a chain in poset $L \backslash F$. So by the Kuratowski's Lemma, there exists a maximal chain $C$ in $L \backslash F$ containing $\{x\}$. Denote $\hat{x} = \bigvee C$, then certainly $\hat{x} \geq x$. If $\hat{x} \notin L \backslash F$, then $\hat{x} \in F$, $\exists i < \omega$ such that $c_{i+1} \ll c_i \leq \hat{x}$, $c_{i+1} \ll$

$\hat{x} = \bigvee C$, $\exists c \in C$, $c \geq c_{i+1}$, $c \in \uparrow c_{i+1} \subset F$. But $c \in C \subset L \backslash F$, this is a contradiction, it follows $\hat{x} \in L\backslash F$. By the maximality of the chain $C$ in $L\backslash F$ and $\hat{x} = \bigvee C$, $\hat{x}$ is a maximal element of $L\backslash F$.

Suppose $\hat{x} \notin pr(L)$, then since $F \in Flt(L)$, $1 \in F$, $\hat{x} \in L\backslash F$, $\hat{x} \neq 1$. So $\exists u, v \in L$ such that $\hat{x} = u \wedge v$ but $\hat{x} \neq u, v$. Hence $\hat{x} < u, v$. If both $u, v \in F$, then by $F \in Flt(L)$, we should have $\hat{x} = u \wedge v \in F$. But this contradicts what we have just proved, so $\{u, v\} \cap (L\backslash F) \neq \emptyset$. But this still contradicts the maximality of $\hat{x}$ in $L\backslash F$. Therefore, $\hat{x}$ can only be irreducible. □

**1.3.22 Theorem**

(i) *For every continuous lattice $L$, $ir(L)$ is a meet-generating set of $L$.*

(ii) *For every distributive continuous lattice $L$, $pr(L)$ is a meet-generating set of $L$.*

**Proof** (i) We prove $ir(L)$ satisfies the condition of Proposition **1.1.26** (iii). $\forall a, b \in L$, $a \not\leq b$. Take $x = b$ in Lemma **1.3.21**, we obtain $c \in ir(L)$ such that $c \geq b$, $c \notin \uparrow a$, i.e. $c \not\geq a$.

(ii) By (i) and Proposition **1.2.13** (iv). □

From Theorem **1.3.10**, reader can find the similarity between completely distributive lattice and continuous lattice. In fact, they are connected with each other in the following

**1.3.23 Theorem** *A distributive complete lattice $L$ is completely distributive if and only if both $L$ and $L^{op}$ are continuous lattices.*

**Proof** (Necessity) By Theorem **1.1.34**, Theorem **1.3.10** (i)$\Longleftrightarrow$(iii) and Theorem **1.3.18** (iv).

(Sufficiency) We prove $L$ satisfies condition (ii) in Theorem **1.3.10**. $\forall a \in L$, denote $D = M(\downarrow a)$. Since $L$ is a continuous lattice, by Theorem **1.3.18** (ii), the following equality holds:

$$D = \{\alpha \in M(L) : \exists c \in L, \alpha \leq c \ll a\} = \bigcup\{M(\Downarrow c) : c \in \downarrow a\}. \tag{1.7}$$

Since $L^{op}$ is a continuous lattice, by Theorem **1.3.22** (i), $M(L) = ir(L^{op})$ is a meet-generating set of $L^{op}$, i.e. a join-generating set of $L$, and hence by equality (1.7) and $L$ being a continuous lattice,

$$\bigvee D = \bigvee(\bigcup\{M(\Downarrow c) : c \in \downarrow a\}) = \bigvee\{\bigvee(M(\Downarrow c)) : c \in \downarrow a\} = \bigvee \downarrow a = a.$$

Now $\forall \alpha \in D$, $\forall S \subset L$, $\bigvee S \geq a$, denote $K = \{\bigvee F : F \in [S]^{<\omega}\}$, then $K$ is an up-directed set in $L$, $\bigvee K = \bigvee S \geq a$. Since $\alpha \ll a$, it follows $\exists c \in K$ such that $c \geq \alpha$, i.e. $\exists F \in [S]^{<\omega}$ such that $\alpha \leq c = \bigvee F$. Since $L$ is distributive, $\alpha \in M(L)$, by Proposition **1.2.13** (v), $\alpha$ is co-prime. Hence by $|F| < \omega$, $\exists s \in F \subset S$ such that $s \geq \alpha$. That is to say, $\alpha \preceq a$, $D \subset \beta(a)$. Combining equation $\bigvee D = a$ proved above, $D$ is a minimal set of $a$. □

**1.3.24 Theorem** *Let $L$ be a completely distributive lattice. Then*

(i) *$\beta : L \to \mathcal{P}(L)$ is a arbitrary join preserving mapping, i.e. for every $A \subset L$,*

$$\beta(\bigvee A) = \bigvee_{a \in A} \beta(a).$$

(ii) *$\beta^* : L \to \mathcal{P}(M(L))$ is a arbitrary join preserving mapping, i.e. for every $A \subset L$,*

$$\beta^*(\bigvee A) = \bigvee_{a\in A} \beta^*(a).$$

**Proof** (i) Suppose $c = \bigvee A$, $B = \bigcup_{a\in A} \beta(a)$, we need only prove that $B$ is the largest minimal set of $c$. If $C \subset L$ is a cover of $c$, then $C$ is a cover of every $a \in A$. So $\beta(a)$ refines $C$ for every $a \in A$, and hence $B$ refines $C$. Since

$$\bigvee B = \bigvee_{a\in A} (\bigvee \beta(a)) = \bigvee_{a\in A} a = c,$$

$B$ is a proper cover of $c$. By Theorem **1.3.4**, $B$ is a minimal set of $c$. If $D \subset L$ is a minimal set of $c$, then $D$ refines the cover $B$ of $c$, i.e. for every $d \in D$, there exists $b_d \in B = \bigcup_{a\in A} \beta(a)$ such that $d \le b_d$. Suppose $a \in A$ such that $b_d \in \beta(a)$, then by Theorem **1.3.6** (iii), $d \in \beta(a) \subset B$. So $D \subset B$, $B$ is the largest minimal set, i.e. $B = \beta(c)$.

(ii) By (i) proved above and Proposition **1.3.2**,

$$\beta^*(\bigvee A) = \beta(\bigvee A) \cap M(L) = (\bigcup_{a\in A} \beta(a)) \cap M(L) = \bigcup_{a\in A} (\beta(a) \cap M(L)) = \bigcup_{a\in A} \beta^*(a). \ \Box$$

To end this section, we introduce another characterization theorem of completely distributive lattice and some relative concepts and results. As for their proofs omitted here, see [44].

**1.3.25 Definition** Let $L$ be a lattice. $L$ is called a *topological lattice*, if $L$ is equipped with a topology such that both the binary meet operation $\wedge:\ L \times L \to L$ and the binary join operation $\vee:\ L \times L \to L$ are continuous with respect to the product topology of $L \times L$.

**1.3.26 Definition** Let $P$ be a poset.

$\forall a, b \in P$, $a \le b$. A *closed interval (determined by a, b) in* $P$ is a set

$$[a, b] = \{x \in P:\ a \le x \le b\}.$$

The *interval topology* on $P$ is the smallest topology on $P$ containing all the closed intervals in $P$ as closed subsets.

Recall the concepts of Hausdorff space and compact space in general topology, we have the following theorems; for their proofs, see [44]:

**1.3.27 Theorem**[44] *The interval topology on a complete lattice is compact.* □

**1.3.28 Theorem**[44] *Let L be a distributive complete lattice, then the following conditions are equivalent:*

(i) *$L$ is a completely distributive lattice.*

(ii) *$L$ is a Hausdorff topological lattice with respect to its interval topology.*

(iii) *There exists a compact Hausdorff topology $\mathcal{T}$ on $L$ such that $L$ is isomorphic to a closed sublattice of some power of* $[0,1]$ *with respect to* $\mathcal{T}$.

(iv) *$L$ is isomorphic to a sub-complete-lattice of some power of* $[0,1]$. □

# Chapter 2

# Fuzzy Topological Spaces

Fuzzy set theory offers us a new angle to observe and investigate the relation between sets and their elements other than traditional "black or white" way. It tells us besides "belonging to" and "not belonging to," more other possibilities exist in the relation between an element and a set emerging in various practical processes. This point of view certainly offers us a new framework of set theory, and then, in this new framework, we face the problems relating to topology, the study on them form the contents of fuzzy topology.

## 2.1 Fuzzy Sets and Mappings

The idea and the concept of fuzzy set were introduced by L.A. Zadeh in 1965,[193] used the unit interval [0,1] to describe and deal with non-crisp phenomena and procedures. Goguen generalized this concept with $L$-fuzzy sets.[26] So in the sequel, we consider fuzzy set based on [0,1] as a special case of general $L$-fuzzy set.

**2.1.1 Definition** Let $X$ be a nonempty ordinary set, $L$ a complete lattice.
An *$L$-fuzzy subset* on $X$ is a mapping $A: X \to L$, i.e. the family of all the $L$-fuzzy subsets on $X$ is just $L^X$ consisting of all the mappings from $X$ to $L$. $L^X$ here is called an *$L$-fuzzy space*, $X$ is called the *carrier domain* of each $L$-fuzzy subset on it, and $L$ is called the *value domain* of each $L$-fuzzy subset on $X$.

$A \in L^X$ is called a *crisp subset* on $X$, if there exists an ordinary subset $U \subset X$ such that $A = \chi_U : X \to \{0,1\} \subset L$, i.e. if $A$ is a charateristic function of some ordinary subset of $X$. For a family $\mathcal{A} \subset L^X$ of $L$-fuzzy subsets, denote the family of all the crisp subsets contained in $\mathcal{A}$ by $crs(\mathcal{A})$, and denote

$$[\mathcal{A}] = \{A \subset X : \chi_A \in crs(\mathcal{A})\}.$$

For every $L$-fuzzy subset $A \in L^X$, define its *support set* by $\{x \in X : A(x) > 0\}$, denoted by $supp(A)$.

An *$L$-fuzzy point* on $X$ is an $L$-fuzzy subset $x_a \in L^X$ defined as

$$\forall y \in X, \quad x_a(y) = \begin{cases} a, & y = x, \\ 0, & y \neq x. \end{cases}$$

For every $\mathcal{A} \subset L^X$, denote the set of all the $L$-fuzzy points on $X$ in $\mathcal{A}$ by $Pt(\mathcal{A})$; especially, the set of all the $L$-fuzzy points on $X$ is denoted by $Pt(L^X)$.

For an $L$-fuzzy point $x_a \in Pt(L^X)$, denote $ht(x_a) = a$, called the *height of $x_a$* or the *value of $x_a$* or the *membership of $x$*.

**2.1.2 Note** (1) Many authors still study fuzziness in the case $L = [0,1]$, and then use word "fuzzy" but not "$L$-fuzzy." So that this word "fuzzy" possesses two level of meanings: one means "[0, 1]-fuzzy" and an other means all the fuzzy cases including both $L$-fuzzy cases and [0, 1]-cases.

In the case that we need to emphasize the value domain $L = [0,1]$, we will often shorten the word "$L$-fuzzy" as "F-". and then "[0, 1]-fuzzy subset," "[0, 1]-fuzzy point" are called "F-subset," "F-point," etc.

(2) For a nonempty ordinary set $X$ and a complete lattice $L$, if equivalently replace the value set $\{0,1\} \subset [0,1]$ of all characteristic functions on $X$ by the two-point set $\{0,1\} \subset L$, we still have a natural one-to-one correspondence between $\mathcal{P}(X)$ and all these characteristic functions. So for convenient, **in the sequel, sometimes we will not distinguish an ordinary subset $A \subset X$ from its characteristic function $\chi_A: X \to \{0,1\} \subset L$, i.e. a crisp subset, if no confusion will be caused.** That is to say, in an investigation of $L$-fuzzy subsets, we will often equivalently consider and deal with an ordinary subset just as its characteristic function.

**2.1.3 Definition** Let $L^X$ be an $L$-fuzzy space, $A \in L^X$, $a \in L$. Define an $L$-fuzzy subset $aA$ by

$$aA(x) = a \wedge A(x),$$

called a *layer* of $A$, or more exactly, the *a-layer* of $A$.

For every ordinary subset $A \subset X$, we also denote $a\chi_A$ by $aA$ for short, unless some confusion will be caused. Particularly, denote $\underline{a} = aX$, or denote $\underline{a}_X = aX$ to emphasize the carrier domain $X$, and call it a *layer* of $X$, or the *a-layer* of $X$.

Since $L^X$ is also a product lattice, by Definition **1.2.31**, we naturally give out the following

**2.1.4 Definition** Let $L^X$ be an $L$-fuzzy space.

Define the partial order $\leq$ in $L^X$ by:

$$\forall A, B \in L^X, \quad A \leq B \Longleftrightarrow \forall x \in X,\ A(x) \leq B(x).$$

Particularly, for an $L$-fuzzy point $p \in Pt(L^X)$ and an $L$-fuzzy subset $A \in L^X$ such that $p \leq A$, we also say "$p$ is in $A$" and denote it as $p \in A$ sometimes.

Then we obtain the following proposition as a corollary of Theorem **1.2.33**:

**2.1.5 Proposition** *Let $L^X$ be an $L$-fuzzy space. Then*

(i) *$L^X$ is a complete lattice, and for every $\mathcal{A} \subset L^X$, the join $\bigvee \mathcal{A}$ and the meet $\bigwedge \mathcal{A}$ in $L^X$ fulfill the following relations:*

$$\forall x \in X, \quad (\bigvee \mathcal{A})(x) = \bigvee_{A \in \mathcal{A}} A(x), \quad (\bigwedge \mathcal{A})(x) = \bigwedge_{A \in \mathcal{A}} A(x).$$

(ii) *$L$ is distributive$\Longleftrightarrow$ $L^X$ is distributive.*

(iii) *$L$ satisfies* (IFD1)$\Longleftrightarrow$ *$L^X$ satisfies* (IFD1).

(iv) *$L$ satisfies* (IFD2)$\Longleftrightarrow$ *$L^X$ satisfies* (IFD2).

(v) *$L$ is completely distributive$\Longleftrightarrow$ $L^X$ is completely distributive.* □

**2.1.6 Definition** Let $L^X$ be an $L$-fuzzy space, $A \in L^X$, $a \in L$. Define the *a-level* (or *a-stratification*) of $A$ as the ordinary set $\{x \in X : A(x) \geq a\}$, denote it by $A_{[a]}$.

Denote

$$A_{(a)} = \{x \in X : A(x) \not\leq a\},$$
$$A^{[a]} = \{x \in X : A(x) \leq a\},$$
$$A^{(a)} = \{x \in X : A(x) \not\geq a\}.$$

Call every $A_{[a]}$ a *level* of $A$.

By this definition, the following conclusion is clear:

**2.1.7 Proposition** *Let $A$ be a L-fuzzy subset on X, then*

$$A = \bigvee_{a \in L} aA_{[a]}. \qquad \square$$

Therefore, an $L$-fuzzy subset can be naturally represented by some crisp sets $A_{[a]}$'s. Because of this reason, the preceding proposition is called "Decomposition Theorem of $L$-fuzzy Subset" by some authors.

Now what is about its converse problem? That is to say, for a family of ordinary sets, what conditions should it satisfy to form a family of levels of an $L$-fuzzy subset?

**2.1.8 Theorem** *Let $L_0$ be a subset of a complete lattice $L$, for every $\alpha \in L_0$, $A_\alpha$ be an ordinary set. Then the following conditions are equivalent:*

(i) *For every $D \subset L_0$ and every $\alpha \in L_0$ such that $\alpha \le \bigvee D$, $\bigcap\{A_\delta : \delta \in D\} \subset A_\alpha$.*

(ii) *For $B = \bigvee\{\alpha A_\alpha : \alpha \in L_0\}$ and every $\alpha \in L_0$, $B_{[\alpha]} = A_\alpha$.*

**Proof** (i)$\Longrightarrow$(ii): Clearly $B_{[\alpha]} \supset A_\alpha$ for every $\alpha \in L_0$. $\forall x \in B_{[\alpha]}$, take $D = \{\delta \in L_0 : x \in A_\delta\}$, then $\alpha \le B(x) = \bigvee D$. By the assumed condition, $x \in \bigcap\{A_\delta : \delta \in D\} \subset A_\alpha$. So $B_{[\alpha]} \subset A_\alpha$.

(ii)$\Longrightarrow$(i): Suppose $D \subset L_0$, $\alpha \in L_0$, $\alpha \le \bigvee D$. Then $\forall x \in \bigcap\{A_\delta : \delta \in D\} = \bigcap\{B_{[\delta]} : \delta \in D\}$, $B(x) \ge \bigvee D \ge \alpha$. So $x \in B_{[\alpha]} = A_\alpha$, $\bigcap\{A_\delta : \delta \in D\} \subset A_\alpha$. $\square$

The preceding result shows the condition for a family of ordinary sets to form levels of an $L$-fuzzy subset. But for an arbitrary family of ordinary sets, what can we do on it? The following lemma will deduce an approach in this aspect:

**2.1.9 Lemma** *Let $L$ be a complete lattice, $a \in L$. Then*

(i) *For every join-generating set $L_0$ of $L$, $\beta(a) \cap L_0$ is a minimal set of $a$ if and only $\beta(a)$ is a minimal set of $a$.*

(ii) *For every strictly join-generating set $L_0$ of $L$, $(\beta(a) \cap L_0)\backslash\{a\}$ is a minimal set of $a$ if and only $\beta(a)$ is a minimal set of $a$.*

**Proof** We only prove (ii), (i) is similar.

The necessity is clear. Suppose $\beta(a)$ is a minimal set of $a$, then since $L_0$ is a strictly join-generating set of $L$, by Theorem **1.3.6** (iii), $\downarrow\beta(a) = \beta(a)$,

$$\begin{aligned}\bigvee((\beta(a) \cap L_0)\backslash\{a\}) &= \bigvee((\downarrow\beta(a) \cap L_0)\backslash\{a\}) \\ &= \bigvee(\bigcup\{\downarrow c \cap L_0 : c \in \beta(a)\}\backslash\{a\}) \\ &\ge \bigvee(\bigcup\{(\downarrow c \cap L_0)\backslash\{c\} : c \in \beta(a)\}) \\ &= \bigvee\{\bigvee((\downarrow c \cap L_0)\backslash\{c\}) : c \in \beta(a)\} \\ &= \bigvee \beta(a) \\ &= a.\end{aligned}$$

By Theorem **1.3.6** (i), $\bigvee((\beta(a) \cap L_0)\backslash\{a\}) \le a$, so $\bigvee((\beta(a) \cap L_0)\backslash\{a\}) = a$, $(\beta(a) \cap L_0)\backslash\{a\}$ is a minimal set of $a$. $\square$

**2.1.10 Theorem** *Let $L$ be a completely distributive lattice, $L_0$ a join-generating set of $L$, for every $\alpha \in L_0$, $A_\alpha$ an ordinary set. If take $B_\gamma = \bigcap\{A_\alpha : \alpha \preceq \gamma, \alpha \in L_0\}$, then $B = \bigvee\{\gamma B_\gamma : \gamma \in L_0\}$ fulfills $B_{[\gamma]} = B_\gamma$, $\forall \gamma \in L_0$; particularly, if $\alpha \preceq \beta \Longrightarrow A_\alpha \supset A_\beta$, then $B = \bigvee\{\alpha A_\alpha : \alpha \in L_0\}$.*

**Proof** We need to prove the family $\{B_\gamma : \ \gamma \in L_0\}$ fulfills the condition (i) of Theorem **2.1.8**.

Suppose $D \subset L_0$, $\alpha \in L_0$, $\alpha \leq \bigvee D = d_0$. $\forall \gamma \in L_0$, denote $\beta_0(\gamma) = \beta(\gamma) \cap L_0$. By Theorem **1.3.10** and Lemma **2.1.9** (i), $\bigvee \beta_0(\gamma) = \gamma$. Since

$$\bigvee\{\delta : \ \delta \in \beta_0(\gamma), \ \gamma \in D\} = \bigvee\{\bigvee \beta_0(\gamma) : \ \gamma \in D\} = \bigvee D = d_0 \geq \alpha,$$

so $\forall \tau \in \beta_0(\alpha)$, $\tau \preceq \alpha$, $\exists \gamma \in D$, $\exists \delta \in \beta_0(\gamma)$ such that $\delta \geq \tau$. Since $\delta \preceq \gamma$, by Theorem **1.3.6** (ii), $\tau \preceq \gamma$, $\tau \in \beta_0(\gamma)$. That is to say,

$$\beta_0(\alpha) \subset \bigcup\{\beta_0(\gamma) : \ \gamma \in D\}. \tag{2.1}$$

By the definition,

$$B_\gamma = \bigcap\{A_\delta : \ \delta \in \beta_0(\gamma)\},$$

$$\bigcap\{B_\gamma : \ \gamma \in D\} = \bigcap\{A_\delta : \ \delta \in \bigcup\{\beta_0(\gamma) : \ \gamma \in D\}\},$$

so by relation (2.1),

$$B_\alpha = \bigcap\{A_\gamma : \ \gamma \in \beta_0(\alpha)\} \supset \bigcap\{A_\delta : \ \delta \in \bigcup\{\beta_0(\gamma) : \ \gamma \in D\}\} = \bigcap\{B_\gamma : \ \gamma \in D\}.$$

Hence the condition (i) of Theorem **2.1.8** is satisfied by $\{B_\gamma : \ \gamma \in L_0\}$, $B_{[\gamma]} = B_\gamma$ for every $\gamma \in L_0$.

Moreover, if we have $\alpha \preceq \beta \Longrightarrow A_\alpha \supset A_\beta$, denote $C = \bigvee\{\alpha A_\alpha : \ \alpha \in L_0\}$, then $\forall \gamma \in L_0$,

$$A_\gamma \subset \bigcap\{A_\alpha : \ \alpha \preceq \gamma, \ \alpha \in L_0\} = B_\gamma,$$

$$C = \bigvee\{\gamma A_\gamma : \ \gamma \in L_0\} \leq \bigvee\{\gamma B_\gamma : \ \gamma \in L_0\} = B.$$

On the other hand, suppose $x \in B_{[\gamma]}$, then by

$$B_{[\gamma]} = B_\gamma = \bigcap\{A_\alpha : \ \alpha \preceq \gamma, \ \alpha \in L_0\} = \bigcap\{A_\alpha : \ \alpha \in \beta_0(\gamma)\},$$

we have $x \in A_\alpha$ for every $\alpha \in \beta_0(\gamma)$. So by Lemma **2.1.9** (i),

$$\begin{aligned} C(x) &= \bigvee\{\alpha : \ \alpha \in L_0, \ x \in A_\alpha\} \\ &\geq \bigvee\{\alpha : \ \alpha \in \beta_0(\gamma), \ x \in A_\alpha\} \\ &= \bigvee \beta_0(\gamma) \\ &= \gamma, \end{aligned}$$

i.e. $x \in C_{[\gamma]}$. Hence $B_{[\gamma]} \subset C_{[\gamma]}$, $B \leq C$. Therefore, $C = B$. □

**2.1.11 Remark** Theorem **2.1.10** offers a method to form a $L$-fuzzy subset, i.e. a lattice-valued mapping, from a family of ordinary sets. Clearly, relation $\preceq$ plays an important role in it. In fact, there exist counterexamples to show that $\preceq$ in the definition of $B_\gamma$ in the preceding proposition cannot be replaced by $\leq$, neither by $<$.

**2.1.12 Definition** Let $L^X$, $L^Y$ be $L$-fuzzy spaces, $f : \ X \to Y$ an ordinary mapping. Based on $f : \ X \to Y$ (and adopting Rodabaugh's symbols[149]), define *$L$-fuzzy mapping* $f^{\to} : \ L^X \to L^Y$ and its *$L$-fuzzy reverse mapping* $f^{\leftarrow} : \ L^Y \to L^X$ by

$$f^{\to} : \ L^X \to L^Y, \ \ f^{\to}(A)(y) = \bigvee\{A(x) : \ x \in X, \ f(x) = y\}, \ \ \forall A \in L^X, \ \ \forall y \in Y,$$

$$f^{\leftarrow} : \ L^Y \to L^X, \ \ f^{\leftarrow}(B)(x) = B(f(x)), \ \ \forall B \in L^Y, \ \ \forall x \in X.$$

In this case, we say the ordindary mapping $f : \ X \to Y$ *produces* the correspondent $L$-fuzzy mapping $f^{\to} : \ L^X \to L^Y$, or say $f^{\to} : \ L^X \to L^Y$ *is induced from* $f : \ X \to Y$.

Symbol $f^{\to}$ and $f^{\leftarrow}$ always mean that $f^{\to}$ is an $L$-fuzzy mapping induced from an ordinary mapping $f$ and $f^{\leftarrow}$ is the $L$-fuzzy reverse mapping of $f^{\to}$.

Let $f : \ L^X \to L^Y$ be an ordinary mapping between two $L$-fuzzy spaces. We say $f$ is *$L$-fuzzy point preserving*, if $f$ maps every $L$-fuzzy point to an $L$-fuzzy point; say $f$ is *$L$-fuzzy point preserving with height*, if $f$ maps every $L$-fuzzy point to an $L$-fuzzy point with the same height; say $f$ is *crisp subset preserving*, if $f$ maps every crisp subset to a crisp subset; say $f$ is *layer preserving*, if $f(\underline{a}) = \underline{a}$ for every $a \in L$.

**2.1.13 Remark** Zadeh is the first author who defined fuzzy mapping in the way shown above,[193] so an $L$-fuzzy mapping in the meaning of Definition **2.1.12** is also called a "*Zadeh type mapping*" or a "*mapping of Zadeh type*". So a "Zadeh type mapping" or a "mapping of Zadeh type" means there exists an ordinary mapping producing it in the way mentioned above.

**2.1.14 Theorem** *Let $L^X$, $L^Y$ be $L$-fuzzy spaces, $f: X \to Y$ an ordinary mapping. Then for every $a \in L$ and every $A \in L^X$, $f^{\to}(aA) = af^{\to}(A)$.*

**Proof** $\forall a \in L$, $\forall A \in L^X$, $\forall y \in Y$, we have

$$\begin{aligned} f^{\to}(aA)(y) &= \bigvee\{(aA)(x) : x \in X, f(x) = y\} \\ &= \bigvee\{a \wedge (A(x)) : x \in X, f(x) = y\} \\ &= a \wedge \bigvee\{A(x) : x \in X, f(x) = y\} \\ &= a \wedge (f^{\to}(A)(y)) \\ &= (af^{\to}(A))(y). \end{aligned}$$

So $f^{\to}(aA) = af^{\to}(A)$. □

**2.1.15 Theorem** *Let $L^X$, $L^Y$ be $L$-fuzzy spaces, $f: X \to Y$ an ordinary mapping. Then*

(i) *$f^{\to}$ is arbitrary join preserving.*
(ii) *$f^{\to}$ is $L$-fuzzy point preserving; exactly, $f^{\to}(x_a) = f(x)_a$.*
(iii) *$f^{\to}$ is crisp subset preserving; exactly, $f^{\to}(\chi_A) = \chi_{f[A]}$, $\forall A \subset X$.*
(iv) *$f^{\to}$ is layer preserving.*
(v) *$\forall A \in L^X$, $f^{\to}(A) = \underline{0} \iff A = \underline{0}$.* □

**2.1.16 Remark** For an ordinary mapping $f: X \to Y$, easy to verify that there uniquely exists $F: \mathcal{P}(X) \to \mathcal{P}(Y)$, so-called the "*lifting of $f$*", such that $F$ is arbitrary union preserving and $F(\{x\}) = f(x)$ for every $x \in X$. In general, the lifting $F$ of $f$ does not preserve even finite intersections. $f^{\to}: L^X \to L^Y$ is equivalent to an ordinary lifting whenever we take $L$ as the two-point lattice $2 = \{0, 1\}$, so $f^{\to}$ is not arbitrary meet preserving, neither be finite meet preserving.

**2.1.17 Theorem** *Let $L^X$, $L^Y$ be $L$-fuzzy spaces, $f: X \to Y$ an ordinary mapping. Then*

(i) *$f^{\leftarrow}$ is arbitrary join preserving.*
(ii) *$f^{\leftarrow}$ is arbitrary meet preserving.*
(iii) *$f^{\leftarrow}$ is crisp subset preserving.*
(iv) *$f^{\leftarrow}$ is layer preserving.* □

**2.1.18 Remark** Note that $f^{\leftarrow}$ is not $L$-fuzzy point preserving whenever $f$ is not injective. In fact, $f^{\leftarrow}(y_b) = bf^{-1}(y)$.

**2.1.19 Proposition** *Let $L^X$, $L^Y$ be $L$-fuzzy spaces, $A \subset X$, $B \subset Y$, $f: X \to Y$ an ordinary mapping. Then*

(i) $f^{\to}(\chi_A) = \chi_{f[A]}$.
(ii) $f^{\leftarrow}(\chi_B) = \chi_{f^{-1}(B)}$. □

A natural question arising now is: "For an ordinary mapping between two $L$-fuzzy spaces, when is it of Zadeh type?" The following two theorems completely answer this question:

**2.1.20 Theorem** *Let $L^X$, $L^Y$ be L-fuzzy spaces, $f: L^X \to L^Y$ an ordinary mapping, then there exists a unique ordinary mapping $f_0: X \to Y$ such that $f = f_0^{\to}$ if and only if $f$ satisfies the following conditions:*

(i) *$f$ is arbitrary join preserving.*

(ii) *$f$ is L-fuzzy point preserving with height.*

**Proof** By Theorem **2.1.15**, the necessity is clear. We prove the sufficiency.

First of all, we prove there exists an ordinary mapping $f_0: X \to Y$ such that $\forall a \in L\backslash\{0\}$, $f(x_a) = (f_0(x))_a$. By (ii), $\forall x \in X$, $\forall a \in L\backslash\{0\}$, $\exists f_{0,a}(x) \in Y$ such that $f(x_a) = (f_{0,a}(x))_a$. But by (i), for a fixed $L$-fuzzy point $x_a$, we have

$$\begin{aligned}(f_{0,a}(x))_a \leq& \bigvee\{(f_{0,c}(x))_c : c \in L\backslash\{0\}\} = \bigvee\{f(x_c) : c \in L\backslash\{0\}\} \\ =& f(\bigvee\{x_c : c \in L\backslash\{0\}\}) = f(x_1) = (f_{0,1}(x))_1.\end{aligned}$$

Note that $(f_{0,1}(x))_1$ here is an $L$-fuzzy point, it turns out $f_{0,a}(x) = f_{0,1}(x) \in Y$, $\forall a \in L\backslash\{0\}$. So take $f_0(x) = f_{0,1}(x)$, we obtain an ordinary mapping $f_0: X \to Y$ such that

$$f(x_a) = (f_0(x))_a, \quad \forall x \in X,\ \forall a \in L\backslash\{0\}. \tag{2.2}$$

$\forall A \in L^X$, $\forall y \in Y$, by the conclusion just proved above and (i),

$$\begin{aligned} f(A)(y) =& f(\bigvee\{x_{A(x)} : x \in X\})(y) = (\bigvee\{f(x_{A(x)}) : x \in X\})(y) \\ =& (\bigvee\{(f_0(x))_{A(x)} : x \in X\})(y) = \bigvee\{((f_0(x))_{A(x)})(y) : x \in X\} \\ =& \bigvee\{A(x) : x \in X,\ f_0(x) = y\} = f_0^{\to}(A)(y), \end{aligned}$$

so $f = f_0^{\to}$, $f$ is induced from $f_0$.

As for the uniqueness of $f_0$, can be clearly deduced from relation (2.2). □

**2.1.21 Theorem** *Let $L^X$, $L^Y$ be L-fuzzy spaces, $g: L^Y \to L^X$ an ordinary mapping, then there exists a unique ordinary mapping $f: X \to Y$ such that $g = f^{\leftarrow}$ if and only if $g$ satisfies the following conditions:*

(i) *$g$ is arbitrary join preserving.*

(ii) *$g$ is finite meet preserving.*

(iii) *$g$ is crisp subset preserving.*

(iv) *$g$ is layer preserving.*

**Proof** Clearly, need only prove the sufficiency.

First of all, we shall prove

(1) $\forall y \in Y$, $\exists X_y \subset X$ such that $\forall a \in L$, $g(y_a) = aX_y$,

(2) $X$ is just the disjoint union of all the $X_y$'s.

By (iii), $\forall y \in Y$, $\exists X_y \subset X$ such that $g(y_1) = \chi_{X_y}$. By (ii) and (iv), $\forall a \in L$, denote the mapping taking consistent value $a$ by $\underline{a}$, we have

$$g(y_a) = g(\underline{a} \wedge y_1) = g(\underline{a}) \wedge g(y_1) = \underline{a} \wedge \chi_{X_y} = aX_y,$$

(1) is proved.

$$\forall y, z \in Y,\ y \neq z, \text{ by (ii) and (i)},$$

$$\chi_{X_y} \wedge \chi_{X_z} = g(y_1) \wedge g(z_1) = g(y_1 \wedge z_1) = g(\underline{0}) = g(\bigvee \emptyset) = \bigvee \emptyset = \underline{0}.$$

Hence $X_y \cap X_z = \emptyset$, $\{X_y : y \in Y\}$ is a disjoint family. On the other hand, by (ii),

$$\bigvee_{y \in Y} \chi_{X_y} = \bigvee_{y \in Y} g(y_1) = g(\bigvee_{y \in Y} y_1) = g(\chi_Y) = g(\bigwedge \emptyset) = \bigwedge \emptyset = \underline{1},$$

so $\bigcup_{y \in Y} X_y = Y$, (2) is true.

Now define $f: X \to Y$ by $f(x) = y$, $\forall x \in X_y$. By (2) proved above, this is well-defined. Therefore, $\forall B \in L^Y$, $\forall x \in X$, suppose $x \in X_y$, for $f^{\leftarrow}: L^Y \to L^X$, by (1), (2) and (i), we have

$$f^{\leftarrow}(B)(x) = B(f(x)) = B(y) = (B(y)\chi_{X_y})(x) = (\bigvee_{z\in Y} B(z)\chi_{X_z})(x)$$
$$= (\bigvee_{z\in Y} g(z_{B(z)}))(x) = g(\bigvee_{z\in Y} z_{B(z)})(x) = g(B)(x).$$

So $f^{\leftarrow} = g$.

If there exists another mapping $h: X \to Y$ such that $g = h^{\leftarrow}$, then $\forall x \in X$, $\forall y \in Y$, $\forall a \in L\backslash\{0\}$,

$$y_a(h(x)) = h^{\leftarrow}(y_a))(x) = (g(y_a))(x) = (f^{\leftarrow}(y_a))(x) = y_a(f(x)).$$

So $h^{-1}[\{y\}] = f^{-1}[\{y\}]$, $h = f$. □

**2.1.22 Theorem** *Let $L^X$ and $L^Y$ be L-fuzzy spaces, $f: X \to Y$ an ordinary mapping. Then*

(i) *$f^{\rightarrow}$ is injective if and only if $f$ is injective.*
(ii) *$f^{\rightarrow}$ is injective if and only if $f^{\leftarrow}\circ f^{\rightarrow} = id_{L^X}: L^X \to L^X$.*
(iii) *$f^{\rightarrow}$ is surjective if and only if $f$ is surjective.*
(iv) *$f^{\rightarrow}$ is surjective if and only if $f^{\rightarrow}\circ f^{\leftarrow} = id_{L^Y}: L^Y \to L^Y$.*
(v) *$f^{\rightarrow}$ is bijective if and only if $f$ is bijective.*
(vi) *$f^{\leftarrow}$ is bijective if and only if $f^{\leftarrow}\circ f^{\rightarrow} = id_{L^X}$, $f^{\rightarrow}\circ f^{\leftarrow} = id_{L^Y}$, i.e. $f^{\leftarrow}$ is just the reverse of $f^{\rightarrow}$, $(f^{\rightarrow})^{-1} = f^{\leftarrow}$.*

**Proof** (i) If $f^{\rightarrow}$ is injective, $\forall x, y \in X$, by Theorem **2.1.15** (ii),

$$f(x) = f(y) \Longrightarrow f(x)_1 = f(y)_1 \Longrightarrow f(x_1) = f(y_1) \Longrightarrow x_1 = y_1 \Longrightarrow x = y,$$

$f$ is injective.

Suppose $f$ is injective. $\forall A, B \in L^X$, if $f^{\rightarrow}(A) = f^{\rightarrow}(B)$, then $\forall x \in X$,

$$A(x) = f^{\rightarrow}(A)(f(x)) = f^{\rightarrow}(B)(f(x)) = B(x),$$

so $A = B$, $f^{\rightarrow}$ is injective.

(ii) The sufficiency is clear. Suppose $f^{\rightarrow}$ is injective, then by (i), $f$ is injective, $\forall A \in L^X$, $\forall x \in X$,

$$(f^{\leftarrow}\circ f^{\rightarrow})(A)(x) = f^{\rightarrow}(A)(f(x)) = \bigvee\{A(z): z \in X, f(z) = f(x)\} = A(x),$$
$$f^{\leftarrow}\circ f^{\rightarrow}(A) = A, \quad f^{\leftarrow}\circ f^{\rightarrow} = id_{L^X}.$$

(iii) If $f^{\rightarrow}$ is surjective, $\forall y \in Y$, then $\exists A \in L^X$ such that $f^{\rightarrow}(A) = y_1$. By Theorem **2.1.15** (i), (ii),

$$\bigvee_{x\in X}(f(x))_{A(x)} = \bigvee_{x\in X} f^{\rightarrow}(x_{A(x)}) = f^{\rightarrow}(\bigvee_{x\in X} x_{A(x)}) = f^{\rightarrow}(A) = y_1.$$

So $\exists x \in X$ such that $f(x) = y$, $f$ is surjective.

Suppose $f$ is surjective, $B \in L^Y$. Take $A = \bigvee_{x\in X} x_{B(f(x))} \in L^X$, then by Theorem **2.1.15** (i), (ii) and $f$ being surjective,

$$f^{\rightarrow}(A) = \bigvee_{x\in X}(f(x))_{B(f(x))} = \bigvee_{y\in Y} y_{B(y)} = B,$$

$f^{\rightarrow}$ is surjective.

(iv) The sufficiency is clear. Suppose $f^{\rightarrow}$ is surjective, then by (iii), $f$ is surjective, $\forall B \in L^Y$, $\forall y \in Y$,

$$(f^{\rightarrow}\circ f^{\leftarrow})(B)(y) = \bigvee\{f^{\leftarrow}(B)(x): x \in X, f(x) = y\} = \bigvee\{B(f(x)): x \in X, f(x) = y\} = B(y),$$
$$f^{\rightarrow}\circ f^{\leftarrow}(B) = B, \quad f^{\rightarrow}\circ f^{\leftarrow} = id_{L^Y}.$$

(v) An implication of (i) and (iii).

(vi) An implication of (ii) and (iv). □

**2.1.23 Theorem** *Let $L^X$, $L^Y$ and $L^Z$ be L-fuzzy spaces, $f: X \to Y$ and $g: Y \to Z$ be orindary mappings. Then*

(i) $g^{\to} f^{\to} = (gf)^{\to}$.

(ii) $f^{\leftarrow} g^{\leftarrow} = (gf)^{\leftarrow}$.

**Proof** (i) $\forall A \in L^X$, $\forall z \in Z$, then

$$\begin{aligned} g^{\to} f^{\to}(A)(z) &= \bigvee\{f^{\to}(A)(y) : y \in Y, g(y) = z\} \\ &= \bigvee\{\bigvee\{A(x) : x \in X, f(x) = y\} : y \in Y, g(y) = z\} \\ &= \bigvee\{A(x) : x \in X, gf(x) = z\} \\ &= (gf)^{\to}(A)(z), \end{aligned}$$

so (i) is true.

(ii) $\forall C \in L^Z$, $\forall x \in X$, then

$$f^{\leftarrow} g^{\leftarrow}(C)(x) = (g^{\leftarrow}(C))(f(x)) = C(g(f(x))) = C((gf)(x)) = (gf)^{\leftarrow}(C)(x),$$

so (ii) is true. □

Is an $L$-fuzzy reverse mapping of an $L$-fuzzy mapping a Zadeh type one? In general, it is not, except in a very special case:

**2.1.24 Theorem** *Let $L^X$, $L^Y$ be L-fuzzy spaces, $f: X \to Y$. Then there exists an ordinary mapping $g: Y \to X$ such that $f^{\leftarrow} = g^{\to}$ if and only if $f$ is bijective and $g = f^{-1}$.*

**Proof** (Sufficiency) Suppose $f$ is bijective and $g = f^{-1}$, then by Theorem **2.1.23** (i),

$$g^{\to} f^{\to} = (gf)^{\to} = (id_X)^{\to} = id_{L^X},$$
$$f^{\to} g^{\to} = (fg)^{\to} = (id_Y)^{\to} = id_{L^Y}.$$

So by Theorem **2.1.22** (v) and (vi), $g^{\to} = (f^{\to})^{-1} = f^{\leftarrow}$.

(Necessity) Suppose there exists $g: Y \to X$ such that $f^{\leftarrow} = g^{\to}$. $\forall x \in X$, for the $L$-fuzzy point $f(x)_1 \in L^Y$, we should have

$$\bigvee\{f(x)_1(y) : y \in Y, g(y) = x\} = g^{\to}(f(x)_1)(x) = f^{\leftarrow}(f(x)_1)(x) = f(x)_1(f(x)) = 1.$$

So it must be $g(f(x)) = x$, otherwise $\bigvee\{f(x)_1(y) : y \in Y, g(y) = x\} = 0$, a contradiction. To show that $g$ maps only $f(x)$ to $x$, take $(f(x)_1)' \in L^Y$, then

$$\begin{aligned} \bigvee\{(f(x)_1)'(y) : y \in Y, g(y) = x\} &= g^{\to}(f(x)_1)'(x) = f^{\leftarrow}(f(x)_1)'(x) \\ &= (f(x)_1)'(f(x)) = 0. \end{aligned}$$

So for any point $y \in Y$, $y \neq f(x)$, $g(y) \neq x$, otherwise $\bigvee\{(f(x)_1)'(y) : y \in Y, g(y) = x\} = 1$, will cause a contradiction. □

Moreover, consider posets $L^X$ and $L^Y$ as categories, $f^{\to}$ and $f^{\leftarrow}$ just constitute a so-called "*adjunction*" as follows (for this categorical concept, see [33] or [1]):

**2.1.25 Theorem** *Let $L^X$, $L^Y$ be L-fuzzy spaces, $f: X \to Y$ an ordinary mapping. Then*

(i) $\forall A \in L^X$, $f^{\leftarrow} f^{\to}(A) \geq A$.

(ii) $\forall B \in L^Y$, $f^{\to} f^{\leftarrow}(B) \leq B$.

(iii) $\forall A \in L^X$, $f^{\to} f^{\leftarrow} f^{\to}(A) = f^{\to}(A)$.

(iv) $\forall B \in L^Y$, $f^{\leftarrow} f^{\to} f^{\leftarrow}(B) = f^{\leftarrow}(B)$. □

## 2.2 Fuzzy Topological Spaces

**2.2.1 Definition** A completely distributive lattice $L$ is called a *F-lattice*, if $L$ has an order-reversing involution $': L \to L$.

Let $X$ be nonempty ordinary set, $L$ a F-lattice, $'$ the order-reversing involution on $L$. $\forall A \in L^X$, $\forall \mathcal{B} \subset L^X$, use the order-reversing involution $'$ to define an operation $'$ on $L^X$ by:

$$A'(x) = (A(x))', \quad \forall x \in X;$$

also define:

$$\mathcal{B}' = \{B' : B \in \mathcal{B}\}.$$

Call $': L^X \to L^X$ the *pseudo-complementary operation* on $L^X$, $A'$ the *pseudo-complementary set of* $A$ *in* $L^X$.

**2.2.2 Proposition** *Let $X$ be a nonempty ordinary set, $L$ a F-lattice, then the pseudo-complementary operation $': L^X \to L^X$ is an order-reversing involution.* □

**2.2.3 Remark** By the preceding proposition, propositions **2.1.5** and **1.2.16**, the pseudo-complementary operation satifies the De Morgan's Law (DM1) and (DM2). But, in general, an order-reversing involution $'$ on a F-lattice $L$ needs not be a complementary operation defined in Definition **1.2.14**, and therefore, the correspondent pseudo-complementary operation needs not be a complementary operation. That is to say, maybe $A \wedge A' \neq \underline{0}$, $A \vee A' \neq \underline{1}$.

**2.2.4 Proposition** *Let $X$, $Y$ be ordinary sets, $L$ a F-lattice, $f: X \to Y$ an ordinary mapping, then for every $A \in L^X$, $f^{\to}(A') \geq (f^{\to}(B))'$.*

**Proof** Since $\forall y \in Y$,

$f(A')(y) = \bigvee\{A'(x): f(x) = y\} \geq \bigwedge\{A'(x): f(x) = y\} = (\bigvee\{A(x): f(x) = y\})' = f(A)'(y)$,

so $f^{\to}(A') \geq (f^{\to}(B))'$. □

**2.2.5 Proposition** *Let $X$, $Y$ be ordinary sets, $L$ a F-lattice, $f: X \to Y$ an ordinary mapping, then $f^{\leftarrow}: L^Y \to L^X$ preserves the pseudo-complementary operation, i.e. $\forall B \in L^Y$, $f^{\leftarrow}(B') = (f^{\leftarrow}(B))'$.* □

**2.2.6 Definition** Let $X$ be a nonempty ordinary set, $L$ a F-lattice, $\delta \subset L^X$. $\delta$ is called a *L-fuzzy topology* on $X$, and $(L^X, \delta)$ is called an *L-fuzzy topological space*, or *L-fts* for short, if $\delta$ satisfies the following three conditions:

(LFT1) $\underline{0}, \underline{1} \in \delta$;

(LFT2) $\forall \mathcal{A} \subset \delta$, $\bigvee \mathcal{A} \in \delta$;

(LFT3) $\forall U, V \in \delta$, $U \wedge V \in \delta$.

Particularly, when $L = [0,1]$, call an $L$-fuzzy topological space $(L^X, \delta)$ a *F-topological space* or a *F-ts* for short, and simply denote it by $(X, \delta)$.

Every element in $\delta$ is called an *open subset* in $L^X$, every pseudo-complementary set of an open subset, i.e. every element in $\delta'$ is called a *closed subset* in $L^X$.

Sometimes, we need to compaire more than one $L$-fuzzy topologies on a common carrier domain $X$, so we define as follows:

**2.2.7 Definition** Let $X$ be a nonempty ordinary set, $L$ a F-lattice, $\delta_0, \delta_1$ two $L$-fuzzy topologies on $X$. We say $\delta_0$ is *coarser* than $\delta_1$, or say $\delta_1$ is finer than $\delta_0$, if $\delta_0 \subset \delta_1$.

Certainly, for a family $\mathcal{T}$ of $L$-fuzzy topologies on $X$ and $\delta_0 \in \mathcal{T}$, call $\delta_0$ the *coarsest* $L$-fuzzy topology in $\mathcal{T}$ if $\delta_0$ is coarser than any other one in $\mathcal{T}$; call $\delta_0$ is the *finest* $L$-fuzzy topology in $\mathcal{T}$ if $\delta_0$ is finer than any other one in $\mathcal{T}$.

**2.2.8 Example** Let $X$ be a nonempty ordinary set, $L$ a F-lattice.

(i) Take $\delta = \{\underline{0}, \underline{1}\} \subset L^X$, then $\delta$ is clearly an $L$-fuzzy topology on $X$. We call this $L$-fuzzy topology $\delta$ the *trivial $L$-fuzzy topology* on $X$, and call the correspondent $L$-fts $(L^X, \delta)$ a *trivial $L$-fts.*

Obviously, the trivial $L$-fuzzy topology on $X$ is the coarsest one.

(ii) Take $\delta = L^X$, then $\delta$ is an $L$-fuzzy topology on $X$, called the *discrete $L$-fuzzy topology* on $X$, and call the correspondent $L$-fts $(L^X, \delta)$ a *discrete $L$-fts.*

By (LFT2), every element of the discrete $L$-fuzzy topology on $X$ is a join of elements in the family $Pt(L^X)$ of all the $L$-fuzzy points on $X$.

The discrete $L$-fuzzy topology on $X$ is clearly the finest one.

(iii) Take $\delta = \{\underline{a} : \ a \in L\} \subset L^X$, then $\delta$ is an $L$-fuzzy topology on $X$.

(iv) Suppose $\mathcal{T}$ is an ordinary topology on $X$, then $\delta = \{\chi_U : \ U \in \mathcal{T}\} \subset L^X$ is an $L$-fuzzy topology on $X$.

Usually, $L$-fts' are defined with open subsets; but dually, we can also define $L$-fts' with closed subsets as follows:

**2.2.9 Definition** Let $X$ be a nonempty ordinary set, $L$ a F-lattice, $\eta \subset L^X$.

$\eta$ is called a *$L$-fuzzy co-topology* on $X$, and $(L^X, \eta)$ is called an *$L$-fuzzy topological space*, or *$L$-fts* for short, if $\eta$ satisfies the following three conditions:

(LFT1$'$) $\underline{0}, \ \underline{1} \in \eta$;

(LFT2$'$) $\forall \mathcal{A} \subset \eta, \ \bigwedge \mathcal{A} \in \eta$;

(LFT3$'$) $\forall P, Q \in \eta, \ P \vee Q \in \eta$.

Every element in $\eta$ is called an *closed subset* in $L^X$, every pseudo-complementary set of an closed subset, i.e. every element in $\eta'$ is called a *open subset* in $L^X$.

**2.2.10 Note** Obviously, since open subsets and closed subsets are correspondent to each other via the order-reversing involution on a F-lattice, definitions of $L$-fts with topology and co-topology are equivalent. In the sequel, "$L$-fts $(L^X, \delta)$" or other similar words without special illustration will always mean that $\delta$ is an $L$-fuzzy topology on $X$.

In the case that $L$-fts $(L^X, \eta)$ defined with $L$-fuzzy co-topology $\eta$, we will emphatically indicate that $\eta$ in $(L^X, \eta)$ is an $L$-fuzzy co-topology. Then it naturally means $\eta'$ is the $L$-fuzzy topology of $(L^X, \eta)$.

**2.2.11 Remark** (1) Take $L$ as the two-point lattice $\{0, 1\}$, then in the view point of lattice theory, $L^X \cong 2^X \cong \mathcal{P}(X)$, so ordinary subsets can be equivalently considered as a special case of $L$-fuzzy subsets. Therefore, ordinary topologies and ordinary topological spaces become special cases of $L$-fuzzy topologies and $L$-fuzzy topological spaces respectively, $L$-fuzzy topological space is a generalization of ordinary topological space.

(2) The preceding definition of $L$-fts, of course, including F-ts, bases on C.L. Chang's definition[15] and is widely adopted, because considering the definition of ordinary topological space, it is natural. However, R. Lowen has another different

definition of F-ts in his a series of deep investigations on F-ts:[92]–[103] For a nonempty ordinary set $X$, $\delta \subset [0,1]^X$ is called a fuzzy topology on $X$, if $\delta$ satisfies previous conditions (LFT2), (LFT3) and the following

(i′) $\forall a \in [0,1]$, $\underline{a} \in \delta$, (FT1′)

i.e. a fuzzy topology on $X$ must contains every constant mapping from $X$ to $[0,1]$, or in another word, all the layers of $X$. His main consideration for this definition of fuzzy topology is to make category of F-ts and continuous fuzzy mappings satisfy the *Terminal Separator Property* of a originally defined topological category,[33] i.e. to make every singleton $\{x\}$ can only be defined with exactly one fuzzy topology.

This Terminal Separator Property has been removed from the recent definition of topological cateory.[1]

(3) There is another interesting definition of fuzzy topology in the literature. The basic idea of which is to assign to every fuzzy set a "degree of openness." Briefly say, fuzzy topology on $X$ is a mapping from $[0,1]^X$ to $[0,1]$ fulfilling certain conditions.[151, 190, 191, 192]

**2.2.12 Proposition** *Let $(L^X, \delta)$ be an L-fts. Then*

(i) *$crs(\delta)$ is an L-fuzzy topology on $X$.*

(ii) *$[\delta]$ is an ordinary topology on $X$.* □

**2.2.13 Definition** Let $(L^X, \delta)$ be an $L$-fts, $\delta_0 \subset \delta$.

$\delta_0$ is called a *base* of $\delta$, if

$$\delta = \{\bigvee \mathcal{A} : \ \mathcal{A} \subset \delta_0\}.$$

$\delta_0$ is called a *subbase* of $\delta$, if the family

$$\{\bigwedge \mathcal{B} : \ \mathcal{B} \in [\delta_0]^{<\omega} \backslash \{\emptyset\}\}$$

is a base of $\delta$.

Say $\delta$ is *generated* by the $\delta_0$, if $\delta_0$ is a base or subbase of $\delta$.

For a nonempty ordinary set $X$, a F-lattice $L$ and a family $\mathcal{B} \subset L^X$, denote by $Grt(\mathcal{B})$ the $L$-fuzzy topology on $X$ generated by $\mathcal{B}$, i.e.

$$Grt(\mathcal{B}) = \{\bigvee\{\bigwedge \mathcal{F} : \ \mathcal{F} \in \mathcal{G}\} : \ \mathcal{G} \subset [\mathcal{B}]^{<\omega} \backslash \{\emptyset\}\}.$$

Dually define "*base,*" "*subbase,*" "*generated*" for $L$-fuzzy co-topology.

Also call a base or a subbase of an $L$-fuzzy topology a *open base* or a *open subbase* respectively; call a base or a subbase of an $L$-fuzzy co-topology a *closed base* or a *closed subbase* respectively.

A very natural quetion is: "How to judge some $L$-fuzzy subsets just form a base or a subbase of some $L$-fuzzy topology?" We have the following simple rules:

**2.2.14 Theorem** *Let $X$ be a nonempty ordinary set, $L$ a F-lattice, $\mathcal{A} \subset L^X$. Then*

(i) *$\mathcal{A}$ is a base of just one L-fuzzy topology on $X$ if and only if $\bigvee \mathcal{A} = \underline{1}$ and $\mathcal{A}$ is closed under binary meets.*

(ii) *$\mathcal{A}$ is a subbase of just one L-fuzzy topology on $X$ if and only if $\bigvee \mathcal{A} = \underline{1}$.*

**Proof** (i) Since a base is always a part of a topology, so the necessity is clear.

Suppose $\mathcal{A}$ satisfies the given conditions, denote $\delta = \{\bigvee \mathcal{C} : \ \mathcal{C} \subset \mathcal{A}\}$, then by $\bigvee \mathcal{A} = \underline{1}$ we have $\underline{1} \in \delta$. Certainly, take $\mathcal{C} = \emptyset \subset \mathcal{A}$, we have $\underline{0} = \bigvee \mathcal{C} \in \delta$. Clearly, $\delta$ is closed under arbitrary joins. Since $L$ and then $L^X$ are completely distributive, so infinitely distributive, by the given condition that $\mathcal{A}$ is closed under binary meets, one can easily check that $\delta$ is closed under nonempty meets. Therefore, $\delta$ is an $L$-fuzzy topology on $X$ generated by the base $\mathcal{A}$. The uniqueness of generated topology is obvious.

(ii) By (i). □

**2.2.15 Definition** Let $(L^X, \delta)$ be an $L$-fts, $A \in L^X$.

Define the *interior of* $A$ as the join of all the open subsets contained in $A$, denote it by $int(A)$ or $A^\circ$, or $int_\delta(A)$ to emphasize the $L$-fuzzy topology $\delta$ in which this interior is produced.

Define the *closure of* $A$ as the meet of all the closed subsets containing $A$, denote it by $cl(A)$ or $A^-$, or $cl_\delta(A)$ to emphasize the $L$-fuzzy topology $\delta$ in which this closure is produced.

**2.2.16 Remark** By condition (LFT2) of an $L$-fuzzy topology, the interior of an $L$-fuzzy set $A \in L^X$ is just the largest open subset contained in $A$, and by the dual of (LFT2), the closure of $A \in L^X$ is just the smallest closed subsets containing $A$. So $A$ is open if and only if $A^\circ = A$, $A$ is closed if and only if $A^- = A$.

**2.2.17 Theorem** *Let* $(L^X, \delta)$ *be an L-fts. Then*

(i) $\underline{0}^\circ = \underline{0}$, $\underline{1}^\circ = \underline{1}$.

(ii) $\forall A \in L^X$, $A^\circ \leq A$.

(iii) $\forall A \in L^X$, $A^{\circ\circ} = A^\circ$.

(iv) $\forall A, B \in L^X$, $A \leq B \Longrightarrow A^\circ \leq B^\circ$.

(v) $\forall A, B \in L^X$, $(A \wedge B)^\circ = A^\circ \wedge B^\circ$.

**Proof** (i) to (iv): Directly from the definition of interior.

(v): By (iv), $(A \wedge B)^\circ \leq A^\circ, B^\circ$, so $(A \wedge B)^\circ \leq A^\circ \wedge B^\circ$. Reversely, by Theorem **2.2.17** (ii), $A^\circ \wedge B^\circ \leq A \wedge B$; by Theorem **2.2.17** (iv) and (iii), $A^\circ \wedge B^\circ = (A^\circ \wedge B^\circ)^\circ \leq (A \wedge B)^\circ$. □

Dually produce the following theorem:

**2.2.18 Theorem** *Let* $(L^X, \delta)$ *be an L-fts. Then*

(i) $\underline{0}^- = \underline{0}$, $\underline{1}^- = \underline{1}$.

(ii) $\forall A \in L^X$, $A \leq A^-$.

(iii) $\forall A \in L^X$, $A^{--} = A^-$.

(iv) $\forall A, B \in L^X$, $A \leq B \Longrightarrow A^- \leq B^-$.

(v) $\forall A, B \in L^X$, $(A \vee B)^- = A^- \wedge B^-$. □

In an $L$-fts, open subsets and closed subsets are correspondent to each other family via the pseudo-complementary operation; moreover, pseudo-complementary operation can be "commuted" with interior operator and closure operator in the following sense:

**2.2.19 Theorem** *Let* $(L^X, \delta)$ *be an L-fts. Then*

(i) $\forall A \in L^X$, $A^{\circ\prime} = A^{\prime -}$.

(ii) $\forall A \in L^X$, $A^{\prime\circ} = A^{-\prime}$.

(iii) $\forall A \in L^X$, $A^\circ = A^{\prime -\prime}$.

(iv) $\forall A \in L^X$, $A^- = A^{\prime\circ\prime}$.

**Proof** (i): Certainly $A^\circ \leq A$, so by Proposition **2.2.2**, $A' \leq A^{\circ\prime}$. Since $A^{\circ\prime}$ is closed, so by Theorem **2.2.18** (iv), $A^{\prime -} \leq A^{\circ\prime -} = A^{\circ\prime}$. Reversely, by Theorem **2.2.18** (ii),

$A' \le A'^-$, by Proposition **2.2.2**, $A'^{-\prime} \le A'' = A$. Since $A'^-$ is closed, so $A'^{-\prime}$ is open, by the definition of interior, $A'^{-\prime} \le A^\circ$. Using Proposition **2.2.2** again, we get $A^{\circ\prime} \le A'^{-\prime\prime} = A'^-$.

(ii) - (iv): By Proposition **2.2.2**, take the pseudo-complements of the both sides of (i), we get (iii). Replace $A$ in (iii) with $A'$, we get (ii). Replace $A$ in (i) with $A'$, we get (iv). □

To an $L$-fuzzy set $A$ in an $L$-fts $(L^X, \delta)$, repeatedly map it in finit number of times with interior operator, closure operator and pseudo-complementary operator, a series of various $L$-fuzzy subsets will be produced. But will this series contain infinite number of different elements if we infinitely last this procedure? In general topology, we know the answer is: "There will be at most 14 different subsets can be produced by repeatedly acting a subset in finite number of times with interior operator, closure operator and pseudo-complementary operator." Now this conclusion is still true in far more general $L$-fuzzy topological spaces:

**2.2.20 Theorem** (Kuratowski's 14 Sets Theorem) *Let $(L^X, \delta)$ be an L-fts, $A \in L^X$, $R$ the set of all the finite compositions of interior operator, closure operator and pseudo-complementary operator, where empty composition, as a finite one, is defined as the identity operator. Then $|\{A^r : r \in R\}| \le 14$.*

**Proof** By Theorem **2.2.19** (iii), interior operator can be replaced by a finite composition of closure operator and pseudo-complementary operator. On the other hand, by Proposition **2.2.2**, pseudo-complementary operator composed with itself is the identity. So take $S$ as the set of all the operators on $L^X$ which is alternately composed in finite number of times with closure operator and pseudo-complementary operator, and also defined the empty composition as the identity operator, we have

$$\{A^s : s \in S\} = \{A^r : r \in R\}.$$

Now we prove

$$A^{-\circ-\circ} = A^{-\circ}, \quad A^{\circ-\circ-} = A^{\circ-}. \tag{2.3}$$

Since $A^{-\circ-} \ge A^{-\circ}$, so $A^{-\circ-\circ} \ge A^{-\circ}$. For another direction, we have $A^{-\circ} \le A^-$, so $A^{-\circ-} \le A^{--} = A^-$, and hence $A^{-\circ-\circ} \le A^{-\circ}$, it follows $A^{-\circ-\circ} = A^{-\circ}$. Similarly prove the second equation.

From the definition of $S$, all the elements in $\{A^s : s \in S\}$ can be listed as follows:

$$A,$$

$$A',\ A'^-,\ A'^{-\prime},\ A'^{-\prime-},\ A'^{-\prime-\prime},\ A'^{-\prime-\prime-},\ A'^{-\prime-\prime-\prime},\ \cdots, \tag{2.4}$$

$$A^-,\ A^{-\prime},\ A^{-\prime-},\ A^{-\prime-\prime},\ A^{-\prime-\prime-},\ A^{-\prime-\prime-\prime},\ \cdots. \tag{2.5}$$

There have been 14 sets here. We will prove that any other set appearing in the series later than these 14 sets must be one of them. In series (2.4), the 8th set is $A'^{-\prime-\prime-\prime-}$, by Theorem **2.2.19** (iii) and equalities (2.3),

$$A'^{-\prime-\prime-\prime-} = (((A'^{-\prime})^-)'^{-\prime})^- = A^{\circ-\circ-} = A^{\circ-} = A'^{-\prime-},$$

that is the 4th set in series (2.4). Therefore, its succesor, the 9th set in series (2.4), must be the 5th set in series (2.4), the next one, the 10th, must be the 6th set in series (2.4), $\cdots$, and so on. So series (2.4) contains at most 7 different sets as it has been shown.

As for series (2.5), the 7th set should be $A^{-\prime-\prime-\prime-}$. So by Theorem **2.2.19** (iii),

$$A^{-\prime-\prime-\prime-}=((((A^{-})^{\prime-\prime})^{-})^{\prime-\prime})^{\prime} = A^{-\circ-\circ\prime} = A^{-\circ\prime} = A^{-\prime-\prime\prime} = A^{-\prime-},$$

just be the 3th set in series (2.5). Similar to the preceding investigation, we can prove that series (2.5) contains at most 6 different sets as it has shown. □

Theorems **2.2.17** and **2.2.18** show the main properties interior operator and closure operator possessing. Reversely, some of these properties can also describe these two operators:

**2.2.21 Theorem** *Let $X$ be a nonempty ordinary set, $L$ a F-lattice, $i: L^X \to L^X$ an operator on $L^X$ satisfying the following conditions:*

(i) $i(\underline{1}) = \underline{1}$,

(ii) $\forall A \in L^X,\ i(A) \leq A$,

(iii) $\forall A, B \in L^X,\ i(A \wedge B) = i(A) \wedge i(B)$,

*then*

$$\delta = \{A \in L^X :\ i(A) = A\}$$

*is an L-fuzzy topology on $X$. Moreover, if the operator $i$ also fulfills*

(iv) $\forall A \in L^X,\ i(i(A)) = i(A)$,

*then with the L-fuzzy topology $\delta$ defined above, in L-fts $(L^X, \delta)$, $A^\circ = i(A)$ for every $A \in L^X$.*

**Proof** By (i), $\underline{1} \in \delta$. By (ii), $i(\underline{0}) \leq \underline{0}$, so $i(\underline{0}) = \underline{0}$, $\underline{0} \in \delta$, $\delta$ fulfills (LFT1).

$\forall \{A_t :\ t \in T\} \subset \delta$. By (iii), $i$ is order preserving (see Proposition **1.1.20** (iii)), so by (ii) and $A_s \leq \bigvee_{t \in T} A_t$ for every $s \in T$,

$$i(\bigvee_{t\in T} A_t) \leq \bigvee_{t\in T} A_t = \bigvee_{s\in T} i(A_s) \leq i(\bigvee_{t\in T} A_t),$$

$$i(\bigvee_{t\in T} A_t) = \bigvee_{t\in T} A_t,\quad \bigvee_{t\in T} A_t \in \delta,$$

$\delta$ fulfills (LFT2).

(LFT3) is guaranteed by (iii).

Suppose $i$ also fulfills (iv). $\forall A \in L^X$, since $A^\circ \in \delta = \{C \in L^X :\ i(C) = C\}$, so $i(A^\circ) = A^\circ$. By (iii), $i$ is order preserving, so $A^\circ = i(A^\circ) \leq i(A)$. Reversely, by (iv), $i(i(A)) = i(A)$, so

$$i(A) \leq \bigvee\{C \in L^X :\ i(C) = C \leq A\} = A^\circ,$$

$$A^\circ = i(A).$$ □

Similarly we have

**2.2.22 Theorem** *Let $X$ be a nonempty ordinary set, $L$ a F-lattice, $c: L^X \to L^X$ an operator on $L^X$ satisfying the following conditions:*

(i) $c(\underline{0}) = \underline{0}$,

(ii) $\forall A \in L^X,\ A \leq c(A)$,

(iii) $\forall A, B \in L^X,\ c(A \vee B) = c(A) \vee c(B)$,

*then*

$$\delta = \{A \in L^X :\ c(A') = A'\}$$

*is an L-fuzzy topology on $X$. Moreover, if the operator $c$ also fulfills*

(iv) $\forall A \in L^X,\ c(c(A)) = c(A)$,

*then with the L-fuzzy topology $\delta$ defined above, in L-fts $(L^X, \delta)$, $A^- = c(A)$ for every $A \in L^X$.* □

By these two theorems **2.2.21** and **2.2.22**, we can introduce the following notions:

**2.2.23 Definition** Let $X$ be a nonempty ordinary set, $L$ a F-lattice, $i, c: L^X \to L^X$ mappings on $L^X$.

$i$ is called an *interior operator* on $L^X$, if it fulfills the following conditions:

(IO1) $i(\underline{1}) = \underline{1}$.

(IO2) $\forall A \in L^X,\ i(A) \leq A$.

(IO3) $\forall A, B \in L^X,\ i(A \wedge B) = i(A) \wedge i(B)$.

(IO4) $\forall A \in L^X,\ i(i(A)) = i(A)$.

For an interior operator $i$ on $L^X$, define the $L$-fuzzy topology *generated by* $i$ as

$$\delta = \{A \in L^X : \ i(A) = A\}.$$

$c$ is called a *closure operator* on $L^X$, if it fulfills the following conditions:

(CO1) $c(\underline{0}) = \underline{0}$.

(CO2) $\forall A \in L^X,\ A \leq c(A)$.

(CO3) $\forall A, B \in L^X,\ c(A \vee B) = c(A) \vee c(B)$.

(CO4) $\forall A \in L^X,\ c(c(A)) = c(A)$.

For a closure operator $c$ on $L^X$, define the $L$-fuzzy topology *generated by* $c$ as

$$\delta = \{A \in L^X : \ c(A') = A'\}.$$

Then, with these definitions, by theorems **2.2.17**, **2.2.18**, **2.2.21** and **2.2.22**, we obtain the following two theorems:

**2.2.24 Theorem** *Let $X$ be a nonempty ordinary set, $L$ a F-lattice, $\mathcal{I}$ the family of all the interior operators on $L^X$, $\mathcal{T}$ the family of all the L-fuzzy topologies on $X$. Then*

$$f: \mathcal{I} \to \mathcal{T}, \quad f(i) = \{A \in L^X : \ i(A) = A\}$$

*is a bijection, and its reverse is just*

$$f^{-1}: \mathcal{T} \to \mathcal{I}, \quad f^{-1}(\delta) = int_\delta. \qquad \square$$

**2.2.25 Theorem** *Let $X$ be a nonempty ordinary set, $L$ a F-lattice, $\mathcal{C}$ the family of all the closure operators on $L^X$, $\mathcal{T}$ the family of all the L-fuzzy topologies on $X$. Then*

$$f: \mathcal{C} \to \mathcal{T}, \quad f(c) = \{A \in L^X : \ c(A') = A'\}$$

*is a bijection, and its reverse is just*

$$f^{-1}: \mathcal{T} \to \mathcal{I}, \quad f^{-1}(\delta) = cl_\delta. \qquad \square$$

## 2.3 Multiple Choice Principle and Neighborhood Structure

**2.3.1 Definition** Let $(L^X, \delta)$ be an $L$-fts. $\forall x_a \in Pt(L^X)$, $\forall A, B \in L^X$.

We say $x_a$ *quasi-coincides with* $A$, or say $x_a$ *is quasi-coincident with* $A$, denoted by $x_a \lessdot A$, if $x_a \not\leq A'$ or $a \not\leq A(x)'$; say $A$ *quasi-coincides with* $B$ *at* $x$, or say $A$ *is quasi-coincident with* $B$ *at* $x$, if $A(x) \not\leq B'(x)$; say $A$ *quasi-coincides with* $B$, or say $A$ *is quasi-coincident with* $B$, if $A$ quasi-coincides with $B$ at some point $x \in X$.

Let $\hat{q}$ denote the relation "quasi-coincides with" or "is quasi-coincident with." Then "$A$ quasi-coincides with $B$ at $x$" or "$A$ is quasi-coincident with $B$ at $x$" is denoted by "$A\hat{q}B$ at $x$," and "$A$ quasi-coincides with $B$" or "$A$ is quasi-coincident with $B$" is denoted by "$A\hat{q}B$".

Relation "does not quasi-coincide with" or "is not quasi-coincident with" is denoted by $\neg\hat{q}$.

Denote the set of all the points in $X$, at which $A\hat{q}B$, by $A\mathbin{⩕}B$, i.e.

$$A\mathbin{⩕}B = \{x \in X : \ x_{A(x)} \lessdot B\}.$$

**2.3.2 Remark** Certainly, in the preceding definition, just as between two $L$-fuzzy subsets, we can also denote "$x_a$ quasi-coincides with $A$" or "$x_a$ is quasi-coincident with $A$" by $x_a\hat{q}A$. But we still use the special symbol $\lessdot$ to denote this kind of relation to emphasize that $x_a$ is an $L$-fuzzy point.

**2.3.3 Proposition** *Let $(L^X, \delta)$ be an L-fts, $A, B, C \in L^X$, $\{A_t : \ t \in T\} \subset L^X$, $x \in X$, $a \in L\backslash\{0\}$. Then*

(i) $A\mathbin{⩕}B = B\mathbin{⩕}A$
$= \{x \in X : \ x_{A(x)} \lessdot B\}$
$= \{x \in X : \ x_{B(x)} \lessdot A\}$
$= \{x \in X : \ A(x) \not\leq B(x)'\}$
$= \{x \in X : \ B(x) \not\leq A(x)'\}.$

(ii) $A\hat{q}B$ *at* $x \Longleftrightarrow B\hat{q}A$ *at* $x \Longleftrightarrow x \in A\mathbin{⩕}B \Longleftrightarrow x \in B\mathbin{⩕}A.$

(iii) $A\hat{q}B \Longleftrightarrow B\hat{q}A \Longleftrightarrow A\mathbin{⩕}B \neq \emptyset \Longleftrightarrow B\mathbin{⩕}A \neq \emptyset \Longleftrightarrow A \not\leq B' \Longleftrightarrow B \not\leq A'.$

(iv) $A \leq B \Longrightarrow A\mathbin{⩕}C \subset B\mathbin{⩕}C.$

(v) $A\mathbin{⩕} \bigvee_{t\in T} A_t = \bigcup_{t\in T} A\mathbin{⩕}A_t.$

(vi) $A\hat{q} \bigvee_{t\in T} A_t \Longleftrightarrow \exists t \in T, \quad A\hat{q}A_t.$

(vii) $A \leq B, \ C\hat{q}A \Longrightarrow C\hat{q}B.$

(viii) $A \leq B, \ x_a \lessdot A \Longrightarrow x_a \lessdot B.$ □

**2.3.4 Proposition** *Let $(L^X, \delta)$, $(L^Y, \mu)$ be L-fts', $A, B \in L^X$, $C, D \in L^Y$, $f : X \to Y$ an ordinary mapping. Then*

(i) $A\hat{q}f^{\leftarrow}(C) \Longleftrightarrow f^{\rightarrow}(A)\hat{q}C.$

(ii) $A\hat{q}B \Longrightarrow f^{\rightarrow}(A)\hat{q}f^{\rightarrow}(B).$

(iii) $f^{\leftarrow}(C)\hat{q}f^{\leftarrow}(D) \Longrightarrow C\hat{q}D.$

**Proof** (i) By Proposition **2.2.5** and Theorem **2.1.25** (ii),

$$\begin{aligned} A\neg\hat{q}f^{\leftarrow}(C) &\Longrightarrow A \leq f^{\leftarrow}(C)' = f^{\leftarrow}(C') \\ &\Longrightarrow f^{\rightarrow}(A) \leq f^{\rightarrow}f^{\leftarrow}(C') \leq C' \\ &\Longrightarrow f^{\rightarrow}(A)\neg\hat{q}C. \end{aligned}$$

By Theorem **2.1.25** (i) and Proposition **2.2.5**,

$$\begin{aligned} f^{\rightarrow}(A)\neg\hat{q}C &\Longrightarrow f^{\rightarrow}(A) \leq C' \\ &\Longrightarrow A \leq f^{\leftarrow}f^{\rightarrow}(A) \leq f^{\leftarrow}(C') = f^{\leftarrow}(C)' \\ &\Longrightarrow A\neg\hat{q}f^{\leftarrow}(C). \end{aligned}$$

So $A\hat{q}f^{\leftarrow}(C) \Longleftrightarrow f^{\rightarrow}(A)\hat{q}C.$

(ii): By the definition of $f^{\rightarrow}$,

$$\begin{aligned} A\hat{q}B &\Longrightarrow \exists x \in X, \ A(x) \not\leq B(x)' \\ &\Longrightarrow \exists x \in X, \ f^{\rightarrow}(A)(f(x)) = \bigvee\{A(z) : z \in X, f(z) = f(x)\} \\ &\qquad\qquad \not\leq \bigwedge\{B(z)' : z \in X, f(z) = f(x)\} \\ &\qquad\qquad = (\bigvee\{B(z) : z \in X, f(z) = f(x)\})' \\ &\qquad\qquad = f^{\rightarrow}(B)(f(x))' \\ &\Longrightarrow f^{\rightarrow}(A)\hat{q}f^{\rightarrow}(B). \end{aligned}$$

(iii): By the definition of $f^{\leftarrow}$,

$$\begin{aligned} f^{\leftarrow}(C)\hat{q}f^{\leftarrow}(D) &\Longrightarrow \exists x \in X,\ C(f(x)) = f^{\leftarrow}(C)(x) \not\leq f^{\leftarrow}(D)(x)' = D(f(x))' \\ &\Longrightarrow C \not\leq D' \\ &\Longrightarrow C\hat{q}D. \end{aligned}$$

□

It's not hard to find that (ii) and (iii) mentioned above cannot be reversed, because they will not be true even in ordinary set theory. However, we still give out the following examples to show this point:

**2.3.5 Example** (i) Take $X = \{x^0, x^1\}$, $Y = \{y\}$, $L = [0,1]$, $f : X \to Y$, $f(x^0) = f(x^1) = y$, $A = x^0{}_1$, $B = x^1{}_1$. Then $f^{\to}(A) = f^{\to}(B) = y_1$, $f^{\to}(A)\hat{q}f^{\to}(B)$, but $A\neg\hat{q}B$.

(ii) Take $X = \{x^0, x^1\}$, $Y = \{y^0, y^1, y^2\}$, $L = [0,1]$, $C = y^0{}_1 \vee y^1{}_1$, $D = y^1{}_1 \vee y^2{}_1$, $f : X \to Y$, $f(x^0) = y^0$, $f(x^1) = y^2$. Then $f^{\leftarrow}(C) = x^0{}_1$, $f^{\leftarrow}(D) = x^1{}_1$, $C\hat{q}D$, but $f^{\leftarrow}(C)\neg\hat{q}f^{\leftarrow}(D)$.

**2.3.6 Definition** Let $(L^X, \delta)$ be an $L$-fts, $\eta$ an $L$-fuzzy co-topology on $X$, $x_a \in Pt(L^X)$.

$U \in \delta$ is called a *neighborhood* of $x_a$ in $(L^X, \delta)$, if $x_a$ belongs to $U$, i.e. $x_a \in U$ or $x_a \leq U$. The family of all the neighborhoods of $x_a$ in $(L^X, \delta)$ is called the *neighborhood system* of $x_a$, denoted by $\mathcal{N}_\delta(x_a)$ or $\mathcal{N}(x_a)$ for short.

$U \in \delta$ is called a *quasi-coincident neighborhood* of $x_a$ in $(L^X, \delta)$, shortened as *Q-neighborhood,* if $x_a \lessdot U$, i.e. $x_a$ quasi-coincides with $U$. The family of all the Q-neighborhoods of $x_a$ in $(L^X, \delta)$ is called the *Q-neighborhood system* of $x_a$, denoted by $\mathcal{Q}_\delta(x_a)$ or $\mathcal{Q}(x_a)$ for short.

In $L$-fts $(L^X, \eta)$ with $L$-fuzzy co-topology $\eta$, $P \in \eta$ is called a *remote neighborhood* of $x_a$, shortened as *R-neighborhood,* if $x_a \not\leq P$, i.e. $x_a$ is not contained in $P$. The family of all the R-neighborhoods of $x_a$ in $(L^X, \eta)$ is called the *R-neighborhood system* of $x_a$, denoted by $\mathcal{R}_\eta(x_a)$ or $\mathcal{R}(x_a)$ for short.

In $L$-fts $(L^X, \delta)$ with $L$-fuzzy topology $\delta$, we call $P \in L^X$ a R-neighborhood of $x_a$ if $P$ is a R-neighborhood of $x_a$ in the $L$-fts $(L^X, \delta')$ with $L$-fuzzy co-topology $\delta'$, i.e. $P \in \mathcal{R}_{\delta'}(x_a)$. Therefore, in $L$-fts $(L^X, \delta)$ with $L$-fuzzy topology $\delta$, the shortened symbol $\mathcal{R}(x_a)$ will denote the family of all the R-neighborhoods of $x_a$ in $L$-fts $(L^X, \delta')$ with $L$-fuzzy co-topology $\delta'$, i.e. $\mathcal{R}(x_a) = \mathcal{R}_{\delta'}(x_a)$.

**2.3.7 Remark** In general topology, the neighborhood structure of points can be decomposed as

"*Structure of open sets + Membership relation between points and sets.*"

In $L$-fuzzy topology, if we regard the ordinary "belonging to" as the membership relation, we obtain the neighborhood structure called "neighborhood system" mentioned above. This structure apeared very early in the study of $L$-fuzzy topology,[177] and encountered many difficulties. For example, the notions of accumulation point, derived set, net and compactness based on this membership relation will cause some preposterous results, such as "every closed subset must be crisp," and so on. [80, 168] In some years, this problem was an obstacle in the study of fuzzy topology. It was breaken by Ying-ming Liu in 1977. In the case of $L = [0,1]$, he introduced the concept of quasi-coincide neighborhood which overcomes these problems well and provides a stable fundamental neighborhood structure to fuzzy topology. [137]

In fact, the cuase of these shortages is the failure of the following fundamental "multiple choice principle" on this membership relation:

**Multiple Choice Principle:** *If a point has relation $\lhd$ with the union of a family of sets then it has also $\lhd$ with one of these sets.*

This principle is a very simple and fundamental fact in set theory. But in $L$-fuzzy set theory, it does not hold for the relation "belonging to." For example, $\forall x \in X$, $\forall n = 2, 3, 4, \cdots$, $x_{1-\frac{1}{n}} \in I^X$, $x_1 \in x_1 = \bigvee\{x_{1-\frac{1}{n}} : n \geq 2\}$, but $\forall n \geq 2$, $x_1 \notin x_{1-\frac{1}{n}}$. By this example we know that it is just the level structure of $L$-fuzzy subsets that violates the multiple choice principle for the membership relation "belonging to."

Does there exist a reasonable membership relation $\lhd$ between $L$-fuzzy points and $L$-fuzzy subsets? What conditions should be fulfilled? The following *Memship Relation Determination Principles:*[73] are produced as a group of standards:

(1) **Extension Principle:** *Restricted to the ordinary set theory, membership $\lhd$ will become the ordinary belonging relation. Precisely, if $p$ and $A$ are an ordinary point and an ordinary subset respectively, then $p \lhd A$ iff $p \in A$.*

(2) **Range Determination Principle:** *The fact that $x_\lambda \lhd A$ or not is completely determined by a system of formula about $\lambda$ and $A(x)$ expressed in terms of the order relation and involution in $L$.*

(3) **Maximum and Minimum Principle:** *For every fuzzy point $p$, $p \lhd \underline{0}$ and $p \lhd \underline{1}$; where $\underline{0}$ and $\underline{1}$ are the smallest $L$-fuzzy subset and the largest one on $X$ respectively.*

(4) **Multiple-choice Principle:** *For every family $\{A_t : t \in T\}$ of $L$-fuzzy subsets on $X$, the "Multiple Choice Principle" holds, i.e. the following implication holds:*

$$x_\lambda \lhd \bigvee\{A_t : t \in T\} \Longrightarrow \exists t \in T,\ x_\lambda \lhd A_t.$$

In Theorem **2.3.8**,[73] Ying-ming Liu proved that quasi-coincidence relation is just the unique membership relation satisfying all these four principles.

**2.3.8 Theorem** *For every F-lattice $L$, the membership relation satisfying the "Membership Relation Determination Principles" uniquely exists and is just the quasi-coincidence relation $\lessdot$.*

**Proof** By the Range Determination Principle, the reasonale membership relation, if exists, must be one of the following eight relations:

$$\left.\begin{array}{llll} (1)\ \lambda \leq A(x), & (2)\ \lambda \not\leq A(x), & (3)\ \lambda \leq A(x)', & (4)\ \lambda \not\leq A(x)', \\ (5)\ \lambda \geq A(x), & (6)\ \lambda \not\geq A(x), & (7)\ \lambda \geq A(x)', & (8)\ \lambda \not\geq A(x)'. \end{array}\right\} \quad (2.6)$$

By the Extension Principle, relations (2), (3), (6) and (8) in relation group (2.6) should be excluded. By the Multiple Choice Principle, relation (1) in group (2.6) must be excluded. In the Maximum and Minimum Principle, take $p = x_{\frac{1}{2}}$, then if we take $\lhd$ as relation (5) in group (2.6), $p \lhd \underline{1}$ will be false; if we take $\lhd$ as relation (7) in group (2.6), $p \lhd \underline{0}$ will not hold. So the only possible reasonable membership relation is relation (4) in group (2.6).

It can be verified that this relation satisfies the Membership Relation Determination Principles indeed and, it is just the quasi-coincidence relation. □

**2.3.9 Remark** (1) A quasi-coincident neighborhood of an $L$-fuzzy point is not necessary to contain the $L$-fuzzy point itself. A similar neighborhood structure which may excludes the corresponding point was investigated by M.Frechet in 1916, and was summarized to the (V)-space theory by W.Sierpinski.[154] But a set does not intersect with its complement there, which as a basic fact cannot be preserved in $L$-fuzzy sets.

(2) The notion of R-neighborhood was introduced by Guo-jun Wang.[161] Obviously, for a F-lattice $L$, R-neighborhoods are just the pseudo-complements of Q-neighborhoods. But since the definition of R-neighborhood does not involve any involution, it can be applied to more general situations. For example, Wang has developed a theory of pointed topology on completely distributive lattices making use of R-neighborhoods.[160] The concerning discussion on the theory can also see Bibliography [31, 32].

Recall the definition of "molecule": in Definition **1.2.12**, the set of all the molecules in a subset $A$ of a lattice $L$ is defined as the set of all the join-irreducible elements of $L$ in $A$ and denoted by $M(A)$. Following this definition, we can describe the set of all the molecules of $L^X$ in a family $\mathcal{A} \subset L^X$ as follows:

**2.3.10 Theorem** *Let $L^X$ be an L-fuzzy space, $\mathcal{A} \subset L^X$, then*

$$M(\mathcal{A}) = \{x_\lambda \in Pt(\mathcal{A}) : \lambda \in M(L)\}. \qquad \square$$

By this theorem, one can directly deduce the following conclusion from Theorem **1.2.33** (viii) and Corollary **1.3.12**:

**2.3.11 Corollary** *Let $X$ be a nonempty ordinary set, $L$ a completely distributive lattice. Then*

(i) *$M(L^X)$ is a join-generating set of $L^X$, i.e. $\bigvee M(\downarrow A) = A$ for every $A \in L^X$.*

(ii) *$\forall A, B \in L^X$, $A \leq B$ if and only if $M(\downarrow A) \subset M(\downarrow B)$.* $\square$

**2.3.12 Theorem** *Let $(L^X, \delta)$ be an L-fts. Then*

(i) *$\forall x_a \in Pt(L^X)$, $\mathcal{N}(x_a)$ is a down-directed in $L^X$ and $\underline{0} \notin \mathcal{N}(x_a)$.*

(ii) *$\forall x_\lambda \in M(L^X)$, $\mathcal{Q}(x_\lambda)$ is a down-directed set in $L^X$ and $\underline{0} \notin \mathcal{Q}(x_a)$.*

(iii) *$\forall x_\lambda \in M(L^X)$, $\mathcal{R}(x_\lambda)$ is a directed set in $L^X$ and $\underline{1} \notin \mathcal{R}(x_a)$.*

**Proof** (i) is clear, (iii) is dual to (ii), so we need only prove (ii).

$\forall U, V \in \mathcal{Q}(x_\lambda)$, then $x_\lambda \not\leq U', V'$. If $x_\lambda \leq U' \vee V'$, then by $x_\lambda \in M(L^X)$, we should have $x_\lambda \leq U'$ or $x_\lambda \leq V'$, contradicts with $U, V \in \mathcal{Q}(x_\lambda)$. So $x_\lambda \not\leq U' \vee V' = (U \wedge V)'$, $U \wedge V \in \mathcal{Q}(x_\lambda)$, $\mathcal{Q}(x_\lambda)$ is a down-directed set in $L^X$.

Since $x_\lambda \not\leq \underline{0}$, so $\underline{0} \notin \mathcal{Q}(x_\lambda)$. $\square$

The definition of base describes it with its global property. Now with local structures in an $L$-fts, we can view it from the angle of neighborhood structures:

**2.3.13 Theorem** *Let $(L^X, \delta)$ be an L-fts, $\mathcal{A} \subset \delta$. Then the following conditions are equivalent:*

(i) *$\mathcal{A}$ is a base of $\delta$.*

(ii) $\forall x_a \in Pt(L^X)$, $\forall U \in \mathcal{N}_\delta(x_a)$, $\forall \lambda \in \beta^*(a)$, $\exists A \in \mathcal{A} \cap \mathcal{N}_\delta(x_\lambda)$, $A \leq U$.

(iii) $\forall x_\lambda \in M(L^X)$, $\forall U \in \mathcal{Q}_\delta(x_\lambda)$, $\exists A \in \mathcal{A} \cap \mathcal{Q}_\delta(x_\lambda)$, $A \leq U$.

(iv) $\forall x_\lambda \in M(L^X)$, $\forall P \in \mathcal{R}_\delta(x_\lambda)$, $\exists Q \in \mathcal{A}' \cap \mathcal{R}_\delta(x_\lambda)$, $Q \geq P$.

**Proof** (i)$\Longrightarrow$(ii): $\forall x_a \in Pt(L^X), \forall U \in \mathcal{N}_\delta(x_a)$, $\forall \lambda \in \beta^*(a)$, then by (i), $\exists \mathcal{A}_0 \subset \mathcal{A}$ such that $U = \bigvee \mathcal{A}_0$. Since $U \in \mathcal{N}_\delta(x_a)$, so $\lambda \leq a \leq U(x) = \bigvee\{A(x) : A \in \mathcal{A}\}$. By Theorem **1.3.10** (iv) and $\lambda \in \beta^*(a)$, $\exists A \in \mathcal{A}_0$ such that $\lambda \leq A(x)$. That is to say, $A \in \mathcal{A} \cap \mathcal{N}_\delta(x_\lambda)$. By $\bigvee \mathcal{A}_0 = U$ and $A \in \mathcal{A}_0$, $A \leq U$.

(ii)$\Longrightarrow$(iii): $\forall x_\lambda \in M(L^X)$, $\forall U \in \mathcal{Q}_\delta(x_\lambda)$, take $a = U(x)$, then $U \in \mathcal{N}_\delta(x_a)$. By (ii), $\forall \mu \in \beta^*(a)$, $\exists A_\mu \in \mathcal{A} \cap \mathcal{N}_\delta(x_\mu)$ such that $A_\mu \leq U$. By Theorem **1.3.10** (iv), $\bigvee \delta^*(a) = a = U(x)$, so $\bigvee\{A_\mu(x) : \mu \in \beta^*(a)\} = U(x)$. By Proposition **2.2.2** and De Morgan's Law (DM1) and (DM2), since $U \in \mathcal{Q}_\delta(x_\lambda)$, $\lambda \not\leq U(x)' = \bigwedge\{A_\mu(x)' : \mu \in \beta^*(a)\}$. So $\exists \mu \in \beta^*(a)$ such that $\lambda \not\leq A_\mu(x)'$. That is to say, $A_\mu \in \mathcal{A} \cap \mathcal{Q}_\delta(x_\lambda)$ and $A_\mu \leq U$.

(iii)$\Longrightarrow$(iv): Every R-neighborhood is just a pseudo-complementary set of a Q-neighborhood in $(L^X, \delta)$.

(iv)$\Longrightarrow$(i): $\forall U \in \delta$. Denote $\mathcal{A}_0 = \{A \in \mathcal{A} : A \leq U\}$, then we need only prove $\bigvee \mathcal{A}_0 = U$. Suppose this is not true, then $\exists x \in X$ such that $(\bigvee \mathcal{A}_0)(x) < U(x)$. Denote $a = (\bigvee \mathcal{A}_0)(x)$, then by Proposition **2.2.2** and De Morgan's Law, $a' > U(x)'$. By Corollary **1.3.12** (i), $\bigvee M(\downarrow a') = a'$, $\exists \lambda \in M(\downarrow a')$ such that $\lambda \not\leq U(x)'$, $U' \in \mathcal{R}_\delta(x_\lambda)$. By (iv), $\exists Q \in \mathcal{A}' \cap \mathcal{R}_\delta(x_\lambda)$ such that $Q \geq U'$. So $Q' \leq U$, $Q' \in \mathcal{A}_0$, $Q \in (\mathcal{A}_0)'$. But $Q \in \mathcal{R}_\delta(x_\lambda)$, $x_\lambda \not\leq Q(x)$, this contradicts with $\lambda \leq a' = (\bigvee \mathcal{A}_0)(x)' = (\bigwedge(\mathcal{A}_0)')(x) \leq Q(x)$. □

We can also consider structures similar to base in a "local area," so we make

**2.3.14 Definition** Let $(L^X, \delta)$ be an $L$-fts, $p \in Pt(L^X)$.

A subfamily $\mathcal{A} \subset \mathcal{N}(p)$ is called a *neighborhood base* of $p$, if for every $U \in \mathcal{N}(p)$, there exists $V \in \mathcal{A}$ such that $V \leq U$.

A subfamily $\mathcal{A} \subset \mathcal{Q}(p)$ is called a *Q-neighborhood base* of $p$, if for every $U \in \mathcal{Q}(p)$, there exists $V \in \mathcal{A}$ such that $V \leq U$.

A subfamily $\mathcal{A} \subset \mathcal{R}(p)$ is called a *R-neighborhood base* of $p$, if for every $P \in \mathcal{R}(p)$, there exists $Q \in \mathcal{A}$ such that $Q \geq P$.

Call every one of these three kinds of bases a *local base*.

Following conclusions show relations among a base and local bases:

**2.3.15 Theorem** *Let $(L^X, \delta)$ be an L-fts, $\delta_0$ a base of $\delta$. Then*

(i) $\forall e \in M(L^X)$, $\mathcal{A} = \{U \in \delta_0 : e \not\leq U'\}$ *is a Q-neighborhood base of* $e$.

(ii) $\forall e \in M(L^X)$, $\mathcal{A} = \{P \in \delta_0' : e \not\leq P\}$ *is a R-neighborhood base of* $e$. □

**2.3.16 Remark** Note that we cannot construct a neighborhood base in an $L$-fts in the way shown in the preceding theorem. In fact, if we take $\mathcal{A} = \{U \in \delta_0 : p \in U\}$ in $L$-fts $(L^X, \delta)$ for a base $\delta_0$ and $p \in Pt(L^X)$, then for an arbitrary $U \in \mathcal{N}(p)$, although there exists $\mathcal{C} \subset \delta_0$ such that $p \in U = \bigvee \mathcal{C}$, but $\mathcal{C}$ is not necessary to contain an element $V$ such that $p \in V$. So we cannot affirm $\mathcal{A}$ is a neighborhood base of $p$. The cause here is just that the "Multiple Choice Principle" mentioned in Remark **2.3.7** is not satisfied by the relation "$\in$," i.e. "$\leq$" in $L$-fuzzy spaces. The next theorem has the same problem.

If we have a base in an $L$-fts, then we can naturally use it to construct a local base and can hope it has less elements:

**2.3.17 Theorem** *Let $(L^X, \delta)$ be an L-fts, $\delta_0$ a base of $\delta$. Then*

(i) $e \in M(L^X)$, *$\xi$ is a Q-neighborhood base of e $\Longrightarrow$ e has a Q-neighborhood base $\zeta \subset \delta_0$ such that $|\zeta| \leq |\xi|$.*

(ii) $e \in M(L^X)$, *$\xi$ is a R-neighborhood base of e $\Longrightarrow$ e has a R-neighborhood base $\zeta \subset \delta_0{}'$ such that $|\zeta| \leq |\xi|$.*

**Proof** Only prove (i), (ii) is dual to (i).

$\forall V \in \xi$, since $\delta_0$ is a base, $\exists \mathcal{A}_V \subset \delta_0$ such that $\bigvee \mathcal{A}_V = V$. Since $\xi$ is a Q-neighborhood base of e, by De Morgan's Law, $e \not\leq V' = \bigwedge\{U' : U \in \mathcal{A}_V\}$. So $\exists U_V \in \mathcal{A}_V$ such that $e \not\leq U_V$, $U_V \in \mathcal{Q}(e)$. Take $\zeta = \{U_V : V \in \xi\} \subset \bigcup_{V\in\xi} \mathcal{A}_V \subset \delta_0$, then since $\xi$ is a Q-neighborhood base of $e$ and $U_V \leq V$, $\zeta$ is also a Q-neighborhood base of $e$. Certainly $|\zeta| \leq |\xi|$. □

**2.3.18 Definition** Let $L^X$ be an $L$-fuzzy space, $A, B \in L^X$. Define the *quasi-difference* of $A$ and $B$, denoted by $A\backslash\!\backslash B$, as

$$A\backslash\!\backslash B = \bigvee\{x_\lambda \in M(\downarrow A) : B(x) = 0\} \vee \bigvee\{x_\lambda \in M(\downarrow A) : \lambda \not\geq B(x) > 0\};$$

particularly, $\forall x_a \in Pt(L^X)$,

$$A\backslash\!\backslash x_a = \bigvee\{y_\lambda \in M(\downarrow A) : y \neq x\} \vee \bigvee\{x_\lambda \in M(\downarrow A) : \lambda \not\geq a\}.$$

**2.3.19 Proposition** *Let $L^X$ be an L-fuzzy space, $A, B, C \in L^X$, $\{A_t : t \in T\} \subset L^X$, $x_a \in Pt(L^X)$. Then the following conclusions hold:*

(i) $A\backslash\!\backslash B \leq A$.

(ii) $A\backslash\!\backslash \underline{0} = A$.

(iii) $1_L \notin M(L) \Longrightarrow A\backslash\!\backslash \underline{1} = A$.

(iv) $A \leq B \Longrightarrow A\backslash\!\backslash C \leq B\backslash\!\backslash C$.

(v) $(\bigvee_{t\in T} A_t)\backslash\!\backslash C = \bigvee_{t\in T}(A_t\backslash\!\backslash C)$.

(vi) $\forall x \in supp(B)$, $B(x) \not\leq A(x) \Longrightarrow A\backslash\!\backslash B = A$.

(vii) $A \wedge B = \underline{0} \Longrightarrow A\backslash\!\backslash B = A$.

(viii) $x_a \not\leq A \Longrightarrow A\backslash\!\backslash x_a = A$.

(ix) $supp(B) = supp(C)$, $B \leq C \Longrightarrow A\backslash\!\backslash B \leq A\backslash\!\backslash C$.

**Proof** (i) By the definition.

(ii) By Corollary **2.3.11** (i).

(iii) Since $1_L \notin M(L)$, $\forall x_\lambda \in A$, $\lambda \not\geq 1_L = \underline{1}(x) > 0_L$ always hold, hence by Corollary **2.3.11** (ii), $A \leq A\backslash\!\backslash\underline{1}$. By (i), $A\backslash\!\backslash\underline{1} = A$.

(iv) By the definition of quasi-difference and Corollary **2.3.11** (ii).

(v) By (iv), $\bigvee_{t\in T}(A_t\backslash\!\backslash C) \leq (\bigvee_{t\in T} A_t)\backslash\!\backslash C$.

Simply denote the condition "$C(x) = 0$ or $\lambda \not\geq C(x) > 0$" by "$P(x_\lambda) > 0$". Suppose $x_\lambda \in M(\downarrow \bigvee_{t\in T} A_t)$ and $P(x_\lambda) > 0$.

If $C(x) = 0$, by Corollary **2.3.11** (i),

$$\begin{aligned}(\bigvee_{t\in T}(A_t\backslash\!\backslash C))(x) &= \bigvee_{t\in T} \bigvee\{\mu \in M(\downarrow A_t(x)) : P(x_\mu) > 0\} = \bigvee_{t\in T} \bigvee M(\downarrow A_t(x)) \\ &= \bigvee_{t\in T} A_t(x) \geq \lambda.\end{aligned}$$

If $\lambda \not\geq C(x) > 0$, $\forall t \in T$, denote $a_t = \lambda \wedge A_t(x)$. Since $\lambda \not\geq C(x)$, $\forall t \in T$, $\mu \in M(\downarrow a_t) \Longrightarrow \mu \not\geq C(x)$. So by Corollary **2.3.11** (i),

$$\begin{aligned}(\bigvee_{t\in T}(A_t\backslash\!\backslash C))(x) &= \bigvee_{t\in T} \bigvee\{\mu \in M(\downarrow A_t(x)) : P(x_\mu) > 0\} \\ &= \bigvee_{t\in T} \bigvee\{\mu \in M(\downarrow A_t(x)) : \mu \not\geq C(x) > 0\}\end{aligned}$$

$$\geq \bigvee_{t\in T} \bigvee\{\mu \in M(\downarrow a_t) : \mu \not\geq C(x) > 0\}$$
$$= \bigvee_{t\in T} \bigvee M(\downarrow a_t) = \bigvee_{t\in T} a_t = \bigvee_{t\in T} (\lambda \wedge A_t(x)) = \lambda \wedge \bigvee_{t\in T} A_t(x) = \lambda.$$

So we always have $(\bigvee_{t\in T}(A_t\backslash C))(x) \geq \lambda$. Since the join of all these $x_\lambda$'s is just $(\bigvee_{t\in T} A_t)\backslash C$, we get $\bigvee_{t\in T}(A_t\backslash C) \geq (\bigvee_{t\in T} A_t)\backslash C$.

(vi) Suppose $B(x) \not\leq A(x)$ for every $x \in supp(B)$. $\forall x_\lambda \in A$. If $B(x) = 0$, we have $x_\lambda \in A\backslash B$. If $B(x) > 0$, then we cannot have $\lambda \geq B(x)$, otherwise we should have $A(x) \geq \lambda \geq B(x) > 0$, i.e. $x \in supp(B)$ and $B(x) \leq A(x)$, this contradicts with the supposition. So $\lambda \not\geq B(x) > 0$, $x_\lambda \in A\backslash B$. By (i) and Corollary **2.3.11** (ii), $A\backslash B = A$.

(vii),(viii) Both are straightforwardly deduced from (vi).

(ix) Directly prove. □

**2.3.20 Remark** In the definition of quasi-difference of $L$-fuzzy sets, take $L = \{0, 1\}$, then $A, B \in L^X$ and $x_\lambda \in M(L^X)$ become crisp subsets and crisp point respectively. One can easily verify that in this case $A\backslash B$ and $A\backslash x_\lambda$ just become $A\backslash B$ and $A\backslash\{x\}$ respectively. On the other hand, however, from the preceding proposition, we can also find that the quasi-difference defined above has properties very different from the ordinary one, e.g. Proposition **2.3.19** (iii), or one can deduce from (vi) that in some cases $A \leq B$ holds but $A\backslash B = A$.

**2.3.21 Definition** Let $(L^X, \delta)$ be an $L$-fts, $A \in L^X$, $x_\lambda \in M(L^X)$.

$x_\lambda$ is called an *adherent point* of $A$, if for every $U \in \mathcal{Q}(x_\lambda)$, $U$ quasi-coincides with $A$, i.e. $U\hat{q}A$.

$x_\lambda$ is called an *accumulation point* of $A$, if $x_\lambda$ is an adherent point of $A\backslash x_\lambda$; that is to say, for every $U \in \mathcal{Q}(x_\lambda)$, $U$ quasi-coincides with the quasi-difference of $A$ and $x_\lambda$, i.e. $U\hat{q}(A\backslash x_\lambda)$.

Denote the set of all the accumulation points of $A$ in $(L^X, \delta)$ by $Acu(A)$ or $Acu_\delta(A)$ to emphasize the concerning topology $\delta$.

Define the *derived set* of $A$ as $\bigvee Acu(A)$, denoted by $A^d$.

**2.3.22 Remark** (1) Note that by Proposition **2.3.19** (i) and Proposition **2.3.3** (iv) every accumulation point $x_\lambda$ of $A$ is an adherent point of $A$.

(2) Since every molecule in an $L$-fts is certainly an adherent point of every molecule which has the same support point and a greater height, naturally, when the notion of accumulation point is defined, the higher molecules with same support point should be removed off. But whether or not the lower molecules at the same support point (they are still distinguished from the considered molecule as well as ones with distinguished support points) "approximate" the considered molecule reflects the structure of the value domain, so they have enough reason to be preserved. This is just the motivation of defining quasi-difference and then defining accumulation point. Certainly, under this definition, a molecule can be an accumulation point of itself if it is the join of the lower molecules, for example, if $L = [0, 1]$; but this exactly reflects the existence of the special ordered struture — the value domain. Since $\{0, 1\}$ has no this property, the preceding definition returns the ordinary correspondent one when $L = \{0, 1\}$.

**2.3.23 Theorem** *Let $(L^X, \delta)$ be an L-fts, $A, B \in L^X$. Then*

(i) *$M(\downarrow A^-)$ is just the set of all the adherent points of $A$.*
(ii) *$A^- = \bigvee\{x_\lambda \in M(L^X):\ x_\lambda$ is an adherent point of $A\}$.*
(iii) *$A \le B \Longrightarrow$ every adherent point of $A$ is an adherent point of $B$.*

**Proof** (i) Suppose $x_\lambda \in M(\downarrow A^-)$. $\forall U \in \mathcal{Q}(x_\lambda)$. If $A \le U'$, then $x_\lambda \le A^- \le U'^- = U'$, $x_\lambda \le U'$, contradicts with $U \in \mathcal{Q}(x_\lambda)$. So $A \not\le U'$, $x_\lambda$ is an adherent point of $A$.

Suppose $x_\lambda$ is an adherent point of $A$. If $x_\lambda \not\le A^-$, then $A^{-\prime} \in \mathcal{Q}(x_\lambda)$. Since $x_\lambda$ is an adherent point of $A$, $A \not\le A^-$. But this is a contradiction. So $x_\lambda \le A^-$.

(ii) By Corollary **2.3.11** (i) and the conclusion proved above,

$$A^- = \bigvee M(\downarrow A^-) = \bigvee\{x_\lambda \in M(L^X):\ x_\lambda \text{ is an adherent point of } A\}.$$

(iii) By (i) and Theorem **2.2.18** (iv). □

**2.3.24 Theorem** *Let $(L^X, \delta)$ be an L-fts, $A, B \in L^X$, $\{A_t:\ t \in T\} \subset L^X$. Then*
(i) $A^- = A \vee A^d$.
(ii) $A \le B \Longrightarrow A^d \le B^d$.
(iii) $(A \vee B)^d = A^d \vee B^d$.
(iv) $\bigvee_{t\in T} {A_t}^d \le (\bigvee_{t\in T} A_t)^d$.

**Proof** (i) By Remark **2.3.22** and Theorem **2.3.23** (ii), $A \vee A^d \le A^-$. $\forall x_\lambda \in M(\downarrow A^-)$. If $x_\lambda \in A$, then we have had $x_\lambda \le A \vee A^d$ already. If $x_\lambda \notin A$, by Theorem **2.3.23** (i), $x_\lambda$ is an adherent point of $A$, by Proposition **2.3.19** (viii), $\forall U \in \mathcal{Q}(x_\lambda)$, $U \mathbin{\bar\wedge} (A \setminus x_\lambda) = U \mathbin{\bar\wedge} A \ne \emptyset$, $x_\lambda$ is an accumulation point of $A$, we still have $x_\lambda \le A^d \le A \vee A^d$. By Corollary **2.3.11**, $A^- = A \vee A^d$.

(ii) Suppose $x_\lambda$ is an accumulation point of $A$, then $\forall U \in \mathcal{Q}(x_\lambda)$, by Proposition **2.3.19** (iv) and Proposition **2.3.3** (iv),

$$U \mathbin{\bar\wedge} (B \setminus x_\lambda) \supset U \mathbin{\bar\wedge} (A \setminus x_\lambda) \ne \emptyset,$$

$x_\lambda$ is also an accumulation point of $B$, $x_\lambda \in B^d$. By Corollary **2.3.11** (ii), $A^d \le B^d$.

(iii) By (ii), $A^d \vee B^d \le (A \vee B)^d$. Suppose $x_\lambda$ is an accumulation point of $A \vee B$. If $x_\lambda$ is not an accumulation point of either $A$ or $B$, then $\exists U_0, U_1 \in \mathcal{Q}(x_\lambda)$ such that $U_0 \mathbin{\bar\wedge} (A \setminus x_\lambda) = U_1 \mathbin{\bar\wedge} (B \setminus x_\lambda) = \emptyset$. Take $U = U_0 \wedge U_1 \in \mathcal{Q}(x_\lambda)$, since $x_\lambda$ is an accumulation point of $A \vee B$,

$$\begin{aligned}
&(U_0 \mathbin{\bar\wedge} (A \setminus x_\lambda)) \cup (U_1 \mathbin{\bar\wedge} (B \setminus x_\lambda) \\
\supset&(U \mathbin{\bar\wedge} (A \setminus x_\lambda)) \cup (U \mathbin{\bar\wedge} (B \setminus x_\lambda)) && \text{(by Proposition 2.3.3 (iv))}\\
=&U \mathbin{\bar\wedge} ((A \setminus x_\lambda) \vee (B \setminus x_\lambda)) && \text{(by Proposition 2.3.3 (v))}\\
=&(U \mathbin{\bar\wedge} ((A \vee B) \setminus x_\lambda) && \text{(by Proposition 2.3.19 (v))}\\
\ne&\emptyset.
\end{aligned}$$

This is a contradiction.

(iv) By (ii). □

**2.3.25 Remark** In ordinary set theory, point is the smallest element of set. Contrasting with this situation, in fuzzy set theory, a fuzzy point may contain other smaller fuzzy points. Just because of this difference, a join of some fuzzy points may contain some other fuzzy points different from those original ones. Therefore, it is natural that a molecule — an $L$-fuzzy point $x_\lambda$ contained in the derived set $A^d$ of an $L$-fuzzy set $A$ perhaps is not an accumulation point of $A$. For example, take $X$ as a singleton $\{x\}$, $L = \{0, a', a, 1\}$ be a chain, $0 < a' < a < 1$, let $\underline{0}$ be mapped to $\underline{1}$ and $a$ to $a'$, we obtain an order-reversing involution $'$ and then a F-lattice $L$. Define

$\delta = \{\underline{0}, x_a, \underline{1}\}$, $A = x_a$, then $x_1 \in Acu(A)$, and hence $A^d = x_1$. But $x_a \in A^d$ is not an accumulation point of $A$.

In general topology, there is a theorem as follows: In a topological space, if the derived set of each singleton is closed, then so is the derived set of each subset. This theorem is called C.T. Yang's Theorem, not difficult to be proved. It still hold for $L$-fuzzy point in $L$-fuzzy topology:

**2.3.26 Theorem**(C.T. Yang's Theorem) *Let $(L^X, \delta)$ be an L-fts. If the derived set of each L-fuzzy point in $(L^X, \delta)$ is a closed subset, then the derived set of each L-fuzzy subset of $(L^X, \delta)$ is a closed subset.*

**Proof** Suppose the derived set of each $L$-fuzzy point is closed, $A \in L^X$. To prove $A^d$ is closed, we need only prove $A^{dd} \leq A^d$, because in this case by Theorem **2.3.24** (i) we shall have $(A^d)^- = A^d \vee A^{dd} = A^d$. So let $x_\lambda \in Acu(A^d)$, we need only prove $x_\lambda \leq A^d$.

Suppose it is not true, i.e. $x_\lambda \not\leq A^d$. Denote $A(x) = a$, by Theorem **2.3.24** (ii), ${x_a}^d \leq A^d$. By $x_\lambda \not\leq A^d$, $x_\lambda \not\leq {x_a}^d$. $\forall U \in \mathcal{Q}(x_\lambda)$, since ${x_a}^d$ is a closed subset, by Theorem **2.3.12** (ii) and $x_\lambda \not\leq {x_a}^d$, $V = (U' \vee {x_a}^d)' = U \wedge ({x_a}^d)' \in \mathcal{Q}(x_\lambda)$. By Corollary **2.3.11** (i),

$$\begin{aligned}(A\backslash x_\lambda) \vee x_a =& \bigvee\{y_\mu \in M(\downarrow A) : \ y \neq x\} \vee \\ & \bigvee\{x_\mu : \ \mu \in M(\downarrow A(x)), \ \mu \not\geq \lambda\} \vee x_{A(x)} \\ =& A. \end{aligned} \tag{2.7}$$

By $x_\lambda \not\leq A^d$ and Proposition **2.3.19** (viii), $A^d\backslash x_\lambda = A^d$, by $V \in \mathcal{Q}(x_\lambda)$ and $x_\lambda \in Acu(A^d)$, $A^d\backslash x_\lambda \not\leq V'$. So by Theorem **2.3.24** (iii) and equality (2.7),

$$(A\backslash x_\lambda)^d \vee {x_a}^d = ((A\backslash x_\lambda) \vee x_a)^d = A^d = A^d\backslash x_\lambda \not\leq V' = U' \vee {x_a}^d,$$
$$(A\backslash x_\lambda)^d \not\leq U',$$

$\exists y_\mu \in Acu(A\backslash x_\lambda)$ such that $y_\mu \not\leq U'$, $U \in \mathcal{Q}(y_\mu)$. So by $y_\mu \in Acu(A\backslash x_\lambda)$, $(A\backslash x_\lambda)\backslash y_\mu \not\leq U'$, by Proposition **2.3.19** (i), $A\backslash x_\lambda \not\leq U'$. Since $U$ was arbitrarily taken out from $\mathcal{Q}(x_\lambda)$, it means $x_\lambda \in Acu(A)$, $x_\lambda \leq A^d$, this is a contradiction. □

**2.3.27 Open Question** Is the C.T. Yang's Theorem still true if the word "point" in its condition "the derived set of each $L$-fuzzy point is closed" is replaced by "molecule?"

Although we have not had yet a perfect answer to the preceding question, but we can prove that it is true for some kind of $L$'s, this is just the Theorem **2.3.30** in the sequel.

**2.3.28 Lemma** *Let L be a complete lattice, then the join of every directed set of molecules in L is still a molecule.*

**Proof** Suppose $D \subset M(L)$ is a directed set in $L$, denote $\zeta = \bigvee D$. If $\zeta \notin M(L)$, then $\exists a, b < \zeta$ such that $a \vee b = \zeta$. Since $\zeta = \bigvee D$, $\exists \alpha, \beta \in D$ such that $\alpha \not\leq a$, $\beta \not\leq b$. Since $D$ is directed, $\exists \gamma \in D$ such that $\gamma \geq \alpha, \beta$. So $\gamma \not\leq a, b$. But $\gamma \in M(L)$, so $\gamma \not\leq a \vee b = \zeta = \bigvee D$, this is a contradiction. □

Since a chain in a complete lattice is naturally a directed set, so by Zorn's Lemma, the following corollary holds:

**2.3.29 Corollary** *Let L be a complete lattice, $a \in L$, $\lambda \in M(L)$, $\lambda \leq a$. Then*

*$M(\uparrow\lambda \cap \downarrow a)$ contains at least one maximal element.* □

**2.3.30 Theorem** *Let $L$ be a F-lattice such that for every $a \in L$ and every molecule $\lambda \leq a$, $M(\uparrow\lambda\cap\downarrow a)$ contains at most finite number of maximal elements, $(L^X, \delta)$ be an L-fts. If the derived set of each L-fuzzy molecule in $(L^X, \delta)$ is a closed subset, then the derived set of each L-fuzzy subset of $(L^X, \delta)$ is a closed subset.*

**Proof** By Theorem **2.3.26**, we need only prove that under the given condition, the derived set of each $L$-fuzzy point is closed.

$\forall x_a \in Pt(L^X)$. For every $\lambda \in M(\downarrow a)$, denote $K_\lambda = M(\uparrow\lambda \cap \downarrow a)$, suppose the set of all the maximal elements of $K_\lambda$ is $F_\lambda$. Note that $F_\lambda^x = \{x_\mu : \mu \in F_\lambda\}$ is just the set of all the maximal elements of $K_\lambda^x = \{x_\mu : \mu \in K_\lambda\}$ and $\bigvee K_\lambda^x = \bigvee F_\lambda^x$, since $F_\lambda$ is a finite set, by Theorem **2.3.24** (iii),

$$(\bigvee K_\lambda^x)^d = (\bigvee F_\lambda^x)^d = (\bigvee\{x_\mu : \mu \in F_\lambda\})^d = \bigvee\{{x_\mu}^d : \mu \in F_\lambda\}. \tag{2.8}$$

Now $\forall y_\lambda \in M(L^X)$ such that $y_\lambda \not\leq {x_a}^d$, we shall prove $\exists V_\lambda \in \mathcal{Q}(y_\lambda)$ such that $V_\lambda' \geq {x_a}^d$. If $y_\lambda \not\leq x_a$, then take $V_\lambda = (x_a \vee {x_a}^d)' = {x_a}^{-\prime} \in \delta$, by $y_\lambda \not\leq {x_a}^d$ and $y_\lambda \in M(L^X)$, we have $y_\lambda \not\leq x_a \vee {x_a}^d = V_\lambda'$, $V_\lambda \in \mathcal{Q}(y_\lambda)$ and, certainly, $V_\lambda' \geq {x_a}^d$. If $y_\lambda \leq x_a$, then $y = x$, $y_\lambda = x_\lambda$. Since $x_\lambda = y_\lambda \not\leq {x_a}^d$, $\exists U_\lambda \in \mathcal{Q}(x_\lambda)$ such that $U_\lambda ⩕ (x_a \backslash x_\lambda) = \emptyset$, i.e. $x_a \backslash x_\lambda \leq U_\lambda'$. Using the previous symbols, we have

$$\begin{aligned}
(x_a\backslash x_\lambda) \vee \bigvee K_\lambda^x &= \bigvee\{x_\mu : \mu \in M(\downarrow a),\ \mu \not\geq \lambda\} \vee \bigvee\{x_\mu : \mu \in M(\downarrow a),\ \mu \geq \lambda\} \\
&= x_a, && \text{(by Corollary } \mathbf{2.3.11}\text{ (i))} \\
{x_a}^d &= ((x_a\backslash x_\lambda) \vee \bigvee K_\lambda^x)^d \\
&= (x_a\backslash x_\lambda)^d \vee (\bigvee K_\lambda^x)^d && \text{(by Theorem } \mathbf{2.3.24}\text{ (iii))} \\
&\leq (U_\lambda')^d \vee (\bigvee K_\lambda^x)^d \\
&\leq (U_\lambda')^- \vee (\bigvee K_\lambda^x)^d && \text{(by Theorem } \mathbf{2.3.24}\text{ (i))} \\
&= U_\lambda' \vee \bigvee\{{x_\mu}^d : \mu \in F_\lambda\}. && \text{(by equality (2.8))}
\end{aligned}$$

Then take $V_\lambda = (U_\lambda' \vee \bigvee\{{x_\mu}^d : \mu \in F_\lambda\})'$, since $F_\lambda$ is a finite set and the derived set of each molecule is closed, $V_\lambda \in \delta$. Since $x_\lambda \not\leq {x_a}^d$, by Theorem **2.3.24** (ii), $\forall \mu \in F_\lambda$, $x_\lambda \not\leq {x_\mu}^d$, and hence $x_\lambda \not\leq \bigvee\{{x_\mu}^d : \mu \in F_\lambda\}$ by $x_\lambda \in M(L^X)$ and the finiteness of the set $F_\lambda$. Since $U_\lambda \in \mathcal{Q}(x_\lambda)$, we still obtain $V_\lambda \in \mathcal{Q}(x_\lambda) = \mathcal{Q}(y_\lambda)$ and $V_\lambda' \geq {x_a}^d$.

At last, $\forall y_\lambda \in M(L^X)$ such that $y_\lambda \not\leq {x_a}^d$, as proved above, we can take $V_\lambda^y \in \mathcal{Q}(y_\lambda)$ such that $(V_\lambda^y)' \geq {x_a}^d$. Denote $C = \bigwedge\{(V_\lambda^y)' : y_\lambda \in M(L^X),\ y_\lambda \not\leq {x_a}^d\}$, then $C$ is a closed subset and clearly $C \geq {x_a}^d$. If $C \not\leq {x_a}^d$, then by Corollary **2.3.11** (i), $\exists y_\lambda \in M(\downarrow C)$, $y_\lambda \not\leq {x_a}^d$. But $y_\lambda \not\leq (V_\lambda^y)'$, and hence $y_\lambda \not\leq C$, this is a contradiction. So ${x_a}^d = C$ is a closed subset. □

**2.3.31 Remark** (1) F-lattice which is a product of a family of chains, for instance, $[0,1]$ or even $[0,1]^\kappa$, is an example of the lattice involved in Theorem **2.3.30**.

(2) The definition of "accumulation point" is different from the one defined previously:[137, 168]

*In an L-fts $(L^X, \eta)$ with L-fuzzy co-topology $\eta$, $x_\lambda \in M(L^X)$ is called an accumulation point of $A \in L^X$, if*

(i) *$x_\lambda$ is an adherent point of $A$, i.e. $x_\lambda \leq A^-$, and*

(ii) *$x_\lambda \not\leq A$, or $x_\lambda \leq A$ and $A \not\leq U' \vee x_\mu$ for every $U \in \mathcal{Q}(x_\lambda)$ and every $x_\mu \in M(\uparrow x_\lambda \cap \downarrow A)$.*

Under this definition, property (i) in Theorem **2.3.24** holds, but (ii) to (iv) not.

(3) Wang [168] introduced the notion of "standard lattice" $L$: $\forall a \in L$, $\forall \lambda \in M(\downarrow a)$, $M(\uparrow\lambda \cap \downarrow a)$ contains just one maximal element; and, replacing the lattice involved in Theorem **2.3.30** with a standard lattice, deduced the correspondent result in the previous sense of "accumulation point" above mentioned in this remark.

## 2.4 Continuous Mappings

Introduce the following

**2.4.1 Definition** Let $(L^X, \delta)$, $(L^Y, \mu)$ be $L$-fts', $f^\rightarrow : L^X \to L^Y$ an $L$-fuzzy mapping. We say $f^\rightarrow$ is an *L-fuzzy continuous mapping* from $(L^X, \delta)$ to $(L^Y, \mu)$ or call $f^\rightarrow$ *continuous* for short, if its $L$-fuzzy reverse mapping $f^\leftarrow : L^Y \to L^X$ maps every open subset in $(L^Y, \mu)$ as an open one in $(L^X, \delta)$, i.e. $\forall V \in \mu$, $f^\leftarrow(V) \in \delta$.

Also denote an $L$-fuzzy continuous mapping as $f^\rightarrow : (L^X, \delta) \to (L^Y, \mu)$ to indicate the $L$-fuzzy topologies concerning.

**2.4.2 Remark** (1) The preceding concept of continuity of $L$-fuzzy mapping bases on the earliest C.L. Chang's definition for the case $L = [0, 1]$.[15]

(2) Recall Remark **2.2.11** (2). Requiring a fuzzy topology (or an $L$-fuzzy topology) containing every constant mapping $\underline{a}$ as Lowen did has an other advantage: It makes every $L$-fuzzy mapping induced from an ordinary constant mapping continuous just as the case of ordinary topology.

Contrasted with the preceding "global continuity," one can also define "local continuity" of a mapping (see [137] and [168]):

**2.4.3 Definition** Let $(L^X, \delta)$, $(L^Y, \mu)$ be $L$-fts'. An $L$-fuzzy mapping $f^\rightarrow : L^X \to L^Y$ is called *continuous at molecule* $e \in M(L^X)$, if $\forall V \in \mathcal{Q}(f(e))$, $f^\leftarrow(V) \in \mathcal{Q}(e)$.

By Proposition **2.3.4** (i), in the definition mentioned above of continuity at a molecule, for every $V \in \mathcal{Q}(f(e))$, $e \lessdot f^\leftarrow(V)$ is always true. So

**2.4.4 Proposition** Let $(L^X, \delta)$, $(L^Y, \mu)$ be $L$-fts'. An $L$-fuzzy mapping $f^\rightarrow : L^X \to L^Y$ is continuous at molecule $e \in M(L^X)$ if and only if $f^\leftarrow(V) \in \delta$ for every $V \in \mathcal{Q}(f^\rightarrow(e))$. □.

**2.4.5 Theorem** *Let $(L^X, \delta)$, $(L^Y, \mu)$ be L-fts', $f^\rightarrow : L^X \to L^Y$ an L-fuzzy mapping. Then the following conditions are equivalent:*

(i) *$f^\rightarrow$ is continuous.*

(ii) $\forall Q \in \mu'$, $f^\leftarrow(Q) \in \delta'$.

(iii) *$f^\rightarrow$ is continuous at every molecule in $(L^X, \delta)$.*

(iv) $\forall e \in M(L^X)$, $\forall Q \in \mathcal{R}(f(e))$, $f^\leftarrow(Q) \in \mathcal{R}(e)$.

(v) *$\mu$ has a subbase $\mu_0$ such that* $\forall V \in \mu_0$, $f^\leftarrow(V) \in \delta$.

(vi) *$\mu$ has a subbase $\mu_0$ such that* $\forall Q \in \mu_0'$, $f^\leftarrow(Q) \in \delta'$.

(vii) $\forall A \in L^X$, $f^\rightarrow(A^-) \le (f^\rightarrow(A))^-$.

(viii) $\forall B \in L^Y$, $(f^\leftarrow(B))^- \le f^\leftarrow(B^-)$.

(ix) $\forall B \in L^Y$, $f^\leftarrow(B^\circ) \le (f^\leftarrow(B))^\circ$.

**Proof** (i)$\Longrightarrow$(ii): By Proposition **2.2.5**.

(ii)$\Longrightarrow$(iii): $\forall e \in M(L^X)$, $\forall V \in \mathcal{Q}(f^\rightarrow(e))$. By Proposition **2.2.5** and the preceding condition (ii), $(f^\leftarrow(V))' = f^\leftarrow(V') \in \delta'$, so $f^\leftarrow(V) \in \delta$. Therefore, if

$f^{\leftarrow}(V) \notin \mathcal{Q}(e)$, then $e \leq (f^{\leftarrow}(V))' = f^{\leftarrow}(V')$, by Theorem **2.1.25** (ii), $f^{\rightarrow}(e) \leq f^{\rightarrow}f^{\leftarrow}(V') \leq V'$, contradicts with $V \in \mathcal{Q}(f^{\rightarrow}(e))$.

(iii)$\Longrightarrow$(iv): By Proposition **2.2.5**.

(iv)$\Longrightarrow$(v): $\forall V \in \mu$. If $\exists e \in M(L^X)$ such that $f^{\rightarrow}(e) \not\leq V'$, then $V' \in \mathcal{R}(f^{\rightarrow}(e))$, by (iv) and Proposition **2.2.5**, $(f^{\leftarrow}(V))' = f^{\leftarrow}(V') \in \mathcal{R}(e) \subset \delta'$, $f^{\leftarrow}(V) \in \delta$. If $\forall e \in M(L^X)$, $f^{\rightarrow}(e) \leq V'$, then by Corollary **2.3.11** (i) and Theorem **2.1.15** (i), for $\underline{1} \in \delta$,

$$f^{\rightarrow}(\underline{1}) = f^{\rightarrow}(\bigvee M(L^X)) = \bigvee\{f^{\rightarrow}(e) : e \in M(L^X)\} \leq V'.$$

By Theorem **2.1.25** (i) and Proposition **2.2.5**,

$$\underline{1} \leq f^{\leftarrow}f^{\rightarrow}(\underline{1}) \leq f^{\leftarrow}(V') = (f^{\leftarrow}(V))', \qquad f^{\leftarrow}(V) = \underline{0} \in \delta.$$

So take $\mu_0 = \mu$ and then (v) holds.

(v)$\Longrightarrow$(vi): By Proposition **2.2.5**.

(vi)$\Longrightarrow$(vii): By (vi) and Theorem **2.1.17** (i), (ii), $f^{\leftarrow}$ maps every element of $\mu'$ into $\delta'$. So $\forall A \in L^X$, by Theorem **2.1.25** (i),

$$A \leq f^{\leftarrow}f^{\rightarrow}(A) \leq f^{\leftarrow}((f^{\rightarrow}(A))^-) \in \delta',$$
$$A^- \leq f^{\leftarrow}((f^{\rightarrow}(A))^-).$$

By Theorem **2.1.25** (ii),

$$f^{\rightarrow}(A^-) \leq f^{\rightarrow}f^{\leftarrow}((f^{\rightarrow}(A))^-) \leq (f^{\rightarrow}(A))^-.$$

(vii)$\Longrightarrow$(viii): Take $A = f^{\leftarrow}(B) \in L^X$, by (vii) and Theorem **2.1.25** (ii),

$$f^{\rightarrow}((f^{\leftarrow}(B))^-) \leq (f^{\rightarrow}f^{\leftarrow}(B))^- \leq B^-,$$

by Theorem **2.1.25** (i),

$$(f^{\leftarrow}(B))^- \leq f^{\leftarrow}f^{\rightarrow}((f^{\leftarrow}(B))^-) \leq f^{\leftarrow}(B^-).$$

(viii)$\Longrightarrow$(ix): By (viii),

$$(f^{\leftarrow}(B'))^- \leq f^{\leftarrow}(B'^-),$$

by Theorem **2.2.19** (i) and Proposition **2.2.5**,

$$(f^{\leftarrow}(B))^{\circ\prime} = (f^{\leftarrow}(B))'^- = (f^{\leftarrow}(B'))^- \leq f^{\leftarrow}(B'^-) = f^{\leftarrow}(B^{\circ\prime}) = (f^{\leftarrow}(B^{\circ}))',$$
$$(f^{\leftarrow}(B))^{\circ} \geq f^{\leftarrow}(B^{\circ}).$$

(ix)$\Longrightarrow$(i): $\forall V \in \mu$, then by (ix),

$$f^{\leftarrow}(V) = f^{\leftarrow}(V^{\circ}) \leq (f^{\leftarrow}(V))^{\circ} \leq f^{\leftarrow}(V).$$

So $f^{\leftarrow}(V) = (f^{\leftarrow}(V))^{\circ} \in \delta$, $f$ is continuous. □

**2.4.6 Definition** Let $(L^X, \delta)$, $(L^Y, \mu)$ be $L$-fts', $f^{\rightarrow} : L^X \rightarrow L^Y$ an $L$-fuzzy mapping.

$f^{\rightarrow} : (L^X, \delta) \rightarrow (L^Y, \mu)$ is called *open*, if it maps every open subset in $(L^X, \delta)$ as an open one in $(L^Y, \mu)$, i.e. $\forall U \in \delta$, $f(U) \in \mu$.

$f^{\rightarrow} : (L^X, \delta) \rightarrow (L^Y, \mu)$ is called *closed*, if it maps every closed subset in $(L^X, \delta)$ as a closed one in $(L^Y, \mu)$, i.e. $\forall F \in \delta'$, $f(F) \in \mu'$.

$f^{\rightarrow} : (L^X, \delta) \rightarrow (L^Y, \mu)$ is called an *L-fuzzy homeomorphism*, if it is bijective, continuous and open.

By the conclusions of theorems **2.1.22** and **2.1.24**, the following results are obvious:

**2.4.7 Proposition** *Let $(L^X, \delta)$ and $(L^Y, \mu)$ be L-fts', $f : X \rightarrow Y$ an ordinary mapping. Then*

(i) *$f^{\rightarrow}$ is open if and only if for every base $\delta_0$ of $\delta$ and every $U \in \delta_0$, $f^{\rightarrow}(U) \in \mu$.*

(ii) *$f^{\rightarrow}$ is an L-fuzzy homeomorphism if and only if $f$ is bijective, $f^{\rightarrow}$ is continuous and open.*

(iii) *$f^{\rightarrow}$ is an L-fuzzy homeomorphism if and only if $f$ is bijective, $f^{\rightarrow}$ is continuous and closed.*

(iv) *$f^{\rightarrow}$ is an L-fuzzy homeomorphism if and only if $f^{\leftarrow}$ is a Zadeh type mapping and an L-fuzzy homeomorphism.* □

**2.4.8 Proposition** *Let $(L^X, \delta)$, $(L^Y, \mu)$ and $(L^Z, \zeta)$ be L-fts', $f^{\rightarrow}: L^X \rightarrow L^Y$, $g^{\rightarrow}: L^Y \rightarrow L^Z$ be L-fuzzy mappings. Then*

(i) *$f^{\rightarrow}$, $g^{\rightarrow}$ are continuous $\Longrightarrow$ $g^{\rightarrow}f^{\rightarrow}$ is continuous.*

(ii) *$f^{\rightarrow}$, $g^{\rightarrow}$ are open $\Longrightarrow$ $g^{\rightarrow}f^{\rightarrow}$ is open.*

(iii) *$f^{\rightarrow}$, $g^{\rightarrow}$ are closed $\Longrightarrow$ $g^{\rightarrow}f^{\rightarrow}$ is closed.*

(iv) *$f^{\rightarrow}$, $g^{\rightarrow}$ are L-fuzzy homeomorphisms $\Longrightarrow$ $g^{\rightarrow}f^{\rightarrow}$ is an L-fuzzy homeomorphism.* □

By Theorem **2.1.22** (vi), we can define two *L*-fts' are homeomorphic to each other as follows:

**2.4.9 Definition** We say *L*-fts' $(L^X, \delta)$ and $(L^Y, \mu)$ are *homeomorphic*, or $(L^X, \delta)$ is *homeomorphic to* $(L^Y, \mu)$, or $(L^Y, \mu)$ is *homeomorphic to* $(L^X, \delta)$, if there exists an *L*-fuzzy homeomorphism $f^{\rightarrow}: (L^X, \delta) \rightarrow (L^Y, \mu)$.

# Chapter 3

# Operations on Fuzzy Topological Spaces

From some given $L$-fts', we can construct some new ones in various ways. In other words, if an $L$-fts is produced from some known $L$-fts' in a certain procedure, then it can help us to increase understanding on this $L$-fts. Therefore, in this chapter, some basic operations to produce new $L$-fts' are introduced and some relative problems are discussed.

## 3.1 Subspaces

Surely, the most natural way to produce new $L$-fts' is considering their "children" — the various parts of these global ones. This is just the idea of "subspace" introduced in this section:

**3.1.1 Definition** Let $L^X$ be an $L$-fuzzy space, $Y \subset X$, $Y \neq \emptyset$, $\mathcal{A} \subset L^X$. Denote

$$\mathcal{A}|_Y = \{A|_Y : A \in \mathcal{A}\},$$

where restriction $A|_Y$ is defined as in Definition **1.1.1**. Particularly, for every $L$-fts $(L^X, \delta)$, $\delta|_Y = \{U|_Y : U \in \delta\}$.

**3.1.2 Proposition** *Let $(L^X, \delta)$ be an $L$-fts, $Z \subset Y \subset X$, $Y \neq \emptyset$, $Z \neq \emptyset$, $\{A_t : t \in T\} \subset L^X$, $A \in L^X$. Then*

(i) $(\bigvee_{t\in T} A_t)|_Y = \bigvee_{t\in T}(A_t|_Y)$.

(ii) $(\bigwedge_{t\in T} A_t)|_Y = \bigwedge_{t\in T}(A_t|_Y)$.

(iii) $A'|_Y = (A|_Y)'$.

(iv) $(A|_Y)|_Z = A|_Z$. □

By the previous conclusions, the following results are natural:

**3.1.3 Corollary** *Let $(L^X, \delta)$ be an $L$-fts, $Z \subset Y \subset X$, $Y \neq \emptyset$, $Z \neq \emptyset$. Then*

(i) *$\delta|_Y$ is an $L$-fuzzy topology on $Y$.*

(ii) $(\delta|_Y)|_Z = \delta|_Z$. □

The previous results guarantee the reasonableness of the following definition:

**3.1.4 Definition** Let $(L^X, \delta)$ be an $L$-fts, $Y \subset X$, $Y \neq \emptyset$.

Call $\delta|_Y$ the *relative topology* of $\delta$ on $Y$ or the *subspace topology* of $Y$, call $(L^Y, \delta|_Y)$ an *$L$-fuzzy subspace* of $(L^X, \delta)$, or a *subspace* for short.

Call $(L^Y, \delta|_Y)$ an *open subspace* of $(L^X, \delta)$, if $\chi_Y \in \delta$.

Call $(L^Y, \delta|_Y)$ a *closed subspace* of $(L^X, \delta)$, if $\chi_Y \in \delta'$.

Call $(L^Y, \delta|_Y)$ a *dense subspace* of $(L^X, \delta)$, if $cl_\delta(\chi_Y) = \chi_X$.

By Proposition **3.1.2** (i), (ii) and relative definitions, the following conclusions are clear and useful:

**3.1.5 Proposition** *Let $(L^X, \delta)$ be an $L$-fts. Then*

(i) *Every subspace $(L^Y, \delta|_Y)$ of $(L^X, \delta)$ is an $L$-fts.*

(ii) *If* $(L^Y, \delta|_Y)$ *is a subspace of* $(L^X, \delta)$, $(L^Z, (\delta|_Y)|_Z)$ *is a subspace of* $(L^Y, \delta|_Y)$, *then* $(L^Z, (\delta|_Y)|_Z)$ *is also a subspace of* $(L^X, \delta)$.

(iii) *The family of all the closed subsets in* $(L^Y, \delta|_Y)$ *consists of the restrictions of all the closed subsets in* $(L^X, \delta)$, *i.e.* $(\delta|_Y)' = \delta'|_Y$.

(iv) *If* $\mathcal{B}$ *is a base of* $\delta$, $(L^Y, \delta|_Y)$ *a subspace of* $(L^X, \delta)$, *then* $\mathcal{B}|_Y$ *is a base of* $\delta|_Y$.

(v) *If* $\mathcal{S}$ *is a subbase of* $\delta$, $(L^Y, \delta|_Y)$ *a subspace of* $(L^X, \delta)$, *then* $\mathcal{B}|_Y$ *is a subbase of* $\delta|_Y$. □

**3.1.6 Theorem** *Let* $(L^X, \delta)$, $(L^Y, \mu)$ *be L-fts', * $f^\to : (L^X, \delta) \to (L^Y, \mu)$ *be an L-fuzzy mapping,* $X_0 \subset X$, $Y_0 \subset Y$. *Then*

(i) $f^\to : (L^X, \delta) \to (L^Y, \mu)$ *is continuous and* $f[X_0] \subset Y_0 \implies (f|_{X_0}^{Y_0})^\to : (L^{X_0}, \delta|_{X_0}) \to (L^{Y_0}, \mu|_{Y_0})$ *is continuous.*

(ii) $f^\to : (L^X, \delta) \to (L^Y, \mu)$ *is open and* $f[X] \subset Y_0 \implies (f|^{Y_0})^\to : (L^X, \delta) \to (L^{Y_0}, \mu|_{Y_0})$ *is open.*

(iii) $f^\to : (L^X, \delta) \to (L^Y, \mu)$ *is closed and* $f[X_0] \subset Y_0 \implies (f|^{Y_0})^\to : (L^X, \delta) \to (L^{Y_0}, \mu|_{Y_0})$ *is closed.*

**Proof** (i) $\forall U \in \mu|_{Y_0}$, $\exists V \in \mu$ such that $U = V|_{Y_0}$. Since $f[X_0] \subset Y_0$ and $f^\to$ is continuous, so

$$(f|_{X_0}^{Y_0})^\leftarrow(U) = U\circ(f|_{X_0}^{Y_0}) = (V|_{Y_0})\circ(f|_{X_0}^{Y_0}) = V\circ(f|_{X_0}) = (V\circ f)|_{X_0} = f^\leftarrow(V)|_{X_0} \in \delta_{X_0},$$

$(f|_{X_0}^{Y_0})^\to$ is continuous.

(ii) $\forall U \in \delta$, $\forall y \in Y_0$,

$$\begin{aligned}(f|^{Y_0})^\to(U)(y) &= \bigvee\{U(x) : x \in X,\ f|^{Y_0}(x) = y\}\\ &= \bigvee\{U(x) : x \in X,\ f(x) = y\}\\ &= (f^\to(U)|_{Y_0})(y).\end{aligned}$$

Since $f^\to$ is open, $(f|^{Y_0})^\to(U) = f^\to(U)|_{Y_0} \in \mu|_{Y_0}$, $(f|^{Y_0})^\to$ is open.

(iii) Similar to (ii). □

**3.1.7 Remark** In ordinary topology, every nonempty subset of a space forms a subspace, or in another expression, every subspace is just a part of its original space. This is natural, because every ordinary topological space is a set itself indeed. But this cannot be still true in fuzzy topology. The essential cause here is that a fuzzy subset, in general, does not possess a form of $L^X$, an $L$-fuzzy space. Sure, one can also define an $L$-fuzzy subspace as the family of all the $L$-fuzzy subsets contained in an arbitrary $L$-fuzzy subset — they still form a completely distributive lattice but without an order-reversing involution in general — and then induce the correspondent concepts and results. But this will unavoidably induce a conclusion that a $L$-fuzzy subspace probably is not an $L$-fts, the relative investigations on this aspect will not be involved in this book.

By the definition, an $L$-fuzzy subspace is not a subfamily of an $L$-fts except the case that it is the $L$-fts itself. But we can still investigate the relation between $L$-fuzzy subspaces and $L$-fuzzy mappings. For this target, introduce the the following definitions:

**3.1.8 Definition** Let $L^X$ be an $L$-fuzzy space, $Y \subset X$, $Y \neq \emptyset$. For every $A \in L^Y$, define its *extension on* $X$ as an $L$-fuzzy subset $A^*$ on $X$ as follows:

$$A^*(x) = \begin{cases} A(x), & x \in Y, \\ 0, & x \notin Y. \end{cases}$$

For every subfamily $\mathcal{A} \subset L^Y$, denote

$\mathcal{A}^* = \{A^* : A \in \mathcal{A}\}$;

particularly, for an $L$-fts $(L^X, \delta)$ and its subspace $(L^Y, \delta|_Y)$, denote

$(\delta|_Y)^* = \{V^* : V \in \delta|_Y\}$.

**3.1.9 Definition** Let $(L^X, \delta)$ be an $L$-fts. For every $\mathcal{A} \subset L^X$ and $\mathcal{B} \subset \mathcal{A}$, call $(\mathcal{A}, \mathcal{B})$ a *quasi-subspace* of $(L^X, \delta)$, if there exists a subspace $(L^Y, \delta|_Y)$ of $(L^X, \delta)$ such that $\mathcal{A} = (L^Y)^*$, $\mathcal{B} = (\delta|_Y)^*$.

The follow conclusion is clear:

**3.1.10 Proposition** *Let $(L^X, \delta)$ be an L-fts, $(\mathcal{A}, \mathcal{B})$ a quasi-subspace of $(L^X, \delta)$. Then the subspace $(L^Y, \delta|_Y)$ of $(L^X, \delta)$ such that $(L^Y)^* = \mathcal{A}$, $(\delta|_Y)^* = \mathcal{B}$ is unique and $Y = supp(\bigvee \mathcal{A})$.* □

Since it is obvious that a subspace $(L^Y, \delta|_Y)$ uniquely produces a quasi-subspace $((L^Y)^*, (\delta|_Y)^*)$, by the preceding proposition, we can introduce the following

**3.1.11 Definition** Let $(L^X, \delta)$ be an $L$-fts. For every subspace $(L^Y, \delta|_Y)$ of $(L^X, \delta)$, $((L^Y)^*, (\delta|_Y)^*)$ is called the *correspondent quasi-subspace* of $(L^Y, \delta|_Y)$; for every quasi-subspace $(\mathcal{A}, \mathcal{B})$ of $(L^X, \delta)$, the subspace $(L^Y, \delta|_Y)$ such that $(L^Y)^* = \mathcal{A}$, $(\delta|_Y)^* = \mathcal{B}$ is called the *correspondent subspace* of $(\mathcal{A}, \mathcal{B})$; and call the one-to-one correspondence which maps every quasi-subspace of $(L^X, \delta)$ to its correspondent subspace the *subspace correspondence on* $(L^X, \delta)$.

**3.1.12 Definition** Let $(L^X, \delta)$, $(L^Y, \mu)$ be $L$-fts', $f^\rightarrow : (L^X, \delta) \to (L^Y, \mu)$ an $L$-fuzzy continuous mapping. $f^\rightarrow$ is called an *embedding*, if $f^\rightarrow$ is injective and $(f^\rightarrow[L^X], f^\rightarrow[\delta])$ is just a quasi-subspace of $(L^Y, \mu)$.

**3.1.13 Lemma** *Let $L^X$, $L^Y$ be L-fuzzy spaces, $f^\rightarrow : L^X \to L^Y$ an L-fuzzy mapping, then $f^\rightarrow[L^X] = (L^{f[X]})^*$.*

**Proof** $\forall A \in L^X$, $f^\rightarrow(A) \leq f^\rightarrow(\chi_X) = \chi_{f[X]}$, so $f^\rightarrow[L^X] \subset (L^{f[X]})^*$. On the other hand, $\forall B \in (L^{f[X]})^*$, $supp(B) \subset f[X]$, then $y \notin f[X] \Longrightarrow y \notin supp(B)$. So take $A = f^\leftarrow(B) \in L^X$, $\forall y \in Y$, no matter $y \in f[X]$ or not, we always have

$$\begin{aligned} f^\rightarrow(A)(y) &= \bigvee\{A(x) : x \in X, f(x) = y\} \\ &= \bigvee\{f^\leftarrow(B)(x) : x \in X, f(x) = y\} \\ &= \bigvee\{B(f(x)) : x \in X, f(x) = y\} \\ &= B(y), \end{aligned}$$

$$B = f^\rightarrow(A) \in f^\rightarrow[L^X], \quad (L^{f[X]})^* \subset f^\rightarrow[L^X].$$

So $f^\rightarrow[L^X] = (L^{f[X]})^*$. □

The following conclusion gives out simple rules to check if an $L$-fuzzy continuous mapping is an embedding:

**3.1.14 Theorem** *Let $(L^X, \delta)$, $(L^Y, \mu)$ be L-fts', $f^\rightarrow : (L^X, \delta) \to (L^Y, \mu)$ an L-fuzzy continuous mapping. Then the following conditions are equivalent:*

(i) $f^{\rightarrow}$ *is an embedding.*

(ii) $(f|^{f[X]})^{\rightarrow}: (L^X, \delta) \rightarrow (L^{f[X]}, \mu|_{f[X]})$ *is an L-fuzzy homeomorphism.*

(iii) $f^{\rightarrow}$ *is injective and* $f^{\rightarrow}[\delta] = (\mu|_{f[X]})^*$;

(iv) $f^{\rightarrow}$ *is injective and* $f^{\rightarrow}[\delta] \subset (\mu|_{f[X]})^*$.

**Proof** (i)$\Longrightarrow$(iii): Since $f^{\rightarrow}$ is an embedding, we can suppose $\exists Y \subset X$, $Y \neq \emptyset$ such that $f^{\rightarrow}[\delta] = (\mu|_Y)^*$. $\forall U \in \delta$, $f^{\rightarrow}(U) \leq f^{\rightarrow}(X) = \chi_{f[X]}$, so $(\mu|_Y)^* = f^{\rightarrow}[\delta] \subset (\mu|_{f[X]})^*$, $Y \subset f[X]$. On the other hand, $\chi_{f[X]} = f^{\rightarrow}(X) \in f^{\rightarrow}[\delta] = (\mu|_Y)^*$, $f[X] \subset Y$, so $Y = f[X]$, $f^{\rightarrow}[\delta] = (\mu|_{f[X]})^*$.

(iii)$\Longrightarrow$(iv): Clear.

(iv)$\Longrightarrow$(i): $\forall V \in (\mu|_{f[X]})^*$, then $\exists W \in \mu$ such that $V = (W|_{f[X]})^*$. Since

$$f^{\leftarrow}(W) = W \circ f = (W|_{f[X]})^* \circ f = V \circ f = f^{\leftarrow}(V),$$

so by $V \in (\mu|_f[X])^* \subset (L^{f[X]})^*$, $\forall y \in Y$, no matter $y \in f[X]$ or not, we always have

$$\begin{aligned} f^{\rightarrow}(f^{\leftarrow}(W))(y) &= f^{\rightarrow}(f^{\leftarrow}(V))(y) \\ &= \bigvee\{f^{\leftarrow}(V)(x) :\ x \in X,\ f(x) = y\} \\ &= \bigvee\{V(f(x)) :\ x \in X,\ f(x) = y\} \\ &= V(y), \end{aligned}$$

$$f^{\rightarrow}(f^{\leftarrow}(W)) = V.$$

By the continuity of $f^{\rightarrow}$, $f^{\leftarrow}(W) \in \delta$, so $V = f^{\rightarrow}(f^{\leftarrow}(W)) \in f^{\rightarrow}[\delta]$. By the arbitrariness of $V \in (\mu|_{f[X]})^*$, $(\mu|_{f[X]})^* \subset f^{\rightarrow}[\delta]$. Then by the assumed condition, $f^{\rightarrow}[\delta] = (\mu|_{f[X]})^*$

At last, by Lemma **3.1.13**, $f^{\rightarrow}[L^X] = (L^{f[X]})^*$, so $(f^{\rightarrow}[L^X], f^{\rightarrow}[\delta])$ is just a quasi-subspace of $(L^Y, \mu)$. Since $f^{\rightarrow}$ is injective, it is an embedding.

(iv)$\Longrightarrow$(ii): Since $f^{\rightarrow}$ is injective. By Theorem **2.1.22** (i), $f|^{f[X]} : X \rightarrow f[X]$ is bijective; by (v) in the same theorem, $(f|^{f[X]})^{\rightarrow}$ is bijective.

$\forall V \in \mu|_{f[X]}$, $\forall x \in X$, $\exists W \in \mu$ such that $V = W|_{f[X]}$. Then

$$(f|^{f[X]})^{\leftarrow}(V)(x) = V \circ (f|^{f[X]})(x) = V \circ f(x) = W \circ f(x) = f^{\leftarrow}(W)(x).$$

Since $f^{\rightarrow}$ is continuous, so $(f|^{f[X]})^{\rightarrow}(V) = f^{\leftarrow}(W) \in \delta$, $(f|^{f[X]})^{\rightarrow}$ is continuous.

$\forall U \in \delta$, $\forall y \in Y$. If $y \in f[X]$,

$$\begin{aligned} (f|^{f[X]})^{\rightarrow}(U)(y) &= \bigvee\{U(x) :\ x \in X,\ f|^{f[X]}(x) = y\} \\ &= \bigvee\{U(x) :\ f(x) = y\} \\ &= f^{\rightarrow}(U)(y); \end{aligned}$$

if $y \notin f[X]$, then

$$\begin{aligned} (f|^{f[X]})^{\rightarrow}(U)(y) &= \bigvee\{U(x) :\ x \in X,\ f|^{f[X]}(x) = y\} \\ &= \bigvee\{U(x) :\ x \in X,\ f(x) = y\} \\ &= 0. \end{aligned}$$

So by (iv), $(f|^{f[X]})^{\rightarrow}(U) = f^{\rightarrow}(U)|_{f[X]} \in \mu|_{f[X]}$, $(f|^{f[X]})^{\rightarrow}$ is open and hence an $L$-fuzzy homeomorphism.

(ii)$\Longrightarrow$(iv): Since $(f|^{f[X]})^{\rightarrow}$ is an $L$-fuzzy homeomorphism, by Theorem **2.1.22** (i), $f|^{f[X]}$ and then $f$ is injective. By the same reason, $f^{\rightarrow}$ is injective.

$\forall U \in \delta$, since $(f|^{f[X]})^{\rightarrow}$ is open, as proved in (iv)$\Longrightarrow$(ii), we have $f^{\rightarrow}(U)|_{f[X]} = (f|^{f[X]})^{\rightarrow}(U) \in \mu|_{f[X]}$. Suppose $y \in Y \backslash f[X]$, then $f^{\rightarrow}(U)(y) = 0$, so $f^{\rightarrow}(U) \in (\mu|_{f[X]})^*$, $f^{\rightarrow}[\delta] \subset (\mu|_{f[X]})^*$. □

If $f$ is surjective, then $f[X] = Y$, so by (i)$\Longleftrightarrow$(ii) in the preceding theorem, the following conclusion holds:

**3.1.15 Corollary** *Let $f^{\rightarrow}: (L^X, \delta) \to (L^Y, \mu)$ be an L-fuzzy mapping, then $f^{\rightarrow}$ is an L-fuzzy homeomorphism if and only if $f^{\rightarrow}$ is a surjective embedding.* □

For a subspace, the basic operations in an $L$-fts have closed relations with the original space, just as follows:

**3.1.16 Theorem** *Let $(L^X, \delta)$ be an L-fts, $(L^Y, \delta|_Y)$ its a subspace, $e \in M(L^Y)$, $A, B \in L^Y$, $\{A_t : t \in T\} \subset L^Y$, $C \in L^X$. Then*

(i) $M(\downarrow A) = \{p \in Pt(L^Y) : p^* \in M(\downarrow A^*)\}$.
(ii) $A \leq B \Longrightarrow A^* \leq B^*$.
(iii) $(\bigvee_{t\in T} A_t)^* = \bigvee_{t\in T}(A_t)^*$.
(iv) $(\bigwedge_{t\in T} A_t)^* = \bigwedge_{t\in T}(A_t)^*$.
(v) $(C|_Y)' = C'|_Y$.
(vi) $A' = (A^*)'|_Y$.
(vii) $A \pitchfork B = A^* \pitchfork B^*$.
(viii) $A \backslash B = (A^* \backslash B^*)|_Y$.
(ix) $(A \backslash B)^* = A^* \backslash B^*$.
(x) $\mathcal{Q}(e) = \mathcal{Q}(e^*)|_Y$.
(xi) $\mathcal{R}(e) = \mathcal{R}(e^*)|_Y$.
(xii) $Acu(A) = (Acu(A^*)|_Y) \backslash \{\underline{0}\}$.
(xiii) $A^d = (A^*)^d|_Y$.
(xiv) $A^- = (A^*)^-|_Y$.
(xv) $A^\circ \geq (A^*)^\circ|_Y$.

**Proof** (i) Obvious.

(ii) - (vi) By Proposition **3.1.2** (i) - (iii).

(vii) By Proposition **2.3.3** (i), note that $A, B \in L^Y$,

$$A^* \pitchfork B^* = \{x \in X : A(x) \not\leq B^*(x)'\} = \{y \in Y : A(y) \not\leq B(y)'\} = A \pitchfork B.$$

(viii) Since

$$x_\lambda \in M(\downarrow A^*) \Longleftrightarrow A^*(x) \geq \lambda > 0 \Longleftrightarrow x \in Y,\ x_\lambda|_Y \leq A.$$

So note that $A, B \in L^Y$, we have

$$\begin{aligned} A \backslash B =& \bigvee\{y_\lambda \in M(\downarrow A) : B(y) = 0\} \vee \bigvee\{y_\lambda \in M(\downarrow A) : \lambda \not\geq B(y) > 0\} \\ =& (\bigvee\{x_\lambda \in M(\downarrow A^*) : B^*(x) = 0\} \vee \bigvee\{x_\lambda \in M(\downarrow A^*) : \lambda \not\geq B^*(x) > 0\})|_Y \\ =& (A^* \backslash B^*)|_Y. \end{aligned}$$

(ix) By Proposition **2.3.19** (i), $A^* \backslash B^* \leq A^*$, so $(A^* \backslash B^*)(x) = 0$ for every $x \in X \backslash Y$. By (vi) proved above, $(A^* \backslash B^*)|_Y = A \backslash B$, so $A^* \backslash B^* = (A \backslash B)^*$.

(x) Since $e^*$ takes value $0_L$ outside $Y$, so

$$\begin{aligned} U \in \mathcal{Q}(e) &\Longleftrightarrow \exists V \in \delta,\ U = V|_Y,\ e \not\leq U' = (V|_Y)' = V'|_Y \\ &\Longleftrightarrow \exists V \in \delta,\ U = V|_Y,\ e^* \not\leq V' \\ &\Longleftrightarrow \exists V \in \mathcal{Q}(e^*),\ U = V|_Y \\ &\Longleftrightarrow U \in \mathcal{Q}(e^*)|_Y. \end{aligned}$$

(xi) Since in the present case $L$ is a F-lattice, a R-neighborhood is just the pseudo-complement of a Q-neighborhood, so the conclusion holds.

$$\begin{aligned}
\text{(xii) } y_\lambda \in Acu(A) &\Longleftrightarrow \forall U \in \mathcal{Q}(y_\lambda),\ A\backslash y_\lambda \not\leq U' \\
&\Longleftrightarrow \forall V \in \mathcal{Q}(y_\lambda{}^*),\ A\backslash y_\lambda \not\leq (V|_Y)' = V'|_Y \quad \text{(by (x) and (v))} \\
&\Longleftrightarrow \forall V \in \mathcal{Q}(y_\lambda{}^*),\ A^*\backslash y_\lambda{}^* = (A\backslash y_\lambda)^* \not\leq V' \quad \text{(by (ix))} \\
&\Longleftrightarrow y_\lambda{}^* \in Acu(A^*) \\
&\Longleftrightarrow y_\lambda \in (Acu(A^*)|_Y.
\end{aligned}$$

So the molecules in $Acu(A^*)$ other than the preceding ones, if exist, can only possess support points in $X\backslash Y$, and hence the conclusion holds.

(xiii) By the definition of join and (xii) proved above,

$$\begin{aligned}
A^d &= \bigvee Acu(A) = \bigvee((Acu(A^*)|_Y)\backslash\{\underline{0}\}) = \bigvee(Acu(A^*)|_Y) \\
&= (\bigvee Acu(A^*))|_Y = (A^*)^d|_Y.
\end{aligned}$$

(xiv) By Theorem **2.3.24** (i) and the preceding (xiii),

$$(A^*)^-|_Y = (A^* \vee (A^*)^d)|_Y = A^*|_Y \vee (A^*)^d|_Y = A \vee A^d = A^-.$$

(xv) Since $A^*|_Y = A$, so $(A^*)^\circ|_Y \leq A$. By the definition of subspace topology, it is also an open subset in $(L^Y, \delta|_Y)$, so $(A^*)^\circ|_Y \leq A^*$. □

## 3.2 Product Spaces

If consider an $L$-fts as a model of a practical system, then in many situations we need to face more than one $L$-fts simultaneously. Composing them into one structure but still keeping their own main properties, this is the motivation of constructing product spaces.

**3.2.1 Definition** Let $\mathcal{S} = \{(L^{X_t}, \delta_t) : t \in T\}$ be a family of $L$-fts', $\mathcal{A} = \{A_t : t \in T\}$ a family of $L$-fuzzy subsets where $A_t \in L^{X_t}$ for every $t \in T$. Denote $X = \prod_{t\in T} X_t$.

For every $t \in T$, suppose $p_t : X \to X_t$ is the ordinary projection, define the *projection* from $L$-fuzzy space $L^X$ to $L$-fuzzy space $L^{X_t}$ as

$$p_t{}^{\to}\colon\ L^X \to L^{X_t}.$$

Define the *product topology* of $L$-fuzzy topologies $\{\delta_t : t \in T\}$ on $X$, denoted by $\prod_{t\in T} \delta_t$, as the $L$-fuzzy topology $\delta$ on $X$ generated by the subbase

$$\{p_t{}^{\leftarrow}(U_t) : U_t \in \delta_t,\ t \in T\},$$

and call the $L$-fts $(L^X, \delta)$ the *product space* of $L$-fts' $\{(L^{X_t}, \delta_t) : t \in T\}$, or an *$L$-fuzzy product space*, denote it by $\prod \mathcal{S}$ or $\prod_{t\in T}(L^{X_t}, \delta_t)$. For every $t \in T$, call $(L^{X_t}, \delta_t)$ a *coordinate space* or a *component* of the product space $\prod \mathcal{S}$; some times also call $(L^{X_t}, \delta_t)$ the *t'th coordinate space* to emphasize its index $t$.

Define the *product* of $L$-fuzzy subsets $\mathcal{A} = \{A_t : t \in T\}$, denoted by $\prod \mathcal{A}$ or $\prod_{t\in T} A_t$, as

$$\prod \mathcal{A} = \prod_{t\in T} A_t = \bigwedge\{p_t{}^{\leftarrow}(A_t) : t \in T\}.$$

For $x \in X$, $s \in T$ and the slice $sl(x,s)$ in $\prod_{t\in T} X_t$ at point $x$ parallel to $X_s$, $L$-fuzzy subspace $(L^{sl(x,s)}, \delta|_{sl(x,s)})$ of $\prod_{t\in T}(L^{X_t}, \delta_t)$ is called the *$L$-fuzzy slice* in $\prod_{t\in T}(L^{X_t}, \delta_t)$ *at $L$-fuzzy point $x_1$ parallel* to $(L^{X_s}, \delta_s)$.

**3.2.2 Remark** In the preceding definition, since every $p_t{}^{\leftarrow}(\underline{1})$ is just the largest element of $L^X$, by Theorem **2.2.14** (ii), the family there is indeed a subbase of some $L$-fuzzy topology on $X$, the definition is reasonable.

Directly follow the definition, we get the following

**3.2.3 Proposition** *Let $\mathcal{S} = \{(L^{X_t}, \delta_t) : t \in T\}$ be a family of L-fts', $(L^X, \delta)$ be their product space. Then*

$$\{\bigwedge_{t \in F} p_t^{\leftarrow}(U_t) : F \in [T]^{<\omega}, \forall t \in F, U_t \in \delta_t\}$$

*is a base of the product topology $\delta$.* □

Obviously, the subbase and base mentioned in Definition **3.2.1** have their special importance, so we give them some special names:

**3.2.4 Definition** Let $\mathcal{S} = \{(L^{X_t}, \delta_t) : t \in T\}$ be a family of $L$-fts'. Call the family

$$\{p_t^{\leftarrow}(U_t) : U_t \in \delta_t, t \in T\},$$

and the family

$$\{\bigwedge_{t \in F} p_t^{\leftarrow}(U_t) : F \in [T]^{<\omega}, \forall t \in F, U_t \in \delta_t\}$$

the *canonical subbase* and the *canonical base* of the product topology $\prod_{t \in T} \delta_t$ respectively.

**3.2.5 Theorem** *Let $\mathcal{S} = \{(L^{X_t}, \delta_t) : t \in T\}$ be a family of L-fts', $(L^X, \delta)$ be their product space. Then*

(i) *For every $t \in T$, projection $p_t^{\rightarrow} : (L^X, \delta) \to (L^{X_t}, \delta_t)$ is continuous.*

(ii) *The product topology $\delta$ is just the coarsest L-fuzzy on X which makes every projection $p_t^{\rightarrow}$ continuous.*

**Proof** (i) $\forall U_t \in \delta_t$, $p_t^{\rightarrow}(U_t)$ is just an element of the canonical base of the product space, so $p_t^{\rightarrow}$ is continuous.

(ii) If $\mu$ is an $L$-fuzzy topology on $X$ making every projection continuous, then $\forall t \in T$, $\forall U_t \in \delta_t$, $\mu$ must contain $p_t^{\leftarrow}(U_t)$, and then their finite meets. So $\mu$ contains the canonical subbase of $\delta$ and then $\delta$ itself, $\delta$ is coarser than any $L$-fuzzy topology on $X$ which makes every projection continuous. By (i), the conclusion holds. □

Projections in an $L$-fuzzy product space connect the product to its components, makes a possibility to reflect the properties of each side to another. The next is a typical conclusion:

**3.2.6 Theorem** *Let $\mathcal{S} = \{(L^{X_t}, \delta_t) : t \in T\}$ be a family of L-fts', $(L^X, \delta)$ be their product space. Then for every L-fts $(L^Y, \mu)$ and every L-fuzzy mapping $f^{\rightarrow} : (L^Y, \mu) \to (L^X, \delta)$, $f^{\rightarrow}$ is continuous if and only if the composition $p_t^{\rightarrow} f^{\rightarrow} : (L^Y, \mu) \to (L^{X_t}, \delta_t)$ is continuous for every $t \in T$.*

**Proof** By Proposition **2.4.8** (i) and preceding theorem, the necessity is clear. We prove the sufficiency.

Suppose $p_t^{\rightarrow} f^{\rightarrow}$ is continuous for every $t \in T$, then by Theorem **2.1.23** (i), so is $(p_t f)^{\rightarrow}$. $\forall t \in T$, $\forall U_t \in \delta_t$, by Theorem **2.1.23** (ii),

$$f^{\leftarrow}(p_t^{\leftarrow}(U_t)) = (p_t f)^{\leftarrow}(U_t) \in \mu.$$

By the structure of the canonical subbase in $(L^X, \delta)$ and Theorem **2.4.5** (v), $f^{\rightarrow}$ is continuous. □

Every projection is an open mapping is another familiar property of projections in ordinary topology. But, in general, it is not still true in $L$-fuzzy product spaces. This point is shown in the following example:

**3.2.7 Example** Take $X_0 = X_1 = \{x, y\}$, $L = \{0, \lambda, 1\}$ is a chain, $\delta_0 = \{\underline{0}, \underline{1}\}$, $\delta_1 =$

$\{\underline{0}, x_\lambda, \underline{1}\}$, then $(L^{X_0}, \delta_0)$ and $(L^{X_1}, \delta_1)$ be $L$-fts'. Denote their product space by $(L^X, \delta)$, then $U = p_1{}^{\leftarrow}(x_\lambda)$ is an open subset in $(L^X, \delta)$. For the two points $x$, $y$ in $X_0 = \{x, y\}$,

$$\begin{aligned} p_0{}^{\rightarrow}(p_1{}^{\leftarrow}(x_\lambda))(x) &= \bigvee\{x_\lambda(p_1(z)) :\ z \in X_0 \times X_1,\ p_0(z) = x\} \\ &= x_\lambda(x) \vee x_\lambda(y) = \lambda, \\ p_0{}^{\rightarrow}(p_1{}^{\leftarrow}(x_\lambda))(y) &= \bigvee\{x_\lambda(p_1(z)) :\ z \in X_0 \times X_1,\ p_0(z) = y\} \\ &= x_\lambda(x) \vee x_\lambda(y) = \lambda. \end{aligned}$$

So we get $p_0{}^{\rightarrow}(p_1{}^{\leftarrow}(x_\lambda)) = \underline{\lambda} \notin \delta_0$, projection $p_0{}^{\rightarrow} :\ (L^X, \delta) \to (L^{X_0}, \delta_0)$ is not open.

The cause of the failure of this property in $L$-fuzzy topology is that $L$-fuzzy subsets have level structures, or in another word, have stratifications. Theorem **3.2.9** in the following shows this point in some degree.

**3.2.8 Lemma** *Let $\{(L^{X_t}, \delta_t) :\ t \in T\}$ be a family of L-fts', $(L^X, \delta)$ their product space. Then*

(i) $A \in L^X,\ s \in T,\ A_s \in L^{X_s} \Longrightarrow p_s{}^{\rightarrow}(p_s{}^{\leftarrow}(A_s) \wedge A) = A_s \wedge p_s{}^{\rightarrow}(A)$.

(ii) $s \in T,\ F \subset T \backslash \{s\},\ \forall t \in F,\ A_t \in L^{X_t},\ \bigvee_{x \in X}(\bigwedge_{t \in T} p_t{}^{\leftarrow}(A_t))(x) = a \Longrightarrow p_s{}^{\rightarrow}(\bigwedge_{t \in T} p_t{}^{\leftarrow}(A_t)) = \underline{a}$.

**Proof** (i) $\forall x^s \in X_s$, since $L$ is a completely distributive lattice and then is infinitely distributive,

$$\begin{aligned} (p_s{}^{\rightarrow}(p_s{}^{\leftarrow}(A_s) \wedge A))(x^s) &= \bigvee\{(p_s{}^{\leftarrow}(A_s) \wedge A)(x) :\ x \in X,\ p_s(x) = x^s\} \\ &= \bigvee\{p_s{}^{\leftarrow}(A_s)(x) \wedge A(x) :\ x \in X,\ p_s(x) = x^s\} \\ &= \bigvee\{A_s(p_s(x)) \wedge A(x) :\ x \in X,\ p_s(x) = x^s\} \\ &= \bigvee\{A_s(x^s) \wedge A(x) :\ x \in X,\ p_s(x) = x^s\} \\ &= A_s(x^s) \wedge \bigvee\{A(x) :\ x \in X,\ p_s(x) = x^s\} \\ &= A_s(x^s) \wedge p_s{}^{\rightarrow}(A)(x^s) \\ &= (A_s \wedge p_s{}^{\rightarrow}(A))(x^s), \end{aligned}$$

so $p_s{}^{\rightarrow}(p_s{}^{\leftarrow} \wedge A) = A_s \wedge p_s{}^{\rightarrow}(A)$.

(ii) Denote $A = \bigwedge_{t \in F} p_t{}^{\leftarrow}(A_t)$, we shall prove $p_s{}^{\rightarrow}(A) = \underline{a}$. $\forall x^s \in X_s$, note that $s \notin F$,

$$\begin{aligned} p_s{}^{\rightarrow}(A)(x^s) &= \bigvee\{(\bigwedge_{t \in F} p_t{}^{\leftarrow}(A_t))(x) :\ x \in X,\ p_s(x) = x^s\} \\ &= \bigvee\{\bigwedge_{t \in F} A_t(p_t(x)) :\ x \in X,\ p_s(x) = x^s\} \\ &= \bigvee\{\bigwedge_{t \in F} A_t(x^t) :\ \forall t \in F,\ x^t \in X_t\} \\ &= \bigvee\{\bigwedge_{t \in F} A_t(p_t(x)) :\ x \in X\} \\ &= \bigvee_{x \in X}(\bigwedge_{t \in F} p_t{}^{\leftarrow}(A_t))(x) \\ &= a, \end{aligned}$$

so $p_s{}^{\rightarrow}(A) = \underline{a}$. □

**3.2.9 Theorem** *Let $\{(L^{X_t}, \delta_t) :\ t \in T\}$ be a family of L-fts', $(L^X, \delta)$ be their product space, $s \in T$. Then the following conditions are equivalent:*

(i) *Projection $p_s :\ (L^X, \delta) \to (L^{X_s}, \delta_s)$ is an open mapping.*

(ii) *For every element $U$ of the canonical base of $(L^X, \delta)$ which takes a constant value on each slice parallel to $X_s$, and for $a = \bigvee_{x \in X} U(x)$, the layer $aX_s$ is an open subset in $(L^{X_s}, \delta_s)$.*

(iii) $F \in [T\backslash\{s\}]^{<\omega}$, $\forall t \in F$, $U_t \in \delta_t$, $\bigvee_{x\in X}(\bigwedge_{t\in F} p_t{}^{\leftarrow}(U_t))(x) = a \implies aX_s \in \delta_s$.

**Proof** (i)$\Longrightarrow$(ii): Suppose $U$ is an element of the canonical base of $X$, and on each slice in $X$ parallel to $X_s$, $U$ takes a constant value. Denote $a = \bigvee_{x\in X} U(x)$.

First of all, note that we can always represent $U$ as

$$U = p_s{}^{\leftarrow}(U_s) \wedge V, \quad V = \bigwedge_{t\in F} p_t{}^{\leftarrow}(U_t), \tag{3.1}$$

where $U_s \in \delta_s$, $F \in [T\backslash\{s\}]^{<\omega}$, $\forall t \in F$, $U_t \in \delta_t$. This is possible because if $U = \bigwedge_{t\in F} p_t{}^{\leftarrow}(U_t)$ but $s \notin F$, we can always take $F' = \{s\} \cup F$, $U_s = \underline{1} \in \delta_s$, then $U$ possesses the form of equalities (3.1).

$\forall u^s, v^s \in X_s$, $\forall x \in X$, for the slice $sl(x,s)$ in $X$ at $x$ parallel to $X_s$, take two points $y', y'' \in sl(x,s)$ such that $p_s(y') = u^s$, $p_s(y'') = v^s$, we have

$$U_s(u^s) \wedge V(x) = U_s(p_s(y')) \wedge V(y') = p_s{}^{\leftarrow}(U_s)(y') \wedge V(y') = U(y').$$

Similarly, $U_s(v^s) \wedge V(x) = U(y'')$. Since both $y'$ and $y''$ are in the same slice parallel to $X_s$, so

$$U_s(u^s) \wedge V(x) = U(y') = U(y'') = U_s(v^s) \wedge V(x). \tag{3.2}$$

Denote $a_0 = \bigvee_{x\in X} V(x)$. Since $u^s, v^s \in X_s$ and $x \in X$ in equalities (3.2) are arbitrary, so

$$\begin{aligned} a = \bigvee_{x\in X} U(x) &= \bigvee\{p_s{}^{\leftarrow}(U_s)(x) \wedge V(x) :\ x \in X\} \\ &= \bigvee\{U_s(p_s(x)) \wedge V(x) :\ x \in X\} = \bigvee\{U_s(u^s) \wedge V(x) :\ x \in X\} \\ &= U_s(u^s) \wedge \bigvee\{V(x) :\ x \in X\} = U_s(u^s) \wedge a_0. \end{aligned}$$

By the arbitrariness of $u^s \in X_s$, this is to say, $a_0U_s = aX_s$. Therefore, by Lemma **3.2.8** and (i),

$$aX_s = a_0U_s = p_s{}^{\rightarrow}(U) \in \delta_s.$$

(ii)$\Longrightarrow$(iii): Suppose $F \in [T\backslash\{s\}]^{<\omega}$, $U_t \in \delta_t$ for every $t \in F$, $U = \bigwedge_{t\in F} p_t{}^{\leftarrow}(U_t)$ and $a = \bigvee_{x\in X} U(x)$. Then since $s \notin F$, $U$ satisfies the condition of (ii), so $aX_s \in \delta_s$.

(iii)$\Longrightarrow$(i): By Proposition **2.4.7** (i), it is sufficient to prove that $p_s{}^{\rightarrow}$ maps every element of the canonical base of $\delta$ into $\delta_s$.

Note that every element $U$ of the canonical base of $\delta$ can be represented as $U = p_s{}^{\leftarrow}(U_s) \wedge \bigwedge_{t\in F} p_t{}^{\leftarrow}(U_t)$, where $U_s \in \delta_s$, $F \in [T\backslash\{s\}]^{<\omega}$, $U_t \in \delta_t$ for every $t \in F$, because we can take $U_s = 1X_s$ if necessary. Therefore, by Lemma **3.2.8**, we always have

$$p_s{}^{\rightarrow}(U) = aU_s = aX_s \wedge U_s\delta_s,$$

$p_s{}^{\rightarrow} :\ (L^X, \delta) \to (L^{X_s}, \delta_s)$ is an open mapping. □

Clearly, by (ii)$\Longrightarrow$(i) or (iii)$\Longrightarrow$(i) in the preceding theorem, if the $L$-fuzzy topology of a coordinate space of a $L$-fuzzy product space contains every layer in this space, then the projection from the product to this space is certainly open. This is just the following definition and corollary will declare:

**3.2.10 Definition** Let $(L^X, \delta)$ be an $L$-fts. $\delta$ is called *stratified*, if $\delta$ contains every layer of $X$, i.e. for every $a \in L$, $\underline{a} \in \delta$. $(L^X, \delta)$ is called *stratified*, if $\delta$ is stratified.

**3.2.11 Corollary** *Let $\{(L^{X_t}, \delta_t) :\ t \in T\}$ be a family of L-fts', $(L^X, \delta)$ be their product space, $s \in T$. If $(L^{X_s}, \delta_s)$ is stratified, then the projection $p_s{}^{\rightarrow} :\ (L^X, \delta) \to (L^{X_s}, \delta_s)$ is an open mapping.* □

In general topology, every slice is homeomorphic to the coordinate space which is parallel to via the correspondent projection; but from the Theorem **3.2.9** we can find that in general this property are not still possessed by projections in $L$-fuzzy product spaces. The correspondent results are shown in the following:

**3.2.12 Theorem** *Let $\{(L^{X_t}, \delta_t) : t \in T\}$ be a family L-fts', $(L^X, \delta)$ their product space, $s \in T$, $(L^{X_s}, \delta_s)$ be stratified. Then*

(i) *For every slice $\tilde{X}_s$ in $X$ parallel to $X_s$, $(p_s|_{\tilde{X}_s})^{\rightarrow} : (L^{\tilde{X}_s}, \delta|_{\tilde{X}_s}) \rightarrow (L^{X_s}, \delta_s)$ is an L-fuzzy homeomorphism.*

(ii) *For every slice $\tilde{X}_s$ in $X$ parallel to $X_s$ and the inclusion mapping $i_{\tilde{X}_s} : \tilde{X}_s \rightarrow X$, $(i_{\tilde{X}_s})^{\rightarrow}(p_s|_{\tilde{X}_s})^{\leftarrow} : (L^{X_s}, \delta_s) \rightarrow (L^X, \delta)$ is an embedding.*

**Proof** (i) Since $p_s|_{\tilde{X}_s} : \tilde{X}_s \rightarrow X_s$ is bijective, by Theorem **2.1.22** (v), $(p_s|_{\tilde{X}_s})^{\rightarrow} : L^{\tilde{X}_s} \rightarrow L^{X_s}$ is bijective, too. By Theorem **3.1.6** (i), $(p_s|_{\tilde{X}_s})^{\rightarrow}$ is continuous.

Now prove $(p_s|_{\tilde{X}_s})^{\rightarrow}$ is open. Denote the canonical base of $\delta$ by $\mathcal{B}$, then by propositions **2.4.7** (i) and **3.1.5** (iv), it is sufficient to prove that $(p_s|_{\tilde{X}_s})^{\rightarrow}$ maps every element of $\mathcal{B}|_{\tilde{X}_s}$ into $\delta_s$. $\forall U \in \mathcal{B}|_{\tilde{X}_s}$, $\exists V \in \mathcal{B}$, $\exists U_s \in \delta_s$, $\exists F \in [T \backslash \{s\}]^{<\omega}$ and $U_t \in \delta_t$ for every $t \in F$ such that $U = V|_{\tilde{X}_s}$, $V = p_s^{\leftarrow}(U_s) \wedge \bigwedge_{t \in F} p_t^{\leftarrow}(U_t)$, where we declare that the meet of $V$ can always contain a $U_s$ because we can take $U_s = \underline{1}$ if it is necessary. Take a point $z \in \tilde{X}_s$. Denote $a = \bigwedge_{t \in F} U_t(z^t)$ and $p_t(z) = z^t$ for every $t \in T$, then

$$\begin{aligned}
(p_s|_{\tilde{X}_s})^{\rightarrow}(U)(x^s) &= \bigvee\{(V|_{\tilde{X}_s})(x) : x \in \tilde{X}_s,\ p_s|_{\tilde{X}_s}(x) = x^s\} \\
&= \bigvee\{(p_s^{\leftarrow}(U_s) \wedge \bigwedge_{t \in F} p_t^{\leftarrow}(U_t))|_{\tilde{X}_s}(x) : x \in \tilde{X}_s,\ p_s(x) = x^s\} \\
&= \bigvee\{U_s(p_s(x)) \wedge \bigwedge_{t \in F} U_t(p_t(x)) : x \in \tilde{X}_s,\ p_s(x) = x^s\} \\
&= \bigvee\{U_s(x^s) \wedge \bigwedge_{t \in F} U_t(z^t) : x \in \tilde{X}_s,\ p_s(x) = x^s\} \\
&= U_s(x^s) \wedge \bigwedge_{t \in F} U_t(z^s) \\
&= (aU_s)(x^s).
\end{aligned}$$

Since $(L^{X_s}, \delta_s)$ is stratified, $(p_s|_{\tilde{X}_s})^{\rightarrow}(U)(x^s) = aU_s = \underline{a} \wedge U_s \in \delta_s$, $(p_s|_{\tilde{X}_s})^{\rightarrow}$ is open. So it is an $L$-fuzzy homeomorphism.

(ii) By (i) proved above and Theorem **2.1.24**, $(p_s|_{\tilde{X}_s})^{\leftarrow}$ here is a Zadeh type mapping and $(p_s|_{\tilde{X}_s})^{\leftarrow} = ((p_s|_{\tilde{X}_s})^{-1})^{\rightarrow}$, by Proposition **2.4.7** (iv), $((p_s|_{\tilde{X}_s})^{-1})^{\rightarrow}$ is an $L$-fuzzy homeomorphism. Denote $f = i_{\tilde{X}_s} \circ (p_s|_{\tilde{X}_s})^{-1} : X_s \rightarrow X$, then $f[X_s] = \tilde{X}_s$, $f|^{f[X_s]} = (p_s|_{\tilde{X}_s})^{-1}$, $(f|^{f[X_s]})^{\rightarrow} = ((p_s|_{\tilde{X}_s})^{-1})^{\rightarrow}$ is an $L$-fuzzy homeomorphism. By Theorem **2.1.23** (i) and the implication (ii)$\Longrightarrow$(i) in Theorem **3.1.14**,

$$(i_{\tilde{X}_s})^{\rightarrow}(p_s|_{\tilde{X}_s})^{\leftarrow} = (i_{\tilde{X}_s})^{\rightarrow}((p_s|_{\tilde{X}_s})^{-1})^{\rightarrow} = (i_{\tilde{X}_s} \circ (p_s|_{\tilde{X}_s})^{-1})^{\rightarrow} = f^{\rightarrow}$$

is an embedding. □

**3.2.13 Corollary** *Let $\{(L^{X_t}, \delta_t) : t \in T\}$ be a family of L-fts', $(L^X, \delta)$ their product space, $s \in T$. If $(L^{X_s}, \delta_s)$ is stratified, then $(L^{X_s}, \delta_s)$ is homeomorphic to every L-fuzzy slice in $(L^X, \delta)$ parallel to $(L^{X_s}, \delta_s)$.* □

Even in general topology, we know that a cube produced with a family of open subsets in a product space need not be open, so it does not hold in $L$-fuzzy product space, too. But, however, note that every projection $p_t^{\rightarrow}$ is continuous and then

$p_t^{\leftarrow}$ maps closed subsets to closed ones, we still has the correspondent positive result about closed subsets in $L$-fuzzy product space:

**3.2.14 Theorem** *Let $\{(L^{X_t}, \delta_t) : t \in T\}$ be a family of L-fts', $(L^X, \delta)$ their product space, $A_t$ is a closed subset in $(L^{X_s}, \delta_s)$ for every $t \in T$. Then $\prod_{t \in T} A_t$ is a closed subset in $(L^X, \delta)$.* □

## 3.3 Sum Spaces

In Section **3.1**, we consider some part of an $L$-fts as an $L$-fts. In fact, we can also make some $L$-fts' along a reverse direction: Combining some $L$-fts' into a "bigger" one. This is just the motivation of sum space.

By Proposition **3.1.2** (i) and (ii), the following proposition can be easily verified:

**3.3.1 Proposition** *Let $\{(L^{X_t}, \delta_t) : t \in T\}$ be a family of L-fts', for different $t, t' \in T$, $X_t$ and $X_{t'}$ be disjoint, $X = \bigcup_{t \in T} X_t$. If define a subfamily $\bigoplus_{t \in T} \delta_t \subset L^X$ as follows:*

$$\forall U \in L^X, \quad U \in \bigoplus_{t \in T} \delta_t \iff \forall t \in T,\ U|_{X_t} \in \delta_t,$$

*then $\bigoplus_{t \in T} \delta_t$ is an L-fuzzy topology on X.* □

So we can give out the following

**3.3.2 Definition** Let $\{(L^{X_t}, \delta_t) : t \in T\}$ be a family of $L$-fts', different $X_t$'s be disjoint, $X = \bigcup_{t \in T} X_t$. Define the *sum topology* of $\{\delta_t : t \in T\}$ on $L^X$, denoted by $\bigoplus_{t \in T} \delta_t$, as follows:

$$\forall U \in L^X, \quad U \in \bigoplus_{t \in T} \delta_t \iff \forall t \in T,\ U|_{X_t} \in \delta_t.$$

Call the $L$-fts $(L^X, \bigoplus_{t \in T} \delta_t)$ the *sum space* of $\{(L^{X_t}, \delta_t) : t \in T\}$, denoted by $\bigoplus_{t \in T}(L^{X_t}, \delta_t)$.

In a situation of finite number of $L$-fts', also denote a sum topology and a sum space as $\delta_0 \oplus \cdots \oplus \delta_n$ and $(L^{X_0}, \delta_0) \oplus \cdots \oplus (L^{X_n}, \delta_n)$ respectively.

**3.3.3 Remark** The preceding definition requires that $X_t$'s must be disjoint. This requirement will not limit us seriously, because for every $t \in T$, take an $L$-fts $(L^{Y_t}, \mu_t)$ as follows:

$$Y_t = X_t \times \{t\},$$
$$\forall A \in L^{X_t},\ A \times \{t\} \in L^{Y_t},$$
$$\forall x \in X_t, \quad (A \times \{t\})(x, t) = A(x),$$
$$\mu_t = \{U \times \{t\} : U \in \delta_t\},$$

then one easily verify that $(L^{Y_t}, \mu_t)$ is an $L$-fts and is homeomorphic to $(L^{X_t}, \delta_t)$, and different $Y_t$'s are disjoint. Therefore, if necessary, we can always use these $(L^{Y_t}, \mu_t)$'s to replace $(L^{X_t}, \delta_t)$ without effecting relative results in topological sense.

By Proposition **3.1.2** (iii), the following results hold:

**3.3.4 Theorem** *Let $\{(L^{X_t}, \delta_t) : t \in T\}$ be a family of L-fts', different $X_t$'s be disjoint, $(L^X, \delta) = \bigoplus_{t \in T}(L^{X_t}, \delta_t)$. Then for every $A \in L^X$, A is a closed subset in $(L^X, \delta)$ if and only if $A|_{X_t}$ is a closed subset in $(L^{X_t}, \delta_t)$ for every $t \in T$.* □

Recall the definition of extension of an $L$-fuzzy subspace (Definition **3.1.8**), we have

**3.3.5 Theorem** *Let $\{(L^{X_t}, \delta_t) : t \in T\}$ be a family of L-fts', different $X_t$'s be disjoint, $(L^X, \delta) = \bigoplus_{t\in T}(L^{X_t}, \delta_t)$. Then for every $t \in T$, $(L^{X_t}, \delta_t)$ is an open-and-closed L-fuzzy subspace of $(L^X, \delta)$ and for the inclusion $i_t : X_t \to X$, $i_t^{\to} : (L^{X_t}, \delta_t) \to (L^X, \delta)$ is an embedding.* □

**3.3.6 Theorem** *Let $\{(L^{X_t}, \delta_t) : t \in T\}$ be a family of L-fts', different $X_t$'s be disjoint, $(L^X, \delta) = \bigoplus_{t\in T}(L^{X_t}, \delta_t)$, $(L^Y, \mu)$ an L-fts, $f^{\to} : L^X \to L^Y$ an L-fuzzy mapping. Then $f^{\to}$ is continuous if and only if for every $t \in T$ and the inclusion $i_t : X_t \to X$, $f^{\to}i_t^{\to} : (L^{X_t}, \delta_t) \to (L^Y, \mu)$ is continuous.* □

**3.3.7 Theorem** *Let $(L^X, \delta)$ be an L-fts, X can be represented as the disjoint union of a family $\{X_t : t \in T\}$ of nonempty subsets of X. Then $(L^X, \delta)$ can be represented as the sum space of the family $\{(L^{X_t}, \delta|_{X_t}) : t \in T\}$ if and only if $\chi_{X_t} \in \delta$ for every $t \in T$.* □

## 3.4 Quotient Spaces

In a set of objects to be investigated, people often divide them into some subsets according to their similar properties, and then considering every these subset as a basic element, investigate the properties they displaying together.

Similar to various quotient objects in many other fields, this is also the background of the idea of quotient space in *L*-fuzzy topology. Let us consider a surjective *L*-fuzzy mapping $f^{\to} : L^X \to L^Y$. In fact, $M(L^X)$ and $M(L^Y)$ are generating sets of $L^X$ and $L^Y$ respectively, and for every $y_\lambda \in M(L^Y)$, $f^{\to}$ makes the subset $\{x_\lambda \in M(L^X) : f(x) = y\}$ of $M(L^X)$ as a single element $y_\lambda$ of $M(L^Y)$. So we can produce quotient space in this way:

**3.4.1 Definition** Let $(L^X, \delta)$ be an *L*-fts, $L^Y$ an *L*-fuzzy space, $f^{\to} : L^X \to L^Y$ a surjective *L*-fuzzy mapping. Define the *L-fuzzy quotient topology* $\delta/f^{\to}$ of $\delta$ with respect to $f^{\to}$ by

$$\delta/f^{\to} = \{V \in L^Y : f^{\leftarrow}(V) \in \delta\}.$$

Call $(L^Y, \delta/f^{\to})$ the *L-fuzzy quotient space* of $(L^X, \delta)$ with respect to $f^{\to}$. Call $f^{\to} : (L^X, \delta) \to (L^Y, \delta/f^{\to})$ an *L-fuzzy quotient mapping.*

For convenience, we also define a symbol for ordinary quotient topology:

**3.4.2 Definition** Let $(X, \mathcal{T})$ be an ordinary topological space, $Y$ a nonempty ordinary set, $f : X \to Y$ an ordinary surjective mapping. Define the ordinary quotient topology on $Y$ with respect to $f$ by $\mathcal{T}/f$

Is an *L*-fuzzy quotient topology an *L*-fuzzy topology indeed? Certainly. Note that every *L*-fuzzy reverse mapping preserves arbitrary joins and meets (showing in Theorem **2.1.17**), we have

**3.4.3 Proposition** *Let $(L^X, \delta)$ be an L-fts, $L^Y$ an L-fuzzy space, $f^{\to} : L^X \to L^Y$ an surjective L-fuzzy mapping. Then $\delta/f^{\to}$ is an L-fuzzy topology on Y, $(L^Y, \delta/f^{\to})$ is an L-fts.* □

Easy to find the following conclusion from the definition of *L*-fuzzy quotient space:

**3.4.4 Theorem** *Let $(L^X, \delta)$ be an L-fts, $L^Y$ an L-fuzzy space, $f^{\rightarrow} : L^X \rightarrow L^Y$ an surjective L-fuzzy mapping. Then $\delta / f^{\rightarrow}$ is the finest L-fuzzy topology on Y making $f^{\rightarrow}$ continuous.* □

**3.4.5 Theorem** *Let $(L^X, \delta)$, $(L^Z, \xi)$ be L-fts', $L^Y$ an L-fuzzy space, $f^{\rightarrow} : L^X \rightarrow L^Y$ an surjective L-fuzzy mapping, $g^{\rightarrow} : L^Y \rightarrow L^Z$ an L-fuzzy mapping. Then $g^{\rightarrow} : (L^Y, \delta/f^{\rightarrow}) \rightarrow (L^Z, \xi)$ is continuous if and only if $g^{\rightarrow} \circ f^{\rightarrow} : (L^X, \delta) \rightarrow (L^Z, \xi)$ is continuous.*

**Proof** The necessity is clear. We prove the sufficiency. Suppose $g^{\rightarrow} \circ f^{\rightarrow}$ is continuous. By Theorem **2.1.23** (i),

$$g^{\rightarrow} \circ f^{\rightarrow} = (g \circ f)^{\rightarrow},$$

so $\forall W \in \xi$, since $g^{\rightarrow} \circ f^{\rightarrow}$ is continuous, by Theorem **2.1.23** (ii),

$$f^{\leftarrow}(g^{\leftarrow}(W)) = (gf)^{\leftarrow}(W) \in \delta.$$

But $f^{\rightarrow}$ is an $L$-fuzzy quotient mapping, this just show $g^{\leftarrow}(W) \in \delta/f^{\rightarrow}$. So $g^{\rightarrow}$ is continuous. □

For any surjective $L$-fuzzy continuous mappings, surely we can verify if it is an $L$-fuzzy quotient mapping by checking if the correspondent $L$-fuzzy topology is finest to make this mapping continuous; but we have two other ways to do this thing:

**3.4.6 Theorem** *Let $(L^X, \delta)$, $(L^Y, \mu)$ be L-fts', $f^{\rightarrow} : (L^X, \delta) \rightarrow (L^Y, \mu)$ a surjective L-fuzzy continuous mapping. Then*

(i) *$f^{\rightarrow}$ is an open mapping $\Longrightarrow$ $f^{\rightarrow}$ is an L-fuzzy quotient mapping.*

(ii) *$f^{\rightarrow}$ is a closed mapping $\Longrightarrow$ $f^{\rightarrow}$ is an L-fuzzy quotient mapping.*

**Proof** (i) Since $f^{\rightarrow}$ is continuous, $\mu \subset \delta/f^{\rightarrow}$. $\forall V \in \delta/f^{\rightarrow}$, since $f^{\rightarrow}$ is surjective and open, by Theorem **2.1.22** (iv), $V = f^{\rightarrow}(f^{\leftarrow}(V)) \in \mu$, so we also have $\delta/f^{\rightarrow} \subset \mu$, $\mu = \delta/f^{\rightarrow}$, $f^{\rightarrow}$ is a quotient mapping.

(ii) Note that every $L$-fuzzy reverse mapping preserves pseudo-complementary operation $'$, the proof is similar to (i). □

**3.4.7 Corollary** *Let $\{(L^{X_t}, \delta_t) : t \in T\}$ be a family of L-fts', $(L^X, \delta)$ their product space, $s \in T$. Then*

(i) *$(L^{X_s}, \delta_s)$ satisfies condition* (ii) *or* (iii) *of Theorem* **3.2.9** $\Longrightarrow$ *$p_s^{\rightarrow}$ is an L-fuzzy quotient mapping.*

(ii) *$(L^{X_s}, \delta_s)$ is stratified $\Longrightarrow$ $p_s^{\rightarrow}$ : is an L-fuzzy quotient mapping.* □

# Chapter 4

# $L$-valued Stratification Spaces

Except constructing new $L$-fts' from ready $L$-fts' just as investigated in the last chapter, we can also produce $L$-fts' from ordinary topological spaces. Moreover, these special $L$-fts' play important roles in the investigation on the relation between $L$-fts' and ordinary topological spaces. In this chapter, after some further investigations on stratified $L$-fts', we introduce the most two kinds of these spaces: weakly induced spaces and induced spaces; these three kinds of $L$-fts' are uniformly called $L$-valued stratification spaces.

## 4.1 Stratified Spaces and Stratifications

In Chapter 3, we defined the concept of stratified $L$-fts. In fact, for an ordinary topological space $(X, \mathcal{T})$ and a F-lattice $L$, $\delta = \{\chi_U : U \in \mathcal{T}\}$ is certainly an $L$-fuzzy topology on $X$; if we add some layers into $\delta$ and consider it as a subbase, then we get a new $L$-fuzzy topology different from the original one. Espeically, if we add all the layers into it, we get a stratified $L$-fuzzy topology. So we give out the following

**4.1.1 Definition** Let $(L^X, \delta)$ be an $L$-fts, $\mu$ be the $L$-fuzzy topology on $X$ generated by $\delta \cup \{\underline{a} : a \in L\}$, then call $\mu$ the *stratification* of $\delta$, and call $(L^X, \mu)$ the *stratification* of $(L^X, \delta)$.

**4.1.2 Proposition** *Let $(L^X, \delta)$ be an L-fts, $\mu$ the stratification of $\delta$. Then*

$$\mu = \{\bigvee\{aU : (a, U) \in \mathcal{A}\} : \mathcal{A} \subset L \times \delta\}. \qquad \square$$

**4.1.3 Proposition** *Let $(L^X, \delta)$, $(L^Y, \mu)$ be L-fts', $(L^Y, \mu)$ be stratified, $f^{\rightarrow} : (L^X, \delta) \to (L^Y, \mu)$ continuous. Then $(L^X, \delta)$ is stratified.* $\square$

**4.1.4 Theorem** *Let $(L^X, \delta)$, $(L^Y, \mu)$ be L-fts', $f^{\rightarrow} : (L^X, \delta) \to (L^Y, \mu)$ an L-fuzzy continuous mapping, $\delta_0$ and $\mu_0$ be respectively the stratifications of $\delta$ and $\mu$. Then $f^{\rightarrow} : (L^X, \delta_0) \to (L^Y, \mu_0)$ is continuous.*

**Proof** $\forall V \in \mu_0$, by Proposition **4.1.2**, $\exists \mathcal{A} \subset L \times \mu$ such that $V = \bigvee\{aW : (a, W) \in \mathcal{A}\}$. Since $f^{\rightarrow} : (L^X, \delta) \to (L^Y, \mu)$ is continuous,

$$\begin{aligned} f^{\leftarrow}(V) &= f^{\leftarrow}(\bigvee\{aW : (a, W) \in \mathcal{A}\}) \\ &= \bigvee\{f^{\leftarrow}(aW) : (a, W) \in \mathcal{A}\} && \text{(by Theorem 2.1.17 (i))} \\ &= \bigvee\{af^{\leftarrow}(W) : (a, W) \in \mathcal{A}\} && \text{(by Theorem 2.1.17 (ii))} \\ &\in \delta_0. \end{aligned}$$

So $f^{\rightarrow} : (L^X, \delta_0) \to (L^Y, \mu_0)$ is continuous. $\square$

Then on the $L$-fts' investigated in Chapter 3, we have

**4.1.5 Theorem** *Let $(L^X, \delta)$ be an L-fts, $Y \subset X$ nonempty, $\mu$ the stratification of $\delta$. Then $\mu|_Y$ is just the stratification of $\delta|_Y$.* $\square$

In Chapter 3, we discussed several kinds of operations on $L$-fts'. Then a natural need is to investigate that in various properties of $L$-fts', which one can be preserved under these operations. So we introduce the following notions:

**4.1.6 Definition** Let $\mathcal{P}$ be a property of $L$-fts', $\mathcal{A}$ a kind of $L$-fuzzy subspaces, $\kappa$ a cardinal number, $\mathcal{M}$ a kind of $L$-fuzzy mappings.

$\mathcal{P}$ is called *hereditary*, if for every $L$-fts $(L^X, \delta)$ which has property $\mathcal{P}$ every $L$-fuzzy subspace of $(L^X, \delta)$ has property $\mathcal{P}$; called *hereditary with respect to* $\mathcal{A}$, if for every $L$-fts $(L^X, \delta)$ which has property $\mathcal{P}$ every $L$-fuzzy subspace of $(L^X, \delta)$ in $\mathcal{A}$ has property $\mathcal{P}$.

$\mathcal{P}$ is called *multiplicative*, if for every family $\{(L^{X_t}, \delta_t) : t \in T\}$ of $L$-fts' which have property $\mathcal{P}$ the $L$-fuzzy product space $\prod_{t\in T}(L^{X_t}, \delta_t)$ has property $\mathcal{P}$; called $\kappa$-*multiplicative*, if for every family $\{(L^{X_t}, \delta_t) : t \in T\}$ of $L$-fts' with $|T| \leq \kappa$ which have property $\mathcal{P}$ the $L$-fuzzy product space $\prod_{t\in T}(L^{X_t}, \delta_t)$ has property $\mathcal{P}$; called *strongly multiplicative*, if for every family $\{(L^{X_t}, \delta_t) : t \in T\}$ of $L$-fts', the $L$-fuzzy product space $\prod_{t\in T}(L^{X_t}, \delta_t)$ has property $\mathcal{P}$ if and only if $(L^{X_t}, \delta_t)$ has property $\mathcal{P}$ for every $t \in T$.

$\mathcal{P}$ is called *preserved by mappings* $\mathcal{M}$, if for every $L$-fts $(L^X, \delta)$ which has property $\mathcal{P}$, every $L$-fts $(L^Y, \mu)$ which is the range space of an $L$-fuzzy mapping $f: (L^X, \delta) \to (L^Y, \mu)$ in $\mathcal{M}$, has property $\mathcal{P}$.

Obviously strong multiplicative property is strictly stronger than multiplicative property. But we have the following

**4.1.7 Theorem** *Let $\{(L^{X_t}, \delta_t) : t \in T\}$ be a family of L-fts', $s \in T$, $(L^{X_s}, \delta_s)$ be stratified, $\mathcal{P}$ a hereditary property of L-fuzzy topological spaces preserved by L-fuzzy homeomorphisms. If $\prod_{t\in T}(L^{X_t}, \delta_t)$ has property $\mathcal{P}$, then $(L^{X_s}, \delta_s)$ has property $\mathcal{P}$.*

**Proof** By Theorem **3.2.12** (i), $(L^{X_s}, \delta_s)$ is homeomorphic to an $L$-fuzzy subspace of $\prod_{t\in T}(L^{X_t}, \delta_t)$. Since $\mathcal{P}$ is hereditary and is preserved by $L$-fuzzy homeomorphisms, $(L^{X_s}, \delta_s)$ has property $\mathcal{P}$. □

**4.1.8 Theorem** *Stratified property is hereditary.* □

**4.1.9 Theorem** *Let $\{(L^{X_t}, \delta_t) : t \in T\}$ be a family of L-fts'. Then*

(i) *If there exists $t \in T$ such that $(L^{X_t}, \delta_t)$ is stratified, then $\prod_{t\in T}(L^{X_t}, \delta_t)$ is stratified.*

(ii) *If $\mu$ is the stratification of $\prod_{t\in T}\delta_t$ and there exists nonempty $S \subset T$ such that for every $t \in S$, $\mu_t$ is the stratification of $\delta_t$, for every $t \in T \backslash S$, $\mu_t = \delta_t$, then $\mu = \prod_{t\in T}\mu_t$.*

**Proof** Denote $(L^X, \delta) = \prod_{t\in T}(L^{X_t}, \delta_t)$.

(i) Suppose $s \in T$, $(L^{X_s}, \delta_s)$ is stratified. Since $\forall a \in L$, $aX = p_s^{\leftarrow}(aX_s) \in \delta$, so $\delta$ is also stratified.

(ii) Denote $\zeta = \prod_{t\in T}\mu_t$. By (i), $\mu \subset \zeta$. $\forall t \in T$, $\forall U \in \mu_t = \delta_t$, certainly we have $p_t^{\leftarrow}(U) \in \delta \subset \mu$. $\forall t \in S$, $\forall U \in \mu_t$, by Proposition **4.1.2**, $\exists \mathcal{A} \subset L \times \delta_t$ such that $U = \bigvee\{aV : (a, V) \in \mathcal{A}\}$. So by Theorem **2.1.17** (i), (ii), we still have

$$\begin{aligned} p_t^{\leftarrow}(U) &= \bigvee\{p_t^{\leftarrow}(\underline{a}) \wedge p_t^{\leftarrow}(V) : (V, a) \in \mathcal{A}\} \\ &= \bigvee\{aX \wedge p_t^{\leftarrow}(V) : (V, a) \in \mathcal{A}\} \in \mu. \end{aligned}$$

Since the family of all the $p_t^{\leftarrow}(U)$'s for every $t \in T$ and every $U \in \mu_t$ is just the canonical subbase of $\zeta$, $\zeta \subset \mu$. □

**4.1.10 Remark** Using the diamond-type lattice, it is easy to construct a counterexample to show that a stratified $L$-fts can be a product of a family of $L$-fts' in which every one is not stratified. In fact, the $L$-fuzzy product topology will always include all the layers contained in the $L$-fuzzy topologies of its coordinate spaces. This makes many conclusions on the relations among an $L$-fuzzy product space and its coordinate spaces involve the property of $L$.

**4.1.11 Theorem** *Let $\{(L^{X_t}, \delta_t) : t \in T\}$ be a family of L-fts', different $X_t$'s be disjoint, $(L^X, \delta) = \bigoplus_{t\in T}(L^{X_t}, \delta_t)$, $\mu_t$ be the stratification of $\delta_t$ for every $t \in T$, $\mu$ the stratification of $\delta$. Then $\mu = \bigoplus_{t\in T}\mu_t$.* □

**4.1.12 Theorem** *Let $\{(L^{X_t}, \delta_t) : t \in T\}$ be a family of L-fts', different $X_t$'s be disjoint, $(L^X, \delta) = \bigoplus_{t\in T}(L^{X_t}, \delta_t)$. Then $\bigoplus_{t\in T}(L^{X_t}, \delta_t)$ is stratified if and only if $(L^{X_t}, \delta_t)$ is stratified for every $t \in T$.* □

**4.1.13 Theorem** *Let $(L^X, \delta)$ be an L-fts, $(L^Y, \delta/f^{\rightarrow})$ the L-fuzzy quotient space of $(L^X, \delta)$ with respect to the L-fuzzy surjective mapping $f^{\rightarrow} : L^X \to L^Y$. Then $(L^X, \delta)$ is stratified if and only if $(L^Y, \delta/f^{\rightarrow})$ is stratified.* □

But for stratifications of quotient spaces, the conclusion parallel to the preceding relative results does not hold. The following counterexample shows this point:

**4.1.14 Example** Take $X$, $Y$ and $f : X \to Y$ as the left one of the following two diagrams, and $L$ as the right one:

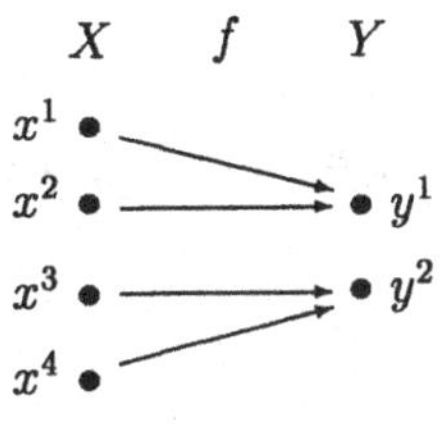


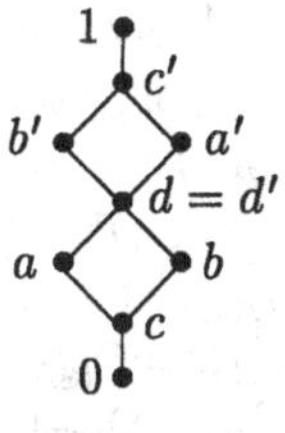


Take $\mathcal{B} \subset L^X$ as $\mathcal{B} = \{\underline{0}, x^1{}_a, x^2{}_b, x^3{}_{a'}, x^4{}_{b'}, \underline{1}\}$, then by Theorem **2.2.14**, $\mathcal{B}$ is a base of an $L$-fuzzy topology $\delta$ on $X$. $\forall V \in \delta/f^{\rightarrow}$, $f^{\leftarrow}(V) \in \delta$ and

$$f^{\leftarrow}(V)(x^1) = V(f(x^1)) = V(y^1) = V(f(x^2)) = f^{\leftarrow}(V)(x^2).$$

But clearly there is no $U \in \delta$ such that $U(x^1) = U(x^2)$ except $U \in \{\underline{0}, \underline{1}\}$. So $f^{\leftarrow}(V) \in \{\underline{0}, \underline{1}\}$. By Theorem **2.1.22** (iv), $V = f^{\rightarrow}f^{\leftarrow}(V) \in f^{\rightarrow}[\{\underline{0}, \underline{1}\}] = \{\underline{0}, \underline{1}\}$, $\delta/f^{\rightarrow} = \{\underline{0}, \underline{1}\}$.

Let $\delta_0$ and $\mu_0$ denote respectively the stratifications of $\delta$ and $\delta/f^{\rightarrow}$, then $\mu_0 = \{\underline{a} : a \in L\}$. For $V = y^1{}_c \vee y^2{}_d \in L^Y$,

$$f^{\leftarrow}(V) = x^1{}_c \vee x^2{}_c \vee x^3{}_d \vee x^4{}_d = bx^1{}_a \vee ax^2{}_b \vee b'x^3{}_{a'} \vee a'x^4{}_{b'} \in \delta_0,$$

so $V \in \delta_0/f^{\rightarrow} \backslash \mu_0$, $\mu_0 \neq \delta_0/f^{\rightarrow}$.

## 4.2 Weakly Induced Spaces

For every $L$-fuzzy topology $\delta$ on a nonempty ordinary set $X$, the family $[\delta]$ of support sets of all the crisp subsets in $\delta$ is certainly an ordinary topology on $X$. Theory of weakly induced spaces or weakly induced topologies is just about the relation between $L$-fuzzy topology and this kind of ordinary topologies. The concept of weakly induced space in the case of $L = [0,1]$ was introduced by H.W. Martin.[116] Liu and Luo generalized this notion to the case that $L$ is an arbitrary F-lattice and introduced the notion of induced $L$-fts.[83] Many nice properties of this kind of spaces were revealed by Chinese scholars.[111, 112, 113, 84, 114, 88, 89, 194, 189, 196] From the ready results in this aspects, we can find that weakly induced spaces play indeed an important role in fuzzy topology.

Since for every $L$-fts $(L^X, \delta)$, $(X, [\delta])$ is clearly a topological space, so we can give out the following

**4.2.1 Definition** Let $(L^X, \delta)$ be an $L$-fts. Call the topological space $(X, [\delta])$ the *background space* of $(L^X, \delta)$.

**4.2.2 Proposition** *Let $(L^X, \delta)$, $(L^Y, \mu)$ be L-fts', $f^{\to}: (L^X, \delta) \to (L^Y, \mu)$ an L-fuzzy continuous mapping. Then $f: (X, [\delta]) \to (Y, [\mu])$ is continuous.* □

Recall the definitions of $A_{(a)}$ and $A_{[a]}$ in Definition **2.1.6**, the following conclusions can be easily proved:

**4.2.3 Proposition** *Let $L^X$ be an L-fuzzy space, $Y \subset X$, $a \in L$, $A \in L^X$, $B \subset L$. Then*

(i) $(A|_Y)_{(a)} = A_{(a)} \cap Y$.

(ii) $(A|_Y)_{[a]} = A_{[a]} \cap Y$.

(iii) $A_{(\bigwedge B)} = \bigcup\{A_{(b)}: b \in B\}$.

(iv) $A_{[\bigvee B]} = \bigcap\{A_{[b]}: b \in B\}$. □

**4.2.4 Proposition** *Let $L^X$ be an L-fuzzy space, $\{A_t : t \in T\} \subset L^X$, $a \in L$, $\gamma \in pr(L)$, $\lambda \in M(L)$, $F \in [T]^{<\omega}$. Then*

(i) $(\bigvee_{t\in T} A_t)_{(a)} = \bigcup_{t\in T}(A_t)_{(a)}$.

(ii) $(\bigwedge_{t\in F} A_t)_{(\gamma)} = \bigcap_{t\in F}(A_t)_{(\gamma)}$.

(iii) $(\bigwedge_{t\in T} A_t)_{[a]} = \bigcap_{t\in T}(A_t)_{[a]}$.

(iv) $(\bigvee_{t\in F} A_t)_{[\lambda]} = \bigcup_{t\in F}(A_t)_{[\lambda]}$. □

**4.2.5 Proposition** *Let $L^X$, $L^Y$ be L-fuzzy spaces, $f: L^X \to L^Y$ an L-fuzzy mapping, $B \in L^Y$, $a \in L$. Then*

(i) $f^{\leftarrow}(B)_{(a)} = f^{-1}(B_{(a)})$.

(ii) $f^{\leftarrow}(B)_{[a]} = f^{-1}(B_{[a]})$. □

**4.2.6 Proposition** *Let $L^X$ be a L-fuzzy space, L a F-lattice, $A \in L^X$, $a \in L$. Then*

(i) $A_{(a)} = X \backslash A'_{[a']}$.

(ii) $A_{[a]} = X \backslash A'_{(a')}$. □

**4.2.7 Proposition** *Let $\{(L^{X_t}, \delta_t) : t \in T\}$ be a family of L-fts', $(L^X, \delta) = \bigoplus_{t\in T}(L^{X_t}, \delta_t)$. Then $(X, [\delta])$ is the ordinary sum space of $\{(X, [\delta_t]) : t \in T\}$.* □

**4.2.8 Note** About $L$-fuzzy subspaces, it is easy to find that the similar conclusion is not true.

**4.2.9 Proposition** *Let $\{(L^{X_t}, \delta_t) : t \in T\}$ be a family of L-fts', $(L^X, \delta) = \prod_{t\in T}(L^{X_t}, \delta_t)$. Then $(X, [\delta]) = \prod_{t\in T}(X, [\delta_t])$.*

**Proof** Let $\mathcal{T}$ denote the ordinary product topology of $\{[\delta_t] : t \in T\}$, then clearly $\mathcal{T} \subset [\delta]$. $\forall U \in [\delta]$, $\forall \gamma \in pr(L)$, then $\exists \mathcal{A} \subset [T]^{<\omega}$, and for every $F \in \mathcal{F}$, every $t \in F$, $\exists U_{F,t} \in \delta_t$ such that $\chi_U = \bigvee_{F\in\mathcal{A}} \bigwedge_{t\in F} U_{F,t}$. By Proposition **4.2.4** (i), (ii),

$$U = (\chi_U)_{(\gamma)} = \bigcup_{F\in\mathcal{A}} \bigcap_{t\in F} (U_{F,t})_{(\gamma)} \in \mathcal{T}.$$

So $[\delta] \subset \mathcal{T}$, $[\delta] = \mathcal{T}$. □

**4.2.10 Proposition** *Let $(L^X, \delta)$ be a weakly induced L-fts, $(L^Y, \delta/f^{\rightarrow})$ the L-fuzzy quotient space of $(L^X, \delta)$ with respect to L-fuzzy surjective mapping $f^{\rightarrow} : L^X \to L^Y$. Then $(Y, [\delta/f^{\rightarrow}])$ is the ordinary quotient space of $(X, [\delta])$ with respect to $f : X \to Y$.*

**Proof** By Theorem **2.1.22** (iii), $f : X \to Y$ is surjective. So the ordinary quotient topology $[\delta]/f$ of $[\delta]$ exists. $\forall V \in [\delta/f^{\rightarrow}]$, $\chi_V \in \delta/f^{\rightarrow}$, since $f^{\rightarrow}$ is the $L$-fuzzy quotient mapping, $\chi_{f^{-1}(V)} = f^{\leftarrow}(\chi_V) \in \delta$, $f^{-1}(V) \in [\delta]$. So $V \in [\delta]/f$, $[\delta/f^{\rightarrow}] \subset [\delta]/f$.

$\forall V \in [\delta]/f$, then $f^{-1}(V) \in [\delta]$. By Proposition **2.1.19** (ii), $f^{\leftarrow}(\chi_V) = \chi_{f^{-1}(V)} \in \delta$. Since $f^{\rightarrow}$ is the $L$-fuzzy quotient mapping, $\chi_V \in \delta/f^{\rightarrow}$, $V \in [\delta/f^{\rightarrow}]$. So $[\delta]/f \subset [\delta/f^{\rightarrow}]$, $[\delta]/f = [\delta/f^{\rightarrow}]$. □

**4.2.11 Definition** Let $L$ be a complete lattice. The co-topologies on $L$ generated respectively by subbases $\{\uparrow a : a \in L\}$, $\{\downarrow a : a \in L\}$ and $\{[a, b] : a, b \in L, a \leq b\}$ are called respectively the *upper co-topology*, the *lower co-topology* and the *interval co-topology* of L respectively denote them by $\underline{\Omega}^*(L)$, $\underline{\Omega}_*(L)$ and $\underline{\Omega}(L)$. The correspondent topologies of $\underline{\Omega}^*(L)$, $\underline{\Omega}_*(L)$ and $\underline{\Omega}(L)$ are respectively called the *upper topology*, the *lower topology* and the *interval topology* of $L$ respectively denote them by $\Omega^*(L)$, $\Omega_*(L)$ and $\Omega(L)$. Also denote respectively these co-topologies and topologies on $L$ by $\underline{\Omega}^*$, $\underline{\Omega}_*$, $\underline{\Omega}$, $\Omega^*$, $\Omega_*$ and $\Omega$ for short.

**4.2.12 Definition** Let $(X, \mathcal{T})$ be an ordinary topological space, $L$ a complete lattice. A mapping $f : X \to L$ is called respcetively *upper semicontinuous*, *lower semicontinuous* and *continuous*, if $f$ is respectively continuous for the topologies $\Omega^*$, $\Omega_*$ and $\Omega$.

**4.2.13 Proposition** *Let $(X, \mathcal{T})$ be a topological space, $L$ a complete lattice. Then a mapping $f : X \to L$ is continuous if and only if $f$ is both upper semicontinuous and lower semicontinuous.* □

**4.2.14 Proposition** *Let $(L^X, \delta)$ is an L-fts. Then every crisp subset in $\delta$ is a lower semicontinuous mapping from $(X, [\delta])$ to $L$.* □

**4.2.15 Definition** Let $(L^X, \delta)$ be an $L$-fts. $\delta$ is called *lattice-valued weakly induced*, or *L-valued weakly induced*, or *weakly induced* for short, if every $U \in \delta$ is a lower semicontinuous mapping from the background space $(X, [\delta])$ to $L$; or equivalently, if every $P \in \delta'$ is an upper semicontinuous mapping from $(X, [\delta])$ to $L$. $(L^X, \delta)$ is called *lattice-valued weakly induced*, or *L-valued weakly induced*, or *weakly induced* for short, if $\delta$ is weakly induced.

By Proposition **4.2.3** (i), (ii), we have

**4.2.16 Theorem** *Let $(X, \mathcal{T})$ be an ordinary topological space, $L$ a complete lattice, $f : X \to L$ an ordinary mapping, $L_0$ a meet-generating set of $L$, $L_1$ a join-generating set of $L$. Then*

(i) *$f$ is lower semicontinuous if and only if for every $a \in L_0$, $f_{(a)}$ is open in $(X, \mathcal{T})$.*

(ii) *$f$ is lower semicontinuous if and only if for every $a \in L_0$, $f^{[a]}$ is closed in $(X,\mathcal{T})$.*

(iii) *$f$ is upper semicontinuous if and only if for every $a \in L_1$, $f_{[a]}$ is closed in $(X,\mathcal{T})$.*

(iv) *$f$ is upper semicontinuous if and only if for every $a \in L_1$, $f^{(a)}$ is open in $(X,\mathcal{T})$.* □

**4.2.17 Theorem** *Let $(L^X,\delta)$ be an L-fts, $L_0$ a meet-generating set of $L$, $L_1$ a join-generating set of $L$. Then the following conditions are equivalent:*

(i) *$(L^X,\delta)$ is weakly induced.*

(ii) *For every $U \in \delta$ and every $a \in L$, $U_{(a)} \in [\delta]$*

(iii) *For every $U \in \delta$ and every $a \in L_0$, $U_{(a)} \in [\delta]$*

(iv) *For every $P \in \delta'$ and every $a \in L$, $P_{[a]} \in [\delta']$.*

(v) *For every $P \in \delta'$ and every $a \in L_1$, $P_{[a]} \in [\delta']$.* □

Particularly, for $L$-fts, we have the following

**4.2.18 Theorem** *Let $(L^X,\delta)$ be an L-fts, $\mathcal{B}$ a subbase of $\delta$. Then the following conditions are equivalent:*

(i) *$(L^X,\delta)$ is weakly induced.*

(ii) *For every $U \in \mathcal{B}$ and every $\gamma \in pr(L)$, $U_{(\gamma)} \in [\delta]$.*

(iii) *For every $P \in \mathcal{B}'$ and every $\lambda \in M(L)$, $P_{[\lambda]} \in [\delta']$.*

**Proof** (i)$\Longrightarrow$(ii): By Theorem **4.2.17** (i)$\Longrightarrow$(ii).

(ii)$\Longrightarrow$(iii): By De Morgan's Law.

(iii)$\Longrightarrow$(i): $\forall P \in \delta'$, then by De Morgan's Law, $\exists \mathcal{A} \subset [\mathcal{B}']^{<\omega}$ such that

$$P = \bigwedge\{\bigvee\{Q : Q \in \mathcal{F}\} : \mathcal{F} \in \mathcal{A}\}.$$

$\forall \lambda \in M(L)$, by Proposition **4.2.4** (iii), (iv),

$$P_{[\lambda]} = \bigcap\{\bigcup\{Q_{[\lambda]} : Q \in \mathcal{F}\} : \mathcal{F} \in \mathcal{A}\} \in [\delta'].$$

By Corollary **1.3.12** (i) and Theorem **4.2.17** (v)$\Longrightarrow$(i), $(L^X,\delta)$ is weakly induced. □

**4.2.19 Theorem** *Weakly induced property is hereditary.*

**Proof** By Proposition **4.2.3** (i). □

**4.2.20 Theorem** *Let $\{(L^{X_t},\delta_t) : t \in T\}$ be a family of L-fts'. Then $\bigoplus_{t\in T}(L^{X_t},\delta_t)$ is weakly induced if and only if for every $t \in T$, $(L^{X_t},\delta_t)$ is weakly induced.*

**Proof** By Proposition **4.2.3** (i). □

**4.2.21 Theorem** *Weakly induced property is strongly multiplicative.*

**Proof** Suppose $\{(L^{X_t},\delta\}$ be a family of $L$-fts', denote $(L^X,\delta) = \prod_{t\in T}(L^{X_t},\delta_t)$.

(Sufficiency) Take arbitrarily an element $P = \bigvee_{t\in F} p_t^{\leftarrow}(P_t)$ of the canonical subbase of the $L$-fuzzy co-topology $\delta'$, where $F \in [T]^{<\omega}$, $P_t \in \delta_t{}'$ for every $t \in F$. $\forall \lambda \in M(L)$, then by propositions **4.2.4** (iv), **4.2.5** (ii) and **4.2.2**,

$$P_{[\lambda]} = \bigcup_{t\in F}(p_t^{\leftarrow}(P_t))_{[\lambda]} = \bigcup_{t\in F} p_t^{-1}((P_t)_{[\lambda]}) \in [\delta'].$$

So by Theorem **4.2.18** (iii)$\Longrightarrow$(i), $(L^X,\delta)$ is weakly induced.

(Necessity) $\forall s \in T$, $\forall U_s \in \delta_s$, $\forall a \in L$. Since $(L^X,\delta)$ is weakly induced, by Proposition **4.2.5** (i),

$$p_s^{-1}((U_s)_{(a)}) = (p_s^{\leftarrow}(U_s))_{(a)} \in [\delta].$$

By Proposition **4.2.9**, $p_s : (X,[\delta]) \to (X_s,[\delta_s])$ is just the $s$'th projection of the ordinary product space $\prod_{t\in T}(X_t, [\delta_t])$, so it is an open surjective mapping and hence

$$(U_s)_{(a)} = p_s(p_s^{-1}((U_s)_{(a)})) \in [\delta_s].$$

By Theorem **4.2.17** (ii)$\Longrightarrow$(i), $(L^{X_s}, \delta_s)$ is weakly induced. □

**4.2.22 Theorem** *Let $(L^X, \delta)$ be a weakly induced L-fts, $(L^Y, \delta/f^{\to})$ the L-fuzzy quotient space of $(L^X, \delta)$ with respect to L-fuzzy surjective mapping $f^{\to}: L^X \to L^Y$. Then $(L^Y, \delta/f^{\to})$ is weakly induced.*

**Proof** $\forall V \in \delta/f^{\to}$, $\forall a \in L$, sicne $(L^X, \delta)$ is weakly induced, by Proposition **4.2.5** (i),

$$f^{-1}(V_{(a)}) = (f^{\leftarrow}(V))_{(a)} \in [\delta].$$

By Proposition **4.2.10**, $V_{(a)} \in [\delta/f^{\to}]$. By Theorem **4.2.17** (ii)$\Longrightarrow$(i), $(L^Y, \delta/f^{\to})$ is weakly induced. □

In [196], De-xue Zhang and Ying-ming Liu introduced the concept of "weakly induced modification of an $L$-fuzzy topological space," and obtained a series of results on their categorical properties which we put in the last part of this book. Here we introduce the concept:

**4.2.23 Definition** Let $(L^X, \delta)$ be an $L$-fts. Denote the family of all the lower semicontinuous mappings in $\delta$ from $(X, [\delta])$ to $L$ by $wi^*(\delta)$, and denote the $L$-fuzzy topology on $X$ generated by the subbase $\delta \cup \{\chi_{U_{(a)}} : U \in \delta, a \in L\}$ by $wi(\delta)$. Call $wi(\delta)$ the *weakly induced modification* of $\delta$, and $(L^X, wi(\delta))$ the *weakly induced modification* of $(L^X, \delta)$.

**4.2.24 Proposition** *Let $(L^X, \delta)$ be an L-fts. Then $wi^*(\delta)$ is an L-fuzzy topology on X.* □

**Proof** Certainly $\underline{0}, \underline{1} \in wi^*(\delta)$. $\forall \mathcal{A} \subset wi^*(\delta)$, $\forall \mathcal{B} \in [wi^*(\delta)]^{<\omega}$, $\forall \gamma \in pr(L)$. By Proposition **4.2.4** (i), (ii),

$$(\bigvee_{U\in\mathcal{A}} U)_{(\gamma)} = \bigcup_{U\in\mathcal{A}} U_{(\gamma)} \in [\delta],$$

$$(\bigwedge_{U\in\mathcal{B}} U)_{(\gamma)} = \bigcap_{U\in\mathcal{B}} U_{(\gamma)} \in [\delta].$$

By Corollary **1.3.13**, $pr(L)$ is a meet-generating set of $L$. By Theorem **4.2.16** (i), $\bigvee\mathcal{A}, \bigwedge\mathcal{B} \in wi^*(\delta)$. □

**4.2.25 Proposition** *Let $(L^X, \delta)$ be an L-fts. Then*

(i) *$wi^*(\delta)$ is the finest weakly induced topology on X contained in $\delta$.*

(ii) *$wi(\delta)$ is the coarest weakly induced topology on X containing $\delta$.*

(iii) *If $(L^X, \delta)$ is weakly induced, then $wi^*(\delta) = wi(\delta) = \delta$.* □

**4.2.26 Theorem** *Let $(L^X, \delta)$ be an L-fts, $Y \subset X$, $Y \neq \emptyset$. Then*

$$wi(\delta|_Y) = wi(\delta)|_Y.$$ □

**4.2.27 Theorem** *Let $\{(L^{X_t}, \delta_t) : t \in T\}$ be a family of L-fts'. Then*

$$wi(\bigoplus_{t\in T} \delta_t) = \bigoplus_{t\in T} wi(\delta_t).$$ □

**4.2.28 Proposition** *Let $(L^X, \delta)$ be an L-fts, $A \in L^X$, $\mathcal{B}$ a subbase of $\delta$. Then*

$$\mathcal{B} \cup \{\chi_{U_{(\gamma)}} : U \in \mathcal{B}, \gamma \in pr(L)\}$$

*is a subbase of $wi(\delta)$.* □

**4.2.29 Theorem** *Let $(L^X, \delta)$, $(L^Y, \mu)$ be L-fts, $f^{\to}: (L^X, \delta) \to (L^Y, \mu)$ continuous. Then $f^{\to}: (L^X, wi(\delta)) \to (L^Y, wi(\mu))$ is continuous.*

**Proof** $\forall U \in wi(\mu)$, then there exists $\mathcal{A} \subset \mu \times [\{\chi_{V_{(a)}} : V \in \mu,\ a \in L\}]^{<\omega}$ such that $U = \bigvee\{V \wedge \bigwedge \mathcal{F} : (V, \mathcal{F}) \in \mathcal{A}\}$. Then

$$\begin{aligned} f^{\leftarrow}(U) &= \bigvee\{f^{\leftarrow}(V) \wedge \bigwedge\{f^{\leftarrow}(\chi_{W_{(a)}}) : W_{(a)} \in \mathcal{F}\} : (V, \mathcal{F}) \in \mathcal{A}\} \\ &\qquad\qquad \text{(by Theorem } \mathbf{2.1.17} \text{ (i), (ii))} \\ &= \bigvee\{f^{\leftarrow}(V) \wedge \bigwedge\{\chi_{f^{-1}(W_{(a)})} : W_{(a)} \in \mathcal{F}\} : (V, \mathcal{F}) \in \mathcal{A}\} \\ &\qquad\qquad \text{(by Proposition } \mathbf{2.1.19} \text{ (ii))} \\ &= \bigvee\{f^{\leftarrow}(V) \wedge \bigwedge\{\chi_{f^{\leftarrow}(W)_{(a)}} : W_{(a)} \in \mathcal{F}\} : (V, \mathcal{F}) \in \mathcal{A}\}. \\ &\qquad\qquad \text{(by Proposition } \mathbf{4.2.5} \text{ (i))} \end{aligned}$$

Since $f^{\rightarrow} : (L^X, \delta) \to (L^Y, \mu)$ is continuous, every $f^{\leftarrow}(V)$ and $f^{\leftarrow}(W)$ in the equations listed above are elements of $\delta$. So $f^{\leftarrow}(U) \in wi(\delta)$, $f^{\rightarrow} : (L^X, wi(\delta)) \to (L^Y, wi(\mu))$ is continuous. □

**4.2.30 Theorem** *Let $\{(L^{X_t}, \delta_t) : t \in T\}$ be a family of L-fts'. Then*

$$wi(\prod_{t\in T} \delta_t) = \prod_{t\in T} wi(\delta_t).$$

**Proof** Denote $\prod_{t\in T}(L^{X_t}, \delta_t)$ by $(L^X, \delta)$. By Proposition **4.2.25** (ii) and Theorem **4.2.21**, $wi(\delta) \subset \prod_{t\in T} wi(\delta_t)$. On the other hand, by Theorem **4.2.29**, $\forall s \in T$, $p_s^{\rightarrow} : (L^X, wi(\delta)) \to (L^{X_s}, wi(\delta_s))$ is continuous, so the canonical subbase of $\prod_{t\in T} wi(\delta_t)$ is contained in $wi(\delta)$, $\prod_{t\in T} wi(\delta_t) \subset wi(\delta)$. □

Similar to the case of stratification of $L$-fuzzy quotient space, the parallel conclusion for weakly induced modification of $L$-fuzzy quotient space is not true either:

**4.2.31 Example** Take $(L^X, \delta)$, $L^Y$ and $f : X \to Y$ as taken in Example **4.1.14**, then it has been proved there $\delta/f^{\rightarrow} = \{\underline{0}, \underline{1}\}$. So $wi(\delta/f^{\rightarrow}) = \delta/f^{\rightarrow}$. Then

$$\forall i \le 2,\ \forall {x^i}_u \in \{{x^1}_a, {x^2}_b\} \subset \delta,\quad {x^i}_1 = \chi_{({x^i}_u)_{(c)}} \in wi(\delta),$$
$$f^{\leftarrow}({y^1}_1) = {x^1}_1 \vee {x^2}_1 \in wi(\delta).$$

So ${y^1}_1$ is an element of $wi(\delta)/f^{\rightarrow}$. But ${y^1}_1 \notin \{\underline{0}, \underline{1}\} = wi(\delta/f^{\rightarrow})$.

## 4.3 Induced Spaces

Between ordinary topological spaces and fuzzy topological spaces, people considered a kind of special fuzzy topological spaces — induced spaces, as a connection between these two categories. Early in [93], R. Lowen studied this kind of fuzzy topological spaces from the angle of operator. In his later monograph [101], where induced spaces are called "topological" to distinguish from the "non-induced" called "nontopological" ones, he deepened the research in this aspect. He even introduced a standard "good extension" via induced spaces to judge whether or not a property of fuzzy topological spaces is "good," which were widely adopted later in fuzzy topology. Ge-ping Wang and Lan-fang Hu studied operating properties of induced spaces in 1985.[158] In [83], replacing value domain $[0, 1]$ and an external ordinary topological space, the study on induced spaces was extended into the case of lattice from the angle of "internal ordinary topology." They were connected with stratified spaces and weakly induced spaces, and the descriptions of the interior operator and the closure operator in these three kinds of spaces were established.[88, 114]

**4.3.1 Definition** For a category **C**, denote the class of all the objects of **C** by $Ob(\mathbf{C})$.

For every two objects $X, Y \in Ob(\mathbf{C})$, denote the set of all the morphisms from $X$ to $Y$ by $hom(X,Y)$, or $hom_{\mathbf{C}}(X,Y)$ to emphasize the category $\mathbf{C}$ to which the morphisms belonging.

Denote the category of topological spaces and continuous mappings by **Top**.

Shorten respectively the phrases "upper semicontinuous," "lower semicontinuous" and "continuous" by "u.s.c.", "l.s.c." and "c.". For a complete lattice $L$, let (US), (LS) and (CT) denote the following three conditions respectively:

$(X,\mathcal{T}) \in Ob(\mathbf{Top})$, $A \in L^X$ is u.s.c. $\Longrightarrow$ For every $x \in X$ and every neighborhood base $\mathcal{B}$ of $x$,
$$A(x) = \bigwedge_{U\in\mathcal{B}} \bigvee_{y\in U} A(y);$$

$(X,\mathcal{T}) \in Ob(\mathbf{Top})$, $A \in L^X$ is l.s.c. $\Longrightarrow$ For every $x \in X$ and every neighborhood base $\mathcal{B}$ of $x$,
$$A(x) = \bigvee_{U\in\mathcal{B}} \bigwedge_{y\in U} A(y);$$

$(X,\mathcal{T}) \in Ob(\mathbf{Top})$, $A \in L^X$ is c. $\Longrightarrow$ For every $x \in X$ and every neighborhood base $\mathcal{B}$ of $x$,
$$A(x) = \bigwedge_{U\in\mathcal{B}} \bigvee_{y\in U} A(y) = \bigvee_{U\in\mathcal{B}} \bigwedge_{y\in U} A(y).$$

Clearly, (US)+(LS)$\Longrightarrow$(CT), but the converse is in general false.

**4.3.2 Theorem** *Every completely distributive lattice $L$ satisfies both conditions* (US) *and* (LS).

**Proof** Suppose $(X,\mathcal{T}) \in Ob(\mathbf{Top})$, $A: X \to L$ is an upper semicontinuous mapping, $x \in X$, $\mathcal{B}$ a neighborhood base of $x$ in $(X,\mathcal{T})$. Let $b = \bigwedge_{U\in\mathcal{B}} \bigvee_{y\in U} A(y)$, then since $L$ is completely distributive,
$$b = \bigvee\{\bigwedge_{U\in\mathcal{B}} A(\varphi(U)) : \varphi \in \prod_{U\in\mathcal{B}} U\}.$$
Let $a_\varphi = \bigwedge_{U\in\mathcal{B}} A(\varphi(U))$ for every $\varphi \in \prod_{U\in\mathcal{B}} U$. Since $A$ is upper semicontinuous, by Theorem **4.2.16** (iii), every $A_{[a_\varphi]}$ is closed in $(X,\mathcal{T})$. Moreover, for every $U \in \mathcal{B}$, we have $\varphi(U) \in U$, $A(\varphi(U)) \geq a_\varphi$, i.e. $\varphi(U) \in A_{[a_\varphi]}$, $U \cap A_{[a_\varphi]} \neq \emptyset$. Hence $x \in (A_{[a_\varphi]})^- = A_{[a_\varphi]}$. Namely, for every $\varphi \in \prod_{U\in\mathcal{B}} U$, we have $x_{a_\varphi} \leq A$, and thus $x_b \leq A$, $A(x) \geq b$. On the other hand, obviously $A(x) \leq b$, so $A(x) = b$, (US) is true.

Replace $L$ by $L^{op}$, by Theorem **1.3.23**, $L^{op}$ is still completely distributive, we get (LS). □

**4.3.3 Theorem** *Let $(X,\mathcal{T}) \in Ob(\mathbf{Top})$, $L$ a complete lattice. Then for every two mappings $A, B: X \to L$ the following conclusions hold:*

(i) *If for every $x \in X$ there exists a family $\mathcal{B}(x)$ of neighborhoods (need not be a neighborhood base) of $x$ such that $A(x) = \bigwedge_{U\in\mathcal{B}(x)} \bigvee_{y\in U} A(y)$, then $A$ is upper semicontinuous.*

(ii) *If for every $x \in X$ there exists a family $\mathcal{B}(x)$ of neighborhoods (need not be a neighborhood base) of $x$ such that $A(x) = \bigvee_{U\in\mathcal{B}(x)} \bigwedge_{y\in U} A(y)$, then $A$ is lower semicontinuous.*

(iii) *If for every $x \in X$ there exists a neighborhood base $\mathcal{B}(x)$ of $x$ such that $B(x) = \bigwedge_{U\in\mathcal{B}_x} \bigvee_{y\in U} A(y)$, then $B$ is upper semicontinuous.*

(iv) *If for every $x \in X$ there exists a neighborhood base $\mathcal{B}(x)$ of $x$ such that $B(x) = \bigvee_{U\in\mathcal{B}_x} \bigwedge_{y\in U} A(y)$, then $B$ is lower semicontinuous.*

**Proof** (i) $\forall a \in L$, we want to prove that $X\backslash A_{[a]}$ is open in $(X, \mathcal{T})$. $\forall x \in X\backslash A_{[a]}$, $\bigwedge_{U\in\mathcal{B}(x)} \bigvee_{y\in U} A(y) = A(x) \not\geq a$. So $\exists U \in \mathcal{B}(x)$ such that $\bigvee_{y\in U} A(y) \not\geq a$, and hence $\forall y \in U$, $A(y) \not\geq a$. That is to say $U \subset X\backslash A_{[a]}$. Thus $X\backslash A_{[a]}$ is an open subset in $(X, \mathcal{T})$. By Theorem **4.2.16** (iii), $A$ is upper semicontinuous.

(ii) Similar to (i).

(iii) $\forall a \in L$, $\forall x \in X\backslash B_{[a]}$, we have $\bigwedge_{U\in\mathcal{B}(x)} \bigvee_{y\in U} A(y) \not\geq a$. So $\exists U \in \mathcal{B}(x)$ such that $\bigvee_{y\in U} A(y) \not\geq a$. $\forall y \in U$, since $\mathcal{B}_y$ is a neighborhood of $y$ in $X$, $\exists V \in \mathcal{B}_y$ such that $V \subset U$. So $\bigvee_{z\in V} A(z) \leq \bigvee_{z\in U} A(z)$, $\bigvee_{z\in V} A(z) \not\geq a$. Hence $B(y) = \bigwedge_{W\in\mathcal{B}_y} \bigvee_{z\in W} A(z) \leq \bigvee_{z\in V} A(z)$, $B(y) \not\geq a$. That is to say, for $x \in X\backslash B_{[a]}$, $\exists U \in \mathcal{N}(x)$ such that $U \subset X\backslash B_{[a]}$, $X\backslash B_{[a]} \in \mathcal{T}$. By Theorem **4.2.16** (iii), $B$ is upper semicontinuous.

(iv) Similar to (iii). □

**4.3.4 Theorem** *Let $(L^X, \delta)$ be an L-fts. Then $(L^X, \delta)$ is stratified if and only if $\delta$ contains all the lower semicontinuous mapping from $(X, [\delta])$ to $L$.*

**Proof** (Necessity) Let $A \in L^X$ be a lower semicontinuous mapping from $(X, [\delta])$ to $L$, then $A'$ is upper semicontinuous, we need only prove $A' \in \delta'$. If it is not true, then there exists an $L$-fuzzy point $x_a \leq A'^-$ but $x_a \not\leq A'$. Since $x_a \not\leq A'$, $A'$ is upper semicontinuous, by Theorem **4.3.2**, there exists a neighborhood $U$ of $x$ in $(X, [\delta])$ such that $\bigvee_{y\in U} A'(y) \not\geq a$. Suppose $b = \bigvee_{y\in U} A'(y)$. Since $\delta$ is stratified, $\underline{b} \in \delta'$, $P = \underline{b} \vee \chi_{X\backslash U} \in \delta'$. Obviously $A' \leq P$, $A'^- \leq P$ and $a \not\leq b = P(x)$. But this contradicts with $x_a \leq A'^-$.

(Sufficiency) Every layer $\underline{a}$ is clearly a lower semicontinuous mappings from $(X, [\delta])$ to $L$. □

**4.3.5 Theorem** *Let $(L^X, \delta)$ be a weakly induced L-fts, $A \subset X$. Then for the interior $A^\circ$ and the closure $A^-$ of $A$ in $(X, [\delta])$, the following equations hold:*

(i) $(\chi_A)^\circ = \chi_{A^\circ}$.

(ii) $(\chi_A)^- = \chi_{A^-}$.

**Proof** (i) Since $A^\circ$ is open in $(X, [\delta])$ and $\chi_{A^\circ} \leq \chi_A$, so $\chi_{A^\circ} \leq (\chi_A)^\circ$. On the other hand, since $(L^X, \delta)$ is weakly induced,

$$((\chi_A)^\circ)_{(0)} = supp((\chi_A)^\circ) \subset supp(\chi_A) = A,$$
$$supp((\chi_A)^\circ) = ((\chi_A)^\circ)_{(0)} = (((\chi_A)^\circ)_{(0)})^\circ \subset A^\circ,$$
$$(\chi_A)^\circ \leq \chi_{A^\circ}.$$

(ii) Replace $A$ in (i) by $A'$, we get

$$(\chi_A)^{-\prime} = (\chi_A)^{\prime\circ} = (\chi_{A'})^\circ = \chi_{A'^\circ} = \chi_{A^{-\prime}} = (\chi_{A^-})'.$$

Take their pseudo-complementary sets,

$$(\chi_A)^- = \chi_{A^-}.$$ □

**4.3.6 Definition** Let $(L^X, \delta)$ be an $L$-fts. $\delta$ is called *lattice-valued induced*, or *L-valued induced*, or *induced* for short, if $\delta$ is exactly the family of all the lower semicontinuous mappings from $(X, [\delta])$ to $L$. $(L^X, \delta)$ is called *lattice-valued induced*, or *L-valued induced*, or *induced* for short, if $\delta$ is induced.

Since stratified spaces, weakly induced spaces and induced spaces play special roles in fuzzy topology, we often need to deal with them uniformly, we give them a general name as follows:

An $L$-fts $(L^X, \delta)$ is called a *stratification L-fts*, or an *lattice-valued stratification space*, or an *L-valued stratification space*, or an *L-valued S-space* for short, if $(L^X, \delta)$ is stratified, or weakly induced, or induced.

By Theorem **4.3.4**, we obtain immediately the following

**4.3.7 Theorem** *Let $(L^X, \delta)$ be an L-fts. Then $(L^X, \delta)$ is induced if and only if $(L^X, \delta)$ is both stratified and weakly induced.* □

**4.3.8 Theorem** *Let $(L^X, \delta)$ be an L-fts. Then*

(i) *$(L^X, \delta)$ is stratified if and only if for every $A \in L^X$, every $a \in L$ and every minimal set $D$ of $a$, $(A^-)_{[a]} \subset \bigcap_{\alpha\in D}(A_{[\alpha]})^-$ in $(X, [\delta])$.*

(ii) *$(L^X, \delta)$ is weakly induced if and only if for every $A \in L^X$, every $a \in L$ and every minimal set $D$ of $a$, $(A^-)_{[a]} \supset \bigcap_{\alpha\in D}(A_{[\alpha]})^-$ in $(X, [\delta])$.*

(iii) *$(L^X, \delta)$ is induced if and only if for every $A \in L^X$, every $a \in L$ and every minimal set $D$ of $a$, $(A^-)_{[a]} = \bigcap_{\alpha\in D}(A_{[\alpha]})^-$ in $(X, [\delta])$.*

**Proof** (i)(Sufficiency) $\forall a \in L$, denote $A = \underline{a}$. $\forall b \not\leq a$, by Theorem **1.3.10**, $b$ has a minimal set $D$, $\bigvee D = b \not\leq a$, $\exists \gamma \in D$ such that $\alpha \not\leq a$, $A_{[\gamma]} = (\underline{a})_{[\gamma]} = \emptyset$. So

$$(A^-)_{[b]} \subset \bigcap_{\alpha\in D}(A_{[\alpha]})^- = \emptyset.$$

That is to say, $\forall x \in X$, $A^-(x) \leq a$, $A^- \leq A$, $A^- = A$. Thus $\underline{a} \in \delta'$. Since $L$ is a F-lattice, we have $\underline{a} \in \delta$ for every $a \in L$, $\delta$ is stratified.

(Necessity) If it is not true, there exist $A \in L^X$, $a \in L$, a minimal set $D$ of $a$ and $\gamma \in D$ such that $x \in (A^-)_{[a]}$ but $x \notin (A_{[\gamma]})^-$. Then there exists a neighborhood $U$ of $x$ in $(X, [\delta])$ such that $U \cap (A_{[\gamma]})^- = \emptyset$. Hence $\forall y \in U$, $A(y) \not\geq \gamma$. Since $D$ is a minimal set of $a$, $\gamma \in D$, we have $b = \bigvee\{A(y) : y \in U\} \not\geq a$ (otherwise we should have $y \in U$ such that $A(y) \geq \gamma$). Let $P = \underline{b} \vee \chi_{X\setminus U}$. Since $(L^X, \delta)$ is stratified and $U \in [\delta]$, we have $P \in \delta'$ and $P(x) = b \not\geq a$. Since $A \leq P$, $A^- \leq P^- = P$, so $A^-(x) \not\geq a$. But $x \in (A^-)_{[a]}$, $A^-(x) \geq a$, this is a contradictions.

(ii)(Sufficiency) $\forall P \in \delta'$, $\forall a \in L$, by Theorem **1.3.10**, we can take a minimal set $D$ of $a$. Then

$$P_{[a]} = (P^-)_{[a]} \supset \bigcap_{\alpha\in D}(P_{[\alpha]})^- \supset (P_{[a]})^-.$$

So $P_{[a]} \in [\delta']$, by Theorem **4.2.17** (v)$\Longrightarrow$(i), $(L^X, \delta)$ is weakly induced.

(Necessity) Since $(L^X, \delta)$ is weakly induced, $\forall \alpha \in D$, $(A^-)_{[\alpha]} \in [\delta']$. So

$$(A_{[\alpha]})^- \subset ((A^-)_{[\alpha]})^- = (A^-)_{[\alpha]}.$$

By $a = \bigvee D$ we have

$$\bigcap_{\alpha\in D}(A_{[\alpha]})^- \subset \bigvee_{\alpha\in D}(A^-)_{[\alpha]} = (A^-)_{[a]}.$$

(iii) By (i), (ii) and Theorem **4.3.7**. □

To establish the parallel conclusions for interior in $L$-fts, we need the following lemma:

**4.3.9 Lemma** *Let $(L^X, \delta)$ be an L-fts, $A \in L^X$, $a \in L$. Then*

(i) *$(L^X, \delta)$ is stratified $\Longrightarrow$ $((A^\circ)_{[a]})^\circ = (A_{[a]})^\circ$.*

(ii) *$(L^X, \delta)$ is weakly induced $\Longrightarrow$ $\forall \alpha \in \beta(a)$, $(A^\circ)_{[a]} \subset (A_{[\alpha]})^\circ$.*

**Proof** (i) Let $U = (A_{[a]})^\circ$. Since $U \in [\delta]$, $(L^X, \delta)$ is stratified, $aU \leq A$ is open in $(L^X, \delta)$. So $aU = (aU)^\circ \leq A^\circ$, $U \subset (A^\circ)_{[a]}$, $U = U^\circ \subset ((A^\circ)_{[a]})^\circ$. The inequality of another direction is obvious.

(ii) $\forall \alpha \in \beta(a)$, $\forall x \in (A^\circ)_{[a]}$. If we can prove that there exists a neighborhood $U \in \mathcal{N}_{[\delta]}(x)$ of $x$ in $(X, [\delta])$ such that $U \subset (A^\circ)_{[\alpha]}$, then $x \in ((A^\circ)_{[\alpha]})^\circ$, the conclusion is then generated by (i). If this is not true, then $\forall U \in \mathcal{N}_{[\delta]}(x)$, $\exists y^U \in U$ such that $A^\circ(y^U) \not\geq \alpha$. Denote $b = \bigvee\{A^\circ(y^U) : U \in \mathcal{N}_{[\delta]}(x)\}$, then since $\alpha \in \beta(a)$ and $\beta(a)$ is a minimal set of $a$, we have $b \not\geq a$. By $A^\circ(x) \geq a$, $b \not\geq A^\circ(x)$. Let $B = \{y \in X : A^\circ(y) \leq b\}$, then $x \notin B$. Since $(L^X, \delta)$ is weakly induced, $B = (A^{\circ\prime})_{[b']} \in [\delta']$, $B^- = B$ in $(X, [\delta])$. But for every $U \in \mathcal{N}_{[\delta]}(x)$, by the choice of $b$ we always have $y^U \in U \cap B$, so $x \in B^- = B$. This contradicts with the result $x \notin B$ proved above. □

**4.3.10 Theorem** *Let $(L^X, \delta)$ be an L-fts. Then*

(i) *$(L^X, \delta)$ is stratified if and only if for every $A \in L^X$, every $a \in L$ and every minimal set $D$ of $a$, $(A^\circ)_{[a]} \supset \bigcap_{\alpha \in D}(A_{[\alpha]})^\circ$ in $(X, [\delta])$.*

(ii) *$(L^X, \delta)$ is weakly induced if and only if for every $A \in L^X$, every $a \in L$ and every minimal set $D$ of $a$, $(A^\circ)_{[a]} \subset \bigcap_{\alpha \in D}(A_{[\alpha]})^\circ$ in $(X, [\delta])$.*

(iii) *$(L^X, \delta)$ is induced if and only if for every $A \in L^X$, every $a \in L$ and every minimal set $D$ of $a$, $(A^\circ)_{[a]} = \bigcap_{\alpha \in D}(A_{[\alpha]})^\circ$ in $(X, [\delta])$.*

**Proof** (i)(Sufficiency) $\forall a \in L$, denote $A = \underline{a}$. Since $A \in \delta$ for $a = 0$, so we assume $a \neq 0$. Then $a$ has a nonempty minimal set $D \subset L$. Since $\forall \alpha \in D$, $\alpha \leq a$, $A = \underline{a}$, we have

$$(A^\circ)_{[a]} \supset \bigcap_{\alpha \in D} (A_{[\alpha]})^\circ = X.$$

So $A^\circ \geq A$, $A^\circ = A$, $\underline{a} = A \in \delta$, $\delta$ is stratified.

(Necessity) By Lemma **4.3.9** (i) and $\bigvee D = a$, we have

$$\bigcap_{\alpha \in D} (A_{[\alpha]})^\circ \subset \bigcap_{\alpha \in D} ((A^\circ)_{[\alpha]})^\circ \subset \bigcap_{\alpha \in D} (A^\circ)_{[\alpha]} = (A^\circ)_{[a]}.$$

(ii)(Sufficiency) $\forall A \in \delta$, $\forall a \in L$, then

$$A_{(a)} = (A^\circ)_{(a)} = \bigcup\{(A^\circ)_{[b]} : b \in L,\ b \not\leq a\} \subset \bigcup\{ \bigcap_{\alpha \in \beta(b)} (A_{[\alpha]})^\circ : b \in L,\ b \not\leq a\}.$$

So $\forall x \in A_{(a)}$, $\exists b \in L$, $b \not\leq a$ such that $x \in \bigcap_{\alpha \in \beta(b)}(A_{[\alpha]})^\circ$. Since $\beta(b)$ is a minimal set of $b$, $\bigvee \beta(b) = b \not\leq a$, $\exists \alpha \in \beta(b)$ such that $\alpha \not\leq a$. Then

$$x \in (A_{[\alpha]})^\circ \subset A_{[\alpha]} \subset A_{(a)}.$$

That is to say, $A_{(a)}$ is an open subset in $(X, [\delta])$. By Theorem **4.2.17**, $(L^X, \delta)$ is weakly induced.

(Necessity) By Lemma **4.3.9** (ii).

(iii) By (i), (ii) and Theorem **4.3.7**. □

In theorems **4.3.8** and **4.3.10**, we use levels $A_{[a]}$'s of an $L$-fuzzy $A$ to describe its interior and closure in stratified spaces, weakly induced spaces and induced spaces. Clearly, we can also use its dual form $A_{(a)}$ to do the same things. With the concept of "maximal set" introduced by Guo-jun Wang and dual to minimal set, it is not hard to obtain those conclusions in the form of $A_{(a)}$'s.

We can concretely describe the structures of $L$-fuzzy topologies in stratified spaces, weakly induced spaces and induced spaces. For this aim, we define a "standard topology" for them:

**4.3.11 Definition** Let $(X,\mathcal{T})$ be an ordinary topological space, $L$ a complete lattice. For every $a\in L$, $U\in\mathcal{T}$, denote

$$F_*(a,U)=aU,\quad F^*(a,U)=\underline{a}\vee\chi_{X\backslash U},$$
$$stb_{*L}(\mathcal{T})=\{F_*(a,U):\ a\in L,\ U\in\mathcal{T}\}\subset L^X,$$
$$stb^*{}_L(\mathcal{T})=\{F^*(a,U):\ a\in L,\ U\in\mathcal{T}\}\subset L^X,$$
$$stt_{*L}(\mathcal{T})=\{\bigvee\mathcal{A}:\ \mathcal{A}\subset stb_{*L}(\mathcal{T})\}\subset L^X,$$
$$stt^*{}_L(\mathcal{T})=\{\bigwedge\mathcal{A}:\ \mathcal{A}\subset stb^*{}_L(\mathcal{T})\}\subset L^X.$$

Call $stb_{*L}(\mathcal{T})$ the *staircase base* associated with $\mathcal{T}$, $stb^*{}_L(\mathcal{T})$ the *staircase co-base* associated with $\mathcal{T}$, $stt_{*L}(\mathcal{T})$ the *staircase topology* associated with $\mathcal{T}$, $stt^*{}_L(\mathcal{T})$ the *staircase co-topology* associated with $\mathcal{T}$.

Denote the family of all the lower semicontinuous mappings from $(X,\mathcal{T})$ to $L$ by $lc_L(\mathcal{T})$, the family of all the upper semicontinuous mappings from $(X,\mathcal{T})$ to $L$ by $uc_L(\mathcal{T})$.

**4.3.12 Proposition** *Let $(X,\mathcal{T})$ be an ordinary topological space, L a complete lattice, $a,b\in L$, $U,V\in\mathcal{T}$. Then*

(i) $a\le b \Longrightarrow F_*(a,U)_{(b)}=\emptyset$,
$a\not\le b \Longrightarrow F_*(a,U)_{(b)}=U$.

(ii) $a\le b \Longrightarrow F_*(a,U)^{[b]}=X$,
$a\not\le b \Longrightarrow F_*(a,U)^{[b]}=X\backslash U$.

(iii) $a\ge b \Longrightarrow F^*(a,U)_{[b]}=X$,
$a\not\ge b \Longrightarrow F^*(a,U)_{[b]}=X\backslash U$.

(iv) $a\ge b \Longrightarrow F^*(a,U)^{(b)}=\emptyset$,
$a\not\ge b \Longrightarrow F^*(a,U)^{(b)}=U$.

(v) $F_*(a,U)\wedge F_*(b,V)=F_*(a\wedge b,U\cap V)$.

(vi) $F^*(a,U)\vee F^*(b,V)=F^*(a\vee b,U\cap V)$. □

By the Proposition **4.3.12**, we have immediately the following

**4.3.13 Proposition** *Let $(X,\mathcal{T})$ be an ordinary topological space, L an infinitely distributive lattice. Then*

(i) $stt_{*L}(\mathcal{T})$ *fulfills* (LFT1), (LFT2) *and* (LFT3).

(ii) $stt^*{}_L(\mathcal{T})$ *fulfills* (LFT1′), (LFT2′) *and* (LFT3′). □

**4.3.14 Proposition** *Let $(X,\mathcal{T})$ be an ordinary topological space, L a complete lattice. Then*

(i) $stt_{*L}(\mathcal{T})\subset lc_L(\mathcal{T})$.

(ii) $stt^*{}_L(\mathcal{T})\subset uc_L(\mathcal{T})$.

**Proof** (i) $\forall A\in stt_{*L}(\mathcal{T})$, $\exists\mathcal{A}\subset L\times\mathcal{T}$ such that $A=\bigvee\{F_*(b,U):\ (b,U)\in\mathcal{A}\}$. $\forall a\in L$, by Proposition **4.2.4** (i) and Proposition **4.3.12** (i),

$$A_{(a)}=\bigcup\{F_*(b,U)_{(a)}:\ (b,U)\in\mathcal{A}\}=\bigcup\{U:\ \exists b\in L,\ b\not\le a,\ (b,U)\in\mathcal{A}\}\in\mathcal{T}.$$

By Theorem **4.2.16** (i), $A\in lc_L(\mathcal{T})$, $stt_{*L}(\mathcal{T})\subset lc_L(\mathcal{T})$.

(ii) Similarly prove. □

**4.3.15 Proposition** *Let $(X,\mathcal{T})$ be an ordinary topological space, L a complete lattice, $x\in X$, $\mathcal{B}$ a neighborhood base of x in X. Then*

(i) $A\in stt_{*L}(\mathcal{T})\Longrightarrow A(x)=\bigvee_{U\in\mathcal{B}}\bigwedge_{y\in U}A(y)$.

(ii) $A \in stt^*_L(\mathcal{T}) \Longrightarrow A(x) = \bigwedge_{U\in\mathcal{B}} \bigvee_{y\in U} A(y)$.

**Proof** We only prove (i). (ii) is similar.

Since $\forall U \in \mathcal{B}$, $x \in U$, so $A(x) \geq \bigvee_{U\in\mathcal{B}} \bigwedge_{y\in U} A(y)$. By $A \in stt_{*L}(\mathcal{T})$, $\exists \mathcal{A} \subset L\times\mathcal{T}$ such that $A = \bigvee\{F_*(a,U) : (a,U) \in \mathcal{A}\}$. Denote $\mathcal{A}_x = \{(a,U) \in \mathcal{A} : x \in U\}$. If $\mathcal{A}_x = \emptyset$, then $\forall (a,U) \in \mathcal{A}$, $x \notin U$, $F_*(a,U)(x) = 0$, and hence $A(x) = 0 \leq \bigvee_{U\in\mathcal{B}} \bigwedge_{y\in U} A(y)$. If $\mathcal{A}_x \neq \emptyset$, then $\forall (a,U) \in \mathcal{A}_x$, since $\mathcal{B}$ is a neighborhood base of $x$, $\exists V_{a,U} \in \mathcal{B}$ such that $V_{a,U} \subset U$. Then $\forall (a,U) \in \mathcal{A}_x$,

$$\bigwedge_{y\in V_{a,U}} A(y) = \bigwedge_{y\in V_{a,U}} \bigvee_{(b,W)\in\mathcal{A}} F_*(b,W)(y) \geq \bigwedge_{y\in V_{a,U}} F_*(a,U)(y) = a,$$

$$\bigvee_{U\in\mathcal{B}} \bigwedge_{y\in U} A(y) \geq \bigvee_{(a,U)\in\mathcal{A}_x} \bigwedge_{y\in V_{a,U}} A(y) \geq \bigvee_{(a,U)\in\mathcal{A}_x} a = A(x).$$

So $A(x) = \bigvee_{U\in\mathcal{B}} \bigwedge_{y\in U} A(y)$. □

**4.3.16 Theorem** *Let $(X,\mathcal{T})$ be an ordinary topological space, $L$ a completely distributive lattice. Then*

(i) $[stt_{*L}(\mathcal{T})] = \mathcal{T}$.

(ii) $stt_{*L}(\mathcal{T}) = lc_L(\mathcal{T})$.

(iii) $stt^*_L(\mathcal{T}) = uc_L(\mathcal{T})$.

**Proof** (i) $\mathcal{T} \subset [stt_{*L}(\mathcal{T})]$ is obvious. Let $U \in [stt_{*L}(\mathcal{T})]$, then $\chi_U \in stt_{*L}(\mathcal{T})$. By Proposition **4.3.14** (i), $\chi_U : (X,\mathcal{T}) \to L$ is lower semicontinuous. Then by Theorem **4.2.16** (i), $U = (\chi_U)_{(0)} \in \mathcal{T}$. So $[stt_{*L}(\mathcal{T})] \subset \mathcal{T}$.

(ii) Suppose $A \in lc_L(\mathcal{T})$. $\forall U \in \mathcal{T}$, denote

$$a_U = \bigwedge_{y\in U} A(y), \qquad B = \bigvee_{U\in\mathcal{T}} F_*(a_U, U),$$

then $\forall x \in X$, by Theorem **4.3.2**, $L$ satisfies (LS),

$$B(x) = \bigvee_{U\in\mathcal{T}} (a_U \wedge \chi_U)(x) = \bigvee_{U\in\mathcal{N}(x)} a_U = \bigvee_{U\in\mathcal{N}(x)} \bigwedge_{y\in U} A(y) = A(x).$$

So $A = B = \bigvee_{U\in\mathcal{T}} F_*(a_U, U) \in stt_{*L}(\mathcal{T})$, $lc_L(\mathcal{T}) \subset stt_{*L}(\mathcal{T})$. By Proposition **4.3.14** (i), the proof is completed.

(iii) Similarly prove. □

By the conclusions proved above, we have immediately the following:

**4.3.17 Theorem** *Let $(L^X,\delta)$ be an L-fts. Then*

(i) *$(L^X,\delta)$ is stratified if and only if $\delta \supset stt_{*L}([\delta])$.*

(ii) *$(L^X,\delta)$ is weakly induced if and only if $\delta \subset stt_{*L}([\delta])$.*

(iii) *$(L^X,\delta)$ is induced if and only if $\delta = stt_{*L}([\delta])$.* □

**4.3.18 Proposition** *Let $(L^X,\delta)$ be an L-fts. Then $stt_{*L}([\delta])$ is just the stratification of the L-fuzzy topology $crs(\delta)$ on $X$.* □

**4.3.19 Corollary** *Let $(L^X,\delta)$ be an L-fts. Then*

(i) *$(L^X,\delta)$ is stratified if and only if $\delta$ contains the stratification of $crs(\delta)$.*

(ii) *$(L^X,\delta)$ is weakly induced if and only if $\delta$ is contained in the stratification of $crs(\delta)$.*

(iii) *$(L^X,\delta)$ is induced if and only if $\delta$ is the stratification of $crs(\delta)$.* □

Using results obtained above, we can also establish some other characterizing theorems for stratified spaces, weakly induced spaces and induced spaces.

**4.3.20 Theorem** *Let $(L^X, \delta)$ be an L-fts. For an arbitrary mapping $\mathcal{B}: X \to \mathcal{P}([\delta])$ such that $\mathcal{B}(x)$ is a neighborhood base of $x$ in $(X, [\delta])$ for every $x \in X$, define an operator $i: L^X \to L^X$ as follows:*

$$\forall A \in L^X,\ \forall x \in X, \quad i(A)(x) = \bigvee_{U \in \mathcal{B}(x)} \bigwedge_{y \in U} A(y).$$

*Then*

(i) *$i$ is an interior operator on $L^X$.*

(ii) *The L-fuzzy topology on $X$ generated by $i$ is exactly the family of all the lower semicontinuous mappings from $(X, [\delta])$ to $L$.*

**Proof** (i) (IO1) and (IO2) are obvious. Since $L$ is completely distributive and then is infinitely distributive, (IO3) is also satisfied. $\forall A \in L^X$, by Theorem **4.3.3** (iv), $i(A)$ is lower semicontinuous. By Theorem **4.3.2**, $i(i(A)) = i(A)$, $i$ satisfies (IO4) as well. So $i$ is an interior operator on $L^X$.

(ii) Let $\mu$ denote the $L$-fuzzy topology on $X$ generated by $i$, i.e. $\mu = \{A \in L^X : A = i(A)\}$, then by Theorem **4.3.3** (iv), all the elements in $\mu$ are lower semicontinuous. On the other hand, by Theorem **4.3.2**, every lower semicontinuous mapping $A$ from $(X, [\delta])$ to $L$ satisfies $A = i(A)$, so $A \in \mu$. □

Dually, we have the following

**4.3.21 Theorem** *Let $(L^X, \delta)$ be an L-fts. For an arbitrary mapping $\mathcal{B}: X \to \mathcal{P}([\delta])$ such that $\mathcal{B}(x)$ is a neighborhood base of $x$ in $(X, [\delta])$ for every $x \in X$, define an operator $c: L^X \to L^X$ as follows:*

$$\forall A \in L^X,\ \forall x \in X, \quad c(A)(x) = \bigwedge_{U \in \mathcal{B}(x)} \bigvee_{y \in U} A(y).$$

*Then*

(i) *$c$ is a closure operator on $L^X$.*

(ii) *The L-fuzzy topology on $X$ generated by $c$ is exactly the family of all the lower semicontinuous mappings from $(X, [\delta])$ to $L$.* □

**4.3.22 Theorem** *Let $(L^X, \delta)$ be an L-fts. For an arbitrary mapping $\mathcal{B}: X \to \mathcal{P}([\delta])$ such that $\mathcal{B}(x)$ is a neighborhood base of $x$ in $(X, [\delta])$ for every $x \in X$, define an operator $i: L^X \to L^X$ as follows:*

$$\forall A \in L^X,\ \forall x \in X, \quad i(A)(x) = \bigvee_{U \in \mathcal{B}(x)} \bigwedge_{y \in U} A(y).$$

*Denote $\mu = \{A \in L^X : A = i(A)\}$. Then*

(i) *$(L^X, \delta)$ is stratified if and only if $\delta \supset \mu$.*

(ii) *$(L^X, \delta)$ is weakly induced if and only if $\delta \subset \mu$.*

(iii) *$(L^X, \delta)$ is induced if and only if $\delta = \mu$.*

(iv) *$(L^X, \delta)$ is induced if and only if $A^\circ = i(A)$ for every $A \in L^X$.*

**Proof** (i) - (iii): By Theorem **4.3.20** (ii).

(iv)(Necessity) $\forall A \in L^X$. By Theorem **4.3.20** (ii), $i(A)$ is lower semicontinuous, so $i(A) \in \delta$. By Theorem **4.3.20** (i), $i$ is an interior operator, so $i(A) \leq A$. Then by $i(A) \in \delta$, $i(A) = (i(A))^\circ \leq A^\circ$. On the other hand, since $A^\circ \in \delta, A^\circ$ is lower semicontinuous. By Theorem **4.3.2**, $A^\circ = i(A^\circ) \leq i(A)$.

(Sufficiency) By (iii). □

**4.3.23 Theorem** *Let $(L^X, \delta)$ be an L-fts. For an arbitrary mapping $\mathcal{B}: X \to \mathcal{P}([\delta])$*

*such that $\mathcal{B}(x)$ is a neighborhood base of $x$ in $(X,[\delta])$ for every $x \in X$, define an operator $c: L^X \to L^X$ as follows:*

$$\forall A \in L^X,\ \forall x \in X, \quad c(A)(x) = \bigwedge_{U\in\mathcal{B}(x)} \bigvee_{y\in U} A(y).$$

*Denote $\eta = \{A \in L^X : A = c(A)\}$. Then*

(i) *$(L^X,\delta)$ is stratified if and only if $\delta \supset \eta'$.*
(ii) *$(L^X,\delta)$ is weakly induced if and only if $\delta \subset \eta'$.*
(iii) *$(L^X,\delta)$ is induced if and only if $\delta = \eta'$.*
(iv) *$(L^X,\delta)$ is induced if and only if $A^- = c(A)$ for every $A \in L^X$.* □

Opposite to Proposition **4.2.2**, by the results prove above, we have the following theorem "lifting the continuity" in background spaces to $L$-fuzzy topological spaces on them (proof is left as an exercise):

**4.3.24 Lemma** (Lifting Lemma) *Let $(L^X,\delta)$, $(L^Y,\mu)$ be L-fts', $f: (X,[\delta]) \to (Y,[\mu])$ a continuous mapping. If $(L^X,\delta)$ is stratified, $(L^Y,\mu)$ is weakly induced, then $f^\to: (L^X,\delta) \to (L^Y,\mu)$ is continuous.* □

Combining the related results in the previous sections of this chapter, it is easy to find the following conclusions:

**4.3.25 Theorem** *Induced property is hereditary.* □

**4.3.26 Theorem** *Let $\{(L^{X_t},\delta_t) : t \in T\}$ be a family of L-fts'. Then $\bigoplus_{t\in T}(L^{X_t},\delta_t)$ is induced if and only if $(L^{X_t},\delta_t)$ is induced for every $t \in T$.* □

**4.3.27 Theorem** *Let $\{(L^{X_t},\delta_t) : t \in T\}$ be a family of weakly induced L-fts' and at least one of them is stratified. Then $\prod_{t\in T}(L^{X_t},\delta_t)$ is induced.* □

**4.3.28 Theorem** *Let $(L^X,\delta)$ be an induced L-fts, $(L^Y,\delta/f^\to)$ the L-fuzzy quotient space of $(L^X,\delta)$ with respect to the L-fuzzy surjective mapping $f^\to: L^X \to L^Y$. Then $(L^Y,\delta/f^\to)$ is induced.* □

Induced spaces have very strong properties. Theorem **4.3.22** (iv) and Theorem **4.3.23** (iv) provide two useful ways to concretely construct the interior and closure of an $L$-fuzzy subset with the algebraic properties of itself and the open subsets in the background space. In a different way, we can also use the nice properties of induced spaces to characterize interior operator and closure operator in an $L$-fts by interior and closure of stratifications in the background space. Then the procedure of producing interior and closure in $L$-fts is transformed into the parallel but simpler problem in an orindary topological space.

By Theorem **4.3.10** (iii) and Theorem **4.3.8** (iii), the following results is clear:

**4.3.29 Theorem** *Let $(L^X,\delta)$ be an induced L-fts, $A \in L^X$, $L_0$ a generating set of $L$, $D(a)$ a minimal set of for every $a \in L_0$. Then*

$$A^\circ = \bigvee\{a\chi_{\bigcap\{(A_{[\alpha]})^\circ :\ \alpha\in D(a)\}} : a \in L_0\}.$$ □

**4.3.30 Theorem** *Let $(L^X,\delta)$ be an induced L-fts, $A \in L^X$, $L_0$ a generating set of $L$, $D(a)$ a minimal set of for every $a \in L_0$. Then*

$$A^- = \bigvee\{a\chi_{\bigcap\{(A_{[\alpha]})^- :\ \alpha\in D(a)\}} : a \in L_0\}.$$ □

Combining Theorem **2.1.10** with Theorem **4.3.29** and Theorem **4.3.30**, we get immediately

**4.3.31 Theorem** *Let* $(L^X, \delta)$ *be an induced L-fts,* $L_0$ *a join-generating set of* $L$, $A \in L^X$. *Then*

$$A^\circ = \bigvee_{a \in L_0} aA_{[a]}{}^\circ. \qquad \square$$

**4.3.32 Theorem** *Let* $(L^X, \delta)$ *be an induced L-fts,* $L_0$ *a join-generating set of* $L$, $A \in L^X$. *Then*

$$A^- = \bigvee_{a \in L_0} aA_{[a]}{}^-. \qquad \square$$

**4.3.33 Remark** Theorems **4.3.31** and **4.3.32** are clearly very useful. They were proved respectively by Yun Yi[188] and De-xue Zhang.[194] Here we can find that these two conclusions are in fact two direct corollaries of results existed already.

Since an $L$-fuzzy induced topology just consists of all the lower semicontinuous mappings from the backgound space to $L$, note that in Theorem **4.3.31**, the right side of the equation there involves only the topology of the background space, so the family $\{\bigvee_{a \in L} aA_{[a]}{}^\circ : A \in L^X\} = lc_L([\delta])$, i.e. just the family of all the lower semicontinuous from $(X, [\delta])$ to $L$. Therefore, we have another characterization theorem for $L$-valued S-spaces:

**4.3.34 Theorem** *Let* $(L^X, \delta)$ *be an L-fts. Then*

(i) $(L^X, \delta)$ *is stratified if and only if*
$$\delta \supset \{\bigvee_{a \in L} aA_{[a]}{}^\circ : A \in L^X\}.$$

(ii) $(L^X, \delta)$ *is weakly induced if and only if*
$$\delta \subset \{\bigvee_{a \in L} aA_{[a]}{}^\circ : A \in L^X\}.$$

(iii) $(L^X, \delta)$ *is induced if and only if*
$$\delta = \{\bigvee_{a \in L} aA_{[a]}{}^\circ : A \in L^X\}. \qquad \square$$

Similarly, by Theorem **4.3.32**, the following conclusions hold:

**4.3.35 Theorem** *Let* $(L^X, \delta)$ *be an L-fts. Then*

(i) $(L^X, \delta)$ *is stratified if and only if*
$$\delta' \supset \{\bigvee_{a \in L} aA_{[a]}{}^- : A \in L^X\}.$$

(ii) $(L^X, \delta)$ *is weakly induced if and only if*
$$\delta' \subset \{\bigvee_{a \in L} aA_{[a]}{}^- : A \in L^X\}.$$

(iii) $(L^X, \delta)$ *is induced if and only if*
$$\delta' = \{\bigvee_{a \in L} aA_{[a]}{}^- : A \in L^X\}. \qquad \square$$

## 4.4 Functors $\omega_L$ and $\iota_L$

We have investigated $L$-valued stratification spaces starting from $L$-fts' themselves. In the case of F-ts', authors studied the relations between F-ts' and ordinary topological spaces from the angle of ordinary topological spaces. In [93], R. Lowen, as the first author, systematically studied this kind of problems in the case of $L = [0, 1]$. Moreover, S.E. Rodabaugh deepened the study on these aspects in the view point of category theory in a series of papers.[142, 143, 145, 146, 147] In Guo-jun Wang's monograph [168], he gave out many interesting results on this aspect in the case $L$ is a completely distributive lattice. For the contents of this section, we refer the reader the references mentioned above.

In this section, we use the commonly used symbol $\omega_L$ to match the discussion here, although we have had another symbol $lc_L$ with the same meaning.

**4.4.1 Lemma** *Let L be a F-lattice. All L-fuzzy topological spaces and all L-fuzzy continuous mappings form a category.* □

**4.4.2 Definition** Let $L$ be a F-lattice. Denote the category of all $L$-fuzzy topological spaces and all $L$-fuzzy continuous mappings by $L$-**FTS**.

**4.4.3 Definition** Let $L$ be a F-lattice.

For every $(X, \mathcal{T}) \in Ob(\mathbf{Top})$, let $\omega_L(\mathcal{T})$ denote the family of all the lower semi-continuous mappings from $(X, \mathcal{T})$ to $L$, i.e. $\omega_L(\mathcal{T}) = lc_L(\mathcal{T})$.

For every $(L^X, \delta) \in Ob(L\text{-}\mathbf{FTS})$, every $\mathcal{A} \subset L^X$ and every $a \in L$, denote

$$\iota_a(\mathcal{A}) = \{A_{(a)} : A \in \mathcal{A}\}.$$

Let $\iota_L(\delta)$ denote the ordinary topology on $X$ generated by the subbase $\bigcup_{a \in L} \iota_a(\delta)$.

By the results proved in the previous several sections, reader can easily verify the following conclusions:

**4.4.4 Proposition** *Let L be a F-lattice. Then*

(i) *For every $(X, \mathcal{T}) \in Ob(\mathbf{Top})$, $\omega_L(\mathcal{T})$ is an induced L-fuzzy topology on X and $[\omega_L(\mathcal{T})] = \mathcal{T}$.*

(ii) *For every $(L^X, \delta) \in Ob(L\text{-}\mathbf{FTS})$ and every $\gamma \in pr(L)$, $\iota_\gamma(\delta)$ is an ordinary topology on X.*

(iii) *For every $(L^X, \delta) \in Ob(L\text{-}\mathbf{FTS})$, $\bigcup_{\gamma \in pr(L)} \iota_\gamma(\delta)$ is a subbase of $\iota_L(\delta)$.* □

**4.4.5 Definition** Let $X$ be a nonempty ordinary set, $L$ a F-lattice.

Denote the family of all the ordinary topologies on $X$ by $Top(X)$.

Denote the family of all the $L$-fuzzy topologies on $X$ by $FT_L(X)$.

Still use the symbols $\omega_L$ and $\iota_L$ to denote respectively their restrictions on $Top(X)$ and $FT_L(X)$; i.e. use them to denote mappings

$$\omega_L : Top(X) \to FT_L(X),$$
$$\iota_L : FT_L(X) \to Top(X).$$

**4.4.6 Proposition** *Let X be a nonempty ordinary set, L a F-lattice. Then both $Top(X)$ and $FT_L(X)$ are complete lattices with inclusion orders.* □

**4.4.7 Exercise** (1) $\omega_L : Top(X) \to FT_L(X)$ preserves arbitrary meets and nonempty joins.

(2) $\omega_L[Top(X)]$ is a complete sublattice of $FT_L(X)$.

The following theorem was proved by R. Lowen in the case $L = [0, 1]$.[93] Wang proved it for completely distributive lattice $L$.[168] We leave its proof to reader as an exercise.

**4.4.8 Theorem** *Let X be a nonempty ordinary set, L a F-lattice. Then*

(i) *For every $\mathcal{T} \in Top(X)$, $\iota_L \circ \omega_L(\mathcal{T}) = \mathcal{T}$.*

(ii) *For every $\delta \in FT_L(X)$, $\omega_L \circ \iota_L(\delta) \supset \delta$.* □

**4.4.9 Corollary** *Let $X$ be a nonempty ordinary, $L$ a F-lattice. Then both*

$$\omega_L : \ Top(X) \to \omega_L[Top(X)] \quad \text{and} \quad \iota_L : \ \omega_L[Top(X)] \to Top(X)$$

*are complete lattice isomorphisms.* □

**4.4.10 Remark** (1) From Theorem **4.4.8** (ii) one can find that every $L$-fuzzy topology $\delta$ on $X$ consists of some $L$-valued lower semicontinuous mappings from $(X, \iota_L(\delta))$ to $L$. This is a reason why $L$-valued weakly induced spaces and $L$-valued induced spaces possess special importance in the study of $L$-fuzzy topology.

(2) Problems studied with $\omega_L$ are in fact same with $L$-valued induced spaces discussed in the previous sections. $\omega_L$ starts from an ordinary topology on $X$ to generates an induced topology, but with the notion "induced topology" defined in the preceding section, people deal with the same problems from $L$-fts' themselves without introducing an "external" topological space.

We leave the proofs of the following conclusions as exercises. With results in the previous sections in this chapter, reader can easily prove them.

**4.4.11 Theorem** *Let $(X, \mathcal{T}) \in Ob(\mathbf{Top})$, $Y \subset X$ nonempty, $\mathcal{T}|_Y$ the subspace topology of $Y$ in $X$, $L$ a F-lattice. Then*

$$\omega_L(\mathcal{T}|_Y) = \omega_L(\mathcal{T})|_Y. \quad □$$

**4.4.12 Theorem** *Let $(X, \mathcal{T}), (Y, \mathcal{S}) \in Ob(\mathbf{Top})$, $f : \ X \to Y$ an ordinary mapping, $L$ a F-lattice. Then the following conditions are equivalent:*

(i) $f \in hom_{\mathbf{Top}}((X, \mathcal{T}), (Y, \mathcal{S}))$.

(ii) $f^{\to} \in hom_{L\textbf{-FTS}}((L^X, \omega_L(\mathcal{T})), (L^Y, \omega_L(\mathcal{S}))$. □

**4.4.13 Theorem** *Let $\{(X_t, \mathcal{T}_t) : \ t \in T\}$ be a family of ordinary topological spaces, $\omega_L(\bigoplus_{t \in T} \mathcal{T}_t)$ the sum topology on $\omega_L(\bigoplus_{t \in T} X_t)$, $L$ a F-lattice. Then*

$$\omega_L(\bigoplus_{t \in T} \mathcal{T}_t) = \bigoplus_{t \in T} \omega_L(\mathcal{T}_t). \quad □$$

**4.4.14 Theorem** *Let $\{(X_t, \mathcal{T}_t) : \ t \in T\}$ be a family of ordinary topological spaces, $(X, \mathcal{T})$ be their product space, $(L^X, \delta)$ be $L$-fuzzy product space of $\{(L^{X_t}, \omega_L(\mathcal{T}_t)) : \ t \in T\}$, $L$ a F-lattice. Then $\omega_L(\mathcal{T}) = \delta$.* □

**4.4.15 Theorem** *Let $(X, \mathcal{T})$ be an ordinary topological space, $(Y, \mathcal{T}/f)$ the ordinary quotient space with respect to the surjective mapping $f : \ X \to Y$, $L$ a F-lattice. Then*

$$\omega_L(\mathcal{T}/f) = \omega_L(\mathcal{T})/f^{\to}. \quad □$$

**4.4.16 Theorem** *Let $X$ be a nonempty ordinary set, $(Y, \mathcal{T})$ an ordinary topological space, $f : \ X \to Y$ an ordinary mapping, $L$ a F-lattice. Then*

$$f^{\leftarrow}[\omega_L(\mathcal{T})] = \omega_L(f^{-1}[\mathcal{T}]). \quad □$$

By the preceding proposition, we can reasonably give out the following definitions:

**4.4.17 Definition** Let $L$ be a F-lattice.

For every $(X, \mathcal{T}) \in Ob(\mathbf{Top})$, call $\omega_L(\mathcal{T})$ the *induced topology* of $\mathcal{T}$, denote $\omega_L(X, \mathcal{T})= (L^X, \omega_L(\mathcal{T}))$, called the *induced L-fts* of $(X, \mathcal{T})$.

For every two $(X, \mathcal{T})$, $(Y, \mathcal{S}) \in Ob(\mathbf{Top})$ and every $f \in hom_{\mathbf{Top}}((X, \mathcal{T}), (Y, \mathcal{S}))$, denote

$$\omega_L(f)= f^{\to} : \ (L^X, \omega_L(\delta)) \to (L^Y, \omega_L(\mu)).$$

For every $(L^X, \delta) \in Ob(L\text{-}\mathbf{FTS})$ and every $\gamma \in pr(L)$, call $\iota_\gamma(\delta)$ the *$\gamma$-level topology* of $\delta$; call $\iota_L(\delta)$ the *topological modification* of $\delta$; denote $\iota_L(L^X, \delta) = (X, \iota_L(\delta))$.

For every two $(L^X, \delta), (L^Y, \mu) \in Ob(L\text{-}\mathbf{FTS})$ and every

$$f^{\to} \in hom_{L\text{-}\mathbf{FTS}}((L^X, \delta), (L^Y, \mu)),$$

denote

$$\iota_L(f^{\to}) = f: \ (X, \iota_L(\delta)) \to (Y, \iota_L(\mu)).$$

Then one can easily verify the following conclusion:

**4.4.18 Proposition** *Let $L$ be a F-lattice. Then*

$$\omega_L: \mathbf{Top} \to L\text{-}\mathbf{FTS} \quad \text{and} \quad \iota_L: L\text{-}\mathbf{FTS} \to \mathbf{Top}$$

*are functors.* □

## 4.5 Analytic and Topological Characterizations of Completely Distributive Law

In Section **1.3**, we established several characterizations of completely distributive law from the pure lattice-theoretic angle, or in another word, from the view of point of algebra. But complete distributivity has still its own analytic property and topological property, they can be still used to describe its distributivity. This is just what we want to do in this section.

**4.5.1 Theorem** *Let $L$ be a distributive complete lattice. Then the following conditions are equvalent:*

(i) *$L$ is completely distributive.*

(ii) *$L$ satisfies both conditions* (US) *and* (LS).

**Proof** By Theorem **4.3.2**, (i)$\Longrightarrow$(ii). So we need only prove (ii)$\Longrightarrow$(i). By Theorem **1.3.28** (ii)$\Longrightarrow$(i), we need only prove that $L$ is a Hausdorff topological lattice with respect to its interval $\Omega(L)$.

(1) $L$ is a topological lattice. Denote the product topology of $(L, \Omega(L)) \times (L, \Omega(L))$ by $\mathcal{T}$, define $g, h: \ L \times L \to L$ by $g(x, y) = x \wedge y$, $h(x, y) = x \vee y$ for every $(x, y) \in L \times L$, then we need to prove that both

$$g: \ (L \times L, \mathcal{T}) \to (L, \Omega(L)) \text{ and } h: \ (L \times L, \mathcal{T}) \to (L, \Omega(L))$$

are continuous.

For $g$, by Proposition **4.2.13**, we need only prove that $g$ is both upper semicontinuous and lower semicontinuous. $\forall a \in L$, $g_{[a]} = (\uparrow a) \times (\uparrow a)$ is closed in $(L \times L, \mathcal{T})$. By Theorem **4.2.16** (iii), $g$ is upper semicontinuous. Then what we need is only to prove that $g: \ (L \times L, \mathcal{T}) \to L$ is lower semicontinuous, i.e. to prove the continuity of

$$g: \ (L \times L, \mathcal{T}) \to (L, \Omega_*(L)).$$

$\forall (x, y) \in L \times L$, $\forall V \in \mathcal{N}_{\Omega_*(L)}(g(x, y))$, by the definition of lower topology, $\exists F \in [L]^{<\omega}$ such that $g(x, y) \in \bigcap\{L \backslash \downarrow b: \ b \in F\} \subset V$. So $\forall b \in F$, $x \wedge y = g(x, y) \not\leq b$. Since $id: \ (L, \Omega_*(L)) \to (L, \Omega_*(L))$ is continuous, i.e. $id: \ (L, \Omega_*(L)) \to L$ is lower semicontinuous, by (ii), (LS) holds for $L$, $\forall b \in F$, $\bigvee\{\bigwedge U: \ U \in \mathcal{N}_{\Omega_*(L)}(x \wedge y)\} = x \wedge y \not\leq b$. So $\exists U_b \in \mathcal{N}_{\Omega_*(L)}(x \wedge y)$ such that $\bigwedge U_b \not\leq b$. By the structure of lower topology $\Omega_*(L)$, $\exists G_b \in [L]^{<\omega}$ such that $x \wedge y \in \bigcap\{L \backslash \downarrow c: \ c \in G_b\} \subset U_b$ and hence $x, y \in \bigcap\{L \backslash \downarrow c: \ c \in G_b\}$ for every $b \in F$. Take $W = \bigcap\{\bigcap\{L \backslash \downarrow c: \ c \in G_b\}: \ b \in F\}$, then $x, y \in W$. Since every $G_b$ is finite, $\bigcap\{L\backslash\Downarrow c: c \in G_b\} \in \Omega_*(L)$; since $F$ is finite, $W \in \Omega_*(L)$. So $W \in \mathcal{N}_{\Omega_*(L)}(x)$, $W \in \mathcal{N}_{\Omega_*(L)}(y)$, $W \times W$ is a neighborhood of $(x, y)$ in $(L, \Omega_*(L)) \times (L, \Omega_*(L))$, and hence a neighborhood of $(x, y)$ in $(L \times L, \mathcal{T})$. Now $\forall (u, v) \in W \times W$, since $W \times W \subset \bigcap\{U_b \times U_b: b \in F\}$ and $\forall b \in F$, $\bigwedge U_b \not\leq b$, we have $g(u, v) = u \wedge v \geq \bigwedge U_b \wedge \bigwedge U_b = \bigwedge U_b$, $g(u, v) \not\leq b$ for every $b \in F$. So $g(u, v) \in \bigcap\{L\backslash\Downarrow b: b \in F\} \subset V$, $g[W \times W] \subset V$. That is to say, $g: (L \times L, \mathcal{T}) \to (L, \Omega_*(L))$ is continuous.

(2) $L$ is a Hausdorff space. $\forall x, y \in L$, since $id: (L, \Omega(L)) \to L$ is continuous and hence both upper semicontinuous and lower semicontinuous, by (ii),

$$x = \bigwedge\{\bigvee U: U \in \mathcal{N}_{\Omega(L)}(x)\}, \quad y = \bigvee\{\bigwedge V: V \in \mathcal{N}_{\Omega(L)}(y)\}. \tag{4.1}$$

If $U \cap V \neq \emptyset$ for every $U \in \mathcal{N}_{\Omega(L)}(x)$ and every $V \in \mathcal{N}_{\Omega(L)}(y)$, then $\exists z_{UV} \in U \cap V$, $\bigvee U \geq z_{UV} \geq \bigwedge V$. By equalities (4.1), $x \geq y$. Similarly prove $x \leq y$. So $x = y$, $L$ is Hausdorff. □

Theorem **4.5.1** characterizes completely distributive law from an angle of analysis. Recall Proposition **4.3.13**, we can consider the following conclusion as a topological way to do the same thing:

**4.5.2 Theorem** *Let $L$ be a distributive complete lattice. Then the following conditions are equivalent:*

(i) *$L$ is completely distributive.*

(ii) *For every ordinary topological space $(X, \mathcal{T})$, $lc_L(\mathcal{T}) \subset stt_{*L}(\mathcal{T})$, $uc_L(\mathcal{T}) \subset stt^*{}_L(\mathcal{T})$.*

**Proof** (i)$\Longrightarrow$(ii): By Theorem **4.3.16** (ii), (iii).

(ii)$\Longrightarrow$(i): Let $(X, \mathcal{T})$ be an ordinary topological space, $A: X \to L$ a lower semicontinuous mapping, $x \in X$, $\mathcal{B}$ a neighborhood base of $x$ in $X$, then $A \in lc_L(\mathcal{T}) \subset stt_{*L}(\mathcal{T})$. By Proposition **4.3.15**, $A(x) = \bigvee_{U \in \mathcal{B}} \bigwedge_{y \in U} A(y)$. So (LS) is true for $L$. Similarly, $uc_L(\mathcal{T}) \subset stt^*{}_L(\mathcal{T})$ implies (US). By Theorem **4.5.1** (ii)$\Longrightarrow$(i), $L$ is completely distributive. □

**4.5.3 Remark** In real analysis, as well-known, semicontinuous functions can be approximated by staircase functions. Theorem **4.5.2** extends this result into the case of lattice. Moreover, this theorem tells us: As a value domain, a distributive complete lattice is completely distributive if and only if every semicontinuous mappings into it can be approximated by staircase mappings. So Theorem **4.5.2** shows the special importance of completely distributive lattice in topology on lattice.

# Chapter 5

# Convergence Theory

To describe trends of some variations in $L$-fuzzy topological spaces, we need some tools to deal with various convergence processes. In this chapter, we investigate three main convergence modes and their relations. In the last section, a way to describe an $L$-fts using net convergence will be studied.

## 5.1 Net Convergence Theory

Certainly, the simplest convergence is sequential convergence which is familiar in Euclidean space $\mathbf{R}^n$. The concept and the method of net convergence is just a generalization of sequential convergence.

**5.1.1 Definition** Let $X$ be a nonempty ordinary set, $D$ a directed set. Then call every mapping $S:\ D \to X$ a *net* in $X$, and $D$ the *index set* of $S$.

In an $L$-fuzzy space $L^X$, call a net $S:\ D \to Pt(L^X)$ a *net* in $L^X$. Especially, call a net $S:\ D \to M(L^X)$ a *molecule net* in $L^X$.

For a subset $A \subset Pt(L^X)$ and a net $S = \{S(n), n \in D\}$ such that $S(n) \in A$ for every $n \in D$, we say $S$ *consists of points in* $A$.

A net $S$ in $L^X$ with index set $D$ is also denoted by $S:\ D \to (L^X, \delta)$ or $S = \{S(n), n \in D\}$.

For a net $S = \{S(n), n \in D\}$ in $L^X$ and an $L$-fuzzy subset $A \in L^X$, We say $S$ is a *net in* $A$, if $S(n) \leq A$ for every $n \in D$.

For a net $S = \{S(n), n \in D\}$ in $L^X$ and $e \in Pt(L^X)$, $S$ is called a *constant net with value* $e$, if $S(n) = e$ for every $n \in D$.

For an $L$-fts $(L^X, \delta)$ and a net or a molecule net $S$ in $L^X$, we say $S$ is a net or a molecule net in $(L^X, \delta)$ respectively.

Clearly, a sequence $x_{n_0}$, $x_{n_1}$, $\cdots$ in a set $X$ is in fact a net with an index set $D = \{n_0, n_1, \cdots\} \subset \omega$. So the concept of net is a generalization of the one of sequence.

**5.1.2 Definition** Let $(L^X, \delta)$ be an $L$-fts, $S = \{S(n), n \in D\}$ a net in $(L^X, \delta)$, $P$ a property, $e \in Pt(L^X)$.

Call $S$ *eventually* possesses the property $P$, if there exists $n_0 \in D$ such that for every $n \in D$, $n \geq n_0$, $S(n)$ always possess the property $P$. Call $S$ *frequently* possesses the property $P$, if for every $n \in D$, there always exist $n_0 \in D$ such that $n_0 \geq n$ and $S(n_0)$ possesses the property $P$.

Call $e$ a *cluster point* of $S$, denoted by (adopting Guo-jun Wang's symbol[168]) $S \infty e$, if for every Q-neighborhood $U$ of $e$, $S$ frequently quasi-coincides with $U$. Call $e$ a *limit point* or a *limit* for short, denoted by $S \to e$, if for every Q-neighborhood $U$ of $e$, $S$ eventually quasi-coincides with $U$; in this case we also say $S$ *converges* to $e$, or say $S$ is *convergent* to $e$.

Denote the join of all the cluster points of $S$ by clu$S$, the join of all the limit points of net $S$ by $\lim S$.

**5.1.3 Theorem** *Let* $(L^X, \delta)$ *be an L-fts,* $S = \{S(n), n \in D\}$ *a net in* $(L^X, \delta)$, $e \in Pt(L^X)$. *Then*

(i) $S \to e \implies S \infty e$.

(ii) $\lim S \leq \mathrm{clu} S$.

(iii) $S \infty e \geq d \implies S \infty d$.

(iv) $S \to e \geq d \implies S \to d$.

(v) $S \infty e \iff e \leq \mathrm{clu} S$.

(vi) $S \to e \iff e \leq \lim S$.

**Proof** (i) Clear.

(ii) By (i).

(iii) Since $e \geq d$, so by the definition of Q-neighborhood, $\mathcal{Q}(d) \subset \mathcal{Q}(e)$. Hence $S \infty d$ by $S \infty e$.

(iv) Similar to (iii).

(v) ($\Longrightarrow$) By the definition of clu$S$.

($\Longleftarrow$) Suppose $e \leq \mathrm{clu} S$. $\forall U \in \mathcal{Q}(e)$, then $e \not\leq U'$, $\mathrm{clu} S \not\leq U'$, there exists a cluster point $d$ of $S$ such that $d \not\leq U'$, $U \in \mathcal{Q}(d)$. So $S$ frequently quasi-coincides with $U$, $S \infty e$.

(vi) Similar to (v). □

The concept of subsequence is familiar to us. Similar to it, we can define the concept of subnet in the following:

**5.1.4 Definition** Let $(L^X, \delta)$ be an $L$-fts, $S: D \to (L^X, \delta)$, $T: E \to (L^X, \delta)$ be two nets in $(L^X, \delta)$. Call $T$ is a *subnet* of $S$, or call $S$ a *parental net* of $T$, if there exists a mapping $N: E \to D$, called a *cofinal selection* on $S$, such that

(i) $T = S \circ N$;

(ii) For every $n_0 \in D$, there exists $m_0 \in E$ such that $N(m) \geq n_0$ for $m \geq m_0$.

**5.1.5 Remark** (1) Suppose $\{x_0, x_1, \cdots\}$ is a sequence in a set $X$, then it can be represented as a net $S: \omega \to X$ with $S(n) = x_n$, and every its subsequence $\{x_{n_0}, x_{n_1}, \cdots\}$ can be represented as a subnet $T: \omega \to X$ with the cofinal selection $N: \omega \to \omega$, $N(m) = n_m$. So a subsequence is naturally a subnet. But need to note that for every subsequence of a sequence, its index set, in fact, is still $\omega$ or equivalently, **N**, just as the index set of the sequence itself. As for nets, being a directed set is the unique limitation for the index set of a subnet, so its structure may be far more complicated than the one of its parental net.

(2) In the two conditions of a subnet, the meaning of the first one is clear, it asks that a subnet can only take values from the image of its parental net, i.e. should be "a part of its parent." The second condition means that this "part" should be cofinal in the parental net according to the two index sets connected by the cofinal selection.

**5.1.6 Example** (i) Let $X = (0, 1]$, $L = [0, 1]$ and $\delta$ be the $L$-fuzzy topology on $X$ generated by the base $\{a\chi_{(x,1]} : x \in X,\ a \in L\}$, then $(L^X, \delta)$ is an $L$-fts.

Let $D$ be the set of all the rational numbers in $(0, 1)$ with the relative order in $[0, 1]$. Since every rational number $r \in (0, 1)$ can be uniquely represented as a reduced proper fraction $r = \frac{m}{n}$, so we can take $S(r) = r_{\frac{n}{n+1}}$ for every $r \in D$ and its reduced proper fraction representation $r = \frac{m}{n}$. Then $S: D \to Pt(L^X)$ is a net in $(L^X, \delta)$ and $S \to 1_1$.

Take $E = (0,1) \times \mathbf{N}$ equipped with the product partial order and $N: E \to D$ as

$$N(t,n) = min\{\frac{m}{n+1} : m \in \{1, \cdots, n\}, |t - \frac{m}{n+1}| \leq \frac{1}{n+1}\}, \ \forall (t,n) \in E,$$

then $E$ is a directed set, $N$ is a cofinal selection, $T = S \circ N$ is a subnet of $S$, and $T$ still converges to $1_1$.

(ii) Take $(L^X, \delta)$ and $S$ as (i), but $E = \mathbf{N}$. Define $N: E \to D$ by $N(n) = \frac{n}{n+1}$ for every $n \in \mathbf{N}$, then $N$ is a cofinal selection, $T = S \circ N: E \to (L^X, \delta)$ is a subnet of $S$ and $T$ still converges to $1_1$. In fact $T(n) = (\frac{n}{n+1})_{\frac{n}{n+1}} \in Pt(L^X)$ for every $n \in \mathbf{N}$.

In the preceding two examples, both subnets converge to the same limit point of their parent net. It is not accidental, this point is shown in the following results:

**5.1.7 Theorem** *Let $(L^X, \delta)$, $(L^Y, \mu)$ be an L-fts, $S$ a net in $(L^X, \delta)$, $T$ a subnet of $S$, $e \in Pt(L^X)$. Then*

(i) $S \to e \implies T \to e$.

(ii) $\lim S \leq \lim T$.

(iii) $T \infty e \implies S \infty e$.

(iv) $\mathrm{clu} T \leq \mathrm{clu} S$. □

**5.1.8 Definition** Let $A$ be a nonempty set, $\rho$ an ordinal number. By Zermelo's Theorem on well-ordering, there exists a relation $<_w$ well-ordering $A$. Define relation $\leq_w$ on $A$ by

$$\forall a, b \in A, \quad a \leq_w b \iff a <_w b \text{ or } a = b$$

and relation $\leq$ on $\rho \times A$ by

$$\forall (\alpha, a), (\beta, b) \in \rho \times A, \quad (\alpha, a) \leq (\beta, b) \iff \alpha < \beta, \text{ or } \alpha = \beta, \ a \leq_w b.$$

Call the set $\rho \times A$ equipped with this relation $\leq$ the *well-ordered $\rho$-copy* of $A$.

**5.1.9 Proposition** *Let $A$ be a nonempty set, $\rho$ an ordinal number. Then the relation $\leq$ on the well-ordered $\rho$-copy $\rho \times A$ of $A$ well-orders $\rho \times A$; i.e. $\rho \times A$ equipped with $\leq$ is a well-ordered set.* □

**5.1.10 Theorem** *Let $(L^X, \delta)$ be an L-fts, $S$ a net in $(L^X, \delta)$, $e \in Pt(L^X)$. Then $S \to e$ if and only if $T \infty e$ for every subnet $T$ of $S$.*

**Proof** (Necessity) By theorems **5.1.7** (i) and **5.1.3** (i).

(Sufficiency) Suppose $S = \{S(n), n \in D\}$, $S$ does not converge to $e$. Then $\exists U_0 \in \mathcal{Q}(e)$ such that $S$ frequently does not quasi-coincide with $U_0$. That is to say, $\exists U_0 \in \mathcal{Q}(e)$, $\forall n \in D$, $\exists N(n) \in D$ such that $N(n) \geq n$, $S(N(n)) \not\prec U_0$. So we get a cofinal selection $N: D \to D$ and then a subnet $T = S \circ N$ of $S$. Since for $U_0 \in \mathcal{Q}(e)$ and every $n \in D$, $T(n) = S(N(n)) \not\prec U_0$, $e$ is not a cluster point of $T$. □

About a cluster point of a net, we have furthermore the following important Theorem:

**5.1.11 Theorem** *Let $(L^X, \delta)$ be an L-fts, $e \in Pt(L^X)$. Then the following conditions are equivalent:*

(i) *$\mathcal{Q}(e)$ is down-directed.*

(ii) *$\mathcal{Q}(e)$ has a down-directed down-cofinal subset.*

(iii) *For every net $S$ in $(L^X, \delta)$, $S\infty e$ if and only if $S$ has a subnet $T \to e$.*

(iv) *For every net $S$ in $(L^X, \delta)$ such that $S\infty e$, $S$ has a subnet $T \to e$.*

(v) *For every molecule net $S$ in $(L^X, \delta)$, $S\infty e$ if and only if $S$ has a subnet $T \to e$.*

(vi) *For every molecule net $S$ in $(L^X, \delta)$ such that $S\infty e$, $S$ has a subnet $T \to e$.*

**Proof** (i)$\Longrightarrow$(ii): Clear.

(ii)$\Longrightarrow$(iii): By theorems **5.1.3** (i) and **5.1.7** (iii), "$S$ has a subnet $T \to e$" $\Longrightarrow$ $S\infty e$. So we need only prove that if (ii) holds and $S\infty e$, then $S$ has a subnet $T \to e$.

Suppose $D$ is the index set of $S$, $\mathcal{A}$ is a down-directed down-cofinal subset of $\mathcal{Q}(e)$. Take $E = D \times \mathcal{A}$, define a relation $\leq$ on $E$ by

$$\forall (m, U), (n, V) \in E, \quad (m, U) \leq (n, V) \Longleftrightarrow m \leq n, \ U \geq V,$$

then $\leq$ is a directed preorder (in fact, a directed partial order) on $E$, $E$ equipped with $\leq$ is a directed set.

$\forall (n, U) \in E$, since $U \in \mathcal{Q}(e)$, $S\infty e$, $\exists N(n, U) \in D$ such that $N(n, U) \geq n$, $S(N(n, U)) \lessdot U$. So we get a mapping $N: E \to D$. $\forall n_0 \in D$, since $\mathcal{A}$ is a down-directed set, $\mathcal{A} \neq \emptyset$, $\exists U_0 \in \mathcal{A}$. So $(n_0, U_0) \in E$. Then $\forall (n, U) \in E$ such that $(n, U) \geq (n_0, U_0)$, we have $N(n, U) \geq n \geq n_0$. That is to say $N: E \to D$ is a cofinal selection, $T = S \circ N$ is a subnet of $S$.

$\forall U \in \mathcal{Q}(e)$, since $\mathcal{A}$ is down-directed in $\mathcal{Q}(e)$, $\exists U_0 \in \mathcal{A}$ such that $U_0 \leq U$. Since $S\infty e$, $\exists n_0 \in D$ such that $S(n_0) \lessdot U_0$. Then $(n_0, U_0) \in E$. $\forall (n, V) \in E$ such that $(n, V) \geq (n_0, U_0)$, we have $T(n, V) = S(N(n, V)) \lessdot V \leq U_0$. By Proposition **2.3.3** (viii), $T(n, V) \lessdot U_0$. So $T$ eventually quasi-coincides with $U_0$, $T \to e$.

(iii)$\Longrightarrow$(iv): Obvious.

(iv)$\Longrightarrow$(v): Need only prove "a molecule net $S$ has a subnet $T \to e$ $\Longrightarrow$ $S\infty$." But by theorems **5.1.3** (i) and **5.1.7** (iii), this is obvious.

(v)$\Longrightarrow$(vi): Obvious.

(vi)$\Longrightarrow$(i): Suppose $\mathcal{Q}(e)$ is not down-directed, then $\exists U_0, U_1 \in \mathcal{Q}(e)$ such that $U_0 \wedge U_1 \notin \mathcal{Q}(e)$.

Take $D$ as the well-ordered $\omega$-copy $\omega \times \mathcal{Q}(e)$ of $\mathcal{Q}(e)$, then by Proposition **5.1.9**, $D$ is a well-ordered set and then clearly a directed set.

Suppose $e = x_a$. $\forall (n, U) \in D$, there totally exist the following five possibilities:

Case (1): $U_0 \wedge U \in \mathcal{Q}(x_a)$, $U_1 \wedge U \notin \mathcal{Q}(x_a)$;

Case (2): $U_0 \wedge U \notin \mathcal{Q}(x_a)$, $U_1 \wedge U \in \mathcal{Q}(x_a)$;

Case (3): $U_0 \wedge U \notin \mathcal{Q}(x_a)$, $U_1 \wedge U \notin \mathcal{Q}(x_a)$;

Case (4): $U_0 \wedge U \in \mathcal{Q}(x_a)$, $U_1 \wedge U \in \mathcal{Q}(x_a)$, $\exists m < \omega$, $n = 2m$;

Case (5): $U_0 \wedge U \in \mathcal{Q}(x_a)$, $U_1 \wedge U \in \mathcal{Q}(x_a)$, $\exists m < \omega$, $n = 2m + 1$,

We shall take $S(n, U) \in M(L^X)$ such that

$$\left.\begin{array}{lll}\text{In Case (1)}: & S(n,U) \lessdot U, & S(n,U) \not\lessdot U_1; \\ \text{In Case (2)}: & S(n,U) \lessdot U, & S(n,U) \not\lessdot U_0; \\ \text{In Case (3)}: & S(n,U) \lessdot U, & S(n,U) \not\lessdot U_0, \quad S(n,U) \not\lessdot U_1; \\ \text{In Case (4)}: & S(n,U) \lessdot (U_0 \wedge U), & S(n,U) \not\lessdot U_1; \\ \text{In Case (5)}: & S(n,U) \lessdot (U_1 \wedge U), & S(n,U) \not\lessdot U_0. \end{array}\right\} \quad (5.1)$$

For Case (1), since $U \in \mathcal{Q}(x_a)$, $a \not\leq U(x)'$. By Corollary **1.3.12** (i), $\bigvee M(\downarrow a) = a$, so $\exists \lambda \in M(\downarrow a)$ such that $\lambda \not\leq U(x)'$. Since $U_1 \wedge U \notin \mathcal{Q}(x_a)$, $\lambda \leq a \leq (U_1 \wedge U)'(x) = U_1(x)' \vee U(x)'$. By $\lambda \in M(L)$ and $\lambda \not\leq U(x)'$, we have $\lambda \leq U_1(x)'$. So we take $S(n,U) = x_\lambda \in M(L^X)$, then $S(n,U) \lessdot U$, $S(n,U) \not\lessdot U_1$.

For Case (2), similar to the Case (1), we can take $S(n,U) \in M(L^X)$ such that $S(n,U) \lessdot U$, $S(n,U) \not\lessdot U_0$.

For Case (3), similar to the Case (1), we can take $S(n,U) \in M(L^X)$ such that $S(n,U) \lessdot U$, $S(n,U) \not\lessdot U_0, U_1$.

For Case (4), since $U_0 \wedge U \in \mathcal{Q}(x_a)$, $a \not\leq (U_0 \wedge U)'(x)$. By Corollary **1.3.12** (i), $\exists \lambda \in M(\downarrow a)$ such that $\lambda \not\leq (U_0 \wedge U)'(x)$. By the supposition at the beginning of this proof of (vii)$\Longrightarrow$(i), $U_0 \wedge U_1 \notin \mathcal{Q}(x_a)$, so

$$\begin{aligned}\lambda \leq a \leq (U_0 \wedge U_1)'(x) &= (U_0' \vee U_1')(x) \leq ((U_0 \wedge U)' \vee U_1')(x) \\ &= (U_0 \wedge U)'(x) \vee U_1'(x).\end{aligned}$$

By $\lambda \in M(L)$ and $\lambda \not\leq (U_0 \wedge U)'(x)$, we have $\lambda \leq U_1'(x)$. Take $S(n,U) = x_\lambda$, then $S(n,U) \lessdot (U_0 \wedge U)$ but $S(n,U) \not\lessdot U_1$.

For Case (5), similar to the Case (4), we can take $S(n,U) \in M(L^X)$ such that $S(n,U) \lessdot (U_1 \wedge U)$ but $S(n,U) \not\lessdot U_0$.

Now $S: D \to M(L^X)$ is a molecule net in $(L^X, \delta)$ satisfying relations in group (5.1).

$\forall U \in \mathcal{Q}(e)$, $\forall (n,V) \in D$, then $(2(n+1),U) \in D$, $(2(n+1),U) \geq (n,V)$ and by relations (5.1), $S(2(n+1),U) \lessdot U$ (in Case (1) to (3)), or $S(2(n+1),U) \lessdot (U_0 \wedge U)$ (in Case (4)) and then $S(2(n+1),U) \lessdot U$, or $S(2(n+1),U) \lessdot (U_1 \wedge U)$ (in Case (5)) and then $S(2(n+1),U) \lessdot U$. So $S \infty e$.

If $T: E \to M(L^X)$ is a subnet of $S$ such that $T \to e$. Then for $U_0, U_1 \in \mathcal{Q}(e)$, there should exist $m_0, m_1 \in E$ such that $\forall m \in E$, $m \geq m_0 \Longrightarrow T(m) \lessdot U_0$, $m \geq m_1 \Longrightarrow T(m) \lessdot U_1$. Since $E$ is directed, $\exists m_2 \in E$ such that $m_2 \geq m_0, m_1$. So $T(m) \lessdot U_0, U_1$ for every $m \in E$, $m \geq m_2$. But $T$ takes values in $img(S)$, $\exists (n,U) \in D$ such that $T(m) = S(n,U)$. By relations (5.1), there is no $S(n,U)$ can quasi-coincide with both $U_0$ and $U_1$ simultaneously. So $T$ cannot be convergent to $e$, $S$ has no subnet which converges to $e$. □

By Theorem **2.3.12** (ii), we obtain the following

**5.1.12 Corollary** *Let $(L^X, \delta)$ be an L-fts, $S$ a net in $(L^X, \delta)$, $e \in M(L^X)$. Then $S \infty e$ if and only if $S$ has a subnet $T \to e$.* □

**5.1.13 Remark** With its convergence theory, net is a powerful tool in $L$-fuzzy topology as well as in general topology. It can be used to describe closedness, compactness, separations, etc. E.H. Moore and H.L. Smith are the earliest authors to introduce and systematically study net convergence theory.[118] So net convergence theory in topology is also called "Moore-Smith Convergence Theory." As for fuzzy topology, Ying-ming Liu is the first author who successfully introduced and systematically studied Moore-Smith Convergence Theory in topological spaces with the Quasi-coincident Neighborhood Theory he introduced.[137]

**5.1.14 Theorem** *Let $(L^X, \delta)$ be an L-fts, $e, A \in L^X$.*

*If $e \in Pt(L^X)$, then the following conditions are equivalent:*

(i) $e \leq A^-$.

(ii) *There exists a net $S$ in $A$ such that $S \infty e$.*

(iii) *There exists a molecule net $S$ in $A$ such that $S \infty e$.*

*If $e \in M(L^X)$, then the following conditions are equivalent to condition* (i) *mentioned above:*

(iv) *$e$ is an adherent point of $A$.*

(v) *There exists a net $S$ in $A$ such that $S \to e$.*

(vi) *There exists a molecule net $S$ in $A$ such that $S \to e$.*

**Proof** Suppose $e = x_a \in Pt(L^X)$.

(i)$\Longrightarrow$(iii): Take $D$ as the well-ordered $\omega$-copy $\omega \times \mathcal{Q}(e)$ of $\mathcal{Q}(e)$. $\forall (n, U) \in D$, then $U \in \mathcal{Q}(x_a)$, by Corollary **1.3.12** (i), $\bigvee M(\downarrow a) = a \not\leq U(x)'$, $\exists \lambda \in M(\downarrow a)$, $\lambda \not\leq U(x)'$, $U \in \mathcal{Q}(x_\lambda)$. Since $x_\lambda \leq x_a \leq A^-$, by Theorem **2.3.23** (i), $x_\lambda$ is an adherent point of $A$, for $U \in \mathcal{Q}(x_\lambda)$, $A \not\leq U'$. By Corollary **2.3.11** (i), $\exists y_\gamma \in M(\downarrow A)$ such that $y_\gamma \not\leq U'$. Denote $S(n, U) = y_\gamma$, we have $S(n, U) \leq A$, $S(n, U) \lessdot U$. So we have obtained a molecule net $S = \{S(n, U), (n, U) \in D\}$ in $A$. $\forall U \in \mathcal{Q}(e)$, $\forall (n, V) \in D$, $(n+1, U) \in D$, $(n+1, U) \geq (n, V)$, $S(n+1, U) \lessdot U$. So $S$ frequently quasi-coincides with $U$. Hence $S \infty e$.

(iii)$\Longrightarrow$(ii): Obvious.

(ii)$\Longrightarrow$(i): If there exists a net $S = \{S(n), n \in D\}$ in $A$ such that $S \infty x_a$, then $\forall \lambda \in M(\downarrow a)$, $\forall U \in \mathcal{Q}(x_\lambda)$, $U \in \mathcal{Q}(x_a)$. $\forall n \in D$, by $S \infty x_a$, $\exists n_0 \in D$ such that $n_0 \geq n$, $S(n_0) \lessdot U$, $S(n_0) \not\leq U'$. Since $S$ is in $A$, $S(n_0) \leq A$, so $A \not\leq U'$, $U \hat{q} A$. By the arbitrariness of $U \in \mathcal{Q}(x_\lambda)$, $x_\lambda$ is an adherent point of $A$. By Theorem **2.3.23** (ii), $x_\lambda \leq A^-$, $\lambda \leq A^-(x)$. By the arbitrariness of $\lambda \in M(\downarrow a)$ and Corollary **1.3.12** (i), $a = \bigvee M(\downarrow a) \leq A^-(x)$, $x_a \leq A^-$.

Now suppose $e \in M(L^X)$.

(vi)$\Longrightarrow$(v): Obvious.

(v)$\Longrightarrow$(iv): By Theorem **5.1.3** (i) and what are proved above, (v)$\Longrightarrow$(ii)$\Longrightarrow$(i). Then since $e \in M(L^X)$, (i)$\Longrightarrow$(iv) by Theorem **2.3.23** (i).

(iv)$\Longrightarrow$(i): By Theorem **2.3.23** (i).

(i)$\Longrightarrow$(vi): We have had (i)$\Longrightarrow$(iii). On the other hand, since $e \in M(L^X)$, by Corollary **5.1.12**, (iii)$\Longrightarrow$(iv). □

**5.1.15 Corollary** *Let $(L^X, \delta)$ be an L-fts, $A \in L^X$. Then the following conditions are equivalent:*

(i) $A = A^-$.

(ii) *For every net $S$ in $A$ and every $e \in Pt(L^X)$, $S \infty e \Longrightarrow e \leq A$.*

(iii) *For every molecule net $S$ in $A$ and every $e \in Pt(L^X)$, $S \infty e \Longrightarrow e \leq A$.*

(iv) *For every net $S$ in $A$ and every $e \in M(L^X)$, $S \to e \Longrightarrow e \leq A$.*

(v) *For every molecule net* $S$ *in* $A$ *and every* $e \in M(L^X)$, $S \to e \implies e \leq A$.□

Preceding theorem shows how nets in an $L$-fts characterize adherent points and closed subsets. Nets can also be used to characterize accumulation points. A molecule $e$ is an accumulation point of an $L$-fuzzy subset $A$ means $e$ is an adherent point of $A \backslash e$, so by Theorem **5.1.14** we have the following result:

**5.1.16 Theorem** *Let* $(L^X, \delta)$ *be an L-fts,* $A \in L^X$, $e \in M(L^X)$. *Then the following conditions are equivalent:*

(i) $e$ *is an accumulation point of* $A$.

(ii) *There exists a net* $S$ *in* $A \backslash e$ *such that* $S \infty e$.

(iii) *There exists a net* $S$ *in* $A \backslash e$ *such that* $S \to e$.

(iv) *There exists a molecule net* $S$ *in* $A \backslash e$ *such that* $S \infty e$.

(v) *There exists a molecule net* $S$ *in* $A \backslash e$ *such that* $S \to e$. □

As the closedness in $L$-fts can be described by nets, the continuity of $L$-fuzzy can also be represented by nets. This point shown in the next result:

**5.1.17 Theorem** *Let* $(L^X, \delta)$, $(L^Y, \mu)$ *be L-fts',* $f^{\to}: L^X \to L^Y$ *an L-fuzzy mapping. Then the following conditions are equivalent:*

(i) $f^{\to}$ *is continuous.*

(ii) *For every net* $S$ *in* $(L^X, \delta)$ *and every* $e \in M(L^X)$, $S \infty e \implies f^{\to} \circ S \infty f^{\to}(e)$.

(iii) *For every net* $S$ *in* $(L^X, \delta)$ *and every* $e \in M(L^X)$, $S \to e \implies f^{\to} \circ S \to f^{\to}(e)$.

(iv) *For every molecule net* $S$ *in* $(L^X, \delta)$ *and every* $e \in M(L^X)$, $S \infty e \implies f^{\to} \circ S \infty f^{\to}(e)$.

(v) *For every molecule net* $S$ *in* $(L^X, \delta)$ *and every* $e \in M(L^X)$, $S \to e \implies f^{\to} \circ S \to f^{\to}(e)$.

**Proof** (i)$\implies$(ii): Suppose $S = \{S(n), n \in D\}$ is a net in $(L^X, \delta)$, $S \infty e \in M(L^X)$. $\forall V \in \mathcal{Q}(f^{\to}(e))$, since $f^{\to}$ is continuous, so continuous at $e$, $f^{\leftarrow}(V) \in \mathcal{Q}(e)$. So $\forall n \in D$, $\exists n_0 \in D$, $n_0 \geq n$ such that $S(n_0) \lessdot f^{\leftarrow}(V)$. By Proposition **2.3.4** (i), $f^{\to} \circ S(n_0) \lessdot V$. Hence $f^{\to} \circ S$ frequently quasi-coincides with $V$, $f^{\to} \circ S \infty e$.

(ii)$\implies$(iii): Suppose $S \to e \in M(L^X)$.

Let $K = \{K(m), m \in E\}$ be a subnet of $f^{\to} \circ S$, $N : E \to D$ is the cofinal selection, then $K = (f^{\to} \circ S) \circ N = f^{\to} \circ (S \circ N)$. Denote $T = S \circ N$, then $T$ is a subnet of $S$. By Theorem **5.1.7** (i), $T \to e$. By Theorem **5.1.3** (i), $T \infty e$. By (ii), $K = f^{\to} \circ T \infty f^{\to}(e)$. By Theorem **5.1.10**, $f^{\to} \circ S \to f^{\to}(e)$.

(iii)$\implies$(iv): Suppose $S$ is a molecule net in $(L^X, \delta)$ and $S \infty e$. Since $e \in M(L^X)$, by Corollary **5.1.12**, $S$ has a subnet $T \to e$. Then by (iii), $f^{\to} \circ T \to f^{\to}(e)$. Since $f^{\to} \circ T$ is clearly a subnet of $f^{\to} \circ S$, using Corollary **5.1.12** again, we get $f^{\to} \circ S \infty f^{\to}(e)$.

(iv)$\implies$(v): Similar to (ii)$\implies$(iii).

(v)$\implies$(i): Suppose $f^{\to}$ is not continuous, by Theorem **2.4.5** (i)$\iff$(iii), $f^{\to}$ is not continuous at a molecule $e \in M(L^X)$. By Proposition **2.4.4**, $\exists V_0 \in \mathcal{Q}(f^{\to}(e))$ such that $f^{\leftarrow}(V_0) \notin \delta$, $f^{\leftarrow}(V_0)'$ is not closed, i.e. $f^{\leftarrow}(V_0)' < f^{\leftarrow}(V_0)'^{-}$. By Theorem **2.3.23** (ii), there exists an adherent point $d$ of $f^{\leftarrow}(V_0)'$ such that $d \not\leq f^{\leftarrow}(V_0)'$.

Denote $D = \mathcal{Q}(d)$, define a relation $\leq_D$ on $D$ by:

$$\forall U, V \in D, \quad U \leq_D V \iff U \geq V.$$

Since $d \in M(L^X)$, by Theorem **2.3.12** (ii), $\leq_D$ is a directed preorder on $D$, $D$

equipped with $\leq_D$ is a directed set. $\forall U \in D$, since $d$ is an adherent point of $f^{\leftarrow}(V_0)'$, $U\hat{q}f^{\leftarrow}(V_0)'$, $U \not\leq (f^{\leftarrow}(V_0)')' = f^{\leftarrow}(V_0)$, $f^{\leftarrow}(V_0)' \not\leq U'$. By Corollary **2.3.11** (i), $\exists S(U) \in M(\downarrow f^{\leftarrow}(V_0)')$ such that $S(U) \not\leq U'$. Then we get a molecule net $S = \{S(n), n \in D\}$ in $(L^X, \delta)$ such that $S(U) \leq f^{\leftarrow}(V_0)'$, $S(U) \not\leq U'$ for every $U \in D = \mathcal{Q}(d)$.

$\forall U \in \mathcal{Q}(d)$, then $U \in D$. Arbitrarily fixed a $U_0 \in D$ such that $U_0 \geq_D U$, then $\forall V \in D$, $V \geq_D U_0$ implies $V \leq U_0 \leq U$. Since $S(V) \lessdot V$, so $S(V) \lessdot U$. Hence we have proved $S \to d$.

Since $d \not\leq f^{\leftarrow}(V_0)'$, i.e. $d \lessdot f^{\leftarrow}(V_0)$. By Proposition **2.3.4** (i), $f^{\rightarrow}(d) \lessdot V_0$, $V_0 \in \mathcal{Q}(f^{\rightarrow}(d))$. Now $\forall U \in D$, $S(U) \leq f^{\leftarrow}(V_0)'$, by Proposition **2.2.5** and Theorem **2.1.25** (ii),

$$f^{\rightarrow}\circ S(U) \leq f^{\rightarrow}(f^{\leftarrow}(V_0)') = f^{\rightarrow}f^{\leftarrow}(V_0') \leq V_0'.$$

So $f^{\rightarrow}\circ S(U) \not\lessdot V_0$. By $V_0 \in \mathcal{Q}(f^{\rightarrow}(d))$ and the arbitrariness of $U \in D$, $f^{\rightarrow}\circ S$ does not converge to $f^{\rightarrow}(d)$. But $S \to d$, so (v) does not hold. □

Reader can easily verify the following conclusions by the preceding theorem:

**5.1.18 Corollary** *Let $(L^X, \delta)$, $(L^Y, \mu)$ be L-fts', $f^{\rightarrow}: L^X \to L^Y$ an L-fuzzy mapping. Then the following conditions are equivalent:*

(i) *$f^{\rightarrow}$ is continuous.*

(ii) *For every net $S$ in $(L^X, \delta)$, $f^{\rightarrow}(\operatorname{clu} S) \leq \operatorname{clu} f^{\rightarrow}\circ S$.*

(iii) *For every net $S$ in $(L^X, \delta)$, $f^{\rightarrow}(\lim S) \leq \lim f^{\rightarrow}\circ S$.*

(iv) *For every molecule net $S$ in $(L^X, \delta)$, $f^{\rightarrow}(\operatorname{clu} S) \leq \operatorname{clu} f^{\rightarrow}\circ S$.*

(v) *For every molecule net $S$ in $(L^X, \delta)$, $f^{\rightarrow}(\lim S) \leq \lim f^{\rightarrow}\circ S$.* □

**5.1.19 Remark** From results in this section, reader can easily find that in general we cannot substitute the notion of net by the one of sequence. The later one cannot describe major and basic properties of *L*-fts'. The related counterexample is not difficult to construct. In fact, for example, even in ordinary topology, we know the existence of this kind of examples.

Using nets, we can provide some new descriptions of upper semicontinuity and lower semicontinuity:[128]

**5.1.20 Theorem** *Let $(X, \mathcal{T})$ be an ordinary topological space, $L$ a completely distributive lattice, $f: X \to L$ an ordinary mapping. Then the following conditions are equivalent:*

(i) *$f$ is lower semicontinuous.*

(ii) *If a net $S = \{S(n), n \in D\}$ in $X$ which converges to $x \in X$, then $f(x) \leq \bigvee_{n \in D} \bigwedge_{m \geq n} f(S(m))$.*

**Proof** (i)$\Longrightarrow$(ii): We use maximal sets defined in Definition **1.3.5** to prove this implication.

Suppose $S = \{S(n), n \in D\}$ is a net in $X$ converging to $x \in X$, denote $a = \bigvee_{n \in D} \bigwedge_{m \geq n} f(S(m))$. $\forall \gamma \in \beta^*(a)$, then $\forall n \in D$, $\bigwedge_{m \geq n} f(S(m)) \leq a$. By Theorem **1.3.11**, $\exists c(n) \in D$ such that $f(S(c(n))) \leq \gamma$. Denote $E = \{c(n) : n \in D\} \subset D$. $\forall i, j \in E$, since $D$ is directed, $\exists n \in D$ such that $n \geq i, j$, then $c(n) \in E$ and

$c(n) \geq n \geq i, j$. So $E$ is also a directed set. Take $N: E \to D$ as $N(m) = m$, then $N$ is obviously a confinal selection on $S$, $T = \{S(m), m \in E\}$ is a subnet of $S$. Since $S \to x$, $x$ is also a limit point of $T$. Then $f(S(c(n))) \leq \gamma$ for every $n \in D$ means $S(m) \in f^{[\gamma]}$ for every $m \in E$. By Theorem **4.2.16** (ii), $f^{[\gamma]}$ is closed in $(X, \mathcal{T})$, so $x \in f^{[\gamma]}$ by the virtue of $T \to x$. Hence $f(x) \leq \gamma$, by Theorem **1.3.11** we have

$$f(x) \leq \bigwedge\{\gamma: \gamma \in \underline{\beta}^*(a)\} = a = \bigvee_{n \in D} \bigwedge_{m \geq n} f(S(m)).$$

(ii)$\Longrightarrow$(i): Suppose $a \in L$, $S = \{S(n), n \in D\}$ is a net in $f^{[a]}$ which converges to $x \in X$, then $\forall n \in D$, $f(S(n)) \leq a$. It follows from (i),

$$f(x) \leq \bigvee_{n \in D} \bigwedge_{m \geq n} f(S(m)) \leq a.$$

So $x \in f^{[a]}$, $f^{[a]}$ is closed in $(X, \mathcal{T})$. By Theorem **4.2.16** (ii), $f$ is lower semicontinuous. □

Dually we have

**5.1.21 Theorem** *Let $(X, \mathcal{T})$ be an ordinary topological space, $L$ a completely distributive lattice, $f: X \to L$ an ordinary mapping. Then the following conditions are equivalent:*

(i) *$f$ is upper semicontinuous.*

(ii) *If a net $S = \{S(n), n \in D\}$ in $X$ which converges to $x \in X$, then $f(x) \geq \bigwedge_{n \in D} \bigvee_{m \geq n} f(S(m))$.* □

## 5.2 Filter Convergence Theory

In posets, concepts of filter and ideal are dual to each other, and the both structures are powerful in studying problems concerning ordered structure. An $L$-fts is in fact an ordered structure, using these two tools are also a natural need. In general topology, the neighborhood system of every point in a topological space naturally forms a down-directed set according to the inclusion order and then it can be formed as a filter simply. In $L$-fts', this is not always true for quasi-coincident neighborhood system. But, however, by Theorem **2.3.12** (ii), this is still true for ones of molecules. We have known all the molecules in an $L$-fts generates every its $L$-fuzzy subset. So in most cases, except some special cases such as C.T. Yang's Theorem we investigated in Chapter 2, molecules are enough. Theorem **5.1.11** shows us some equivalent characterizations of this properties with net.

**5.2.1 Definition** Let $(L^X, \delta)$ be an $L$-fts, $e \in Pt(L^X)$, $\mathcal{F} \subset L^X$ is a proper filter in $L^X$. $e$ is called a *cluster point* of $\mathcal{F}$, denoted by $\mathcal{F} \infty e$, if for every $U \in \mathcal{Q}(e)$ and every $A \in \mathcal{F}$, $U \wedge A \neq \underline{0}$. $e$ is called a *limit point* of $\mathcal{F}$, denoted by $\mathcal{F} \to e$, if $\mathcal{Q}(e) \subset \mathcal{F}$.

Denote the join of all the cluster points of $\mathcal{F}$ by $\operatorname{clu} \mathcal{F}$, the join of all the limit points of $\mathcal{F}$ by $\lim \mathcal{F}$.

For a filter $\mathcal{F}$ in $L^X$, we also say "$\mathcal{F}$ is a filter in $(L^X, \delta)$."

Similar to Theorem **5.1.3**, it is not difficult to verify the following

**5.2.2 Theorem** *Let $(L^X, \delta)$ be an $L$-fts, $\mathcal{F}$ a proper filter in $(L^X, \delta)$, $e \in Pt(L^X)$. Then*

(i) $\mathcal{F}\to e \implies \mathcal{F}\infty e.$
(ii) $\lim \mathcal{F} \le \mathrm{clu}\mathcal{F}.$
(iii) $\mathcal{F}\infty e \ge d \implies \mathcal{F}\infty d.$
(iv) $\mathcal{F}\to e \ge d \implies \mathcal{F}\to d.$
(v) $\mathcal{F}\infty e \iff e \le \mathrm{clu}\mathcal{F}.$
(vi) $\mathcal{F}\to e \iff e \le \lim \mathcal{F}.$ □

Comparing with the concepts of parent net and subnet, we have the concept of *coarser filter* and *finer filter* as follows:

**5.2.3 Definition** Let $(L^X, \delta)$ be an $L$-fts, $\mathcal{F}$, $\mathcal{G}$ be proper filters in $(L^X, \delta)$. Say $\mathcal{G}$ is *finer* than $\mathcal{F}$, or say $\mathcal{F}$ is *coarser* than $\mathcal{G}$, if $\mathcal{F} \subset \mathcal{G}$.

Then similar to Theorem **5.1.7**, we have

**5.2.4 Theorem** *Let $(L^X, \delta)$ be an L-fts, $\mathcal{F}$, $\mathcal{G}$ be proper filters in $(L^X, \delta)$, $\mathcal{F}$ be coarser than $\mathcal{G}$, $e \in Pt(L^X)$. Then*
(i) $\mathcal{F}\to e \implies \mathcal{G}\to e.$
(ii) $\lim \mathcal{F} \le \lim \mathcal{G}.$
(iii) $\mathcal{G}\infty e \implies \mathcal{F}\infty e.$
(iv) $\mathrm{clu}\mathcal{G} \le \mathrm{clu}\mathcal{F}.$ □

From conclusions mentioned above, we can find that it is different from the case of net and subnet, filter corresponding to parent net is smaller than its "children," but one corresponding to subnet is not "a part of its parent." This point can also be found in a correspondence between filters and nets in an $L$-fts described in the follows:

**5.2.5 Definition** Let $(L^X, \delta)$ be an $L$-fts, $S$ a molecule net in $(L^X, \delta)$, $\mathcal{F}$ a proper filter in $(L^X, \delta)$.

For $S$, define the *filter associated with* the net $S$ as the family $\mathcal{F}(S)$ of all the $L$-fuzzy subsets on $X$ which the molecule net $S$ eventually quasi-coincides with.

For $\mathcal{F}$, let

$$D(\mathcal{F}) = \{(e, A) \in M(L^X) \times \mathcal{F} :\ e \lessdot A \in \mathcal{F}\}$$

and equip it with a relation $\le$ on it as

$$\forall (e, A), (d, B) \in D(\mathcal{F}), \quad (e, A) \le (d, B) \iff A \ge B.$$

Define the *net associated with* the filter $\mathcal{F}$ as the mapping

$$S(\mathcal{F}):\ D(\mathcal{F}) \to M(L^X), \quad S(\mathcal{F})(e, A) = e, \quad \forall (e, A) \in D(\mathcal{F}).$$

**5.2.6 Proposition** *Let $(L^X, \delta)$ be an L-fts. Then*
(i) *For every molecule net $S$ in $(L^X, \delta)$, the filter $\mathcal{F}(S)$ associated with $S$ is a proper filter in $(L^X, \delta)$.*
(ii) *For every proper filter $\mathcal{F}$ in $(L^X, \delta)$, the relation $\le$ on $D(\mathcal{F})$ is a directed preorder, $D(\mathcal{F})$ equipped with $\le$ is a directed set.*
(iii) *For every proper filter $\mathcal{F}$ in $(L^X, \delta)$, the net $S(\mathcal{F})$ associated with $\mathcal{F}$ is a molecule net in $(L^X, \delta)$.* □

For a molecule net $S$ and a proper filter $\mathcal{F}$ in an $L$-fts, their correspondent ones defined above with the terminology "associated with" have indeed closed connections with the original ones as shown in the following theorems. This shown that the used terminology "associated with" is reasonable.

**5.2.7 Theorem** *Let $(L^X, \delta)$ be an L-fts, S a molecule net in $(L^X, \delta)$, $\mathcal{F}$ a proper filter in $(L^X, \delta)$, $e \in Pt(L^X)$. Then*

(i) $S \to e \iff \mathcal{F}(S) \to e$.
(ii) $\mathcal{F} \to e \iff S(\mathcal{F}) \to e$.
(iii) $\mathcal{F} \infty e \iff S(\mathcal{F}) \infty e$.
(iv) $S \infty e \implies \mathcal{F}(S) \infty e$.

**Proof** (i)($\iff$) By the relative definitions.

(ii)($\implies$) Suppose $\mathcal{F} \to e = x_a \in Pt(L^X)$, $U \in \mathcal{Q}(e)$. Then $x_a \not\leq U'$, $a \not\leq U'(x)$, by Corollary **1.3.12** (i), $\exists \lambda \in M(\downarrow a)$ such that $\lambda \not\leq U'(x)$, $x_\lambda \not\leq U'$, $x_\lambda \lessdot U$. By Theorem **5.2.2** (iv), $\mathcal{F} \to x_\lambda \leq x_a = e$, $U \in \mathcal{Q}(x_\lambda) \subset \mathcal{F}$. So $(x_\lambda, U) \in D(\mathcal{F})$. $\forall (d, A) \in D(\mathcal{F})$ such that $(d, A) \geq (x_\lambda, U)$, then $d \lessdot A \leq U$. By Proposition **2.3.3** (viii), $S(\mathcal{F})(d, A) = d \lessdot U$, $S(\mathcal{F})$ eventually quasi-coincides with $U$. By the arbitrariness of $U \in \mathcal{Q}(e)$, $S(\mathcal{F}) \to e$.

($\impliedby$) Suppose $S(\mathcal{F}) \to e$, $U \in \mathcal{Q}(e)$. Then $S(\mathcal{F})$ eventually quasi-coincides with $U$, $\exists (d_0, A_0) \in D(\mathcal{F})$ such that $\forall (d, A) \geq (d_0, A_0)$, $d = S(\mathcal{F})(d, A) \lessdot U$. So $\forall d \in M(L^X)$ such that $d \lessdot A_0$, we have $(d, A_0) \in D(\mathcal{F})$, $(d, A_0) \geq (d_0, A_0)$, and hence $d \lessdot U$. That is to say $\forall d \in M(L^X)$, $d \lessdot A_0 \implies d \lessdot U$, so $d \leq U' \implies d \leq A_0'$. By Corollary **2.3.11** (ii), $U' \leq A_0'$, $U \geq A_0$. Since $A_0 \in \mathcal{F}$, $\mathcal{F}$ is a filter, so $U \in \mathcal{F}$. By the arbitrariness of $U \in \mathcal{Q}(e)$, $\mathcal{Q}(e) \subset \mathcal{F}$, $\mathcal{F} \to e$.

(iii)($\implies$) Suppose $\mathcal{F} \infty e$, $U \in \mathcal{Q}(e)$, $(d_0, A_0) \in D(\mathcal{F})$. We need to find a $(d, A) \in D(\mathcal{F})$ such that $(d, A) \geq (d_0, A_0), S(d, A) \lessdot U$. Since $\mathcal{F} \infty e$ and $A_0 \in \mathcal{F}$, $A_0 \wedge U \neq \underline{0}$, $A_0' \vee U' \neq \underline{1}$. By Corollary **2.3.11** (i), $\exists d \in M(L^X)$ such that $d \not\leq A_0' \vee U'$. So $d \lessdot A_0 \wedge U$. By Proposition **2.3.3** (viii), $d \lessdot A_0$, by $(d_0, A_0) \in D(\mathcal{F})$, $A_0 \in \mathcal{F}$, so $(d, A_0) \in D(\mathcal{F})$, $(d, A_0) \geq (d_0, A_0)$. Using $d \lessdot A_0 \wedge U$ and Proposition **2.3.3** (viii) again, $S(d, A_0) = d \lessdot U$, this is what we need to prove.

($\impliedby$) Suppose $S(\mathcal{F}) \infty e$, $U \in \mathcal{Q}(e)$, $A \in \mathcal{F}$. We need to show $A \wedge U \neq \underline{0}$. Since $A \in \mathcal{F}$ and $\mathcal{F}$ is a proper filter in $(L^X, \delta)$, $A \neq \underline{0}, A' \neq \underline{1}$. By Corollary **2.3.11** (i), $\exists d \in M(L^X)$ such that $d \not\leq A'$, $d \lessdot A$, $(d, A) \in D(\mathcal{F})$. Since $S(\mathcal{F}) \infty e$, $\exists (d_0, A_0) \in D(\mathcal{F})$ such that $(d_0, A_0) \geq (d, A)$, $d_0 = S(d_0, A_0) \lessdot U$. So $d_0 \not\leq A_0'$, $d_0 \not\leq U'$. By $d \in M(L^X)$, $d_0 \not\leq A_0' \vee U' = (A_0 \wedge U)'$, $(A_0 \wedge U)' \neq \underline{1}$, $A_0 \wedge U \neq \underline{0}$. Since $(d_0, A_0) \geq (d, A)$, $A_0 \leq A$, so $A \wedge U \neq \underline{0}$.

(iv) Suppose $S = \{S(n), n \in D\}$, $A \in \mathcal{F}(S)$, $U \in \mathcal{Q}(e)$, going to show $A \wedge U \neq \underline{0}$. Since $A \in \mathcal{F}(S)$, $\exists n_0 \in D$ such that $\forall n \geq n_0$, $S(n) \lessdot A$. Since $U \in \mathcal{Q}(e)$, $S \infty e$, $\exists n_1 \in D$, $n_1 \geq n_0$ such that $S(n_1) \lessdot U$. So $S(n_1) \not\leq A', U'$. But $S(n_1) \in M(L^X)$, so $S(n_1) \not\leq A' \vee U' = (A \wedge U)'$, $(A \wedge U)' \neq \underline{1}$, $A \wedge U \neq \underline{0}$. □

As for the case $\mathcal{F}(S) \infty e \implies S \infty e$, the situation has some trouble, it can be false for even $L = [0, 1]$ just as the following example shows:

**5.2.8 Example** Take $X$ as an arbitrary singleton $\{x\}$, $L = [0, 1]$, $\delta = \{\underline{0}, x_{\frac{1}{2}}, \underline{1}\}$, $D = \{0\}$, $S: D \to M(L^X)$, $S(0) = x_{\frac{1}{3}}$, then $(L^X, \delta)$ is an $L$-fts, $D$ is a directed set, $S$ is a molecule net in $(L^X, \delta)$, $x_{\frac{2}{3}} \in M(L^X)$, $\mathcal{Q}(x_{\frac{2}{3}}) = \{x_{\frac{1}{2}}, \underline{1}\}$, $\mathcal{F}(S) = \{x_t : t \in$

$(\frac{2}{3},1]\}\infty x_{\frac{2}{3}}$. But $S(0) = x_{\frac{1}{3}} \not\leqslant x_{\frac{1}{2}} \in \mathcal{Q}(x_{\frac{2}{3}})$, so $x_{\frac{2}{3}}$ is not a cluster point of $S$.

But, however, we still have a equivalent characterization for this implication $\mathcal{F}(S)\infty e \Longrightarrow S\infty e$, although it is not very straightforward:

**5.2.9 Theorem** *Let $(L^X, \delta)$ be an L-fts, $e \in Pt(L^X)$. Then the following conditions are equivalent:*

(i) *For every $U \in \mathcal{Q}(e)$, there exists $V \in \mathcal{Q}(e)$ such that for every $u \in M(L^X)$, if*

$$\forall v \in M(L^X), \quad v\hat{q}u \Longrightarrow v \wedge V \neq \underline{0},$$

*then $u \lessdot U$.*

(ii) *For every molecule net $S$ in $(L^X, \delta)$, $\mathcal{F}(S)\infty e \Longrightarrow S\infty e$.*

**Proof** (i)$\Longrightarrow$(ii): Suppose (i) is true and $e$ is not a cluster point of $S = \{S(n), n \in D\}$, going to show $e$ is not a cluster point of $\mathcal{F}(S)$, either.

Since $e$ is not a cluster point of $S$, $\exists U_0 \in \mathcal{Q}(e)$ and $n_0 \in D$ such that $S(n) \not\lessdot U_0$ for every $n \geq n_0$. $\forall n \in D$, suppose $S(n) = (x^n)_{\lambda_n}$. By (i), for $U_0 \in \mathcal{Q}$, $\exists V \in \mathcal{Q}(e)$ satisfying condition stated in (i). Since $\forall n \geq n_0$, $S(n) \in M(L^X)$, $S(n) \not\lessdot U_0$, so by (i), $\exists (y^n)_{\xi_n} \in M(L^X)$ such that $(y^n)_{\xi_n}\hat{q}S(n) = (x^n)_{\lambda_n}$, $(y^n)_{\xi_n} \wedge V = \underline{0}$. So $y^n = x^n$, $\lambda_n \not\leq \xi_n{}'$, $\xi_n \wedge V(x^n) = 0$.

Take $A = \bigvee\{(x^n)_{\xi_n} : n \geq n_0\}$, then $\forall n \geq n_0$, $A(x^n) \geq \xi_n$, $A(x^n)' \leq \xi_n{}'$. Since $\lambda_n \not\leq \xi_n{}'$, so $\lambda_n \not\leq A(x^n)'$, $S(n) = (x^n)_{\lambda_n} \not\leq A'$, $S(n) \lessdot A$. That is to say, $A \in \mathcal{F}(S)$.

$\forall x \in X$. If $\forall m \geq n_0$, $x \neq x^m$, we have

$$(A \wedge V)(x) = A(x) \wedge V(x) = 0 \wedge V(x) = 0.$$

If $\exists m \geq n_0$ such that $x = x^m$, then

$$\begin{aligned}(A \wedge V)(x) &= A(x) \wedge V(x)\\ &= V(x^m) \wedge \bigvee\{\xi_n : n \geq n_0, x^n = x^m\}\\ &= \bigvee\{\xi_n \wedge V(x^m) : n \geq n_0, x^n = x^m\}\\ &= \bigvee\{\xi_n \wedge V(x^n) : n \geq n_0, x^n = x^m\}\\ &= \bigvee\{0 : n \geq n_0, x^n = x^m\}\\ &= 0.\end{aligned}$$

So $A \wedge V = \underline{0}$, $e$ is not a cluster point of $\mathcal{F}(S)$.

(ii)$\Longrightarrow$(i): Suppose (i) is not true, going prove (ii) is not true, either.

Since (i) is not true, $\exists U_0 \in \mathcal{Q}(e)$ such that $\forall V \in \mathcal{Q}(e)$, $\exists (x^V)_{\lambda_V} \in M(L^X)$, $(x^V)_{\lambda_V} \not\lessdot U_0$ but $\forall v \in M(L^X)$, $v\hat{q}(x^V)_{\lambda_V} \Longrightarrow v \wedge V \neq \underline{0}$. Take $D$ as the well-ordered $\omega$-copy $\omega \times \mathcal{Q}(e)$ of $\mathcal{Q}(e)$, then by Proposition **5.1.9**, $D$ with the relation $\leq$ on it is a well-ordered set and hence a directed set. $\forall (n, V) \in D$, let $S(n, V) = (x^V)_{\lambda_V}$, then $S : D \to Pt(L^X)$ is a molecule net in $(L^X, \delta)$. Since $\forall (n, V) \in D$, $S(n, V) = (x^V)_{\lambda_V} \not\lessdot U_0 \in \mathcal{Q}(e)$, $e$ is not a cluster point of $S$.

$\forall A \in \mathcal{F}(S)$, $\forall V \in \mathcal{Q}(e)$, suppose $(n_1, V_1) \in D$ such that $S(n, V) \lessdot A$ for every $(n, V) \geq (n_1, V_1)$, then $(n_1 + 1, V) \geq (n_1, V_1)$ and hence

$$\lambda_V = ht(S(n_1 + 1, V)) \not\leq A(supp(S(n_1 + 1, V)))' = A(x^V)'.$$

That is to say, $A(x^V) \not\leq (\lambda_V)'$. By Corollary **1.3.12** (ii), $\exists \xi_V \in M(\downarrow A(x^V))$ such that $\xi_V \not\leq (\lambda_V)'$. Hence

$$(x^V)_{\xi_V}\hat{q}(x^V)_{\lambda_V},$$

and by what mentioned above,

$(x^V)_{\xi_V} \wedge V \neq \underline{0}$,
$(A \wedge V)(x^V) = A(x^V) \wedge V(x^V) \geq \xi_V \wedge V(x^V) \neq 0$,
$A \wedge V \neq \underline{0}$.

Thus we have proved $\mathcal{F}(S) \infty e$, (ii) is not true. □

If the largest element $1_L$ of $L$ is a molecule, the situation can be simpler:

**5.2.10 Corollary** *Let $(L^X, \delta)$ be an L-fts, $e \in Pt(L^X)$. If $1_L \in M(L)$, then the following conditions are equivalent:*

(i) *For every $U \in \mathcal{Q}(e)$, there exists $V \in \mathcal{Q}(e)$ such that*
$$\forall x \in X, \quad U(x) \neq 1 \Longrightarrow V(x) = 0.$$
(ii) *For every molecule net $S$ in $(L^X, \delta)$, $\mathcal{F}(S) \infty e \Longrightarrow S \infty e$.*

**Proof** (i)$\Longrightarrow$(ii): We prove (i) implying Theorem **5.2.9** (i).

Suppose $U \in \mathcal{Q}(e)$, then $\exists V \in \mathcal{Q}(x)$ satisfying "$\forall x \in X, \ U(x) \neq 1 \Longrightarrow V(x) = 0$". $\forall u \in M(L^X)$, if
$$\forall v \in M(L^X), \quad v\hat{q}u \Longrightarrow v \wedge V \neq \underline{0}, \tag{5.2}$$
then especially for $v = (supp(u))_1 \in M(L^X)$, $v\hat{q}u$, we have $V(supp(u)) = 1 \wedge V(supp(u)) = (v \wedge V)(supp(u)) \neq 0$. Hence by (i), $U(supp(u)) = 1, u \lessdot U$. So Theorem **5.2.9** (i) is true.

(ii)$\Longrightarrow$(i): We prove that Theorem **5.2.9** (i) implying the condition (i) mentioned above.

Suppose Theorem **5.2.9** (i) is true, $U \in \mathcal{Q}(e)$, then $\exists V \in \mathcal{Q}(e)$ such that $\forall u \in M(L^X)$, if condition (5.2) holds then $u \lessdot U$. Take $x \in X$ arbitrarily. If $U(x) \neq 1$, then $U(x)' \neq 0$, by Theorem **1.3.12** (i), $\exists \lambda \in M(\downarrow U(x)')$, $u = x_\lambda \in M(L^X)$, $x_\lambda \not\lessdot U$. So $\exists v \in M(L^X)$, $v\hat{q}u$ but $v \wedge V = \underline{0}$, $ht(v) \wedge V(supp(v)) = 0$. Since $v\hat{q}u$, $supp(v) = supp(u) = x$. Since $1 \in M(L)$, $0$ is a prime element of $L$, by $ht(v) > 0$, $V(x) = V(supp(v)) = 0$. Hence (i) is true. □

**5.2.11 Corollary** *Let $(L^X, \delta)$ be an L-fts, $e \in M(L^X)$. If $1_L \in M(L)$ and every $U \in \mathcal{Q}(e)$ is crisp, then for every molecule net $S$ in $(L^X, \delta)$, $\mathcal{F}(S) \infty e \Longrightarrow S \infty e$.*

**Proof** Need only to show (i) implying Corollary **5.2.10** (i). $\forall U \in \mathcal{Q}(e)$, $U$ is crisp, so take $V = U \in \mathcal{Q}(e)$, we have
$$\forall x \in X, \quad U(x) \neq 1 \Longrightarrow V(x) = U(x) = 0.$$ □

**5.2.12 Corollary** *Let $(L^X, \delta)$ be an L-fts, $e \in M(L^X)$. If $1_L \in M(L)$ and for every molecule net $S$ in $(L^X, \delta)$, $\mathcal{F}(S) \infty e \Longrightarrow S \infty e$, then $U(supp(e)) = 1$ for every $U \in \mathcal{Q}(e)$.*

**Proof** By Corollary **5.2.10**, the condition (i) there should be true. Suppose $U \in \mathcal{Q}(e)$, $\exists V \in \mathcal{Q}(e)$ satisfying condition in (i). If $U(supp(e)) \neq 1$, then $V(supp(e)) = 0$. But this contradicts with $V \in \mathcal{Q}(e)$. □

For a molecule net $S$ and a proper filter $\mathcal{F}$, what are the relations between $S$ and $S(\mathcal{F}(S))$, $\mathcal{F}$ and $\mathcal{F}(S(\mathcal{F}))$, between parent nets with subnets and coarser filters with finer filters respectively? The following theorem lists all those relations which can be shown:

**5.2.13 Theorem** *Let $(L^X, \delta)$ be an L-fts, $S$ a molecule net, $T$ a subnet of $S$, $\mathcal{F}$ a proper filter in $(L^X, \delta)$. Then*

(i) $\mathcal{F}(S) \subset \mathcal{F}(T)$.
(ii) $\mathcal{F}(S(\mathcal{F})) = \mathcal{F}$.

**Proof** (i) Follow the relative definitions.

(ii) $\forall C \in \mathcal{F}(S(\mathcal{F}))$, then $S(\mathcal{F})$ eventually quasi-coincides with $C$, $\exists (e, A) \in D(\mathcal{F}) = \{(d, B) : d \in M(L^X), d \lessdot B \in \mathcal{F}\}$ such that $\forall (d, B) \geq (e, A)$, $d = S(\mathcal{F})(d, B) \lessdot C$. If $C \not\geq A$, then $C' \not\leq A'$. By Corollary **2.3.11** (ii), $\exists d \in M(\downarrow C')$ such that $d \not\leq A'$, $d \lessdot A$, $(d, A) \in D(\mathcal{F})$, $(d, A) \geq (e, A)$, $d \lessdot C$. But $d \leq C'$, this is a contradiction. So $C \geq A$. Since $A \in \mathcal{F}$, $\mathcal{F}$ is a filter, $C \in \mathcal{F}$. Hence $\mathcal{F}(S(\mathcal{F})) \subset \mathcal{F}$.

$\forall A \in \mathcal{F}$. Since $\mathcal{F}$ is a proper filter, $A \neq \underline{0}$, $A' \neq \underline{1}$, $\exists e \in M(L^X)$ such that $e \lessdot A$, $(e, A) \in D(\mathcal{F}) = \{(d, B) : d \in M(L^X), d \lessdot B \in \mathcal{F}\}$. Then $\forall (d, B) \in D(\mathcal{F})$ such that $(d, B) \geq (e, A)$, we have $S(\mathcal{F})(d, B) = d \lessdot B \leq A$. By Proposition **2.3.3** (viii), $S(\mathcal{F})(d, B) \lessdot A$. That is to say, $S(\mathcal{F})$ eventually quasi-coincides with $A$, $A \in \mathcal{F}(S(\mathcal{F}))$. Therefore, $\mathcal{F} \subset \mathcal{F}(S(\mathcal{F}))$ and hence $\mathcal{F}(S(\mathcal{F})) = \mathcal{F}$. □

But we cannot claim that $S(\mathcal{G})$ is a subnet of $S(\mathcal{F})$ for arbitrary filters $\mathcal{F} \subset \mathcal{G}$. The following example shows this point:

**5.2.14 Example** Take $X$ as a singleton $\{x\}$, $L = [0, 1]$ and $\delta$ an arbitrary subfamily of $L^X$ (in fact, $L$-fuzzy topology here will not be concerned with the subject), then we get an $L$-fts $(L^X, \delta)$ with $1 \in M(L)$. Take

$$\mathcal{F} = \{x_s : s \in (\frac{2}{3}, 1]\}, \quad \mathcal{G} = \{x_t : t \in [\frac{1}{3}, 1]\},$$

then both $\mathcal{F}$ and $\mathcal{G}$ are (molecule) filters and $\mathcal{F} \subset \mathcal{G}$. The index sets of $S(\mathcal{F})$ and $S(\mathcal{G})$ are

$$D(\mathcal{F}) = \{(e, A) : e \in M(L^X), e \lessdot A \in \mathcal{F}\} = \{(x_s, x_t) : t > \frac{2}{3}, s > 1 - t\},$$

$$E(\mathcal{G}) = \{(d, B) : d \in M(L^X), d \lessdot B \in \mathcal{G}\} = \{(x_s, x_t) : t \geq \frac{1}{3}, s > 1 - t\},$$

respectively. If $S(\mathcal{G})$ is a subnet of $S(\mathcal{F})$, then there exists a cofinal selection $N : E \to D$. For $(x_1, x_{\frac{1}{3}}) \in E$, suppose $N(x_1, x_{\frac{1}{3}}) = (x_{s_0}, x_{t_0})$, then take $t_1 \in (\frac{2}{3}, t_0)$, for $(x_1, x_{t_1}) \in D$, there should exist a $(x_{s_2}, x_{t_2}) \in E$ such that $N(x_s, x_t) \geq (x_1, x_{t_1})$ for every $(x_s, x_t) \in D$, $(x_s, x_t) \geq (x_{s_2}, x_{t_2})$. But $(x_1, x_{\frac{1}{3}}) \geq (x_{s_2}, x_{t_2})$, so

$$N(x_1, x_{\frac{1}{3}}) \geq (x_1, x_{t_1}) > (x_{s_0}, x_{t_0}) = N(x_1, x_{\frac{1}{3}}).$$

This is a contradiction.

We cannot either claim anything about the relation between $S$ and $S(\mathcal{F}(S))$ for a molecule net $S$:

**5.2.15 Example** Take $X$ as a singleton $\{x\}$. Denote $P_0 = \{\frac{1}{n+2} : n < \omega\}$. For the first uncountable ordinal number $\omega_1$, take $L_0$ and a relation $\leq$ on it as

$$L_0 = \omega_1 \cup P_0, \quad \forall a, b \in L_0, \quad a \leq b \iff \begin{array}{ll} & a \in \omega_1, \quad b \in \omega_1, \quad a \leq b, \\ \text{or} & a \in \omega_1, \quad b \in P_0, \\ \text{or} & a \in P_0, \quad b \in P_0, \quad a \leq b, \end{array}$$

then $L_0$ with $\leq$ is a complete chain. Take $L$ and a relation $\leq$ on it as

$$L = 2 \times L_0, \quad \forall (i,a),(j,b) \in L, \quad (i,a) \leq (j,b) \iff \begin{array}{l} i=0,\ j=0,\ a \leq b, \\ \text{or}\ \ i=0,\ j=1, \\ \text{or}\ \ i=1,\ j=1,\ a \geq b. \end{array}$$

Define an operation $'$ on $L$ as

$$\forall (i,a) \in L, \quad (i,a)' = (1-i, a),$$

then $'$ is an order-reversing involution on $L$, $L$ is a F-lattice. Take an $L$-fuzzy topology $\delta$ on $X$ arbitrarily, then $(L^X, \delta)$ is an $L$-fts.

Let $D = \omega$. Define $S:\ D \to M(L^X)$ by $S(n) = x_{(0,\frac{1}{n+2})}$ for every $n \in D$, then $S$ is a molecule net in $(L^X, \delta)$, and

$$\begin{aligned} \mathcal{F}(S) &= \{A \in L^X :\ S \text{ eventually quasi-coincides with } A\} \\ &= \{x_{(1,u)} :\ u \in \omega_1\}, \\ D(\mathcal{F}(S)) &= \{(e,A) :\ e \in M(L^X),\ e \lessdot A \in \mathcal{F}(S)\} \\ &= \{(x_{(i,a)}, x_{(1,u)}) :\ u \in \omega_1,\ (i,a) > (0,u)\}, \\ S(\mathcal{F}(S))(x_{(i,a)}, x_{(1,u)}) &= x_{(i,a)}, \quad \forall (x_{(i,a)}, x_{(1,u)}) \in D(\mathcal{F}(S)). \end{aligned}$$

If $S$ is a subnet of $S(\mathcal{F}(S))$, then there exists a cofinal selection $N:\ D \to D(\mathcal{F}(S))$. Suppose

$$\forall n \in D, \quad N(n) = (x_{(i_n,a_n)}, x_{(1,u_n)}),$$

denote $v = \bigvee_{n<\omega} u_n$, then since $u_n < \omega_1$ for every $n \in D = \omega$, $v < \omega_1$. Now for $(x_{(1,0)}, x_{(1,v+1)}) \in D(\mathcal{F}(S))$, as $N$ is a cofinal selection, $\exists m \in D$ such that $\forall n \in D,\ n \geq m$, we always have

$$\begin{aligned} &(x_{(i_n,a_n)}, x_{(1,u_n)}) = N(n) \geq (x_{(1,0)}, x_{(1,v+1)}), \\ &x_{(1,u_n)} \leq x_{(1,v+1)}, \\ &(1,u_n) \leq (1, v+1), \\ &u_n \geq v+1 = (\bigvee_{k<\omega} u_k) + 1, \end{aligned}$$

this is a contradiction. So there is no such a cofinal selection, $S$ is not a subnet of $S(\mathcal{F}(S))$.

If $S\mathcal{F}(S))$ is a subnet of $S$, then there exists a cofinal selection $N:\ D(\mathcal{F}(S)) \to D$. $\forall n \in D$, $\exists d_n \in D(\mathcal{F}(S))$ such that

$$d \in D(\mathcal{F}(S)),\ d \geq d_n \Longrightarrow N(d) \geq n.$$

Suppose $d_n = (x_{(i_n,a_n)}, x_{(1,u_n)})$, $v = \bigvee_{n<\omega} u_n$, then since $u_n < \omega_1$ for every $n \in D = \omega$, $v < \omega_1$, $(x_{(1,0)}, x_{(1,v)}) \in D(\mathcal{F}(S))$. Suppose $N(x_{(1,0)}, x_{(1,v)}) = m \in D = \omega$, then by $(x_{(1,0)}, x_{(1,v)}) \geq (x_{(i_{m+1},a_{m+1})}, x_{(1,u_{m+1})})$,

$$m = N(x_{(1,0)}, x_{(1,v)}) \geq m+1,$$

a contradiction. So there is no such a cofinal selection, $S(\mathcal{F}(S))$ is not a subset of $S$.

**5.2.16 Theorem** *Let $(L^X, \delta)$ be an L-fts, $e \in Pt(L^X)$. If $\mathcal{Q}(e)$ is a down-directed set, then for every proper filter $\mathcal{F}$ in $(L^X, \delta)$, $\mathcal{F} \infty e$ if and only if $\mathcal{F}$ has a finer proper filter $\mathcal{G} \to e$.*

**Proof** By Theorem **5.2.2** (i) and Theorem **5.2.4** (iii), we have "$\mathcal{F}$ has a finer proper filter $\mathcal{G} \to e$" $\Longrightarrow$ $\mathcal{F} \infty e$, so we need only prove that if $\mathcal{Q}(e)$ is down-directed and $\mathcal{F} \infty e$, then there exists a proper filter $\mathcal{G} \supset \mathcal{F}$ such that $\mathcal{G} \to e$.

By Theorem **5.2.7** (iii), $S(\mathcal{F})\infty e$. By Theorem **5.1.11** (ii)$\Longrightarrow$(iii), $S(\mathcal{F})$ has a subnet $T\to e$. Take $\mathcal{G}=\mathcal{F}(T)$, then by Proposition **5.2.6** (i), $\mathcal{G}$ is a proper filter. By Theorem **5.2.13** (i), $\mathcal{G}\supset\mathcal{F}$. By Theorem **5.2.7** (i), we get $\mathcal{G}\to e$. □

**5.2.17 Corollary** *Let $(L^X,\delta)$ be an L-fts, $e\in M(L^X)$. Then for every proper filter $\mathcal{F}$ in $(L^X,\delta)$, $\mathcal{F}\infty e$ if and only if $\mathcal{F}$ has a finer proper filter $\mathcal{G}\to e$.*

**Proof** By Theorem **2.3.12** (ii), $\mathcal{Q}(e)$ is a down-directed set, so obtain the conclusion from Theorem **5.2.16**. □

**5.2.18 Theorem** *Let $(L^X,\delta)$ be an L-fts, $A\in L^X$, $e\in M(L^X)$. Then $e$ is an adherent point of $A$ if and only if there exists a proper filter $\mathcal{F}$ in $(L^X,\delta)$ such that $A'\notin\mathcal{F}$, $\mathcal{F}\to e$.*

**Proof** (Necessity) By Theorem **5.1.14** (i)$\Longrightarrow$(vi), there exists a molecule net $S$ in $A$ such that $S\to e$. By Theorem **5.2.7** (i), we get a proper filter $\mathcal{F}(S)\to e$. If $A'\in\mathcal{F}(S)$, then $S$ eventually quasi-coincides with $A'$, i.e. is eventually not in $A$, this contradicts with the fact that $S$ is a net in $A$. So $A'\notin\mathcal{F}(S)$.

(Sufficiency) Since $\mathcal{F}$ is a filter and $A'\notin\mathcal{F}$, $A^{-\prime}=A^{\prime\circ}\notin\mathcal{F}$. Since $\mathcal{F}\to e$, $\mathcal{Q}(e)\subset\mathcal{F}$, $A^{-\prime}\notin\mathcal{Q}(e)$, $e\le(A^{-\prime})'=A^-$. By Theorem **2.3.23** (i), $e$ is an adherent point of $A$. □

**5.2.19 Theorem** *Let $(L^X,\delta)$ be an L-fts, $A\in L^X$. Then the following conditions are equivalent:*

(i) $A=A^-$.

(ii) *For every proper filter $\mathcal{F}$ in $(L^X,\delta)$ and every $e\in M(L^X)$ such that $A'\notin\mathcal{F}$, $\mathcal{F}\to e\Longrightarrow e\le A$.*

**Proof** (i)$\Longrightarrow$(ii): Suppose $\mathcal{F}$ is a proper filter in $(L^X,\delta)$ and $e\in M(L^X)$ such that $A'\notin\mathcal{F}$, $\mathcal{F}\to e$, going to prove $e\le A$. By Theorem **5.2.7** (ii), the molecule net $S(\mathcal{F})\to e$. Since $A'\notin\mathcal{F}$, by Theorem **5.2.13** (ii), $A'\notin\mathcal{F}(S(\mathcal{F}))$, $S(\mathcal{F})$ frequently does not quasi-coincide with $A'$, i.e. frequently is in $A$. So there exists a subnet $T$ of $S(\mathcal{F})$ such that $T$ is in $A$ and $T\to e$. By Corollary **5.1.15** (i)$\Longrightarrow$(v), $e\le A$.

(ii)$\Longrightarrow$(i): Suppose (ii) holds, we prove Corollary **5.1.15** (v) is true.

Let $S=\{S(n),n\in D\}$ be a molecule net in $A$, $e\in M(L^X)$ and $S\to e$. By Theorem **5.2.7** (i), the proper filter $\mathcal{F}(S)\to e$. Since $S$ is in $A$, $S(n)\le A=(A')'$, $S(n)\not\prec A'$ for every $n\in D$, so $A'\notin\mathcal{F}(S)$. Hence $e\le A$. □

Similarly we have

**5.2.20 Theorem** *Let $(L^X,\delta)$ be an L-fts, $A\in L^X$, $e\in M(L^X)$. Then $e$ is an accumulation point of $A$ if and only if there exists a proper filter $\mathcal{F}$ in $(L^X,\delta)$ such that $(A\backslash e)'\notin\mathcal{F}$, $\mathcal{F}\to e$.* □

We can also use filters to describe continuities of $L$-fuzzy mappings. For this aim, we need a

**5.2.21 Definition** Let $(L^X,\delta)$ be an $L$-fts. A subfamily $\mathcal{A}\subset L^X$ is called a *filter base* in $(L^X,\delta)$, if $\underline{0}\notin\mathcal{A}$ and $\mathcal{A}$ is a down-directed set in $L^X$.

For a filter base $\mathcal{A}$ in $(L^X,\delta)$, define the *filter generated by* $\mathcal{A}$ as $\uparrow\mathcal{A}$.

With this definition, we can restate Theorem **2.3.12** (ii) as follows:

**5.2.22 Theorem** *Let $(L^X, \delta)$ be an L-fts. Then for every $e \in M(L^X)$, $\mathcal{Q}(e)$ is a filter base in $(L^X, \delta)$.* □

**5.2.23 Proposition** *Let $(L^X, \delta)$ be an L-fts, $\mathcal{A}$ a filter base in $(L^X, \delta)$. Then the filter $\uparrow\mathcal{A}$ generated by $\mathcal{A}$ is a proper filter in $(L^X, \delta)$.* □

**5.2.24 Definition** Let $(L^X, \delta)$ be an $L$-fts, $\mathcal{A}$ a filter base in $(L^X, \delta)$. An $L$-fuzzy point $e \in Pt(L^X)$ is called an *cluster point* of $\mathcal{A}$, denoted by $\mathcal{A} \infty e$, if $\uparrow\mathcal{A} \infty e$; called a *limit point* of $\mathcal{A}$, denoted by $\mathcal{A} \to e$, if $\uparrow\mathcal{A} \to e$.

**5.2.25 Proposition** *Let $(L^X, \delta)$ be an L-fts, $\mathcal{A}$ a filter base in $(L^X, \delta)$, $e \in Pt(L^X)$. Then*

(i) $\mathcal{A} \infty e \iff \forall A \in \mathcal{A},\ \forall U \in \mathcal{Q}(e),\ A \wedge U \neq \underline{0}$
$\iff \uparrow\mathcal{A} \infty e$
$\iff e \leq \mathrm{clu}(\uparrow\mathcal{A})$.

(ii) $\mathcal{A} \to e \iff \forall U \in \mathcal{Q}(e),\ \exists A \in \mathcal{A},\ A \leq U$
$\iff \uparrow\mathcal{A} \to e$
$\iff e \leq \lim(\uparrow\mathcal{A})$. □

**5.2.26 Proposition** *Let $(L^X, \delta)$, $(L^Y, \mu)$ be L-fts', $f^{\to}: L^X \to L^Y$ an L-fuzzy mapping, $\mathcal{A}$ a filter base in $(L^X, \delta)$. Then $f^{\to}[\mathcal{A}]$ is a filter base in $(L^Y, \mu)$.*

**Proof** $\forall A, B \in \mathcal{A}$, then

$$f^{\to}(A) \wedge f^{\to}(B) \geq f^{\to}(A \wedge B) \in f^{\to}[\mathcal{A}].$$

Combining $\underline{1} = f^{\to}(\underline{1}) \in f^{\to}[\mathcal{A}]$, we know $f^{\to}[\mathcal{A}]$ is a down-directed set. Since $\underline{0} \notin \mathcal{A}$, by Theorem **2.1.15** (v), $\underline{0} \notin f^{\to}[\mathcal{A}]$. Hence $f^{\to}[\mathcal{A}]$ is a filter base in $(L^Y, \mu)$. □

Then by Proposiont **5.2.26**, we can state the following

**5.2.27 Theorem** *Let $(L^X, \delta)$, $(L^Y, \mu)$ be L-fts', $f^{\to}: L^X \to L^Y$ an L-fuzzy mapping. Then the following conditions are equivalent:*

(i) *$f^{\to}$ is continuous.*

(ii) *For every filter base $\mathcal{A}$ in $(L^X, \delta)$ and every $e \in M(L^X)$,*
$$\mathcal{A} \to e \Longrightarrow f^{\to}[\mathcal{A}] \to f^{\to}(e).$$

(iii) *For every filter base $\mathcal{A}$ in $(L^X, \delta)$,*
$$f^{\to}(\lim(\uparrow\mathcal{A})) \leq \lim(\uparrow f^{\to}[\mathcal{A}]).$$

**Proof** (i)$\Longrightarrow$(ii): Suppose $\mathcal{A}$ is a filter base in $(L^X, \delta)$, $e \in M(L^X)$, $\mathcal{A} \to e$, then $\mathcal{Q}(e) \subset \uparrow\mathcal{A}$. $\forall V \in \mathcal{Q}(f^{\to}(e))$, since $f^{\to}$ is continuous, $f^{\leftarrow}(V) \in \mathcal{Q}(e)$, $\exists A \in \mathcal{A}$ such that $f^{\leftarrow}(V) \geq A$. By Theorem **2.1.25** (ii), $V \geq f^{\to}f^{\leftarrow}(V) \geq f^{\to}(A) \in f^{\to}[\mathcal{A}]$, $V \in \uparrow f^{\to}[\mathcal{A}]$, $f^{\to}[\mathcal{A}] \to f^{\to}(e)$.

(ii)$\Longrightarrow$(i): Suppose $f^{\to}$ is not continuous, then $\exists V \in \mu$ such that $f^{\leftarrow}(V) \notin \delta$. By Theorem **5.2.19**, there exist $x_\lambda \in M(L^X)$ and a proper filter $\mathcal{F}$ in $(L^X, \delta)$ such that $f^{\leftarrow}(V) = (f^{\leftarrow}(V)')' \notin \mathcal{F}$, $\mathcal{F} \to x_\lambda$ but $x_\lambda \not\leq f^{\leftarrow}(V)'$. By Proposition **2.2.5**, $x_\lambda \not\leq f^{\leftarrow}(V'), \lambda \not\leq f^{\leftarrow}(V')(x) = V'(f(x))$. By Theorem **2.1.15** (ii), $f^{\to}(x_\lambda) = f(x)_\lambda \in M(L^Y)$, $f^{\to}(x_\lambda) = f(x)_\lambda \not\leq V'$.

If $V \in \uparrow f^{\to}[\mathcal{F}]$, then $\exists A \in \mathcal{F}$ such that $V \geq f^{\to}(A)$. By Theorem **2.1.25** (i), $f^{\leftarrow}(V) \geq f^{\leftarrow}f^{\to}(A) \geq A$. Since $\mathcal{F}$ is a filter, so is a upper set, $f^{\leftarrow}(V) \in \mathcal{F}$. But by the preceding discussion, this is impossible. So $V \notin \uparrow f^{\to}[\mathcal{F}]$.

$\mathcal{F}$ is a filter and then is certainly a filter base in $(L^X, \delta)$, by Proposition **5.2.23**, $\uparrow f^{\to}[\mathcal{F}]$ is a proper filter in $(L^Y, \mu)$. If (ii) is true, then for $V' = (V')^-$ in $(L^Y, \mu)$, we have found $f^{\rightarrow}(x_\lambda) \in M(L^Y)$ and a proper filter $\uparrow f^{\rightarrow}[\mathcal{F}]$ in $(L^Y, \mu)$ such that

$$(V')' = V \not\in \uparrow f^{\rightarrow}[\mathcal{F}] \rightarrow f^{\rightarrow}(x_\lambda) \not\leq V',$$

this contradicts with Theorem **5.2.19** (i)$\Longrightarrow$(ii). So the conclusion (ii) is not true whenever (i) is not true.

(ii)$\Longleftrightarrow$(iii): By Proposition **5.2.25** (ii). □

## 5.3 Convergence Classes

J.L. Kelley gave out a way to describe an arbitrary topology on a set with a collection of nets and associated "limit points" and proved its correctness in 1950.[49, 50] Then this method became one of the typical ways of generating topologies on a given set. In $L$-fuzzy topology, the original conditions are not enough because every $L$-fuzzy space $L^X$ has one more dimension — stratifications of $L$ which cannot be well reflected by those conditions only for describing "convergence" in $X$. In fact, in Ref.[137], Liu and Pu introduced the notion of fuzzy convergence classes, and proved that the fuzzy topologies and fuzzy convergence classes were completely determined each other. But it was pointed[79] out that the characterization of fuzzy convergence classes given in Ref.[137] was not perfect, and a correct condition was given at the same time. This new condition is just for reflecting the stratification structure of the value domain $L = [0, 1]$ there.

By the Theorem **1.2.32** (i), the following definition is reasonable:

**5.3.1 Definition** Let $L^X$ be an $L$-fuzzy space, $D$ a directed set, $\{E^n : n \in D\}$ a family of directed sets, $S^n = \{S^n(m), m \in E^n\}$ a net in $Pt(L^X)$ for every $n \in D$. Then for the product directed set $\vec{D} = D \times \prod_{n \in D} E^n$, the net $\vec{S} : \vec{D} \rightarrow L^X$ defined as

$$\forall (n, f) \in \vec{D}, \quad \vec{S}(n, f) = S^n(f(n))$$

is called the *induced net* of the net family $\{S^n : n \in D\}$.

**5.3.2 Definition** Let $X$ be a nonempty ordinary set, $L$ a F-lattice, $\mathcal{A} \subset L^X$.

Denote the class of all the nets $S = \{S(n), n \in D\}$ such that $S(n) \in M(\mathcal{A})$ for every $n \in D$ by $\mathcal{S}_M(\mathcal{A})$.

Let $\mathcal{C} \subset \mathcal{S}_M(L^X) \times M(L^X)$, $(S, e) \in \mathcal{S}_M(L^X) \times M(L^X)$. Say $S$ *$\mathcal{C}$-converges to e*, denoted by $S \rightarrow_{\mathcal{C}} e$, if $(S, e) \in \mathcal{C}$. Denote the case that $S$ does not $\mathcal{C}$-converge to $e$ by $S \not\rightarrow_{\mathcal{C}} e$.

$\mathcal{C}$ is called an *L-fuzzy convergence class* on $L^X$, if it fulfills the following five conditions:

(CC1) If $S \in \mathcal{S}_M(L^X)$ is a constant net with value $e \in M(L^X)$, then $S \rightarrow_{\mathcal{C}} e$.

(CC2) If $S \rightarrow_{\mathcal{C}} e$ and $T$ is a subnet of $S$, then $T \rightarrow_{\mathcal{C}} e$.

(CC3) For every $(S, e) \in \mathcal{S}_M(L^X) \times M(L^X)$, if $S \not\rightarrow_{\mathcal{C}} e$, then there exists a subnet $T$ of $S$ such that for every subnet $R$ of $T$, $R \not\rightarrow_{\mathcal{C}} e$.

(CC4) For every directed set $D$ and every $\{S^n : n \in D\} \subset \mathcal{S}_M(L^X)$, where $S^n = \{S^n(m), m \in E^n\}$ and $S^n \rightarrow_{\mathcal{C}} S(n)$ for every $n \in D$, if $S \rightarrow_{\mathcal{C}} e$ for the obtained molecule net $S = \{S(n), n \in D\}$, then for the induced net $\vec{S}$ of $\{S^n : n \in D\}$, $\vec{S} \rightarrow_{\mathcal{C}} e$.

(CC5) For every $\mathcal{A} \subset M(L^X)$ and every $e \in M(\downarrow \bigvee \mathcal{A})$, there exists $S \in \mathcal{S}_M(\mathcal{A})$ such that $S \rightarrow_{\mathcal{C}} e$.

**5.3.3 Definition** Let $(L^X, \delta)$ be an $L$-fts. Denote

$$\varphi(\delta) = \{(S, e) \in \mathcal{S}_M(L^X) \times M(L^X) : S \to e\}.$$

Call $\varphi(\delta)$ the *L-fuzzy convergence class on* $L^X$ *generated by* $\delta$.

**5.3.4 Theorem** *Let* $(L^X, \delta)$ *be an L-fts, then* $\varphi(\delta)$ *is an L-fuzzy convergence class on* $L^X$.

**Proof** Verifications are straightforward. We verify (CC5) as an example. Suppose $\mathcal{A} \subset M(L^X)$, $e \in M(\downarrow \bigvee \mathcal{A})$. $\forall U \in \mathcal{Q}(e)$, then $e \leq \bigvee \mathcal{A}$, $e \not\leq U'$, $\bigvee \mathcal{A} \not\leq U'$, $\exists S(U) \in \mathcal{A}$ such that $S(U) \not\leq U'$, $S(U) \lessdot U$. Since $e \in M(L^X)$, by Theorem **2.3.12** (ii), $D = \mathcal{Q}(e)$ equipped with the partial order $\leq_D$ defined as

$$\forall U, V \in D, \quad U \leq_D V \Longleftrightarrow U \geq V$$

is a directed set. So we obtain a molecule net $S = \{S(U), U \in D\} \in \mathcal{S}_M(\mathcal{A})$ and clearly $S \to e$. □

**5.3.5 Lemma** *Let X be a nonempty ordinary set, L a F-lattice,* $\mathcal{C} \subset \mathcal{S}_M(L^X) \times M(L^X)$ *fulfill the conditions* (CC4) *and* (CC5), $\{(S^t, e^t) : t \in T\} \subset \mathcal{C}$, $e \in M(\downarrow \bigvee_{t \in T} e^t)$. *Then there exists* $\vec{S} \in \mathcal{S}_M(\bigcup_{t \in T} img(S^t))$ *such that* $\vec{S} \to_{\mathcal{C}} e$.

**Proof** By (CC5), $\exists S = \{S(n), n \in D\} \in \mathcal{S}_M(\{e^t : t \in T\})$ such that $S \to_{\mathcal{C}} e$. $\forall n \in D$, $\exists t(n) \in T$ such that $S^{t(n)} \to_{\mathcal{C}} e^{t(n)} = S(n)$. Then by (CC4), the induced net $\vec{S}$ of $\{S^{t(n)} : n \in D\}$ $\mathcal{C}$-converges to $e$. Clearly $\vec{S} \in \mathcal{S}_M(\bigcup_{t \in T} img(S^t))$. □

**5.3.6 Lemma** *Let X be a nonempty ordinary set, L a F-lattice,* $\mathcal{C} \subset \mathcal{S}_M(L^X) \times M(L^X)$ *fulfill the conditions* (CC4) *and* (CC5), $\mathcal{A} \subset M(L^X)$, $S \in \mathcal{S}_M(\downarrow \bigvee \mathcal{A})$, $S \to_{\mathcal{C}} e$. *Then there exists* $\vec{S} \in \mathcal{S}_M(\mathcal{A})$ *such that* $\vec{S} \to_{\mathcal{C}} e$.

**Proof** Suppose the index set of $S$ is $D$. $\forall n \in D$, since $S(n) \in M(\downarrow \bigvee \mathcal{A})$, by (CC5), $\exists T^n \in \mathcal{S}_M(\mathcal{A})$ such that $T^n \to_{\mathcal{C}} S(n)$. Denote the induced net of $\{T^n : n \in D\}$ by $\vec{S}$, then $\vec{S} \in \mathcal{S}_M(\mathcal{A})$ and by (CC4), $\vec{S} \to_{\mathcal{C}} e$. □

**5.3.7 Definition** Let $X$ be a nonempty ordinary set, $L$ a F-lattice, $\mathcal{C} \subset \mathcal{S}_M(L^X) \times M(L^X)$.

For every $A \in L^X$, denote

$$\mathrm{clu}_{\mathcal{C}}(A) = \{e \in M(L^X) : \exists S \in \mathcal{S}_M(\downarrow A),\ S \to_{\mathcal{C}} e\}.$$

Define an operator $c$ on $L^X$, called the *closure operator generated by* $\mathcal{C}$, as follows:

$$c : L^X \to L^X, \quad c(A) = \bigvee \mathrm{clu}_{\mathcal{C}}(A), \quad \forall A \in L^X.$$

**5.3.8 Theorem** *Let X be a nonempty ordinary set, L a F-lattice,* $\mathcal{C}$ *an L-fuzzy convergence class on* $L^X$. *Then the closure operator c generated by* $\mathcal{C}$ *is a closure operator on* $L^X$.

**Proof** Verify the conditions (CO1) – (CO4) of a closure operator.

(CO1): Clear.

(CO2): By (CC1).

(CO3): Let $A, B \in L^X$. By the definition of $c$, clearly $A \leq B \Longrightarrow c(A) \leq c(B)$, so we need only prove $\mathrm{clu}_{\mathcal{C}}(A \vee B) \subset \mathrm{clu}_{\mathcal{C}}(A) \cup \mathrm{clu}_{\mathcal{C}}(B)$.

$\forall e \in \mathrm{clu}_{\mathcal{C}}(A \vee B)$, then $\exists S \in \mathcal{S}_M(A \vee B)$ such that $S \to_{\mathcal{C}} e$. Suppose the index set of $S$ is $D$, denote $D_A = \{n \in D : S(n) \leq A\}$, $D_B = \{n \in D : S(n) \leq B\}$. $\forall n \in D$, $S(n) \leq A \vee B$. Since $S(n) \in M(L^X)$, so $S(n) \leq A$ or $S(n) \leq B$, $D = D_A \cup D_B$. Thus clearly there is at least one of $D_A$ and $D_B$ must be a cofinal subset of $D$, suppose it is $D_A$. $\forall n_1, n_2 \in D_A \subset D$, $\exists n_3 \in D$ such that $n_3 \geq n_1, n_2$. Since $D_A$ is a cofinal subset of $D$, $\exists n_0 \in D_A$ such that $n_0 \geq n_3$. So $n_0 \geq n_1, n_2$, $D_A$ is a directed set. Then we get a subnet $T \in \mathcal{S}_M(\downarrow A)$ of $S$ with the cofinal selection $N: D_A \to D$, $N(n) = n$. By (CC2), $T \to_{\mathcal{C}} e$, $e \in \text{clu}_{\mathcal{C}}(A)$.

(CO4): Let $A \in L^X$. By (CO2) proved above, $c(A) \leq c(c(A))$. So we need only prove $\text{clu}_{\mathcal{C}}(c(A)) \subset \text{clu}_{\mathcal{C}}(A)$.

Suppose $\mathcal{A} = \{S \in \mathcal{S}_M(\downarrow A) : \exists e \in M(L^X), (S, e) \in \mathcal{C}\}$. $\forall e \in \text{clu}_{\mathcal{C}}(c(A))$, $\exists S = \{S(n), n \in D\} \in \mathcal{S}_M(\downarrow c(A))$ such that $S \to_{\mathcal{C}} e$. $\forall n \in D$, since $S(n) \in M(\downarrow c(A)) = M(\downarrow \bigvee \text{clu}_{\mathcal{C}}(A))$, by Lemma **5.3.5**, $\exists S^n \in \mathcal{S}_M(\bigcup_{T \in \mathcal{A}} img(T)) \subset \mathcal{S}_M(\downarrow A)$ such that $S^n \to_{\mathcal{C}} S(n)$. Denote the induced net of $\{S^n : n \in D\}$ by $\vec{S}$, then $\vec{S} \in \mathcal{S}_M(\downarrow A)$ and by (CC4), $\vec{S} \to_{\mathcal{C}} e$. By the definition of $\text{clu}_{\mathcal{C}}(A)$, $e \in \text{clu}_{\mathcal{C}}(A)$, $\text{clu}_{\mathcal{C}}(c(A)) \subset \text{clu}_{\mathcal{C}}(A)$. □

By the theorem proved above and Theorem **2.2.25**, we can introduce the following

**5.3.9 Definition** Let $X$ be a nonempty ordinary set, $L$ a F-lattice, $\mathcal{C}$ an $L$-fuzzy convergence class on $L^X$. Denote the $L$-fuzzy topology on $X$ generated by the closure operator on $L^X$ generated by $\mathcal{C}$ as $\psi(\mathcal{C})$.

Then we can establish the following results on the correspondence between $L$-fuzzy topologies and $L$-fuzzy convergence classes:

**5.3.10 Theorem** *Let $X$ be a nonempty ordinary, $L$ a F-lattice. Then*

(i) *For every $L$-fuzzy topology $\delta$ on $X$, $\psi(\varphi(\delta)) = \delta$.*

(ii) *For every $L$-fuzzy convergence class $\mathcal{C}$ on $L^X$, $\varphi(\psi(\mathcal{C})) = \mathcal{C}$.*

(iii) *For every pair $\delta$, $\mu$ of $L$-fuzzy topologies on $L^X$ such that $\delta \subset \mu$, $\varphi(\delta) \supset \varphi(\mu)$.*

**Proof** (i) Suppose the closure operator generated by $\varphi(\delta)$ as $c$, still denote the closure of $A \in L^X$ in $(L^X, \delta)$ as $A^-$. $\forall U \in \delta$, $(U')^- = U'$. Since $\varphi(\delta)$ is an $L$-fuzzy convergence class on $L^X$, by Theorem **5.3.8**, $c$ is a closure operator on $L^X$. So by (CO1), $U' \leq c(U')$. $\forall e \in M(\downarrow c(U'))$, since $\psi(\delta)$ is an $L$-fuzzy convergence class, by Lemma **5.3.5**, $\exists S \in \mathcal{S}_M(\downarrow U')$ such that $S \to_{\varphi(\delta)} e$. But this just means $S \to e$ in $(L^X, \delta)$, so by Theorem **5.1.14** (vi)$\Longrightarrow$(i), $e \leq (U')^- = U'$. Therefore, $c(U') \leq U'$, $c(U') = U'$, $U \in \psi(\varphi(\delta))$, $\delta \subset \psi(\varphi(\delta))$.

$\forall U \in \psi(\varphi(\delta))$, $c(U') = U'$. Let $S$ be a molecule net in $U'$, $S \to e \in M(L^X)$. Then $S \to_{\varphi(\delta)} e$, $e \in \text{clu}_{\varphi(\delta)}(U')$, $e \leq \bigvee \text{clu}_{\varphi(\delta)}(U') = c(U') = U'$. By Corollary **5.1.15** (v)$\Longrightarrow$(i), $(U')^- = U'$, $U \in \delta$. Hence $\psi(\varphi(\delta)) \subset \delta$, $\psi(\varphi(\delta)) = \delta$.

(ii) Denote the closure operator generated by $\mathcal{C}$ as $c$. $\forall (S, e) \in \mathcal{C}$. If $S$ does not converge to $e$ in $(L^X, \psi(\mathcal{C}))$, then $\exists U \in \mathcal{Q}_{\psi(\mathcal{C})}(e)$ such that $S$ frequently does not quasi-coincide with $U$, i.e. $S$ is frequently in $U'$. So $S$ has a subnet $T \in \mathcal{S}_M(\downarrow U')$. In view of (CC2), $T \to_{\mathcal{C}} e$. Hence $e \in \text{clu}_{\mathcal{C}}(U')$, $e \leq c(U') = cl_{\psi(\mathcal{C})}(U') = U'$, this is a contradiction. So $S \to e$ in $(L^X, \psi(\mathcal{C}))$, $(S, e) \in \varphi(\psi(\mathcal{C}))$. This proves $\mathcal{C} \subset \varphi(\psi(\mathcal{C}))$.

$\forall (S, e) \in \varphi(\psi(\mathcal{C}))$, then $S \to e$ in $(L^X, \psi(\mathcal{C}))$. We have to prove $S \to_{\mathcal{C}} e$. Suppose this is not true, then by (CC3), $S$ has a subnet $T = \{T(m), m \in E\}$ such that no subnet of $T$ which $\mathcal{C}$-converges to $e$. By Theorem **5.1.7** (i), $T \to e$ in $(L^X, \psi(\mathcal{C}))$. $\forall m \in$

$E$, denote $E_m = \{k \in E : k \geq m\}$, then $E_m$ is a directed set, $T_m = \{T(k), k \in E_m\}$ is a subnet of $T$. By $T \to e$ in $(L^X, \psi(\mathcal{C}))$ and Theorem **5.1.7** (i), every $T_m \to e$ in $(L^X, \psi(\mathcal{C}))$. By Theorem **5.1.14** (vi)$\Longrightarrow$(i),

$$e \leq cl_{\psi(\mathcal{C})}(\bigvee img(T_m)) = c(\bigvee img(T_m)) = \bigvee \mathrm{clu}_{\mathcal{C}}(\bigvee img(T_m)).$$

By Lemma **5.3.5**, $\exists R_m \in \mathcal{S}_M(\downarrow \bigvee img(T_m))$ such that $R_m \to_{\mathcal{C}} e$. By Lemma **5.3.6**, $\exists \vec{R}_m \in \mathcal{S}_M(img(T_m))$ and $\vec{R}_m \to_{\mathcal{C}} e$. Take $K(m) = e$ for every $m \in E$, then by (CC1), $K = \{K(m), m \in E\} \to_{\mathcal{C}} e$, $\forall m \in E$, $\vec{R}_m \to_{\mathcal{C}} K(m)$. By (CC4), the induced net $\vec{K}$ of $\{\vec{R}_m : m \in E\}$ $\mathcal{C}$-converges to $e$.

For the index set $E \times \prod_{m \in E} D_m$ of $\vec{K}$ and arbitrary $(m, f) \in E \times \prod_{m \in E} D_m$, since $\vec{K}(m, f) = \vec{R}_m(f(m)) \in img(T_m)$, $\exists N(m, f) \in E$, $N(m, f) \geq m$ such that $\vec{K}(m, f) = T(N(m, f))$. So we obtain a mapping $N : E \times \prod_{m \in E} D_m \to E$ such that $\vec{K} = T \circ N$. $\forall m_0 \in E$, arbitrarily fix $f_0 \in \prod_{m \in E} D_m$, then $\forall (m, f) \in E \times \prod_{m \in E} D_m$ such that $(m, f) \geq (m_0, f_0)$, $N(m, f) \geq m \geq m_0$. Therefore, $N$ is a cofinal selection on $T$, $\vec{K}$ is a subnet of $T$. But we have proved above that $\vec{K} \to_{\mathcal{C}} e$, this contradicts with the property of $T$ that it has no subnet which $\mathcal{C}$-converges to $e$.

(iii) Suppose $\delta$, $mu$ are two $L$-fuzzy topologies on $L^X$ and $\delta \subset \mu$. If $(S, e) \in \varphi(\mu)$, then $S \to e$ in $(L^X, \mu)$, i.e. $S$ eventually quasi-coincides with element of $\mathcal{Q}_\mu(e)$. But $\delta \subset \mu$, so $\mathcal{Q}_\mu(e) \supset \mathcal{Q}_\delta(e)$. Hence $S$ eventually quasi-coincides with every element of $\mathcal{Q}_\delta(e)$, $S \to e$ in $(L^X, \delta)$, $(S, e) \in \varphi(\delta)$, $\varphi(\mu) \subset \varphi(\delta)$. □

**5.3.11 Remark** Comparing the conditions of the $L$-fuzzy convergence class with the ones in general topology, one can find that the condition (CC5) is new. It is not difficult to show that (CC5) can be equivalently replaced by

(CC5′) For every $x \in X$ and every $A \subset M(L)$ and every $\lambda \in M(\downarrow \bigvee A)$, there exists $S \in \mathcal{S}_M(\{x_\xi : \xi \in A\})$ such that $S \to_{\mathcal{C}} x_\lambda$.

So it is in fact a requirement to $\mathcal{C}$-convergence on the stratifications of $L$, which is just the essential difference between $L$-fuzzy topological spaces and general topological spaces. For the reason, (CC5) cannot be omitted from the definition of $L$-fuzzy convergence class (Example **5.3.12**), although it will become a trivial condition in general topology.

**5.3.12 Example** Take $X$ be a singleton $\{x\}$, $L = [0, 1]$. For every net $S = \{S(n), n \in D\}$ on $L^X$, denote $F(S) = \{ht(S(n)), n \in D\}$, then $F(S)$ is a net in $[0, 1]$. Using $S$ to denote a net in $L^X$, we take

$$\begin{aligned} \mathcal{C} = \{(S, x_\lambda) &: \lambda \in [0,1] \backslash \{\tfrac{2}{3}, 1\},\ F(S) \text{ eventually equals to } \lambda\} \cup \\ \{(S, x_{\frac{2}{3}}) &: \forall \varepsilon > 0,\ F(S) \text{ eventually is in } (\tfrac{1}{3} - \varepsilon, \tfrac{1}{3}) \cup \{\tfrac{2}{3}\}\} \cup \\ \{(S, x_1) &: \forall \varepsilon > 0,\ F(S) \text{ eventually is in } (\tfrac{2}{3} - \varepsilon, \tfrac{2}{3}) \cup \{1\}\}, \end{aligned}$$

then $\mathcal{C} \subset \mathcal{S}_M(L^X) \times M(L^X)$ and clearly satisfies (CC1) – (CC3).

Suppose $S = \{S(n), n \in D\}$ is a net in $L^X$, $S \to_{\mathcal{C}} x_\lambda$, $\forall n \in D$, $T^n$ is a net in $L^X$, $T^n \to_{\mathcal{C}} S(n)$.

If $\lambda \in [0, 1] \backslash \{\frac{2}{3}, 1\}$, then $F(S)$ eventually equals to $\lambda \in [1, 1] \backslash \{\frac{2}{3}, 1\}$, so for every

$n \in D$, $F(T^n)$ eventually equals to $\lambda$. Hence the induced net of $\{T^n : n \in D\}$ $\mathcal{C}$-converges to $x_\lambda$.

If $\lambda = \frac{2}{3}$, then for every $\varepsilon > 0$, $F(S)$ eventually is in $(\frac{1}{3} - \varepsilon, \frac{1}{3}) \cup \{\frac{2}{3}\}$. $\forall n \in D$,

(1) If $F(S)(n) \in (\frac{1}{3} - \varepsilon, \frac{1}{3})$, then $F(T^n)$ eventually equals to $F(S)(n)$;

(2) If $F(S)(n) = \frac{2}{3}$, then $F(T^n)$ eventually is in $(\frac{1}{3} - \varepsilon, \frac{1}{3}) \cup \{\frac{2}{3}\}$.

So for every $n \in D$, $F(T^n)$ always eventually is in $(\frac{1}{3} - \varepsilon, \frac{1}{3}) \cup \{\frac{2}{3}\}$, this implies that the induced net of $\{T^n : n \in D\}$ $\mathcal{C}$-converges to $x_{\frac{2}{3}} = x_\lambda$.

If $\lambda = 1$, the discussion is similar.

So we have proved that the induced net of $\{T^n : n \in D\}$ always $\mathcal{C}$-converges to $x_\lambda$, (CC4) is true for $\mathcal{C}$.

Now for the closure operator $c$ generated by $\mathcal{C}$, clearly $c(x_{\frac{1}{3}}) = x_{\frac{2}{3}}$, $cc(x_{\frac{1}{3}}) = c(x_{\frac{2}{3}}) = x_1$, the condition (CO4) is not satisfied, $c$ is not a closure operator on $L^X$.

# Chapter 6

# Connectedness

In contrast with $L$-fts's consisting of many pieces separated from one another such as sum spaces defined in Chapter 3, we shall introduce a kind of $L$-fts's in this chapter each of which consists of, figuratively speaking, "one single piece."

## 6.1 Connectedness

Recall the definition of sum spaces, it seems that we can define an $L$-fuzzy connected space as one which cannot be represented as a sum space. In general topology, it is just done as this. But we cannot do the same thing in $L$-fuzzy topology because subsets and subspaces of a topological space in general topology are essentially same, but the situation is completely different in $L$-fuzzy topology: Only a few of $L$-fuzzy subsets can be considered as $L$-fuzzy subspaces. If we follow the definition in the form of sum space, we shall have to give up discussion on this topic for many $L$-fuzzy subsets. So we still adopt the following definition which is a equivalent one of connected space in general topology:

**6.1.1 Definition** Let $(L^X, \delta)$ be an $L$-fts, $A, B \in L^X$.
$A$ and $B$ are called *separated*, if

$$A^- \wedge B = A \wedge B^- = \underline{0}.$$

$A$ is called *connected*, if there not exist separated $C, D \in L^X \backslash \{\underline{0}\}$ such that $A = C \vee D$. Call $(L^X, \delta)$ is *connected*, if the largest $L$-fuzzy subset $\underline{1}$ is connected.

**6.1.2 Theorem** *Let $(L^X, \delta)$ be an $L$-fts. Then the following conditions are equivalent:*

(i) $(L^X, \delta)$ *is connected.*

(ii) $A, B \in \delta,\ A \vee B = \underline{1},\ A \wedge B = \underline{0} \implies \underline{0} \in \{A, B\}$.

(iii) $A, B \in \delta',\ A \vee B = \underline{1},\ A \wedge B = \underline{0} \implies \underline{0} \in \{A, B\}$.

**Proof** (i)$\Longrightarrow$(ii): If (ii) is not true, then $\exists A, B \in \delta \backslash \{\underline{0}\}$ such that $A \vee B = \underline{1},\ A \wedge B = \underline{0}$. By the De Morgan's Law,

$$A' \wedge B' = \underline{0}, \quad A' \vee B' = \underline{1},$$

$$(A')^- \wedge B' = A' \wedge B' = \underline{0}, \quad A' \wedge (B')^- = A' \wedge B' = \underline{0}.$$

Since no one of $A$ and $B$ is $\underline{0}$, by $A \wedge B = \underline{0}$, both $A$ and $B$ are not $\underline{1}$ and hence no one of $A'$ and $B'$ is $\underline{0}$. So $A'$ and $B'$ are separated $L$-fuzzy sets. By $A' \vee B' = \underline{1}$, $(L^X, \delta)$ is not connected.

(ii)$\Longrightarrow$(iii): By the De Morgan's Law.

(iii)$\Longrightarrow$(i): If (i) is not true, then there exist separated $A, B \in \delta' \backslash \{0\}$ such that $A \vee B = \underline{1}$. This just means that (iii) is not true. □

**6.1.3 Remark** In general topology, a topological space $X$ is connected if and only if $X$ itself is the unique nonempty open-and-closed subset in $X$. This is true because any other nonempty open-and-closed subset will generate a pair of separated subsets with itself and its complementary set and their union is just $X$. But in $L$-fuzzy topology, the meet of a subset and its "dual form" — pseudo-complementary set need not be $\underline{0}$. So the parallel conclusion does not hold in $L$-fuzzy topology.

On the other hand, for two $L$-fuzzy subsets $A$ and $B$ in an $L$-fts, it is possible that $supp(A) \cup supp(B) \neq \emptyset$ but $A \wedge B = \underline{0}$, so we cannot either define connected $L$-fts with only crisp subsets just as discussed in the beginning of this section. Guo-jun Wang gave out an example to show this point:[168]

**6.1.4 Example** Let $L$ be the diamond-type lattice, i.e. $L = \{0, a, b, 1\}$, where $0 < a < 1$, $0 < b < 1$, $a \perp b$, $0' = 1$, $1' = 0$, $a' = b$, $b' = a$. Take $X$ be a singleton $\{x\}$, $\delta = L^X$, $A = x_a$, $B = x_b$, then both of $A$ and $B$ are open subsets and no one of them is $\underline{0}$, $(L^X, \delta)$ is not connected. But $\underline{1}$ cannot be represented as a join of two non-zero crisp open subsets, so $(L^X, \delta)$ should be connected if we adopt the definition of connectedness in crisp form.

**6.1.5 Lemma** *Let $(L^X, \delta)$ be an L-fts, $A, B, C \in L^X$, $B$ and $C$ be separated. Then $A \wedge B$ and $A \wedge C$ are separated.*

**Proof** Since $B$ and $C$ are separated, so

$$(A \wedge B)^- \wedge (A \wedge C) \leq B^- \wedge C = \underline{0}.$$

Similarly, $(A \wedge B) \wedge (A \wedge C)^- = \underline{0}$, $A \wedge B$ and $A \wedge C$ are separated. □

**6.1.6 Theorem** *Let $(L^X, \delta)$ be an L-fts, $A \in L^X$. Then the following conditions are equivalent:*

(i) *$A$ is connected.*

(ii) *$B, C \in L^X$ are separated, $A \leq B \vee C \Longrightarrow A \leq B$ or $A \leq C$.*

(iii) *$B, C \in L^X$ are separated, $A \leq B \vee C \Longrightarrow A \wedge B = \underline{0}$ or $A \wedge C = \underline{0}$.*

**Proof** (i)$\Longrightarrow$(iii): By Lemma **6.1.5**, $A \wedge B$ and $A \wedge C$ are separated. Since $A$ is connected and $A = A \wedge (B \vee C) = (A \wedge B) \vee (A \wedge C)$, one of $A \wedge B$ and $A \wedge C$ equals to $\underline{0}$.

(iii)$\Longrightarrow$(ii): Suppose $A \wedge B = \underline{0}$, then

$$A = A \wedge (B \vee C) = (A \wedge B) \vee (A \wedge C) = \underline{0} \vee (A \wedge C) = A \wedge C.$$

So $A \leq C$. Similarly, $A \wedge C = \underline{0} \Longrightarrow A \leq B$.

(ii)$\Longrightarrow$(i): Suppose $B, C \in L^X$ are separated and $A = B \vee C$, by (ii), $A \leq B$ or $A \leq C$. If $A \leq B$, then since $B$, $C$ are separated,

$$C = C \wedge A \leq C \wedge B \leq C \wedge B^- = \underline{0}.$$

If $A \leq C$, we can similarly have $B = \underline{0}$. So $A$ can be represented as the join of two separated non-zero subsets, $A$ is continuous. □

**6.1.7 Theorem** *Let $(L^X, \delta)$ be an L-fts, $A \in L^X$ be connected, $B \in L^X$, $A \leq B \leq A^-$. Then $B$ is connected.*

**Proof** If $B = C \vee D$, $C$ and $D$ are separated. Denote $F = A \wedge C$, $G = A \wedge D$, then by Lemma **6.1.5**, $F$ and $G$ are separated. Since $A$ is connected and

$$F \vee G = (A \wedge C) \vee (A \wedge D) = A \wedge (C \vee D) = A \wedge B = A,$$

one of $F$ and $G$, say $F$, equals to $\underline{0}$. Then

$$A = F \vee G = G = A \wedge D,$$

$$A \leq D, \quad B \leq A^- \leq D^-,$$

$$C = C \wedge B \leq C \wedge A^- \leq C \wedge D^- = \underline{0}.$$

By Theorem **6.1.6** (iii)$\Longrightarrow$(i), $B$ is connected. □

**6.1.8 Theorem** *Let $(L^X,\delta)$ be an L-fts, $\{A_t:\ t\in T\}\subset L^X$, $s\in T$ such that $A_t$ be connected and not separated from $A_s$ for every $t\in T$. Then $\bigvee_{t\in T}A_t$ is connected.*
**Proof** Denote $A=\bigvee_{t\in T}A_t$. Suppose $B,C\in L^X$ are separated and $A=B\vee C$. Since $A_s$ is connected, by Theorem **6.1.6** (i)$\Longrightarrow$(ii), $A_s\le B$ or $A_s\le C$, say $A_s\le B$. $\forall t\in T$, since $A_t$ is also connected and is not separated from $A_s$, by Lemma **6.1.5** and Theorem **6.1.6** (i)$\Longrightarrow$(ii), $A_t\le B$. So $A\le B$. By Theorem **6.1.6** (ii)$\Longrightarrow$(i), $A$ is connected.□
**6.1.9 Corollary** *Let $(L^X,\delta)$ be an L-fts, $\mathcal{A}$ be a family of connected L-fuzzy subsets on $X$, $\bigwedge\mathcal{A}\ne\underline{0}$. Then $\bigvee\mathcal{A}$ is connected.* □
**6.1.10 Corollary** *Let $(L^X,\delta)$ be an L-fts. Then $(L^X,\delta)$ is connected if and only if every pair of molecules in $(L^X,\delta)$ are contained in a connected L-fuzzy subset on $X$.*□
**6.1.11 Definition** Let $(L^X,\delta)$ be an L-fts, $A\in L^X$. $A$ is called a *connected component* of $(L^X,\delta)$, if $A$ is a maximal connected subset in $(L^X,\delta)$, i.e.

$$B\in L^X,\ B \text{ is connected},\ B\ge A\ \Longrightarrow\ B=A.$$

**6.1.12 Lemma** *Let $(L^X,\delta)$ be an L-fts, $e\in M(L^X)$. Then $e$ is a connected subset in $(L^X,\delta)$.*
**Proof** By the definition of molecule and Theorem **6.1.6** (ii)$\Longrightarrow$(i). □
**6.1.13 Theorem** *Let $(L^X,\delta)$ be an L-fts. Then*

(i) *Every molecule in $(L^X,\delta)$ is contained in a connected component of $(L^X,\delta)$.*
(ii) *The join of all the connected components of $(L^X,\delta)$ equals to $\underline{1}$.*
(iii) *Each connected component of $(L^X,\delta)$ is a closed subset.*
(iv) *Different connected compnents of $(L^X,\delta)$ are separated.*

**Proof** (i) $\forall e\in M(L^X)$, denote the family of all the connected subsets in $(L^X,\delta)$ which contains $e$ by $\mathcal{A}$, then by Lemma **6.1.12**, $\bigwedge\mathcal{A}\ne\underline{0}$. By Corollary **6.1.9**, $A=\bigvee\mathcal{A}$ is connected. Clearly $A$ is a connected component.

(ii) By (i) and Corollary **2.3.11** (i).

(iii) By Theorem **6.1.7**.

(iv) By Corollary **6.1.9** and (iii). □

**6.1.14 Theorem** *Let $(L^X,\delta)$ be an L-fts. Then*

(i) *If $A\subset M(L)$ is a pairwisely disjoint set, $x\in X$, $\lambda\in A$, $C$ is the connected component in $(L^X,\delta)$ containing $x_\lambda$, then $x_{\bigvee A}\le C$.*
(ii) *If $1\in M(L)$, then every connected component in $(L^X,\delta)$ is a crisp subset.*
(iii) *If $1\in M(L)$, $A\subset L$, $\mu$ is the L-fuzzy topology on $X$ generated by $\delta\cup\{\underline{a}:\ a\in A\}$, $C\in L^X$, then $C$ is a connected component in $(L^X,\delta)$ if and only if $C$ is a connected component in $(L^X,\mu)$.*

**Proof** (i) $\forall\xi\in A$, by Theorem **6.1.13** (i), $x_\xi$ is contained in a connected component $D$. If $D\ne C$, then by Theorem **6.1.13** (iv), $x_\xi\wedge x_\lambda\le D\wedge C=\underline{0}$, $\xi\wedge\lambda=0$. But $A$ is pairwisely disjoint, that is impossible. So $x_\xi\le D=C$, $x_{\bigvee A}\le C$.

(ii) Let $C$ be a connected component in $(L^X,\delta)$, $x\in$ *supp*$(C)$. By $1\in M(L)$ and Theorem **6.1.13** (i), $x_1\in M(L^X)$ must be contained in a connected component $D$. Since $C\wedge D\ge x_{C(x)}\wedge x_1=x_{C(x)}\ne\underline{0}$, by Theorem **6.1.13** (iv), $C=D\ge x_1$. By the arbitrariness of $x\in$ *supp*$(C)$, $C$ is crisp.

(iii) Clearly, without loss of generality, we can assume $0\notin A$, $1\in A$.

Suppose $D \in L^X$ is not connected in $(L^X, \delta)$, then there exist $B, C \in L^X \backslash \{\underline{0}\}$ separated in $(L^X, \delta)$ such that $D = B \vee C$. Since $\delta \subset \mu$, so $\delta' \subset \mu'$, $B$ and $C$ are also separated in $(L^X, \mu)$, $D$ is not connected in $(L^X, \mu)$. Hence every connected component in $(L^X, \mu)$ is connected in $(L^X, \delta)$.

Let $C \in L^X$ be a connected component in $(L^X, \delta)$. If $C$ is not connected in $(L^X, \mu)$, then $\exists D, E \in L^X \backslash \{\underline{0}\}$ separated in $(L^X, \mu)$ such that $C = D \vee E$.

Deote $W = \{\bigwedge F : \ F \in [A]^{<\omega}\}$. Since $\delta \cup \{\underline{a} : \ a \in A\}$ is a subbase of $\mu$ and $1 \in A$,

$$\begin{aligned}\mu &= \{\bigvee\{\bigwedge\{aU : \ (a, U) \in \mathcal{F}\} : \ \mathcal{F} \in \mathcal{A}\} : \ \mathcal{A} \subset [A \times \delta]^{<\omega}\} \\ &= \{\bigvee\{aU : \ (a, U) \in \mathcal{A}\} : \ \mathcal{A} \subset W \times \delta\}. \end{aligned} \tag{6.1}$$

By Theorem **6.1.13** (iii), $C$ is closed in $(L^X, \delta)$. By $\delta' \subset \mu'$, $C$ is also closed in $(L^X, \mu)$. Hence by $C = D \vee E$, we have $cl_\mu(D) \leq C$, $C = cl_\mu(D) \vee E$. $\forall x \in supp(cl_\mu(D))$. If $cl_\mu(D)(x) \neq 1$, then by (ii), $x_1 \leq C = cl_\mu(D) \vee E$. By $1 \in M(L)$, $x_1 \leq E$. But then

$$(cl_\mu(D) \wedge E)(x) = cl_\mu(D)(x) \wedge E(x) = cl_\mu(D)(x) \wedge 1 = cl_\mu(D)(x) \neq 0,$$

this contradicts with that $D$ and $E$ are separated in $(L^X, \mu)$. Hence $cl_\mu(D)(x) = 1$. By the arbitrariness of $x \in supp(cl_\mu(D))$, $cl_\mu(D)$ is crisp. By equality (6.1), $\exists \mathcal{A} \subset W \times \delta$ such that

$$cl_\mu(D) = \bigwedge\{\underline{a}' \times U' : \ (U, a) \in \mathcal{A}\} = \bigwedge\{\underline{a}' \times U' : \ (a, U) \in \mathcal{A}\}.$$

So $\forall x \in supp(D)$, $\forall (a, U) \in \mathcal{A}$, $a' \vee U'(x) = 1$. But $1 \in M(L)$, so $U'(x) = 1$ or $a' = 1$. If $a' = 1$, by $(a, U) \in \mathcal{A} \subset W \times \delta$, $a \in W = \{\bigwedge F : \ F \in [A]^{<\omega}\}$, $\exists F \in [A]^{<\omega}$ such that $\bigvee\{u' : \ u \in F\} = a' = 1 \in M(L)$. So $\exists u \in F$ such that $u' = 1$, $0 = u \in F \subset A$, this is a contradiction. So $a' \neq 1$, $U'(x) = 1$,

$$cl_\mu(D)(x) = \bigwedge\{a' \vee U'(x) : \ (a, U) \in \mathcal{A}\} = \bigwedge\{U'(x) : \ \exists a \in W, \ (a, U) \in \mathcal{A}\},$$
$$cl_\mu(D) = \bigwedge\{U' : \ \exists a \in W, \ (a, U) \in \mathcal{A}\} \in \delta',$$

$cl_\mu(D)$ is also closed in $(L^X, \delta)$. Similarly prove $cl_\mu(E)$ is closed in $(L^X, \delta)$. Then we obtain $D, E \in L^X \backslash \{\underline{0}\}$ separated in $(L^X, \delta)$ such that $C = D \vee E$, this contradicts with that $C$ is a connected component in $(L^X, \delta)$. Therefore, $C$ is also connected in $(L^X, \mu)$.

Now suppose $C$ is a connected component in $(L^X, \delta)$, since it is connected in $(L^X, \mu)$, there exists a connected component $D$ in $(L^X, \mu)$ containing $C$. But $D$ is also connected in $(L^X, \delta)$, so $D = C$, $C$ is also a connected component in $(L^X, \mu)$. Suppose $D$ is a connected component in $(L^X, \mu)$, $D$ is connected in $(L^X, \delta)$, there exists a connected component $C$ in $(L^X, \delta)$ containing $D$. Since $C$ is connected in $(L^X, \mu)$, so $C = D$, $D$ is also a connected component in $(L^X, \delta)$. □

**6.1.15 Remark** The condition "$1 \in M(L)$" cannot be omitted from the conclusions (ii) and (iii) in Theorem **6.1.14**. Take $X$ and $L$ as shown in Example **6.1.4**, $\delta$ as the trivial $L$-fuzzy topology on $X$, i.e. $\delta = \{\underline{0}, \underline{1}\}$, and $\mu = \delta \cup \{\underline{a}, \underline{b}\}$, then $\mu$ is the discreate $L$-fuzzy topology on $X$, $\underline{a}$ and $\underline{b}$ are connected components in $(L^X, \mu)$ but not crisp. $(L^X, \delta)$ is connected but $(L^X, \mu)$ not.

**6.1.16 Corollary** *Let $(L^X, \delta)$ be an L-fts, $1 \in M(L)$, $A \subset L$, $\mu$ be the L-fuzzy topology on $X$ generated by $\delta \cup \{\underline{a} : \ a \in A\}$. Then $(L^X, \delta)$ is connected if and only if $(L^X, \mu)$ is connected.* □

**6.1.17 Corollary** *Let $(X, \delta)$ be a F-ts. Then*

(i) *Every connected component in $(X,\delta)$ is a crisp subset.*
(ii) *For every $A \subset [0,1]$ and the fuzzy topology $\mu$ on $X$ generated by $\delta \cup \{\underline{a} : a \in A\}$, a fuzzy subset on $X$ is a connected component in $(X,\delta)$ if and only if it is a connected component in $(X,\mu)$.* □

**6.1.18 Theorem** *Let $(L^X,\delta)$, $(L^Y,\mu)$ be L-fts', $f^{\rightarrow} : (L^X,\delta) \to (L^Y,\mu)$ be an L-fuzzy continuous mapping, $A \in L^X$ be connected. Then $f^{\rightarrow}(A)$ is connected.*

**Proof** Suppose $B, C \in L^Y$ are separated and $f^{\rightarrow}(A) = B \vee C$, then by Theorem **2.1.25** (i),

$$A \leq f^{\leftarrow}f^{\rightarrow}(A) = f^{\leftarrow}(B \vee C) = f^{\leftarrow}(B) \vee f^{\leftarrow}(C).$$

Since $f^{\rightarrow}$ is continuous, by Theorem **2.4.5** (i)$\Longrightarrow$(viii),

$$(f^{\leftarrow}(B))^- \wedge f^{\leftarrow}(C) \leq f^{\leftarrow}(B^-) \wedge f^{\leftarrow}(C) = f^{\leftarrow}(B^- \wedge C) = f^{\leftarrow}(\underline{0}) = \underline{0}.$$

Similarly, $f^{\leftarrow}(B) \wedge (f^{\leftarrow}(C))^- = \underline{0}$. So $f^{\leftarrow}(B)$ and $f^{\leftarrow}(C)$ are separated. By Theorem **6.1.6** (i)$\Longrightarrow$(ii), $A \leq f^{\leftarrow}(B)$ or $A \leq f^{\leftarrow}(C)$, say $A \leq f^{\leftarrow}(B)$. Then by Theorem **2.1.25** (ii), $f^{\rightarrow}(A) \leq f^{\rightarrow}f^{\leftarrow}(B) \leq B$. By Theorem **6.1.6** (ii)$\Longrightarrow$(i), $f^{\rightarrow}(A)$ is connected. □

**6.1.19 Definition** Let $\{X_t : t \in T\}$ be a family of nonempty sets, $X = \prod_{t\in T} X_t$, $z \in X$, $S \subset T$. Then denote

$$hpl(z,S) = \{x \in X : \forall t \in T \backslash S,\ p_t(x) = p_t(z)\},$$

call $hpl(z,S)$ the *hyperplane* in $X$ at point $z$ with *free range* $S$.

Let $\{(L^{X_t},\delta_t) : t \in T\}$ be an family of L-fts', $(L^X,\delta) = \prod_{t\in T}(L^{X_t},\delta_t)$, $z \in X$, $S \subset T$. Then the L-fuzzy subset $hpl_L(z,S) = \chi_{hpl(z,S)} \in L^X$ is called the *L-fuzzy hyperplane* in $(L^X,\delta)$ at L-fuzzy point $z_1$ with *free range* $S$.

**6.1.20 Remark** Clearly, in the preceding definition, $hpl(z,\emptyset) = \{z\}$, for every singleton $\{s\} \subset T$, the hyperplane $hpl(z,\{s\})$ is just the slice $sl(z,s)$ in $X$ at $z$ parallel to $X_s$ and $hpl(z,T) = X$, $hpl_L(z,T) = \chi_X$.

**6.1.21 Lemma** *Let $\{(L^{X_t},\delta_t) : t \in T\}$ be a family of L-fts', $(L^X,\delta) = \prod_{t\in T}(L^{X_t},\delta_t)$. If every L-fuzzy slice in $(L^X,\delta)$ is a connected L-fts, then $(L^X,\delta)$ is connected.*

**Proof** Arbitrarily fix a point $x \in X$.

$\forall S \subset T$, $\forall t \in T$. If $\forall z \in sl(x,t)$, $hpl_L(z,S)$ is connected in $(L^X,\delta)$, then since $\chi_{sl(x,t)}$ is also connected in $(L^X,\delta)$ and $hpl_L(z,S) \wedge \chi_{sl(x,t)} \geq z_1 \neq \underline{0}$, by Corollary **6.1.9**, $hpl_L(z,S) \vee \chi_{sl(x,t)}$ is connected. Since

$$\bigwedge\{hpl_L(z,S) \vee \chi_{sl(x,t)} : z \in sl(x,t)\} \geq \chi_{sl(x,t)} \neq \underline{0},$$

use Corollary **6.1.9** again,

$$hpl_L(x, S \cup \{t\}) = \bigvee\{hpl_L(z,S) \vee \chi_{sl(x,t)} : z \in sl(x,t)\}$$

is connected.

By the given condition and what we proved above, $\forall F \in [T]^{<\omega}$, $hpl_L(x,F)$ is connected. Denote $A = \bigvee\{hpl_L(x,F) : F \in [T]^{<\omega}\}$. Suppose $P$ is a closed subset in $(L^X,\delta)$ such that $A \leq P$, then $\exists \mathcal{P} \subset [T]^{<\omega}$ such that $\forall F \in \mathcal{P}$, $\forall t \in F$, $\exists P_{F,t} \in \delta_t'$, $P = \bigwedge_{F\in\mathcal{P}} \bigvee_{t\in F} p_t^{\leftarrow}(P_{F,t})$. So $\forall F \in \mathcal{P}$, $\forall z \in X$, take $y \in hpl(x,F)$ such that $\forall t \in F$, $p_t(y) = p_t(z)$, $\forall t \in T\backslash F$, $p_t(y) = p_t(x)$, we have

$$\bigvee_{t\in F} p_t^{\leftarrow}(P_{F,t}) \geq hpl_L(x,F),$$

$$\Big(\bigvee_{t\in F} p_t^{\leftarrow}(P_{F,t})\Big)(z) = \bigvee_{t\in F} P_{F,t}(p_t(z)) = \bigvee_{t\in F} P_{F,t}(p_t(y)) = \Big(\bigvee_{t\in F} p_t^{\leftarrow}(P_{F,t})\Big)(y)$$
$$\geq hpl_L(x,F)(y) = 1.$$

So $\bigvee_{t\in F} p_t^{\leftarrow}(P_{F,t}) = \underline{1}$, $P = \underline{1}$, $A^- = \underline{1}$. By Theorem **6.1.7**, $\chi_X$ is connected, $(L^X, \delta)$ is connected. □

**6.1.22 Theorem** *Let $\{(L^{X_t}, \delta_t) : t \in T\}$ be a family of stratified L-fts', $(L^X, \delta) = \prod_{t\in T}(L^{X_t}, \delta_t)$. Then $(L^X, \delta)$ is connected if and only if $(L^{X_t}, \delta_t)$ is connected for every $t \in T$.*

**Proof** (Necessity) $\forall t \in T$. Since the projection $p_t^{\rightarrow} : (L^X, \delta) \to (L^{X_t}, \delta_t)$ is continuous and surjective, by Theorem **6.1.18**, $(L^{X_t}, \delta_t)$ is connected.

(Sufficiency) By Corollary **3.2.13** and Theorem **6.1.18**, every $L$-fuzzy slice in $(L^X, \delta)$ is connected. By Lemma **6.1.21**, $(L^X, \delta)$ is connected. □

**6.1.23 Theorem** *Let $\{(L^{X_t}, \delta_t) : t \in T\}$ be a family of L-fts', $(L^X, \delta) = \prod_{t\in T}(L^{X_t}, \delta_t)$, $1 \in M(L)$. Then $(L^X, \delta)$ is connected if and only if $(L^{X_t}, \delta_t)$ is connected for every $t \in T$.*

**Proof** (Necessity) Same with the correspondent part in the proof of Theorem **6.1.22**.

(Sufficiency) Let $\mu$ be the stratification of $\delta$, and let $\mu_t$ denote the stratification of $\delta_t$ for every $t \in T$, then by Theorem **6.1.14** (iii), $(L^{X_t}, \mu_t)$ is also connected. By Theorem **6.1.22** and Theorem **4.1.9**, $(L^X, \mu) = \prod_{t\in T}(L^{X_t}, \mu_t)$ is connected. Using Theorem **6.1.14** (iii) again, we know that $(L^X, \delta)$ is connected. □

**6.1.24 Corollary** *Let $\{(X_t, \delta_t) : t \in T\}$ be a family of F-ts', $(X, \delta) = \prod_{t\in T}(X_t, \delta_t)$. Then $(X, \delta)$ is connected if and only if $(X_t, \delta_t)$ is connected for every $t \in T$.* □

Diferent from the correspondent case in general topology, the product $L$-fts of a family of connected $L$-fts can be a non-connected one:

**6.1.25 Example** Let $X$ be a singleton $\{x\}$, $L$ be the diamond-type lattice $\{0, a, b, 1\}$ as mentioned in Example **6.1.4**. Take $\delta_0 = \{\underline{0}, \underline{a}, \underline{1}\}$, $\delta_1 = \{\underline{0}, \underline{b}, \underline{1}\}$, then both $(L^X, \delta_0)$ and $(L^X, \delta_1)$ are connected $L$-fts'. Denote $Z = X \times X$, $\delta = \delta_0 \times \delta_1$, then

$$aZ = p_0^{\leftarrow}(aX) \in \delta, \quad bZ = p_1^{\leftarrow}(bX) \in \delta,$$

so $\delta = \{\underline{0}, \underline{a}, \underline{b}, \underline{1}\}$ is the discrete $L$-fuzzy topology on $Z$, $(L^Z, \delta)$ is clearly not connected.

In [168], Guo-jun Wang gave out an example to show a product of two connected $L$-fuzzy subsets is not connected as follows:

**6.1.26 Example** Let $X = \{x, y\}$, $L = \{0, a, b, 1\}$ be the diamond-type lattice. Take $A, B \in L^X$ such that

$$A(x) = a, \quad A(y) = b, \quad B(x) = b, \quad B(y) = a,$$

and $C \in L^X$ as

$$C(x) = 1, \quad C(y) = 0,$$

then $\delta = \{0, C, 1\}$ is an $L$-fuzzy topology on $X$. It can be easily verified that both $A$ and $B$ are connected subsets in $(L^X, \delta)$. In fact, for example, there is only a unique way to represent $A$ as a join of two disjoint non-zero $L$-fuzzy subsets:

$$A = P \vee Q, \quad P \wedge Q = \underline{0},$$

where $P(x) = a$, $P(y) = 0$, $Q(x) = 0$, $Q(y) = b$. Since $P^- = 1$, so $P^- \wedge Q \neq 0$, i.e. $P$ and $Q$ are not separated, $A$ is connected. Similarly, $B$ is connected. But $A \times B$ is not connected in the $L$-fuzzy product space $(L^X, \delta) \times (L^X, \delta) = (L^{X\times X}, \delta \times \delta)$. In fact, let $K = A \times B$, then

$$K(x, x) = a \wedge b = 0, \quad K(x, y) = a \wedge a = a,$$

$$K(y,x) = b \wedge b = b, \quad K(y,y) = b \wedge a = 0.$$

Take $M, N, R, S \in L^{X \times X}$ as follows:

$$M(x,y) = a, \quad M(x,x) = M(y,x) = M(y,y) = 0,$$
$$N(y,x) = b, \quad N(x,x) = N(x,y) = N(y,y) = 0,$$
$$R(x,x) = R(x,y) = 0, \quad R(y,x) = R(y,y) = 1,$$
$$S(x,x) = S(y,x) = 0, \quad S(x,y) = S(y,y) = 1,$$

then it is easy to verify

$$K = M \vee N, \quad M^- \wedge N = S \wedge N = 0, \quad M \wedge N^- = M \wedge R = 0.$$

So $K = A \times B$ is not connected.

## 6.2 Connectedness of $L$-valued Stratification Spaces

In the last section, we have faced the problem of connectedness in some stratified spaces — stratifizations of $L$-fts'. In this section, we shall discuss the relation among stratified spaces, weakly induced spaces, induced spaces and their background spaces.

By Theorem **6.1.14** (iii), the following conclusion is obvious:

**6.2.1 Theorem** *Let $(L^X, \delta)$ be an L-fts, $1 \in M(L)$. Then $(L^X, \delta)$ is connected if and only if the stratification of $(L^X, \delta)$ is connected.* □

Certainly, in general, we cannot assert any thing about the connectedness of a stratified $L$-fts only from the connectedness of its background space, e.g. discrete $L$-fuzzy topology is also stratified. But for weakly induced $L$-fts', especially for induced $L$-fts', we have still some useful information obtained from their background spaces. To discuss this problem, we need to introduce a kind of lattices:

**6.2.2 Definition** A lattice $L$ is called *anti-diamond-type*, if there not exists a sub-lattice of $L$ which is isomorphic to the diamond-type lattice; i.e. there not exist $a, b \in L \backslash \{0, 1\}$ such that $a \wedge b = 0, \ a \vee b = 1$.

**6.2.3 Example** (i) Every totally ordered lattice is anti-diamond-type.

(ii) Every lattice $L$ with $0_L \in pr(L)$ is anti-diamond-type.

(iii) Every lattice $L$ with $1_L \in M(L)$ is anti-diamond-type.

(iv) The topology of an ordinary connected topological space, with the inclusion order, is an anti-diamond-type lattice.

Then we have the following results:

**6.2.4 Theorem** *Let L be a F-lattice. Then the following conditions are equivalent:*

(i) *L is anti-diamond-type.*

(ii) *For every weakly induced L-fts $(L^X, \delta)$, $(L^X, \delta)$ is connected if and only if $(X, [\delta])$ is connected.*

(iii) *For every induced L-fts $(L^X, \delta)$, $(L^X, \delta)$ is connected if and only if $(X, [\delta])$ is connected.*

**Proof** (i)$\Longrightarrow$(ii): "$(L^X, \delta)$ is connected $\Longrightarrow$ $(X, [\delta])$ is connected" is clearly always true. Suppose $(L^X, \delta)$ is weakly induced, $(X, [\delta])$ is connected. If $(L^X, \delta)$ is not connected, then $\exists A, B \in L^X \backslash \{\underline{0}\}$ separated and $A \wedge B = \underline{0}, \ A \vee B = \underline{1}$. If $A$ is not crisp, then $\exists x \in supp(A)$ such that $A(x) \notin \{0, 1\}$. Denote $a = A(x), \ b = B(x)$, then by $A \wedge B = \underline{0}, \ A \vee B = \underline{1}$ we have $b \notin \{0, 1\}, \ a \wedge b = 0, \ a \vee b = 1$. But this contradicts with (i), so $A$ is crisp. Similarly, $B$ is also crisp, $\exists C, D \subset X$ such that $C, D \neq \emptyset$, $A = \chi_C$, $B = \chi_D$. So $C \cup D = X$. By Proposition **4.2.4** (iii) and $A^- \wedge B = \underline{0}$, we have

$$(A^-)_{[1]} \cap D = (A^-)_{[1]} \cap B_{[1]} = (A^- \wedge B)_{[1]} = \underline{0}_{[1]} = \emptyset.$$

Similarly, $C \cap (B^-)_{[1]} = \emptyset$. Since $(L^X, \delta)$ is weakly induced, by Theorem **4.2.16** (iii), both $(A^-)_{[1]}$ and $(B^-)_{[1]}$ are closed in $(X, [\delta])$. Then by $C = A_{[1]} \subset (A^-)_{[1]}$, $D = B_{[1]} \subset (B^-)_{[1]}$ we have $C^- \subset (A^-)_{[1]}$, $D \subset (B^-)_{[1]}$, and hence

$$C^- \cap D \subset (A^-)_{[1]} \cap D = \emptyset, \quad C \cap D^- \subset C \cap (B^-)_{[1]} = \emptyset.$$

But $C \cup D = X$, so $(X, [\delta])$ is not connected.

(ii)$\Longrightarrow$(iii): Obvious.

(iii)$\Longrightarrow$(i): Suppose $L$ is not anti-diamond-type, then $\exists a, b \in L \backslash \{0, 1\}$ such that $a \wedge b = 0$, $a \vee b = 1$. Take $X$ as a singleton $\{x\}$, $\delta$ as the discrete $L$-fuzzy topology on $X$, then $(L^X, \delta)$ is certainly induced and $(X, [\delta])$ is connected. But for $A = x_a$, $B = x_b$ we have $A^- \wedge B = A \wedge B^- = \underline{0}$, $A \vee B = \underline{1}$, $(L^X, \delta)$ is not connected. □

Obviously, for an anti-diamond-type lattice $L$, it is not necessary $0 \in pr(L)$ or $1 \in M(L)$. So with the concept of anti-diamond-type lattice, we can enhance Theorem **6.1.14** (ii) as follows in a little bit. The proof is omitted off.

**6.2.5 Theorem** *Let $(L^X, \delta)$ be an L-fts, L be anti-diamond-type. Then every connected component of $(L^X, \delta)$ is crisp.* □

# Chapter 7

# Some Properties Related to Cardinals

Some topics in $L$-fuzzy topology naturally involve the problem of "how many" or "how much," then are connected with problems concerning cardinals. Certainly, the simplest infinite cardinal number is $\omega$, so most of investigations are concertrated on this case.

## 7.1 Weight, Characteristic and Density

By Corollary **2.3.11**, every $L$-fuzzy subset in an $L$-fts can be represented as a join of all the molecules contained in it, so we can introduce

**7.1.1 Definition** Let $(L^X, \delta)$ be an $L$-fts, $A \in L^X$, $x_\lambda \in M(L^X)$.

Define the *weight* $w(\delta)$ of $(L^X, \delta)$ by:

$$w(\delta) = min\{|\mathcal{B}| : \mathcal{B} \text{ is a base of } \delta\}.$$

Define the *local characteristic* $chr_\delta(x_\lambda)$ of $x_\lambda$ in $(L^X, \delta)$ by:

$$chr_\delta(x_\lambda) = min\{|\mathcal{B}| : \mathcal{B} \text{ is a Q-neighborhood base of } x_\lambda \text{ in } (L^X, \delta)\},$$

and the *characteritic* $chr(\delta)$ of $(L^X, \delta)$ by:

$$chr(\delta) = \bigvee\{chr_\delta(x_\lambda) : x_\lambda \in M(L^X)\}.$$

Define the *power* $|A|$ of $A$ by

$$|A| = min\{|C| : C \subset M(L^X), \ \bigvee C = A\}.$$

Define the *density* $dn(\delta)$ of $(L^X, \delta)$ by

$$dn(\delta) = min\{|A| : A \in L^X, \ A^- = \underline{1}\}.$$

First of all, we have the following relations for every $L$-fts:

**7.1.2 Theorem** *Let $(L^X, \delta)$ be an L-fts. Then*

$$chr(\delta) \le w(\delta) \ge dn(\delta).$$

**Proof** The first inequality is obvious. As for the second one, since every set of cardinals is well-ordered, there exists a base $\mathcal{B}$ of $\delta$ such that $|\mathcal{B}| = w(\delta)$. Certainly we can assume $\underline{0} \notin \mathcal{B}$, so by Corollary **2.3.11**, $\forall U \in \mathcal{B}$, $\exists e^U \in M(L^X)$ such that $e^U \lessdot U$. Now denote $A = \bigvee\{e^U : U \in \mathcal{B}\}$, then $\forall e \in M(L^X)$, $\forall U \in \mathcal{Q}_\delta(e)$, since $\mathcal{B}$ is a base of $\delta$, $\exists V \in \mathcal{B}$ such that $V \subset U$, $V \in \mathcal{Q}_\delta(e)$. By $e^V \le A$, $e^V \lessdot V$, we have $V \pitchfork A$, and hence $U \pitchfork A$, $e$ is an adherent point of $A$. By Theorem **2.3.23**, $e \le A^-$. By the arbitrariness of $e \in M(L^X)$, $A^- = \underline{1}$. So

$$dn(\delta) \le |A| \le |\{e^U : U \in \mathcal{B}\}| \le w(\delta). \qquad \square$$

The following conclusions are obvious:

**7.1.3 Theorem** *Let $(L^X, \delta)$ be an L-fts, $\kappa \ge \omega$. Then $w(\delta) \le \kappa$ if and only if $\delta$ has a subbase $\mathcal{B}$ such that $|\mathcal{B}| \le \kappa$.* $\square$

**7.1.4 Theorem** *Let $(L^X, \delta)$ be an L-fts, $Y \subset X$ nonempty. Then*

(i) $w(\delta|_Y) \le w(\delta)$.

(ii) $chr(\delta|_Y) \leq chr(\delta)$. □

**7.1.5 Theorem** *Let $(L^X, \delta)$, $(L^Y, \mu)$ be L-fts', $f^{\rightarrow}: (L^X, \delta) \rightarrow (L^Y, \mu)$ an L-fuzzy continuous open surjection. Then*

(i) $w(\mu) \leq w(\delta)$.

(ii) $chr(\mu) \leq chr(\delta)$.

**Proof** (i) Clear.

(ii) $\forall y_\lambda \in M(L^Y)$. Since $f^{\rightarrow}$ is surjective, so is $f: X \rightarrow Y$, $\exists x^y \in X$ such that $f(x^y) = y$. Suppose $\mathcal{B}$ is a Q-neighborhood base of $x^y{}_\lambda$, $V \in \mathcal{Q}_\mu(y_\lambda)$, then by the continuity of $f^{\rightarrow}$, $f^{\rightarrow}(V) \in \mathcal{Q}_\delta(x^y{}_\lambda)$, $\exists U \in \mathcal{B}$ such that $U \leq f^{\rightarrow}(V)$. So

$$\lambda \not\leq U'(x^y), \quad \lambda \not\leq \bigwedge\{U'(u): u \in X, f(u) = y\} = f^{\rightarrow}(U)'(y).$$

Since $f^{\rightarrow}$ is open and surjective, $f^{\rightarrow}(U) \in \mathcal{Q}_\mu(y_\lambda)$, $f^{\rightarrow} \leq f^{\rightarrow}f^{\leftarrow}(V) = V$. That is to say, $\{f^{\rightarrow}(U) : U \in \mathcal{B}\}$ is a Q-neighborhood base of $y_\lambda$ in $(L^Y, \mu)$, $chr_\mu(y_\lambda) \leq chr_\delta(x^y{}_\lambda)$. Therefore,

$$chr(\mu) = \bigvee\{chr_\mu(y_\lambda): y_\lambda \in M(L^Y)\} \leq \bigvee\{chr_\delta(x^y{}_\lambda): y_\lambda \in M(L^Y)\}$$
$$\leq \bigvee\{chr_\delta(e): e \in M(L^X)\} = chr(\delta).$$ □

**7.1.6 Corollary** *Let $(L^X, \delta)$, $(L^Y, \mu)$ be isomorphic L-fts'. Then*

(i) $w(\mu) = w(\delta)$.

(ii) $chr(\mu) = chr(\delta)$. □

**7.1.7 Theorem** *Let $(L^X, \delta)$, $(L^Y, \mu)$ be L-fts', $f^{\rightarrow}: (L^X, \delta) \rightarrow (L^Y, \mu)$ an L-fuzzy continuous surjective mapping. Then $dn(\mu) \leq dn(\delta)$.*

**Proof** Let $A \subset M(L^X)$ such that $(\bigvee A)^- = \underline{1}$. Denote $B = \{f^{\rightarrow}(e) : e \in A\}$, then by Theorem **2.1.15** (ii), $B \subset M(L^Y)$. $\forall y_\lambda \in M(L^Y)$, $\forall V \in \mathcal{Q}_\mu(y_\lambda)$, since $f^{\rightarrow}$ is surjective and so is $f$, we can take a point $x \in X$ such that $f(x) = y$, then $x_\lambda \in M(L^X)$. Since $f^{\rightarrow}$ is continuous, by $y_\lambda \lessdot V$, we have $f^{\leftarrow}(V) \in \mathcal{Q}_\delta(x_\lambda)$. So by $(\bigvee A)^- = \underline{1}$, $\exists x^1{}_\gamma \in A$ such that $x^1{}_\gamma \lessdot f^{\leftarrow}(V)$. That is to say,

$$\gamma \not\leq (f^{\leftarrow}(V))'(x^1) = f^{\leftarrow}(V')(x^1) = V'(f(x^1)).$$

So $f(x^1)_\gamma = f^{\rightarrow}(x^1{}_\gamma) \in B$, $f(x^1)_\gamma \lessdot V$, $y_\lambda$ is an adherent point of $\bigvee B$. Hence $(\bigvee B) = \underline{1}$ and $|B| \leq |A|$. This clearly completes the proof. □

**7.1.8 Exercise** *Let $\{(L^{X_t}, \delta_t) : t \in T\}$ be a family of L-fts'. Then*

(i) *If $w(\delta_t) \leq \kappa$ for every $t \in T$, then $w(\bigoplus_{t\in T} \delta_t) \leq \kappa \cdot |T|$.*

(ii) *If $chr(\delta_t) \leq \kappa$ for every $t \in T$, then $chr(\bigoplus_{t\in T} \delta_t) \leq \kappa$.*

(iii) *If $dn(\delta_t) \leq \kappa$ for every $t \in T$, then $dn(\bigoplus_{t\in T} \delta_t) \leq \kappa \cdot |T|$.*

**7.1.9 Theorem** *Let $\{(L^{X_t}, \delta_t) : t \in T\}$ be a family of L-fts', $(L^X, \delta)$ the L-fuzzy product space of them, $\kappa \geq \omega$. Then*

(i) *If $w(\delta_t) \leq \kappa$ for every $t \in T$, then $w(\delta) \leq \kappa \cdot |T|$.*

(ii) *If $w(\delta) \leq \kappa$, $t \in T$, $(L^{X_t}, \delta_t)$ is stratified, then $w(\delta_t) \leq \kappa$.*

(iii) *If $chr(\delta_t) \leq \kappa$ for every $t \in T$, then $chr(\delta) \leq \kappa \cdot |T|$.*

(iv) *If $chr(\delta) \leq \kappa$, $t \in T$, $(L^{X_t}, \delta_t)$ is stratified, then $chr(\delta_t) \leq \kappa$.*

**Proof** (i) $\forall t \in T$, suppose $\mathcal{B}_t$ is a subbase of $\delta_t$, then $\mathcal{B} = \bigcup_{t\in T}\{p_t{}^{\leftarrow}(U) : U \in \mathcal{B}_t\}$ is a subbase of $\delta$. Clearly $|\mathcal{B}| \leq \kappa \cdot |T|$, by Theorem **7.1.3**, $w(\delta) \leq \kappa \cdot |T|$.

(ii) By Corollary **3.2.11**, $p_t{}^{\rightarrow}: (L^X, \delta) \rightarrow (L^{X_t}, \delta_t)$ is open and surjective. The conclusion is obtained by Theorem **7.1.5** (i).

(iii) $\forall z_\lambda \in M(L^X)$, suppose $\mathcal{B}$ is a Q-neighborhood base of $z_\lambda$ in $(L^X, \delta)$. Certainly, furthermore, we can assume $\mathcal{B}$ is a subset of the canonical base of $\delta$. $\forall \in T$, suppose $\mathcal{B}_t$ is a Q-neighborhood base of $p_t(z)_\lambda$ in $(L^{X_t}, \delta_t)$ such that $|\mathcal{B}_t| = chr_{\delta_t}(p_t(z)_\lambda)$. $\forall V \in \mathcal{B}$, then $\exists F \in [T]^{<\omega}$ such that $\forall t \in F$, $\exists U_t \in \delta_t$, $V = \bigwedge_{t\in T} p_t{}^{\leftarrow}(U_t)$. Then $\forall t \in F$, $z_\lambda \lessdot p_t{}^{\leftarrow}(U_t)$. That is to say, $\forall t \in T$, $p_t(z)_\lambda \lessdot U_t$. Since $\mathcal{B}_t$ is a Q-neighborhood base of $p_t(z)_\lambda$, $\exists W_t \in \mathcal{B}_t$ such that $W_t \le U_t$. Since $F$ is finite, $\lambda \in M(L)$, $W = \bigwedge_{t\in F} p_t{}^{\leftarrow}(W_t) \in \mathcal{Q}_\delta(z_\lambda)$ and $W \le V$. By $|\mathcal{B}_t| = chr_{\delta_t}(p_t(z)_\lambda) \le chr(\delta_t) \le \kappa$ and $\kappa \ge \omega$, we are informed that there exists a Q-neighborhood base of $z_\lambda$ with power not greater than $\kappa\cdot|[T]^{<\omega}| = \kappa\cdot|T|$. So $chr(\delta) \le \kappa\cdot|T|$.

(iv) By Corollary **3.2.11**, $p_t{}^{\rightarrow} : (L^X, \delta) \to (L^{X_t}, \delta_t)$ is open and surjective. The conclusion is obtained by Theorem **7.1.5** (ii). □

As for the density of an $L$-fuzzy product space, the situation is not so clear. We begin from a set theoretical lemma:

**7.1.10 Lemma** *Let $\{A_t : t \in T\}$ be a family nonempty sets, $\kappa \ge \omega$, $|A_t| \le \kappa$ for every $t \in T$ and $|T| \le 2^\kappa$. Then there exists $A \subset \prod_{t\in T} A_t$ satisfying the following conditions:*

(i) $|A| \le \kappa$.

(ii) *For every $x \in \prod_{t\in T} A_t$ and every finite subset $C \subset T$, there exists $a \in A$ such that $p_t(a) = p_t(x)$ for every $t \in C$.*

**Proof** Take a set $S$ such that $|S| = \kappa$. Since $|T| \le 2^\kappa$, we can assume $T \subset 2^S$. Denote $\mathcal{F} = \{M : \exists F \in [S]^{<\omega},\ M \subset 2^F\}$, then by $\kappa \ge \omega$, $|\mathcal{F}| = |S| = \kappa$. $\forall M \in \mathcal{F}$, $\forall f \in M$, $\forall t \in T$, we denote $f \le t$ if $f(s) = t(s)$ for every $s \in dom(f)$.

Suppose $t_0, \cdots, t_n \in T$, $t_i \ne t_j$ whenever $i \ne j$, then $\exists s_{ij} \in S$ such that $t_i(s_{ij}) \ne t_j(s_{ij})$ for every two distinguished $i, j \le n$. Take $F = \{s_{ij} : i, j \le n\}$. For every $i \le n$, define $f_i \in 2^F$ as $f_i(s) = t_i(s)$ for every $s \in F$, then $f_i \le t_i$ for every $i \le n$ and $M = \{f_i : i \le n\} \subset 2^F$, $M \in \mathcal{F}$. Suppose $f_i, f_j \in M$, $f_i \ne f_j$, $r, t \in T$, $f_i \le r$, $f_j \le t$, then we have

$$r(s_{ij}) = f_i(s_{ij}) = t_i(s_{ij}) \ne t_j(s_{ij}) = f_j(s_{ij}) = t(s_{ij}),$$

so $r \ne t$. That is to say, $\forall C \in [T]^{<\omega}$, $\exists M \in \mathcal{F}$ such that

(1) There exists bijective mapping $\varphi_C : M \to C$ such that $\forall f \in M$, $f \le \varphi_C(f)$;

(2) $r, t \in C$, $f, g \in M$, $f \ne g$, $f \le r$, $g \le t \implies r \ne t$.

For every $t \in T$, since $|A_t| \le \kappa$, there exists a surjective mapping $h_t : S \to A_t$. $\forall M \in \mathcal{F}$, take $A_M \subset \prod_{t\in T} A_t$ as

$$a \in A_M \iff \exists s_a \in S,\ \forall f \in M,\ \exists s_f \in S,\ p_t(a) = \begin{cases} h_t(s_f), & \exists f \in M,\ f \le t, \\ h_t(s_a), & \text{otherwise.} \end{cases}$$

By (2), the definition is reasonable. By $|M| < \omega$, $|A_M| \le |[S]^{<\omega}| = |S| = \kappa$. Take $A = \bigcup_{M\in\mathcal{F}} A_M$, by $|\mathcal{F}| = \kappa$, we have $|A| \le \kappa\cdot\kappa = \kappa$, (i) is satisfied.

$\forall x \in \prod_{t\in T} A_t$, $\forall C \in [T]^{<\omega}$, $\exists M \in \mathcal{F}$ satisfying conditions (1), (2). Take $a \in \prod_{t\in T} A_t$ as follows: Fix a $s_a \in S$. $\forall f \in M$, by (1), $f \le \varphi_C(f) \in C$. Since $h_{\varphi_C(f)} : S \to A_{\varphi_C(f)}$ is surjective, $\exists s_f \in S$ such that $h_{\varphi_C(f)}(s_f) = p_{\varphi_C(f)}(x)$. $\forall t \in T$, if $\exists f \in M$ such that $f \le t$, take $p_t(a) = h_t(s_f)$; otherwise, take $p_t(a) = h_t(s_a)$. So we get $a \in A_M \subset A$, and $\forall r \in C$, by (1) and (2), $\exists f = \varphi_C{}^{-1}(r) \in M$, $f \le r$,

$$p_r(a) = h_r(s_f) = h_{\varphi_C(f)}(s_f) = p_{\varphi_C(f)}(x) = p_r(x).$$

(ii) is also proved. □

The relation between the densities of an $L$-fuzzy product space and its coordinate spaces is tightly combined with the property of their value domain lattice just as the following theorem shows. In Theorem **7.1.15** we shall find that in the following theorem, there is no possibility to remove the limitation added on the value domain off, or even only replace it with a weaker one.

**7.1.11 Theorem** *Let $\{(L^{X_t}, \delta_t) : t \in T\}$ be a family of L-fts', $(L^X, \delta)$ their L-fuzzy product space, $1 \in M(L)$, $\kappa \geq \omega$. Then*

(i) *If $dn(\delta_t) \leq \kappa$ for every $t \in T$, $|T| \leq 2^\kappa$, then $dn(\delta) \leq \kappa$.*

(ii) *If $dn(\delta_t) \leq \kappa$ for every $t \in T$, $|T| > 2^\kappa$, then $dn(\delta) \leq |T|$.*

**Proof** (i) $\forall t \in T$, take $C_t \subset M(L^{X_t})$ such that $|C_t| = dn(\delta_t)$, $(\bigvee C_t)^- = \underline{1}$. Clearly, by $1 \in M(L)$, we can assume $\forall e \in C_t$, $ht(e) = 1$. Denote $A_t = \{supp(e) : e \in C_t\}$, then $A_t \subset X_t$, $|A_t| \leq |C_t| \leq dn(\delta_t) \leq \kappa$. By Lemma **7.1.10**, $\exists A \subset \prod_{t\in T} A_t \subset \prod_{t\in T} X_t$ satisfying conditions (i), (ii) in Lemma **7.1.10**. Let $A^* = \{x_1 : x \in A\} \subset M(L^X)$, then $|A^*| = |A| \leq \kappa$. So we need only prove $(\bigvee A^*)^- = \underline{1}$. $\forall z_\lambda \in M(L^X)$, $\forall W \in \mathcal{Q}_\delta(z_\lambda)$, then $\exists F \in [T]^{<\omega}$ such that $\forall t \in F$, $\exists U_t \in \delta_t$, $V = \bigwedge_{t\in F} p_t^{\leftarrow}(U_t) \leq W$, $V \in \mathcal{Q}_\delta(z_\lambda)$. So $\forall t \in F$, $U_t \in \mathcal{Q}_{\delta_t}(p_t(z)_\lambda)$. Since $(\bigvee C_t)^- = \underline{1}$, $U_t \hat{q} \bigvee C_t$. By Proposition **2.3.3** (vi), $\exists {x^t}_1 \in C_t$ such that ${x^t}_1 \lessdot U_t$, $1 \not\leq U_t(x^t)'$. By Lemma **7.1.10**, $\exists a \in A$ such that $\forall t \in F$, $p_t(a) = p_t(x^t)$. Since $1 \not\leq U_t(x^t)'$ for every $t \in F$ and $1 \in M(L)$,

$$1 \not\leq \bigvee_{t\in T} U_t(x^t)' = (\bigwedge_{t\in T} U_t)'(a) = V'(a).$$

Then $a_1 \lessdot V$. Since $a_1 \in A^*$, by the arbitrariness of $z_\lambda \in M(L^X)$, $(\bigvee A^*)^- = \underline{1}$, $dn(\delta) \leq |A^*| \leq \kappa$.

(ii) Take $C_t$ and $A_t$ for every $t \in T$ as in the proof of (i). Fix an ordinary set $S$ such that $|S| = \kappa$, then since $|A_t| \leq \kappa$ for every $t \in T$, there exists a surjective mapping $h_t : S \to A_t$ for every $t \in T$. Let

$$A = \{a \in X : \exists F \in [T]^{<\omega}, \exists s \in S, \forall t \in F, p_t(a) \in A_t, \forall t \in T\backslash F, p_t(a) = h_t(s)\},$$

then by $|T| > 2^\kappa$, $|A| \leq \kappa \cdot |[T]^{<\omega}| = |T|$. Denote $A^* = \{a_1 : a \in A\}$, then $|A^*| = |A| \leq |T|$. Just as the proof of (i), one can easily find $(\bigvee A^*)^- = \underline{1}$. So the conculusion holds. □

**7.1.12 Exercises** Let $\{(L^{X_t}, \delta_t) : t \in T\}$ be a family of $L$-fts', $(L^X, \delta)$ their $L$-fuzzy product space, $\kappa \geq \omega$. Then for every $t \in T$, $dn(\delta_t) \leq dn(\delta)$.

Since the following investigation involves classes, we need to generalize the concept of mapping. Sure, it is just a parallel generalization:

**7.1.13 Definition** Let $\mathcal{A}$, $\mathcal{B}$ be classes. Denote the class of all the ordered pairs $(a, b)$, where $a \in \mathcal{A}$, $b \in \mathcal{B}$, by $\mathcal{A} \times \mathcal{B}$. Similarly define $\mathcal{A}_1 \times \cdots \times \mathcal{A}_n$ for finite number of classes $\mathcal{A}_1, \cdots, \mathcal{A}_n$. If $\mathcal{A}_i = \mathcal{A}$ for $i = 1, \cdots, n$, denote $\mathcal{A}_1 \times \cdots \times \mathcal{A}_n$ by $\mathcal{A}^n$. A subclass $f$ of $\mathcal{A} \times \mathcal{B}$ is called a *general mapping* from $\mathcal{A}$ to $\mathcal{B}$, denoted by $f : \mathcal{A} \to \mathcal{B}$, if for every $a \in \mathcal{A}$, there exists exactly one $b \in \mathcal{B}$ such that $(a, b) \in f$. For every general mapping $f$ from $\mathcal{A}$ to $\mathcal{B}$ and every $a \in \mathcal{A}$, denote $f(a) = b$ if and only if $(a, b) \in f$.

**7.1.14 Definition** Denote the category of all cardinal numbers and all the order preserving mappings among them by **Card**.

Denote the category of all completely distributive lattices and all the complete lattice homomorphisms among them by **CDL**.

**7.1.15 Theorem** *Let $L$ be a F-lattice. Then the following conditions are equivalent:*

(i) $1 \in M(L)$.

(ii) *There exist general mappings*

$$l: Ob(\mathbf{CDL}) \to Ob(\mathbf{Card}), \quad f: Ob(\mathbf{Card})^3 \to Ob(\mathbf{Card})$$

*such that for every $\kappa \geq \omega$, every family $\{(L^{X_t}, \delta_t): t \in T\}$ of L-fts' with property $dn(\delta_t) \leq \kappa$ for every $t \in T$, and their L-fuzzy product space $(L^X, \delta)$, $dn(\delta) \leq f(\kappa, |T|, l(L))$.*

**Proof** (i)$\Longrightarrow$(ii): This has been proved in Theorem **7.1.11**.

(ii)$\Longrightarrow$(i): Suppose $1 \notin M(L)$, then $\exists a, b \in L$ such that $a, b < 1$, $a \vee b = 1$. Take a cardinal number $\kappa \geq \omega$ such that $|M(L)| \leq \kappa$.

We shall show: For every cardinal number $\sigma$, there exist two $L$-fts' $(L^{X_s}, \delta_s)$, $(L^{X_t}, \delta_t)$ such that $dn(\delta_s), dn(\delta_t) \leq \kappa$; but for their $L$-fuzzy product space $(L^X, \delta)$, $dn(\delta) > \sigma$. Then it completes the proof.

Take ordinary set $X_s = \{u\}$ as a singleton, $X_t$ as an ordinary set such that $u \in X_t$ and $|X_t| > \sigma^+$, where $\sigma^+$ is the successor cardinal number of $\sigma$. Denote $X = X_s \times X_t$. Define $\delta_s \subset L^{X_s}$, $\delta_t \subset L^{X_t}$ as follows:

$$\delta_s = \{\underline{0}, u_{a'}, \underline{1}\},$$

$$\delta_t = \{\underline{0}, \underline{1}\} \cup \{u_{b'} \vee \chi_A: A \subset X_s \backslash \{u\}\},$$

then $\delta_s$ and $\delta_t$ are clearly $L$-fuzzy topologies on $X_s$ and $X_t$ respectively. For $C = \{u_\lambda: \lambda \in M(L)\} \subset M(L^{X_s}), M(L^{X_t})$, we have $(\bigvee C)^- = (u_1)^- = \underline{1}$ in both $(L^{X_s}, \delta_s)$ and $(L^{X_t}, \delta_t)$. So $dn(\delta_s), dn(\delta_t) \leq |C| \leq |M(L)| \leq \kappa$.

We now prove the following implying relation:

$$C \subset M(L^X),\ (\bigvee C)^- = \underline{1} \Longrightarrow \{supp(e): e \in C\} \supset X \backslash \{(u, u)\}, \qquad (7.1)$$

then $|C| \geq |X_t \backslash \{u\}| \geq \sigma^+$, it clearly completes the proof.

Suppose $C \subset M(L^X)$ such that $\{supp(e): e \in C\} \not\supset X \backslash \{(u, u)\}$, then $\exists v \in X_t \backslash \{u\}$ such that $supp(e) \neq (u, v)$ for every $e \in C$. By $v \neq u$, we have

$$U = u_{a'} \in \delta_s, \quad V = u_{b'} \vee v_1 \in \delta_t.$$

Since $a \vee b = 1$, $a' \wedge b' = 0$, so by $a < 1$,

$$W = p_s^{\leftarrow}(U) \wedge p_t^{\leftarrow}(V) = (u, v)_{a'},$$

$$W' = (u, v)_a \vee \chi_{X \backslash \{(u,v)\}} < \underline{1}.$$

Since there not exists $e \in C$ such that $supp(e) = (u, v)$, we have $\bigvee C \leq W' \in \delta'$, $(\bigvee C)^- \leq (W')^- = W' < \underline{1}$. So relation (7.1) is true. □

**7.1.16 Remark** (1) The preceding theorem is interesting. It tells us, without some certain lattice theoretical property of the value domain, it is even impossible to find any mathematical result – no matter what method will be used – on some relations, although in the crisp case they all hold.

(2) By Lemma **2.3.28**, the condition "$1 \in M(L)$" in Theorem **7.1.11** and Theorem **7.1.15** can be equivalently replaced by "There exists directed subset $A \subset M(L)$ such that $\bigvee A = 1$." Some authors just used this equivalent condition to state and prove Theorem **7.1.11** (i).[168]

## 7.2 Countability

As well known, in the case of infinity, countable infinity is the simplest one and has many special nice property. So people often particularly investigate this case. In this section, we shall discuss some conclusions about weight, characteristic and density in the case of countability.

**7.2.1 Definition** Let $(L^X, \delta)$ be an $L$-fts. $(L^X, \delta)$ is called *firstly countable* or $C_{\mathrm{I}}$, if $chr(\delta) \le \omega$; called *secondly countable* or $C_{\mathrm{II}}$, if $w(\delta) \le \omega$; called *separable*, if $dn(\delta) \le \omega$.

By the results in the last section, we have the following conclusions:

**7.2.2 Corollary** *Let $(L^X, \delta)$ be a $C_{\mathrm{II}}$ L-fts. Then $(L^X, \delta)$ is both $C_{\mathrm{I}}$ and separable.* □

**7.2.3 Corollary** *Let $(L^X, \delta)$ be an L-fts. Then $(L^X, \delta)$ is $C_{\mathrm{II}}$ if and only if $\delta$ has a countable subbase.* □

**7.2.4 Corollary** *Let $(L^X, \delta)$ be an L-fts. Then*

(i) *$(L^X, \delta)$ is $C_{\mathrm{II}}$ if and only if every L-fuzzy subspace of $(L^X, \delta)$ is $C_{\mathrm{II}}$.*

(ii) *$(L^X, \delta)$ is $C_{\mathrm{I}}$ if and only if every L-fuzzy subspace of $(L^X, \delta)$ is $C_{\mathrm{I}}$.* □

**7.2.5 Corollary** *Let $(L^X, \delta)$, $(L^Y, \mu)$ be L-fts', $f^{\to}: (L^X, \delta) \to (L^Y, \mu)$ an L-fuzzy continuous open surjection. Then*

(i) *If $(L^X, \delta)$ is $C_{\mathrm{II}}$ then so is $(L^Y, \mu)$.*

(ii) *If $(L^X, \delta)$ is $C_{\mathrm{I}}$ then so is $(L^Y, \mu)$.* □

**7.2.6 Corollary** *Let $\{(L^{X_t}, \delta_t) : t \in T\}$ be a countable family of L-fts'. Then*

(i) *If $(L^{X_t}, \delta_t)$ is $C_{\mathrm{II}}$ for every $t \in T$, then so is $\bigoplus_{t\in T}(L^{X_t}, \delta_t)$.*

(ii) *If $(L^{X_t}, \delta_t)$ is $C_{\mathrm{I}}$ for every $t \in T$, then so is $\bigoplus_{t\in T}(L^{X_t}, \delta_t)$.*

(iii) *If $(L^{X_t}, \delta_t)$ is separable for every $t \in T$, then so is $\bigoplus_{t\in T}(L^{X_t}, \delta_t)$.* □

**7.2.7 Corollary** *Let $\{(L^{X_t}, \delta_t) : t \in T\}$ be a countable family of L-fts'. Then*

(i) *If $(L^{X_t}, \delta_t)$ is $C_{\mathrm{II}}$ for every $t \in T$, then so is $\prod_{t\in T}(L^{X_t}, \delta_t)$.*

(ii) *If $\prod_{t\in T}(L^{X_t}, \delta_t)$ is $C_{\mathrm{II}}$, $t \in T$, $(L^{X_t}, \delta_t)$ is stratified, then $(L^{X_t}, \delta_t)$ is $C_{\mathrm{II}}$.*

(iii) *If $(L^{X_t}, \delta_t)$ is $C_{\mathrm{I}}$ for every $t \in T$, then so is $\prod_{t\in T}(L^{X_t}, \delta_t)$.*

(iv) *If $\prod_{t\in T}(L^{X_t}, \delta_t)$ is $C_{\mathrm{I}}$, $t \in T$, $(L^{X_t}, \delta_t)$ is stratified, then $(L^{X_t}, \delta_t)$ is $C_{\mathrm{I}}$.* □

**7.2.8 Corollary** *Let $\{(L^{X_t}, \delta_t) : t \in T\}$ be a family of L-fts', $|T| \le 2^{\omega}$, $1 \in M(L)$. Then $\prod_{t\in T}(L^{X_t}, \delta_t)$ is separable if and only if $(L^{X_t}, \delta_t)$ is separable for every $t \in T$.* □

In Chapter 5, we discussed various properties of net convergence in $L$-fts' in detail, and pointed out there that the notion of net cannot be simplified as the one of sequence in general. But in $C_{\mathrm{I}}$ $L$-fts', we can replace net with simpler sequence in many situations. For this aim, we still formlly define the notion of sequence in $L$-fts':

**7.2.9 Definition** Call every net with index set $D = \mathbf{N}$ a *sequence.*

**7.2.10 Theorem** *Let $(L^X, \delta)$ be a $C_{\mathrm{I}}$ L-fts. Then every molecule in $(L^X, \delta)$ has a countable Q-neighborhood base $\{U_0, U_1, \cdots\}$ such that $U_n \ge U_{n+1}$ for every $n < \omega$.* □

The following results follow from Theorem **7.2.10** and Theorem **5.1.14**:

**7.2.11 Corollary** *Let $(L^X, \delta)$ be a $C_{\mathrm{I}}$ L-fts, $e \in M(L^X)$, $A \in L^X$. Then the following conditions are equivalent:*

(i) $e \le A^-$.

(ii) *$e$ is an adherent point of $A$.*

(iii) *There exists a sequence $S$ in $A$ such that $S \rightarrow e$.*
(iv) *There exists a molecule sequence $S$ in $A$ such that $S \rightarrow e$.* □

By Theorem **7.2.10** and Corollary **5.1.15**, we have

**7.2.12 Corollary** *Let $(L^X, \delta)$ be a $C_{\mathrm{I}}$ L-fts, $A \in L^X$. Then the following conditions are equivalent:*

(i) $A = A^-$.
(ii) *For every sequence $S$ in $A$ and every $e \in M(L^X)$, $S \rightarrow e \Longrightarrow e \leq A$.*
(iii) *For every molecule sequence $S$ in $A$ and every $e \in M(L^X)$, $S \rightarrow e \Longrightarrow e \leq A$.* □

Theorem **7.2.10** and Theorem **5.1.16** imply the following results:

**7.2.13 Corollary** *Let $(L^X, \delta)$ be a $C_{\mathrm{I}}$ L-fts, $A \in L^X$, $e \in M(L^X)$. Then the following conditions are equivalent:*

(i) *$e$ is an accumulation point of $A$.*
(ii) *There exists a sequence $S$ in $A \setminus e$ such that $S \infty e$.*
(iii) *There exists a sequence $S$ in $A \setminus e$ such that $S \rightarrow e$.*
(iv) *There exists a molecule sequence $S$ in $A \setminus e$ such that $S \infty e$.*
(v) *There exists a molecule sequence $S$ in $A \setminus e$ such that $S \rightarrow e$.* □

**7.2.14 Exercise** *Let $(L^X, \delta)$ be a $C_{\mathrm{I}}$ L-fts, $(L^Y, \mu)$ an L-fts', $f^{\rightarrow} : L^X \rightarrow L^Y$ an L-fuzzy mapping. Then the following conditions are equivalent:*

(i) *$f^{\rightarrow}$ is continuous.*
(ii) *For every sequence $S$ in $(L^X, \delta)$ and every $e \in M(L^X)$,*
$$S \infty e \Longrightarrow f^{\rightarrow} \circ S \infty f^{\rightarrow}(e).$$
(iii) *For every sequence $S$ in $(L^X, \delta)$ and every $e \in M(L^X)$,*
$$S \rightarrow e \Longrightarrow f^{\rightarrow} \circ S \rightarrow f^{\rightarrow}(e).$$
(iv) *For every molecule sequence $S$ in $(L^X, \delta)$ and every $e \in M(L^X)$,*
$$S \infty e \Longrightarrow f^{\rightarrow} \circ S \infty f^{\rightarrow}(e).$$
(v) *For every molecule sequence $S$ in $(L^X, \delta)$ and every $e \in M(L^X)$,*
$$S \rightarrow e \Longrightarrow f^{\rightarrow} \circ S \rightarrow f^{\rightarrow}(e).$$

## 7.3 On $L$-valued Weakly Induced Spaces

In weakly induced spaces, the properties related to cardinal numbers can be described by the cardinal number of their background spaces in some way. We discuss these problems in this section.

**7.3.1 Definition** Let $(X, \mathcal{T})$ be an ordinary topological space.

Define the *weight* $w(\mathcal{T})$ of $(X, \mathcal{T})$ by:
$$w(\mathcal{T}) = min\{|\mathcal{B}| : \mathcal{B} \text{ is a base of } \mathcal{T}\}.$$
For a point $x \in X$, define the *local characteristic* $chr_{\mathcal{T}}(x)$ of $x$ in $(X, \mathcal{T})$ by:
$$chr_{\mathcal{T}}(x) = min\{|\mathcal{B}| : \mathcal{B} \text{ is a neighborhood base of } x \text{ in } (X, \mathcal{T})\},$$
and the *characteritic* $chr(\mathcal{T})$ of $(X, \mathcal{T})$ by:
$$chr(\mathcal{T}) = \bigvee\{chr_{\mathcal{T}}(x) : x \in X\}.$$
Define the *density* $dn(\mathcal{T})$ of $(X, \mathcal{T})$ by
$$dn(\mathcal{T}) = min\{|A| : A \subset X, A^- = X\}.$$

**7.3.2 Definition** Let $L$ be a completely distributive lattice. Define the *mass* of $L$, denoted by $ms(L)$, as

$$ms(L) = \min\{|A| : A \subset M(L),\ A \text{ is a join-generating set of } L\}.$$

**7.3.3 Proposition** *Let $(L^X, \delta)$ be an L-fts. Then*

$$dn([\delta]) \leq dn(\delta). \qquad \square$$

**7.3.4 Problem** Let $(L^X, \delta)$ be an $L$-fts.

(1) When does the inequality $w([\delta]) \leq w(\delta)$ hold?

(2) When does the inequality $chr([\delta]) \leq chr(\delta)$ hold?

The proof of the following theorem is left to reader as an exercise:

**7.3.5 Theorem** *Let $(L^X, \delta)$ be a weakly induced L-fts. Then*

(i)

$$w([\delta]) \leq w(\delta) \begin{cases} < \omega, & ms(L)\cdot w([\delta]) < \omega, \\ \leq ms(L)\cdot w([\delta]), & ms(L)\cdot w([\delta]) \geq \omega. \end{cases}$$

(ii)

$$chr([\delta]) \leq chr(\delta) \begin{cases} < \omega, & ms(L)\cdot chr([\delta]) < \omega, \\ \leq ms(L)\cdot chr([\delta]), & ms(L)\cdot chr([\delta]) \geq \omega. \end{cases}$$

(iii)

$$dn([\delta]) \leq dn(\delta) \begin{cases} < \omega, & ms(L)\cdot dn([\delta]) < \omega, \\ \leq ms(L)\cdot dn([\delta]), & ms(L)\cdot dn([\delta]) \geq \omega. \end{cases}$$

In Theorem **7.1.11**, we limited the value domain $L$ with $1 \in M(L)$. Furthermore, we proved in Theorem **7.1.15** that this limitation on $L$ cannot been substituted by any non-equivalent one. But for weakly induced $L$-fts', this limitation can be removed off. However, because of the generality the value domain now, it is impossible to avoid involving its some cardinal number property:

**7.3.6 Definition** Let $L$ be a completely distributive lattice, $a \in L$. Define the *mass* of $a$ in $L$, denoted by $ms_L(a)$, as

$$ms_L(a) = min\{|A| : A \subset M(L),\ \bigvee A = a\}.$$

Since every element in a completely distributive lattice can be represented as a join of molecules, the preceding definition is reasonable. Certainly we have the following conclusion:

**7.3.7 Proposition** *Let $L$ be a completely distributive lattice, $a \in L$. Then*

$$ms_L(a) \leq ms(L). \qquad \square.$$

Then we can prove the following theorem:

**7.3.8 Theorem** *Let $\{(L^{X_t}, \delta_t) : t \in T\}$ be a family of weakly induced L-fts', $(L^X, \delta)$ their product L-fts, $\kappa \geq \omega$. Then*

(i) *If $dn(\delta_t) \leq \kappa$ for every $t \in T$, $|T| \leq 2^\kappa$, then $dn(\delta) \leq \kappa\cdot ms_L(1)$.*

(ii) *If $dn(\delta_t) \leq \kappa$ for every $t \in T$, $|T| > 2^\kappa$, then $dn(\delta) \leq |T|\cdot ms_L(1)$.*

**Proof** (i) $\forall t \in T$, take $C_t \subset M(L^{X_t})$ such that $|C_t| = dn(\delta_t)$, $(\bigvee C_t)^- = \underline{1}$. Denote $A_t = \{supp(e) : e \in C_t\}$, then $A_t \subset X_t$, $|A_t| \leq |C_t| \leq dn(\delta_t) \leq \kappa$. Take $M \subset M(L)$

such that $\bigvee M = 1$, $|M| = ms_L(1)$. By Lemma **7.1.10**, $\exists A \subset \prod_{t\in T} A_t \subset \prod_{t\in T} X_t$ satisfying conditions (i), (ii) in Lemma **7.1.10**. Let $A^* = \{u_\gamma : u \in A, \gamma \in M\} \subset M(L^X)$, then $|A^*| \le |A|\cdot|M| \le \kappa\cdot ms_L(1)$. So we need only prove $(\bigvee A^*)^- = \underline{1}$.

$\forall z_\lambda \in M(L^X)$, $\forall W \in \mathcal{Q}_\delta(z_\lambda)$. Without loss of generality, we can assume $W$ is an element of the canonical base of $(L^X, \delta)$, then $\exists F \in [T]^{<\omega}$ such that $\forall t \in F$, $\exists U_t \in \delta_t$, $W = \bigwedge_{t\in F} p_t^{\leftarrow}(U_t)$. So $\forall t \in F$, $U_t \in \mathcal{Q}_{\delta_t}(p_t(z)_\lambda)$. Since $(L^{X_t}, \delta_t)$ is weakly induced, note that $aA$ for $a \in L$ and $A \subset X$ means $\underline{a} \wedge \chi_A$, by Theorem **4.3.34** (ii),

$$\lambda \not\le U_t'(p_t(z)) = (\bigvee_{a\in L} a(U_t)_{[a]}{}^\circ)'(p_t(z)) = \bigwedge_{a\in L} (a(U_t)_{[a]}{}^\circ)'(p_t(z)).$$

So $\exists a(t) \in L$ such that

$$\lambda \not\le (a(t)(U_t)_{[a(t)]}{}^\circ)'(p_t(z)). \tag{7.2}$$

Then

$$\forall t \in F, \quad \lambda \not\le a(t)'. \tag{7.3}$$

Take $V_t = U_t \wedge \chi_{((U_t)_{[a(t)]})^\circ} \in \delta_t$ for evey $t \in F$, we have

$$\forall t \in F, \quad V_t \le U_t, \tag{7.4}$$

$$\forall t \in F, \quad x \in supp(V_t) \implies V_t(x) = U_t(x) \ge a(t). \tag{7.5}$$

Use inequality (7.2) again, $\forall t \in F$, we have

$$p_t(z) \in ((U_t)_{[a(t)]})^\circ \subset (U_t)_{[a(t)]},$$

$$(V_t)'(p_t(z)) = (U_t)'(p_t(z)) \le a(t)'.$$

So by inequality (7.3), $\lambda \not\le (V_t)'(p_t(z))$ and hence

$$\forall t \in F, \quad V_t \in \mathcal{Q}_{\delta_t}(p_t(z)_\lambda). \tag{7.6}$$

Since $(\bigvee C_t)^- = \underline{1}$, by relation (7.6), $V_t \hat{q} \bigvee C_t$. By Proposition **2.3.3** (vi), $\exists {x^t}_{\lambda_t} \in C_t$ such that ${x^t}_{\lambda_t} \lessdot V_t$. Then $x^t \in supp(V_t)$, by relation (7.5),

$$\forall t \in F, \quad V_t(x^t) \ge a(t). \tag{7.7}$$

By $\lambda \in M(L)$ and inequality (7.3), $\lambda \not\le \bigvee_{t\in F} a(t)'$, so $\bigvee_{t\in F} a(t)' \ne 1$. By $\bigvee M = 1$, $\exists \gamma \in M$ such that $\gamma \not\le \bigvee_{t\in F} a(t)'$. By the way we took $A$, $\exists u \in A$ such that $\forall t \in F$, $p_t(u) = x^t$. Then $u_\gamma \in A^*$, by ineqality (7.7),

$$(\bigwedge_{t\in F} p_t^{\leftarrow}(V_t))'(u) = \bigvee_{t\in F} V_t'(p_t(u)) = \bigvee_{t\in F} V_t'(x^t) \le \bigvee_{t\in F} a(t)'.$$

Since $\gamma \not\le \bigvee_{t\in F} a(t)'$, so $\gamma \not\le (\bigwedge_{t\in F} p_t^{\leftarrow}(V_t))'(u)$, $u_\gamma \lessdot \bigwedge_{t\in F} p_t^{\leftarrow}(V_t)$. By inequality (7.4),

$$u_\gamma \lessdot \bigwedge_{t\in F} p_t^{\leftarrow}(U_t) = W.$$

Since $u_\gamma \in A^*$, by the arbitrariness of $z_\lambda \in M(L^X)$ and $W \in \mathcal{Q}_\delta(z_\lambda)$, $(\bigvee A^*)^- = \underline{1}$.

(ii) Take $M$, $C_t$ and $A_t$ for every $t \in T$ as in the proof of (i). Fix an ordinary set $S$ such that $|S| = \kappa$, then since $|A_t| \le \kappa$ for every $t \in T$, there exists a surjective mapping $h_t : S \to A_t$ for every $t \in T$. Let

$$A = \{u \in X : \exists F \in [T]^{<\omega}, \exists s \in S, \forall t \in F, p_t(u) \in A_t, \forall t \in T\backslash F, p_t(u) = h_t(s)\},$$

then by $|T| > 2^\kappa$, $|A| \le \kappa\cdot|[T]^{<\omega}| = |T|$. Denote $A^* = \{u_\gamma : u \in A, \gamma \in M\}$, then

$$|A^*| \le |A|\cdot|M| \le |T|\cdot ms_L(1).$$

Just as the proof of (i), one can similarly show $(\bigvee A^*)^- = \underline{1}$. So the conculusion holds. □

Then we have immediately the following conclusions. Reader can compare it with Corollary **7.2.8**.

**7.3.9 Corollary** *Let $\{(L^{X_t}, \delta_t) : t \in T\}$ be a family of weakly induced L-fts', $|T| \leq 2^\omega$, $ms_L(1) \leq \omega$. Then $\prod_{t\in T}(L^{X_t}, \delta_t)$ is separable if and only if $(L^{X_t}, \delta_t)$ is separable for every $t \in T$.* □

**7.3.10 Exercise** (1)Use another characterization of *L*-valued S-spaces (other than Theorem **4.3.32**) to prove Theorem **7.3.8**.

(2) Construct a counterexample to show the cardinal number $ms_L(1)$ in Theorem **7.3.8** cannot be removed off.

As direct corollaries of Theorem **7.3.5**, we have immediately the following results on countabilities of weakly induced *L*-fts' and then induced ones:

**7.3.11 Corollary** *Let $(L^X, \delta)$ be a weakly induced L-fts. Then*

(i) *If $(L^X, \delta)$ is $C_{\text{II}}$, then $(X, [\delta])$ is $C_{\text{II}}$.*
(ii) *If $(X, [\delta])$ is $C_{\text{II}}$, $ms(L) \leq \omega$, then $(L^X, \delta)$ is $C_{\text{II}}$.*
(iii) *If $(L^X, \delta)$ is $C_{\text{I}}$, then $(X, [\delta])$ is $C_{\text{I}}$.*
(iv) *If $(X, [\delta])$ is $C_{\text{I}}$, $ms(L) \leq \omega$, then $(L^X, \delta)$ is $C_{\text{I}}$.*
(v) *If $(L^X, \delta)$ is separable, then $(X, [\delta])$ is separable.*
(vi) *If $(X, [\delta])$ is separable, $ms(L) \leq \omega$, then $(L^X, \delta)$ is separable.* □

# Chapter 8

# Separation (I)

On separation, we consider how molecules, or more generally, how points in $L$-fts' can be "separated" by open subsets. Since $L$-fuzzy topological spaces contain a lattice in itself, this kind of problems unavoidably involve the property and the structure of the lattice – the value domain. As a natural result, comparing with the correspondent ones in ordinary topology, these problems have many new characteristics and new colors of lattice theory.

## 8.1 Quasi-$T_0$-, Sub-$T_0$-, $T_0$- and $T_1$-separations

Different from the situation in ordinary topology, besides the "separating degree" of points in background, we must also consider the the same problem for points or molecules in different layers at same time. This make the problems more colorful but also more complicated.

**8.1.1 Definition** Let $(L^X, \delta)$ be an $L$-fts.

$(L^X, \delta)$ is called *quasi-$T_0$*, if for every two distinguished molecules $x_\lambda$ and $x_\gamma$ in $(L^X, \delta)$ with same support point $x$, there exists $U \in \mathcal{Q}_\delta(x_\lambda)$ such that $x_\gamma \not\leq U$, or there exists $V \in \mathcal{Q}_\delta(x_\gamma)$ such that $x_\lambda \not\leq V$.

$(L^X, \delta)$ is called *sub-$T_0$*, if for every two distinguished ordinary points $x$, $y$ in $X$, there exists $\lambda \in M(L)$ such that there exists $U \in \mathcal{Q}_\delta(x_\lambda)$, $y_\lambda \not\leq U$, or there exists $V \in \mathcal{Q}_\delta(y_\lambda)$, $x_\lambda \not\leq V$.

$(L^X, \delta)$ is called $T_0$, if for every two distinguished molecules $x_\lambda$, $y_\gamma$ in $(L^X, \delta)$, there exists $U \in \mathcal{Q}_\delta(x_\lambda)$ such that $y_\gamma \not\leq U$, or there exists $V \in \mathcal{Q}_\delta(y_\gamma)$ such that $x_\lambda \not\leq V$.

Clearly, quasi-$T_0$ separation and sub-$T_0$ separation are in fact respectively the special restrictions of $T_0$ separation in "vertical direction" – in the value domain, and in "horizontal level" – in the layers. We have already known in previous chapters that the structure of the value domain must be considered in many investigation in $L$-fuzzy topology, so these distinguishments are clearly necessary.

**8.1.2 Example** (i) Every stratified $L$-fts is quasi-$T_0$, but clearly is not necessary to be $T_0$ or even only sub-$T_0$.

(ii) For an ordinary set $X$ such that $|X| \geq 2$ and a F-lattice $L$ with $|L| > 2$, take $\delta \subset L^X$ as $\delta = \{\chi_A : A \subset X\}$, then obviously $(L^X, \delta)$ is a sub-$T_0$ $L$-fts, but not $T_0$ or even only quasi-$T_0$.

**8.1.3 Theorem** *Let $(L^X, \delta)$ be an L-fts. Then*

(i) *$(L^X, \delta)$ is quasi-$T_0$ if and only if for every two distinguished molecules $e, d \in M(L^X)$ with same support point, $\mathcal{Q}(e) \neq \mathcal{Q}(d)$.*

(ii) *$(L^X, \delta)$ is quasi-$T_0$ if and only if for every two distinguished molecules $e, d \in M(L^X)$ with same support point, $e \not\leq d^-$ or $d \not\leq e^-$.*

(iii) *$(L^X, \delta)$ is sub-$T_0$ if and only if for every two distinguished points $x, y \in X$, there exists $\lambda \in M(L)$ such that $\mathcal{Q}(x_\lambda) \neq \mathcal{Q}(y_\lambda)$.*

(iv) *$(L^X, \delta)$ is sub-$T_0$ if and only if for every two distinguished points $x, y \in X$, there exists $\lambda \in M(L)$ such that $x_\lambda \not\leq y_\lambda^-$ or $y_\lambda \not\leq x_\lambda^-$.*

(v) *$(L^X, \delta)$ is $T_0$ if and only if for every two distinguished molecules $e, d \in M(L^X)$, $\mathcal{Q}(e) \neq \mathcal{Q}(d)$.*

(vi) *$(L^X, \delta)$ is $T_0$ if and only if for every two distinguished molecules $e, d \in M(L^X)$, $e \not\leq d^-$ or $d \not\leq e^-$.* □

By the relative definitions, we have immediately the following

**8.1.4 Proposition** *Every $T_0$ L-fts is both quasi-$T_0$ and sub-$T_0$.* □

**8.1.5 Proposition** *Every stratified L-fts is quasi-$T_0$.* □

But quasi-$T_0$ + sub-$T_0$ cannot imply $T_0$:

**8.1.6 Example** Take $X = \{x, y\}$, $L$ as the diamond-type lattice $\{0, a, b, 1\}$. Define an order-reversing involution $' : L \to L$ by $a' = a$, $b' = b$, then $L$ with this involution is a F-lattice, $M(L) = \{a, b\}$. Let

$$\delta = \{\underline{0},\ x_a \vee y_b\ , x_b \vee y_a,\ \underline{1}\},$$

then $\delta$ is an $L$-fuzzy topology on $X$.

For elements in $M(L^X) = \{x_a, x_b, y_a, y_b\}$, we have

$$x_a \lessdot x_b \vee y_a \in \delta,\quad x_b \not\lessdot x_b \vee y_a,$$

$$y_a \lessdot x_a \vee y_b \in \delta,\quad y_b \not\lessdot x_a \vee y_b,$$

so $(L^X, \delta)$ is quasi-$T_0$.

For distinguished $x, y \in X$, we have $a \in M(L)$ and

$$x_a \lessdot x_b \vee y_a,\quad y_a \not\lessdot x_b \vee y_a,$$

so $(L^X, \delta)$ is sub-$T_0$.

But for distinguished $x_a, y_b \in M(L^X)$, we have

$$\mathcal{Q}(x_a) = \{x_b \vee y_a,\ \underline{1}\} = \mathcal{Q}(y_b),$$

so $(L^X, \delta)$ is not $T_0$.

**8.1.7 Theorem** *Quasi-$T_0$ property, sub-$T_0$ property and $T_0$ property are hereditary.* □

**8.1.8 Theorem** *Quasi-$T_0$ property, sub-$T_0$ property and $T_0$ property are preserved by L-fuzzy homeomorphisms.* □

**8.1.9 Theorem** *Let $\{(L^{X_t}, \delta_t) : t \in T\}$ be a family of L-fts', $(L^X, \delta)$ their product L-fts. If there exists $s \in T$ such that $(L^{X_s}, \delta_s)$ is quasi-$T_0$, then $(L^X, \delta)$ is quasi-$T_0$.*

**Proof** $\forall x \in X$, $\forall \lambda, \gamma \in M(L)$, $\lambda \neq \gamma$. Since $(L^{X_s}, \delta_s)$ is quasi-$T_0$, without loss of generality, we can assume $\exists U \in \mathcal{Q}_{\delta_s}(p_t(x)_\lambda)$ such that $p_t(x)_\gamma \not\lessdot U$. Then $V = p_t^{\leftarrow}(U) \in \mathcal{Q}_\delta(x_\lambda)$ but $x_\gamma \not\lessdot V$. So $(L^X, \delta)$ is quasi-$T_0$. □

**8.1.10 Corollary** *Quasi-$T_0$ property is multiplicative.* □

**8.1.11 Theorem** *$T_0$ property are multiplicative.*

**Proof** Let $\{(L^{X_t}, \delta_t) : t \in T\}$ be a family of $T_0$ $L$-fts', $(L^X, \delta)$ their $L$-fuzzy product space, $x_\lambda, y_\gamma \in M(L^X)$ are distinguished. Then $x \neq y$ or $\lambda \neq \gamma$. By Theorem **2.1.15** (ii), $\exists t \in T$ such that $p_t(x)_\lambda \neq p_t(y)_\gamma$. Since $(L^{X_t}, \delta_t)$ is $T_0$, without loss of generality, we can assume $\exists U \in \mathcal{Q}_{\delta_t}(p_t(x)_\lambda)$ such that $p_t(y)_\gamma \notin U$. Then $V = p_t^{\leftarrow}(U) \in \mathcal{Q}_\delta(x_\lambda)$ but $y_\gamma \not\lessdot V$. □

But both quasi-$T_0$ and $T_0$ properties are not strongly mutiplicative:

**8.1.12 Example** Take $X = \{x\}$, $Y = \{y\}$, $Z = \{z\}$, $L = \{0,1\}^3$. Recall that $\underline{a}$ for an element $a$ of the value domain always denotes the $a$-layer, take

$$\delta_1 = \{\underline{(0,0,0)}, \underline{(1,0,0)}, \underline{(1,1,1)}\},$$
$$\delta_2 = \{\underline{(0,0,0)}, \underline{(0,1,0)}, \underline{(1,1,1)}\},$$
$$\delta_3 = \{\underline{(0,0,0)}, \underline{(0,0,1)}, \underline{(1,1,1)}\},$$

then $\delta_1$, $\delta_2$ and $\delta_3$ are respectively $L$-fuzzy topologies on $X$, $Y$ and $Z$. Clearly the $L$-fuzzy product topology $\delta$ of $\delta_1$, $\delta_2$ and $\delta_3$ exactly consists of all the layers on $X \times Y \times Z = \{(x,y,z)\}$. Since $X \times Y \times Z$ is a singleton, so the product $L$-fts $(L^{X\times Y\times Z}, \delta)$ is certainly $T_0$; but clearly every one of $(L^X, \delta_1)$, $(L^Y, \delta_2)$ and $(L^Z, \delta_3)$ is not even quasi-$T_0$.

As for sub-$T_0$ property, we have the following stronger result:

**8.1.13 Theorem** *Sut-$T_0$ property is strongly multiplicative.*

**Proof** Suppose $\{(L^{X_t}, \delta_t) : t \in T\}$ is a family of $L$-fts', $(L^X, \delta)$ is their product $L$-fts.

Suppose $(L^{X_t}, \delta_t)$ is sub-$T_0$ for every $t \in T$. Let $x, y \in X$, $x \neq y$, then $\exists s \in T$ such that $p_s(x) \neq p_s(y)$. Since $(L^{X_s}, \delta_s)$ is sub-$T_0$, without loss of generality, we can assume $\exists \lambda \in M(L)$, $\exists U \in \mathcal{Q}_{\delta_s}(p_s(x)_\lambda)$ such that $p_s(y)_\lambda \not\leqslant U$. Then $V = p_s^{\leftarrow}(U) \in \mathcal{Q}_\delta(x_\lambda)$ but $y_\lambda \not\leqslant V$. So $(L^X, \delta)$ is sub-$T_0$.

Suppose $(L^X, \delta)$ is sub-$T_0$, $s \in T$. Let $x^s, y^s \in X_s$, $x^s \neq y^s$. $\forall t \in T \backslash \{s\}$, since $X_t$ is nonempty, $\exists x^t \in X_t$. Take $x, y \in X$ such that

$$p_t(x) = \begin{cases} x^s, & t = s, \\ x^t, & t \in T\backslash\{s\}, \end{cases} \qquad p_t(y) = \begin{cases} y^s, & t = s, \\ x^t, & t \in T\backslash\{s\}, \end{cases}$$

then $x \neq y$. Since $(L^X, \delta)$, we can assume that there exsit $\lambda \in M(L)$ and an element $U$ of the canonical base of $\delta$ such that $U \in \mathcal{B}_\delta(x_\lambda)$, $y_\lambda \not\leqslant U$. Then $\exists F \in [T]^{<\omega}$, $\forall t \in F$, $\exists U_t \in \delta_t$ such that $U = \bigwedge_{t\in T} p_t^{\leftarrow}(U_t)$. Since $x$ and $y$ is different at only their $s$'th coordinates, by $x_\lambda \leqslant U$ but $y_\lambda \not\leqslant U$, $s \in F$. Denote $F_0 = F \backslash \{s\}$, then we have

$$\lambda \not\leq U'(x) = U_s{}'(x^s) \vee \bigvee_{t\in F_0} U_t{}'(x^t),$$
$$\lambda \leq U'(y) = U_s{}'(y^s) \vee \bigvee_{t\in F_0} U_t{}'(x^t).$$

So $\lambda \not\leq U_s{}'(x^s)$, $x^s{}_\lambda \leqslant U_s$. Moreover, $\lambda \not\leq \bigvee_{t\in F_0} U_t{}'(x^t)$. Since $\lambda \in M(L)$, $F$ is finite, by $\lambda \not\leq \bigvee_{t\in F_0} U_t{}'(x^t)$, it must be $\lambda \leq U_s{}'(y^s)$. So $y^s{}_\lambda \not\leqslant U_s$, $(L^{X_s}, \delta_s)$ is sub-$T_0$. □

**8.1.14 Remark** With their style of pointless, Hutton and Reilly introduced a kind of $T_0$ property not involving points.[41] Liu introduced sub-$T_0$ property for the need in embedding theory.[71] Wuyts and Lowen, moreover, gave out a series of extensions of $T_0$ property for various purposes.[179]

**8.1.15 Theorem** *Let $(L^X, \delta)$ be an $L$-fts. Then*

(i) *If $(X, [\delta])$ is $T_0$, then $(L^X, \delta)$ is sub-$T_0$.*

(ii) *If $(L^X, \delta)$ is weakly induced and sub-$T_0$, then $(X, [\delta])$ is $T_0$.*

**Proof** (i) $\forall x, y \in X$, $x \neq y$, then since $(X, [\delta])$ is $T_0$, without loss of generality, we can assume $\exists U \in \mathcal{N}_{[\delta]}(x)$ such that $y \notin U$. Then for every $\lambda \in M(L)$, we have $\chi_U \in \mathcal{Q}_\delta(x_\lambda)$ but $y_\lambda \not\leqslant \chi_U$. So $(L^X, \delta)$ is sub-$T_0$.

(ii) $\forall x, y \in X$, $x \neq y$, then since $(L^X, \delta)$ is sub-$T_0$, without loss of generality, we can assume $\exists \lambda \in M(L)$, $U \in \mathcal{Q}(x_\lambda)$ such that $y_\lambda \not\leq U$. Since $(L^X, \delta)$ is weakly induced, $V = X \backslash U'_{[\lambda]} \in [\delta]$. Then $x \in V$ but $y \notin V$, $(X, [\delta])$ is $T_0$. □

**8.1.16 Theorem** *Let $(L^X, \delta)$ be an induced L-fts. Then the following conditions are equivalent:*

(i) $(L^X, \delta)$ *is* $T_0$.

(ii) $(L^X, \delta)$ *is both quasi-*$T_0$ *and sub-*$T_0$.

(iii) $(L^X, \delta)$ *is sub-*$T_0$.

(iv) $(X, [\delta])$ *is* $T_0$.

**Proof** (i)$\Longrightarrow$(ii)$\Longrightarrow$(iii): Clear.

(iii)$\Longrightarrow$(iv): By Theorem **8.1.15** (ii).

(iv)$\Longrightarrow$(i): Suppose $x_\lambda, y_\gamma \in M(L^X)$ are distinguished. If $x \neq y$, since $(X, [\delta])$ is $T_0$, nothing need to be proved. If $x = y$, then $\lambda \neq \gamma$. By Proposition **8.1.5**, proof is completed. □

By Theorem **4.1.7** and Theorem **8.1.11**, we have the following

**8.1.17 Theorem** *Let $\{(L^{X_t}, \delta_t) : t \in T\}$ be a family of L-fts', $s \in T$, $(L^{X_s}, \delta_s)$ is $T_0$. If $\prod_{t \in T}\{(L^{X_t}, \delta_t) : t \in T\}$ is $T_0$, then $(L^{X_s}, \delta_s)$ is $T_0$.* □

Since $T_0$ property make different molecules in an $L$-fts can be distinguished by the $L$-fuzzy topology and then naturally, by its a base, so it is reasonable to consider the possibility of estimating $|M(L^X)|$ by the weight. Similarly, we can consider the problems of estimating $|X|$ and $|M(L)|$. The following theorems show these points. We only prove one of them, the others are left as exercises.

**8.1.18 Theorem** *Let $(L^X, \delta)$ be a sub-$T_0$ L-fts. Then*

$$|X| \leq 2^{|w(\delta)| \cdot |M(L)|}. \qquad \square$$

**Proof** Suppose $\mathcal{B}$ is a base of $\delta$ such that $|\mathcal{B}| = w(\delta)$. Define a mapping $F : X \to \mathcal{P}(\mathcal{B})^{M(L)}$ as follows:

$$\forall x \in X,\ \forall \lambda \in M(L), \quad F(x)(\lambda) = \mathcal{Q}_\delta(x_\lambda) \cap \mathcal{B}.$$

Suppose $x, y \in X$, $x \neq y$, then $\exists \lambda \in M(L)$ such that $\exists U \in \mathcal{Q}_\delta(x_\lambda)$, $y_\lambda \not\leq U$, or $\exists V \in \mathcal{Q}_\delta(y_\lambda)$, $x_\lambda \not\leq V$. Since $\mathcal{B}$ is a base of $\delta$, we can assume that $U$ or $V$ where is an element of $\mathcal{B}$, then $U \in \mathcal{Q}_\delta(x_\lambda) \cap \mathcal{B}$ but $U \notin \mathcal{Q}_\delta(y_\lambda) \cap \mathcal{B}$, or $V \in \mathcal{Q}_\delta(y_\lambda) \cap \mathcal{B}$ but $V \notin \mathcal{Q}_\delta(x_\lambda) \cap \mathcal{B}$. So $F(x)(\lambda) \neq F(y)(\lambda)$, and hence $F(x) \neq F(y)$. Thus, $F$ is injective, $|X| \leq |\mathcal{P}(\mathcal{B})^{M(L)}| = 2^{|w(\delta)| \cdot |M(L)|}$. □

**8.1.19 Theorem** *Let $(L^X, \delta)$ be a quasi-$T_0$ L-fts. Then*

$$|M(L)| \leq 2^{|w(\delta)|}. \qquad \square$$

**8.1.20 Theorem** *Let $(L^X, \delta)$ be a $T_0$ L-fts. Then*

$$|M(L^X)| \leq 2^{|w(\delta)|}. \qquad \square$$

**8.1.21 Definition** Let $(L^X, \delta)$ be an $L$-fts. $(L^X, \delta)$ is called $T_1$, if for every two distinguished molecules $e$ and $d$ in $(L^X, \delta)$ such that $e \not\leq d$, there exists $U \in \mathcal{Q}_\delta(e)$ such that $d \not\leq U$.

Clearly we have

**8.1.22 Proposition** *Every $T_1$ L-fts is $T_0$.* □

**8.1.23 Theorem** *Let $(L^X, \delta)$ be an L-fts. Then $(L^X, \delta)$ is $T_1$ if and only if every molecule in $(L^X, \delta)$ is a closed subset.*

**Proof** (Necessity) $\forall e \in M(L^X)$. Suppose $d \in M(L^X)$ such that $d \leq e^-$. If $d \not\leq e$, then since $(L^X, \delta)$ is $T_1$, $\exists U \in \mathcal{Q}(d)$ such that $e \not\leq U$. But this means $d$ is not an adherent point of $e$, $d \not\leq e^-$, a contradiction.

(Sufficiency) $\forall e, d \in M(L^X)$, $e \not\leq d$. Since $d$ is closed in $(L^X, \delta)$, $e \not\leq d^-$, $e$ is not an adherent point of $d$. So $\exists U \in \mathcal{Q}(e)$ such that $d \not\leq U$, $(L^X, \delta)$ is $T_1$. □

In fact, some authors just use the preceding equivalent property of $T_1$ property as its definition. Strengthening this property, we get the following

**8.1.24 Definition** Let $(L^X, \delta)$ be an $L$-fts. $(L^X, \delta)$ is called *typically* $T_1$ or *t*-$T_1$ for short, if every $L$-fuzzy point in $(L^X, \delta)$ is a closed subset.

**8.1.25 Theorem** *Let $(L^X, \delta)$ be an L-fts. Then $(L^X, \delta)$ is t-$T_1$ if and only if for every $p \in Pt(L^X)$ and every $e \in M(L^X)$ such that $e \not\leq p$, there exists $U \in \mathcal{Q}(e)$ such that $p \not\leq U$.*

**Proof** (Necessity) Suppose $p \in Pt(L^X)$, $e \in M(L^X)$, $e \not\leq p$. Then by the t-$T_1$ property of $(L^X, \delta)$, $\exists U \in \mathcal{Q}(e)$ such that $p \not\leq U$. So $U \neg \hat{q} p$, $e$ is not an adherent point of $p$, $e \not\leq p^-$. This means $p^- = p$, $p$ is a closed subset in $(L^X, \delta)$.

(Sufficiency) Suppose $p \in Pt(L^X)$, $e \in M(L^X)$, $e \not\leq p$. Since $p$ is a closed subset in $(L^X, \delta)$, $e \not\leq p = p^-$, $e$ is not an adherent point of $p$. So $\exists U \in \mathcal{Q}(e)$ such that $U \neg \hat{q} p$, $p \not\leq U$. Hence $(L^X, \delta)$ is t-$T_1$. □

S-$T_1$ property is obviously stronger than $T_1$ property. But the converse is not true. The following example constructed in [168] shows this point:

**8.1.26 Example** Take $X = [0,1]$, $L = [0,1]^{[0,1]}$. Define $A \in L^X$ is a closed subset if $A = \underline{0}$, or $A = \underline{1}$, or $|supp(A)| < \omega$ and $\forall x \in supp(A)$, $|supp(A(x))| < \omega$ (considering $A(x) \in L = [0,1]^{[0,1]}$ as a $[0,1]$-fuzzy subset on $[0,1]$). Let $\delta'$ denote all this kind of closed subsets, then $\delta'$ is closed to finite joins and arbitrary meets, so $\delta$ is a $L$-fuzzy topology on $X$. $\forall \lambda \in M(L) = M([0,1]^{[0,1]})$, $\lambda$ is a $[0,1]$-fuzzy point on $[0,1]$, so every $x_\lambda \in M(L^X)$ is a closed subset in $(L^X, \delta)$, $(L^X, \delta)$ is $T_1$. But for every $x \in X$ and the largest element $1_L = \chi_{[0,1]}$ of $L$, $x_{1_L}$ is not a closed subset in $(L^X, \delta)$, so $(L^X, \delta)$ is not t-$T_1$.

But, however, in $L$-valued induced spaces, these two kinds of $T_1$ properties are equivalent:

**8.1.27 Theorem** *Let $(L^X, \delta)$ be an L-fts. Then*

(i) $(L^X, \delta)$ *is t-$T_1$* $\Longrightarrow$ $(L^X, \delta)$ *is $T_1$.*

(ii) $(L^X, \delta)$ *is weakly induced and $T_1$* $\Longrightarrow$ $(X, [\delta])$ *is $T_1$.*

(iii) $(L^X, \delta)$ *is stratified, $(X, [\delta])$ is $T_1$* $\Longrightarrow$ $(L^X, \delta)$ *is t-$T_1$.* □

**8.1.28 Corollary** *Let $(L^X, \delta)$ be an induced L-fts. Then the following conditions are equivalent:*

(i) $(L^X, \delta)$ *is t-$T_1$.*

(ii) $(L^X, \delta)$ *is $T_1$.*

(iii) $(X, [\delta])$ *is $T_1$.* □

The proofs of the following theorems are left as exercises:

**8.1.29 Theorem** *t-$T_1$ property and $T_1$ property are hereditary.* □

**8.1.30 Theorem** *t-$T_1$ property and $T_1$ property are preserved by L-fuzzy homeomorphisms.* □

**8.1.31 Theorem** *t-$T_1$ property and $T_1$ property are multiplicative.* □

**8.1.32 Theorem** *Let $\{(L^{X_t}, \delta_t) : t \in T\}$ be a family of L-fts', $s \in T$, $(L^{X_s}, \delta_s)$ is stratified. Then*

(i) *If $\prod_{t \in T}(L^{X_t}, \delta_t)$ is t-$T_1$, then $(L^{X_s}, \delta_s)$ is t-$T_1$.*

(ii) *If $\prod_{t \in T}(L^{X_t}, \delta_t)$ is $T_1$, then $(L^{X_s}, \delta_s)$ is $T_1$.* □

## 8.2 $T_2$-separation

For convenience, just as "disjoint" in ordinary sets, we introduce the following terminology to shorten statements:

**8.2.1 Definition** Let $L$ be a lattice. $a, b \in L$ is called *zero-meet*, if $a \wedge b = 0$.

Since every $L$-fuzzy space $L^X$ is also a lattice, so every two $A, B \in L^X$ is called *zero-meet* if $A \wedge B = \underline{0}$.

**8.2.2 Definition** Let $(L^X, \delta)$ be an $L$-fts, $\alpha \in M(L)$.

$(L^X, \delta)$ is called $T_2$, if for every two molecules $x_\lambda$ and $y_\gamma$ in $(L^X, \delta)$ with distinguished support points $x \neq y$, there exist $U \in \mathcal{Q}_\delta(x_\lambda)$, $V \in \mathcal{Q}_\delta(y_\gamma)$ such that $U \wedge V = \underline{0}$.

$(L^X, \delta)$ is called *strongly* $T_2$, or called *s*-$T_2$ for short,if for every two molecules $x_\lambda$ and $y_\gamma$ in $(L^X, \delta)$ with distinguished supporing points $x \neq y$, there exist $U \in \mathcal{Q}_\delta(x_\lambda)$, $V \in \mathcal{Q}_\delta(y_\gamma)$ such that $U_{(0)} \cap V_{(0)} = \emptyset$.

$(L^X, \delta)$ is called *purely* $T_2$, or called *p*-$T_2$ for short, if for every two zero-meet molecules $x_\lambda$ and $y_\gamma$ in $(L^X, \delta)$, there exist $U \in \mathcal{Q}_\delta(x_\lambda)$, $V \in \mathcal{Q}_\gamma(y_\gamma)$ such that $U \wedge V = \underline{0}$.

$(L^X, \delta)$ is called $\alpha$-$T_2$, if for every distinguished points $x, y \in X$, there exist $U \in \mathcal{Q}(x_\alpha)$, $V \in \mathcal{Q}(y_\alpha)$ such that $U \wedge V = \underline{0}$.

$(L^X, \delta)$ is called *level*-$T_2$, if $(L^X, \delta)$ is $\alpha$-$T_2$ for every $\alpha \in M(L)$.

**8.2.3 Proposition** *Let $(L^X, \delta)$ be an L-fts, $\alpha \in M(L)$. Then*

(i) *s-$T_2$ property $\Longrightarrow$ $T_2$ property.*

(ii) *p-$T_0$ property $\Longrightarrow$ $T_2$ property.*

(iii) *$T_2$ property $\Longrightarrow$ level-$T_2$ property $\Longrightarrow$ $\alpha$-$T_2$ property.* □

Easy to find that the converse of the preceding implications are not true. Moreover, each one of s-$T_2$ property and p-$T_2$ property cannot also imply the other (as two exercises, reader can make correspondent counterexamples). But for the case $1 \in M(L)$, we have the following conclusion:

**8.2.4 Lemma** *Let $(L^X, \delta)$ be an L-fts, $1 \in M(L)$. Then for every two $A, B \in L^X$,*

$$supp(A) \cap supp(B) \iff A \wedge B = \underline{0} \implies A \neg \hat{q} B.$$

**Proof** "$supp(A) \cap supp(B) \iff A \wedge B = \underline{0}$" is clear. We prove "$A \wedge B = \underline{0} \implies A \neg \hat{q} B$". If $A \hat{q} B$, then $\exists x \in X$ such that $A(x) \not\leq B'(x)$. So $A(x) > 0$, $B'(x) < 1$.

Then $B(x) > 0$. By $1 \in M(L)$, $A(x) \wedge B(x) > 0$, $A \wedge B \neq \underline{0}$. So the implication is true. □

**8.2.5 Proposition** *Let $(L^X, \delta)$ be an L-fts, $1 \in M(L)$. Then the following conditions are equivalent:*

(i) *$(L^X, \delta)$ is s-$T_2$.*

(ii) *$(L^X, \delta)$ is p-$T_2$.*

(iii) *$(L^X, \delta)$ is $T_2$.* □

The following results are left as exercises:

**8.2.6 Theorem** *S-$T_2$ property, p-$T_2$ property, $T_2$ property, level-$T_2$ and $\alpha$-$T_2$ property for every $\alpha \in M(L)$ are hereditary.* □

**8.2.7 Theorem** *S-$T_2$ property, p-$T_2$ property, $T_2$ property, level-$T_2$ and $\alpha$-$T_2$ property for every $\alpha \in M(L)$ are preserved by L-fuzzy homeomorphisms.* □

**8.2.8 Theorem** *S-$T_2$ property, p-$T_2$ property, $T_2$ property, level-$T_2$ and $\alpha$-$T_2$ property for every $\alpha \in M(L)$ are multiplicative.* □

**8.2.9 Theorem** *Let $\{(L^{X_t}, \delta_t) : t \in T\}$ be a family of L-fts', $s \in T$, $(L^{X_s}, \delta_s)$ is stratified. Then*

(i) *If $\prod_{t \in T}(L^{X_t}, \delta_t)$ is s-$T_2$, then $(L^{X_s}, \delta_s)$ is s-$T_2$.*

(ii) *If $\prod_{t \in T}(L^{X_t}, \delta_t)$ is p-$T_2$, then $(L^{X_s}, \delta_s)$ is p-$T_2$.*

(iii) *If $\prod_{t \in T}(L^{X_t}, \delta_t)$ is $T_2$, then $(L^{X_s}, \delta_s)$ is $T_2$.*

(iv) *If $\prod_{t \in T}(L^{X_t}, \delta_t)$ is level-$T_2$, then $(L^{X_s}, \delta_s)$ is level-$T_2$.*

(v) *If $\alpha \in M(L)$, $\prod_{t \in T}(L^{X_t}, \delta_t)$ is $\alpha$-$T_2$, then $(L^{X_s}, \delta_s)$ is $\alpha$-$T_2$.* □

**8.2.10 Proposition** *Every level-$T_2$ L-fts is sub-$T_0$.* □

**8.2.11 Remark** Since $T_2$ property only "separates" molecules in "horizontal level" – molecules with distinguished support points, but provides nothing for the difference in "vertical direction" – even the difference of zero-meet elements in the value domain. So it is natural that $T_2$ property cannot imply $T_0$ even only quasi-$T_0$. In fact, every $L$-fts with a $T_2$ background space is certainly $T_2$, but clearly need not be quasi-$T_0$.

Why do not define a $T_2$ property implying $T_1$ and so on, just as in ordinary topology? The key is that an $L$-fuzzy topology is not only a topological object, it is in fact a fusion of a topological structure and an ordered structure – a lattice. Elements in an $L$-fuzzy space have had some ordered relations already before defining an $L$-fuzzy topology for them, then a topological notion introduced is certainly affected by them. If define the notion of $T_2$ property completely as in ordinary topology, many unnatural results will emerge.

**8.2.12 Theorem** *Let $(L^X, \delta)$ be an L-fts. Then the following conditions are equivalent:*

(i) *$(L^X, \delta)$ is $T_2$.*

(ii) *For every molecule net $S$ in $(L^X, \delta)$, $|supp(\lim S)| \leq 1$.*

(iii) *For every proper filter $\mathcal{F}$ in $(L^X, \delta)$, $|supp(\lim \mathcal{F})| \leq 1$.*

**Proof** (i)$\Longrightarrow$(i): Suppose $S = \{S(n), n \in D\}$ is a molecule net in $(L^X, \delta)$, $x_a, y_b \leq \lim S$, $x \neq y$. Since $(L^X, \delta)$ is $T_2$, $\exists U \in \mathcal{Q}(x_a)$, $\exists V \in \mathcal{Q}(y_b)$ such that $U \wedge V = \underline{0}$. By Theorem **5.1.3** (vi), both $x_a$ and $y_b$ are limit points of $S$. Then $\exists n_1, n_2 \in D$ such that $\forall m, n \in D$ with $m \geq n_1$, $n \geq n_2$, we have $S(m) \lessdot U$, $S(n) \lessdot V$. Since $D$ is a directed set, $\exists n_0 \in D$ such that $n_0 \geq n_1, n_2$. So for every $n \in D$ with $n \geq n_0$ we have $S(n) \lessdot U, V$, $ht(S(n)) \not\leq U'(supp(S(n))), V'(supp(S(n)))$. Since $S$ is a molecule net, $ht(S(n)) \in M(L)$, we have

$$ht(S(n)) \not\leq U'(supp(S(n))) \vee V'(supp(S(n))) = (U \wedge V)'(supp(S(n))) = \underline{1}(supp(S(n))) = 1,$$

this is a contradiction.

(ii)$\Longrightarrow$(iii): Suppose $\mathcal{F}$ a proper filter in $(L^X, \delta)$, then $S(\mathcal{F})$ a molecule net in $(L^X, \delta)$. By Theorem **5.2.7** (ii), $\lim \mathcal{F} = \lim S(\mathcal{F})$, so by (i),

$$|supp(\lim \mathcal{F})| = |supp(\lim S(\mathcal{F}))| \leq 1.$$

(iii)$\Longrightarrow$(i): Suppose (i) is not true, then $\exists x_\lambda, y_\gamma \in M(L^X)$ such that $x \neq y$, $\forall U \in \mathcal{Q}(x_\lambda)$, $\forall V \in \mathcal{Q}(y_\gamma)$, $U \wedge V \neq \underline{0}$. Take

$$\mathcal{F} = \{A \in L^X : \exists U \in \mathcal{Q}(x_\lambda), \exists V \in \mathcal{Q}(y_\gamma), A \geq U \wedge V\}.$$

Then $\mathcal{F}$ is an upper set in $L^X$, $\forall A \in \mathcal{F}$, $A \neq \underline{0}$. By Theorem **2.3.12** (ii), $\mathcal{F}$ is a proper filter in $(L^X, \delta)$. Clearly $\mathcal{Q}(x_\lambda), \mathcal{Q}(y_\gamma) \subset \mathcal{F}$, so $\mathcal{F} \to x_\lambda, y_\gamma$. But $x \neq y$, contradicts with the condition (iii). □

**8.2.13 Definition** Let $L^X$ be an $L$-fuzzy space, $a \in L$, $S = \{S(n), n \in D\}$ a net in $L^X$. $S$ is called an *$a$-net*, if $ht(S(n)) = a$ for every $n \in D$.

Similar to the case of $T_2$ property, for s-$T_2$ property, p-$T_2$ property and $\alpha$-$T_2$ property, we have the following characterizations. Their proofs are left as exercises.

**8.2.14 Definition** Let $L^X$ be an $L$-fuzzy space. A filter $\mathcal{F}$ in $L^X$ is called *crisp*, if every element of $\mathcal{F}$ is a crisp subset in $L^X$.

**8.2.15 Theorem** *Let $(L^X, \delta)$ be an L-fts. Then the following conditions are equivalent:*

(i) *$(L^X, \delta)$ is s-$T_2$.*
(ii) *For every 1-net $S$ in $(L^X, \delta)$, $|supp(\lim S)| \leq 1$.*
(iii) *For every proper crisp filter $\mathcal{F}$ in $(L^X, \delta)$, there exists at most one point $x \in X$ such that $\{U_{(0)} : U \in \mathcal{Q}(x_1)\} \subset \mathcal{F}$.* □

**8.2.16 Theorem** *Let $(L^X, \delta)$ be an L-fts. Then the following conditions are equivalent:*

(i) *$(L^X, \delta)$ is p-$T_2$.*
(ii) *Every molecule net $S$ in $(L^X, \delta)$ has no zero-meet limit points.*
(iii) *Every proper filter $\mathcal{F}$ in $(L^X, \delta)$ has no zero-meet limit points.* □

**8.2.17 Theorem** *Let $(L^X, \delta)$ be an L-fts, $\alpha \in M(L)$. Then the following conditions are equivalent:*

(i) *$(L^X, \delta)$ is $\alpha$-$T_2$.*
(ii) *Every molecule net $S$ in $(L^X, \delta)$ has at most one limit point with height $\alpha$.*
(iii) *Every proper filter $\mathcal{F}$ in $(L^X, \delta)$ has at most one limit point with height $\alpha$.* □

Level-$T_2$ has another "geometrical" characterization as follows:

**8.2.18 Definition** Let $(L^X, \delta)$ be an $L$-fts. Call the $L$-fuzzy subset $\Delta_L(X) \in L^{X \times X}$ defined as follows by the *diagonal* of $(L^X, \delta)^2 = (L^X, \delta) \times (L^X, \delta)$:

$$\forall (x, y) \in X \times X, \quad \Delta_L(X)(x, y) = \begin{cases} 1, & x = y, \\ 0, & x \neq y. \end{cases}$$

**8.2.19 Theorem** *Let $(L^X, \delta)$ be an L-fts. Then $(L^X, \delta)$ is level-$T_2$ if and only if the diagonal of $(L^X, \delta)^2$ is a closed subset of $(L^X, \delta)^2$.*

**Proof** (Necessity) Suppose $(x, y)_\lambda \in M(L^{X\times X})$, $(x, y)_\lambda \not\leq \Delta_L(X)$. We try to prove $(x, y)_\lambda \not\leq \Delta_L(X)^-$, then $\Delta_L(X)$ is closed in $(L^X, \delta)^2$. Since $(x, y)_\lambda \not\leq \Delta_L(X)$, $x \neq y$. By the level-$T_2$ property of $(L^X, \delta)$, $\exists U \in \mathcal{Q}(x_\lambda)$, $\exists V \in \mathcal{Q}(y_\lambda)$ such that $U \wedge V = \underline{0}$. Take $W = p_1{}^{\leftarrow}(U) \wedge p_2{}^{\leftarrow}(V)$, then $W$ is open in $(L^X, \delta)^2$ and by $U \in \mathcal{Q}(x_\lambda)$, $V \in \mathcal{Q}(y_\lambda)$, $\lambda \in M(L)$, we have

$$\lambda \not\leq U'(x) \vee V'(y) = W'(x, y),$$

$$W \in \mathcal{Q}((x, y)_\lambda).$$

$\forall (x, x) \in X \times X$, since

$$W(x, x) = p_1{}^{\leftarrow}(U)(x, x) \wedge p_2{}^{\leftarrow}(V)(x, x) = U(x) \wedge V(x) = 0,$$

so $\forall (u, v) \in supp(W)$, $u \neq v$, $W \leq \Delta_L(X)'$, $W \neg\hat{q}\, \Delta_L(X)$. Hence $(x, y)_\lambda$ is not an adherent point of $\Delta_L(X)$, $(x, y)_\lambda \not\leq \Delta_L(X)^-$, $\Delta_L(X)$ is closed.

(ii)$\Longrightarrow$(i): Suppose $x, y \in X$, $x \neq y$, $\alpha \in M(L)$, denote $W = \Delta_L(X)'$. Since $\Delta_L(X)$ is closed, $W$ is open. Since $x \neq y$, $(x, y) \in supp(W)$. Since $W$ is crisp, $(x, y)_\alpha \lessdot W$. Then $\exists U, V \in \delta$ such that

$$(x, y)_\alpha \lessdot p_1{}^{\leftarrow}(U) \wedge p_2{}^{\leftarrow}(V) \leq W.$$

So $U \in \mathcal{Q}(x_\alpha)$, $V \in \mathcal{Q}(y_\alpha)$. $\forall x \in X$,

$$(U \wedge V)(x) = U(x) \wedge V(x) = p_1{}^{\leftarrow}(U)(x, x) \wedge p_2{}^{\leftarrow}(V)(x, x)$$
$$= (p_1{}^{\leftarrow}(U) \wedge p_2{}^{\leftarrow}(V))(x, x) \leq W(x, x) = 0,$$

so $U \wedge V = \underline{0}$. □

**8.2.20 Theorem** *Let $(L^X, \delta)$ be a weakly induced L-fts. Then the following conditions are equivalent:*

(i) *$(L^X, \delta)$ is s-$T_2$.*
(ii) *$(L^X, \delta)$ is $T_2$.*
(iii) *$(L^X, \delta)$ is level-$T_2$.*
(iv) *There exists $\alpha \in M(L)$ such tht $(L^X, \delta)$ is $\alpha$-$T_2$.*
(v) *$(X, [\delta])$ is $T_2$.*

**Proof** (i)$\Longrightarrow$(ii)$\Longrightarrow$(iii) $\Longrightarrow$(iv): Clear.

(iv)$\Longrightarrow$(v): Suppose $\alpha \in M(L)$ such that $(L^X, \delta)$ is $\alpha$-$T_2$, $x_\lambda, y_\gamma \in M(L^X)$, $x \neq y$. By $\alpha$-$T_2$ property, $\exists U \in \mathcal{Q}(x_\alpha)$, $\exists V \in \mathcal{Q}(y_\alpha)$ such that $U \wedge V = \underline{0}$. Since $\alpha \in M(L)$, by Proposition **4.2.4** (ii), $U_{(\alpha)} \cap V_{(\alpha)} = \underline{0}_{(\alpha)} = \emptyset$. Since $(L^X, \delta)$ is weakly induced, $U_{(\alpha)}, V_{(\alpha)} \in [\delta]$. So $(X, [\delta])$ is $T_2$.

(v)$\Longrightarrow$(i): Suppose $x_\lambda, y_\gamma \in M(L^X)$, $x \neq y$. Since $(X, [\delta])$ is $T_2$, there exist $W_1, W_2 \in [\delta]$ such that $x \in W_1$, $y \in W_2$, $W_1 \cap W_2 = \emptyset$. Take $U = \chi_{W_1}$, $V = \chi_{W_2}$, then $U \in \mathcal{Q}_\delta(x_\lambda)$, $V \in \mathcal{Q}_\delta(y_\gamma)$, $U_{(0)} \cap V_{(0)} = W_1 \cap W_2 = \emptyset$, $(L^X, \delta)$ is s-$T_2$. □

Combining Proposition **8.2.5**, the following conclusions are natural:

**8.2.21 Corollary** *Let $(L^X, \delta)$ is a weakly induced L-fts. Then the following conditions are equivalent:*

(i) *$(L^X, \delta)$ is s-$T_2$.*
(ii) *$(L^X, \delta)$ is $T_2$.*
(iii) *$(L^X, \delta)$ is level-$T_2$.*
(iv) *There exists $\alpha \in M(L)$ such that $(L^X, \delta)$ is $\alpha$-$T_2$.*

(v) $(X, [\delta])$ *is* $T_2$.

*Moreover, if* $1 \in M(L)$, *then the following condition is equivalent to all conditions listed above:*

(vi) $(L^X, \delta)$ *is p-*$T_2$. □

**8.2.22 Theorem** *Let* $(L^X, \delta)$ *be an L-fts,* $\alpha, 1 \in M(L)$, $(L^X, \delta)$ *is* $\alpha$-$T_2$. *Then*

$$|X| \leq 2^{2^{dn(\delta)}}.$$

**Proof** Suppose $A \subset M(L^X)$ such that $(\bigvee A)^- = \underline{1}$, $|A| = dn(\delta)$, denote

$$D = \{z_1 : \exists \xi \in M(L),\ z_\xi \in A\},$$

then $|D| \leq |A| = dn(\delta)$. $\forall x \in X$, $\forall U \in \mathcal{Q}(x_\alpha)$, take

$$W(x, U) = \{z_1 \in D : z \in supp(U)\},$$

$$F : X \to \mathcal{P}(\mathcal{P}(D)), \quad F(x) = \{W(x, U) : U \in \mathcal{Q}(x_\alpha)\}.$$

We shall prove that $F$ is injective, then $|X| \leq |\mathcal{P}(\mathcal{P}(D))| = 2^{2^D} \leq 2^{2^{dn(\delta)}}$. For this aim, we need only prove that for every two distinguished $x, y \in X$, the following two formulas hold:

$$x_\alpha \leq \bigwedge\{(\bigvee W(x, U))^- : U \in \mathcal{Q}(x_\alpha)\}, \tag{8.1}$$

$$y_\alpha \not\leq \bigwedge\{(\bigvee W(x, U))^- : U \in \mathcal{Q}(x_\alpha)\}, \tag{8.2}$$

since then we have $F(x) \neq F(y)$. $\forall x, y \in X$, $x \neq y$. $\forall U, V \in \mathcal{Q}(x_\alpha)$. Since $\alpha \in M(L)$, by Theorem **2.3.12** (ii), $U \wedge V \in \mathcal{Q}(x_\alpha)$. By $(\bigvee A)^- = \underline{1}$, $x_\alpha$ is an adherent point of $\bigvee A$, $\exists z_\xi \in A$ such that $z_\xi \lessdot U \wedge V$ and hence $z \in supp(U)$, $z_1 \lessdot V$. Then $z_1 \in W(x, U)$, $V \hat{q} \bigvee W(x, U)$. By the arbitrariness of $V \in \mathcal{Q}(x_\alpha)$, $x_\alpha \leq (\bigvee W(x, U))^-$, inequality (8.1) holds. Since $x \neq y$, by the $\alpha$-$T_2$ property of $(L^X, \delta)$, $\exists U \in \mathcal{Q}(x_\alpha)$, $\exists V \in \mathcal{Q}(y_\alpha)$ such that $U \wedge V = \underline{0}$. By $1 \in M(L)$ and Lemma **8.2.4**, $\forall z_1 \in W(x, U)$, $z \notin supp(V)$, $V \wedge \bigvee W(x, U) = \underline{0}$. Apply Lemma **8.2.4** again, $V \neg\hat{q} \bigvee W(x, U)$. So $y_\alpha \not\leq (\bigvee W(x, U))^-$, inequality (8.2) is true. The proof is completed. □

**8.2.23 Theorem** *Let* $(L^X, \delta)$ *be an L-fts,* $\alpha, 1 \in M(L)$, $(L^X, \delta)$ *is* $\alpha$-$T_2$. *Then*

$$|X| \leq dn(\delta)^{chr(\delta)}.$$

**Proof** If $chr(\delta) < \omega$, then $\forall x \in X$, $\mathcal{Q}(x_\alpha)$ has a finite base $\mathcal{B}_x$. Since $\alpha \in M(L)$, by Theorem **2.3.12** (ii), $\bigwedge \mathcal{B}_x \in \mathcal{Q}(x_\alpha)$ and by the $\alpha$-$T_2$ property, clearly, for every two distinguished points $x, y \in X$, $(\bigwedge \mathcal{B}_x) \wedge (\bigwedge \mathcal{B}_y) = \underline{0}$. Since $1 \in M(L)$, by Lemma **8.2.4**, $supp(\bigwedge \mathcal{B}_x) = x$ for every $x \in X$. So for every $A \subset M(L^X)$ such that $(\bigvee A)^- = \underline{1}$, we have $|X| = |supp(\bigvee A)|$, and hence $|X| \leq dn(\delta)$, the conclusion holds.

Suppose $chr(\delta) \geq \omega$, $A \subset M(L^X)$ such that $(\bigvee A)^- = \underline{1}$, $|A| = dn(\delta)$. Denote

$$D = \{z_1 : \exists \xi \in M(L),\ z_\xi \in A\},$$

then $|D| \leq |A| = dn(\delta)$. $\forall x \in X$, suppose $\mathcal{B}_x$ is a Q-neighborhood base of $\mathcal{Q}(x_\alpha)$ such that $|\mathcal{B}_x| \leq chr(\delta)$. $\forall U, V \in \mathcal{B}_x$, since $\alpha \in M(L)$, by Theorem **2.3.12** (ii), $U \wedge V \in \mathcal{Q}(x_\alpha)$. By $(\bigvee A)^- = \underline{1}$, $\exists z(U, V) \in X$, $\xi \in M(L)$ such that $z(U, V)_\xi \in A$, $z(U, V)_\xi \lessdot U \wedge V$. Then $z(U, V)_1 \in D$. For every $U \in \mathcal{B}_x$, take

$$W(x, U) = \{z(U, V)_1 : V \in \mathcal{B}_x\},$$

$$F : X \to \mathcal{P}(\mathcal{P}(D)), \quad F(x) = \{W(x, U) : U \in \mathcal{B}_x\}.$$

Since $\forall x \in X$, $\forall U \in \mathcal{B}_x$, $|W(x, U)| \leq |\mathcal{B}_x| \leq chr(\delta)$, so $W(x, U) \subset D$ can considered as an image of a mapping from $chr(\delta)$ to $D$, and hence denote

$$\mathcal{A} = \{W(x,U) : x \in X,\ U \in \mathcal{B}_x\}$$

we have

$$|\mathcal{A}| \le |D^{chr(\delta)}| \le dn(\delta)^{chr(\delta)}.$$

Since $|F(x)| \le |\mathcal{B}_x| \le chr(\delta)$ for every $x \in X$, we can considered every $F(x)$ as an image of a mapping from $chr(\delta)$ to $\mathcal{A}$. So by $chr(\delta) \ge \omega$,

$$|img(F)| \le |\mathcal{A}^{chr(\delta)}| \le ((\delta)^{chr(\delta)})^{chr(\delta)} = dn(\delta)^{chr(\delta)}.$$

Therefore, if we can show that $F$ is an injective mapping, then the proof will be completed.

$F$ is certainly a mapping. To prove it is injective, we need only prove the following two formulas hold for every two distinguished points $x, y \in X$:

$$x_\alpha \le \bigwedge\{(\bigvee W(x,U))^- : U \in \mathcal{B}_x\}, \tag{8.3}$$

$$y_\alpha \not\le \bigwedge\{(\bigvee W(x,U))^- : U \in \mathcal{B}_x\}, \tag{8.4}$$

since if $F(x) = F(y)$ then equality (8.4) will generate a contradiction in the case that $x$ is replaced by $y$ in inequality (8.3). Suppose $U \in \mathcal{B}_x$. $\forall V \in \mathcal{Q}(x_\alpha)$, since $\mathcal{B}_x$ is a base of $\mathcal{Q}(x_\alpha)$, $\exists V_0 \in \mathcal{B}_x$ such that $V_0 \le V$. Then $z(U,V_0)_1 \in W(x,U)$, $\exists \xi \in M(L)$ such that $z(U,V_0)_\xi \lessdot U \wedge V_0$, $z(U,V_0)_1 \lessdot V_0$. By $V_0 \le V$, $z(U,V_0)_1 \lessdot V$. By the arbitrariness of $V \in \mathcal{Q}(x_\alpha)$, $x_\alpha$ is an adherent point of $\bigvee W(x,U)$, $x_\alpha \le (\bigvee W(x,U))^-$. So inequality (8.3) holds. As for inequality (8.4), since $y \ne x$, by the $\alpha$-$T_0$ property, since $\mathcal{B}_x$ and $\mathcal{B}_y$ are respectively Q-neighborhood bases of $\mathcal{Q}(x_\alpha)$ and $\mathcal{Q}(y_\alpha)$, $\exists U \in \mathcal{B}_x$, $\exists V \in \mathcal{B}_y$ such that $U \wedge V = \underline{0}$. By $1 \in M(L)$ and Lemma **8.2.4**, $supp(U) \cap supp(V) = \emptyset$. We know $supp(\bigvee W(x,U)) \subset supp(U)$, so $supp(V) \cap supp(\bigvee W(x,U)) = \emptyset$, $V \neg\hat{q} \bigvee W(x,U)$. That is to say, $y_\alpha$ is not an adherent point of $\bigvee W(x,U)$, $y_\alpha \not\le (\bigvee W(x,U))^-$. Hence (8.4) is true. □

**8.2.24 Remark** Since $T_2$ property, introduced in a variant form by F. Hausdorff in original[30] and so is also called *Hausdorff property*, can perfectly separate two distinguished points by the topology, it has essential importance in both ordinary topology and fuzzy topology. Because the situation is more complicated in fuzzy topology, authors introduced various extensions of $T_2$ to its major properties. They can be roughly divided into three kinds:

(1) *To guarantee the uniqueness of limit in some sense.*

If an $L$-fuzzy point $x_\lambda$ is a limit of some $L$-fuzzy net (or filter) in an $L$-fts, then for every $\mu \le \lambda$, $x_\mu$ is also a limit point. Hence by uniqueness of limits we mean the uniqueness of support points of limits. Bearing this mind, [137] and [168] defined s-$T_2$ and $T_2$ just as discussed above. It has been shown above that these $T_2$ properties and the limit uniqueness of net or filter can be characterized with each other in some forms, and, they are equivalent respectively in the cased $1 \in M(L)$ and weakly induced.

Wuyts and Lowen gave out several definitions of $T_2$ property to utilize filter convergence theory.[178]

(2) *From the stand point of level structure*

In order to give separation axioms suiting the $\alpha$-compactness (see below Chapter 10), starting with the level structures in F-ts', Rodabaugh introduced the notion of $\alpha$-Hausdorff property and $\alpha^*$-Hausdorff property and studied their basic properties systematically:[140, 148] (for convenience, we denote $I = [0,1]$ in this remark)

*$\forall \alpha \in I\backslash\{1\}$, $(I^X, \delta)$ is called an $\alpha$-Hausdorff space if for arbitrary $x, y \in X$, $x \neq y$, there exist $U, V \in \delta$ with $U(x) > \alpha$, $V(y) > \alpha$ and $U \wedge V = \underline{0}$; called an $\alpha^*$-Hausdorff space if for arbitrary $x, y \in X$, $x \neq y$, there exist $U, V \in \delta$ with $U(x) \leq \alpha$, $V(y) \leq \alpha$ and $U \wedge V = \underline{0}$.*

$(I^X, \delta)$ is $\alpha$-Hausdorff if and only if $(X, \iota_\alpha(\delta))$ is a $T_2$ space. If $(I^X, \delta)$ is $\alpha$-Hausdorff for every $\alpha \in I\backslash\{1\}$, then $(I^X, \delta)$ is a $T_2$ space.

Then notions of $\alpha$-$T_2$ and level-$T_2$ in this section was introduced by Yu-wei Peng in his study of compactness in $L$-fuzzy topological spaces.

(3) *Degree of Hausdorff Separation*

Under the influence of the point of view that one should no longer decide merely whether a fuzzy space has a given property or not but measure a degree of its having this property, a lot of papers have developed different techniques to do this for the Hausdorff separation. For details we refer the reader to [153, 179] and their bibliography.

## 8.3 Regularity

**8.3.1 Definition** Let $L^X$ be an $L$-fuzzy space, $a \in L\backslash\{0\}$. $A \in L^X$ is called *a-crisp*, if $A_{(0)} = A_{[a]}$; $A$ is called *pseudo-crisp*, if there exists $a \in L\backslash\{0\}$ such that $A$ is $a$-cirsp.

**8.3.2 Proposition** *Let $X$ be a nonempty ordinary set, $L$ a completely distributive lattice. Then every $A \in L^X$ is pseudo-crisp if and only if there exists $\lambda \in M(L)$ such that $A$ is $\lambda$-cirsp.* □

We defined a Q-neighborhood of an $L$-fuzzy point. With the same motivation, we can also introduce the notion of a Q-neighborhood of an $L$-fuzzy subset:

**8.3.3 Definition** Let $(L^X, \delta)$ be an $L$-fts, $A \in L^X$, $U \in \delta$. $U$ is called a *quasi-coincident neighborhood* of $A$, or a *Q-neighborhood* of $A$ for short, if $A$ quasi-coincides with $U$ at every $x \in supp(A)$; i.e. for every $x \in supp(A)$, $x_{A(x)} \lessdot U$.

Denote the family of all the Q-neighborhoods of $A$ in $(L^X, \delta)$ by $\mathcal{Q}_\delta(A)$, or $\mathcal{Q}(A)$ for short.

**8.3.4 Definition** Let $(L^X, \delta)$ be an $L$-fts.

$(L^X, \delta)$ is called *regular*, if for every $U \in \delta$, there exists $\mathcal{V} \subset \delta$ such that $\bigvee \mathcal{V} = U$ and $V^- \leq U$ for every $V \in \mathcal{V}$.

$(L^X, \delta)$ is called *pararegular*, or *p-regular* for short, if for every non-zero pseudo-crisp closed subset $P$ in $(L^X, \delta)$ and every molecule $x_\lambda \in M(L^X)$ such that $x \notin supp(P)$, there exist $U \in \mathcal{Q}(x_\lambda)$ and $V \in \mathcal{Q}(A)$ such that $U \wedge V = \underline{0}$. Call $T_1$ pararegular $L$-fts *para-$T_3$*, or *p-$T_3$* for short.

$(L^X, \delta)$ is called *strongly pararegular*, or called *s-p-regular* for short, if for every non-zero pseudo-crisp closed subset $P$ in $(L^X, \delta)$ and every molecule $x_\lambda \in M(L^X)$ such that $x \notin supp(P)$, there exist $U \in \mathcal{Q}(x_\lambda)$ and $V \in \mathcal{Q}(A)$ such that $U_{(0)} \cap V_{(0)} = \emptyset$. Call $T_1$ strongly pararegular $L$-fts *strongly para-$T_3$*, or *s-p-$T_3$* for short.

**8.3.5 Remark** Regularity was defined by Hutton and Reilly.[41] Pseudo-crispness and paraegularity introduced above was defined by Wang[169, 168] to satisfy the need of stratification structure of $L$-fts.

**8.3.6 Theorem** *Let $(L^X, \delta)$ be an L-fts. Then the following conditions are equivalent:*

(i) *$(L^X, \delta)$ is regular.*

(ii) *For every L-fuzzy point $p \in Pt(L^X)$ and every $U \in \mathcal{Q}(p)$, there exists $V \in \mathcal{Q}(p)$ such that $V^- \leq U$.*

(iii) *For every molecule $e \in M(L^X)$ and every $U \in \mathcal{Q}(e)$, there exists $V \in \mathcal{Q}(e)$ such that $V^- \leq U$.*

**Proof** (i)$\Longrightarrow$(ii): $\forall p \in Pt(L^X)$, $\forall U \in \mathcal{Q}(p)$. By the regularity of $(L^X, \delta)$, $\exists \mathcal{V} \subset \delta$ such that $\bigvee \mathcal{V} = U$ and $V^- \leq U$ for every $V \in \mathcal{V}$. Since $p \lessdot U$, $\exists V \in \mathcal{V}$ such that $p \lessdot V$. So $V \in \mathcal{Q}(p)$, $V^- \leq U$.

(ii)$\Longrightarrow$(iii): Clear.

(iii)$\Longrightarrow$(i): $\forall U \in \delta$. $\forall e \in M(L^X)$ such that $e \lessdot U$, then $\exists V_e \in \mathcal{Q}(e)$ such that $V_e^- \leq U$. Denote $\mathcal{V} = \{V_e : e \in M(L^X), e \lessdot U\}$, then $\bigvee \mathcal{V} \leq U$, we need only prove $\bigvee \mathcal{V} \geq U$. Denote $W = \bigvee \mathcal{V}$. If $W \not\geq U$, $W' \not\leq U'$, $\exists d \in M(\downarrow W')$, $d \not\leq U'$. That is to say, $d \leq W' = \bigwedge \{V_e' : e \in M(L^X), e \lessdot U\}$. So $\forall e \in M(L^X)$, $e \lessdot U \Longrightarrow d \leq V_e'$. Especially, by $d \in M(L^X)$, $d \not\leq U'$, we have $d \lessdot U$, and hence $d \leq V_d'$. But by the definition of $V_d$, $V_d \in \mathcal{Q}(d)$, this is a contradicition. Hence $\bigvee \mathcal{V} = W \geq U$. □

Then we have

**8.3.7 Theorem** *Let $(L^X, \delta)$ be an L-fts, $\mathcal{B}$ a subbase of $\delta$. Then the following conditions are equivalent:*

(i) *$(L^X, \delta)$ is regular.*

(ii) *For every $U \in \mathcal{B}$, there exists $\mathcal{V} \subset \delta$ such that $\bigvee \mathcal{V} = U$ and $V^- \leq U$ for every $V \in \mathcal{V}$.*

(iii) *For every L-fuzzy point $p \in Pt(L^X)$ and every $U \in \mathcal{B}$ such that $p \lessdot U$, there exists $V \in \mathcal{Q}(p)$ such that $V^- \leq U$.*

(iv) *For every molecule $e \in M(L^X)$ and every $U \in \mathcal{B}$ such that $e \lessdot U$, there exists $V \in \mathcal{Q}(e)$ such that $V^- \leq U$.*

**Proof** (i)$\Longrightarrow$(ii): Clear.

(ii)$\Longrightarrow$(iii): $\forall p \in Pt(L^X)$, $\forall U \in \mathcal{B}$, $p \lessdot U$, then by (ii), $\exists \mathcal{V} \subset \delta$ such that $\bigvee \mathcal{V} = U$ and $V^- \leq U$ for every $V \in \mathcal{V}$. Since $p \not\leq U' = \bigwedge_{V \in \mathcal{V}} V'$, $\exists V \in \mathcal{V}$ such that $p \not\leq V'$. So $V \in \mathcal{Q}(p)$ and $V^- \leq U$.

(iii)$\Longrightarrow$(iv): Clear.

(iv)$\Longrightarrow$(i): $\forall U \in \mathcal{B}$, $\forall e \in M(L^X)$ such that $e \lessdot U$, then by (iv), $\exists V_e \in \mathcal{Q}(e)$ such that $V_e^- \leq U$. Denote $\mathcal{V} = \{V_e : e \in M(L^X), e \lessdot U\}$, then $\bigvee \mathcal{V} = U$. Denote $W = \bigvee \mathcal{V}$. If $W \not\geq U$, $W' \not\leq U'$, then $\exists d \in M(\downarrow W')$ such that $d \not\leq U'$. So $d \leq W' = \bigwedge \{V_e' : e \in M(L^X), e \lessdot U\}$. That is to say, $\forall e \in M(L^X)$, $e \lessdot U \Longrightarrow d \leq V_e'$. Especially, by $d \in M(L^X)$, $d \not\leq U'$, we have $d \lessdot U$, and hence $d \leq V_d'$. But by the definition of $V_d$, $V_d \in \mathcal{Q}(d)$, a contradiction. So $W \geq U$, $\bigvee \mathcal{V} = W = U$. That is to say, we have proved

$$U \in \mathcal{B} \Longrightarrow \exists \mathcal{V} \subset \delta,\ \bigvee \mathcal{V} = U,\ \forall V \in \mathcal{V},\ V^- \leq U. \tag{8.5}$$

Now suppose $U \in \delta$, then $\exists \mathcal{A} \subset \mathcal{P}([\mathcal{B}]^{<\omega} \backslash \{\emptyset\})$ such that $U = \bigvee_{\mathcal{F} \in \mathcal{A}} \bigwedge \mathcal{F}$. $\forall \mathcal{F} \in \mathcal{A}$, $\forall V \in \mathcal{F}$, by property (8.5), $\exists \mathcal{W}_V \subset \delta$ such that $\bigvee \mathcal{W}_V = V$ and $W^- \leq V$ for every $W \in \mathcal{W}_V$. Then for

$$\mathcal{W} = \{\bigwedge_{i\leq n} W_i : \exists\{V_i : i \leq n\} \in \mathcal{A},\ \forall i \leq n,\ W_i \in \mathcal{W}_{V_i}\}$$

we have

$$\bigvee \mathcal{W} = \bigvee\{\bigwedge\{\bigvee \mathcal{W}_{V_i} : i \leq n\} : \{V_i : i \leq n\} \in \mathcal{A}\} = \bigvee_{\mathcal{F}\in\mathcal{A}} \bigwedge \mathcal{F} = U$$

and $\forall \bigwedge_{i\leq n} W_i \in \mathcal{W}$,

$$(\bigwedge_{i\leq n} W_i)^- \leq \bigwedge_{i\leq n} W_i^- \leq \bigwedge_{i\leq n} V_i \leq U.$$

This completes the proof of the regularity. □

**8.3.8 Proposition**

(i) *S-p-regularity $\Longrightarrow$ p-regularity.*

(ii) *S-p-$T_3$ property $\Longrightarrow$ s-$T_2$ property.*

(iii) *P-$T_3$ property $\Longrightarrow$ $T_2$ property.* □

Clearly, the converse of the preceding implications are not true, there not exist implications between regularity and p-regularity, regularity and s-p-regularity respectively.

**8.3.9 Theorem** *Let $(L^X, \delta)$ be an L-fts, $1 \in M(L)$. Then*

(i) *$(L^X, \delta)$ is s-p-regular if and only if $(L^X, \delta)$ is p-regular.*

(ii) *$(L^X, \delta)$ is s-p-$T_3$ if and only if $(L^X, \delta)$ is p-$T_3$.* □

**8.3.10 Theorem** *Let $(L^X, \delta)$ be an L-fts. Then*

(i) *$(L^X, \delta)$ is weakly induced and regular $\Longrightarrow$ $(X, [\delta])$ is regular.*

(ii *$(L^X, \delta)$ is induced and $(X, [\delta])$ is regular $\Longrightarrow$ $(L^X, \delta)$ is regular.*

**Proof** (i) Suppose $x \in X$, $U \in \mathcal{N}_{[\delta]}(x)$, then $\chi_U \in \mathcal{Q}_\delta(x_1)$. By Theorem **8.3.6**, $\exists W \in \mathcal{Q}_\delta(x_1)$ such that $W^- \leq \chi_U$. By Theorem **4.3.34**, there exists $A \in L^X$ such that $W = \bigvee_{a\in L} aA_{[a]}{}^\circ$. Since $x_1 \lessdot W$,

$$1 \not\leq W(x)' = \bigwedge_{a\in L} (aA_{[a]}{}^\circ)(x)' = \bigwedge_{a\in L} (a' \vee (\chi_{A_{[a]}{}^\circ})'(x)).$$

So $\exists c \in L$ such that $1 \not\leq c' \vee (\chi_{A_{[c]}{}^\circ})'(x)$, $(\chi_{A_{[c]}{}^\circ})'(x) \neq 1$, $x \in A_{[c]}{}^\circ$. Since

$$y \in A_{[c]}{}^\circ \Longrightarrow W(y) \geq (cA_{[c]}{}^\circ)(y) = c,$$

we have $A_{[c]}{}^\circ \subset W_{[c]}{}^\circ$. Then by $W^- \leq \chi_U$,

$$x \in A_{[c]}{}^\circ \subset W_{[c]}{}^\circ \subset W_{[c]} \subset W^-{}_{[c]} \subset (\chi_U)_{[c]} = U.$$

Since $(L^X, \delta)$ is weakly induced, $W^-{}_{[c]}$ is closed in $(X, [\delta])$. By $W_{[c]}{}^\circ \subset W_{[c]} \subset W^-{}_{[c]}$,

$$(A_{[c]}{}^\circ)^- \subset (W_{[c]}{}^\circ)^- \subset W^-{}_{[c]} \subset U.$$

Take $V = A_{[c]}{}^\circ \in [\delta]$, we have

$$x \in V \subset V^- \subset U,$$

$(X, [\delta])$ is regular.

(ii) Suppose $x_\lambda \in M(L^X)$, $U \in \mathcal{Q}_\delta(x_\lambda)$. Since $(L^X, \delta)$ is induced, by Theorem **4.3.31**, $U = \bigvee_{a\in L} aU_{[a]}{}^\circ$. Since $x_\lambda \lessdot U$,

$$\lambda \not\leq U(x)' = \bigwedge_{a\in L} (aU_{[a]}{}^\circ)(x)' = \bigwedge_{a\in L} (a' \vee (\chi_{U_{[a]}{}^\circ})'(x)).$$

So $\exists c \in L$ such that $\lambda \not\leq c' \vee (\chi_{U_{[c]}{}^\circ})'(x)$, $\lambda \not\leq c'$, $\lambda \not\leq (\chi_{U_{[c]}{}^\circ})'(x)$, $x \in U_{[c]}{}^\circ \in [\delta]$. By the regularity of $(X, [\delta])$, $\exists W \in \mathcal{N}_{[\delta]}(x)$ such that $cl_{[\delta]}(W) \subset U_{[c]}{}^\circ$. Since $(L^X, \delta)$ is stratified, $cW \in \delta$. By $\lambda \not\leq c'$, $cW \in \mathcal{Q}_\delta(x_\lambda)$,

$$(cW)^- = (\underline{c} \wedge \chi_W)^- \leq (\underline{c})^- \wedge (\chi_W)^- \leq \underline{c} \wedge \chi_{cl_{[\delta]}(W)} \leq cU_{[c]}{}^\circ \leq U.$$

So take $V = cW \in \mathcal{Q}_\delta(x_\lambda)$, we have $V^- \leq U$. By Theorem **8.3.6**, $(L^X, \delta)$ is regular. □

The implication in Theorem **8.3.10** (i) cannot be inverted:

**8.3.11 Example** There exists a weakly induced F-ts $(X, \delta)$ which is not regular but its background space $(X, [\delta])$ is $T_4$.

Denote the ordinary topology of the real number space $\mathbf{R}$ by $\Omega(\mathbf{R})$, take $X = \mathbf{R}$, $\mathcal{B} = \{\chi_U, \frac{2}{3}\chi_U : U \in \Omega(\mathbf{R})\}$, denote by $\delta$ the F-topology on $X$ generated by $\mathcal{B}$, then $(X, \delta)$ is weakly induced and $[\delta] = \Omega(\mathbf{R})$. But for every $x \in X$ and the Q-neighborhood $\frac{2}{3}\chi_X$ of $x_{\frac{1}{2}}$, there does not exist any $V \in \mathcal{Q}(x_{\frac{1}{2}})$ such that $V^- \leq \frac{2}{3}\chi_X$.

**8.3.12 Theorem** *Let $(L^X, \delta)$ be an L-fts. Then*

(i) *$(L^X, \delta)$ is induced and p-regular $\Longrightarrow$ $(X, [\delta])$ is regular.*

(ii) *$(L^X, \delta)$ is weakly induced, $(X, [\delta])$ is regular $\Longrightarrow$ $(L^X, \delta)$ is s-p-regular.*

**Proof** (i) Suppose $x \in X$, $P$ is a nonempty closed subset in $(X, [\delta])$ and $x \notin P$. Fix a molecule $\lambda \in M(L)$, then since $(L^X, \delta)$ is stratified, $\lambda P \in \delta'$ and is non-zero pseudo-crisp, and $x_\lambda \in M(L^X)$, $x \notin supp(\lambda P)$. By the p-regularity, $\exists U \in \mathcal{Q}(x_\lambda)$, $\exists V \in \mathcal{Q}(\lambda P)$ such that $U \wedge V = \underline{0}$. By $\bigvee M(\Downarrow\lambda') = \lambda'$, we can take a molecule $\gamma \in M(\Downarrow\lambda')$. Then $U(x) \not\leq \gamma$, $x \in U_{(\gamma)}$. If $y \in P = supp(\lambda P)$, then by $V \in \mathcal{Q}(\lambda P)$, $\lambda \not\leq V(y)'$, $V(y) \not\leq \lambda'$. By $\gamma \leq \lambda'$, $V(y) \not\leq \gamma$, $y \in V_{(\gamma)}$. So $P \subset V_{(\gamma)}$. By the weakly induced property of $(L^X, \delta)$, $U_{(\gamma)}, V_{(\gamma)} \in [\delta]$, so $U_{(\gamma)}$ and $V_{(\gamma)}$ are respectively neighborhoods of $x$ and $P$ in $(X, [\delta])$. At last, since $\gamma \in M(L)$, by $U \wedge V = \underline{0}$ and Proposition **4.2.4** (ii), $U_{(\gamma)} \cap V_{(\gamma)} = \underline{0}_{(\gamma)} = \emptyset$. Hence $(X, [\delta])$ is regular.

(ii) Suppose $P$ is a non-zero pseudo-crisp closed subset in $(L^X, \delta)$, $x_\lambda \in M(L^X)$, $x \notin supp(P)$. Then $\exists a \in L$ such that $P_{(0)} = P_{[a]}$. By the weakly induced property of $(L^X, \delta)$, $P_{(0)} = P_{[a]}$ is closed in $(X, [\delta])$. By the regularity of $(X, [\delta])$, $\exists U \in \mathcal{N}_{[\delta]}(x)$, $\exists V \in \mathcal{N}_{[\delta]}(P_{(0)})$ such that $U \cap V = \emptyset$. Then clearly $\chi_U \in \mathcal{Q}_\delta(x_\lambda)$, $\chi_V \in \mathcal{Q}_\delta(P)$ and

$$(\chi_U)_{(0)} \cap (\chi_V)_{(0)} = U \cap V = \emptyset,$$

$(L^X, \delta)$ is s-p-regular. □

**8.3.13 Corollary** *Let $(L^X, \delta)$ be an induced L-fts. Then the following conditions are equivalent:*

(i) *$(L^X, \delta)$ is regular.*

(ii) *$(L^X, \delta)$ is s-p-regular.*

(iii) *$(L^X, \delta)$ is p-regular.*

(iv) *$(X, [\delta])$ is regular.* □

**8.3.14 Theorem** *Regularity is hereditary.* □

**8.3.15 Theorem** *S-p-regularity and p-regularity are hereditary with respect to closed subspaces.* □

In general, s-p-regularity and p-regularity are not hereditary. The following counterexample constructed by Wang[168] shows this point:

**8.3.16 Example** Take $X = L = [0, 1]$. Define $A \in L^X$ as follows:

$$A(x) = \begin{cases} 2x, & x \in [0, \frac{1}{2}], \\ 1, & x \in (\frac{1}{2}, 1]. \end{cases}$$

Let $\delta = \{\underline{0}, A, \underline{1}\}$, then $\delta$ is an $L$-fuzzy topology on $X$. Since there exists only one non-zero pseudo-crisp closed subset $\underline{1}$ in $(L^X, \delta)$, so $(L^X, \delta)$ is s-p-regular. Take $Y = [0, \frac{1}{4}] \cup [\frac{1}{2}, 1]$, then $B = A'|_Y$ is a non-zero pseudo-crisp closed subset in $(L^Y, \delta|_Y)$, where for every $y \in Y$,

$$B(y) = \begin{cases} 1 - 2y, & y \in [0, \frac{1}{4}], \\ 0, & y \in (\frac{1}{2}, 1]. \end{cases}$$

For an arbitrary point $x \in [\frac{1}{2}, 1]$, $\mathcal{Q}(x_1) = \{B', \underline{1}\}$, $\mathcal{Q}(B) = \{\underline{1}\}$, so $(L^Y, \delta|_Y)$ is not even p-regular.

**8.3.17 Theorem** *Regularity, s-p-regularity and p-regularity are preserved by L-fuzzy homeomorphisms.* □

**8.3.18 Theorem** *Let $\{(L^{X_t}, \delta_t) : t \in T\}$ be a family of L-fts', $(L^X, \delta)$ their L-fuzzy product space. Then*

(i) *If $(L^{X_t}, \delta_t)$ is regular for every $t \in T$, then $(L^X, \delta)$ is regular.*

(ii) *If $(L^X, \delta)$ is regular, $s \in T$, $(L^{X_s}, \delta_s)$ is stratified, then $(L^{X_s}, \delta_s)$ is regular.*

**Proof** (i) Suppose $x_\lambda \in M(L^X)$, $U \in \mathcal{Q}_\delta(x_\lambda)$. Without loss of generality, we can assume that $U$ is an element of the canonical base of $\delta$, $\exists F \in [T]^{<\omega}$, $\forall t \in T$, $\exists U_t \in \delta_t$ such that $U = \bigwedge_{t\in T} p_t^{\leftarrow}(U_t)$. Then by $x_\lambda \lessdot U$,

$$\lambda \not\leq U(x)' = \bigvee_{t\in F} p_t^{\leftarrow}(U_t')(x) = \bigvee_{t\in F} U_t(p_t(x))'.$$

So $\forall t \in F$, $\lambda \not\leq U_t(p_t(x))'$, $U_t \in \mathcal{Q}_{\delta_t}(p_t(x)_\lambda)$. Since every $(L^{X_t}, \delta_t)$ is regular, by Theorem **8.3.6** (iii), $\forall t \in F$, $\exists V_t \in \mathcal{Q}_{\delta_t}(p_t(x)_\lambda)$ such that $V_t^- \leq U_t$. Take $V = \bigwedge_{t\in F} p_t^{\leftarrow}(V_t) \in \delta$, then by $\lambda \in M(L)$, $|F| < \omega$ and $\lambda \not\leq V_t(p_t(x))'$,

$$\lambda \not\leq \bigvee_{t\in F} V_t(p_t(x))' = (\bigwedge_{t\in F} p_t^{\leftarrow}(V_t))'(x) = V'(x).$$

Plus $V \in \delta$, it means $V \in \mathcal{Q}_\delta(x_\lambda)$. Now, by the continuity of each $p_t^{\rightarrow}$,

$$V^- \leq \bigwedge_{t\in F} p_t^{\leftarrow}(V_t)^- \leq \bigwedge_{t\in F} p_t^{\leftarrow}(V_t^-) \leq \bigwedge_{t\in F} p_t^{\leftarrow}(U_t) = U.$$

By Theorem **8.3.6** (iii), $(L^X, \delta)$ is regular.

(ii) Since regularity is hereditary, the conclusion is obtained from Corollary **3.2.13**. □

By Theorem **4.2.21** and Corollary **8.3.13**, we have immediately the following conclusions:

**8.3.19 Theorem** *Let $\{(L^{X_t}, \delta) : t \in T\}$ be a family of induced L-fts'. Then*

(i) *$\prod_{t\in T}(L^{X_t}, \delta_t)$ is s-regular if and only if $(L^{X_t}, \delta_t)$ is s-regular for every $t \in T$.*

(ii) *$\prod_{t\in T}(L^{X_t}, \delta_t)$ is regular if and only if $(L^{X_t}, \delta_t)$ is regular for every $t \in T$.* □

The following example shows that there exists an $L$-fts which is a $L$-fuzzy product space of two s-p-regular $L$-fts' but is not p-regular:

**8.3.20 Example** Let $L = \{0, a, b, 1\}$ be the diamond-type lattice, take $a' = b$, $b' = a$, then $L$ is a F-lattice. Take

$X = \{x\}, \quad Y = \{y, z\},$
$A = x_a \in L^X, \quad B = y_a \vee z_b \in L^Y,$
$\delta = \{\underline{0}, A, \underline{1}\}, \quad \mu = \{\underline{0}, B, \underline{1}\}.$

Then $(L^X, \delta)$, $(L^Y, \mu)$ are $L$-fts'. In both $(L^X, \delta)$ and $(L^Y, \mu)$, since the support set of each non-zero pseudo-crisp closed subset is respectively $X$ and $Y$, so both $(L^X, \delta)$ and $(L^Y, \mu)$ are s-p-regular.

Let $(L^Z, \zeta) = (L^X, \delta) \times (L^Y, \mu)$. Since $a' = a$, $b' = b$, $A$ and $B$ are also respectively closed in $(L^X, \delta)$ and $(L^Y, \mu)$. Therefore, besides $\underline{0}$ and $\underline{1}$, $(L^Z, \zeta)$ has totally the following open subsets $U_1$, $U_2$, $U_3$, $U_4$ and closed subsets $U_1{}'$, $U_2{}'$, $U_3{}'$, $U_4{}'$:

$U_1 = p_1{}^{\leftarrow}(A) = (x, y)_a \vee (x, z)_a,$
$U_2 = p_2{}^{\leftarrow}(B) = (x, y)_a \vee (x, z)_b,$
$U_3 = p_1{}^{\leftarrow}(A) \vee p_2{}^{\leftarrow}(B) = (x, y)_a \vee (x, z)_1,$
$U_4 = p_1{}^{\leftarrow}(A) \wedge p_2{}^{\leftarrow}(B) = (x, y)_a,$
$U_1{}' = U_1, \quad U_2{}' = U_2, \quad U_3{}' = U_4, \quad U_4{}' = U_3.$

Then $(x, y)_a$ is a non-zero pseudo-crisp closed subset in $(L^Z, \zeta)$ and $supp((x, y)_a) = \{(x, y)\} \neq Z$. But $\mathcal{Q}_\zeta((x, y)_a) = \{\underline{1}\}$, $(L^Z, \zeta)$ is not p-regular.

# Chapter 9

# Separation (II)

In separation properties, the problems related to complete regularity and normality are more complicated. The first author studied this kind of problem in fuzzy topology is B. Hutton.[37] To establish the Urysohn Lemma in fuzzy form, he also constructed fuzzy unit interval which is widely used later in fuzzy topology. In 1982, Ying-ming Liu proved the pointwise characterization of $L$-fuzzy complete regularity and, using it as a tool, established the embedding theory.[71] 1983, Rodabaugh raised the problem of $L$-fuzzy form of Tietze Extension Theorem.[148] It was positively solved by Kubiak.[54] On $L$-fuzzy complete regularity, Kubiak also stated and proved a series of results.[55] Another problem concerning normality is insertion theorem. Extending the classical insertion theorem to the case of lattice-valued domain and as a practical application of the results on lattice-valued semicontinuous mappings and $L$-fuzzy normality, the Lattice-valued Hahn-Dieudonné-Tong Insertion Theorem will be proved. Kubiak's Fuzzy Insertion Theorem will also be introduced.

## 9.1 $L$-fuzzy Unit Interval — $I(L)$ and $\tilde{I}(L)$

In $L$-fuzzy topology, we hope to find a standard space which plaies a role just as the ordinary unit interval $[0,1]$ playing in ordinary topology, similarly possesses some nice properties. B. Hutton is the first author to make this effort. He defined $L$-fuzzy unit interval $I(L)$ as follows, which is widely used in $L$-fuzzy topology:

**9.1.1 Definition** Let $L$ be a complete lattice with an order-reversing involution $': L \to L$.

$\lambda \in L^{\mathbf{R}}$ is called *monotonically increasing*, if

$$s, t \in \mathbf{R},\ s \leq t \implies \lambda(s) \leq \lambda(t).$$

$\lambda \in L^{\mathbf{R}}$ is called *monotonically decreasing*, if

$$s, t \in \mathbf{R},\ s \leq t \implies \lambda(s) \geq \lambda(t).$$

Denote the family of all the monotonically decreasing mappings $\lambda \in L^{\mathbf{R}}$ fulfilling the following conditions by $md_{\mathbf{R}}(L)$:

$$\bigvee_{t \in \mathbf{R}} \lambda(t) = 1, \qquad \bigwedge_{t \in \mathbf{R}} \lambda(t) = 0.$$

Denote the family of all the elements in $md_{\mathbf{R}}(L)$ fulfilling the following conditions by $md_I(L)$:

$$t < 0 \Longrightarrow \lambda(t) = 1, \qquad t > 1 \Longrightarrow \lambda(t) = 0.$$

For every $\lambda \in md_{\mathbf{R}}(L)$ and every $t \in \mathbf{R}$, define

$$\lambda(t-) = \bigwedge_{s<t} \lambda(s), \qquad \lambda(t+) = \bigvee_{s>t} \lambda(s).$$

Define an equivalent relation $\sim$ on $md_{\mathbf{R}}(L)$ as follows:

$$\forall \lambda, \mu \in md_{\mathbf{R}}(L), \qquad \lambda \sim \mu \iff \forall t \in \mathbf{R},\ \lambda(t-) = \mu(t-),\ \lambda(t+) = \mu(t+).$$

For every $\lambda \in md_{\mathbf{R}}(L)$, let $[\lambda]$ denote the equivalent class in $md_{\mathbf{R}}(L)$ with respect to $\sim$ and containing $\lambda$; i.e.

$$[\lambda] = \{\mu \in md_{\mathbf{R}}(L) : \ \mu \sim \lambda\}.$$

Denote the family of all the equivalent classes in $md_{\mathbf{R}}(L)$ with respect to $\sim$ by $\mathbf{R}(L)$. For convenience, for every $\lambda \in md_{\mathbf{R}}(L)$, if no confusion will be caused, we identify $\lambda$ with every $\mu$ such that $\mu \sim \lambda$, and still use $\lambda$ to denote $[\lambda]$.

For every $t \in \mathbf{R}$, define $L_t, R_t \in L^{\mathbf{R}[L]}$ as follows:

$$\forall \lambda \in \mathbf{R}[L], \quad L_t(\lambda) = \lambda(t-)', \quad R_t(\lambda) = \lambda(t+).$$

Restrict the equivalent relation $\sim$ defined above on $md_I(L)$, still denote the correspondent equivalent classes in $md_I(L)$ by $[\lambda]$ for every $\lambda \in md_I(L)$, and denote the family of all the correspondent equivalent classes in $md_I(L)$ by $I[L]$. Also denote the restrictions of $L_t, R_t : \mathbf{R}[L] \to L$ on $I[L]$ by $L_t$, $R_t$ for every $t \in \mathbf{R}$ respectively.

Denote

$$\mathcal{S}_L^{\mathbf{R}} = \{L_t, R_t \in L^{\mathbf{R}[L]} : \ t \in \mathbf{R}\}, \qquad \mathcal{S}_L^I = \{L_t, R_t \in L^{I[L]} : \ t \in \mathbf{R}\},$$

$$\mathcal{B}_L^{\mathbf{R}} = \{\bigwedge \mathcal{F} : \ \mathcal{F} \in [\mathcal{S}_L^{\mathbf{R}}]^{<\omega} \backslash \{\emptyset\}\}, \qquad \mathcal{B}_L^I = \{\bigwedge \mathcal{F} : \ \mathcal{F} \in [\mathcal{S}_L^I]^{<\omega} \backslash \{\emptyset\}\},$$

$$\mathcal{T}_L^{\mathbf{R}} = \{\bigvee \mathcal{A} : \ \mathcal{A} \subset \mathcal{B}_L^{\mathbf{R}}\}, \qquad \mathcal{T}_L^I = \{\bigvee \mathcal{A} : \ \mathcal{A} \subset \mathcal{B}_L^I\}.$$

Denote $(L^{\mathbf{R}[L]}, \mathcal{T}_L^{\mathbf{R}})$ by $\mathbf{R}(L)$, called the *L-fuzzy real line*. Call respectively $\mathcal{S}_L^{\mathbf{R}}$, $\mathcal{B}_L^{\mathbf{R}}$, $\mathcal{T}_L^{\mathbf{R}}$ the *standard subbase* of $\mathbf{R}(L)$, the *standard base* of $\mathbf{R}(L)$, the *standard topology* of $\mathbf{R}(L)$.

Denote $(L^{I[L]}, \mathcal{T}_L^I)$ by $I(L)$, called the *L-fuzzy unit interval*. Call respectively $\mathcal{S}_L^I$, $\mathcal{B}_L^I$, $\mathcal{T}_L^I$ the *standard subbase* of $I(L)$, the *standard base* of $I(L)$, the *standard topology* of $I(L)$.

The $L$-fuzzy product space $I(L)^\kappa = \prod_{t\in T} I_t$, where $I_t = I(L)$ for every $t \in T$ and $|T| = \kappa$, is called an *L-fuzzy unit cube* of weght $\kappa$.

**9.1.2 Proposition** *In the standard subbase $\mathcal{S}_L^I$ of $I(L)$,*

(i) *$L_t$ is monotonically increasing with respect to $t \in \mathbf{R}$.*

(ii) *$R_t$ is monotonically decreasing with respect to $t \in \mathbf{R}$.*

(iii) *If $A \subset \mathbf{R}$ is nonempty and upper bounded, then $L_{\bigvee A} = \bigvee_{t\in A} L_t$.*

(iv) *If $A \subset \mathbf{R}$ is nonempty and lower bounded, then $R_{\bigwedge A} = \bigvee_{t\in A} R_t$.*

**Proof** (i) and (ii) are clear.

(iii) By the given condition, we can suppose $\bigvee A = s \in \mathbf{R}$. By (i), $L_s \geq \bigvee_{t\in A} L_t$. $\forall \lambda \in I[L]$, $\forall r < s$, $\exists t_r \in A$ such that $t_r > r$, note that $\lambda$ is monotonically decreasing and $r < \frac{r+t_r}{2} < t_r$, we have

$$L_s(\lambda) = \lambda(s-)' = \bigvee_{r<s} \lambda(r)' \leq \bigvee_{r<s} \lambda(\tfrac{r+t_r}{2})' \leq \bigvee_{r<s} \lambda(t_r-)' \leq \bigvee_{t\in A} \lambda(t-)' = (\bigvee_{t\in A} L_t)(\lambda).$$

So (iii) is true.

(iv) Similar to (iii). □

**9.1.3 Exercise** Find the correspondent counterexamples to show that the conclusions (iii) and (iv) in Proposition **9.1.2** will not hold any longer if exchange "upper" and "lower," joins and meets there.

**9.1.4 Proposition** *Let $L$ be a complete lattice with an order-reversing involution $': L \to L$, $a, b \in \mathbf{R}$, $a < b$. Then in $I(L)$,*

(i) $L_a^- \leq R_a' \leq L_b$.

(ii) $R_b^- \leq L_b' \leq R_a$.

**Proof** (i) $\forall \lambda \in I[L]$,

$$L_a'(\lambda) = \lambda(a-) = \bigwedge_{r<a} \lambda(r) \geq \bigvee_{s>a} \lambda(s) = \lambda(a+) = R_a(\lambda),$$
$$R_a(\lambda) = \lambda(a+) = \bigvee_{a<s<b} \lambda(s) \geq \bigwedge_{a<r<b} \lambda(r) = \lambda(b-) = L_b'(\lambda).$$

So

$$L_a' \geq R_a \geq L_b',$$
$$L_a \leq R_a' \leq L_b.$$

Since $R_a'$ is closed in $I(L)$, (i) is true.

(ii) Similar. □

By Proposition **9.1.2**, note that $L_t = \underline{1}$ for every $t > 1$ and $R_t = \underline{1}$ for every $t < 0$, the following conclusions are natural:

**9.1.5 Proposition** *Let $L$ be a complete lattice with an order-reversing involution $': L \to L$ and*

$$\mathcal{L} = \{L_t \in L^{I[L]} : t \in \mathbf{R}\}, \quad \mathcal{R} = \{R_t \in L^{I[L]} : t \in \mathbf{R}\},$$

*then both $\mathcal{L} \subset L^{I[L]}$ and $\mathcal{R} \subset L^{I[L]}$ fulfil the conditions* (LFT1)–(LFT3). □

**9.1.6 Definition** Let $L$ be a complete lattice with an order-reversing involution $': L \to L$. On $I[L]$, call respectively $\mathcal{L}$ and $\mathcal{R}$ the *left topology* and the *right topology* on $I[L]$.

**9.1.7 Proposition** *Let $L$ be a complete lattice with an order-reversing involution $': L \to L$.*

(i) *$\mathcal{T}_L^I$ fulfills the conditions* (LFT1)–(LFT3).

(ii) *Every element of $\mathcal{T}_L^I$ is a join of the elements in $\mathcal{B}_L^I = \{\underline{1_L}\} \cup \{L_b, R_a \wedge L_b, R_a : 0 \leq a, b \leq 1\}$.*

(iii) *Every element of $\mathcal{B}_L^I$ is a nonempty finite meet of the elements in $\mathcal{S}_L^I = \{\underline{1_L}\} \cup \{L_t, R_t : 0 \leq t \leq 1\}$.*

**Proof** (iii) Denote $\mathcal{A} = \{\underline{1_L}\} \cup \{L_t, R_t : t \in [0,1]\}$, need only prove $\mathcal{S}_L^I = \mathcal{A}$. $\forall t \in (-\infty, 0)$, $\forall \lambda \in I[L]$,

$$L_t(\lambda) = \lambda(t-)' = 1_L' = 0_L = R_1(\lambda),$$
$$R_t(\lambda) = \lambda(t+) = 1_L = \underline{1_L}(\lambda),$$

so $L_t = R_1 \in \mathcal{A}$, $R_t = \underline{1_L} \in \mathcal{A}$. $\forall t \in (1, +\infty)$, $\forall \lambda \in I[L]$,

$$L_t(\lambda) = \lambda(t-)' = 0_L' = 1_L = \underline{1_L}(\lambda),$$
$$R_t(\lambda) = \lambda(t+) = 0_L = 1_L' = L_0(\lambda),$$

so $L_t = \underline{1_L} \in \mathcal{A}$, $R_t = L_0 \in \mathcal{A}$, $\mathcal{S}_L^I \subset \mathcal{A}$. As just proved, $\underline{1_L} = L_t \in \mathcal{S}_L^I$ for every $t \in (1, +\infty)$, so $\mathcal{A} \subset \mathcal{S}_L^I$, $\mathcal{S}_L^I = \mathcal{A}$.

(ii) Denote $\mathcal{A} = \{\underline{1_L}\} \cup \{L_b, R_a \wedge L_b, R_a : 0 \leq a, b \leq 1\}$, we need only prove $\mathcal{B}_L^I = \mathcal{A}$. $\forall \mathcal{F} \in [\mathcal{S}_L^I]^{<\omega} \backslash \{\emptyset\}$, without loss of generality, we can suppose $\underline{1_L} \notin \mathcal{F}$. Then $\exists F_1, F_2 \in [[0,1]]^{<\omega}$ such that $F_1 \cup F_2 \neq \emptyset$ and

$$\mathcal{F} = \{L_s : s \in F_1\} \cup \{R_t : t \in F_2\}.$$

If $F_1, F_2 \neq \emptyset$, take $a = max\ F_2$, $b = min\ F_1$, by Proposition **9.1.2** (i), (ii), we have

$\bigwedge\mathcal{F} = R_a \wedge L_b \in \mathcal{A}$. If $F_1 = \emptyset$, then $F_2 \neq \emptyset$, take $a = max\ F_2$, we have $\bigwedge\mathcal{F} = R_a \in \mathcal{A}$. If $F_2 = \emptyset$, then $F_1 \neq \emptyset$, take $b = min\ F_1$, we have $\bigwedge\mathcal{F} = L_b \in \mathcal{A}$. So we have always $\mathcal{B}_L^I \subset \mathcal{A}$. By (iii) proved above, $\mathcal{A} \subset \mathcal{B}_L^I$, so $\mathcal{B}_L^I = \mathcal{A}$.

(i) By (ii). □

**9.1.8 Remark** (1) Liu and Luo also gave out a homeomorphic form of $I(L)$ by replacing $\mathbf{R}$ in Definition **9.1.1** with $[0, 1]$.[83]

(2) $I(L)$ can be considered as an $L$-fuzzy subspace of $L$-fuzzy real line $\mathbf{R}(L)$.

(3) By Proposition **9.1.2** (i), (ii) and Proposition **9.1.7** (ii), besides $1_L$, $L_1 \vee R_0$ is the largest element in $\mathcal{T}_L^I$. Similarly, besides $0_L$, $L_1{}' \wedge R_0{}'$ is the smallest element in $\mathcal{T}_L^I$. But, in general, $L_1 \vee R_0 \neq \underline{1_L}$, $L_1{}' \wedge R_0{}' \neq \underline{0_L}$; this point can be verified with $L = [0, 1]$ and $\lambda = \underline{a} \in md_I(L)$ for $a = \frac{1}{2}$. That is to say, in $\mathcal{T}_L^I$, there probably exist two gaps between $\underline{1_L}$ and other open subsets, and between $\underline{0_L}$ and other open subsets. That is also the reason why $\underline{1_L}$ must be added into Proposition **9.1.7** (iii) and (ii).

As it was proved in [37], $I(L)$ possesses some properties similar to that of the real unit interval $I$. Especially, for some lattices possess better property, it can has even the same topological structure with $I$. The following Theorem **9.1.12** shows this point:

**9.1.9 Definition** A lattice $L$ is called a *Heyting algebra*, if for every two $a, b \in L$, there exists an element $(a \to b) \in L$ such that

$$\forall c \in L, \quad c \leq (a \to b) \Longleftrightarrow c \wedge a \leq b.$$

For every element $a$ of a Heyting algebra $L$, especially denote $\neg a = (a \to 0)$, call $\neg a$ the *negation* of $a$.

**9.1.10 Proposition** *A complete lattice is a Heyting algebra if and only if it fulfills the 1st infinitely distributive law* (IFD1).

**Proof** (Necessity) Suppose $L$ is a complete Heyting algebra, $a \in L$, $B \subset L$. Denote $d = \bigvee_{b \in B}(a \wedge b)$. We need only prove $a \wedge \bigvee B \leq d$. Since $L$ is a Heyting algebra, $\exists (a \to d) \in L$. Since $\forall b \in B$, $a \wedge b \leq d$, we have $b \leq (a \to d)$, $\bigvee B \leq (a \to d)$. Hence $a \wedge \bigvee B \leq d$.

(Sufficiency) Let $L$ be a complete lattice fulfilling the 1st infinitely distributive law (IFD1), $a, b \in L$. Take $e = \bigvee\{d \in L : \ d \wedge a \leq b\}$, then $\forall c \in L$, by (IFD1),

$$c \leq e \Longrightarrow c \wedge a = (c \wedge e) \wedge a = \bigvee\{c \wedge d \wedge a : \ d \wedge a \leq b\} \leq c \wedge b \leq b,$$
$$c \wedge a \leq b \Longrightarrow c \leq \bigvee\{d \in L : \ d \wedge a \leq b\} = e.$$

So $e = (a \to b)$, $L$ is a Heyting algebra. □

**9.1.11 Definition** Call every complete Heyting algebra a *frame*. Let $L_1$, $L_2$ be frames, $f : \ L_1 \to L_2$ a mapping. Call $f$ a *frame homomorphism*, if $f$ is both arbitrary join preserving and finite meet preserving. Call $f$ a *frame isomorphism*, if $f$ is bijective and both $f$ and $f^{-1}$ are frame homomorphisms.

Denote the category of all the frames and frame homomorphisms by **Frm**.

By Proposition **9.1.10**, every frame can be equivalently considered as a complete lattice fulfilling (IFD1). So the terminology "frame" in the sequel always mean a lattice of this kind.

**9.1.12 Theorem** *Let $L$ be a frame with an order-reversing involution $'$ : $L \to L$.*

*Then the following conditions are equivalent:*

(i) *$L$ is a Boolean algebra.*

(ii) *There exists a frame isomorphism $f: \Omega(I) \to \mathcal{T}_L^I$ from the topology $\Omega(I)$ of the ordinary unit interval $I = [0,1]$ to the standard topology $\mathcal{T}_L^I$ of $I(L)$.*

**Proof** Denote $I = [0,1]$ and

$$\mathcal{B}_0 = \{[0,b), (a,1]: \ 0 < b \leq 1, \ 0 \leq a < 1\},$$
$$\mathcal{B}_1 = \{\emptyset\} \cup \{[0,b), (a,b), (a,1]: \ 0 \leq a < b \leq 1\},$$

then clearly $\mathcal{B}_0$ is a subbase of the topology $\Omega(I)$ of $I$, $\mathcal{B}_1$ is generated by $\mathcal{B}_0$ and hence is a base of $\Omega(I)$. That is to say,

$$\mathcal{B}_1 = \{\bigwedge \mathcal{F}: \ \mathcal{F} \in [\mathcal{B}_0]^{<\omega} \backslash \{\emptyset\}\}. \tag{9.1}$$
$$\Omega(I) = \{\bigvee \mathcal{C}: \ \mathcal{C} \subset \mathcal{B}_1\}. \tag{9.2}$$

(i)$\Longrightarrow$(ii): Denote

$$\mathcal{P} = \{L_b, R_a: \ 0 < b \leq 1, \ 0 \leq a < 1\} \subset \mathcal{S}_L^I,$$
$$\mathcal{B} = \{\underline{0_L}\} \cup \{L_b, R_a \wedge L_b, R_a: \ 0 \leq a < b \leq 1\} \subset \mathcal{B}_L^I.$$

First of all, we prove

$$0 \leq b \leq a \leq 1 \implies L_b \wedge R_a = \underline{0_L}, \tag{9.3}$$
$$L_1 \vee R_0 = \underline{1_L}, \tag{9.4}$$
$$\mathcal{B} = \{\bigwedge \mathcal{F}: \ \mathcal{F} \in [\mathcal{P}]^{<\omega} \backslash \{\emptyset\}\}, \tag{9.5}$$
$$\mathcal{T}_L^I = \{\bigvee \mathcal{C}: \ \mathcal{C} \subset \mathcal{B}\}. \tag{9.6}$$

For relation (9.3), $\forall \lambda \in I[L]$, $\forall s,t \in I$ such that $s < b \leq a < t$, since $\lambda$ is monotonical decreasing, $L$ is a Boolean algebra, we have

$$\lambda(s)' \wedge \lambda(t) \leq \lambda(s)' \wedge \lambda(s) = 0_L.$$

Since $L$ is a frame and hence fulfills the 1st infinitely distributive law, by Proposition **1.1.31** (i),

$$(L_b \wedge R_a)(\lambda) = \lambda(b-)' \wedge \lambda(a+) = (\bigvee_{s<b} \lambda(s)') \wedge (\bigvee_{t>a} \lambda(t))$$
$$= \bigvee_{s<b\leq a<t} (\lambda(s)' \wedge \lambda(t)) \leq 0_L.$$

So relation (9.3) is true.

For equality (9.4), $\forall \lambda \in I[L]$, by the monotonical decreasing property of $\lambda$ and that $L$ is a Boolean algebra,

$$(L_1 \vee R_0)(\lambda) = \lambda(1-)' \vee \lambda(0+) \geq \lambda(1-)' \vee \lambda(1-) = (\lambda(1-) \wedge \lambda(1-)')' = 0_L{}' = 1_L.$$

So equality (9.4) is true.

As for equality (9.5) and equality (9.6), they are implied by relation (9.3), Proposition **9.1.7** (ii) and equality (9.4).

Define a mapping $f_0: \ \mathcal{B}_0 \to \mathcal{P}$ as follows:

$$\forall U \in \mathcal{B}_0, \quad f_0(U) = \begin{cases} L_b, & U = [0,b), \ 0 < b \leq 1, \\ R_a, & U = (a,1], \ 0 \leq a < 1. \end{cases}$$

We shall prove that $f_0$ is a poset isomorphism with respect to the inclusion order on $\mathcal{B}_0$.

$f_0$ is clearly surjective. Suppose $0 < b, c \leq 1$, $b \neq c$. Without loss of generality, we can assume $b < c$. Take $d \in (b,c)$ and $\lambda \in I[L]$ such that $\lambda(t) = 1_L$ for every $t < d$ and $\lambda(t) = 0_L$ for every $t \geq d$, then by $b < d < c$,

$$f_0([0,b))(\lambda) = L_b(\lambda) = \lambda(b-)' = 1_L{}' = 0_L,$$

$f_0([0,c))(\lambda) = L_c(\lambda) = \lambda(c-)' = 0_L{}' = 1_L.$

So $f_0([0,b)) \neq f_0([0,c))$. Similarly prove $(a,1] \neq (d,1] \Longrightarrow f_0((a,1]) \neq f_0((d,1])$. Suppose $[0,b),(a,1] \in \mathcal{P}$, then $0 < b \leq 1,\ 0 \leq a < 1$. Take $\lambda \in I[L]$ such that $\lambda(t) = 1_L$ for every $t < 0$, $\lambda(t) = 0_L$ for every $t \geq 0$, then by $0 < b \leq 1,\ 0 \leq a < 1$,

$f_0([0,b))(\lambda) = L_b(\lambda) = \lambda(b-)' = 0_L{}' = 1_L,$

$f_0((a,1])(\lambda) = R_a(\lambda) = \lambda(a+) = 0_L.$

So $f_0([0,b)) \neq f_0((a,1])$. Therefore, $f_0$ is injective and hence bijective. Moreover, it also shows that $L_b \not\leq R_a$ for all the $a,b \in I$ such that $0 < b \leq 1,\ 0 \leq a < 1$.

By Proposition **9.1.2** (i), (ii), $f_0$ is order preserving. For $f_0{}^{-1}$, if $L_b < L_d$ or $R_a < R_c$, by Proposition **9.1.2** (i), (ii) and the injective property of $f_0$, we have $f_0{}^{-1}(L_b) = b < d = f_0{}^{-1}(L_d)$ or $f_0{}^{-1}(R_a) = a < c = f_0{}^{-1}(R_c)$. We have proved above that $L_b \not\leq R_a$ is always true, similarly we can prove that $R_a \not\leq L_b$ is also always true, so $f_0{}^{-1}$ is also order preserving, $f_0$ is a poset isomorphism.

Define $f_1 : \mathcal{B}_1 \to \mathcal{B}$ as follows:

$$\forall U \in \mathcal{B}_1, \quad f_1(U) = \begin{cases} L_b, & U = [0,b),\ 0 < b \leq 1, \\ R_a \wedge L_b, & U = (a,b),\ 0 \leq a < b \leq 1, \\ R_a, & U = (a,1],\ 0 \leq a < 1. \end{cases}$$

Then $f_1$ is an extension of $f_0$ on $\mathcal{B}_1$. Since $f_0$ is a poset isomorphism, by equality (9.1) and equality (9.5), $f_1$ is a poset isomorphism.

At last, by equality (9.2) and equality (9.6), we can define $f : \Omega(I) \to \mathcal{T}_L^I$ as follows:

$$\forall \mathcal{C} \subset \mathcal{B}_1, \quad f(\bigvee \mathcal{C}) = \bigvee_{U \in \mathcal{C}} f_1(U).$$

Since $f_1$ is a poset isomorphism, by equality (9.2) and equality (9.6), $f$ is also a poset isomorphism. Since joins and meets in a lattice are completely determined by its partial order, $f$ is a frame isomorphism.

(ii)$\Longrightarrow$(i): Suppose $\Omega(I)$ is frame isomorphic to $\mathcal{T}_L^I$, then $\mathcal{T}_L^I \backslash \{\underline{1_L}\}$, just as $\Omega(I)$, has no largest element. By Proposition **9.1.2** (i), (ii), just as mentioned in Remark **9.1.8** (3), $L_1 \vee R_0$ is the largest element of $\mathcal{T}_L^I$ besides $1_L$. So $L_1 \vee R_0 = \underline{1_L}$. Now suppose $a \in L$, take $\lambda \in I[L]$ such that

$$\forall t \in \mathbf{R}, \quad \lambda(t) = \begin{cases} 1_L, & t < 0, \\ a, & 0 \leq t \leq 1, \\ 0_L, & t > 1, \end{cases}$$

then

$a \vee a' = \lambda(1-)' \vee \lambda(0+) = (L_1 \vee R_0)(\lambda) = \underline{1_L}(\lambda) = 1_L,$

$a \wedge a' = 0_L,$

$L$ is a Boolean algebra. □

**9.1.13 Proposition** *Let $L$ be a frame with an order-reversing involution $'$ : $L \to L$. Then*

(i) *Every element of $\mathcal{T}_L^I$ is a join of elements in*

$$\{R_a \wedge L_b : \ a,b \in \mathbf{Q}\}.$$

(ii) *Every element of $\mathcal{T}_L^I$ is a join of elements in*

$$\{\underline{1_L}\} \cup \{L_b, R_a \wedge L_b, R_a : \ a,b \in \mathbf{Q} \cap [0,1]\}.$$

**Proof** (i) Denote $\mathcal{B}_0 = \{R_a \wedge L_b : a, b \in \mathbf{Q}\}$. We need only prove that for every two $s, t \in \mathbf{R}$, $\exists \mathcal{A} \subset \mathcal{B}_0$ such that $\bigvee \mathcal{A} = R_s \wedge L_t$. For $s, t \in \mathbf{R}$, take $A, B \subset \mathbf{Q}$ such that $\bigwedge A = s$, $\bigvee B = t$, then $\mathcal{A} = \{R_a \wedge L_b : a \in A, b \in B\} \subset \mathcal{B}_0$. By Proposition **9.1.2** (iii), (iv) and Proposition **1.1.31** (i),

$$R_s \wedge L_t = R_{\bigwedge A} \wedge L_{\bigvee B} = (\bigvee_{a \in A} R_a) \wedge (\bigvee_{b \in B} L_b)$$
$$= \bigvee\{R_a \wedge L_b : a \in A, b \in B\} = \bigvee \mathcal{A}.$$

(ii) By Proposition **9.1.7** (ii), similar to the preceding proof. □

**9.1.14 Proposition** *Let $L$ be a F-lattice. Then $I(L)$ is a sub-$T_0$ L-fts.*

**Proof** Suppose $[\lambda], [\mu] \in I[L]$, $[\lambda] \neq [\mu]$, then $\exists t \in [0,1]$ such that $\lambda(t-) \neq \mu(t-)$ or $\lambda(t+) \neq \mu(t+)$. If $\lambda(t-) \neq \mu(t-)$, no loss of generality, suppose $\lambda(t-) \not\leq \mu(t-)$, then $\exists \alpha \in M(\Downarrow\lambda(t-))$ such that

$$\alpha \not\leq \mu(t-) = L_t'([\mu]),$$
$$\alpha \leq \lambda(t-) = L_t'([\lambda]).$$

So $L_t \in \mathcal{Q}([\mu]_\alpha)$, $L_t \notin \mathcal{Q}([\lambda]_\alpha)$. Similarly, from the assumption $\lambda(t+) \not\leq \mu(t+)$ we can show that $\exists \alpha \in M(\Downarrow\mu(t+)')$ such that $R_t \in \mathcal{Q}([\lambda]_\alpha)$, $R_t \notin \mathcal{Q}([\mu]_\alpha)$. So $I(L)$ is sub-$T_0$. □

But in general, $I(L)$ is not even quasi-$T_0$:

**9.1.15 Example** Take $L = I = [0,1]$, $x = [\varphi] \in I[I]$ is defined as follows:

$$\varphi(t) = \begin{cases} 1, & t < 0, \\ \frac{1}{2}, & 0 \leq t \leq 1, \\ 0, & t > 1, \end{cases}$$

then it is clear that for every open subset $U$ in $I(I)$, $U(x) \in \{0, \frac{1}{2}, 1\}$. In the definition of quasi-$T_0$ property, take $\lambda = \frac{2}{3}$, $\gamma = \frac{3}{4}$, then the condition there cannot be satisfied by $I(I)$, $I(I)$ is not quasi-$T_0$.

Just as disscussed above, $I(L)$ has some nice properties. But as an $L$-fts, the separation property of $I(L)$ is too weak. For instance, it is not even $T_0$. This is caused because the standard topology $\mathcal{T}_L^I$ of $I(L)$ has not been fine enough. So our target is to add some open subsets into $\mathcal{T}_L^I$ to make it finer in some degree. For this aim, it is helpful to consider the ordered structure of $I[L]$:

**9.1.16 Definition** Let $L$ be a complete lattice with an order-reversing involution $' : L \to L$. Define a relation $\leq$ on $I[L]$ as follows:

$$\forall [\lambda], [\mu] \in I[L], \quad [\lambda] \leq [\mu] \iff \forall t \in \mathbf{R},\ \lambda(t-) \leq \mu(t-),\ \lambda(t+) \leq \mu(t+).$$

By the definition of $I[L]$, the preceding definition is reasonable. In fact, this relation is a partial order. Moreover, it makes $I[L]$ be a completely distributive lattice:

**9.1.17 Proposition** *Let $L$ be a complete lattice with an order-reversing involution $' : L \to L$. Then*

(i) *$\leq$ is a partial order on $I[L]$.*

(ii) *$I[L]$ equipped with $\leq$ is a complete lattice and for every $\mathcal{A} \subset I[L]$,*

$$\bigvee_{[\lambda] \in \mathcal{A}} [\lambda] = [\bigvee_{[\lambda] \in \mathcal{A}} \lambda], \qquad \bigwedge_{[\lambda] \in \mathcal{A}} [\lambda] = [\bigwedge_{[\lambda] \in \mathcal{A}} \lambda].$$

**Proof** (i) Certainly, $\leq$ is reflexive and transitive. If $[\lambda] \leq [\mu] \leq [\lambda]$, then $\forall t \in \mathbf{R}$,

$$\lambda(t-) \leq \mu(t-) \leq \lambda(t-), \quad \lambda(t+) \leq \mu(t+) \leq \lambda(t+).$$

But this just means $\lambda(t-) = \mu(t-)$, $\lambda(t+) = \mu(t+)$ for every $t \in \mathbf{R}$, so $[\lambda] = [\mu]$, $\leq$ is antisymmetric, and hence a partial order.

(ii) First of all, we prove that $I[L]$ is a complete lattice. Take $\lambda_0 \in md_I(L)$ which takes $1_L \in L$ on $(-\infty, 0)$ and $0_L \in L$ on $[0, +\infty)$, then clearly $\lambda_0$ is the smallest element in $md_I(L)$, and hence $[\lambda_0]$ is the smallest element in $I[L]$. By Proposition **1.1.24** (ii), to prove the completeness of $L$, need only prove the smallest upper bound of each nonempty subfamily of $I[L]$ exists in $I[L]$. Suppose $\mathcal{A} \subset I[L]$, $\mathcal{A} \neq \emptyset$, then $\forall [\lambda] \in \mathcal{A}$, $\lambda \in md_I(L)$. Denote $\gamma = \bigvee_{[\lambda]\in\mathcal{A}} \lambda$, clearly $\gamma \in md_I(L)$, $[\gamma] \in I[L]$, $[\gamma]$ is an upper bound of $\mathcal{A}$ in $I[L]$. Suppose $[\zeta] \in I[L]$ is also an upper bound of $\mathcal{A}$ in $I[L]$, then

$$\forall [\lambda] \in \mathcal{A},\ \forall t \in \mathbf{R}, \quad \lambda(t-) \leq \zeta(t-), \quad \lambda(t+) \leq \zeta(t+). \tag{9.7}$$

Suppose $\forall s, t \in \mathbf{R}$, $s < t$, denote

$$f(s,t) = s + \tfrac{1}{3}(t-s), \quad g(s,t) = s + \tfrac{2}{3}(t-s).$$

then $s < f(s,t) < g(s,t) < t$. $\forall s, t \in \mathbf{R}$ such that $s < t$, by inequalities (9.7), we have

$$\zeta(s) \geq \zeta(f(s,t)-) \geq \bigvee_{[\lambda]\in\mathcal{A}} \lambda(f(s,t)-) \geq \bigvee_{[\lambda]\in\mathcal{A}} \lambda(g(s,t)),$$

$$\zeta(t-) = \bigwedge_{s<t} \zeta(s) \geq \bigwedge_{s<t} \bigvee_{[\lambda]\in\mathcal{A}} \lambda(g(s,t)) = \bigwedge_{r<t} \bigvee_{[\lambda]\in\mathcal{A}} \lambda(r) = \bigwedge_{r<t} \gamma(r) = \gamma(t-).$$

On the other hand,

$$\zeta(t+) \geq \bigvee_{[\lambda]\in\mathcal{A}} \lambda(t+) = \bigvee_{[\lambda]\in\mathcal{A}} \bigvee_{s>t} \lambda(s) = \bigvee_{s>t} \bigvee_{[\lambda]\in\mathcal{A}} \lambda(s) = \bigvee_{s>t} \gamma(s) = \gamma(t+).$$

So

$$[\zeta] \geq [\gamma] = [\bigvee_{[\lambda]\in\mathcal{A}} \lambda],$$

$$\bigvee_{[\lambda]\in\mathcal{A}} [\lambda] = [\gamma] = [\bigvee_{[\lambda]\in\mathcal{A}} \lambda].$$

This completes the proof of completeness of $I[L]$.

Similarly prove $\bigwedge_{[\lambda]\in\mathcal{A}} [\lambda] = [\bigwedge_{[\lambda]\in\mathcal{A}} \lambda]$. □

The proof of the following corollary is left as an exercise:

**9.1.18 Corollary** *Let $L$ be a complete lattice with an order-reversing involution $' : L \to L$. Then*

(i) *$L$ is distributive if and only if $I[L]$ is distributive.*

(ii) *$L$ fulfills the 1st infinitely distributive law* (IFD1) *if and only if $I[L]$ fulfills the 1st infinitely distributive law* (IFD1).

(iii) *$L$ fulfills the 2nd infinitely distributive law* (IFD2) *if and only if $I[L]$ fulfills the 2nd infinitely distributive law* (IFD2).

(iv) *$L$ is completely distributive if and only $I[L]$ is completely distributive.* □

Then by theorems **1.3.27** and **1.3.28**, the following conclusion holds:

**9.1.19 Theorem** *Let $L$ be a F-lattice. Then $I[L]$ equipped with the interval topology $\Omega(I[L])$ on it is a compact Hausdorff space.* □

**9.1.20 Definition** Let $L$ be a F-lattice. Define the *enhanced topology* $\mathcal{T}_L^{\mathcal{I}}$ on $L^{I[L]}$

as the $L$-fuzzy topology generated by

$$\mathcal{T}_L^I \cup \{\chi_U : U \in \Omega(I[L])\}.$$

Denote the $L$-fts $(L^{I[L]}, \mathcal{T}_L^{\tilde{I}})$ by $\tilde{I}(L)$, call $\tilde{I}(L)$ the *$L$-fuzzy enhanced unit interval.*

The $L$-fuzzy product space $\tilde{I}(L)^\kappa = \prod_{t\in T} I_t$, where $I_t = \tilde{I}(L)$ for every $t \in T$ and $|T| = \kappa$, is called an *$L$-fuzzy enhanced unit cube* of weght $\kappa$.

**9.1.21 Remark** In order to construct a standard space in $L$-fuzzy topology, Liu and Luo [83, 85] ever introduced the $L$-valued induced space of $[0,1]$ to play this role. But its properties are too strong to fit some situations. Wang and Xu [170] constructed an associate weakly induced space of $I(L)$ based on Lawson topology to improve these efforts, which is just the weakly induced modification of $I(L)$, and hence by Theorem **9.1.30** in the sequel, it is just $\tilde{I}(L)$ here.

**9.1.22 Proposition** *Let $L$ be a F-lattice. Then*

(i) $\mathcal{S}_L^I \cup \{\chi_{I[L]\setminus\Uparrow[\lambda]}, \chi_{I[L]\setminus\Downarrow[\mu]} : [\lambda], [\mu] \in I[L]\}$ *is a subbase of* $\mathcal{T}_L^{\tilde{I}}$.

(ii) $\{L_t, R_s \in L^{I[L]} : s,t \in [0,1]\} \cup \{\chi_{I[L]\setminus\Uparrow[\lambda]}, \chi_{I[L]\setminus\Downarrow[\mu]} : [\lambda], [\mu] \in I[L]\}$ *is a subbase of* $\mathcal{T}_L^{\tilde{I}}$.

(iii) $\{L_t, R_s \in L^{I[L]} : s,t \in \mathbf{Q}\} \cup \{\chi_{I[L]\setminus\Uparrow[\lambda]}, \chi_{I[L]\setminus\Downarrow[\mu]} : [\lambda], [\mu] \in I[L]\}$ *is a subbase of* $\mathcal{T}_L^{\tilde{I}}$.

(iv) $\{L_t, R_s \in L^{I[L]} : s,t \in \mathbf{Q} \cap [0,1]\} \cup \{\chi_{I[L]\setminus\Uparrow[\lambda]}, \chi_{I[L]\setminus\Downarrow[\mu]} : [\lambda], [\mu] \in I[L]\}$ *is a subbase of* $\mathcal{T}_L^{\tilde{I}}$. □

By Theorem **9.1.19**, we have straightforwardly the following conclusion:

**9.1.23 Theorem** *Let $L$ be a F-lattice. Then $\tilde{I}(L)$ is a s-$T_2$ $L$-fts.* □

**9.1.24 Definition** Let $L$ be a complete lattice with an order-reversing involution. For every two $s, t \in [0,1]$ such that $s \leq t$ and every $a \in L$, denote the monotonical decreasing mapping defined as follows by $seg(s,t,a)$, call it a *segment* in $I[L]$:

$$\forall r \in \mathbf{R}, \quad seg(s,t,a)(r) = \begin{cases} 1_L, & r < s, \\ a, & s \leq r \leq t, \\ 0_L, & r > t. \end{cases}$$

By Proposition **9.1.17**, the following definition is natural:

**9.1.25 Definition** Let $L$ be a complete lattice with an order-reversing involution $': L \to L$. Denote respectively the smallest element $[seg(0,1,0_L)]$ and the largest element $[seg(0,1,1_L)]$ of $I[L]$ by $\tilde{0}$ and $\tilde{1}$.

The verifications of the following conclusions are straightforward, leave them as exercises:

**9.1.26 Proposition** *Let $L$ be a complete lattice with an order-reversing involution $': L \to L$, $\lambda \in md_I(L)$. Then*

$$\lambda = \bigvee_{0\leq t\leq 1} seg(0,t,\lambda(t)) = \bigwedge_{0\leq t\leq 1} seg(t,1,\lambda(t)). \qquad \square$$

**9.1.27 Proposition** *Let $L$ be a complete lattice with an order-reversing involution $': L \to L$, $t \in [0,1]$, $a \in L$, $\lambda \in md_I(L)$. Then*

(i) $\lambda(t-) \geq a \iff [\lambda] \geq [seg(0,t,a)]$.
(ii) $\lambda(t+) \leq a \iff [\lambda] \leq [seg(t,1,a)]$. □

**9.1.28 Proposition** *Let $L$ be a complete lattice with an order-reversing involution $': L \to L$, $t \in [0,1]$, $a \in L$. Then in $I[L]$,*
(i) $(L_t)_{(a)} = I[L] \setminus \uparrow[seg(0,t,a')]$.
(ii) $(L_t{}')_{[a]} = \uparrow[seg(0,t,a)]$.
(iii) $(R_t)_{(a)} = I[L] \setminus \downarrow[seg(t,1,a)]$.
(iv) $(R_t{}')_{[a]} = \downarrow[seg(t,1,a')]$. □

**9.1.29 Proposition** *Let $L$ be a complete lattice with an order-reversing involution $': L \to L$, $[\lambda] \in I[L]$. Then in $I[L]$,*
(i) $\uparrow[\lambda] = \bigcap_{0\leq t\leq 1}(L_t{}')_{[\lambda(t)]}$.
(ii) $\downarrow[\lambda] = \bigcap_{0\leq t\leq 1}(R_t{}')_{[\lambda(t)']}$.
(iii) $I[L] \setminus \uparrow[\lambda] = \bigcup_{0\leq t\leq 1}(L_t)_{(\lambda(t)')}$.
(iv) $I[L] \setminus \downarrow[\lambda] = \bigcup_{0\leq t\leq 1}(R_t)_{(\lambda(t))}$. □

**9.1.30 Theorem** *Let $L$ be a F-lattice. Then on $I[L]$, the enhanced topology $\mathcal{T}_L^{\tilde{I}}$ is the weakly induced modification $wi(\mathcal{T}_L^I)$ of the standard topology $\mathcal{T}_L^I$.*

**Proof** By Proposition **9.1.28**, $\tilde{I}(L)$ is weakly induced, so $\mathcal{T}_L^{\tilde{I}} \supset wi(\mathcal{T}_L^I)$. On the other hand, $\forall [\lambda] \in I[L]$, by Proposition **9.1.29**, both $\uparrow[\lambda]$ and $\downarrow[\lambda]$ are closed subsets in $(L^{I[L]}, wi(\mathcal{T}_L^I))$, so $\mathcal{T}_L^{\tilde{I}} \subset wi(\mathcal{T}_L^I)$. □

**9.1.31 Proposition** *Let $L$ be a F-lattice. Then $[\mathcal{T}_L^{\tilde{I}}] = \Omega(I[L])$.*

**Proof** Need only prove $[\mathcal{T}_L^{\tilde{I}}] \subset \Omega(I[L])$. Suppose $U \in [\mathcal{T}_L^{\tilde{I}}]$, then by the definition of $\mathcal{T}_L^{\tilde{I}}$, $\exists \{U_t : t \in T\} \subset \mathcal{T}_L^I$, $\exists \{V_t : t \in T\} \subset \Omega(I[L])$ such that $\chi_U = \bigvee_{t\in T} U_t \wedge \chi_{V_t}$. Since $L$ is completely distributive, by Corollary **1.3.13**, $pr(L) \neq \emptyset$, $\exists \gamma \in pr(L)$. Then by Proposition **4.2.4**,

$$U = (\chi_U)_{(\gamma)} = \bigcup_{t\in T} (U_t)_{(\gamma)} \cap (\chi_{V_t})_{(\gamma)} = \bigcup_{t\in T} (U_t)_{(\gamma)} \cap V_t.$$

Since every $V_t \in \mathcal{T}_L^I$, by propositions **4.2.4** and **9.1.28**, every $(U_t)_{(\gamma)} \in \Omega(I[L])$. So $U \in \Omega(I[L])$. □

**9.1.32 Proposition** *If $L \neq \{0,1\}$, then $\tilde{I}(L)$ is not induced.*

**Proof** At points $\tilde{0}, \tilde{1} \in I[L]$ (see Definition **9.1.25**), every $R_s$ and $L_t$ take values $0_L$ and $1_L$, so does every open subset in $\mathcal{T}_L^I$. For every $U \in \Omega(I[L])$, $\chi_U$ is always crisp. So every open subset in $\mathcal{T}_L^{\tilde{I}}$ takes values from $\{0_L, 1_L\}$ at points $\tilde{0}$ and $\tilde{1}$. Hence $\tilde{I}(L)$ is not stratified if $L \neq \{0,1\}$. □

Moreover, $I[L]$ has another nice property:[170]

**9.1.33 Theorem** *$(I[L], \Omega(I[L]))$ is a connected space.*

**Proof** $\forall z \in I[L] \setminus \{\tilde{0}\}$, it is sufficient to prove that both $z$ and $\tilde{0}$ are in the same connected component. By the Kuratowski's Lemma, $\{\tilde{0}, z\}$ is contained in a maximal chain $M$ in $I[L]$, and

(i) $M = \bigcap\{(\uparrow x) \cup (\downarrow x) : x \in M\}$. In fact, denote $C = \bigcap\{(\uparrow x) \cup (\downarrow x) : x \in M\}$, then every element in $C$ is comparable with every element in $M$. Since $M$ is a maximal chain in $I[L]$, $C \subset M$. Clearly, $M \subset C$, so $M = C$.

(ii) $M$ is a compact and $T_2$ subspace of $I[L]$. This is true because every $\uparrow x$ and every $\downarrow x$ are closed subsets in compact and $T_2$ space $I[L]$.

(iii) The subspace topology of $M$ in $I[L]$ is just the interval topology of $M$. This is true because $I[L]$ has a subbase $\mathcal{B} = \{\uparrow_P x, \downarrow_P y : x, y \in I[L]\}$, and by Proposition **1.1.12**, its restriction

$$\mathcal{B}_M = \{(\uparrow_P u) \cap M, (\downarrow_P v) \cap M : u, v \in M\} = \{\uparrow_M u, \downarrow_M v : u, v \in M\},$$

is just a subbase of the interval topology of $M$.

(iv) $M$ is order dense. If it is false, the $\exists u, v \in M$ such that there is no $w$ in $M$ such that $u < w < v$. Suppose $u = [\lambda]$, $v = [\mu]$. Since $u < v$, $\exists t \in [0,1]$ such that $\lambda(t+) \leq \mu(t+)$, $\lambda(t-) \leq \mu(t-)$, and one of them is a strict inequality. Not loss any generality, suppose $\lambda(t+) < \mu(t+)$. Then $t < 1$, and easy to show that $\exists s > t$ such that $\lambda(s+) \leq \mu(s+)$. Take $\rho : \mathbf{R} \to L$ as

$$\rho(r) = \begin{cases} \mu(r), & r \in (-\infty, s), \\ \lambda(r), & r \in [s, +\infty), \end{cases}$$

then $w = [\rho] \in I[L]$ and $u < w < v$. Since $M$ is a maximal chain, $w \in M$, this is a contradiction.

(v) $M$ is a sub-complete-lattice of $I[L]$. In fact, since $M$ is a maximal chain in $I[L]$, $\tilde{0}, \tilde{1} \in M$. $\forall A \subset M$, let $u = \sup_{I[L]} A$, $M_1 = \{x \in M : \forall y \in A,\ y \leq x\}$, $M_2 = M \backslash M_1 = \{x \in M : \exists y \in A,\ x < y\}$, then $\forall x \in M_1$, $u \leq x$ and $\forall x \in M_2$, $x < n$. So $u$ is comparable with every element of $M$. Since $M$ is a maximal chain, $u \in M$. Dually $inf_{I[L]}\ A \in M$ can be proved.

(vi) $M$ is connected space. By (iii)-(v), $M$ is an order dense complete chain, so $M$ is connected space with respect to its interval topology. □

**9.1.34 Theorem** *Let $L$ be a F-lattice and* $1 \in M(L)$, *then $\tilde{I}(L)$ is a connected L-fts.*

**Proof** If $\tilde{I}(L)$ is not connected, then $\exists A, B \in \mathcal{T}_L^{\tilde{I}} \backslash \{\underline{0}\}$ such that $A \wedge B = \underline{0}$ and $A \vee B = \underline{1}$. Since $1 \in M(L)$, $\forall u \in I[L]$, by $A(u) \vee B(u) = 1$, we have $A(u) = 1$ or $B(u) = 1$. Then $B(u) = 0$ or $A(u) = 0$. So both $A$ and $B$ are crisp. So by Theorem **9.1.30** and Proposition **9.1.31**, $supp(A) = A_{(0)}, supp(B) = B_{(0)} \in [\mathcal{T}_L^{\tilde{I}}] = \Omega(I[L])$, and we have $supp(A) \cap supp(B) = \emptyset$, $supp(A) \cup supp(B) = I[L]$, $(I[L], \Omega(I[L]))$ is not connected. But this contradicts with Theorem **9.1.33**. □

## 9.2 Complete Regularity, Normality and Embedding Theory

Before $\tilde{I}(L)$ was introduced, Liu and Luo [88] constructed a new kind of $L$-fuzzy unit interval in order to strengthen the weak properties of $I(L)$:

**9.2.1 Definition** Let $L$ be a F-lattice, $\Omega(I)$ be the ordinary topology, i.e. the interval topology of unit interval $I = [0,1]$. Denote the $L$-valued induced space $(L^I, \omega_L(\Omega(I)))$ of the ordinary topological space $(I, \Omega(I))$ by $I^*(L)$, call $I^*(L)$ the *L-fuzzy $*$-unit interval.*

The $L$-fuzzy product space $I^*(L)^\kappa = \prod_{t \in T} I_t$, where $I_t = I^*(L)$ for every $t \in T$ and $|T| = \kappa$, is called an *L-fuzzy $*$-unit cube* of weght $\kappa$.

**9.2.2 Definition** Let $(L^X, \delta)$ be an $L$-fts.

$(L^X, \delta)$ is called *completely regular*, if for every $U \in \delta$, there exists a family $\{U_t : t \in T\} \subset \delta$ such that $\bigvee_{t \in T} U_t = U$, and for every $t \in T$, there exists a $L$-fuzzy continuous mapping $f_t^{\to} : (L^X, \delta) \to I(L)$ such that

(CR1) $\quad U_t \leq f_t^{\leftarrow}(L_1{}') \leq f_t^{\leftarrow}(R_0) \leq U.$

$(L^X, \delta)$ is called *completely pararegular*, or *completely p-regular* for short, if for every non-zero pseudo-crisp closed subset $P$ in $(L^X, \delta)$ and every molecule $x_\alpha \in M(L^X)$ such that $x \notin supp(P)$, there exists a $L$-fuzzy continuous mapping $f^{\to} : (L^X, \delta) \to I^*(L)$ such that

(CPR) $\quad x_\alpha \leq f^{\leftarrow}(0_{1_L}), \quad P \leq f^{\leftarrow}(1_{1_L}).$

Call $T_1$ completely pararegular $L$-fts *para-*$T_{3\frac{1}{2}}$, or *p-*$T_{3\frac{1}{2}}$ for short.

$(L^X, \delta)$ is called *completely enhanced-regular*, or *completely e-regular* for short, if for every $U \in \delta$, there exists a family $\{U_t : t \in T\} \subset \delta$ such that $\bigvee_{t \in T} U_t = U$, and for every $t \in T$, there exists a $L$-fuzzy continuous mapping $f_t^{\to} : (L^X, \delta) \to \tilde{I}(L)$ such that

(CER) $\quad U_t \leq f_t^{\leftarrow}(L_1{}') \leq f_t^{\leftarrow}(R_0) \leq U.$

$(L^X, \delta)$ is called *normal*, if for every $L$-fuzzy closed subset $P$ and every $L$-fuzzy open subset $U$ in $(L^X, \delta)$ such that $P \leq U$, there exists an $L$-fuzzy open subset $V$ in $(L^X, \delta)$ such that

$$P \leq V \leq V^- \leq U.$$

$(L^X, \delta)$ is called *paranormal*, or *p-normal* for short, if for every two non-zero pseudo-crisp closed subsets $P$ and $Q$ in $(L^X, \delta)$ such that $supp(P) \cap supp(Q) = \emptyset$, there exist $U \in \mathcal{Q}(P)$, $V \in \mathcal{Q}(Q)$ such that $U \wedge V = \underline{0}$. Call $T_1$ paranormal $L$-fts *para-*$T_4$, or *p-*$T_4$ for short.

**9.2.3 Remark** (1) The condition (CR1) is equivalent to the following one introduced by Wang:[168]

(CR2) $\quad f_t(W_t) \leq L_1{}', \quad f_t(U') \leq R_0{}'.$

(2) Complete regularity was defined by B. Hutton in [38]. The equivalent condition of (CR1) he used there is

(CR3) $\quad \forall x \in X, \quad W_t(x) \leq f_t(x)(1-) \leq f_t(x)(0+) \leq U(x),$

where, just as mentioned in Definition **9.1.1**, identify a equivalent class $[\lambda] \in I[L]$ with its a member $\lambda$. Liu introduced the clearer one (CR1) to replace it in [71].

(3) Normality was also defined by B. Hutton.[37] For establishing the fuzzy form of Urysohn Lemma, he constructed the $L$-fuzzy unit interval meanwhile. P-normality was defined by Wang,[169, 168] completely e-regularity was defined by Wang and Xu.[170] Liu and Luo [83, 85] defined completely p-regularity to overcome the difficulty of spatial property in compatifications, and investigated various relations among previously defined separation properties. [68, 71, 83, 84, 85, 86, 87]

The definition of complete regularity follows the idea of neighborhoods of points in an open subsets. Liu proved its pointwise characterization as follows:[71]

**9.2.4 Theorem** *Let $(L^X, \delta)$ be an $L$-fts. Then the following conditions are equivalent:*

(i) *$(L^X, \delta)$ is completely regular.*

(ii) *For every $p \in Pt(L^X)$ and every $U \in \mathcal{Q}(p)$, there exist an $L$-fuzzy continuous mapping $f^{\to} : (L^X, \delta) \to I(L)$ and $V \in \mathcal{Q}(p)$ such that*

$$V \leq f^{\leftarrow}(L_1{}') \leq f^{\leftarrow}(R_0) \leq U. \tag{9.8}$$

(iii) *For every* $e \in M(L^X)$ *and every* $U \in \mathcal{Q}(e)$, *there exist an* $L$*-fuzzy continuous mapping* $f^{\rightarrow} : (L^X, \delta) \to I(L)$ *and* $V \in \mathcal{Q}(e)$ *such that*

$$V \leq f^{\leftarrow}(L_1{}') \leq f^{\leftarrow}(R_0) \leq U.$$

**Proof** (i)$\Longrightarrow$(ii): By the complete regularity, $\exists\{U_t : t \in T\} \subset \delta$ such that $\bigvee_{t\in T} U_t = U$ and $\forall t \in T$, there exists an $L$-fuzzy continuous mapping $f_t{}^{\rightarrow} : (L^X, \delta) \to I(L)$ such that

$$U_t \leq f_t{}^{\leftarrow}(L_1{}') \leq f_t{}^{\leftarrow}(R_0) \leq U.$$

Since $U \in \mathcal{Q}(x_a)$, $a \not\leq U' = \bigwedge_{t\in T} U_t{}'$. So $\exists t_0 \in T$ such that $a \not\leq U_{t_0}{}'$. Take $V = U_{t_0} \in \delta$, $f = f_{t_0}$, then $V \in \mathcal{Q}(x_a)$, inequalities (9.8) holds.

(ii)$\Longrightarrow$(iii): Clear.

(iii)$\Longrightarrow$(i): $\forall U \in \delta$. If $U = \underline{0_L}$, take $\lambda \in I[L]$ as follows:

$$\forall t \in \mathbf{R}, \quad \lambda(t) = \begin{cases} 1_L, & t \leq 0, \\ 0_L, & t > 0, \end{cases}$$

and take an ordinary mapping $f : X \to I[L]$ such that $f(x) = \lambda$ for every $x \in X$. Then $\forall s, t \in [0,1]$, $\forall x \in X$,

$$f^{\leftarrow}(L_s)(x) = L_s(f(x)) = L_s(\lambda) = \lambda(s-)', \tag{9.9}$$

$$f^{\leftarrow}(R_t)(x) = R_t(f(x)) = R_t(\lambda) = \lambda(t+). \tag{9.10}$$

By the definition of $\lambda \in I[L]$, $f^{\leftarrow}(L_s), f^{\leftarrow}(R_t) \in \{\underline{0_L}, \underline{1_L}\} \subset \mathcal{T}_L^I$. By Proposition **9.1.7** (iii), $f^{\rightarrow}$ is continuous. Now take $s = 1$ in equalities (9.9) and $t = 0$ in equalities (9.10), we get $f^{\leftarrow}(L_1{}') = \underline{0_L}$, $f^{\leftarrow}(R_0) = \underline{0_L}$. Let $\{U_t : t \in T\} = \{\underline{0_L}\}$ and $f_t{}^{\rightarrow} = f^{\rightarrow}$ for every $t \in T$, we have

$$U_t = \underline{0_L} = f^{\leftarrow}(L_1{}') = f^{\leftarrow}(R_0) = \underline{0_L} = U,$$

(CR1) holds.

Suppose $U \neq \underline{0_L}$, then $U' \neq \underline{1_L}$, $\exists e \in M(L^X)$ such that $e \lessdot U$, $U \in \mathcal{Q}(e)$. By (iii), there exists an $L$-fuzzy continuous mapping $f_e{}^{\rightarrow} : (L^X, \delta) \to I(L)$ and $U_e \in \mathcal{Q}(e)$ such that

$$U_e \leq f_e{}^{\leftarrow}(L_1{}') \leq f_e{}^{\leftarrow}(R_0) \leq U. \tag{9.11}$$

Denote $W = \bigvee\{U_e : e \in M(L^X),\ e \lessdot U\}$, then $W \leq U$. If $W \neq U$, then $U \not\leq W$, $W' \not\leq U'$, $\exists d \in M(\downarrow W')$, $d \not\leq U'$. That is to say, $d \leq W' = \bigwedge\{U_e{}' : e \in M(L^X),\ e \lessdot U\}$, $\forall e \in M(L^X)$, $e \lessdot U \Longrightarrow d \leq U_e{}'$. Especially, by $d \in M(L^X)$, $d \not\leq U'$, we have $d \lessdot U$, and then $d \leq U_d{}'$. But by the definition of $U_d$, $U_d \in \mathcal{Q}(d)$, this is a contradicition. Hence $\bigvee\{U_e : e \in M(L^X),\ e \lessdot U\} = W = U$. By (9.11), (i) is true. □

**9.2.5 Theorem** *Complete regularity and complete regularity are hereditary.* □

**9.2.6 Theorem** *Normality, complete p-regularity and p-normality are hereditary with respect to closed subspaces.* □

**9.2.7 Theorem** *Complete regularity, complete p-regularity, complete e-regularity, normality and p-normality are preserved by L-fuzzy homeomorphisms.* □

**9.2.8 Proposition** *Complete e-regularity $\Longrightarrow$ cmplete regularity $\Longrightarrow$ regularity.*

**Proof** Since the standard topology $\mathcal{T}_L^I$ is contained in the topology $\mathcal{T}_L^{\tilde{I}}$ of the $L$-fuzzy enhanced unit interval $\tilde{I}(L)$, the first implication is clear.

Let $(L^X, \delta)$ be a completely regular $L$-fts, $U \in \delta$, then by the complete regularity of $(L^X, \delta)$, $\exists\{U_t : t \in T\} \subset \delta$ such that $\bigvee_{t\in T} U_t = U$, $\forall t \in T$, there exists continuous

$f_t^{\rightarrow}: (L^X, \delta) \to I(L)$ such that (CR1) holds. $\forall t \in T$, since $f_t^{\rightarrow}$ is continuous, $L_1{}'$ is closed in $I(L)$, $f_t^{\leftarrow}(L_1{}')$ is closed in $(L^X, \delta)$. So $U_t^- \leq f_t^{\leftarrow}(L_1{}')^- \leq f_t^{\leftarrow}(L_1{}') \leq U$, $(L^X, \delta)$ is regular. □

**9.2.9 Proposition** *Complete p-regularity $\Longrightarrow$ s-p-regularity.* □

**9.2.10 Proposition** *Every completely p-regular L-fts is stratified.* □

The definition of normality is in the form of neighborhood. But it can also be characterized by the terminology of Q-neighborhood:

**9.2.11 Throrem** *Let $(L^X, \delta)$ be an L-fts. Then $(L^X, \delta)$ is normal if and only if for every two closed subsets $P$ and $Q$ in $(L^X, \delta)$ such that $P$ does not quasi-coincide with $Q$, there exists open subsets $U$ and $V$ in $(L^X, \delta)$ such that $P \leq U$, $Q \leq V$ and $U$ does not quasi-coincide with $V$.*

**Proof** (Necessity) Suppose $P, Q \in \delta'$, $P \neg\hat{q} Q$, then $P \leq Q' \in \delta$. By the normality of $(L^X, \delta)$, $\exists U \in \delta$ such that $P \leq U \leq U^- \leq Q'$. Take $V = U^{-\prime} \in \delta$, then $Q \leq U^{-\prime} = V$, $U \leq U^- = V'$, $U \neg\hat{q} V$.

(Sufficiency) Suppose $P \in \delta'$, $U \in \delta$, $P \leq U$. Take $Q = U'$, then $P \leq U = Q'$, $P \neg\hat{q} Q$. So $\exists V, W \in \delta$, $P \leq V$, $Q \leq W$, $V \neg\hat{q} W$, $V \leq W' \in \delta'$. So

$$V^- \leq W'^- = W' \leq Q' = U.$$

That is just $P \leq V \leq V^- \leq U$, $(L^X, \delta)$ is normal. □

**9.2.12 Theorem** *Let $(L^X, \delta)$ be an L-fts, $P_0$ a closed subset and $U_0$ an open subset in $(L^X, \delta)$ such that $P_0 \leq U_0$. Then the following conditions are equivalent:*

(i) *For every closed subset $P$ and every open subset $U$ in $(L^X, \delta)$ such that $P_0 \leq P \leq U \leq U_0$, there exists an open subset $V$ such that*
$$P \leq V \leq V^- \leq U.$$

(ii) *For every closed subset $P$ and every open subset $U$ in $(L^X, \delta)$ such that $P_0 \leq P \leq U \leq U_0$, there exists the family $\{V_t : 0 < t \leq 1\}$ of open subsets in $(L^X, \delta)$ such that*
$$0 < s < t \leq 1 \Longrightarrow P \leq V_s^- \leq V_t \leq U \tag{9.12}$$

(iii) *For every closed subset $P$ and every open subset $U$ in $(L^X, \delta)$ such that $P_0 \leq P \leq U \leq U_0$, there exists an L-fuzzy continuous mapping $f: (L^X, \delta) \to I(L)$ such that*
$$P \leq f^{\leftarrow}(L_1{}') \leq f^{\leftarrow}(R_0) \leq U. \tag{9.13}$$

**Proof** (i)$\Longrightarrow$(ii): Since the set of all the rational numbers in $(0,1]$ is countable, so we can assume $\mathbf{Q} \cap (0,1] = \{r_n : n < \omega\}$. With the inductive method, by (i), one can easily find a family $\{V_{r_n} : n < \omega\} \subset \delta$ such that

$$r_m, r_n \in \mathbf{Q} \cap (0,1], \ r_m < r_n \Longrightarrow P \leq V_{r_m}^- \leq V_{r_n} \leq U. \tag{9.14}$$

$\forall t \in (0,1]$, take

$$V_t = \bigvee\{V_{r_n} : \exists r_n \in \mathbf{Q} \cap (0,1], \ r_n \leq t\}, \tag{9.15}$$

then for every two $s, t \in (0,1]$ such that $s < t$, $\exists r_m, r_n \in \mathbf{Q} \cap (0,1]$ such that $s < r_m < r_n < t$. By definition (9.15) and condition (9.14),

$$V_s \leq V_{r_m} \leq V_{r_m}^-,$$
$$P \leq V_s \leq V_s^- \leq V_{r_m}^- \leq V_{r_n} \leq V_t \leq U.$$

So (ii) is proved.

(ii)$\Longrightarrow$(iii): $\forall x \in X$, take $f(x) \in L^{\mathbf{R}}$ as follows:

$$\forall t \in \mathbf{R}, \quad f(x)(t) = \begin{cases} 1_L, & t < 0, \\ V_{1-t}(x), & 0 \leq t \leq 1, \\ 0_L, & t > 1, \end{cases} \tag{9.16}$$

then by (ii), $\forall x \in X$, $f(x) \in I[L]$, $f: X \to I[L]$ is an ordinary mapping. Now prove the $L$-fuzzy continuity of $f^{\to}$. By Proposition **9.1.7** (iii) and Proposition **9.1.2** (iii) and (iv), we need only prove $f^{\leftarrow}(L_s{}') \in \delta'$, $f^{\leftarrow}(R_t) \in \delta$ for every $s \in (0,1]$ and every $t \in [0,1)$. $\forall x \in X$, $\forall s \in (0,1]$, $\forall t \in [0,1)$, by definition (9.16) and relation (9.12) in (ii), we have

$$\begin{aligned} f^{\leftarrow}(L_s{}')(x) &= L_s{}'(f(x)) = f(x)(s-) = \bigwedge_{0<r<s} f(x)(r) = \bigwedge_{0<r<s} V_{1-r}(x) \\ &= (\bigwedge_{0<r<s} V_{1-r})(x) = (\bigwedge_{0<r<s} V_{1-r}{}^{-})(x), \end{aligned} \tag{9.17}$$

$$\begin{aligned} f^{\leftarrow}(R_t)(x) &= R_t(f(x)) = f(x)(t+) = \bigvee_{t<r<1} f(x)(r) = \bigvee_{t<r<1} V_{1-r}(x) \\ &= (\bigvee_{t<r<1} V_{1-r})(x). \end{aligned} \tag{9.18}$$

So

$$\begin{aligned} &\forall s \in (0,1], \quad f^{\leftarrow}(L_s{}') = \bigwedge_{0<r<s} V_{1-r}{}^{-} \in \delta', \\ &\forall t \in [0,1), \quad f^{\leftarrow}(R_t) = \bigvee_{t<r<1} V_{1-r} \in \delta, \end{aligned} \tag{9.19}$$

$f^{\to}$ is continuous.

Now for the inequality (9.13) in (iii), by (ii) and equality (9.19),

$$P \leq \bigwedge_{0<r<1} V_{1-r}{}^{-} = f^{\leftarrow}(L_1{}') \leq f^{\leftarrow}(R_0) = \bigvee_{t<r<1} V_{1-r} \leq U,$$

(iii) is true.

(iii)$\Longrightarrow$(i): Suppose $P \in \delta'$, $U \in \delta$, $P_0 \leq P \leq U \leq U_0$, then by (iii), there exists an $L$-fuzzy continuous mapping $f^{\to}: (L^X, \delta) \to I(L)$ fulfilling inequalities (9.13). Then by Proposition **9.1.4**,

$$P \leq f^{\leftarrow}(L_1{}') \leq f^{\leftarrow}(R_{0.5}) \leq f^{\leftarrow}(R_{0.5})^{-} \leq f^{\leftarrow}(R_{0.5}{}^{-}) \leq f^{\leftarrow}(R_0) \leq U.$$

Take $V = f^{\leftarrow}(R_{0.5})$, we get (i). □

Then we obtain Hutton's $L$-fuzzy form of the Urysohn's Lemma:[37]

**9.2.13 Theorem** (Urysohn's Lemma for Normality) *Let $(L^X, \delta)$ be an L-fts. Then $(L^X, \delta)$ is normal if and only if for every closed subset $P$ and every open subset $U$ in $(L^X, \delta)$ such that $P \leq U$, there exists a continuous L-fuzzy mapping $f^{\to}: (L^X, \delta) \to I(L)$ such that*

$$P \leq f^{\leftarrow}(L_1{}') \leq f^{\leftarrow}(R_0) \leq U.$$

**Proof** (Necessity) By Theorem **9.2.12** (i)$\Longrightarrow$(ii).

(Sufficiency) Suppose $P$ is a closed subset and $U$ an open subset in $(L^X, \delta)$ such that $P \leq U$, then by $L_1{}' \leq R_{0.5} \leq L_{0.5}{}' \leq R_0$ and $L_{0.5}{}' \in (\mathcal{T}_L^I)'$, take $V = f^{\leftarrow}(R_{0.5})$, we have

$$\begin{aligned} &V^{-} \leq f^{\leftarrow}(L_{0.5}{}'^{-}) = f^{\leftarrow}(L_{0.5}{}'), \\ &P \leq f^{\leftarrow}(L_1{}') \leq V \leq V^{-} \leq f^{\leftarrow}(L_{0.5}{}') \leq f^{\leftarrow}(R_0) \leq U, \end{aligned}$$

$(L^X, \delta)$ is normal. □

By the Urysohn's Lemma, the following result is easy to be proved:

**9.2.14 Theorem** *Every $T_1$ normal L-fts is completely regular.* □

Also easy to find

**9.2.15 Theorem** *Every regular and normal L-fts is completely regular.* □

By the Urysohn's Lemma for p-normality, we also have the following conclusion:

**9.2.16 Proposition** $P\text{-}T_4 \implies p\text{-}T_{3\frac{1}{2}}$. □

T. Kubiak proved the L-fuzzy form of the Tietze Theorem as follows (for its proof, see [54]):

**9.2.17 Theorem** (Fuzzy Tietze Theorem) *Let $(L^X,\delta)$ be an L-fts, $(L^Y,\delta|_Y)$ a closed subspace of $(L^X,\delta)$, $f^{\rightarrow}: (L^Y,\delta|_Y) \to I(L)$ an L-fuzzy continuous mapping. Then there exists an L-fuzzy continuous mapping $g^{\rightarrow}: (L^X,\delta) \to I(L)$ such that $g|_Y = f$.* □

**9.2.18 Definition** Let $(L^X,\delta)$ be an L-fts, $A \in L^X$.

$A$ is called a *$F_\sigma$-set* in $(L^X,\delta)$, if $A$ is a join of countable number of closed subsets in $(L^X,\delta)$; called a *$G_\delta$-set* in $(L^X,\delta)$, if $A$ is a meet of countable number of open subsets in $(L^X,\delta)$.

**9.2.19 Definition** Let $(L^X,\delta)$ be an L-fts, $A, B \in L^X$.

Say $A$ is *absolutely contained in* $B$, denoted by $A \lessdot B$, if

$$A \le B^\circ, \quad A^- \le B.$$

**9.2.20 Remark** Easy to find that the notion "absolutely containing" follows the idea of "separated sets." In fact, in an L-fts, if the value-domain $L$ is Boolean, then $A$ is absolutely contained in $B$ if and only if $A$, $B'$ are separated from each other.

**9.2.21 Proposition** *Let $(L^X,\delta)$ be an L-fts, $A, B \in L^X$. Then*

$$A \lessdot B \iff B' \lessdot A'.$$ □

Then we can prove a useful property of normality in L-fts':

**9.2.22 Theorem** *Let $(L^X,\delta)$ be a normal L-fts, $A$ a $F_\sigma$-set and $B$ a $G_\delta$-set in $(L^X,\delta)$, $A \lessdot B$. Then there exists an open subset $U$ such that*

$$A \le U \le U^- \le B.$$

**Proof** Suppose $\{P_i : i < \omega\} \subset \delta'$, $\{U_j : j < \omega\} \subset \delta$ such that $A = \bigvee_{i<\omega} P_i$, $B = \bigwedge_{j<\omega} U_j$. We are going to use inductive method to construct two countable families

$$\mathcal{V} = \{V_i : i < \omega\} \subset \delta, \qquad \mathcal{Q} = \{Q_j : j < \omega\} \subset \delta'$$

such that $\forall k < \omega$, the following conditions are satisfied:

(1) $\forall i, j \le k, \quad V_i \le Q_j$.

(2) $\forall i, j \le k, \quad V_i^- \le B^\circ, \ A^- \le Q_j{}^\circ$.

(3) $\bigvee_{i\le k} P_i \le V_k, \quad Q_k \le \bigwedge_{j\le k} U_j$.

Since $A \lessdot B$,

$$\bigvee_{i<\omega} P_i = A \le B = \bigwedge_{j<\omega} U_j. \tag{9.20}$$

So $P_0 \le U_0$. By the normality of $(L^X,\delta)$, $\exists O_0 \in \delta$ such that $P_0 \le O_0 \le O_0{}^- \le U_0$. Since $A \lessdot B$, $P_0 \le A \le B^\circ$, $P_0 \le B^\circ \wedge O_0$, $\exists S_0 \in \delta$ such that $P_0 \le S_0 \le S_0{}^- \le B^\circ \wedge O_0$. Since $A \lessdot B$, $A^- \le B \le U_0$, $A^- \vee O_0{}^- \le U_0$, $\exists T_0 \in \delta$ such that $A^- \vee O_0{}^- \le T_0 \le T_0{}^- \le U_0$. Take $V_0 = S_0$, $Q_0 = T_0{}^-$, we have

$$P_0 \le V_0 \le V_0{}^- \le B^\circ \wedge V_0 \le V_0 \le V_0{}^- \le A^- \vee V_0{}^- \le Q_0{}^\circ \le Q_0 \le U_0.$$

So conditions (1) – (3) are satisfied for $k = 0$. Suppose conditions (1) – (3) are satisfied for $k < \omega$. By inequalities (9.20), $\bigvee_{i\le k+1} P_i \le \bigwedge_{j\le k+1} U_j$. Since $\bigvee_{i\le k+1} P_i \in \delta'$, $\bigwedge_{j\le k+1} U_j \in \delta$, by the normality of $(L^X, \delta)$, $\exists O_{k+1} \in \delta$ such that

$$\bigvee_{i\le k+1} P_i \le O_{k+1} \le O_{k+1}^- \le \bigwedge_{j\le k+1} U_j.$$

Since $A \le B$, we have

$$\bigvee_{i\le k+1} P_i \le A \le B^\circ, \quad A^- \le B \le \bigwedge_{j\le k+1} U_j,$$

so $\exists S_{k+1}, T_{k+1} \in \delta$ such that

$$\bigvee_{i\le k+1} P_i \le S_{k+1} \le S_{k+1}^- \le B^\circ \wedge O_{k+1}, \tag{9.21}$$

$$A^- \vee O_{k+1}^- \le T_{k+1} \le T_{k+1}^- \le \bigwedge_{j\le k+1} U_j. \tag{9.22}$$

Take

$$V_{k+1} = S_{k+1} \wedge \bigwedge_{j\le k} Q_j^\circ, \quad Q_{k+1} = T_{k+1}^- \vee \bigvee_{i\le k} V_i^-. \tag{9.23}$$

By inductive assumption (2), we have

$$\bigvee_{i\le k+1} P_i \le A \le A^- \le \bigwedge_{j\le k} Q_j^\circ, \quad \bigvee_{i\le k} V_i^- \le B^\circ \le B \le \bigwedge_{j\le k+1} U_j,$$

so by inequalities (9.21), (9.22),

$$\bigvee_{i\le k+1} P_i \le S_{k+1} \wedge \bigwedge_{j\le k} Q_j^\circ = V_{k+1}, \quad Q_{k+1} = T_{k+1}^- \vee \bigvee_{i\le k} V_i^- \le \bigwedge_{j\le k+1} U_j,$$

condition (3) holds for $k+1$. By inequalities (9.21), (9.22),

$$V_{k+1}^- = (S_{k+1} \wedge \bigwedge_{j\le k} Q_j^\circ)^- \le S_{k+1}^- \le B^\circ \wedge O_{k+1} \le B^\circ,$$

$$A^- \le A^- \vee O_{k+1}^- \le T_{k+1} \le T_{k+1}^{-\circ} \le (T_{k+1}^- \vee \bigvee_{i\le k} V_i^-)^\circ = Q_{k+1}^\circ,$$

condition (2) holds for $k+1$. As for condition (1), suppose $i, j \le k+1$. If $i, j \le k$, nothing needs to prove. If $i = k+1$, $j \le k$ or $i \le k$, $j = k+1$, (1) holds by equalities (9.23). If $i = j = k+1$, by inequalities (9.21), (9.22) and equalities (9.23),

$$V_{k+1} \le S_{k+1}^- \le B^\circ \wedge O_{k+1} \le A^- \vee O_{k+1}^- \le T_{k+1}^- \le Q_{k+1},$$

condition (1) still holds. So we have constructed $\mathcal{V} = \{V_i : i < \omega\} \subset \delta$ and $\mathcal{Q} = \{Q_j : j < \omega\} \subset \delta'$ satisfying conditions (1) – (3) for every $k < \omega$. Take

$$U = \bigvee \mathcal{V} \in \delta, \quad Q = \bigwedge \mathcal{Q} \in \delta',$$

by conditions (1) – (3) we have

$$A = \bigvee_{i<\omega} P_i \le U \le U^- \le Q^- = Q \le \bigwedge_{j<\omega} U_j = B.$$

This inequality produces what we need. □

**9.2.23 Corollary** *Let $(X, \mathcal{T})$ be a normal topological space, $A$ and $B$ the separated $F_\sigma$-sets in $(X, \mathcal{T})$. Then there exists disjoint open subsets $U$ and $V$ in $(X, \mathcal{T})$ such that $A \subset U$, $B \subset V$.* □

The following definitions belong to Hutton and Reilly:[41]

**9.2.24 Definition** Let $(L^X, \delta)$ be an $L$-fts.

$(L^X, \delta)$ is called *completely normal*, if for every two $A, B \in L^X$ such that $A \le B$, there exists an open subset $U$ in $(L^X, \delta)$ such that $A \le U \le U^- \le B$.

$(L^X, \delta)$ is called *perfectly normal*, if $(L^X, \delta)$ is normal and every open subset in $(L^X, \delta)$ is a $F_\sigma$-set.

The following theorem was proved by Kubiak,[54] but here we give out a direct proof different from his original one:

**9.2.25 Theorem** *Let $(L^X, \delta)$ be a regular L-fts with $w(\delta) \leq \omega$. Then $(L^X, \delta)$ is perfectly normal.*

**Proof** Since $w(\delta) \leq \omega$, by the regularity of $(L^X, \delta)$ and Proposition **1.1.27** (i), we have

$$U \in \delta \implies \exists \{U_i : i < \omega\} \subset \delta,\ U = \bigvee_{i<\omega} U_i = \bigvee_{i<\omega} U_i^-. \tag{9.24}$$

So every open subset in $(L^X, \delta)$ is a $F_\sigma$-set, we need only prove the normality of $(L^X, \delta)$.

Suppose $P \in \delta'$, $U \in \delta$, $P \leq U$. By the virtue of implication (9.24), $\exists \{P_i : i < \omega\} \subset \delta'$, $\exists \{U_j : j < \omega\} \subset \delta$ such that

$$P = \bigwedge_{i<\omega} P_i = \bigwedge_{i<\omega} P_i^\circ, \qquad U = \bigvee_{j<\omega} U_j = \bigvee_{j<\omega} U_j^-. \tag{9.25}$$

For every two $i, j < \omega$, let

$$V_i = U_i \wedge \bigwedge_{k\leq i} P_k^\circ, \qquad Q_j = P_j \vee \bigvee_{k\leq j} U_k^-,$$

then $V_i \in \delta$, $Q_j \in \delta'$. We say the following implication holds:

$$i, j < \omega \implies U_i \leq Q_j. \tag{9.26}$$

In fact, if $i \leq j$,

$$V_i = U_i \wedge \bigwedge_{k\leq i} P_k^\circ \leq U_i^- \leq P_j \vee \bigvee_{k\leq j} U_k^- = P_j;$$

if $i \geq j$,

$$V_i = U_i \wedge \bigwedge_{k\leq i} P_k^\circ \leq \bigwedge_{k\leq i} P_k^\circ \leq P_j^\circ \leq P_j \vee \bigvee_{k\leq j} U_k^- = P_j.$$

So implication (9.26) is always true. Then by $P \leq U$, equalities (9.25) and implication (9.26),

$$P = P \wedge U = \bigvee_{i<\omega} (U_i \wedge P) \leq \bigvee_{i<\omega} (U_i \wedge \bigwedge_{k\leq i} P_k^\circ) = \bigvee_{i<\omega} V_i \leq \bigwedge_{j<\omega} Q_j$$
$$= \bigwedge_{j<\omega} (P_j \vee \bigvee_{k\leq j} U_k^-) \leq \bigwedge_{j<\omega} (P_j \vee U) = (\bigwedge_{j<\omega} P_j) \vee U = P \vee U = U.$$

So take $V = \bigvee_{i<\omega} V_i$, $Q = \bigwedge_{j<\omega} Q_j$, we have $V \in \delta$, $Q \in \delta'$ and

$$P \leq V \leq V^- \leq Q^- = Q \leq U,$$

$(L^X, \delta)$ is normal and hence perfectly normal. □

**9.2.26 Proposition** Let $L$ be a F-lattice. Then $L$-fuzzy unit interval $I(L)$ is regular.

**Proof** By the virtue of Theorem **8.3.7**, we need only to verify the regularity for $U = R_a \wedge L_b$, where $a, b \in \mathbf{R}$. By Proposition **9.1.2** (iii), (iv),

$$R_a = \bigvee_{t>0} R_{a+t}, \qquad L_b = \bigvee_{s>0} L_{b-s}.$$

By the infinite distributivity of $L$,

$$R_a \wedge L_b = \bigvee\{R_{a+t} \wedge L_{b-s} : t > 0,\ s > 0\} \geq \bigvee\{R_{a+t} \wedge L_{b-t} : t > 0\}.$$

On the other hand, for every $R_{a+\varepsilon} \wedge L_{b-\sigma}$, take $t = min\{\varepsilon, \sigma\}$, we have $R_{a+\varepsilon} \wedge L_{b-\sigma} \leq R_{a+t} \wedge L_{b-t}$. So

$$U = R_a \wedge L_b = \bigvee\{R_{a+t} \wedge L_{b-t} : t > 0\},$$

we need only prove $(R_{a+t} \wedge L_{b-t})^- \leq U$ for every $t > 0$. By Proposition **9.1.4**,

$$(R_{a+t} \wedge L_{b-t})^- \leq R_{a+t}^- \wedge L_{b-t}^- \leq R_a \wedge L_b = U.$$

So $I(L)$ is regular. □

Then the following conclusion is produced from Proposition **9.2.26**, Proposition **9.1.13** and Theorem **9.2.25**:

**9.2.27 Corollary** *Let $L$ be a F-lattice. Then $I(L)$ is perfectly normal.* □

Then by Proposition **9.2.26** and Theorem **9.2.15** we get

**9.2.28 Corollary** *Let $L$ be a F-lattice. Then $I(L)$ is completely regular.* □

The family $\{U_t : t \in T\}$ of open subsets in the definition of complete regularity can also be changed to a family of arbitrary subsets or a family of closed subsets just as follows:

**9.2.29 Theorem** *Let $(L^X, \delta)$ is an L-fts. Then the following conditions are equivalent:*

(i) *$(L^X, \delta)$ is completely regular.*

(ii) *For every $U \in \delta$, there exists a family $\{W_t : t \in T\} \subset L^X$ such that $\bigvee_{t\in T} W_t = U$, and for every $t \in T$, there exists a continuous L-fuzzy mapping $f_t : (L^X, \delta) \to I(L)$ such that*

$$W_t \leq f_t^{\leftarrow}(L_1{}') \leq f_t^{\leftarrow}(R_0) \leq U.$$

(iii) *For every $U \in \delta$, there exists a family $\{W_t : t \in T\}$ of closed subsets in $(L^X, \delta)$ such that $\bigvee_{t\in T} W_t = U$, and for every $t \in T$, there exists a continuous L-fuzzy mapping $f_t : (L^X, \delta) \to I(L)$ such that*

$$W_t \leq f_t^{\leftarrow}(L_1{}') \leq f_t^{\leftarrow}(R_0) \leq U.$$

**Proof** (i)$\Longrightarrow$(iii): Proved by replacing $U_t$ in (CR1) with $f_t^{\leftarrow}(L_1{}') \in \delta'$.

(iii)$\Longrightarrow$(ii): Clear.

(ii)$\Longrightarrow$(i): Suppose $U \in \delta$. By (ii), $\exists\{W_t : t \in T\} \subset L^X$ such that $\bigvee_{t\in T} W_t = U$, and for every $t \in T$, there exists a continuous $L$-fuzzy mapping $f_t : (L^X, \delta) \to I(L)$ such that

$$W_t \leq f_t^{\leftarrow}(L_1{}') \leq f_t^{\leftarrow}(R_0) \leq U. \tag{9.27}$$

For a arbitrarily fixed $t \in T$, $\forall s \in [0,1]$, take $V_s = f_t^{\leftarrow}(R_{\frac{1-s}{2}}) \in \delta$. Then for every two $r, s \in [0,1]$ such that $r < s$, by Proposition **9.1.4** (ii), we have $R_{\frac{1-r}{2}}{}^{-} \leq R_{\frac{1-s}{2}}$. So by the continuity of $f_t$,

$$V_r{}^{-} \leq f_t^{\leftarrow}(R_{\frac{1-r}{2}}{}^{-}) \leq f_t^{\leftarrow}(R_{\frac{1-s}{2}}) = V_s.$$

Since

$$f_t^{\leftarrow}(R_{\frac{1}{2}}) \leq f_t^{\leftarrow}(R_{\frac{1-r}{2}}) = V_r,$$

$$V_s = f_t^{\leftarrow}(R_{\frac{1-s}{2}}) \leq f_t^{\leftarrow}(R_0) \leq U,$$

by inequalities (9.27) and Proposition **9.1.4** (ii), we have

$$W_t \leq f_t^{\leftarrow}(L_1{}') \leq f_t^{\leftarrow}(R_{\frac{1}{2}}) \leq V_r{}^{-} \leq V_s \leq U.$$

By Theorem **9.2.12** (ii)$\Longrightarrow$(iii), $\exists g_t{}^{\rightarrow} : (L^X, \delta) \to I(L)$ such that

$$W_t \leq f_t^{\leftarrow}(R_{\frac{1}{2}}) \leq g_t^{\leftarrow}(L_1{}') \leq g_t^{\leftarrow}(R_0) \leq U.$$

Replace $W_t$ with $f_t^{\leftarrow}(R_{\frac{1}{2}})$, and $f_t$ with $g_t$, (i) is proved. □

**9.2.30 Lemma** *Let $L$ be an infinitely distributive lattice with an order-reversing involution $'$ : $L \to L$, $\lambda, \mu \in md_I(L)$. Then*

(i) $\forall t \in \mathbf{R}$, $(\lambda \vee \mu)(t-) = \lambda(t-) \vee \mu(t-)$.

(ii) $\forall t \in \mathbf{R}$, $(\lambda \vee \mu)(t+) = \lambda(t+) \vee \mu(t+)$.

(iii) $\forall t \in \mathbf{R}$, $(\lambda \wedge \mu)(t-) = \lambda(t-) \wedge \mu(t-)$.
(iv) $\forall t \in \mathbf{R}$, $(\lambda \wedge \mu)(t+) = \lambda(t+) \wedge \mu(t+)$.

**Proof** (i) Clearly $(\lambda \vee \mu)(t-) \geq \lambda(t-) \vee \mu(t-)$. By the 2nd infinitely distributive law (IFD2) and $\lambda, \mu \in md_I(L)$,

$$\begin{aligned}\lambda(t-) \vee \mu(t-) &= (\bigwedge_{r<t} \lambda(r)) \vee (\bigwedge_{s<t} \mu(s)) \\ &= \bigwedge_{r,s<t} (\lambda(r) \vee \mu(s)) \\ &= (\bigwedge_{r<s<t} (\lambda(r) \vee \mu(s))) \wedge (\bigwedge_{s\leq r<t} (\lambda(r) \vee \mu(s))) \\ &\geq (\bigwedge_{r<s<t} (\lambda(s) \vee \mu(s))) \wedge (\bigwedge_{s\leq r<t} (\lambda(r) \vee \mu(r))) \\ &= \bigwedge_{s<t} (\lambda(s) \vee \mu(s)) \\ &= \bigwedge_{s<t} (\lambda \vee \mu)(s) \\ &= (\lambda \vee \mu)(t-).\end{aligned}$$

So $(\lambda \vee \mu)(t-) = \lambda(t-) \vee \mu(t-)$.

(ii) $(\lambda \vee \mu)(t+) = \bigvee_{s>t}(\lambda(s) \vee \mu(s)) = (\bigvee_{s>t} \lambda(s)) \vee (\bigvee_{s>t} \mu(s)) = \lambda(t+) \vee \mu(t+)$.

Similarly prove (iii) and (iv). □

**9.2.31 Proposition** *Let $L$ be an infinitely distributive lattice with an order-reversing involution $'$ : $L \to L$. Then*

(i) *For every $t \in \mathbf{R}$, $L_t'$ : $I[L] \to L$ is a lattice homomorphism.*
(ii) *For every $t \in \mathbf{R}$, $R_t$ : $I[L] \to L$ is a lattice homomorphism.*

**Proof** (i) $[seg(0,1,0_L)]$ and $[seg(0,1,1_L)]$ are clearly the smallest element and the largest element in $I[L]$ respectively. Easy to find

$$L_t'([seg(0,1,0_L)]) = 0_L,\ L_t'([seg(0,1,1_L)]) = 1_L,$$

so $L_t'$ preserves the join and the meet of empty set. Suppose $[\lambda], [\mu] \in I[L]$, then by Proposition **9.1.17** and Lemma **9.2.30**,

$$L_t'([\lambda]\vee[\mu]) = L_t'([\lambda\vee\mu]) = (\lambda\vee\mu)(t-) = \lambda(t-)\vee\mu(t-) = L_t'([\lambda])\vee L_t'([\mu]).$$

Similarly prove $L_t'([\lambda] \wedge [\mu]) = L_t'([\lambda]) \wedge L_t'([\mu])$, $L_t'$ is a lattice homomorphism.

(ii) Similar to (i). □

**9.2.32 Proposition** *Let $(L^X, \delta)$ be an L-fts, $f^{\rightarrow}, g^{\rightarrow}$ : $(L^X, \delta) \to I(L)$ be L-fuzzy continuous mappings, $t \in \mathbf{R}$. Then*

(i) $(f \vee g)^{\leftarrow}(L_t) = f^{\leftarrow}(L_t) \wedge g^{\leftarrow}(L_t)$.
(ii) $(f \vee g)^{\leftarrow}(R_t) = f^{\leftarrow}(R_t) \vee g^{\leftarrow}(R_t)$.
(iii) $(f \wedge g)^{\leftarrow}(L_t) = f^{\leftarrow}(L_t) \vee g^{\leftarrow}(L_t)$.
(iv) $(f \wedge g)^{\leftarrow}(R_t) = f^{\leftarrow}(R_t) \wedge g^{\leftarrow}(R_t)$.

**Proof** (i) $\forall x \in X$, by Proposition **9.2.31**,

$$\begin{aligned}(f \vee g)^{\leftarrow}(L_t)(x) &= L_t(f(x) \vee g(x)) \\ &= L_t(f(x)) \wedge L_t(g(x)) \\ &= f^{\leftarrow}(L_t)(x) \wedge g^{\leftarrow}(L_t)(x) \\ &= (f^{\leftarrow}(L_t) \wedge g^{\leftarrow}(L_t))(x).\end{aligned}$$

So $(f \vee g)^{\leftarrow}(L_t) = f^{\leftarrow}(L_t) \wedge g^{\leftarrow}(L_t)$.

(ii), (iii) and (iv) are similar to (i). □

**9.2.33 Corollary** *Let $(L^X, \delta)$ be an L-fts, $f^{\rightarrow}, g^{\rightarrow}$ : $(L^X, \delta) \to I(L)$ be L-fuzzy continuous mappings. Then both $(f \vee g)^{\rightarrow}$ and $(f \wedge g)^{\rightarrow}$ are L-fuzzy continuous* *mappings from* $(L^X, \delta)$ *to* $I(L)$. □

Then not difficult to verify the following conclusions with the preceding corollary:

**9.2.34 Theorem** *Let* $(L^X, \delta)$ *be an L-fts,* $\mathcal{P}$ *a subbase of* $\delta$, $\mathcal{B}$ *a base of* $\delta$. *Then the following conditions are equivalent:*

(i) $(L^X, \delta)$ *is completely regular.*

(ii) *For every* $p \in Pt(L^X)$ *and every* $U \in \mathcal{P}$ *such that* $p \lessdot U$, *there exists an L-fuzzy continuous mapping* $f^{\rightarrow}: (L^X, \delta) \to I(L)$ *and* $V \in \mathcal{Q}(p)$ *such that*
$$V \leq f^{\leftarrow}(L_1{}') \leq f^{\leftarrow}(R_0) \leq U.$$

(iii) *For every* $e \in M(L^X)$ *and every* $U \in \mathcal{B}$ *such that* $p \lessdot U$, *there exists an L-fuzzy continuous mapping* $f^{\rightarrow}: (L^X, \delta) \to I(L)$ *and* $V \in \mathcal{Q}(p)$ *such that*
$$V \leq f^{\leftarrow}(L_1{}') \leq f^{\leftarrow}(R_0) \leq U.$$

(iv) *For every* $U \in \mathcal{B}$, *there exists a family* $\{W_t : t \in T\} \subset L^X$ *such that* $\bigvee_{t\in T} W_t = U$, *and for every* $t \in T$, *there exists a continuous L-fuzzy mapping* $f_t: (L^X, \delta) \to I(L)$ *such that*
$$W_t \leq f_t{}^{\leftarrow}(L_1{}') \leq f_t{}^{\leftarrow}(R_0) \leq U.$$

(v) *For every* $U \in \mathcal{B}$, *there exists a family* $\{W_t : t \in T\}$ *of closed subsets in* $(L^X, \delta)$ *such that* $\bigvee_{t\in T} W_t = U$, *and for every* $t \in T$, *there exists a continuous L-fuzzy mapping* $f_t: (L^X, \delta) \to I(L)$ *such that*
$$W_t \leq f_t{}^{\leftarrow}(L_1{}') \leq f_t{}^{\leftarrow}(R_0) \leq U.$$ □

One can easily verify the complete regularity of an $L$-fts which possesses an $L$-fuzzy topology consisting of just all layers $\underline{a}$'s. Then it follows from Theorem **9.2.34** (i)$\Longleftrightarrow$(iii), we have

**9.2.35 Theorem** *The stratification of a completely regular L-fts is completely regular.* □

The following theorem shows the multiplicative property of complete regularity:

**9.2.36 Theorem** *Let* $\{(L^{X_t}, \delta_t) : t \in T\}$ *be a family of L-fts', $(L^X, \delta)$ their L-fuzzy product space. Then*

(i) *If* $(L^{X_t}, \delta_t)$ *is completely regular for every* $t \in T$, *then* $(L^X, \delta)$ *is completely regular.*

(ii) *If* $(L^X, \delta)$ *is completely regular,* $s \in T$, $(L^{X_s}, \delta_s)$ *is stratified, then* $(L^{X_s}, \delta_s)$ *is completely regular.*

**Proof** (i) Suppose $x_\lambda \in M(L^X)$, $U \in \mathcal{Q}_\delta(x_\lambda)$. Without loss of generality, we can assume that $U$ is an element of the canonical base of $\delta$, $\exists F \in [T]^{<\omega}$, $\forall t \in T$, $\exists U_t \in \delta_t$ such that $U = \bigwedge_{t\in T} p_t{}^{\leftarrow}(U_t)$. Then by $x_\lambda \lessdot U$,
$$\lambda \not\leq U(x)' = \bigvee_{t\in F} p_t{}^{\leftarrow}(U_t{}')(x) = \bigvee_{t\in F} U_t(p_t(x))'.$$
So $\forall t \in F$, $\lambda \not\leq U_t(p_t(x))'$, $U_t \in \mathcal{Q}_{\delta_t}(p_t(x)_\lambda)$. Since every $(L^{X_t}, \delta_t)$ is completely regular, by Theorem **9.2.4**, $\forall t \in F$, $\exists V_t \in \mathcal{Q}_{\delta_t}(p_t(x)_\lambda)$ and there exists an $L$-fuzzy continuous mapping $f_t{}^{\rightarrow}: (L^{X_t}, \delta_t) \to I(L)$
$$V_t \leq f_t{}^{\leftarrow}(L_1{}') \leq f_t{}^{\leftarrow}(R_0) \leq U_t. \tag{9.28}$$
Take $V = \bigwedge_{t\in F} p_t{}^{\leftarrow}(V_t) \in \delta$, then by $\lambda \in M(L)$, $|F| < \omega$ and $\lambda \not\leq V_t(p_t(x))'$,
$$\lambda \not\leq \bigvee_{t\in F} V_t(p_t(x))' = (\bigwedge_{t\in F} p_t{}^{\leftarrow}(V_t))'(x) = V'(x).$$

Plus $V \in \delta$, it means $V \in \mathcal{Q}_\delta(x_\lambda)$. Since for every $t \in T$, we have an ordinary mapping $f_t: X_t \to I[L]$, so by Proposition **9.1.17**, we can define an ordinary mapping $f: X \to I[L]$ as follows:

$$\forall x \in X, \quad f(x) = \bigwedge_{t \in F} f_t \circ p_t(x).$$

Then by Corollary **9.2.33**, $f^{\to}: (L^X, \delta) \to I(L)$ is continuous. $\forall x \in X$, by inequalities (9.28),

$$\begin{aligned} V(x) &= \bigwedge_{t\in F} p_t^{\leftarrow}(V_t)(x) \le \bigwedge_{t\in F} p_t^{\leftarrow}(f_t^{\leftarrow}(L_1{}'))(x) = \bigwedge_{t\in F} L_1{}' \circ f_t \circ p_t(x) \\ &= \bigwedge_{t\in F} f_t \circ p_t(x)(1-)' = f(x)(1-)' = L_1{}'(f(x)) = f^{\leftarrow}(L_1{}')(x) \\ &\le f^{\leftarrow}(R_0)(x) = f(x)(0+) = \bigwedge_{t\in F} f_t \circ p_t(x)(0+) \\ &= \bigwedge_{t\in F} p_t^{\leftarrow}(f_t^{\leftarrow}(R_0))(x) \le \bigwedge_{t\in F} p_t^{\leftarrow}(U_t) = U. \end{aligned}$$

By the arbitrariness of $x \in X$, we have

$$V \le f^{\leftarrow}(L_1{}') \le f^{\leftarrow}(R_0) \le U.$$

By Theorem **9.2.4**, $(L^X, \delta)$ is completely regular.

(ii) Since complete regularity is hereditary, the conclusion is obtained from Corollary **3.2.13**. □

**9.2.37 Corollary** *Every L-fuzzy unit cube and every its L-fuzzy subspace is completely regular.* □

As for complete e-regularity, we have some other characterizations:

**9.2.38 Theorem** *An L-fts $(L^X, \delta)$ is completely e-regular if and only if for every $U \in \delta$, there exists a family $\{W_t: t \in T\} \subset L^X$ such that $\bigvee_{t\in T} W_t = U$, and for every $t \in T$, there exists a L-fuzzy continuous mapping $f_t^{\to}: (L^X, \delta) \to \tilde{I}(L)$ such that*

$$W_t \le f_t^{\leftarrow}(L_1{}') \le f_t^{\leftarrow}(R_0) \le U.$$ □

**9.2.39 Theorem** *An L-fts $(L^X, \delta)$ is compeltely e-regular if and only if for every base $\mathcal{B}$ of $\delta$ and every $U \in \mathcal{B}$, there exists a family $\{W_t: t \in T\} \subset L^X$ such that $\bigvee_{t\in T} W_t = U$, and for every $t \in T$, there exists a L-fuzzy continuous mapping $f_t^{\to}: (L^X, \delta) \to \tilde{I}(L)$ such that*

$$W_t \le f_t^{\leftarrow}(L_1{}') \le f_t^{\leftarrow}(R_0) \le U.$$ □

For weakly induced $L$-fts, the following Urysohn's Lemma holds:

**9.2.40 Theorem** (Urysohn's Lemma for $\tilde{I}(L)$) *Let $(L^X, \delta)$ be a weakly induced L-fts, $K \le U \in \delta$, where $K$ is not necessary to be closed. Then in the following conditions,* (i) *implies* (ii)*:*

(i) *There exists family $\mathcal{V} = \{V_t: 0 < t \le 1\} \subset \delta$ satisfying*

$$0 \le s < t \le 1 \implies V_s{}^- \le V_t, \tag{9.29}$$

*where $V_0{}^- = K$ is only a symbol, $V_1 = U$.*

(ii) *There exists L-fuzzy continuous mapping $f^{\to}: (L^X, \delta) \to \tilde{I}(L)$ such that*

$$K \le f^{\leftarrow}(L_1{}') \le f^{\leftarrow}(R_0) \le U.$$

**Proof** Suppose (i) holds, take $f: X \to I[L]$ as $f(x) = [\lambda_x]$ for every $x \in X$, where

$$\lambda_x(t) = \begin{cases} 1, & t < 0, \\ V_{1-t}(x), & 0 \le t < 1, \\ 0, & t \ge 1, \end{cases}$$

then $[\lambda_x] \in I[L]$. $\forall t \in (0,1]$, then by (i), $\forall x \in X$, we have

$$f^{\leftarrow}(L_t{}')(x) = L_t{}'(f(x)) = \lambda_x(t-) = \bigwedge_{s<t} \lambda_x(s)$$
$$= (\bigwedge_{s<t} V_{1-s})(x) = (\bigwedge_{s<t} V_{1-s}{}^{-})(x).$$

So $f^{\leftarrow}(L_t{}') = \bigwedge_{s<t} V_{1-s}{}^{-}$ is closed in $(L^X, \delta)$, and hence $f^{\leftarrow}(L_t) \in \delta$. Similarly one can prove for every $t \in [0,1)$,

$$f^{\leftarrow}(R_t) = \bigvee_{s>t} V_{1-s} \in \delta.$$

Since $f^{\leftarrow}(L_t), f^{\leftarrow}(R_t) \in \delta$ obviously holds for other $t$'s, so $f^{\leftarrow}(V) \in \delta$ for every $V \in \mathcal{T}_L^I$.

By Theorem **9.1.30**, $\tilde{I}(L)$ is weakly induced, so $\forall t \in \mathbf{R}$, $\forall a \in L$, $(L_t{}')_{[a]}$ and $(R_t{}')_{[a]}$ are closed in $(I[L], [\mathcal{T}_L^{\tilde{I}}])$. But by Proposition **9.1.31**, $(I[L], [\mathcal{T}_L^{\tilde{I}}])$ is just $(I[L], \Omega(I[L]))$. So by Proposition **9.1.29**, $\forall [\lambda] \in I[L]$, $f^{\leftarrow}(\chi_{\Uparrow[\lambda]}), f^{\leftarrow}(\chi_{\Downarrow[\lambda]}) \in \delta'$. Hence for every $U \in \Omega(I[L])$, $f^{\leftarrow}(\chi_U) \in \delta$, and hence $f^{\rightarrow}$ is continuous. □

**9.2.41 Definition** Let $(L^X, \delta)$ be an $L$-fts. An open subset $U \in \delta$ is called *chain representable*, if $U$ fulfils the condition (9.29) for every $K \le U$.

**9.2.42 Lemma** *Let $(L^X, \delta)$ be an L-fts. Then every finite meet of chain representable open subsets in $(L^X, \delta)$ is chain representable.* □

**9.2.43 Theorem** *$\tilde{I}(L)$ is completely e-regular.*

**Proof** By Proposition **9.1.13**, $\mathcal{T}_L^I$ has a base

$$\mathcal{B}_0 = \{R_s \wedge L_t : \ s, t \in \mathbf{R}\}.$$

By Proposition **9.1.29** (iii) and (iv), $\Omega(I[L])$ has a base

$$\mathcal{B}_1 = \{(\bigwedge_{i \le m} (R_{s_i})_{(a_i)}) \wedge (\bigwedge_{j \le n} (L_{t_j})_{(b_j)}) : \ m, n \in \mathbf{N}, \ s_i, t_j \in \mathbf{R}, \ a_i, b_j \in L\}.$$

So $\mathcal{T}_L^{\tilde{I}}$ has a base (for symbol $Chr(A)$ see Definition **1.1.1**):

$$\mathcal{B} = \{R_s \wedge L_t \wedge (\bigwedge_{i \le m} Chr((R_{s_i})_{(a_i)})) \wedge (\bigwedge_{j \le n} Chr((L_{t_j})_{(b_j)})) :$$
$$m, n \in \mathbf{N}, \ s, t, s_i, t_j \in \mathbf{R}, \ a_i, b_j \in L\}.$$

By Theorem **9.1.33** and Lemma **9.2.42**, it sufficient to prove that every one of the four kinds of open subsets: $R_s$, $L_t$, $(R_s)_{(a)}$, $(L_t)_{(b)}$, is chain representable; moreover, by $R_1 = \underline{0}$, $L_1 = \underline{1}$ we can assume $0 \le s < 1$, $0 < t \le 1$. By Proposition **9.1.4** and Proposition **9.1.2**, both $R_s$ and $L_t$ are chain representable. By Theorem **9.1.19**, $(I[L], \Omega(I[L]))$ is a compact Hausdorff space, so it is normal, open subsets $Chr((R_s)_{(a)})$ and $Chr((L_t)_{(b)})$ in $(I[L], \Omega(I[L]))$ are chain representable. Since $\Omega(I[L])$ can be considered as a subfamily of $\mathcal{T}_L^{\tilde{I}}$, $Chr((R_s)_{(a)})$ and $Chr((L_t)_{(b)})$ are also chain representable in $\tilde{I}(L)$. □

**9.2.44 Corollary** *Every L-fuzzy enhanced unit cube and every its L-fuzzy subspace are completely e-regular.* □

Similar to the case of complete regularity, the reader can also prove the following conclusions:

**9.2.45 Theorem** *Let $\{(L^{X_t}, \delta_t) : t \in T\}$ be a family of L-fts', $(L^X, \delta)$ their L-fuzzy product space. Then*

(i) *If $(L^{X_t}, \delta_t)$ is completely e-regular for every $t \in T$, then $(L^X, \delta)$ is completely e-regular.*

(ii) *If $(L^X, \delta)$ is completely e-regular, $s \in T$, $(L^{X_s}, \delta_s)$ is stratified, then $(L^{X_s}, \delta_s)$ is completely e-regular.* □

Since $L$-fuzzy $*$-unit interval $I^*(L)$ is induced, it inherits many very nice properties from $[0,1]$. The reader can verify them with related results in Chapter 4:

**9.2.46 Theorem** *Every L-fuzzy $*$-unit cube $I^*(L)^\kappa$ and every its closed subset is N-compact.* □

**9.2.47 Definition** $(L^X, \delta)$ is called *strongly paranormal*, or *s-p-normal* for short, if for every two non-zero pseudo-crisp closed subsets $P$ and $Q$ in $(L^X, \delta)$ such that $supp(P) \cap supp(Q) = \emptyset$, there exist $U \in \mathcal{Q}(P)$, $V \in \mathcal{Q}(Q)$ such that $U_{(0)} \wedge V_{(0)} = \emptyset$. Call $T_1$ s-p-normal $L$-fts *strongly para-$T_4$*, or *s-p-$T_4$* for short.

**9.2.48 Proposition** *S-p-normality $\Longrightarrow$ p-normality.* □

**9.2.49 Proposition** *S-p-normality is a good extension of ordinaty normality.* □

**9.2.50 Theorem** *Every s-$T_2$ and N-compact L-fts is s-p-normal. In particular, every L-fuzzy $*$-unit cube $I^*(L)^\kappa$ is s-p-$T_4$.* □

**9.2.51 Theorem** *Let $(L^X, \delta)$ be an induced p-$T_4$ L-fts, then for every pair A, B of pseudo-crisp closed subsets in $(L^X, \delta)$ such that $supp(A) \cap supp(B) = \emptyset$, there exists an L-fuzzy continuous mapping $f^{\to} : (L^X, \delta) \to I^*(L)$ such that*

$$A \leq f^{\leftarrow}(0_1), \quad B \leq f^{\leftarrow}(1_1).$$ □

**9.2.52 Theorem** *Every induced N-compact s-$T_2$ L-fts (in particular, $I^*(L)^\kappa$) and every its subspace are p-$T_{3\frac{1}{2}}$.* □

By Theorem **9.1.30** and Theorem **4.2.29**, we have

**9.2.53 Proposition** *Let $(L^X, \delta)$ be a completely regular L-fts. Then for every $p \in Pt(L^X)$ and every $U \in \mathcal{Q}(p)$, there exists an L-fuzzy continuous mapping $f^{\to} : (L^X, wi(\delta)) \to \tilde{I}(L)$ and $V \in \mathcal{Q}(p)$ such that*

$$V \leq f^{\leftarrow}(L_1{}') \leq f^{\leftarrow}(R_0) \leq U.$$ □

By Proposition **4.2.25** (iii), we have

**9.2.54 Corollary** *Let $(L^X, \delta)$ be a weakly indcued completely regular L-fts. Then for every $p \in Pt(L^X)$ and every $U \in \mathcal{Q}(p)$, there exists an L-fuzzy continuous mapping $f^{\to} : (L^X, \delta) \to \tilde{I}(L)$ and $V \in \mathcal{Q}(p)$ such that*

$$V \leq f^{\leftarrow}(L_1{}') \leq f^{\leftarrow}(R_0) \leq U.$$ □

Similar to Theorem **9.2.4**, the following pointwise characterization of complete e-regularity also holds:

**9.2.55 Theorem** *Let $(L^X, \delta)$ be an L-fts. Then the following conditions are equivalent:*

(i) *$(L^X, \delta)$ is completely e-regular.*

(ii) *For every* $p \in Pt(L^X)$ *and every* $U \in \mathcal{Q}(p)$*, there exist an L-fuzzy continuous mapping* $f^{\rightarrow}: (L^X, \delta) \to \tilde{I}(L)$ *and* $V \in \mathcal{Q}(p)$ *such that*

$$V \leq f^{\leftarrow}(L_1{}') \leq f^{\leftarrow}(R_0) \leq U. \tag{9.30}$$

(iii) *For every* $e \in M(L^X)$ *and every* $U \in \mathcal{Q}(e)$*, there exist an L-fuzzy continuous mapping* $f^{\rightarrow}: (L^X, \delta) \to \tilde{I}(L)$ *and* $V \in \mathcal{Q}(e)$ *such that*

$$V \leq f^{\leftarrow}(L_1{}') \leq f^{\leftarrow}(R_0) \leq U.$$

**Proof** (i)$\Longrightarrow$(ii): By the complete e-regularity, there exists $\{U_t: t \in T\} \subset \delta$ such that $\bigvee_{t\in T} U_t = U$ and for every $t \in T$, there exists an $L$-fuzzy continuous mapping $f_t{}^{\rightarrow}: (L^X, \delta) \to \tilde{I}(L)$ such that the inequality (CER) holds. Since $p \lessdot U$, there exists $t_0 \in T$ such that $p \lessdot U_{t_0}$. Take $V = U_{t_0}$, $f = f_{t_0}$, then inequality (9.30) holds.

(ii)$\Longrightarrow$(iii): Obvious.

(iii)$\Longrightarrow$(i): $\forall U \in \delta$. If $U = \underline{0}$, take $\lambda \in I[L]$ as follows:

$$\forall t \in \mathbf{R}, \quad \lambda(t) = \begin{cases} 1_L, & t \leq 0, \\ 0_L, & t > 0, \end{cases}$$

and take an ordinary mapping $f: X \to I[L]$ such that $f(x) = \lambda$ for every $x \in X$. Then $\forall s, t \in [0,1]$, $\forall x \in X$,

$$f^{\leftarrow}(L_s)(x) = L_s(f(x)) = L_s(\lambda) = \lambda(s-)', \tag{9.31}$$

$$f^{\leftarrow}(R_t)(x) = R_t(f(x)) = R_t(\lambda) = \lambda(t+). \tag{9.32}$$

By the definition of $\lambda \in I[L]$, $f^{\leftarrow}(L_s), f^{\leftarrow}(R_t) \in \{\underline{0}, \underline{1}\} \subset \mathcal{T}_L^I \subset \mathcal{T}_L^{\tilde{I}}$. Clearly, for every $W \in \Omega(I[L])$, we also have $f^{\leftarrow}(\chi_W) \in \{\underline{0}, \underline{1}\}$. So $f^{\rightarrow}$ is continuous. Now take $s = 1$ in equalities (9.31) and $t = 0$ in equalities (9.32), we get $f^{\leftarrow}(L_1{}') = \underline{0}$, $f^{\leftarrow}(R_0) = \underline{0}$. Let $\{U_t: t \in T\} = \{\underline{0}\}$ and $f_t{}^{\rightarrow} = f^{\rightarrow}$ for every $t \in T$, we have

$$U_t = \underline{0} = f^{\leftarrow}(L_1{}') = f^{\leftarrow}(R_0) = \underline{0} = U,$$

(CER) holds.

Suppose $U \neq \underline{0}$, then $U' \neq \underline{1}$, $\exists e \in M(L^X)$ such that $e \lessdot U$, $U \in \mathcal{Q}(e)$. By (iii), there exists an $L$-fuzzy continuous mapping $f_e{}^{\rightarrow}: (L^X, \delta) \to \tilde{I}(L)$ and $U_e \in \mathcal{Q}(e)$ such that

$$U_e \leq f_e{}^{\leftarrow}(L_1{}') \leq f_e{}^{\leftarrow}(R_0) \leq U. \tag{9.33}$$

Denote $W = \bigvee\{U_e: e \in M(L^X), e \lessdot U\}$, then $W \leq U$. If $W \neq U$, then $U \not\leq W$, $W' \not\leq U'$, $\exists d \in M(\downarrow W')$, $d \not\leq U'$. That is to say,

$$d \leq W' = \bigwedge\{U_e{}': e \in M(L^X), e \lessdot U\},$$

$$e \in M(L^X), e \lessdot U \Longrightarrow d \leq U_e{}'.$$

Especially, by $d \in M(L^X)$, $d \not\leq U'$, we have $d \lessdot U$, and then $d \leq U_d{}'$. But by the definition of $U_d$, $U_d \in \mathcal{Q}(d)$, this is a contradicition. Hence

$$\bigvee\{U_e: e \in M(L^X), e \lessdot U\} = W = U.$$

By (9.33), (i) is true. □

**9.2.56 Theorem** *Let* $(L^X, \delta)$ *be a weakly induced normal L-fts. Then* $(X, [\delta])$ *is normal.*

**Proof** Suppose $P \in [\delta']$ and $U \in [\delta]$ such that $P \subset U$. By the normality of $(L^X, \delta)$, $\exists V \in \delta$ such that $\chi_P \leq V \leq V^- \leq \chi_U$. Arbitrarily fix a molecule $\alpha \in \beta(1)$, by Theorem **4.3.10** (ii),

$$P = (\chi_P)_{[1]} \subset V_{[1]} \subset V_{[\alpha]}{}^\circ \subset V_{[\alpha]} \subset V^-{}_{[\alpha]} \subset (\chi_U)_{[\alpha]} = U.$$

Since $(L^X, \delta)$ is weakly induced, $V^-{}_{[\alpha]}$ is closed in $(L^X, \delta)$. Then take $W = V_{[\alpha]}{}^\circ$, we have

$$P \subset W \subset W^- \subset (V^-{}_{[\alpha]})^- = V^-{}_{[\alpha]} \subset U,$$

$(X, [\delta])$ is normal. □

But the converse of the preceding proposition, even in the case of induced $L$-fts, is more complicated. We put the discussion into the next section and show the conclusions related to p-normality in the following

**9.2.57 Theorem** *Let $(L^X, \delta)$ be an L-fts. Then*

(i) *$(L^X, \delta)$ is induced and p-normal $\Longrightarrow$ $(X, [\delta])$ is normal.*

(ii) *$(L^X, \delta)$ is weakly induced, $(X, [\delta])$ is normal $\Longrightarrow$ $(L^X, \delta)$ is s-p-normal.*

**Proof** (i) Suppose $A$, $B$ are disjoint nonempty closed subsets in $(X, [\delta])$. Fix a molecule $\lambda \in M(L)$, then since $(L^X, \delta)$ is stratified, $\lambda A, \lambda B \in \delta' \backslash \{\underline{0}\}$, $supp(\lambda A) \cap supp(\lambda B) = \emptyset$ and both of $\lambda A$, $\lambda B$ are pseudo-crisp. By the p-normality, $\exists U \in \mathcal{Q}(\lambda A)$, $\exists V \in \mathcal{Q}(\lambda B)$ such that $U \wedge V = \underline{0}$. By $\bigvee M(\Downarrow\lambda') = \lambda'$, we can take a molecule $\gamma \in M(\Downarrow\lambda')$. Then $\forall x \in A$, $U(x) \not\leq \gamma$, $x \in U_{(\gamma)}$. So $A \subset U_{(\gamma)}$. Similarly, $B \subset V_{(\gamma)}$. By the weakly induced property of $(L^X, \delta)$, $U_{(\gamma)}, V_{(\gamma)} \in [\delta]$, so $U_{(\gamma)}$ and $V_{(\gamma)}$ are respectively neighborhoods of $A$ and $B$ in $(X, [\delta])$. At last, since $\gamma \in M(L)$, by $U \wedge V = \underline{0}$ and Proposition **4.2.4** (ii), $U_{(\gamma)} \cap V_{(\gamma)} = \underline{0}_{(\gamma)} = \emptyset$. Hence $(X, [\delta])$ is normal.

(ii) Suppose $A$, $B$ are non-zero pseudo-crisp closed subsets in $(L^X, \delta)$, $supp(A) \cap supp(B) = \emptyset$. Then $\exists a, b \in L$ such that $A_{(0)} = A_{[a]}$, $B_{(0)} = B_{[b]}$. By the weakly induced property of $(L^X, \delta)$, $A_{(0)} = A_{[a]}$ and $B_{(0)} = B_{[b]}$ are closed in $(X, [\delta])$ and $A_{(0)} \cap B_{(0)} = supp(A) \cap supp(B) = \emptyset$. By the normality of $(X, [\delta])$, $\exists U \in \mathcal{N}_{[\delta]}(A_{(0)})$, $\exists V \in \mathcal{N}_{[\delta]}(B_{(0)})$ such that $U \cap V = \emptyset$. Then clearly $\chi_U \in \mathcal{Q}_\delta(A)$, $\chi_V \in \mathcal{Q}_\delta(B)$ and

$$(\chi_U)_{(0)} \cap (\chi_V)_{(0)} = U \cap V = \emptyset,$$

$(L^X, \delta)$ is s-p-normal. □

Now we consider the Embedding Lemma of $L$-fuzzy completely regular spaces. It was established at first by Liu,[71] then Wang improved the proof.[168] Some generalized results for complete lattices and frames were also investigated by Kubiak.[55] But since we primarily consider completely distributive lattices in this book, those results will not be involved here.

**9.2.58 Definition** Let $\mathcal{F}$ be a family of $L$-fuzzy mappings, for every $f^\rightarrow \in \mathcal{F}$, $f^\rightarrow : (L^X, \delta) \to (L^{Y_f}, \mu_f)$, $(L^Y, \mu) = \prod_{f \in \mathcal{F}} (L^{Y_f}, \delta_f)$.

Define $E: X \to Y$ as follows:

$$\forall x \in X, \ \forall f^\rightarrow \in \mathcal{F}, \quad p_f(E(x)) = f(x).$$

For the mapping $E: X \to Y$ defined above, call the $L$-fuzzy mapping $E^\rightarrow : L^X \to L^Y$ induced from $E$ the *evaluation mapping* determined by $\mathcal{F}$. Also denote $E$ by $\Delta\mathcal{F}$.

Say $\mathcal{F}$ *distinguishes points*, if for every two $L$-fuzzy points $a$, $b$ in $(L^X, \delta)$, there exists $f^\rightarrow \in \mathcal{F}$ such that $f^\rightarrow(a) \neq f^\rightarrow(b)$.

Say $\mathcal{F}$ *distinguishes closed subsets and points*, if for every closed subset $P$ and every $L$-fuzzy point $p$ in $(L^X, \delta)$ such that $p \not\leq P$, there exists $f^\rightarrow \in \mathcal{F}$ such that $f^\rightarrow(p) \not\leq (f^\rightarrow(P))^-$.

For a family $\mathcal{A} = \{f_t : t \in T\}$ of ordinary mappings, where $f_t : X \to Y_t$ for every $t \in T$, we also define the *evaluation mapping* $\Delta\mathcal{A}$ of $\mathcal{A}$ as the ordinary mapping $E$ defined above, i.e. $\Delta\mathcal{A} : X \to \prod_{t\in T} Y_t$, where $(\Delta\mathcal{A})(x) = (f_t(x))_{t\in T}$ for every $x \in X$. We also denote $\Delta\mathcal{A}$ by $\Delta_{t\in T} f_t$ sometimes. Then $(\Delta\mathcal{A})^{\to} : L^X \to L^{\prod_{t\in T} Y_t}$ is just the evaluation mapping determined by the family $\{f_t^{\to} : t \in T\}$ of $L$-fuzzy mappings.

**9.2.59 Lemma** *Let $\mathcal{F}$ be a family of L-fuzzy mappings, for every $f^{\to} \in \mathcal{F}$, $f^{\to} : (L^X, \delta) \to (L^{Y_f}, \mu_f)$ be continuous. Then $\mathcal{F}$ distinguishes closed subsets and points if and only if $P = \bigwedge_{f^{\to}\in\mathcal{F}_0} f^{\leftarrow}((f^{\to}(P))^-)$ for every $P \in \delta'$.*

**Proof** (Necessity) By Theorem **2.4.5** (vii) and Theorem **2.1.25** (i), $\forall P \in \delta'$, $\forall f^{\to} \in \mathcal{F}$,

$$P \le f^{\leftarrow} f^{\to}(P^-) \le f^{\leftarrow}((f^{\to}(P))^-).$$

So $P \le \bigwedge_{f^{\to}\in\mathcal{F}} f^{\leftarrow}((f^{\to}(P))^-)$. Suppose $p \in Pt(L^X)$, $p \not\le P$. Since $\mathcal{F}$ distinguishes closed subsets and points, $\exists f^{\to} \in \mathcal{F}$ such that $f^{\to}(p) \not\le (f^{\to}(P))^-$. If $p \le f^{\leftarrow}((f^{\to}(P))^-)$, then by Theorem **2.1.25** (ii),

$$f^{\to}(p) \le f^{\to} f^{\leftarrow}((f^{\to}(P))^-) \le (f^{\to}(P))^-,$$

a contradiction. So $p \not\le f^{\leftarrow}((f^{\to}(P))^-)$. Then $p \not\le \bigwedge_{f^{\to}\in\mathcal{F}} f^{\leftarrow}((f^{\to}(P))^-)$. Hence we have already proved $p = \bigwedge_{f^{\to}\in\mathcal{F}} f^{\leftarrow}((f^{\to}(P))^-)$.

(Sufficiency) Suppose $P \in \delta'$, $p \in Pt(L^X)$, $p \not\le P$, then by the given condition, $\exists f^{\to} \in \mathcal{F}$ such that $p \not\le f^{\leftarrow}((f^{\to}(P))^-)$. If $f^{\to}(p) \le (f^{\to}(P))^-$, then by Theorem **2.1.25** (i) and Theorem **2.4.5** (vii),

$$p \le f^{\leftarrow} f^{\to}(p) \le f^{\leftarrow}((f^{\to}(P))^-),$$

a contradiction. So $f^{\to}(p) \not\le (f^{\to}(P))^-$, $\mathcal{F}$ distinguishes closed subsets and points. □

**9.2.60 Remark** Alternating the terminology "distinguish closed subsets and points," Kubiak called this kind of family by "completely regular family" and defined them by the preceding equivalent condition "$P = \bigwedge_{f^{\to}\in\mathcal{F}} f^{\leftarrow}((f^{\to}(P))^-)$".[55]

**9.2.61 Lemma** *Let $\mathcal{F}$ be a family of L-fuzzy mappings, for every $f^{\to} \in \mathcal{F}$, $f^{\to} : (L^X, \delta) \to (L^{Y_f}, \mu_f)$ be continuous. If $\mathcal{F}$ distinguishes closed subsets and points, then there exists $\mathcal{F}_0 \subset \mathcal{F}$ such that $|\mathcal{F}_0| \le w(\delta)^2$ distinguishing closed subsets and points.*

**Proof** Suppose $\mathcal{P}$ is a closed base of $\delta'$ such that $|\mathcal{P}| = w(\delta)$. $\forall P \in \mathcal{P}$, by Proposition **1.1.27** (i) and the dual of Lemma **9.2.59**, $\exists \mathcal{F}_P \subset \mathcal{F}$ such that $|\mathcal{F}_P| \le |\mathcal{P}| = w(\delta)$ and $P = \bigwedge_{f^{\to}\in\mathcal{F}_P} f^{\leftarrow}((f^{\to}(P))^-)$. Take $\mathcal{F}_0 = \bigcup_{P\in\mathcal{P}} \mathcal{F}_P$, then $|\mathcal{F}_0| \le w(\delta)^2$. Suppose $A \in \delta'$, $\exists \mathcal{A} \subset \mathcal{P}$ such that $A = \bigwedge \mathcal{A}$. Then $P \ge A$ for every $P \in \mathcal{A}$. By Theorem **2.1.25** (i),

$$\begin{aligned} A &= \bigwedge_{P\in\mathcal{A}} P = \bigwedge_{P\in\mathcal{A}} \bigwedge_{f^{\to}\in\mathcal{F}_P} f^{\leftarrow}((f^{\to}(P))^-) \ge \bigwedge_{f^{\to}\in\mathcal{F}_0} f^{\leftarrow}((f^{\to}(A))^-) \\ &\ge \bigwedge_{f^{\to}\in\mathcal{F}_0} f^{\leftarrow} f^{\to}(A) \ge A, \\ A &= \bigwedge_{f^{\to}\in\mathcal{F}_0} f^{\leftarrow}((f^{\to}(A))^-). \end{aligned}$$

By Lemma **9.2.59**, $\mathcal{F}_0$ distinguishes closed subsets and points. □

**9.2.62 Theorem** (Embedding Lemma) *Let $\mathcal{F}$ be a family of L-fuzzy mappings, for every $f^{\to} \in \mathcal{F}$, $f^{\to} : (L^X, \delta) \to (L^{Y_f}, \mu_f)$ be continuous, $(L^Y, \mu) = \prod_{f\in\mathcal{F}} (L^{Y_f}, \delta_f)$. Then*

(i) *The evaluation mapping* $E^{\rightarrow}: (L^X, \delta) \rightarrow (L^Y, \mu)$ *determined by* $\mathcal{F}$ *is continuous.*

(ii) *If* $\mathcal{F}$ *distinguishes points,* $E^{\rightarrow}: (L^X, \delta) \rightarrow (L^Y, \mu)$ *is injective.*

(iii) *If* $\mathcal{F}$ *distinguishes closed subsets and points,* $E^{\rightarrow}: (L^X, \delta) \rightarrow (L^{E[X]}, \mu|_{E[X]})$ *is open.*

**Proof** (i) For every element $p_f^{\leftarrow}(V_f)$ of the canonical subbase of $\mu$, where $f \in \mathcal{F}$, $V_f \in \mu_f$, by $p_f \circ E = f$,

$$E^{\leftarrow}(p_f^{\leftarrow}(V_f)) = (p_f \circ E)^{\leftarrow}(V_f) = f^{\leftarrow}(V_f) \in \delta.$$

So by Theorem **2.4.5** (v), $E^{\rightarrow}$ is continuous.

(ii) Suppose $x, y \in X$, $x \neq y$, then since $\mathcal{F}$ distinguishes points, by Theorem **2.1.15** (ii), $\exists f^{\rightarrow} \in \mathcal{F}$ such that $f(x) \neq f(y)$, so $E(x) \neq E(y)$, $E: X \rightarrow Y$ is injective. By Theorem **2.1.22** (i), $E^{\rightarrow}: (L^X, \delta) \rightarrow (L^Y, \mu)$ is injective.

(iii) Suppose $E[X] = Y_0 \subset Y$, $A \in \delta$. We need only prove that $(E^{\rightarrow}(A))'|_{Y_0}$ is a closed subset in $(L^{Y_0}, \mu|_{Y_0})$.

Let $y \in Y_0$, $\lambda \in M(L)$ and $y_\lambda \not\leq (E^{\rightarrow}(A))'$, then

$$\lambda \not\leq (E^{\rightarrow}(A))'(y) = \bigwedge\{A'(x) : E(x) = y\}.$$

So $\exists x \in X$ such that $\lambda \not\leq A'(x)$ and $E(x) = y$. Hence $x_\lambda \not\leq A'$. Since $A'$ is closed and $\mathcal{F}$ distinguishes closed subsets and points, $\exists f^{\rightarrow} \in \mathcal{F}$ such that

$$p_f^{\rightarrow}E^{\rightarrow}(x_\lambda) = f^{\rightarrow}(x_\lambda) \not\leq (f^{\rightarrow}(A'))^-,$$
$$y_\lambda = E(x)_\lambda = E^{\rightarrow}(x_\lambda) \not\leq p_f^{\leftarrow}((f^{\rightarrow}(A'))^-) \in \delta'.$$

So denote $U = (p_f^{\leftarrow}((f^{\rightarrow}(A'))^-))'$, we have $U \in \mathcal{Q}_\mu(y_\lambda)$. But by Proposition **2.2.4**,

$$U' = p_f^{\leftarrow}((f^{\rightarrow}(A'))^-) \geq p_f^{\leftarrow}f^{\rightarrow}(A') = p_f^{\leftarrow}p_f^{\rightarrow}E^{\rightarrow}(A')$$
$$= E^{\rightarrow}(A') \geq (E^{\rightarrow}(A))'.$$

So $U \neg \hat{q} (E^{\rightarrow}(A))'$, $y_\lambda \not\leq (E^{\rightarrow}(A))'^-$. Note that $y_\lambda \in M(L^{Y_0})$, $y_\lambda \not\leq (E^{\rightarrow}(A))'$, so we have proved

$$(E^{\rightarrow}(A))'^-|_{Y_0} = (E^{\rightarrow}(A))'|_{Y_0},$$

$(E^{\rightarrow}(A))'$ is a closed subset in $(L^{Y_0}, \mu|_{Y_0})$. □

**9.2.63 Theorem** *Let* $(L^X, \delta)$ *be a completely regular L-fts,* $\mathcal{F}$ *the family of all the L-fuzzy continuous mappings from* $(L^X, \delta)$ *to* $I(L)$*. Then* $\mathcal{F}$ *distinguishes closed subsets and points. If* $(L^X, \delta)$ *is sub-*$T_0$ *in extra, then* $\mathcal{F}$ *even distinguishes points.*

**Proof** Suppose $P \in \delta'$, $x_a \in Pt(L^X)$, $x_a \not\leq P$, then $P' \in \mathcal{Q}(x_a)$. By the complete regularity of $(L^X, \delta)$ and Theorem **9.2.4**, there exists an $L$-fuzzy continuous mapping $f^{\rightarrow}: (L^X, \delta) \rightarrow I(L)$ and $U \in \mathcal{Q}(x_a)$ such that

$$U \leq f^{\leftarrow}(L_1{}') \leq f^{\leftarrow}(R_0) \leq P'.$$

So $P \leq f^{\leftarrow}(R_0{}')$, $f^{\rightarrow}(P) \leq f^{\rightarrow}f^{\leftarrow}(R_0{}') \leq R_0{}'$. Since $R_0{}'$ is closed in $I(L)$, we have $(f^{\rightarrow}(P))^- \leq R_0{}'$. We say $f^{\rightarrow}(x_a) \not\leq R_0{}'$, otherwise

$$x_a \leq f^{\leftarrow}f^{\rightarrow}(x_a) \leq f^{\leftarrow}(R_0{}') \leq f^{\leftarrow}(L_1) \leq U',$$

contradicts with $U \in \mathcal{Q}(x_a)$. So by $(f^{\rightarrow}(P))^- \leq R_0{}'$ we have $f^{\rightarrow}(x_a) \not\leq (f^{\rightarrow}(P))^-$, $\mathcal{F}$ distinguishes closed subsets and points.

Suppose $(L^X, \delta)$ is sub-$T_0$ in extra, $x_a, y_b \in Pt(L^X)$, $x_a \neq y_b$. By Theorem **2.1.15** (ii), we need only prove the conclusion in the case $x \neq y$. Then $\exists \lambda \in M(L)$ such that $x_\lambda \not\leq y_\lambda{}^-$ or $y_\lambda \not\leq x_\lambda{}^-$. Without loss of generality, we can assume $x_\lambda \not\leq y_\lambda{}^-$. Since $\mathcal{F}$ distinguishes closed subsets and points, $\exists f^{\rightarrow} \in \mathcal{F}$ such that $f^{\rightarrow}(x_\lambda) \not\leq (f^{\rightarrow}(y_\lambda{}^-))^-$. By $f^{\rightarrow}(y_\lambda) \leq (f^{\rightarrow}(y_\lambda{}^-))^-$, we have $f^{\rightarrow}(x_\lambda) \not\leq f^{\rightarrow}(y_\lambda)$, $\mathcal{F}$ distinguishes points. □

**9.2.64 Theorem** *Let $(L^X, \delta)$ be an L-fts. Then the following conditions are equivalent:*

(i) *$(L^X, \delta)$ is sub-$T_0$ and completely regular.*

(ii) *$(L^X, \delta)$ is homeomorphic to a subspace of an L-fuzzy unit cube $I(L)^\kappa$.*

(iii) *There exists $\kappa \le w(\delta)^2$ such that $(L^X, \delta)$ is homeomorphic to a subspace of the L-fuzzy unit cube $I(L)^\kappa$.*

**Proof** (i)$\Longrightarrow$(iii): By Theorem **9.2.63**, Theorem **9.2.62** and Lemma **9.2.61**.

(iii)$\Longrightarrow$(ii): Clear.

(ii)$\Longrightarrow$(i): By Corollary **9.2.37**, Proposition **9.1.14**, Theorem **8.1.7** and Theorem **8.1.13**. □

With regard to complete p-regularity, we have the following

**9.2.65 Theorem** *Let $(L^X, \delta)$ be a weakly induced p-$T_{3\frac{1}{2}}$ L-fts, then for every point $x_a \in Pt(L^X)$ and every closed subset $P \in \delta'$ such that $x_a \not\le P$, there exists an L-fuzzy continuous $f^\to: (L^X, \delta) \to I^*(L)$ such that $f^\to(x_a) \not\le f^\to(P)^-$.*

**Proof** Since $x_a \not\le P$, we can take $\alpha \in M(\downarrow a)$ such that $x_\alpha \not\le P$. Then there exists $U \in \mathcal{N}_{[\delta]}(x)$ such that $\alpha \not\le \bigwedge_{y\in U}(\chi_U \wedge P)(y)$. In fact, if it is false, then by Theorem **4.5.1**,

$$\alpha \le \bigwedge_{V\in\mathcal{N}(x)} \bigvee_{y\in V} (\chi_V \wedge P)(y) \le \bigwedge_{V\in\mathcal{N}(x)} \bigvee_{y\in V} P(y) = P(x),$$

contradicts with $x_\alpha \not\le P$. Now take $\gamma \in \beta^*(\alpha)$ and let $P^* = P \wedge \chi_{X\setminus U} \wedge \chi_{P_{[\gamma]}}$, since $(L^X, \delta)$ is weakly induced, $P^*$ is closed and $x \notin supp(P^*)$. By the p-$T_{3\frac{1}{2}}$ property of $(L^X, \delta)$, there exists an $L$-fuzzy continuous mapping $f^\to: (L^X, \delta) \to I^*(L)$ such that $f^\to(x_\alpha) = 0_\alpha$, $P^* \le f^\leftarrow(1_1)$.

Now we prove $f^\to(x_a) \not\le f^\to(P)^-$. Let

$$d = (\bigvee_{y\in U} (\chi_U \wedge P)(y)) \vee (\bigvee(L\setminus\uparrow\gamma)),$$

$$G = \underline{d} \vee 1_1,$$

then for every point $y_b \le P$ but $y_b \not\le P^*$, $y \in U$ or $b \not\ge \gamma$. So $b \le \bigvee_{y\in U}(\chi_U \wedge P)(y)$ or $b \in L\setminus\uparrow\gamma$, $b \le d$, $f^\to(y_b) \le G$. Thus $f^\to(P) \le G$. Since $I^*(L)$ is the $L$-valued induced space of $I$, $G$ is closed, $f^\to(P)^- \le G$. If $\alpha \le \bigvee(L\setminus\uparrow\gamma)$, then by $\gamma \in \beta^*(\alpha)$, there exists $c \in L\setminus\uparrow\gamma$ such that $c \ge \gamma$, this is a contradiction. Hence $\alpha \not\le \bigvee(L\setminus\uparrow\gamma)$. By $\alpha \not\le \bigvee_{y\in U}(\chi_U \wedge P)(y)$ proved above and $\alpha \in M(L)$, $\alpha \not\le d$. Therefore $f(x_\alpha) = 0_\alpha \not\le G$, and by $f^\to(P)^- \le G$ we have $f^\to(x_\alpha) \not\le f^\to(P)^-$. □

By Proposition **9.2.10** and Corollary **8.1.28**, every weakly induced and completely p-regular $L$-fts is t-$T_1$, so by the virtue of Theorem **9.2.65**, the following conclusion holds:

**9.2.66 Corollary** *Let $(L^X, \delta)$ be a weakly induced and completely p-regular L-fts, $\mathcal{F}$ the family of all the L-fuzzy continuous mappings from $(L^X, \delta)$ to $I(L)$. Then $\mathcal{F}$ distinguishes closed subsets and points, and distinguishes points.* □

Since $I^*(L)$ is induced and induced property is multiplicative and hereditary, by the virtue of Embedding Lemma (Theorem **9.2.62**), we have

**9.2.67 Theorem** *Let $(L^X, \delta)$ be an L-fts. Then the following conditions are equivalent:*

(i) *$(L^X, \delta)$ is weakly induced and completely p-regular.*

(ii) *$(L^X, \delta)$ is homeomorphic to a subspace of an L-fuzzy $*$-unit cube $I^*(L)^\kappa$.* □

For complete e-regularity, the conclusion similar to Theorem **9.2.63** also holds as follows and its proof is completely parallel:

**9.2.68 Theorem** *Let $(L^X, \delta)$ be a completely e-regular L-fts, $\mathcal{F}$ the family of all the L-fuzzy continuous mappings from $(L^X, \delta)$ to $\tilde{I}(L)$. Then $\mathcal{F}$ distinguishes closed subsets and points. If $(L^X, \delta)$ is sub-$T_0$ in extra, then $\mathcal{F}$ even distinguishes points.* □

Then it follows from Embedding Lemma (Theorem **9.2.62**), we have

**9.2.69 Theorem** *Let $(L^X, \delta)$ be an L-fts. Then the following conditions are equivalent:*

(i) *$(L^X, \delta)$ is sub-$T_0$ and completely e-regular.*

(ii) *$(L^X, \delta)$ is homeomorphic to a subspace of an L-fuzzy enhanced unit cube $\tilde{I}(L)^\kappa$.* □

By Theorem **9.1.23**, Theorem **8.2.8**, Theorem **8.2.6** and Theorem **9.2.69**, the following conclusion holds:

**9.2.70 Corollary** *Every sub-$T_0$ and completely e-regular L-fts is s-$T_2$.* □

Since weakly induced property is hereditary (Theorem **4.2.19**), by Theorem **9.2.69**, every completely e-regular $L$-fts is weakly induced. Then by Corollary **9.2.54** and Theorem **9.2.55**, we have

**9.2.71 Theorem** *Let $(L^X, \delta)$ be an L-fts. Then the following conditions are equivalent:*

(i) *$(L^X, \delta)$ is completely e-regular.*

(ii) *$(L^X, \delta)$ is weakly induced and completely regular.* □

**9.2.72 Remark** Complete e-regularity was defined by Wang and Xu[170] and was called "$H(\lambda)$ complete regularity" in their original work. In order to embed a sub-$T_0$ and completely e-regular $L$-fts into an $L$-fuzzy enhanced unit cube, Wang and Xu limited their investigation in weakly induced $L$-fts'. But by Theorem **9.2.71**, this limitation is not necessary. In fact, the weakly induced property of completely e-regular $L$-fts' can be also straight proved from the original definition:

Suppose $(L^X, \delta)$ is a completely e-regular $L$-fts, $U \in \delta$, $a \in L$. Then there exists $\{U_t : t \in T\} \subset \delta$ such that $\bigvee_{t\in T} U_t = U$ and for every $t \in T$, there exists an $L$-fuzzy continuous mapping $f_t^{\rightarrow} : (L^X, \delta) \to \tilde{I}(L)$ such that

$$U_t \leq f_t^{\leftarrow}(L_t{}') \leq f_t^{\leftarrow}(R_0) \leq U.$$

By Proposition **4.2.4**,

$$(U_t)_{(a)} \subset (f_t^{\leftarrow}(R_0))_{(a)} \subset U_{(a)},$$

$$U_{(a)} = \bigcup_{t\in T} (U_t)_{(a)} \subset \bigcup_{t\in T} (f_t^{\leftarrow}(R_0))_{(a)} \subset U_{(a)}.$$

Since $\tilde{I}(L)$ is weakly induced, by Proposition **4.2.5** and Proposition **4.2.2**,

$$U_{(a)} = \bigcup_{t\in T}(f_t^{\leftarrow}(R_0))_{(a)} = \bigcup_{t\in T} f_t^{-1}((R_0)_{(a)}) \in [\delta].$$

So $(L^X, \delta)$ is weakly induced.

The previous results related to complete e-regularity can be also proved from another angle with Theorem **9.2.71**.

## 9.3 Insertion Theorem

Let $(X, \mathcal{T})$ be an ordinary topological space, $f, g : X \to \mathbf{R}$ be an upper semicontinuous function and a lower semicontinuous function respectively, and $f \leq g$; then a classical problem is: "Is it possible to insert a continuous function $h : X \to \mathbf{R}$ between $f$ and $g$, i.e. $f \leq h \leq g$?" Hahn,[28] Dieudonné [16] and Tong [157] gave out respectively the insertion functions $h$ for metric spaces, paracompact spaces and normal spaces. In recent several decades, caused by the attention on computer science, the interest in ordered structures is increasing. Then problem of extending the value domain in the insertion problem of semicontinuous mappings emerges naturally. Theorem **9.3.6** shows how the classic insertion theorem is extended to the case of lattice-valued semicontinuous mappings, and Theorem **9.3.11** provides us a solution of fuzzy insertion problem. Just as it will be proved in the sequel, besides the structure of the value domain, this problem is also closely connected with normality in $L$-fuzzy topological spaces.

**9.3.1 Lemma** *Let $(L^X, \delta)$ be a weakly induced L-fts, $F$, $U$ be respectively a closed subset and an open subset in $(L^X, \delta)$ and $F \leq U$, $L_0$ a join-generating set of $L$, $a \in L_0$. Then $(U_{[a]})'$ and $\bigcup\{F_{[b]} : b \in L_0,\ a \preceq b\}$ are separated in $(X, [\delta])$.*

**Proof** Denote these two sets by $A$ and $B$ respectively. If $a \preceq b$, we have $F_{[b]} \subset F_{[a]} \subset U_{[a]}$. Since $(L^X, \delta)$ is weakly induced, $F_{[a]}$ is closed in $(X, [\delta])$, so

$$B^- \cap A \subset F_{[a]}{}^- \cap A = F_{[a]} \cap (U_{[a]})' = \emptyset.$$

As for another equality, if $a \preceq b$, by Theorem **4.3.10** (ii),

$$F_{[b]} \subset U_{[b]} = U^\circ{}_{[b]} \subset U_{[a]}{}^\circ,$$

so $B \subset U_{[a]}{}^\circ$. But $U_{[a]}{}^\circ \subset U_{[a]}$, $A^- = U_{[a]}{}'^- \subset U_{[a]}{}^{\circ\prime}$, so $B \cap A^- = \emptyset$. □

**9.3.2 Theorem** *Let $(L^X, \delta)$ be an induced L-fts, $F$, $U$ be respectively a closed subset and an open subset in $(L^X, \delta)$ and $F \leq U$, $L_0$ a join-generating set of $L$, $A \in L^X$. Then the following conditions are equivalent:*

(i) $F \leq A^\circ \leq A^- \leq U$.

(ii) *For every $a \in L_0$,*

$$\bigcup\{F_{[b]} : b \in L_0,\ a \preceq b\} \subset A_{[a]}{}^\circ \subset A_{[a]}{}^- \subset U_{[a]}.$$

**Proof** (i)$\Longrightarrow$(ii): Suppose $a \preceq b$, by Theorem **4.3.10** (iii), $A^\circ{}_{[b]} \subset A_{[a]}{}^\circ$. Since $(L^X, \delta)$ is induced, $A^-{}_{[a]}$ is closed in $(X, [\delta])$, so by $A_{[a]} \subset A^-{}_{[a]}$ we have $A_{[a]}{}^- \subset A^-{}_{[a]}$. Then by $F \leq A^\circ \leq A^- \leq U$ we have

$$F_{[b]} \subset A^\circ{}_{[b]} \subset A_{[a]}{}^\circ \subset A_{[a]} \subset A_{[a]}{}^- \subset A^-{}_{[a]} \subset U_{[a]}.$$

(ii)$\Longrightarrow$(i): Suppose $b \in L_0$. By Lemma **2.1.9** (i), $D = \{a : a \in L_0,\ a \preceq b\}$ is a minimal set of $b$. By Theorem **4.3.10** (iii), $A^\circ_{[b]} = \bigcap\{A_{[a]}{}^\circ : a \in D\}$. But by the condition (ii), $F_{[b]} \subset A_{[a]}{}^\circ$ for every $a \in D$, so $F_{[b]} \subset A^\circ_{[b]}$ for every $b \in L_0$. Hence $F \leq A^\circ$. On the other hand, by $\bigvee D = b$, $U_{[b]} = \bigcap\{U_{[a]} : a \in D\}$. By Theorem **4.3.8** (iii), $A^-{}_{[b]} = \bigcap\{A_{[a]}{}^- : a \in D\}$. By the condition (ii), $A_{[a]}{}^- \subset U_{[a]}$ for every $a \in D$, so $A^-{}_{[b]} \subset U_{[b]}$, and hence $A^- \leq U$. □

**9.3.3 Lemma** *Let $L$ be a completely distributive lattice, $L_0$ a join-generating set of $L$. Then*

$$\forall a \in L,\ \forall b \in M(L),\ a \neq b,\ a \preceq b \Longrightarrow \exists c \in L_0,\ c \neq a,\ a \preceq c \preceq b.$$

**Proof** Suppose $a \in L$, $b \in M(L)$, $a \neq b$, $a \preceq b$. By Theorem **1.3.14** (ii), $\exists c_0 \in L$ such that $c_0 \neq a$, $a \preceq c_0 \preceq b$. Since $L$ is completely distributive, by Lemma **2.1.9** (i), $\beta(b) \cap L_0$ is a minimal set of $b$, $\bigvee(\beta(b) \cap L_0) = b$. Since by $c_0 \preceq b$ we have $c_0 \leq b$, so $\exists c \in \beta(b) \cap L_0$ such that $c_0 \leq c$. Then by $a \preceq c_0$ and $a \neq c_0$ we have $c \in L_0$, $a < c_0 \leq c$, $a \preceq c \preceq b$. □

**9.3.4 Theorem** *Let $L$ be a F-lattice possessing a countable join-generating set $L_0$, $(L^X, \delta)$ an induced L-fts. Then the following conditions are equivalent:*

(i) *$(L^X, \delta)$ is normal.*

(ii) *$(X, [\delta])$ is normal.*

*Moreover, if $L_0$ is a countable strictly generating set of $L$, then the following condition is equivalent to the conditions* (i) *and* (ii)*:*

(iii) *For every closed subset $F$ and every open subset $U$ in $(L^X, \delta)$ such that $F \leq U$, there exists an open and closed subset $C$ in $(L^X, \delta)$ such that $F \leq C \leq U$.*

**Proof** (i)$\Longrightarrow$(ii): By Theorem **9.2.56**.

(ii)$\Longrightarrow$(i): For convenience, for an ordinary subset $A$ of $X$, we use still the symbol $A'$ to denote the complementary set $X \backslash A$ of $A$ in $X$.

Suppose $F$ and $U$ are respectively a closed subset and an open subset in $(L^X, \delta)$ such that $F \leq U$. By Theorem **1.1.28** (i), we can assume $L_0 \subset M(L)$, $L_0 = \{\alpha_i : i \in \mathbf{N}\}$, where $i \neq j \Longrightarrow \alpha_i \neq \alpha_j$. $\forall i \in \mathbf{N}$, denote $G_i = \bigcup\{F_{[\alpha]} : \alpha \in L_0,\ \alpha_i \preceq \alpha\}$. In the following, we are going to find by inductive method a series $\{A_{\alpha_i} : i \in \mathbf{N}\}$ of ordinary subsets in $X$ such that

(a) $G_i \subset A_{\alpha_i}{}^\circ \subset A_{\alpha_i}{}^- \subset U_{[\alpha_i]}$.

(b) $\alpha_i \preceq \alpha_j$, $\alpha_i < \alpha_j \implies A_{\alpha_i}{}^\circ \supset A_{\alpha_j}{}^-$ ; especially, $A_{\alpha_i} \supset A_{\alpha_j}$.

When $i = 1$, by Theorem **4.3.10** (iii) and Lemma **2.1.9** (i), $U_{[\alpha_1]}{}'$ is a $F_\sigma$-set. Since $(L^X, \delta)$ is weakly induced, every $F_{[\alpha]}$ is a closed subset, so $G_1$ is also a $F_\sigma$-set. By Lemma **9.3.1**, $U_{[\alpha_1]}{}'$ and $G_1$ are separated. By the normality of $(X, [\delta])$ and Corollary **9.2.23**, $\exists A_{\alpha_1}$ satisfying inductive conditions (a) and (b). Now suppose $k \in \mathbf{N}$ and for every $i < k$, we have found $A_{\alpha_i}$ satisfying conditions (a) and (b). Divide these $i$'s into two groups

$$J_0 = \{i : i < k,\ \alpha_k \preceq \alpha_i\}, \quad J_1 = \{j : j < k,\ \alpha_j \preceq \alpha_k\}.$$

Denote

$$E = G_k \cup \bigcup\{A_{\alpha_i}{}^- : i \in J_0\}, \quad F = U_{[\alpha_k]}{}' \cup \bigcup\{A_{\alpha_j}{}^{\circ\prime} : j \in J_1\},$$

then $E$ and $F$ are separated in $(X, [\delta])$. In fact, by Lemma **9.3.1**, $G_k$ and $U_{[\alpha_k]}'$ are separated in $(X, [\delta])$. By condition (b), $\bigcup\{A_{\alpha_i}^- : i \in J_0\}$ and $\bigcup\{A_{\alpha_j}^{\circ\prime} : j \in J_1\}$ are disjoint closed subsets in $(X, [\delta])$. By condition (a), $\forall j \in J_1$,

$$G_k^- \subset F_{[\alpha_k]}^- = F_{[\alpha_k]} \subset G_j \subset A_{\alpha_j}^\circ,$$

so $G_k^- \subset \bigcap\{A_{\alpha_j}^\circ : j \in J_1\}$. That is to say, $G_k$ and $\bigcup\{A_{\alpha_j}^{\circ\prime} : j \in J_1\}$ are separated. At last, $\forall i \in J_0$, by condition (a) and Theorem **4.3.10** (ii), $A_{\alpha_i}^- \subset U_{[\alpha_i]} \subset U_{[\alpha_k]}^\circ$, so $U_{[\alpha_k]}'$ and $\bigcup\{A_{\alpha_i}^- : i \in J_0\}$ are separated. Hence $E$ and $F$ are separated in $(X, [\delta])$. Applying the countability of $L_0$, we know $G_k$ is a $F_\sigma$-set. By Theorem **4.3.10** (iii), $U_{[\alpha_k]}'$ is also a $F_\sigma$-set. So both $E$ and $F$ are $F_\sigma$-sets. By Corollary **9.2.23**, there exists an ordinary set $A_{\alpha_k} \subset X$ such that $E \subset A_{\alpha_k}^\circ \subset A_{\alpha_k}^- \subset F'$ and conditions (a), (b) certainly hold. This completes the induction.

$\forall k \in \mathbf{N}$, let

$$C_{\alpha_k} = \bigcap\{A_{\alpha_j} : \alpha_j \in L_0,\ \alpha_j \preceq \alpha_k\}, \tag{9.34}$$

then by Theorem **2.1.10**, $C = \bigvee\{\alpha C_\alpha : \alpha \in L_0\}$ fulfills $C_{[\alpha_k]} = C_{\alpha_k}$ for every $k < \mathbf{N}$. Now we need to prove $F \leq C^\circ \leq C^- \leq U$. By Theorem **9.3.2**, we need only prove $G_i \subset C_{[\alpha_i]}^\circ \subset C_{[\alpha_i]}^- \subset U_{[\alpha_i]}$ for every $\alpha_i \in L_0$. By condition (b), $\alpha_j \preceq \alpha_k \Longrightarrow A_{\alpha_j} \supset A_{\alpha_k}$, so $A_{\alpha_i} \subset C_{\alpha_i}$ by the definition of $C_\alpha$. Then by condition (a), $G_i \subset A_{\alpha_i}^\circ \subset C_{\alpha_i}^\circ = C_{[\alpha_i]}^\circ$. On the other hand, $C_{[\alpha_i]}^- \subset \bigcap\{A_{\alpha_i}^- : \alpha_j \preceq \alpha_i,\ \alpha_j \in L_0\}$, by Lemma **2.1.9** (i), $D = \{\alpha_j : \alpha_j \in L_0,\ \alpha_j \preceq \alpha_i\}$ is a minimal set of $\alpha_i$, so $U_{[\alpha_i]} = \bigcap\{U_{[\alpha_j]} : \alpha_j \in D\}$. By condition (a), $A_{\alpha_j}^- \subset U_{[\alpha_j]}$, so $C_{[\alpha_i]}^- \subset U_{[\alpha_i]}$, (i) has already been proved.

(iii)$\Longrightarrow$(i): Clear.

(ii)$\Longrightarrow$(iii): Now we use the assumption that $L_0$ is a countable strictly join-generating set of $L$ to prove that the $L$-fuzzy set $C$ constructed in the proof of (ii)$\Longrightarrow$(i) is open and closed in $(L^X, \delta)$. For this aim, it is enough to prove $C^- \leq C^\circ$; or in an alternative way, prove $C^-{}_{[\alpha]} \subset C^\circ{}_{[\alpha]}$ for every $\alpha \in L_0$. Denote

$$H(\alpha) = \beta(\alpha) \cap L_0, \quad H_-(\alpha) = (\beta(\alpha)\backslash\{\alpha\}) \cap L_0.$$

By Lemma **2.1.9**, both $H(\alpha)$ and $H_-(\alpha)$ are minimal sets of $\alpha$. By Lemma **9.3.3**, $\{\sigma : \sigma \in H_-(\theta),\ \theta \in H(\alpha)\} \supset H_-(\alpha)$; by Theorem **1.3.14** (i), its converse holds. So

$$\{\sigma : \sigma \in H_-(\theta),\ \theta \in H(\alpha)\} = H_-(\alpha). \tag{9.35}$$

Then we have

$$\begin{aligned}
C_{[\alpha]}^- &= \bigcap\{C_{[\gamma]}^- : \gamma \in H(\alpha)\} && \text{(Theorem } \mathbf{4.3.8} \text{ (iii))}\\
&\subset \bigcap\{\bigcap\{A_\theta^- : \theta \in H(\gamma)\} : \gamma \in H(\alpha)\} && \text{(Equation (9.34))}\\
&= \bigcap\{A_\theta^- : \theta \in H(\alpha)\} && \text{(Theorem } \mathbf{1.3.14} \text{ (i))}\\
&\subset \bigcap\{\bigcap\{A_\sigma^\circ : \sigma \in H_-(\theta)\} : \theta \in H(\alpha)\} && \text{(Condition (b))}\\
&= \bigcap\{A_\sigma^\circ : \sigma \in H_-(\alpha)\} && \text{(Equation (9.35))}\\
&\subset \bigcap\{C_\sigma^\circ : \sigma \in H_-(\alpha)\} && (A_\sigma \subset C_\sigma)\\
&= C^\circ{}_{[\alpha]}, && \text{(Theorem } \mathbf{4.3.10} \text{ (iii))}
\end{aligned}$$

this completes the proof. □

**9.3.5 Definition** Let $(X, \mathcal{T})$ be an ordinary topological space, $L$ a complete lattice. We say *the insertion theorem holds in* $(X, \mathcal{T})$ *for* $L$, if for every upper semicontinuous mapping $f$ and every lower semicontinuous mapping $g$ from $(X, \mathcal{T})$ to $L$ such that $f \leq g$, there exists a continuous mapping $h$ form $(X, \mathcal{T})$ to $L$ such that $f \leq h \leq g$.

Then by Theorem **9.3.4**, we directly have an extension of Hahn-Dieudonné-Tong Insertion Theorem as follows:

**9.3.6 Theorem** (Lattice-valued Insertion Theorem) *Let $L$ be a F-lattice with a countable strictly join-generating set. Then the insertion theorem holds in a topological space $(X, \mathcal{T})$ if and only if $(X, \mathcal{T})$ is normal.* □

**9.3.7 Remark** (1) In Theorem **9.3.6**, the requirement for the order-reversing involution on $L$ is not necessary; i.e. complete distributivity of $L$ in that theorem is enough but not necessary to require $L$ to be a F-lattice. Reader can trace the proof of Theorem **9.3.4** to prove Theorem **9.3.6** again under the assumtion that $L$ is a completely distributive lattice. Certainly, all the involved propositions need to be restated and proved at first.

(2) The method used in the preceding proof to analysis and to construct lattice-valued mappings with its levels has an evidently different style compared with the classical one, which is called "stratification structure analysis."

In Theorem **9.3.6**, the strictness of the countable generating set of $L$ cannot be removed. The following counterexample shows this point:

**9.3.8 Example** Suppose $L$ be a F-lattice which has not a strictly join-generating set (e.g. every finite lattice). Let $(X, \mathcal{T})$ be a normal space with a closed set $F$ and an open set $U$ such that $F \subset U$ but there does not exist any closed and open set $W$ in it such that $F \subset W \subset U$ (e.g. every connected normal space), let $\delta = \omega(\mathcal{T})$, then $(L^X, \delta)$ is an induced $L$-fts and $[\delta] = \mathcal{T}$. Since $L$ has no strictly generating set, so $\exists a \in L$ such that $b = \bigvee\{c \in L : \ c < a\} < a$, i.e. there is a gap from $b$ to $a$. Denote this gap by $(b, a)$. Take

$$f = \underline{b} \vee aF, \quad g = \underline{b} \vee aU,$$

then, clearly, $f$ and $g$ are respectively closed and open in $(L^X, \delta)$, hence are respectively upper semicontinuous and lower semicontinuous, and $f \leq g$. If the insertion theorem holds in $(X, \mathcal{T})$, then there exists a continuous mapping $h : \ (X, \mathcal{T}) \to L$ such that $f \leq h \leq g$. $\forall x \in X$, we should have $b \leq f(x) \leq h(x) \leq g(x) \leq a$. Since $(b, a)$ is a gap, $h(x)$ must be either $a$ or $b$. By the upper semicontinuity of $h$, $W = h_{[a]}$ is a closed subset in $(X, [\delta]) = (X, \mathcal{T})$. Certainly, $F \subset W \subset U$. Since $h(x)$ is either $a$ or $b$ for every $x \in X$, we have $h(x) \not\geq a \Longleftrightarrow h(x) \leq b$. So $X \backslash W = \{x \in X : \ h(x) \leq b\}$. By the lower semicontinuity of $h$, $X \backslash W$ is closed in $(X, [\delta])$. Hence $W$ is also open in $(X, [\delta]) = (X, \mathcal{T})$. This is a contradiction.

If the value-domain $L$ has no countable generating set, then not only the Lattice-valued Insertion Theorem will not hold any longer, but also the implication (ii)$\Longrightarrow$(i) in Theorem **9.3.4** cannot be true:

**9.3.9 Example** Let $\omega_0$ and $\omega_1$ be the first countably infinite ordinal number and the first uncountable ordinal number respectively, $\omega_0+1=\omega_0\cup\{\omega_0\}$, $\omega_1+1=\omega_1\cup\{\omega_1\}$ be equipped with the interval topology. Let $X=(\omega_0+1)\times(\omega_1+1)$, then the product space $(X,\mathcal{T})$ (i.e. the Tychonoff plane) is a normal compact space. We construct a F-lattice $L$ as follows: Let $\mathcal{P}(\omega_1)$ be the power set of $\omega_1$ equipped with the inclusion order, then it is a completely distributive lattice. Take a copy of it, denote this copy by $\widetilde{\Omega}$. Take the opposite lattice of $\widetilde{\Omega}$, denote it by $\underset{\sim}{\Omega}$. We can suppose $\widetilde{\Omega}$ does not intersect with $\underset{\sim}{\Omega}$. For every $a\in\mathcal{P}(\omega_1)$, denote the correspoindent element in $\widetilde{\Omega}$ by $\tilde{a}$ and the correspondent element in $\underset{\sim}{\Omega}$ by $\underset{\sim}{a}$. Moreover, we suppose that both $\widetilde{\Omega}$ and $\underset{\sim}{\Omega}$ do not intersect with the open interval $(-1,1)$. Now take $L=\underset{\sim}{\Omega}\cup(-1,1)\cup\widetilde{\Omega}$, define a partial order $\leq$ in $L$ as follows: For every two $a,b\in\mathcal{P}(\omega_1)$ and every $t\in(-1,1)$,

$$\underset{\sim}{a}\leq\underset{\sim}{b}\Longleftrightarrow a\supset b\Longleftrightarrow\tilde{b}\leq\tilde{a},\quad \underset{\sim}{a}\leq t\leq\tilde{b};$$

furthermore, the elements in $(-1,1)$ still preserve their order of real numbers. The order-reversing involution on $L$ can be defined by

$$z'=\begin{cases}\tilde{a}, & z=\underset{\sim}{a},\\ -z, & z\in(-1,1),\\ \underset{\sim}{a}, & z=\tilde{a}.\end{cases}$$

Denote the smallest element and the largest one of $L$ by $\check{0}$ and $\check{1}$ respectively. Clearly, $L$ is a F-lattice and has not any countable generating set but has a generating set of power $\omega_1$.

Let $\delta=\omega(\mathcal{T})$, then $(L^X,\delta)$ is an induced $L$-fts, its background space $(X,[\delta])=(X,\mathcal{T})$ is normal. We say $(L^X,\delta)$ is not normal. In fact, take $F,U\in L^X$ as follows:

$$F(x,y)=\begin{cases}\check{0}, & x<\omega_0,\\ \tilde{\emptyset}, & (x,y)=(\omega_0,\omega_1),\\ \widetilde{\{y\}}, & x=\omega_0,\ y<\omega_1,\end{cases}$$

$$U(x,y)=\begin{cases}\check{1}, & y<\omega_1,\\ \tilde{\emptyset}, & (x,y)=(\omega_0,\omega_1),\\ 1-\frac{1}{x+1}, & x<\omega_0,\ y=\omega_1.\end{cases}$$

Since for every $\alpha\in M(L)$, both $F_{[\alpha]}$ and $U'_{[\alpha]}$ are closed in $(X,\mathcal{T})$, so both $F$ and $U'$ are closed subsets in induced $L$-fts $(L^X,\delta)$, and easy to find $F\leq U$. If $(L^X,\delta)$ is normal, then there exists $A\in L^X$ such that $F\leq A^\circ\leq A^-\leq U$. Take $b=\tilde{\emptyset}$, then $\bigcup\{F_{[\alpha]}:\ \alpha\in M(L),\ b\preceq\alpha\}=\{\omega_0\}\times\omega_1$ and $U_{[b]}=X\backslash(\omega_0\times\{\omega_1\})$. By Theorem **9.3.2**,

$$\{\omega_0\}\times\omega_1\subset A_{[b]}{}^\circ\subset A_{[b]}{}^-\subset X\backslash(\omega_0\times\{\omega_1\}).$$

So $\{\omega_0\}\times\omega_1$ and $\omega_0\times\{\omega_1\}$ have disjoint neighborhoods in $(X,\mathcal{T})$. But this is impossible (refer to Question F, Chapter 4 of [50]). Hence $(L^X,\delta)$ is not normal.

The Lattice-valued Insertion Theorem provides an extension of the classical insertion theorem. But will the fuzzy insertion theorem be true? Kubiak gave out the positive answer as follows. About its proof, we refer the reader to [54].

**9.3.10 Definition** Let $(L^X, \delta)$ be an $L$-fts. An $L$-fuzzy mapping $f^{\rightarrow} : (L^X, \delta) \to \mathbf{R}(L)$ is called *lower semicontinuous* (resp. *upper semicontinuous*), if $f^{\leftarrow}(R_t) \in \delta$ (resp. $f^{\leftarrow}(L_t) \in \delta$) for every $t \in \mathbf{R}$.

**9.3.11 Theorem** (Fuzzy Insertion Theorem) *Let $(L^X, \delta)$ be an L-fts. Then the following conditions are equivalent:*

(i) *$(L^X, \delta)$ is normal.*

(ii) *If $f^{\rightarrow}, g^{\rightarrow} : (L^X, \delta) \to \mathbf{R}(L)$ are respectively upper semicontinuous and lower semicontinuous and $f^{\rightarrow} \le g^{\rightarrow}$, then there exists a continuous L-fuzzy mapping $h^{\rightarrow} : (L^X, \delta) \to \mathbf{R}(L)$ such that $f^{\rightarrow} \le h^{\rightarrow} \le g^{\rightarrow}$.* □

# Chapter 10

# Compactness

Compactness is always one of the most important concepts in topology. In fuzzy topology, after the initial work of straight imitation of ordinary compactness in the pattern of covers of a whole space, many authors tried to establish various reasonable compactness with consideration of levels of fuzzy topological spaces, and obtained many important results. Among these notions, based on stratification structures of fuzzy topological spaces, N-compactness looks like to be possessed of more nice properties. Hence it will play a main role in compactness theory in this chapter.

## 10.1 Some Kinds of Compactness in Fuzzy Topological Spaces

Since the level structures, or in the other word, stratifications of lattices, are involved, compactness in fuzzy topological spaces is one of most complicated problems in this field. Many kinds of fuzzy compactnesses were raised, and each of them has its own advantage and shortage. In [94], R. Lowen introduced some notions of compactness in fuzzy topological spaces which were produced by himself and C. L. Chang, T. E. Gantner, R. C. Steinlage, R. H. Warren etc., and compared their properties. All the value domains used in these notions are $[0,1]$'s, i.e. they are all about F-ts'. This section will introduce these investigations.

Based on quasi-coincident neighborhood structure, Ying-Ming Liu also produced Q-compactness in fuzzy topological spaces with some nice properties.[68] It will also be introduced in this section.

**10.1.1 Definition** Let $(L^X, \delta)$ be an $L$-fts, $A \in L^X$, $\mathcal{A}, \mathcal{B} \subset L^X$.

$\mathcal{A}$ is called a *cover* of $A$, if $\bigvee \mathcal{A} \geq A$; particularly, $\mathcal{A}$ is called a cover of $(L^X, \delta)$, if $\mathcal{A}$ is a cover of $\underline{1}$. $\mathcal{A}$ is called an *open cover* of $A$, if $\mathcal{A} \subset \delta$ and $\mathcal{A}$ is a cover of $A$. For a cover $\mathcal{A}$ of $A$, $\mathcal{B}$ is called a *subcover* of $\mathcal{A}$, if $\mathcal{B} \subset \mathcal{A}$ and $\mathcal{B}$ is still a cover of $A$.

**10.1.2 Definition** Let $(X, \delta)$ be a F-ts.

For every $\alpha \in [0,1)$, a family $\mathcal{A} \subset I^X$ is called an *$\alpha$-cover*, if for every $x \in X$, there exists $A \in \mathcal{A}$ such that $A(x) > \alpha$; $\mathcal{A}$ is called an *open $\alpha$-cover*, if $\mathcal{A} \subset \delta$ and $\mathcal{A}$ is an $\alpha$-cover; $\mathcal{A}_0 \subset I^X$ is called a *sub-$\alpha$-cover* of $\mathcal{A}$, if $\mathcal{A}_0 \subset \mathcal{A}$ and $\mathcal{A}_0$ is an $\alpha$-cover; $\mathcal{A}$ is called an *$\alpha^*$-cover*, if for every $x \in X$, there exists $A \in \mathcal{A}$ such that $A(x) \geq \alpha$; $\mathcal{A}$ is called an *open $\alpha^*$-cover*, if $\mathcal{A} \subset \delta$ and $\mathcal{A}$ is an $\alpha^*$-cover; $\mathcal{A}_0 \subset I^X$ is called a *sub-$\alpha^*$-cover* of $\mathcal{A}$, if $\mathcal{A}_0 \subset \mathcal{A}$ and $\mathcal{A}_0$ is an $\alpha^*$-cover.

**10.1.3 Definition** A F-ts $(X, \delta)$ is called *C-compact*, if every open cover of $(X, \delta)$ has a finite subcover.

**10.1.4 Definition** Let $\alpha \in [0,1)$. $(X, \delta)$ is called *$\alpha$-compact*, if every open $\alpha$-cover has a finite sub-$\alpha$-cover.

**10.1.5 Definition** Let $\alpha \in [0,1)$. $(X,\delta)$ is called *$\alpha^*$-compact,* if every open $\alpha^*$-cover has a finite sub-$\alpha^*$-cover.
**10.1.6 Definition** A F-ts $(X,\delta)$ is called *strongly compact,* if it is $\alpha$-compact for every $\alpha \in [0,1)$.

Using the symbol $\iota_L(\delta)$ introduced in Section **4.4** and taking $L = I = [0,1]$, we have
**10.1.7 Definition** A F-ts $(X,\delta)$ is called *ultra-compact,* if $(X, \iota_I(\delta))$ is compact.
**10.1.8 Definition** A F-ts $(X,\delta)$ is called *weakly compact,* if for every $\mathcal{A} \subset \delta$ such that $\bigvee \mathcal{A} = \underline{1}$ and every $\varepsilon > 0$, there exists a finite subfamily $\mathcal{A}_0 \subset \mathcal{A}$ such that $\bigvee \mathcal{A}_0 \geq 1 - \varepsilon$.
**10.1.9 Definition** A F-ts $(X,\delta)$ is called *fuzzy compact,* if for every $\alpha \in [0,1]$, every $\mathcal{A} \subset \delta$ such that $\bigvee \mathcal{A} \geq \underline{\alpha}$ and every $\varepsilon \in (0,\alpha)$, there exists a finite subfamily $\mathcal{A}_0 \subset \mathcal{A}$ such that $\bigvee \mathcal{A}_0 \geq \underline{\alpha - \varepsilon}$.
**10.1.10 Definition** Let $(L^X,\delta)$ be an $L$-fts, $A \in L^X$, $\Phi \subset L^X$. $\Phi$ is called a *Q-cover* of $A$, if for every $x \in supp(A)$, there exists $U \in \Phi$ such that $x_{A(x)} \lessdot U$. $\Phi$ is called an *open Q-cover* of $A$, if $\Phi \subset \delta$ and $\Phi$ is a Q-cover of $A$. $\Phi_0 \subset L^X$ is called a *sub-Q-cover* of $\Phi$, if $\Phi_0 \subset \Phi$ and $\Phi_0$ is also a Q-cover of $A$. $\Phi$ is called a *Q-cover* of $(X,\delta)$, if $\Phi$ is a Q-cover of $\underline{1}$.
**10.1.11 Definition** Let $(X,\delta)$ be a F-ts, $A \in I^X$. $A$ is called *Q-compact,* if every open Q-cover of $A$ has a finite sub-Q-cover. $(X,\delta)$ is called *Q-compact,* if $\underline{1}$ is Q-compact.

Recall Section **4.4**, where we defined a family $\iota_\alpha(\delta)$ and an ordinary topology $\iota_L(\delta)$ on $X$ for an $L$-fts $(X,\delta)$ and $\alpha \in L$. For every $\alpha \in [0,1)$, since $\alpha$ is prime in $[0,1]$, the following proposition is direct by Propositions **4.2.3** and **4.2.4**:
**10.1.12 Proposition** *Let $(X,\delta)$ be a F-ts, $\alpha \in [0,1)$, then $\iota_\alpha(\delta)$ is a topology on $X$.* □

Similarly, we have the following
**10.1.13 Definition** Let $L$ be a F-lattice.
For every $(L^X,\delta) \in Ob(L\text{-}\mathbf{FTS})$ and every $a \in L$, denote

$$\iota_a^-(\delta) = \{U_{[a]} : U \in \delta\}.$$

Let $\iota_a^*(\delta)$ denote the ordinary topology on $X$ generated by the subbase $\iota_a^-(\delta)$, let $\iota_L^*(\delta)$ denote the ordinary topology on $X$ generated by the subbase $\bigcup_{a \in L} \iota_a^-(\delta)$.

Then the following result can be directly obtained:
**10.1.14 Proposition** *Let $(X,\delta)$ be a F-ts. Then*
(i) *For every $\alpha \in [0,1)$, $(X,\delta)$ is $\alpha$-compact if and only $(X, \iota_\alpha(\delta))$ is compact.*
(ii) *For every $\alpha \in (0,1]$, $(X,\delta)$ is $\alpha^*$-compact if and only if $(X, \iota_\alpha^*(\delta))$ is compact.* □

In Section **4.4**, we introduce the functor $\omega_L : \mathbf{Top} \to L\text{-}\mathbf{FTS}$. It is easy to verify that $\omega_L(\mathbf{Top})$ is a full subcategory of $L$-**FTS** and is isomorphic to **Top**, so one can consider **Top** as a full subcategory of $L$-**FTS**. Starting from this point, Lowen introduced the concept "good extension" for F-ts'. For convenience, we prescribe the meanings of notions "good extension" and "good $L$-extension" as follows:

> For a property $P$ of ordinary topological spaces, an property $P^*$ of $L$-fuzzy topological spaces is called a *good $L$-extension* of $P$, if for every ordinary topological space $(X, \mathcal{T})$, $(X, \mathcal{T})$ has property $P$ if and only if $(X, \omega_L(\mathcal{T}))$ has property $P^*$. $P^*$ is called a *good extension* of $P$ for short, if $P^*$ is a good $L$-extension of $P$ and $L = [0, 1]$.

With this definition, the following results can be easily shown:

**10.1.15 Theorem** *$\alpha$-compactness, strong compactness, ultra-compactness, weak compactness, fuzzy compactness and Q-compactness in fuzzy topological spaces are good extensions of compactness in ordinary topological spaces.* □

From the following examples, one can find that the notion C-compactness has many shortages.

**10.1.16 Example** Let $(X, \delta)$ be a stratified F-ts. Then $\mathcal{A} = \{\underline{\alpha} : \alpha \in (0, 1)\}$ is an open cover of $(X, \delta)$, but $\mathcal{A}$ has not any finite subcover. So every stratified F-ts is not C-compact. It follows that C-compactness is not a good extension.

**10.1.17 Example** Let $(X, \delta)$ be a $T_1$ F-ts, $x \in X$. Then $\mathcal{A} = \{(x_{\frac{1}{n}})' : n \in \mathbf{N}\}$ is an open cover of $(X, \delta)$. But it obviously has not a subcover. So every $T_1$ F-ts is not C-compact.

C-compactness is not multiplicative either:

**10.1.18 Example**[168] Let $X$ be a non-empty ordinary set. For every $n \in \mathbf{N}$, denote $a_n = 1 - \frac{1}{n}$, $\delta_n = \{\underline{0}, \underline{1}, \underline{a_n}\}$. Then $(X, \delta_n)$ is a C-compact F-ts for every $n \in \mathbf{N}$. Let

$$(Y, \delta) = \prod_{n \in \mathbf{N}} (X, \delta_n),$$

then

$$\delta = \{\underline{0}, \underline{1}\} \cup \{\underline{a_n} : n \in \mathbf{N}\},$$

and $\mathcal{A} = \{\underline{a_n} : n \in \mathbf{N}\}$ is an open cover of $\prod_{n \in \mathbf{N}} (X, \delta_n)$. But $\mathcal{A}$ has not any finite subcover. Hence $\prod_{n \in \mathbf{N}} (X, \delta_n)$ is not C-compact.

With an example, Lowen[94] showed that $\alpha^*$-compactness is not a good extension. In fact, he showed that for every ordinary $T_1$ topological space $(X, \mathcal{T})$, $(X, \omega(\mathcal{T}))$ is not $\alpha^*$-compact for every $\alpha \in (0, 1]$.

On the multiplicative property of other kinds of compactness in F-ts, Lowen, Gantner, Steinlage and Warren proved the following conclusions for F-ts' based on the definition of Lowen's, i.e. for stratified F-ts' (see [92, 93, 94]):

**10.1.19 Theorem** *$\alpha$-compactness, strong compactness, ultra-compactness and fuzzy compactness in fuzzy topological spaces are strongly multiplicative.* □

Liu proved the multiplicativity of Q-compactness in Ref.[68]:

**10.1.20 Theorem** *Q-compactness for fuzzy topological spaces is multiplicative.* □

## 10.2 N-compactness

N-compactness was introduced by Guo-Jun Wang[162] in 1983, where "N" means "Nice". In Wang's original paper, he defined this kind of compactness for $L = [0, 1]$

with fuzzy nets. Since this kind of compactness possesses many nice properties and is defined for $L = [0,1]$, later it was generalized to the case of general F-lattice and was characterized geometrically by Yu-Wei Peng[126, 127] and Dong-Sheng Zhao.[198]

In the original work of Wang, Peng and Zhao, N-compactness was described via R-neighborhoods. We use Q-neighborhoods to do the same thing in the sequel.

**10.2.1 Definition** Let $(L^X, \delta)$ be an $L$-fts, $A \in L^X$, $C \in \delta$, $\Phi \subset L^X$, $\alpha \in M(L)$. $C$ is called an *α-Q-neighborhood* of $A$, if $C \in \mathcal{Q}(x_\alpha)$ for every $x_\alpha \leq A$. $\Phi$ is called an *α-Q-cover* of $A$, denoted by $\bigvee \Phi \hat{q} A(\alpha)$, if for every $x_\alpha \leq A$, there exists $U \in \Phi$ such that $x_\alpha \lessdot U$. $\Phi$ is called an *open α-Q-cover* of $A$, if $\Phi \subset \delta$ and $\Phi$ is an α-Q-cover of $A$. $\Phi_0 \subset L^X$ is called a *sub-α-Q-cover* of $\Phi$, if $\Phi_0 \subset \Phi$ and $\Phi_0$ is also an α-Q-cover of $A$. $\Phi$ is called an *$\alpha^-$-Q-cover* of $A$, denoted by $\bigvee \Phi \hat{\hat{q}} A(\alpha)$, if there exists $\gamma \in \beta^*(\alpha)$ such that $\Phi$ is a γ-Q-cover of $A$.

Denote by $\bigvee \Phi \neg \hat{q} A(\alpha)$ the statement "$\Phi$ is not an α-Q-cover of $A$", and denote by $\bigvee \Phi \neg \hat{\hat{q}} A(\alpha)$ the statement "$\Phi$ is not an $\alpha^-$-Q-cover of $A$".

Clearly, α-Q-cover and $\alpha^-$-Q-cover will always be non-empty whenever there exists indeed a molecule which is contained in $A$ with height α.

**10.2.2 Proposition** *Let $(L^X, \delta)$ be an L-fts, $\alpha \in M(L)$, $A \in L^X$, $\mathcal{A} \subset L^X$. Then*

(i) *$\mathcal{A}$ is an α-Q-cover of $A$ if and only if $\mathcal{A}$ is a Q-cover of $\alpha A_{[\alpha]}$.*

(ii) *$\mathcal{A}$ is an α-Q-cover of $\underline{1}$ if and only if $\mathcal{A}$ is a Q-cover of $\underline{\alpha}$.* □

**10.2.3 Definition** Let $(L^X, \delta)$ be an $L$-fts, $A \in L^X$. $A$ is called N-compact, if for every $\alpha \in M(L)$, every open α-Q-cover of $A$ has a finite subfamily which is an $\alpha^-$-Q-cover of $A$. $(L^X, \delta)$ is called N-compact, if $\underline{1}$ is N-compact.

**10.2.4 Theorem** *Let $(L^X, \delta)$ be an L-fts, $A \in L^X$. Then $A$ is N-compact if and only if the following conditions hold*

(i) *For every $\alpha \in M(L)$, every open α-Q-cover of $A$ has a finite sub-α-Q-cover.*

(ii) *For every $\alpha \in M(L)$, every open α-Q-cover of $A$ which consists of just one subset is an $\alpha^-$-Q-cover of $A$.*

**Proof** (Necessity) Let $A$ be N-compact, $\alpha \in M(L)$, $\Phi$ is an open α-Q-cover of $A$. Then $\Phi$ has a finite subfamily $\Psi$ such that $\bigvee \Psi \hat{\hat{q}} A(\alpha)$. It follows that $\bigvee \Psi \hat{q} A(\alpha)$, (i) is true. Suppose $U \in \delta$ and suppose that $\Phi = \{U\}$ is an open α-Q-cover of $A$. Then by the N-compactness of $A$, $\Phi$ has a finite subfamily $\Psi$ such that $\Psi$ is an $\alpha^-$-Q-cover of $A$. But clearly we have $\Psi = \Phi$ now, so $\Phi$ is also an $\alpha^-$-Q-cover of $A$, (ii) holds.

(Sufficiency) Suppose the conditions (i) and (ii) has been fulfilled, $\alpha \in M(L)$, $\Phi$ is an open α-Q-cover of $A$. By the virtue of (i), $\Phi$ has a finite sub-α-Q-cover $\Psi$. Take $U = \bigvee \Psi$, then $\{U\}$ is an α-Q-cover of $A$ by Proposition **2.3.3** (vi). By (ii), $\{U\}$ is also an $\alpha^-$-Q-cover of $A$, so there exists $\gamma \in \beta^*(\alpha)$ such that $\gamma \not\leq U(x)' = \bigwedge\{V(x)' : V \in \Psi\}$ for every $x_\gamma \leq A$. Then there exists $V \in \Psi$ such that $\gamma \not\leq V(x)'$. That is to say, $V \in \mathcal{Q}_\delta(x_\gamma)$. So $\Psi$ is an $\alpha^-$-Q-cover of $A$, $A$ is N-compact. □

Just as it was pointed previously, N-compactness was defined via fuzzy nets in the original paper. Now we characterize N-compactness with molecule nets in $L$-fts.

**10.2.5 Definition** Let $(L^X, \delta)$ be an $L$-fts, $S = \{S(n), n \in D\}$ a net in $L^X$, $a \in L$.

Define the *support net* of $S$ as the net $supp(S) = \{supp(S(n)), n \in D\}$ in $(X, [\delta])$.

Define the *height net* of $S$ as the net $ht(S) = \{ht(S(n)), n \in D\}$ in $L$. $S$ is called an *eventual a-net*, if $S$ is a molecule net and for every $\gamma \in \beta^*(a)$, $ht(S)$ is eventually larger than or equal to $\gamma$, i.e. there exists $n_0 \in D$ such that $ht(S(n)) \geq \gamma$ for every $n \in D$, $n \geq n_0$.

**10.2.6 Proposition** *Let $(L^X, \delta)$ be an L-fts, $S = \{S(n), n \in D\}$ a net in $L^X$, $x \in X$, $a \in L$. Then*

(i) $S \to x_a \implies supp(S) \to x$.

(ii) *$S$ is an eventual a-net if and only if*

$$\bigvee_{n\in D}\bigwedge_{m\geq n} ht(S(m)) \geq a.$$

**Proof** (i) By $[\delta] \subset \delta$.

(ii) Suppose $S$ is an eventual $a$-net in $(L^X, \delta)$, then for every $\gamma \in \beta^*(a)$, there exists $n_\gamma \in D$ such that $ht(S(m)) \geq \gamma$ for every $m \geq n_\gamma$. So $\bigwedge\{ht(S(m)) : m \in D, m \geq n_\gamma\} \geq \gamma$. Hence by Theorem **1.3.10**,

$$\bigvee_{n\in D}\bigwedge_{m\geq n} ht(S(m)) \geq \bigvee_{\gamma\in\beta^*(a)}\bigwedge_{m\geq n_\gamma} ht(S(m)) \geq \bigvee_{\gamma\in\beta^*(a)} \gamma = a.$$

Suppose $\bigvee_{n\in D}\bigwedge_{m\geq n} ht(S(m)) \geq a$, $\gamma \in \beta^*(a)$, then by Theorem **1.3.10**, $\beta^*(a)$ is a minimal set of $a$, $\exists n_\gamma \in D$ such that $\gamma \leq \bigwedge_{m\geq n} ht(S(m))$. So $ht(S(m)) \geq \gamma$ for every $m \in D$, $m \geq n_\gamma$, $S$ is an eventual $a$-net. □

**10.2.7 Theorem** *Let $(L^X, \delta)$ be an L-fts, $A \in L^X$. Then $A$ is N-compact if and only if for every $\alpha \in M(L)$, every eventual $\alpha$-net in $A$ has a cluster point in $A$ with height $\alpha$.*

**Proof** Suppose $A$ is N-compact, $S = \{S(n), n \in D\}$ is an eventual $\alpha$-net in $A$. Moreover, suppose $S$ has not any cluster point in $A$ with height $\alpha$. Then for every $x_\alpha \in M(\downarrow A)$, there exists $U_x \in \mathcal{Q}(x_\alpha)$ such that eventually $S$ is not quasi-coincident with $U_x$, i.e. there exists $n_x \in D$ such that for every $n \in D$, $n \geq n_x$, $S(n) \leq (U_x)'$. Take $\Phi = \{U_x : x_\alpha \leq A\}$, then $\Phi$ is an open $\alpha$-Q-cover of $A$. Since $A$ is N-compact, $\Phi$ has a finite subfamily $\Psi = \{U_{x_i} : i = 1, \cdots, k\}$ such that $\bigvee \Psi \hat{q} A(\alpha)$, i.e. there exists $\gamma \in \beta^*(\alpha)$ such that for every $y_\gamma \in M(\downarrow A)$ there exists $i \leq k$ such that $y_\gamma \lessdot U_{x_i}$. Take $U = \bigvee\{U_{x_i} : i = 1, \cdots, k\}$, then for every $y_\gamma \in M(\downarrow A)$, $y_\gamma \lessdot U$, i.e.

$$\forall y_\gamma \leq A, \quad \gamma \not\leq U(y)'. \tag{10.1}$$

Since $D$ is a directed set, there exists $n_0 \in D$ such that $n_0 \geq n_{x_i}$ for $i = 1, \cdots, k$. Then $S(n) \leq (U_{x_i})'$ for every $n \geq n_0$ and $i = 1, \cdots, k$, and hence

$$n \geq n_0 \implies S(n) \leq U'. \tag{10.2}$$

By the virtue of (10.1), (10.2) and $S(n) \leq A$, we have

$$n \geq n_0 \implies ht(S(n)) \not\geq \gamma.$$

But $S$ is an eventual $\alpha$-net in $A$, it is a contradiction. So $S$ has at least one cluster point in $A$ with height $\alpha$.

Reversely, suppose that for every $\alpha \in M(\downarrow A)$, every eventual $\alpha$-net in $A$ has a cluster point in $A$ with height $\alpha$, $\Phi$ is an open $\alpha$-Q-cover of $A$. If every finite subfamily $\Psi$ of $\Phi$ is not an $\alpha^-$-Q-cover of $A$, then $\forall \Psi \in [\Phi]^{<\omega}$, $\forall \gamma \in \beta^*(\alpha)$, $\exists x_\gamma^\Psi \in M(\downarrow A)$ such that $x_\gamma^\Psi \not\lessdot U$ for every $U \in \Psi$, i.e.

$$\Psi \in [\Phi]^{<\omega}, \ \gamma \in \beta^*(\alpha) \implies \exists x_\gamma^\Psi \in M(\downarrow A), \ \forall U \in \Psi, \ x_\gamma^\Psi \not\lessdot U. \tag{10.3}$$

Take $D = \beta^*(\alpha) \times [\Phi]^{<\omega}$, and define a relation $\leq$ on $D$ as follows:

$$\forall(\gamma_1,\Psi_1),(\gamma_2,\Psi_2)\in D,\quad (\gamma_1,\Psi_1)\leq(\gamma_2,\Psi_2)\implies \gamma_1\leq\gamma_2,\ \Psi_1\subset\Psi_2,$$

then it follows from Theorme **1.3.8**, $D$ is a directed set. Take

$$S=\{x_\gamma^\Psi:\ (\gamma,\Psi)\in D\},$$

then $S$ is a molecule net in $A$. $\forall\gamma\in\beta^*(\alpha)$, fix arbitrarily $\Psi_0\in[\Phi]^{<\omega}$, then $ht(x_\lambda^\Psi)\geq\gamma$ whenever $(\lambda,\Psi)\geq(\gamma,\Psi_0)$, so $S$ is an eventual $\alpha$-net in $A$. Let $x_\alpha\in M(\downarrow A)$. Since $\Phi$ is an open $\alpha$-Q-cover of $A$, there exists $U\in\Phi$ such that $U\in\mathcal{Q}(x_\alpha)$. Then $\{U\}\in[\Phi]^{<\omega}$. $\forall\lambda\in\beta^*(\alpha)$, then for every $(\gamma,\Psi)\geq(\lambda,\{U\})$, by (10.3) and $U\in\Psi$, we have $x_\gamma^\Psi\not\leq U$. Thus $x_\alpha$ is not a cluster point of $S$. It follows that $S$ has no cluster point in $A$ with height $\alpha$. This conclusion contradicts the previous assumption that every finite subfamily $\Psi$ of $\Phi$ is not an $\alpha^-$-Q-cover of $A$. □

N-compactness has the following interesting property:

**10.2.8 Theorem** *Let $(L^X,\delta)$ be an L-fts, $A\in L^X$ is N-compact, $\alpha=\bigvee\{A(x):\ x\in X\}\in M(L)$. Then there exists $x\in X$ such that $A(x)=\alpha$.*

**Proof** Since $\beta^*(\alpha)$ is a minimal set of $\alpha=\bigvee\{A(x):\ x\in X\}$, we have

$$\gamma\in\beta^*(\alpha)\implies\exists x^\gamma\in X,\ \gamma\leq A(x^\gamma).\tag{10.4}$$

Take $D=\beta^*(\alpha)$, then by the virtue of $\alpha\in M(L)$ and Theorem **1.3.8**, $D$ is a directed set. Take $S(\gamma)=x^\gamma{}_\gamma$ for every $\gamma\in D$, then by the viertue of (10.4), $S=\{S(\gamma),\gamma\in D\}$ is a molecule net in $A$. In order to show that $S$ is an eventual $\alpha$-net in $A$, take arbitrarily $\lambda\in\beta^*(\alpha)$, then $\lambda\in D$. For every $\gamma\in D$ such that $\gamma\geq\lambda$, we have

$$ht(S(\gamma))=ht(x^\gamma{}_\gamma)=\gamma\geq\lambda.$$

So $S$ is an eventual $\alpha$-net in $A$. Then by the N-compactness of $A$, $S$ has a cluster point $x_\alpha\leq A$. That is to say, $A(x)\geq\alpha=\bigvee\{A(y):\ y\in X\}$, $A(x)=\alpha$. □

Then we obtain directly the following conclusion:

**10.2.9 Corollary** *Let $(X,\delta)$ be a F-ts, $A$ is a N-compact subset in $(X,\delta)$. Then as a function from $X$ to $[0,1]$, $A$ takes its largest value at some point $x$.* □

On the hereditary property of N-compactness, we have the following result:

**10.2.10 Theorem** *Let $(L^X,\delta)$ be an L-fts, $A\in L^X$ be N-compact, $B\in L^X$ be closed. Then $A\wedge B$ is N-compact.*

**Proof** Let $S$ be an eventual $\alpha$-net in $A\wedge B$, then $S$ is also an eventual $\alpha$-net in $A$. Since $A$ is N-compact, $S$ has a cluster point $x_\alpha$ in $A$ with height $\alpha$. But $S$ is also a net in closed subset $B$, by Corollary **5.1.15**, $x_\alpha\leq B$. So $x_\alpha\leq A\wedge B$, i.e $x_\alpha$ is a cluster point of $S$ in $A\wedge B$ with height $\alpha$, $A\wedge B$ is N-compact. □

Peng[128] also shown us the following interesting conclusion:

**10.2.11 Theorem** *Let $(L^X,\delta)$ be a N-compact L-fts, $A:\ (X,[\delta])\to L$ is upper semicontinuous, then $A$ is a N-compact subset in $(L^X,\delta)$.*

**Proof** Suppose $\alpha\in M(L)$, $S=\{S(n),n\in D\}$ is an eventual $\alpha$-net in $A$, then by the N-compactness of $(L^X,\delta)$, $S$ has a cluster point $x_\alpha$ in $(L^X,\delta)$ with height $\alpha$. By Corollary **5.1.12**, $S$ has a subnet $T=\{T(m),m\in E\}$ such that $T\to x_\alpha$. Then $T$ is also an eventual $\alpha$-net in $A$. By Proposition **10.2.6** (i) and (ii), $supp(T)\to x$, $\bigvee_{m\in E}\bigwedge_{k\geq m}ht(T(k))\geq\alpha$. Since the mapping $A:\ (X,[\delta])\to L$ is upper semicontinuous and $T$ is a net in $A$, by Theorem **5.1.21**,

$$A(x) \geq \bigwedge_{m\in E} \bigvee_{k\geq m} A(supp(T(k))) \geq \bigvee_{m\in E} \bigwedge_{k\geq m} ht(T(k)) \geq \alpha.$$

So $S\infty x_\alpha \leq A$. By Theorem **10.2.7**, $A$ is N-compact. □

**10.2.12 Theorem** *Let $(L^X, \delta)$ be an L-fts, $A \in L^X$. If the support set of $A$ is finite, then $A$ is N-compact.*

**Proof** Let $supp(A) = \{x^0, \cdots, x^n\}$, $\alpha \in M(L)$, $\Phi$ is an open $\alpha$-Q-cover of $A$. $\forall i \leq n$, take $U_i \in \Phi$ such that $x^i{}_\alpha \lessdot U_i$. Then $\alpha \not\leq U_i(x^i)'$. Since $\alpha = \bigvee \beta^*(\alpha)$, $\exists \gamma_i \in \beta^*(\alpha)$ such that $\gamma_i \not\leq U_i(x^i)'$. By Theorem **1.3.8**, $\beta^*(\alpha)$ is directed, so there exists $\gamma \in \beta^*(\alpha)$ such that $\gamma \geq \gamma_i$ for every $i \leq n$. Then $\gamma \not\leq U_i(x^i)'$ for every $i \leq n$. Thus the finite subfamily $\Psi = \{U_0, \cdots, U_n\}$ of $\Phi$ is an $\alpha^-$-Q-cover. That means $A$ is N-compact. □

**10.2.13 Theorem** *Let $(L^X, \delta)$ be a weakly induced L-fts. Then $(L^X, \delta)$ is N-compact if and only if its background space $(X, [\delta])$ is compact.*

**Proof** (Necessity) Suppose $(L^X, \delta)$ is N-compact, $\mathcal{U}$ is an open cover of $(X, [\delta])$. Then for every $\alpha \in M(L)$, $\Phi = \{\chi_U : U \in \mathcal{U}\}$ is an open $\alpha$-Q-cover of $\underline{1}$. By the N-compactness of $(L^X, \delta)$, $\Phi$ has a finite subfamily $\Psi = \{\chi_{U_0}, \cdots, \chi_{U_n}\}$ is an $\alpha^-$-Q-cover of $\underline{1}$. It follows that $\mathcal{U}_0 = \{U_0, \cdots, U_n\}$ is a subcover of $\mathcal{U}$, $(X, [\delta])$ is compact.

(Sufficiency) Suppose $(X, [\delta])$ is compact, $\alpha \in M(L)$, $\Phi$ is an open $\alpha$-Q-cover of $\underline{1}$. Then $\forall x \in X$, $\exists U_x \in \Phi$ such that $x_\alpha \lessdot U_x$, i.e. $\alpha \not\leq U_x(x)'$. Since $\bigvee \beta^*(\alpha) = \alpha$, $\exists \gamma(x) \in \beta^*(\alpha)$ such that $\gamma(x) \not\leq U_x(x)'$, $U_x(x) \not\leq \gamma(x)'$, $x \in (U_x)_{(\gamma(x)')}$. Since $(L^X, \delta)$ is weakly induced, by Theorem **4.2.17**, $(U_x)_{(\gamma(x)')} \in [\delta]$, $\mathcal{U} = \{(U_x)_{(\gamma(x)')} : x \in X\}$ is an open cover of $(X, [\delta])$. By the compactness of $(X, [\delta])$, $\mathcal{U}$ has a finite subcover $\mathcal{U}_0 = \{(U_{x^0})_{(\gamma(x^0)')}, \cdots, (U_{x^n})_{(\gamma(x^n)')}\}$. By $\alpha \in M(L)$ and Theorem **1.3.8**, $\beta^*(\alpha)$ is a directed set, $\exists \gamma \in \beta^*(\alpha)$ such that $\gamma \geq \gamma(x^i)$ for $i \leq n$. Then $\forall x \in X$, $\exists (U_{x^i})_{(\gamma(x^i)')} \in \mathcal{U}$, such that $x \in (U_{x^i})_{(\gamma(x^i)')}$, i.e. $\gamma(x^i) \not\leq U_{x^i}(x)'$. Since $\gamma \geq \gamma(x^i)$, we have $\gamma \not\leq U_{x^i}(x)'$, $x_\gamma \lessdot U_{x^i}$. Hence we have proved that the subfamily $\Psi = \{U_{x^i} : i \leq n\}$ of $\Phi$ is an $\alpha^-$-Q-cover of $\underline{1}$, $(L^X, \delta)$ is N-compact. □

Note that for every ordinary topological space $(X, \mathcal{T})$, $(L^X, \omega(\mathcal{T}))$ is an induced $L$-fts, we have immediately the following

**10.2.14 Corollary** *N-compactness is a good L-extension of ordinary compactness.* □

N-compactness is preserved by continuous $L$-fuzzy mappings:

**10.2.15 Theorem** *Let $(L^X, \delta)$, $(L^Y, \mu)$ be L-fts', $f^\rightarrow : (L^X, \delta) \rightarrow (L^Y, \mu)$ a continuous L-fuzzy mapping, $A$ be a N-compact subset in $(L^X, \delta)$. Then $f^\rightarrow(A)$ is a N-compact subset in $(L^Y, \mu)$.*

**Proof** Suppose $\alpha \in M(L)$, $\Phi$ is an open $\alpha$-Q-cover of $f^\rightarrow(A)$, then for every $x_\alpha \in M(\downarrow A)$, $f^\rightarrow(x_\alpha) = f(x)_\alpha \in M(\downarrow f^\rightarrow(A))$, $\exists U \in \Phi$ such that $f(x)_\alpha \lessdot U$. So

$$\alpha \not\leq U'(f(x)) = f^\leftarrow(U')(x) = f^\leftarrow(U)(x)'.$$

That is to say, $x_\alpha \lessdot f^\leftarrow(U)$. Since $f^\rightarrow$ is continuous, $f^\leftarrow(U) \in \delta$, and hence $f^\leftarrow(U) \in \mathcal{Q}(x_\alpha)$, we obtain an open $\alpha$-Q-cover $f^\leftarrow[\Phi]$ of $A$. Since $A$ is N-compact, $\Phi$ has a finite subfamily $\Psi = \{U_i : i \leq n\}$ such that $f^\leftarrow[\Psi]$ is an $\alpha^-$-Q-cover of $A$. Now we are going to prove that $\Psi$ is just an $\alpha^-$-Q-cover of $f^\rightarrow(A)$. Since $\bigvee f^\leftarrow[\Psi]\hat{q}A(\alpha)$, $\exists \gamma \in \beta^*(\alpha)$ such that $\bigvee f^\leftarrow[\Psi]\hat{q}A(\gamma)$. By Theorem **1.3.14** (i), $\exists \lambda \in M(L)$ such that

$\gamma \preceq \lambda \preceq \alpha$, so $\lambda \in \beta^*(\alpha)$. For every $y_\lambda \in M(\downarrow f^\rightarrow(A))$, by the definition of $f^\rightarrow$, we have

$$\lambda \leq f^\rightarrow(A)(y) = \bigvee\{A(x) : x \in X, f(x) = y\}.$$

Hence by $\gamma \preceq \lambda$, $\exists x \in X$ such that $f(x) = y$, $A(x) \geq \gamma$, $x_\gamma \in M(\downarrow A)$. Since $\bigvee f^\leftarrow[\Psi]\hat{q}A(\gamma)$, $\exists i \leq n$ such that $x_\gamma \lessdot f^\leftarrow(U_i)$. That is to say,

$$\gamma \not\leq (f^\leftarrow(U_i))'(x) = f^\leftarrow(U_i{}')(x) = U_i{}'(f(x)) = U_i{}'(y).$$

By $\gamma \preceq \lambda$ and hence $\gamma \leq \lambda$, we have $\lambda \not\leq U_i{}'(y)$, $y_\lambda \lessdot U_i$. So $\Psi$ is an open $\lambda$-Q-cover of $f^\rightarrow(A)$, and hence an $\alpha^-$-Q-cover of $f^\rightarrow(A)$, $f^\rightarrow(A)$ is N-compact. □

**10.2.16 Theorem** *Let $(L^X, \delta)$ be a stratified level-$T_2$ L-fts, A is a N-compact subset in $(L^X, \delta)$. Then A is a closed subset.*

**Proof** Let $x_\alpha \leq A^-$, we need to prove $x_\alpha \leq A$. By Theorem **5.1.14**, there exists a molecule net $S = \{x^n{}_{\lambda(n)}, n \in D\}$ in $A$ such that $S \rightarrow x_\alpha$. Now with $S$ we are going to construct an eventual $\alpha$-net in $A$ as follows. Let

$$\mu = \bigwedge_{m \in D} \bigvee_{n \geq m} \lambda(n),$$

then $\mu \geq \alpha$. Since if it is false, then there exists $m \in D$ such that $\bigvee\{\lambda(n) : n \geq m\} \not\geq \alpha$. Denote $d = \bigvee\{\lambda(n) : n \geq m\}$, then by the stratified property of $(L^X, \delta)$, the $d'$-layer $\underline{d'}$ of $X$ is an open subset in $(L^X, \delta)$ and $x_\alpha \not\leq \underline{d} = (\underline{d'})'$, $\underline{d'} \in \mathcal{Q}(x_\alpha)$. But for every $k \geq m$, we have

$$\lambda(k) \leq \bigvee_{n \geq m} \lambda(n) = d = (d')' = (\underline{d'})'(x^k),$$

so eventually we have $x^k{}_{\lambda(k)} \not\lessdot \underline{d'}$, this contradicts the fact $S \rightarrow x_\alpha$ and $\underline{d'} \in \mathcal{Q}(x_\alpha)$. Thus $\mu \geq \alpha$, $\bigvee\{\lambda(n) : n \geq m\} \geq \alpha$ for every $m \in D$. For every $\gamma \in \beta^*(\alpha)$ and every $m \in D$, by $\bigvee\{\lambda(n) : n \geq m\} \geq \alpha$, there exists $n(\gamma, m) \in D$, $n(\gamma, m) \geq m$ such that $\lambda(n(\gamma, m)) \geq \gamma$. Define a relation on $\beta^*(alpha) \times D$ as follows:

$$(\gamma_1, m_1) \leq (\gamma_2, m_2) \iff \gamma_1 \leq \gamma_2,\ m_1 \leq m_2.$$

Then it follows from Theorem **1.3.8** that $\beta^*(\alpha) \times D$ is a directed set. Take $N : \beta^*(\alpha) \times D \rightarrow D$ as $N(\gamma, m) = n(\gamma, m)$, then $N$ is a cofinal selection on $S$. So let

$$T = \{x^{n(\gamma,m)}{}_{\lambda(n(\gamma,m))}, (\gamma, m) \in \beta^*(\alpha) \times D\},$$

$T = S \circ N$ is a subnet of $S$, and hence it also converges to $x_\alpha$. By the definition of $T$, it is an eventual $\alpha$-net in $A$. Since $A$ is N-compact, $T$ has a cluster point $y_\alpha$ in $A$ with height $\alpha$. By Corollary **5.1.12**, $T$ has a subnet $T_0$ converging to $y_\alpha$. But $T_0$ is also a subnet of $S$, so $T_0 \rightarrow x_\alpha$. By the level-$T_2$ property of $(L^X, \delta)$ and Theorem **8.2.17**, we have $x = y$, and hence $x_\alpha = y_\alpha \leq A$. □

**10.2.17 Remark** The condition "stratified" in Theorem **10.2.16** cannot be omitted. In fact, take $X$ as a singleton, then by Theorem **10.2.12**, for every F-lattice $L$ and every $L$-fuzzy topology $\delta$ on $L^X$, every $L$-fuzzy subset $A$ in $L$-fts $(L^X, \delta)$ is N-compact. Moreover, since $X$ is a singleton, $(L^X, \delta)$ is $T_2$ and hence is level-$T_2$. But for every $L \neq \{0, 1\}$, we can choose an $L$-fuzzy topology $\delta' \neq L^X$. Then there exists an $L$-fuzzy subset $A$ in $(L^X, \delta)$ which is not a closed subset.

The following conclusion is deduced directly from Theorem **10.2.16**:

**10.2.18 Theorem** *Let $(L^X, \delta)$ be a N-compact L-fts, $(L^Y, \mu)$ a stratified level-$T_2$ L-fts, $f^\rightarrow : (L^X, \delta) \rightarrow (L^Y, \mu)$ a continuous bijective L-fuzzy mapping. Then $f^\rightarrow$ is an L-fuzzy homeomorphism.* □

**10.2.19 Remark** In Peng's paper [126] of generalizing N-compactness to the case of general F-lattices, he defined $\alpha$-net as follows:

A molecule net $S = \{S(n), n \in D\}$ in $L$-fts $(L^X, \delta)$ is called an $\alpha$-net for $\alpha \in M(L)$, if

$$\bigwedge_{m \in D} \bigvee_{n \geq m} ht(S(n)) = \alpha.$$

Then Peng used this notion to characterize N-compactness in $L$-fts.

N-compactness can strengthen some kinds of separation in $L$-fts' as the following theorem shows.

**10.2.20 Lemma** *Let $(L^X, \delta)$ be a N-compact $T_2$ L-fts, $A \in \delta'$, $\alpha \in M(L)$, $A_{(0)} = A_{[\alpha]}$, $x_\lambda \in M(L^X)$, $x \notin supp(A)$. Then there exist $U \in \mathcal{Q}(x_\lambda)$ and $V \in \delta$ such that $V \in \mathcal{Q}(z_\alpha)$ for every $z \in supp(A)$, and $U \wedge V = \underline{0}$.*

**Proof** For every $y_\alpha \leq A$, since $x \neq y$ and $(L^X, \delta)$ is $T_2$, $\exists U_y \in \mathcal{Q}(x_\lambda)$, $\exists V_y \in \mathcal{Q}(y_\alpha)$ such that $U_y \wedge V_y = \underline{0}$. Take $\Phi = \{V_y : y_\alpha \leq A\}$, then $\Phi$ is an open $\alpha$-Q-cover of $A$. Since $(L^X, \delta)$ is N-compact, by Theorem **10.2.10**, $A$ is N-compact. So $\Phi$ has a finite subfamily $\Psi = \{V_{y_0}, \cdots, V_{y_n}\}$ which is an $\alpha^-$-Q-cover of $A$, i.e. there exists $\gamma \in \beta^*(\alpha)$ such that $\bigvee \Psi \hat{q} A(\gamma)$. For every $z \in supp(A)$, by $A_{(0)} = A_{[\alpha]}$ and $\gamma \in \beta^*(\alpha)$, we have $\gamma \leq \alpha$, $z_\gamma \leq A$. Since $\bigvee \Psi \hat{q} A(\gamma)$, there exists $i \leq n$ such that $z_\gamma \lessdot V_{y_i}$. Let $V = \bigvee_{i \leq n} V_{y_i}$, then for every $z \in supp(A)$ we have $z_\gamma \lessdot V$. By $\gamma \leq \alpha$, $V \in \mathcal{Q}(z_\alpha)$.

Since $A(z) > 0$ implies $A(z) \geq \alpha$, for every $z \in supp(A)$, we have $A(z) \lessdot V$, $V \in \mathcal{Q}(A)$. Let $U = \bigwedge_{i \leq n} U_{y_i}$, then $U \in \mathcal{Q}(x_\lambda)$, and

$$U \wedge V = (\bigwedge_{i \leq n} U_{y_i}) \wedge (\bigvee_{i \leq n} V_{y_i}) \leq \bigvee_{i \leq n} (U_{y_i} \wedge V_{y_i}) = \underline{0}.$$

So $\gamma$, $U$ and $V$ are exactly what we need. □

**10.2.21 Theorem** *Let $(L^X, \delta)$ be a N-compact and $T_2$ L-fts. Then $(L^X, \delta)$ is p-normal.*

**Proof** Suppose $A$ and $B$ are non-zero pseudo-crisp closed subsets in $(L^X, \delta)$ such that $supp(A) \cap supp(B) = \emptyset$. Then $\exists \lambda, \mu \in M(L)$ such that $A_{(0)} = A_{[\lambda]}$, $B_{(0)} = B_{[\mu]}$. By Lemma **10.2.20**, $\forall y_\mu \leq B$, $\exists \gamma \in \beta^*(\lambda)$, $\exists U_y \in \delta$, $\exists V_y \in \mathcal{Q}(y_\mu)$ such that $U_y \in \mathcal{Q}(z_\gamma)$ for every $z \in supp(A)$, and $U_y \wedge V_y = \underline{0}$. Then $\Phi = \{V_y : y_\mu \leq B\}$ is an open $\mu$-Q-cover of N-compact subset $B$, $\Phi$ has a finite subfamily $\Psi = \{V_{y_i} : i \leq n\}$ which is a $\mu^-$-Q-cover of $B$. Take $U = \bigwedge\{U_{y_i} : i \leq n\}$, $V = \bigvee\{V_{y_i} : i \leq n\}$, then $U \in \mathcal{Q}(A)$, $V \in \mathcal{Q}(B)$ and $U \wedge V = \underline{0}$. So $(L^X, \delta)$ is p-normal. □

Moreover, N-compactness has the following nice properties:

**10.2.22 Proposition** *Every finite join of N-compact subsets is N-compact.* □

**10.2.23 Proposition** *In every stratified level-$T_2$ L-fts, every meet of N-compact subsets is N-compact.* □

**10.2.24 Proposition** *Let $A$ be a N-compact subset in L-fts $(L^X, \delta)$. Then for every $a \in L$, $A \wedge \underline{a}$ is N-compact.* □

## 10.3 Tychonoff Product Theorem

In this section, we investigate the multiplicative property of N-compactness.

**10.3.1 Lemma** (Alexander Subbase Lemma) *Let* $(L^X, \delta)$ *be an L-fts,* $\mathcal{B}$ *is a subbase of* $\delta$, $A \in L^X$. *If for every* $\Phi \subset \mathcal{B}$ *and every* $\alpha \in M(L)$, $\bigvee \Phi \hat{q} A(\alpha)$ *always imply that* $\Phi$ *has a finite subfamily* $\Psi$ *such that* $\bigvee \Psi \hat{q} A(\alpha)$, *then* $A$ *is N-compact.*

**Proof** Let $\Phi$ be an open $\alpha$-Q-cover of $A$. If $\bigvee \Phi \neg \hat{q} A(\alpha)$, take

$$H = \{\Omega : \ \Phi \subset \Omega \subset \delta, \ \forall \Psi \in [\Omega]^{<\omega}, \ \bigvee \Psi \neg \hat{q} A(\alpha)\},$$

then $\Phi \in H$, $H$ is not empty. Equip $H$ with the inclusion order, then one can easily find that every chain in the poset $H$ has an upper bound (e.g., the union of this chain), so it follows from the Zorn Lemma, $H$ has a maximal element $\Omega_0$, which satisfies the following conditions:

(i) $\bigvee \Omega_0 \hat{q} A(\alpha)$.

It holds because $\bigvee \Phi \hat{q} A(\alpha)$ and $\Phi \subset \Omega_0$.

(ii) $V \in \delta, \ V \subset U \in \Omega_0 \implies V \in \Omega_0$.

In fact, if it is false, we can take $\Omega^* = \Omega_0 \cup \{V\}$ and verify that $\Omega^* \in H$. Then $\Omega^*$ is strictly larger than $\Omega_0$, contradicts the maximal property of $\Omega_0$.

(iii) $U, V \in \delta, \ U \wedge V \in \Omega_0 \implies U \in \Omega_0$ or $V \in \Omega_0$.

If $U \notin \Omega_0$, $V \notin \Omega_0$, then since $\Omega_0$ is maximal in $H$, $\Omega_0 \cup \{U\}, \Omega_0 \cup \{V\} \notin H$. Therefore, by the definition of $H$, there exist $\Psi_0, \Psi_1 \in [\Omega_0]^{<\omega}$ such that

$$\bigvee(\Psi_0 \cap \{U\}) \hat{q} A(\alpha), \quad \bigvee(\Psi_1 \cap \{V\}) \hat{q} A(\alpha),$$

i.e. there exist $\gamma_0, \gamma_1 \in \beta^*(\alpha)$ such that

$$\bigvee(\Psi_0 \cap \{U\}) \hat{q} A(\gamma_0), \quad \bigvee(\Psi_1 \cap \{V\}) \hat{q} A(\gamma_1). \tag{10.5}$$

Since $\alpha \in M(L)$, $\beta^*(\alpha)$ is a directed set, so there exist $\gamma \in \beta^*(\alpha)$ such that $\gamma \geq \gamma_0, \gamma_1$. Now we want to prove

$$\bigvee(\Psi_0 \cup \Psi_1 \cup \{U \wedge V\}) \hat{q} A(\gamma). \tag{10.6}$$

For every $x_\gamma \leq A$, if there does not exist a Q-neighborhood of $x_\gamma$ in $\Psi_0 \cup \Psi_1$, then $\Psi_0 \cup \Psi_1$ does not contain any Q-neighborhood of $x_{\gamma_0}$ or $x_{\gamma_1}$. By (10.5), $U \in \mathcal{Q}(x_{\gamma_0})$, $V \in \mathcal{Q}(x_{\gamma_1})$, and hence $U, V \in \mathcal{Q}(x_\gamma)$, $U \wedge V \in \mathcal{Q}(x_\gamma)$. So (10.6) is true. Then since $\Psi_0 \cup \Psi_1 \in [\Phi]^{<\omega}$, by the definition of $\Omega_0$, we have $U \wedge V \notin \Omega_0$. Thus (iii) is true.

By (ii) and (iii), we have

$$\{U_i : \ i \leq n\} \subset \delta, \ \bigwedge_{i \leq n} U_i \leq W \in \Omega_0 \implies \exists i \leq n, \ U_i \in \Omega_0. \tag{10.7}$$

Now return to our subbase $\mathcal{B}$. If $\bigvee(\mathcal{B} \cap \Omega_0) \hat{q} A(\alpha)$, then by the assumption of this lemma $\mathcal{B} \cap \Omega_0$ has a finite subfamily $\Delta$ such that $\bigvee \Delta \hat{q} A(\alpha)$. But $\Delta$ is also a finite subfamily of $\Omega_0$, this contradicts the definition of $\Omega_0$. So $\bigvee(\mathcal{B} \cap \Omega_0) \neg \hat{q} A(\alpha)$, there exists $x_\alpha \leq A$ such that $(\mathcal{B} \cap \Omega_0) \cap \mathcal{Q}(x_\alpha) = \emptyset$. So $x_\alpha \leq \bigwedge(\mathcal{B} \cap \Omega_0)') = \bigwedge(\mathcal{B}' \cap \Omega_0')$. If $x_\alpha \not\leq \bigwedge \Omega_0'$, then $\exists V \in \Omega_0$ such that $x_\alpha \not\leq V'$. Since $\mathcal{B}$ is a subbase of $\delta$, there exists $\{U_{ij} : \ j \in J_i, \ i \in I\} \subset \mathcal{B}$ such that $V = \bigvee_{i \in I} \bigwedge_{j \in J_i} U_{ij}$, where $J_i$ is finite for every $i \in I$. Since $x_\alpha \not\leq V'$, there exists $i \in I$ such that $x_\alpha \not\leq \bigvee_{j \in J_i} U_{ij}'$. Then we have $\bigwedge_{j \in J_i} U_{ij} \leq V \in \Omega_0$. By (10.7), $\exists j \in J_i$ such that $U_{ij} \in \Omega_0$. Thus $U_{ij} \in \mathcal{B} \cap \Omega_0$, $x_\alpha \not\leq U_{ij}'$, $U_{ij} \in \mathcal{Q}(x_\alpha)$. This contradicts $(\mathcal{B} \cap \Omega_0) \cap \mathcal{Q}(x_\alpha) = \emptyset$. So the assumption $x_\alpha \not\leq \bigwedge \Omega_0'$ is not true, we have $x_\alpha \leq \bigwedge \Omega_0'$. But this means $\Omega_0 \cap \mathcal{Q}(x_\alpha) = \emptyset$, it contradicts (i). Therefore, the previous assumption "$\bigvee \Phi \neg \hat{q} A(\alpha)$" is false, $A$ is N-compact. □

**10.3.2 Theorem** *N-compactness of L-fuzzy subsets is multiplicative.*

**Proof** Let $\{(L^{X_t},\delta_t):\ t\in T\}$ be a family of N-compact $L$-fts', $(L^X,\delta)=\prod_{t\in T}(L^{X_t},\delta_t)$ be their product $L$-fts, $A_t$ be a N-compact subset in $(L^{X_t},\delta_t)$ for every $t\in T$, $A=\prod_{t\in T}A_t$.

Let $\alpha\in M(L)$, $\Phi$ be an open $\alpha$-Q-cover of $A$. By the Alexander Subbase Lemma, we can suppose

$$\Phi\subset\{p_t^{\leftarrow}(U_t):\ t\in T,\ U_t\in\delta_t\}. \tag{10.8}$$

Divide the investigation into two cases as follows:

(i) If there exists $t_0\in T$ such that $x^{t_0}{}_\alpha\not\leq A_{t_0}$ for every $x^{t_0}\in X_{t_0}$. If for every $\gamma\in\beta^*(\alpha)$, there exists $S(\gamma)\leq A_{t_0}$, then since $\alpha\in M(L)$, $\beta^*(\alpha)$ is a directed set, $S=\{S(\gamma),\gamma\in\beta^*(\alpha)\}$ is an eventual $\alpha$-net in $A_{t_0}$. By the N-compactness of $A_{t_0}$ and Theorem **10.2.7**, $S$ has a cluster point in $A_{t_0}$ with height $\alpha$. But it is impossible. So there exists $\gamma\in\beta^*(\alpha)$ such that $x^{t_0}{}_\gamma\not\leq A_{t_0}$ for every $x^{t_0}\in X_{t_0}$. Then $\forall x\in X$,

$$A(x)=(\prod_{t\in T}A_t)(x)=(\bigwedge_{t\in T}p_t^{\leftarrow}(A_t))(x)=\bigwedge_{t\in T}A_t(p_t(x))\leq A_{t_0}(p_{t_0}(x))=A_{t_0}(x^{t_0}).$$

It follows that $\gamma\not\leq A(x)$ for every $x\in X$, i.e. $A$ does not contain any molecule with height $\gamma$. Thus every finite subfamily $\Psi$ of $\Phi$ is a $\gamma$-Q-cover of $A$, $\bigvee\Psi\hat{q}A(\alpha)$.

(ii) If for every $t\in T$, there exists $x^t{}_\alpha\leq A_t$. By (10.8), for every $t\in T$, there exists $\mathcal{B}_t\subset\delta_t$ (may be empty subfamily) such that

$$\Phi=\{p_t^{\leftarrow}(U_t):\ t\in T,\ U_t\in\mathcal{B}_t\}. \tag{10.9}$$

If $\forall t\in T$, $\bigvee\mathcal{B}_t\neg\hat{q}A_t(\alpha)$, then $\forall t\in T$, $\exists y^t\in X_t$ such that $y^t{}_\alpha\leq A_t\wedge(\bigwedge\mathcal{B}_t{}')$. Take $z\in X$ such that $\forall t\in T$,

$$p_t(z)=\begin{cases} y^t, & \mathcal{B}_t\neq\emptyset,\\ x^t, & \mathcal{B}_t=\emptyset.\end{cases}$$

Then $\forall t\in T$, $\forall U_t\in\mathcal{B}_t$,

$$\alpha\leq(A_t\wedge(\bigwedge\mathcal{B}_t{}')(y^t)\leq U_t{}'(y^t)=U_t{}'(p_t(z))=p_t^{\leftarrow}(U_t{}')(z)=p_t^{\leftarrow}(U_t)'(z),$$

so $z_\alpha\leq p_t^{\leftarrow}(U_t)'$. By (10.9), we have $\Phi\cap\mathcal{Q}(z_\alpha)=\emptyset$. But $\forall t\in T$, $x^t{}_\alpha, y^t{}_\alpha\leq A_t$, so

$$A(z)=\bigwedge_{t\in T}A_t(p_t(z))=(\bigwedge_{t\in T,\ \mathcal{B}_t\neq\emptyset}A_t(y^t))\wedge(\bigwedge_{t\in T,\ \mathcal{B}_t=\emptyset}A_t(x^t))\geq\alpha.$$

It means $z_\alpha\leq A$. By $\bigvee\Phi\hat{q}A(\alpha)$, $\Phi\cap\mathcal{Q}(z_\alpha)\neq\emptyset$, it contradicts the equality $\Phi\cap\mathcal{Q}(z_\alpha)=\emptyset$ proved above. Therefore, $\exists s\in T$ such that $\bigvee\mathcal{B}_s\hat{q}A_s(\alpha)$.

Since $A_s$ is N-compact, $\mathcal{B}_s$ has a finite subfamily $\mathcal{D}_s$ such that $\bigvee\mathcal{D}_s\hat{q}A_s(\gamma)$ for some $\gamma\in\beta^*(\alpha)$. Take $\Psi=\{p_s^{\leftarrow}(U_s):\ U_s\in\mathcal{D}_s\}$, then $\Psi\in[\Phi]^{<\omega}$. $\forall u_\gamma\leq A$, then

$$A_s(p_s(u))=p_s^{\leftarrow}(A_s)(u)\geq A(u)\geq\gamma,$$

$p_s(u)_\gamma\leq A_s$. By $\bigvee\mathcal{D}_s\hat{q}A_s(\gamma)$, $\exists U_s\in\mathcal{D}_s$ such that $U_s\in\mathcal{Q}(p_s(u)_\gamma)$. Then

$$p_s^{\leftarrow}(U_s)'(u)=U_s{}'(p_s(u))\not\geq\gamma.$$

It follows that $p_s^{\leftarrow}(U_s)\in\mathcal{Q}(u_\gamma)$, $\Psi$ is a $\gamma$-Q-cover of $A$. □

**10.3.3 Theorem** (Tychonoff Product Theorem) *N-compactness of L-fts's is strongly multiplicative.*

**Proof** Let $\{(L^{X_t},\delta_t):\ t\in T\}$ be a family of $L$-fts', $(L^X,\delta)=\prod_{t\in T}(L^{X_t},\delta_t)$ be their product $L$-fts.

If $(L^X, \delta)$ is N-compact, $t \in T$, then since $p_t^{\to} : (L^X, \delta) \to (L^{X_t}, \delta_t)$ is surjective and continuous, then by Theorem **10.2.15**, $(L^{X_t}, \delta_t)$ is N-compact.

The sufficiency has been proved in Theorem **10.3.2**. □

Comparing Theorem **10.3.2** and Theorem **10.3.3**, we can find that the strongly multiplicative property of N-compactness is only guaranteed for $L$-fuzzy topological spaces but not for $L$-fuzzy subsets. In fact, if some $A_s = \underline{0}$ in $\{A_t : t \in T\}$, then $A = \prod_{t\in T} A_t = \underline{0}$ will always be N-compact but no matter whether $A_t$'s other than $A_s$ are N-compact or not. Moreover, it is shown by the following example[168] that the N-compactness of $A = \prod_{t\in T} A_t$ cannot imply the N-compactness of $A_t$'s even $A \neq \underline{0}$:

**10.3.4 Example** Take $Y = \{a_i : i \in \mathbf{N}\}$ as an ordinary countable infinite set. For every $i \in \mathbf{N}$, take

$$X_i = Y, \quad \delta_i = [0,1]^{X_i},$$

$$A_i(a_j) = \begin{cases} 1, & j = 1, \\ \frac{1}{i}, & j \geq 2. \end{cases}$$

Then for every $i \in \mathbf{N}$, since $(L^{X_i}, \delta_i)$ is discrete, $A_i$ is clearly not N-compact.

Take $(L^X, \delta) = \prod_{i\in\mathbf{N}}(L^{X_i}, \delta_i)$, $A = \prod_{i\in\mathbf{N}} A_i$, then since

$$A = \bigwedge_{i\in\mathbf{N}} p_i^{\leftarrow}(A_i) = \bigwedge_{i\in\mathbf{N}} A_i \circ p_i,$$

for every $x \in X$ and every $i \in \mathbf{N}$, denote $p_i(x) = x^i \in X_i$, we have

$$A(x) = \begin{cases} 0, & \bigvee\{i \in \mathbf{N} : x^i \neq a_1\} = \omega, \\ \frac{1}{i_0}, & \bigvee\{i \in \mathbf{N} : x^i \neq a_1\} = i_0 < \omega. \end{cases} \tag{10.10}$$

Then $A \neq \emptyset$. Now suppose $\Phi$ is an open $\alpha$-Q-cover of $A$ for some $\alpha \in (0,1]$. Take $m \in \mathbf{N}$ such that $\frac{1}{m} < \alpha$, then for an arbitrary $x \in X$, by (10.10), in order to make $A(x) \geq \frac{1}{m}$, all the coordinates $x^i$'s of $x$ following the $m$'s coordinate $x^m$ must be $a_1$. So

$$|\{x \in X : A(x) \geq \tfrac{1}{m}\}| \leq 2^m < \omega.$$

Denote $Y = \{x \in X : A(x) \geq \frac{1}{m}\}$. If $x \in Y$ such that $x_\alpha \leq A$, then since $\Phi$ is an $\alpha$-Q-cover of $A$, there exists $U_x \in \Phi$ such that $x_\alpha \lessdot U_x$, i.e. $\alpha > 1 - U_x(x)$. Since $Y$ is a finite set, we can let $\gamma = max\{1 - U_x(x) : x \in Y\}$ and hence $\gamma < \alpha$. Since we also have $\frac{1}{m} < \alpha$, we can take $\gamma_0$ such that

$$max\{\gamma, \tfrac{1}{m}\} < \gamma_0 < \alpha.$$

Then $\gamma_0 \in \beta^*(\alpha)$, $Y_0 = \{x \in X : A(x) \geq \gamma_0\} \subset Y$, $Y_0$ is also a finite set. For every $x \in X$ such that $x_{\gamma_0} \leq A$, we have $x \in Y_0$. By the virtue of $\gamma = max\{1 - U_x(x) : x \in Y\}$, $\gamma_0 > \gamma \geq 1 - U_x(x)$, $x_{\gamma_0} \lessdot U_x \in \Phi$. So $\Psi = \{U_x : x \in Y_0\}$ is both a finite subfamily of $\Phi$ and a $\gamma_0$-Q-cover of $A$, i.e. an $\alpha^-$-Q-cover of $A$, $A$ is N-compact.

The following theorem shows us two kinds of important N-compact $L$-fts':

**10.3.5 Theorem** *For every F-lattice L and every cardinal number $\kappa$, $I(L)^\kappa$ and every its closed subset is N-compact.*

**Proof** As a preparation, we prove the following conclusion in advance:

$$\alpha \in M(L),\ \alpha \leq \bigvee\{a \wedge a' :\ a \in L\} \implies \alpha \leq \alpha'. \tag{10.11}$$

For every $\gamma \in \beta^*(\alpha)$, since $\alpha \leq \bigvee\{a \wedge a' :\ a \in L\}$, there exists $a_\gamma \in L$ such that $\gamma \leq a_\gamma \wedge a_\gamma{}'$. So $\gamma \leq a_\gamma$, $\gamma \leq a_\gamma{}'$, and hence $a_\gamma \leq \gamma'$, $\gamma \leq \gamma'$. Suppose $\gamma, \zeta \in \beta^*(\alpha)$, as a result of Theorem **1.3.8**, there exists $\tau \in \beta^*(\alpha)$ such that $\tau \geq \gamma, \zeta$. It implies $\gamma \leq \tau \leq \tau' \leq \zeta'$. Applying Proposition **1.2.16** on $\gamma \leq \zeta'$, we have

$$\alpha = \bigvee\{\gamma :\ \gamma \in \beta^*(\alpha)\} \leq \bigwedge\{\zeta' :\ \zeta \in \beta^*(\alpha)\} = \alpha'.$$

Now we prove the N-compactness of $I(L)^\kappa$ and its closed subsets. By the Tychonoff Product Theorem for N-compactness (Theorem **10.3.3**) and Theorem **10.2.10**, it is sufficient to prove the N-compactness of $I(L)$. By the Alexander Subbase Lemma (Lemma **10.3.1**) and Proposition **9.1.13**, we need only prove that for every open $\alpha$-Q-cover $\mathcal{U} \subset \{L_b, R_a :\ a, b \in [0,1]\}$ of $I(L)$, there exists a finite subfamily $\mathcal{U}_0 \subset \mathcal{U}$ such that $\mathcal{U}_0$ is an $\alpha^-$-Q-cover of $I(L)$. We divide the investigation into two cases:

(1) If $\alpha \leq \alpha'$, take $\gamma \in I[L]$ such that

$$\gamma(t) = \begin{cases} 1, & t < 0, \\ \alpha, & 0 \leq t \leq 1, \\ 0, & t > 1, \end{cases}$$

then there exists $U \in \mathcal{U}$ such that $\alpha \not\leq U'(\gamma)$. If $U = L_a$, then $a > 1$ by $\alpha \not\leq U'(\gamma) = \gamma(a-)$; if $U = R_a$, then $a < 0$ by $\alpha \not\leq U'(\gamma) = \gamma(a+)'$. So we always have $U = \underline{1}$, we can choose the needed $\mathcal{U}_0$ as $\{U\}$.

(2) If $\alpha \not\leq \alpha'$, then by (10.11), $\alpha \not\leq \bigvee\{a \wedge a' :\ a \in L\}$. For every $r \in [0,1]$, take $\lambda_r \in I[L]$ such that

$$\lambda_r(t) = \begin{cases} 1, & t < r, \\ 0, & t \geq r, \end{cases}$$

then there exists $U \in \mathcal{U}$ such that $\alpha \not\leq U'(\lambda_r)$. If $U = L_b$, then $r \in (-\infty, b)$; if $U = R_a$, then $r \in (a, +\infty)$. So $\{(-\infty, b), (a, +\infty) :\ a, b \in [0,1],\ L_b, R_a \in \mathcal{U}\}$ is an open cover of $[0,1]$, and hence it has a finite subcover $\mathcal{V}$. Take

$$A = \{a :\ (a, +\infty) \in \mathcal{V}\}, \qquad B = \{b :\ (-\infty, b) \in \mathcal{V}\},$$

then clearly we have $A \neq \emptyset$, $B \neq \emptyset$ and both of them are finite. So we can suppose $min\, A = c \geq 0$, $max\, B = d \leq 1$ and we have $c < d$.

Now we affirm that $\mathcal{U}_0 = \{L_d, R_c\}$ is what we need. If it is not try, for every $\sigma \in \beta^*(\alpha)$, there exists $\lambda^\sigma \in I[L]$ such that

$$\sigma \leq (L_d{}' \cap R_c{}')(\lambda^\sigma) = \lambda^\sigma(d-) \wedge \lambda^\sigma(c+)' \leq \lambda^\sigma(c+) \wedge \lambda^\sigma(c+)' \leq \bigvee\{u \wedge u' :\ u \in L\}.$$

So by $\bigvee \beta^*(\alpha) = \alpha$ we have

$$\alpha = \bigvee\{\sigma :\ \sigma \in \beta^*(\alpha)\} \leq \bigvee\{u \wedge u' :\ u \in L\},$$

this is a contradiction. □

**10.3.6 Theorem** *For every F-lattice L and every cardinal number $\kappa$, $\tilde{I}(L)^\kappa$ and every its closed subset is N-compact.*

**Proof** By Theorem **9.1.30**, $\tilde{I}(L)$ is weakly induced. By Theorem **10.2.13**, $\tilde{I}(L)$ is N-compact. By Theorem **10.3.3** and Theorem **10.2.10**, $\tilde{I}(L)^\kappa$ and its closed subsets are N-compact. □

## 10.4 Comparison of Different Compactness in Fuzzy Topological Spaces

In the first section of this chapter, we introduced several kinds of compactness in fuzzy topological spaces. We have also been shown various properties of N-compactness in the last section. In this section, some comparisons will be made between N-compactness and some of these kinds of compactness in $L$-fuzzy topological spaces; the reader can consult Ref. [168] and Ref. [128] for further details.

Just as it was shown in the first section of this chapter, compactness defined there are all for the case $L = [0,1]$. Now we generlize them into the case of F-lattices and then compare them with N-compactness.

**10.4.1 Definition** Let $(L^X, \delta)$ be a $L$-fts, $A \in L^X$, $\gamma \in pr(L)$.

A family $\mathcal{A} \subset L^X$ is called a *$\gamma$-cover* of $A$, if for every $x_\gamma \leq A$, there exists $C \in \mathcal{A}$ such that $C(x) \not\leq \gamma$. A family $\mathcal{A}_0 \subset L^X$ is called a *sub-$\gamma$-cover* of a $\gamma$-cover $\mathcal{A}$ of $A$, if $\mathcal{A}_0 \subset \mathcal{A}$ and $\mathcal{A}_0$ is a $\gamma$-cover of $A$. A $\gamma$-cover of $A$ is called an *open $\gamma$-cover* of $A$, if it consists of open subsets.

$(L^X, \delta)$ is called *$\gamma$-compact*, if every open $\gamma$-cover of $\underline{1}$ has a sub-$\gamma$-cover. $(L^X, \delta)$ is called *strongly compact*, if it is $\gamma$-compact for every $\gamma \in pr(L)$.

Recall the definition of $\alpha$-Q-cover and Proposition **1.2.15**, the following conclusion is obvious:

**10.4.2 Proposition** *Let $(L^X, \delta)$ be an L-fts, Phi $\subset L^X$. Then*

(i) *For every $\alpha \in M(L)$, $\Phi$ is an $\alpha$-Q-cover of $\underline{1}$ if and only if $\Phi$ is an $\alpha'$−cover of $\underline{1}$.*

(ii) *For every $\gamma \in pr(L)$, $\Phi$ is a $\gamma$-cover of $\underline{1}$ if and only if $\Phi$ is a $\gamma'$-Q-cover of $\underline{1}$.* □

Recall the related definitions and results in Section **4.4**, we have

**10.4.3 Theorem** *Let $(L^X, \delta)$ be an L-fts, $\gamma \in pr(L)$. Then $(L^X, \delta)$ is $\gamma$-compact if and only if $(X, \iota_\gamma(\delta))$ is compact.* □

Then the following result is clear by the virtue of Proposition **1.2.15**:

**10.4.4 Theorem** *Let $(L^X, \delta)$ be an L-fts. Then*

(i) *For every $\gamma \in pr(L)$, $(L^X, \delta)$ is $\gamma$-compact if and only if every open $\gamma'$-Q-cover of $\underline{1}$ has a finite sub-$\gamma'$-Q-cover.*

(ii) *$(L^X, \delta)$ is strongly compact if and only if for every $\alpha \in M(L)$, every open*

*$\alpha$-Q-cover $\Phi$ of $\underline{1}$ has a finite sub-$\alpha$-Q-cover.* □

Similar with N-compactness, $\gamma$-compactness and strong compactness can also be characterized with nets in $L$-fuzzy topological spaces:

**10.4.5 Theorem** *Let $(L^X, \delta)$ be an L-fts. Then*

(i) *For every $\gamma \in pr(L)$, $(L^X, \delta)$ is $\gamma$-compact if and only if every $\gamma'$-net in $(L^X, \delta)$ has cluster point with height $\gamma'$.*

(ii) *$(L^X, \delta)$ is strongly compact if and only if for every $\alpha \in M(L)$, every $\alpha$-net in $(L^X, \delta)$ has a cluster point with height $\alpha$.*

**Proof** Clearly, by Proposition **1.2.15**, we need only prove (i).

(Sufficiency) Suppose $\gamma \in pr(L)$ and $(L^X, \delta)$ is not $\gamma$-compact, denote $\alpha = \gamma' \in M(L)$, then by Theorem **10.4.4** (i), there exists an open $\alpha$-Q-cover $\Phi$ of $\underline{1}$ which has not any finite sub-$\alpha$-Q-cover. That is to say, for every $\Psi \in [\Phi]^{<\omega}$, there exists $x(\Psi) \in X$ such that $x(\Psi)_\alpha \not\prec \bigvee \Psi$. Take $D = [\Phi]^{<\omega}$ equipped with inclusion order, then $D$ is a directed set and $S = \{x(\Psi)_\alpha, \Psi \in D\}$ is a $\gamma'$-net in $(L^X, \delta)$. If $x_\alpha$ is a cluster point of $S$ in $(L^X, \delta)$ with height $\alpha$, then since $\Phi$ is an open $\alpha$-Q-cover of $\underline{1}$, there exists $U \in \Phi$ such that $U \in \mathcal{Q}(x_\alpha)$. Take $\Psi_0 = \{U\} \in D$, then by $S \infty x_\alpha$, there exists $\Psi \in D = [\Phi]^{<\omega}$ such that $\Psi \geq \Psi_0 = \{U\}$, $x(\Psi)_\alpha \prec U$. But $U \in \Psi$, so $x(\Psi)_\alpha \prec \bigvee \Psi$, this contradicts the choose of $x(\Psi)_\alpha$, $x_\alpha$ cannot be a cluster point of $S$. Hence $S$ has no cluster point in $(L^X, \delta)$ with height $\alpha = \gamma'$.

(Necessity) Let $\alpha = \gamma'$, $S = \{x(n)_\alpha, n \in D\}$ be a $\gamma'$-net in $(L^X, \delta)$ without any cluster point which possessing height $\gamma'$. Then for every $x \in X$, there exists $U(x) \in \mathcal{Q}(x_\alpha)$ such that eventually $S$ does not quasi-coincide with $U(x)$. Take $\Phi = \{U(x) : x \in X\}$, then $\Phi$ is an open $\alpha$-Q-cover of $\underline{1}$. If $\Phi_0 = \{U(x^0), \cdots, U(x^k)\} \subset \Phi$ is finite and is an $\alpha$-Q-cover of $\underline{1}$, then for every $i \in \{0, \cdots, k\}$, there exists $n_i \in D$ such that $x(n)_\alpha \not\prec U(x^i)$ for every $n \geq n_i$. Since $D$ is directed, we can take $n^* \in D$ such that $n^* \geq n_i$ for every $i \in \{0, \cdots, k\}$. So $x(n^*)_\alpha \not\prec U(x^i)$ for every $i \in \{0, \cdots, k\}$, $\Phi_0$ is not an $\gamma'$-Q-cover of $\underline{1}$. By Theorem **10.4.4** (i), $(L^X, \delta)$ is not $\gamma$-compact. □

Strong compactness was originally defined for spaces. Zhong-Fu Li defined it for subsets in the case of $L = [0, 1]$ by $\alpha$-Q-covers just as the condition in Theorem **10.4.11** in the sequel, and called it "strong Q-compactness".[56] By Theorem **10.4.5**, we can generalize it for subsets and general F-lattices:

**10.4.6 Definition** Let $(L^X, \delta)$ be an $L$-fts, $A \in L^X$. $A$ is called *strongly compact*, if for every $\alpha \in M(L)$, every $\alpha$-net in $A$ has a cluster point in $A$ with height $\alpha$.

It is not very difficult to prove the following conclusions:

**10.4.7 Theorem** *Let $(L^X, \delta)$ be an L-fts, $A \in L^X$ a strongly compact subset, $B \in L^X$ a closed subset, then $A \wedge B$ is strongly compact.* □

**10.4.8 Theorem** *Every strongly compact subset in a stratified $T_2$ L-fts is a closed subset.* □

**10.4.9 Theorem** *Stong compactness for L-fuzzy subsets is multiplicative.* □

But strong compactness for $L$-fuzzy subsets cannot be preserved under continuous $L$-fuzzy mappings, even if $L = [0,1]$:

**10.4.10 Example** Take $X = \bigcup\{(n-1,n) : n \in \mathbf{N}\}$. $\forall n \in \mathbf{N}$, take $A_n \in I^X$ as follows:

$$A_n(x) = \begin{cases} \frac{1}{n+1}, & x \in (n-1,n), \\ 0, & x \notin (n-1,n), \end{cases}$$

and take $\delta = Grt(\{A_n : n \in \mathbf{N}\})$, then $(X,\delta)$ is a F-ts. Let $C \in I^X$ be defined as $C(x) = (x)$ for every $x \in X$, where $(x)$ is the decimal part of $x$, then $C$ is strongly compact in $(X,\delta)$. In fact, suppose $S = \{x^n{}_\alpha, n \in D\}$ is an $\alpha$-net in $C$, then by the structure of $C$, $\alpha < 1$. Take $n \in \mathbf{N}$ large enough such that $\forall x \in (n-1,n)$, $A_n{}'(x) > \alpha$. Arbitrarily fix such a point $x$, then we can show that $x_\alpha$ is a cluster point of $S$: Let $U \in \mathcal{Q}(x_\alpha)$, then $\alpha \not\leq U(x)'$, $\alpha > U(x)'$. But it is easy to verify $U = \underline{1}$ at this time, so for every $U \in \mathcal{Q}(x_\alpha)$, $S$ eventually quasi-coincides with $U$, $S \to x_\alpha$, $x_\alpha$ is a cluster point of $S$. Hence $C$ is a strongly compact subset.

Take $Y = \mathbf{N}$. $\forall n \in \mathbf{N}$, take $B_n \in I^Y$ as follows:

$$B_n(y) = \begin{cases} \frac{1}{y+1}, & y = n, \\ 0, & y \neq n, \end{cases}$$

and take $\mu = Grt(\{B_n : n \in \mathbf{N}\})$, then $(Y,\mu)$ is a F-ts. Take an ordinary mapping $f : X \to Y$ as follows:

$$\forall x \in X, \quad f(x) = [x] + 1,$$

where $[x]$ is the integral part of $x$, then $f[(n-1,n)] = \{n\}$ for every $n \in \mathbf{N}$, and hence $f^\to(C) = \underline{1}$. Since $\forall n \in \mathbf{N}$, $f^\leftarrow(B_n) = A_n \in \delta$, so $f^\to$ is continuous. In order to show that $\underline{1} \in I^Y$ is not strongly compact, take 1-net $T = \{y_1, y \in \mathbf{N}\}$ in $\underline{1}$, $\forall z \in Y$, then $B_z \in \mathcal{Q}(z_1)$ and eventually we have $B_z(y)' = 1$, $y_1 \not\prec B_z$. So $z_1$ is not a cluster point of $T$ in $\underline{1}$, and hence $\underline{1} = f^\to(C)$ is not a strongly compact subset in $(Y,\mu)$.

Moreover, similar to the proof of Theorem **10.4.5**, the reader can verify the following interesting result:

**10.4.11 Theorem** *Let $(L^X,\delta)$ be an L-fts, $A \in L^X$. Then $A$ is strongly compact if and only if for every $\alpha \in M(L)$, every open $\alpha$-Q-cover of $A$ has a finite sub-$\alpha$-Q-cover.* □

In fact, it has been shown that strong compactness is weaker than N-compactness. Guo-Jun Wang also show us another interesting relation between strong compactness and N-compactness as follows:[168]

**10.4.12 Theorem** *Let $(X,\delta)$ be a strongly compact F-ts. Then $(X,\delta)$ is N-compact if and only if every closed subset in $(X,\delta)$, as a function from $X$ to $[0,1]$, takes its largest value at some point in $X$.*

**Proof** By Theorem **10.2.10** and Corollary **10.2.9**, the necessity is obvious. We need only prove the sufficiency. Suppose $(X,\delta)$ is strongly compact but not N-compact, then by Theorem **10.2.7**, there exists an eventual $\alpha$-net $S = \{S(n), n \in D\}$ such that for every $x \in X$, $x_\alpha$ is not a cluster point of $S$. So $\forall x \in X$, $\exists U_x \in \mathcal{Q}(x_\alpha)$ and $n(x) \in D$ such that $S(n) \not\leq U_x$ for every $n \geq n_x$. Let $\Phi = \{U_x : x \in X\}$, then $\Phi$ is an open $\alpha$-Q-cover of $\underline{1}$, and hence it has a finite sub-$\alpha$-Q-cover $\Phi_0 = \{U_{x_0}, \cdots, U_{x_k}\}$. Take

$$A = \bigwedge \Phi_0{}' = U_{x_0}{}' \wedge \cdots \wedge U_{x_k}{}',$$

then $A$ is a closed subset in $(X, \delta)$, it is sufficient to prove that as a function from $X$ to $[0,1]$, $A$ has not the largest value.

Take $n_0 \in D$ such that $n_0 \geq n(x_i)$ for every $i \in \{0, \cdots, k\}$, then $S(n) \leq A$ for every $n \geq n_0$, i.e. $S$ is eventually in $A$. Since $\Phi_0$ is an $\alpha$-Q-cover, we have $A(x) < \alpha$ for every $x \in X$. Since $S$ is an eventual $\alpha$-net, for every $\varepsilon > 0$, the height net $ht(S)$ of $S$ is eventually larger than or equal to $\alpha - \varepsilon$. So there exists $n_1 \in D$ such that $n_1 \geq n_0$ and $ht(S(n_1)) > \alpha - \varepsilon$. But by the choose of $n_0$, we have $S(n_1) \leq A$, so $A(supp(S(n_1))) > \alpha - \varepsilon$. Combining $A(x) < \alpha$ for every $x \in X$ shown previously, we have already proved that as a function, $A$ has not the largest value. □

**10.4.13 Theorem** *Let $(L^X, \delta)$ be a weakly induced L-fts. Then $(L^X, \delta)$ is strongly compact if and only if its background space $(X, [\delta])$ is compact.*

**Proof** (Sufficiency) Suppose $(X, [\delta])$ is compact, then by Theorem **10.2.13**, $(L^X, \delta)$ is N-compact. By Theorem **10.2.7** and Theorem **10.4.5** (ii), $(L^X, \delta)$ is strongly compact.

(Necessity) Suppose $(L^X, \delta)$ is strongly compact, $\mathcal{U}$ is an open cover of $(X, [\delta])$. Since $L$ is competely distributive, $pr(L)$ is not empty, there exists $\gamma \in pr(L)$. Then $\Phi = \{\chi_U : U \in \mathcal{U}\}$ is a $\gamma$-cover of $\underline{1}$. By the strong compactness of $(L^X, \delta)$, $\Phi$ has a finite sub-$\gamma$-cover $\Phi_0 = \{\chi_{U^0}, \cdots, \chi_{U^k}\}$. Since $\gamma < 1$, $\mathcal{U}_0 = \{U^0, \cdots, U^k\} \subset \mathcal{U}$ is a cover of $X$, $(X, [\delta])$ is compact. □

So for the case of F-lattices, we still have

**10.4.14 Corollary** *Strong compactness in L-fuzzy topological spaces is a good L-extension of compactness in ordinary topological spaces.* □

The following conclusion can be easily shown with Theorem **10.4.5**:

**10.4.15 Theorem** *Let $(L^X, \delta)$ be a strongly compact L-fts, $f^{\rightarrow} : (L^X, \delta) \rightarrow (L^Y, \mu)$ a surjective L-fuzzy continuous mapping. Then $(L^Y, \mu)$ is strongly compact.* □

**10.4.16 Remark** The conclusion similar to Theorem **10.4.15** for strongly compact $L$-fuzzy subsets does not hold even if $L = [0,1]$. A related counterexample is provided in Ref. [168].

The symbols used in the following lemma are respectively defind in Definition **1.1.2** and Definition **2.2.13**):

**10.4.17 Lemma** *Let $\{(L^{X_t}, \delta_t) : t \in T\}$ be a family of L-fts', $X$ a non-empty ordinary set, $f_t{}^{\rightarrow} : L^X \rightarrow L^{X_t}$ an L-fuzzy mapping, $\gamma \in pr(L)$. Then*

$$\iota_\gamma(Grt(\bigcup\{f_t{}^{\leftarrow}[\delta_t] : t \in T\})) = Grt(\bigcup\{f_t{}^{-1}[\iota_\gamma(\delta_t)] : t \in T\}).$$

**Proof** By Proposition **4.2.5** (i), the following equation holds for every $t \in T$:

$$\iota_\gamma(f_t^{\leftarrow}[\delta_t]) = f_t^{-1}(\iota_\gamma(\delta_t)).$$

So in order to prove this lemma, it is sufficient to prove that the following equation holds for every $\mathcal{A} \subset L^X$:

$$\iota_\gamma(Grt(\mathcal{A})) = Grt(\iota_\gamma(\mathcal{A})).$$

But by Proposition **4.2.4**, it is true, so it completes the proof. □

**10.4.18 Theorem** *Strong compactness in L-fuzzy topological spaces is strongly multiplicative.*

**Proof** Let $\{(L^{X_t}, \delta_t) : t \in T\}$ be a family of $L$-fts', $(L^X, \delta) = \prod_{t\in T}(L^{X_t}, \delta_t)$.

If $(L^X, \delta)$ is strongly compact, then for every $t \in T$, since the projection $p_t^{\rightarrow} : (L^X, \delta) \to (L^{X_t}, \delta_t)$ is surjective and continuous, by Theorem **10.4.15**, $(L^{X_t}, \delta_t)$ is strongly compact. If $(L^{X_t}, \delta_t)$ is strongly compact for every $t \in T$, $\gamma \in pr(L)$, then by Theorem **10.4.3**, $\iota_\gamma(\delta_t)$ is an ordinary compact topology on $X_t$ for every $t \in T$. Hence their ordinary product topology $Grt(\bigcup\{p_t^{-1}[\iota_\gamma(\delta_t)] : t \in T\})$ is compact. But by Lemma **10.4.17**, it is just the $\gamma$-level topology $\iota_\gamma(Grt(\bigcup\{f_t^{\leftarrow}[\delta_t] : t \in T\}))$ of the product topology $\delta$ of $\{\delta_t : t \in T\}$. Applying Theorem **10.4.3** again, $(L^X, \delta)$ is strongly compact. □

On the other relations between weakly induced property and strong compactness in $L$-fts', Peng[128] revealed the following results:

**10.4.19 Lemma** *Let $(L^X, \delta)$ be a strongly $T_2$ L-fts, $A \in L^X$ be strongly compact. Then $A : (X, [\delta]) \to L$ is upper semicontinuous.*

**Proof** $\forall \alpha \in M(L)$, $\forall x \in X \backslash A_{[\alpha]}$, we need to prove $x \notin A_{[\alpha]}{}^-$. Since $x \notin A_{[\alpha]}$ and $(L^X, \delta)$ is strongly $T_2$, $\forall \lambda \in M(L)$, $\forall y \in A_{[\alpha]}$, $\exists U^x \in \mathcal{Q}(x_\lambda)$, $\exists V^y \in \mathcal{Q}(y_\alpha)$ such that $U^x{}_{(0)} \cap V^y{}_{(0)} = \emptyset$. Then $\{V^y{}_{(0)} : y \in A_{[\alpha]}\}$ is an open $\alpha$-Q-cover of $A$. By Theorem **10.4.11**, $\{V^y{}_{(0)} : y \in A_{[\alpha]}\}$ has a finite sub-$\alpha$-Q-cover $\{V^{y_i}{}_{(0)} : i \leq n\}$. Take

$$U = \bigwedge_{i\leq n} U^{x_i}, \qquad V = \bigvee_{i\leq n} V^{y_i},$$

then $U \in \mathcal{Q}(x_\lambda)$, $V$ is an $\alpha$-Q-neighborhood of $A$ and by Proposition **4.2.4** (i), (ii),

$$U_{(0)} \cap V_{(0)} = (\bigcap_{i\leq n} U^{x_i}{}_{(0)}) \cap (\bigcup_{i\leq n} V^{y_i}{}_{(0)}) \leq \bigcup_{i\leq n} (U^{x_i}{}_{(0)} \cap V^{y_i}{}_{(0)}) = \emptyset.$$

Since $\forall y \in A_{[\alpha]}$, $y_\alpha \lessdot V$, $V(y) \neq 0$, $y \in V_{(0)}$, $y \notin U_{(0)}$, $y \in U_{(0)}{}'$. So $A_{[\alpha]} \subset U_{(0)}{}'$, $\chi_{A_{[\alpha]}} \leq \chi_{U_{(0)}{}'} \leq U'$, $(\chi_{A_{[\alpha]}})^- \leq U'$. By $U \in \mathcal{Q}(x_\lambda)$, we have $x_\lambda \not\leq U'$, so $x_\lambda \not\leq (\chi_{A_{[\alpha]}})^-$. Since $\lambda$ is arbitrarily chosen from $M(L)$, $(\chi_{A_{[\alpha]}})^-(x) = 0$. Therefore, we have actually already proved $\chi_{A_{[\alpha]}}(x) = 0 \Longrightarrow (\chi_{A_{[\alpha]}})^-(x) = 0$, so $(\chi_{A_{[\alpha]}})^- = \chi_{A_{[\alpha]}}$, $\chi_{A_{[\alpha]}}$ is closed in $(L^X, \delta)$, $A_{[\alpha]}$ is closed in $(X, [\delta])$. By Theorem **4.2.17**, $A : (X, [\delta]) \to L$ is upper semicontinuous. □

**10.4.20 Theorem** *Every strongly $T_2$ and strongly compact L-fts is weadly induced.*

**Proof** Let $(L^X, \delta)$ be a strongly $T_2$ and strongly compact $L$-fts, $A \in \delta'$, then $A$ is strongly compact. By Lemma **10.4.19**, $A$ is upper semicontinuous as a mapping from $(X, [\delta])$ to $L$. So $(L^X, \delta)$ is weakly induced. □

Now we turn to fuzzy compactness defined by Lowen.[92] Similar to the investigation for strong compactness, we also generalize fuzzy compactness from F-ts' to $L$-fts' and from spaces to subsets:[168]

**10.4.21 Definition** Let $(L^X, \delta)$ be an $L$-fts, $A \in L^X$. $A$ is called *fuzzy compact*, if for every $\alpha \in M(L)$, every $\alpha$-net $S$ in $A$ and every $\gamma \in \beta^*(\alpha)$, $S$ has a cluster point in $A$ with height $\gamma$. $(L^X, \delta)$ is called *fuzzy compact*, if $\underline{1}$ is fuzzy compact.

**10.4.22 Theorem** *Let $(L^X, \delta)$ be an $L$-fts, $A \in L^X$. Then $A$ is fuzzy compact if and only if for every $\alpha \in M(L)$, every eventual $\alpha$-net $S$ in $A$ and every $\gamma \in \beta^*(\alpha)$, $S$ has a cluster point in $A$ with height $\gamma$.*

**Proof** It is sufficient to show the necessity. Suppose $S = \{S(n), n \in D\}$ is an eventual $\alpha$-net in $A$. By Theorem **1.3.14** (i), $\exists \lambda \in \beta^*(\alpha)$ such that $\gamma \in \beta^*(\lambda)$. Since $S$ is an eventual $\alpha$-net in $A$, $\exists n_0 \in D$ such that $ht(S(n)) \geq \lambda$ for every $n \geq n_0$. Take $D_0 = \{n \in D : \ n \geq n_0\}$, then $D_0$ is a directed set. $\forall n \in D_0$, take $T(n) = (supp(S(n)))_\lambda$, then

$$ht(T(n)) = \lambda \leq ht(S(n)) \leq A(supp(S(n))) = A(supp(T(n))),$$

$T = \{T(n), n \in D_0\}$ is a $\lambda$-net in $A$. By the fuzzy compactness of $A$ and $\gamma \in \beta^*(\lambda)$, $T$ has a cluster point $x_\gamma \leq A$. So $\forall U \in \mathcal{Q}(x_\gamma)$, $\forall n \in D_0$, $\exists n_1 \in D_0$ such that $n_1 \geq n$, $T(n_1) \lessdot U$. Since $supp(T(n_1)) = supp(S(n_1))$, $ht(T(n_1)) \leq ht(S(n_1))$, so $S(n_1) \lessdot U$. Since $D$ is directed, for every $n \in D$, we can take $n_1 \in D$ such that $n_1 \geq n_0$, $n_1 \geq n$ and $S(n_1) \lessdot U$, $x_\gamma$ is also a cluster point of $S$. □

**10.4.23 Theorem** *Let $(L^X, \delta)$ be a weakly induced $L$-fts. Then $(L^X, \delta)$ is fuzzy compact if and only its background space $(X, [\delta])$ is compact.*

**Proof** (Sufficiency) Suppose $(X, [\delta])$ is compact, then by Theorem **10.4.13**, $(L^X, \delta)$ is strongly compact. By Theorem **10.4.5** (ii), $(L^X, \delta)$ is fuzzy compact.

(Necessity) Suppose $(L^X, \delta)$ is fuzzy compact, $S = \{S(n), n \in D\}$ is a net in $(X, [\delta])$. Since $L$ is completely distributive, $M(L) \neq \emptyset$, we can fix a molecule $\alpha \in M(L)$. For every $n \in D$, take $T(n) = S(n)_\alpha$, then $T = \{T(n), n \in D\}$ is an $\alpha$-net in $(L^X, \delta)$. By Theorem **1.3.10**, there exists $\gamma \in \beta^*(\alpha)$. By the fuzzy compactness of $(L^X, \delta)$, $T$ has a cluster point $x_\gamma$. $\forall U \in \mathcal{N}_{[\delta]}(x)$, $\forall n \in D$, then $V = \chi_U \in \mathcal{Q}(x_\gamma)$, $\exists n_0 \in D$ such that $n_0 \geq n$, $T(n) \lessdot V$, $S(n) \in U$. So $x$ is a cluster point of $S$, $(X, [\delta])$ is compact. □

**10.4.24 Corollary** *Fuzzy compactness in L-fuzzy topological spaces is a good L-extension of compactness in ordinary topological spaces.* □

It is not very difficult to prove the following conclusions:

**10.4.25 Theorem** *Let $(L^X, \delta)$ be an $L$-fts, $A \in L^X$ a fuzzy compact subset, $B \in L^X$ a closed subset, then $A \wedge B$ is fuzzy compact.* □

**10.4.26 Theorem** *Every fuzzy compact subset in a stratified $T_2$ L-fts is a closed subset.* □

**10.4.27 Theorem** *Let $(L^X, \delta)$ be a fuzzy compact L-fts, $f^\rightarrow : \ (L^X, \delta) \rightarrow (L^Y, \mu)$ a surjective L-fuzzy continuous mapping. Then $(L^Y, \mu)$ is fuzzy compact.* □

**10.4.28 Theorem** *Let $(L^X, \delta)$ be a $T_2$ L-fts, $A \in L^X$. Then $A$ is strongly compact if and only if $A$ is fuzzy compact.*

**Proof** Suppose $A$ is fuzzy compact, $\alpha \in M(L)$, $S = \{x^n{}_\alpha, n \in D\}$ is an $\alpha$-net in $A$. $\forall \gamma \in \beta^*(\alpha)$, by the fuzzy compactness of $A$, $S$ has a cluster point $x_\gamma \leq A$. By Corollary **5.1.12**, $S$ has a subnet $T \to x_\gamma$, $T$ is still an $\alpha$-net in $A$. By the fuzzy compactness of $A$, $\forall \lambda \in \beta^*(\alpha)$, $T$ has a cluster point $y_\lambda \leq A$, $T$ has a subnet $R \to y_\lambda$. Certainly, $y_\lambda$ is also a cluster point of $S$. Since $R$ is a subnet of $T$ and $T \to x_\gamma$, we have $R \to x_\gamma$ either. By $T_2$ property of $(L^X, \delta)$ and Theorem **8.2.12**, $x = y$. So $S \infty x_\gamma$, $S \infty x_\lambda$, we have already proved

$$\exists x \in X,\ \forall \gamma \in \beta^*(\alpha),\ S \infty x_\gamma. \tag{10.12}$$

$\forall U \in \mathcal{Q}(x_\alpha)$, then $\bigvee \beta^*(\alpha) = \alpha \not\leq U(x)'$, $\exists \gamma \in \beta^*(\alpha)$ such that $\gamma \not\leq U(x)'$. So $U \in \mathcal{Q}(x_\gamma)$. By (10.12), $S$ frequently does not quasi-coincides with $U$. So $S \infty x_\alpha$, $A$ is strongly compact. □

**10.4.29 Remark** (1) It is not hard to verify that in the case of $L = [0,1]$, the definition of fuzzy compactness of spaces in Definition **10.4.21** is coincident with its original definition in Definition **10.1.9**. But it looks like be hard to generalize the original definition to the case of F-lattices in the pattern of some kind of covers or Q-covers, because the order relation used there is $\leq$ and the Multiple Choice Principle (see Section **2.3**) needs the order relation to be $\not\leq$. In F-ts', $\not\leq$ and $>$ are equivalent to each other, so in order to describe some kind of compactness in F-ts' in the pattern of some kind of covers or Q-covers, usually we need the order relation used in the definition to be $>$ or $<$, just as it was done in the original definition of strong compactness.

(2) Lowen[92] also defined fuzzy compactness for fuzzy subsets as follows:

> *A fuzzy subset $A$ in a F-ts $(X, \delta)$ is called fuzzy compact, if for every family $\mathcal{U} \subset \delta$ such that $\bigvee \mathcal{U} \geq A$ and every $\varepsilon > 0$, $\mathcal{U}$ has a finite subfamily $\mathcal{V}$ such that $\bigvee \mathcal{V} \geq A - \varepsilon$.*

But according to this definition of fuzzy compact subsets, fuzzy compactness is not hereditary for closed subsets.[92]

The similar generalizition can also be made for ultra-compactness:

**10.4.30 Definition** An $L$-fts $(L^X, \delta)$ is called *ultra-compact*, if $(X, \iota_L(\delta))$ is compact.

**10.4.31 Lemma** *Let $(L^X, \delta)$ be a weakly induced L-fts, $a \in L \backslash \{1\}$, then*

$$\iota_a(\delta) = \iota_L(\delta) = [\delta].$$

**Proof** Since $(L^X, \delta)$ is weakly induced, by Theorem **4.2.17**, $\iota_a(\delta) \subset [\delta]$. So

$$\iota_a(\delta) \subset \iota_L(\delta) = Grt(\bigcup_{a \in L} \iota_a(\delta)) \subset [\delta].$$

$\forall U \in [\delta]$, then $\chi_U \in \delta$. Since $a \in L \backslash \{1\}$, so

$$U = \{x \in X : \chi_U(x) \not\leq a\} = (\chi_U)_{(a)} \in \iota_a(\delta),$$

i.e. $[\delta] \subset \iota_a(\delta)$. Hence

$$\iota_a(\delta) = \iota_L(\delta) = [\delta].$$ □

By Lemma **10.4.31**, the following theorem is obvious:

**10.4.32 Theorem** *Let $(L^X, \delta)$ be a weakly induced L-fts. Then $(L^X, \delta)$ is ultra-compact if and only its background space $(X, [\delta])$ is compact.* □

**10.4.33 Corollary** *Ultra-compactness in L-fuzzy topological spaces is a good L-extension of compactness in ordinary topological spaces.* □

Then we have the following relations among these kinds of compactness in $L$-fuzzy topological spaces:

**10.4.34 Theorem** *The following implications hold in L-fuzzy topological spaces:*
*Ultra-compactness$\Longrightarrow$N-compactness$\Longrightarrow$strong compactness$\Longrightarrow$fuzzy compactness.*

**Proof** By Theorem **10.2.7** and Theorem **10.4.5**,

N-compactness $\Longrightarrow$ strong compactness $\Longrightarrow$ fuzzy compactness.

So we need only prove Ultra-compactness $\Longrightarrow$ N-compactness. Suppose $(L^X, \delta)$ is an ultra-compact $L$-fts, then $(X, \iota_L(\delta))$ is compact. Since N-compactness is a good $L$-extension, so $(L^X, \omega_L \circ \iota_L(\delta))$ is N-compact. By Theorem **4.4.8** (ii), $\omega_L \circ \iota_L(\delta) \supset \delta$, so $(L^X, \delta)$ is N-compact. □

Moreover, each of the implications listed above is strict just as the following examples show:

**10.4.35 Example** There exists a fuzzy compact F-ts which is not strong compact.

Let $(Y, \mu)$ be defined as in Example **10.4.10**, then it has been shown there that $(Y, \mu)$ is not strongly compact. Suppose $S = \{y^n{}_{\alpha}, n \in D\}$ is an $\alpha$-net in $(Y, \mu)$. If $\alpha < 1$, take $y \in Y = \mathbf{N}$ large enough such that $\frac{1}{y+1} < \alpha'$, then $y_\alpha$ has only one Q-neighborhood $\underline{1}$, $y_\alpha$ is a cluster point of $S$, and hence $y_{\alpha-\varepsilon}$ is also a cluster point of $S$. If $\alpha = 1$, suppose $\varepsilon \in (0,1)$, take $y$ large enough such that $\frac{1}{y+1} < \varepsilon$. Since $\forall n \in \mathbf{N}$, $B_n(y) \leq \frac{1}{y+1}$, $1 - \varepsilon < 1 - \frac{1}{y+1} \leq B_n(y)'$, so $B_n \notin \mathcal{Q}(y_{1-\varepsilon})$. Since $\{B_n : n \in \mathbf{N}\}$ is a base of $\mu$, the unique Q-neighborhood of $y_\varepsilon$ is $\underline{1}$. That means $S \infty y_\varepsilon$. Therefore, $(Y, \mu)$ is fuzzy compact, fuzzy compactness is strictly weaker than strong compactness in $L$-fuzzy topological spaces.

**10.4.36 Example** There exists a strong compact F-ts which is not N-compact.

Take $X = (0,1)$ and $\delta$ as the fuzzy topology on $X$ generated by the family $\{\underline{a} : a \in L\} \cup \{A\}$, where $A \in L^X$ and $A(x) = x$ for every $x \in X$. Clearly, every $\gamma$-cover of $\underline{1}$ must contain a layer $\underline{a}$ such that $a > \gamma$, so $(X, \delta)$ is strongly compact. But, obviously, the closed subset $A'$ cannot reach the largest value, so $(X, \delta)$ is not N-compact by Theorem **10.4.12**. This means that strong compactness is strictly weaker than N-compactness in $L$-fuzzy topological spaces.

**10.4.37 Example** There exists a N-compact F-ts which is not ultra-compact.

Take $X = \mathbf{N}$. For every real number $\alpha$ in $(0,1)$, construct a series of fuzzy subsets on $X$ as follows: Suppose $\frac{n-1}{n} < \alpha \leq \frac{n}{n+1}$. $\forall \alpha_1, \cdots, \alpha_n \in [\frac{n-1}{n}, \alpha)$, define $F(\alpha, \alpha_1, \cdots, \alpha_n) \in I^X$ as

$$F(\alpha, \alpha_1, \cdots, \alpha_n)(x) = \begin{cases} \alpha, & x > n, \\ \alpha_i, & x = i \in \{1, \cdots, n\}. \end{cases} \tag{10.13}$$

Then for every fixed $n \in \mathbf{N}$, we have

$$\{F(\alpha,\alpha_1,\cdots,\alpha_n):\ \alpha \in (\tfrac{n-1}{n},\tfrac{n}{n+1}],\ \alpha_1,\cdots,\alpha_n \in [\tfrac{n-1}{n},\alpha)\} \subset [\tfrac{n-1}{n},\tfrac{n}{n+1}],$$

and it is closed under nonempty finite joins and nonempty arbitrary intersections. So

$$m,n \in \mathbf{N},\ m>n \implies F(\alpha,\alpha_1,\cdots,\alpha_n) \le F(\beta,\beta_1,\cdots,\beta_m).$$

By the virtue of this implication, the family

$$\mathcal{T}_0 = \{F(\alpha,\alpha_1,\cdots,\alpha_n):\ n\in\mathbf{N},\ \alpha \in (\tfrac{n-1}{n},\tfrac{n}{n+1}],\ \alpha_1,\cdots,\alpha_n \in [\tfrac{n-1}{n},\alpha)\}$$

is also closed under nonempty finite joins and nonempty arbitrary intersections. Take $\mu \subset I^X$ as $\mu = \{\underline{0},\underline{1}\} \cup \mathcal{T}_0$, then $\delta = \mu'$ is a fuzzy topology on $X$, $(X,\delta)$ is a fuzzy topological space.

Suppose $m \in X$, take $n \in \mathbf{N}$ such that $n > m$ and take $A = F(\alpha,\alpha_1,\cdots,\alpha_n)$ such that $\alpha = \frac{n}{n+1}$, $\alpha_m = \frac{n-1}{n}$, $\{\alpha_i:\ i\in\{1,\cdots,n\}\backslash\{m\}\} \subset (\alpha_m,\alpha)$. Then $x = m$ is just the unique point in $X$ at which closed subset $A$ takes its smallest value $\frac{n-1}{n}$, so $x = m$ is also the unique point in $X$ at which open subset $A'$ takes its largest value $1-\frac{n-1}{n} = \frac{1}{n}$. Since $A'[X\backslash\{m\}] \subset [\alpha,\frac{1}{n})$ is a finite set, so we can take $\gamma \in \mathbf{R}$ such that $max(A'[X\backslash\{m\}]) < \gamma < \frac{1}{n}$, and hence

$$\iota_\gamma(A') = \{x \in X:\ A'(x) > \gamma\} = \{m\},$$

the singleton $\{m\}$ is an open subset in $(X,\iota_I(\delta))$. So $(X,\iota_I(\delta))$ is discrete and hence is not compact, $(X,\delta)$ is not ultra-compact.

Suppose $S$ is an eventual $\alpha$-net in $(X,\delta)$. If $\alpha<1$, suppose $\frac{n-1}{n} < \alpha \le \frac{n}{n+1}$. $\forall m>n$, $\forall U \in \mathcal{Q}(m_\alpha)$, then $U = \underline{1}$ or $U' \in \mathcal{T}_0$. If $U = \underline{1}$, $S$ eventually quasi-coincides with $U$. If $U' = F(\beta,\beta_1,\cdots,\beta_k) \in \mathcal{T}_0$, then by definition (10.13) and $U \in \mathcal{Q}(m_\alpha)$, we have $U'(m) = \beta < \alpha$ or $U'(m) = \beta_i < \alpha$ for some $i \in \{1,\cdots,k\}$. So by $\frac{k-1}{k} \le \beta_i < \beta \le \frac{k}{k+1}$ and $\frac{n-1}{n} < \alpha \le \frac{n}{n+1}$ we always have $k \le n$. Since $m > n \ge k$, $U'(m) = \beta < \alpha$. Then by $U' = F(\beta,\beta_1,\cdots,\beta_k)$, the eventual $\alpha$-net $S$ eventually quasi-coincides with $U$, $S \to m_\alpha$, $S \infty m_\alpha$. If $\alpha = 1$, then for every $m \in X$ and every $U \in \mathcal{Q}(m_1)$, $max(U'[X]) < 1$, the eventual $\alpha$-net $S$ eventually quasi-coincides with $U$, $S \to m_1$, $S \infty m_1$. So every eventual $\alpha$-net in $(X,\delta)$ has a cluster point with height $\alpha$, $(X,\delta)$ is N-compact.

But in weakly induced $L$-fts', all these kinds of compactness are coincident to each other:

**10.4.38 Theorem** *Let $(L^X,\delta)$ be a weakly induced L-fts, then the following conditions are equivalent:*

(i) *$(L^X,\delta)$ is ultra-compact.*
(ii) *$(L^X,\delta)$ is N-compact.*
(iii) *$(L^X,\delta)$ is strongly compact.*
(iv) *$(L^X,\delta)$ is fuzzy compact.*
(v) *$(X,[\delta])$ is compact.*

**Proof** By theorems **10.4.34**, **10.4.32** and **10.4.23**. □

By Theorem **10.3.6** and Theorem **9.1.30**, we have the following consequence of the above theorem:

**10.4.39 Corollary** *$\tilde{I}(L)$ is ultra-compact.* □

In strongly $T_2$ $L$-fts', these kinds of compactness are also coincides with each other:

**10.4.40 Theorem** *Let $(L^X, \delta)$ be a strongly $T_2$ L-fts, then the following conditions are equivalent:*

(i) *$(L^X, \delta)$ is ultra-compact.*
(ii) *$(L^X, \delta)$ is N-compact.*
(iii) *$(L^X, \delta)$ is strongly compact.*
(iv) *$(L^X, \delta)$ is fuzzy compact.*

**Proof** By Theorem **10.4.34** and Theorem **10.4.38**, we need only prove that every strongly $T_2$ and fuzzy compact $L$-fts $(L^X, \delta)$ is weakly induced. By Proposition **8.2.3** and Theorem **10.4.28**, $(L^X, \delta)$ is strongly compact. By Theorem **10.4.20**, $(L^X, \delta)$ is weakly induced. □

**10.4.41 Corollary** *Let $(L^X, \delta)$ be a $T_2$ L-fts, $1_L \in M(L)$, then the following conditions are equivalent:*

(i) *$(L^X, \delta)$ is ultra-compact.*
(ii) *$(L^X, \delta)$ is N-compact.*
(iii) *$(L^X, \delta)$ is strongly compact.*
(iv) *$(L^X, \delta)$ is fuzzy compact.*

**Proof** By Proposition **8.2.5** and Theorem **10.4.40**. □

**10.4.42 Corollary** *Let $(X, \delta)$ be a $T_2$ F-ts, then the following conditions are equivalent:*

(i) *$(X, \delta)$ is ultra-compact.*
(ii) *$(X, \delta)$ is N-compact.*
(iii) *$(X, \delta)$ is strongly compact.*
(iv) *$(X, \delta)$ is fuzzy compact.* □

# Chapter 11

# Compactification

Since compactness in fuzzy topology can be defined in various forms, correspondingly, compatification in fuzzy topology can also appear in various forms. But, just as it was explained previously, N-compactness possess better properties among these kinds of compactness in fuzzy topology. Hence, compatification theory in this chapter will be based on N-compactness.

## 11.1 Basic Theory of Compactification

A $L$-fuzzy topological space can possess many compatifications, so the investigation on the realtions among these compatifications is required; especially, the existence and the uniqueness of the largest compatification closely relates to the problem of mapping extension, brings us much interest. We introduce a preorder in the family of all compatifications of an $L$-fts and, under this preorder, we can prove that a largest compatification in a arbitrarily given subfamily exists. But since a preorder is not necessary to be antisymetric, so the largest compactification may not be unique. As an example, we also construct an $L$-fts to show the existence of two distinct largest compactifications of a common space. But under a natural assumption of separation, i.e. $T_2$ property, we prove that this preorder is just a partial order and hence guarantee the uniqueness of the largest compatification in this kind of compatifications.

For symbols $\underline{1}_X$, $\underline{1}_Y$ used in the following, see Definition **2.1.3**:

**11.1.1 Definition** Let $(L^X, \delta)$, $(L^Y, \mu)$ be $L$-fts'. $(L^Y, \mu)$ is called a *compactification* of $(L^X, \delta)$, if there exists an embedding $c^{\rightarrow} : (L^X, \delta) \rightarrow (L^Y, \mu)$ such that $c^{\rightarrow}(\underline{1}_X)^-$ is N-compact and $supp(c^{\rightarrow}(\underline{1}_X)^-) = Y$; denote $Y$ and $\mu$ by $cX$ and $c\delta$ respectively. Furthermore, $(L^Y, \mu)$ is called a *space-compatification* of $(L^X, \delta)$, if $c^{\rightarrow}(\underline{1}_X)^- = \underline{1}_Y$.

**11.1.2 Remark** Obviously, a compactification is not necessarily N-compact, and $c^{\rightarrow}(\underline{1}_X)^-$ is not necessarily a crisp subset. In fact, let $Z = c^{\rightarrow}(\underline{1}_X)^-_{[1]}$, then we can find that $Z$ is not necessarily N-compact. This fact shows that it is necessary to consider the supporting set of $c^{\rightarrow}(\underline{1}_X)^-$. For example, let $X = (\frac{1}{2}, 1]$, $Y = [0,1]$, $L = [0,1]$, $c: X \rightarrow Y$ be the inclusion mapping. Defince

$$\mu = \{\underline{0}_Y, \underline{1}_Y\} \cup \{\tfrac{1}{2}\chi_{[0,\frac{1}{2}]}\} \cup \{\tfrac{1}{2}\chi_{[0,t]} \vee \chi_{(t,1]} : t \in [0,1]\},$$

and define $\delta$ as the subspace topology of $X$. Then $C = c^{\rightarrow}(\underline{1}_X)^- = \frac{1}{2}\chi_{[0,\frac{1}{2}]} \vee \chi_{(\frac{1}{2},1]}$ is N-compact, but for $X = c^{\rightarrow}(\underline{1}_X)^-_{[1]}$, $\chi_X$ is not N-compact. In fact,

$$\{\chi_{[0,\frac{1}{2}]} \vee \tfrac{1}{2}\chi_{(\frac{1}{2},t]} \vee \chi_{(t,1]} : \tfrac{1}{2} < t \leq 1\}$$

is an open $\frac{1}{2}$-Q-cover of $\mathcal{X}_X$ in $(L^Y, \mu)$, but clearly it has not any subfamily which is a $(\frac{1}{2})^-$-Q-cover of $\mathcal{X}_X$.

But, however, we have the following

**11.1.3 Proposition** *Let $(L^X, \delta)$ be an L-fts, $(L^{cX}, c\delta)$ be a weakly induced compactification of $(L^X, \delta)$. Then $(L^{cX}, c\delta)$ is a N-compact space-compactification.*

**Proof** Since $(L^{cX}, c\delta)$ is weakly induced, by Theorem **4.3.5** (ii),

$$cX = supp(c^{\rightarrow}(\underline{1}_X)^-) = supp((\mathcal{X}_{c[X]})^-) = supp(\mathcal{X}_{(c[X])^-}) = (c[X])^-.$$

So

$$c^{\rightarrow}(\underline{1}_X)^- = (\mathcal{X}_{c[X]})^- = \mathcal{X}_{(c[X])^-} = \mathcal{X}_{cX} = \underline{1}_{cX},$$

$(L^{cX}, c\delta)$ is a space-compactification. According to the definition of compactification, $c^{\rightarrow}(\underline{1}_X)^-$ is N-compact, hence $\underline{1}_{cX}$ is N-compact, i.e. $(L^{cX}, c\delta)$ is N-compact. □

The rationality of the following definition is guaranteed by Theorem **11.1.5**:

**11.1.4 Definition** Let $(L^X, \delta)$ be an $L$-fts which is not N-compact. The $L$-fts $(L^{X^*}, \delta^*)$ defined as follows is called the *one-point compactification* of $(L^X, \delta)$:

$$X^* = X \cup \{\infty\},$$
$$\delta^* = \{U \in L^{X^*} : U(\infty) = 0,\ U|_X \in \delta\} \cup$$
$$\{V \in L^{X^*} : V(\infty) = 1,\ V'|_X \in \delta' \text{ and is N-compact in } (L^X, \delta)\},$$

where $\infty \notin X$ is any point outside $X$.

**11.1.5 Theorem** *Let $(L^X, \delta)$ be an L-fts which is not N-compact. Then*

(i) *The one-point compactification of $(L^X, \delta)$ is an L-fts.*

(ii) *The one-point compactification of $(L^X, \delta)$ is a space-compactification of $(L^X, \delta)$.*

**Proof** (i) By Proposition **3.1.2** and Proposition **10.2.22**.

(ii) Denote

$$\mathcal{U} = \{U \in L^{X^*} : U(\infty) = 0,\ U|_X \in \delta\},$$
$$\mathcal{V} = \{V \in L^{X^*} : V(\infty) = 1,\ V'|_X \in \delta' \text{ and is N-compact in } (L^X, \delta)\},$$

then $\delta^* = \mathcal{U} \cup \mathcal{V}$. If $\alpha \in M(L)$, $\Phi$ is an open $\alpha$-Q-cover of $(L^{X^*}, \delta^*)$, then by $\delta^* = \mathcal{U} \cup \mathcal{V}$, for $\infty_\alpha \leq \underline{1}_{X^*}$, there exists $V \in \mathcal{V} \cap \Phi$ such that $\infty_\alpha \lessdot V$. Since $\Phi$ is also an open $\alpha$-Q-cover of $V'$, there exist $\lambda \in \delta^*(\alpha)$ and a finite subfamily $\Phi_0 = \{W_1, \cdots, W_n\}$ of $\Phi$ such that $\Phi_0$ is a $\lambda$-Q-cover of $V'$. Let $\Psi = \{V\} \cup \Phi_0$. For every $x \in X^*$, if $x = \infty$, $x_\lambda \lessdot V \in \Psi$; if $x \neq \infty$ and $x_\lambda \not\lessdot V$, then $x_\lambda \leq V'$, then since $\Phi_0$ is a $\lambda$-Q-cover of $V'$, there exists $W_i \in \Phi_0$ such that $x_\lambda \lessdot W_i \in \Phi_0 \subset \Psi$. So $\Psi$ is a finite subfamily of $\Phi$ and is an $\alpha^-$-Q-cover of $(L^{X^*}, \delta^*)$, $(L^{X^*}, \delta^*)$ is N-compact.

Let $i$ denote the inclusion mapping $x \mapsto x : X \to X^*$, then $i^{\rightarrow} : (L^X, \delta) \to (L^{X^*}, \delta^*)$ is an embedding. Since $(L^X, \delta)$ is not N-compact, i.e. $\underline{1}_X = i^{\rightarrow}(\underline{1}_X)|_X$ is not N-compact, we have $i^{\rightarrow}(\underline{1}_X) \notin (\delta^*)'$. Hence $i^{\rightarrow}(\underline{1}_X)^- = \underline{1}_{X^*}$, $(L^{X^*}, \delta^*)$ is a space-compactification of $(L^X, \delta)$. □

Besides one-point compactifications, we can find some other kind of typical compactifications as follows: By Theorem **9.2.64**, an $L$-fts can be embedded into an $L$-fuzzy unit cube if and only if it is sub-$T_0$ and completely regular. By Theorem **10.3.5**, every $L$-fuzzy unit cube and every its closed subset is N-compact. So denote the family of all the $L$-fuzzy continuous mappings from $L$-fts $(L^X, \delta)$ to $I(L)$ by $\mathcal{F}(L^X, \delta)$, we have

**11.1.6 Theorem** *Every sub-$T_0$ and completely regular L-fts $(L^X, \delta)$ has a compactification $(L^{\beta X}, \beta\delta)$ which is a subspace of the L-fuzzy unit cube $I(L)^{\mathcal{F}(L^X,\delta)}$, where $\beta X = supp(F(\underline{1}_X)^-)$, $F = \Delta(\mathcal{F}(L^X, \delta))$, $\beta\delta$ is the subspace topology of $I(L)^{\mathcal{F}(L^X,\delta)}$.* □

**11.1.7 Definition** Let $(L^X, \delta)$ be a sub-$T_0$ and completely regular $L$-fts. Call $(L^{\beta X}, \beta\delta)$ the *Stone-Čech compactification* of $(L^X, \delta)$.

**11.1.8 Definition** Let $(L^X, \delta)$ be an $L$-fts. Denote the family of all the compactifications of $(L^X, \delta)$ by $\mathcal{C}(L^X, \delta)$.

In Section **3.1**, we have defined subspaces of an $L$-fuzzy topological space. But in order to fulfill the need in the investigation of compactifications, we extend the meaning of the notion "subspace" as follows, and, for convenience, the extension will be in the form of "co-topology":

**11.1.9 Definition** Let $(L^X, \delta)$ be an $L$-fts, $A \in L^X$. Denote

$$\mathcal{P}(A) = \{B \in L^X : \ B \leq A\},$$
$$\delta'_A = \{P \wedge A : \ P \in \delta'\},$$

and call $(\mathcal{P}(A), \delta'_A)$ a *pseudo-subspace* of $(L^X, \delta)$ with quasi-co-topology $\delta'_A$. For every $C \in \mathcal{P}(A)$, $C$ is called *closed* or a *closed subset* in $(\mathcal{P}(A), \delta'_A)$, if $C \in \delta'_A$.

Define the *closure* $cl_{\delta_A}(C)$ of $C$ in $(\mathcal{P}(A), \delta_A)$ as follows:

$$cl_{\delta_A}(C) = \bigwedge\{P \in \delta'_A : \ P \geq C\}.$$

Let $(\mathcal{P}(A), \delta'_A)$, $(\mathcal{P}(B), \mu'_B)$ be pseudo-subspaces of $L$-fts' $(L^X, \delta)$ and $(L^Y, \mu)$ respectively, $f : \ \mathcal{P}(A) \to \mathcal{P}(B)$ an ordinary mapping. Define the mapping $f^\vee$ : $\mathcal{P}(B) \to \mathcal{P}(A)$ as follows:

$$f^\vee(D) = \bigvee\{C \in \mathcal{P}(A) : \ f(C) \leq D\}, \quad \forall D \in \mathcal{P}(B).$$

For an $L$-fuzzy mapping $f^\to : \ L^X \to L^Y$ such that $f^\to[\mathcal{P}(A)] \subset \mathcal{P}(B)$, simply denote the restriction $(f^\to)|_{\mathcal{P}(A)}^{\mathcal{P}(B)}$ of $f^\to$ from $\mathcal{P}(A)$ to $\mathcal{P}(B)$ by $\overline{f^\to}_{AB}$, or $\overline{f^\to}$ for short, and call $\overline{f^\to}$ a *subspace-mapping*.

A subspace-mapping $\overline{f^\to}$ : $(\mathcal{P}(A), \delta_A) \to (\mathcal{P}(B), \delta_B)$ is called *continuous*, if $\overline{f^\to}^\vee(Q) \in \delta'_A$ for every $Q \in \mu'_B$. $\overline{f^\to}$ is called *closed*, if $\overline{f^\to}(P) \in \mu'_B$ for every $P \in \delta'_A$. $\overline{f^\to}$ is called a *homeomorphism* from $(\mathcal{P}(A), \delta'_A)$ to $(\mathcal{P}(B), \mu'_B)$, if $\overline{f^\to}$ is bijective, continuous and closed.

**11.1.10 Proposition** *Let $(L^X, \delta)$ be an L-fts, $(\mathcal{P}(A), \delta_A)$ a pseudo-subspace of $(L^X, \delta)$. Then*

(i) *$\mathcal{P}(A)$ is a compaletely distributive lattice with the smallest element $\underline{0}_X$ and the largest element $A$.*

(ii) *$\delta'_A$ is closed under arbitrary meets and finite joins; especially, $\delta'_A$ has the smallest element $\underline{0}_X$ and the largest element $A$.* □

**11.1.11 Proposition** *Let $(L^X,\delta)$ be an L-fts, $(\mathcal{P}(A),\delta'_A)$ a pseudo-subspace of $(L^X,\delta)$. Then*

(i) $C \in \mathcal{P}(A) \Longrightarrow cl_{\delta_A}(C) = C^- \wedge A.$

(ii) $\forall C \in \mathcal{P}(A), \quad C \in \delta'_A \Longleftrightarrow C = cl_{\delta_A}(C).$ □

**11.1.12 Proposition** *Let $\overline{f^\to}: \mathcal{P}(A) \to \mathcal{P}(B)$ be a subspace-mapping, then*

$$\overline{f^\to}^\vee(D) = f^\leftarrow(D) \wedge A$$

*for every $D \in \mathcal{P}(B)$.*

**Proof** For every $C$ satisfying $f^\to(C) \le D$, by Theorem **2.1.25** (i), we have $C \le f^\leftarrow f^\to(C) \le f^\leftarrow(D)$. Hence

$$\overline{f^\to}^\vee(D) = \bigvee\{C \in \mathcal{P}(A) : f^\to(C) \le D\} \le f^\leftarrow(D).$$

On the reverse of this inequality, by Theorem **2.1.25** (ii),

$$f^\to(f^\leftarrow(D) \wedge A) \le f^\to f^\leftarrow(D) \le D,$$

so $f^\leftarrow(D) \wedge A \le \overline{f^\to}^\vee(D)$. □

**11.1.13 Proposition** *Let $\overline{f^\to}: \mathcal{P}(A) \to \mathcal{P}(B)$ be a subspace-mapping. Then*

(i) $\overline{f^\to}$ *is arbitrary join preserving.*

(ii) $\overline{f^\to}^\vee$ *is arbitrary join preserving and arbitrary meet preserving.* □

**11.1.14 Proposition** *Let $\overline{f^\to}: \mathcal{P}(A) \to \mathcal{P}(B)$, $\overline{g^\to}: \mathcal{P}(B) \to \mathcal{P}(C)$ be subspace-mappings. Then*

(i) $\overline{f^\to}\,\overline{g^\to} = \overline{f^\to g^\to}$.

(ii) $(\overline{f^\to}\,\overline{g^\to})^\vee = \overline{g^\to}^\vee\,\overline{f^\to}^\vee$.

(iii) $\overline{f^\to}^\vee\,\overline{f^\to} \ge id_{\mathcal{P}(A)}$.

(iv) $\overline{f^\to}\,\overline{f^\to}^\vee \le id_{\mathcal{P}(B)}$.

(v) $\overline{f^\to}$ *is bijective if and only if*

$$\overline{f^\to}\,\overline{f^\to}^\vee = id_{\mathcal{P}(B)}, \quad \overline{f^\to}^\vee\,\overline{f^\to} = id_{\mathcal{P}(A)}.$$ □

**11.1.15 Proposition** *Let $\overline{f^\to}: (\mathcal{P}(A),\delta'_A) \to (\mathcal{P}(B),\mu'_B)$ be a subspace-mapping. Then $\overline{f^\to}$ is continuous if and only if for every $C \in \mathcal{P}(A)$,*

$$\overline{f^\to}(cl_{\delta_A}(C)) \le cl_{\mu_B}(\overline{f^\to}(C)).$$ □

**11.1.16 Definition** In the family $\mathcal{C}(L^X,\delta)$ of all the compactifications of $L$-fts $(L^X,\delta)$, define a preorder $\le$ as follows:

$(L^{c_1X}, c_1\delta) \le (L^{c_2X}, c_2\delta)$ if and only if there exists a continuous subspace-mapping $\overline{f^\to}: \mathcal{P}(c_2{}^\to(\underline{1}_X)^-) \to \mathcal{P}(c_1{}^\to(\underline{1}_X)^-)$ such that $f^\to c_2{}^\to = c_1{}^\to$;

moreover, if $\overline{f^\to}$ is a homeomorphism, we say that these two compactifications are equivalent to each other.

**11.1.17 Theorem** *Let $(L^X,\delta)$ be an L-fts, $\mathcal{C}_0 \subset \mathcal{C}(L^X,\delta)$ is nonempty. Then $\mathcal{C}_0$ has a supremum in $\mathcal{C}(L^X,\delta)$.*

**Proof** In fact, a stronger conclusion will be proved in the following: The desired supremum $(L^{c_0X}, c_0\delta)$ exists and for every $(L^{cX}, c\delta) \in \mathcal{C}_0$, there exists a $L$-fuzzy continuous mapping (but not only a continuous subspace-mapping) $f_0{}^\to: (L^{c_0X}, c_0\delta) \to (L^{cX}, c\delta)$ such that $f_0{}^\to c_0{}^\to = c^\to$.

Let $\mathcal{C}_0 = \{(L^{c_tX}, c_t\delta) : t \in T\}$. Clearly, the following evaluation mapping defined in Definition **9.2.58** is an embedding:

$$c_0{}^{\rightarrow} = (\mathop{\triangle}_{t\in T} c_t)^{\rightarrow}: (L^X, \delta) \rightarrow \prod_{t\in T}(L^{c_tX}, c_t\delta)$$

Denote $A_t = c_t{}^{\rightarrow}(\underline{1}_X)^-$, $A = \prod_{t\in T} A_t = \bigwedge_{t\in T} p_t{}^{\leftarrow}(A_t)$, where $p_t{}^{\leftarrow}: \prod_{s\in T}(L^{c_sX}, c_s\delta) \rightarrow (L^{c_tX}, c_t\delta)$ is the $t$'s projection. Denote $(L^Y, \mu) = \prod_{s\in T}(L^{c_sX}, c_s\delta)$. By Theorem **10.3.2**, $A$ is a closed and N-compact subset in $(L^Y, \mu)$. Since

$$c_0{}^{\rightarrow}(\underline{1}_X) \le \prod_{t\in T} c_t{}^{\rightarrow}(\underline{1}_X) \le \prod_{t\in T}(c_t{}^{\rightarrow}(\underline{1}_X)^- = A,$$

we have $c_0{}^{\rightarrow}(\underline{1}_X)^- \le A^- = A$, hence $c_0{}^{\rightarrow}(\underline{1}_X)^-$ is a closed N-compact subset in $(L^Y, \mu)$. Let $c_0X = supp(c_0{}^{\rightarrow}(\underline{1}_X)^-)$ and let $c_0\delta$ be the subspace toplogy of $c_0X$, then $(L^{c_0X}, c_0\delta) \in \mathcal{C}(L^X, \delta)$.

$\forall (L^{c_tX}, c_t\delta) \in \mathcal{C}(L^X, \delta)$, let $f_0 = p_t|_{c_0X}: c_0X \rightarrow c_tX$, then clearly $f_0{}^{\rightarrow}: (L^{c_0X}, c_0\delta) \rightarrow (L^{c_tX}, c_t\delta)$ is continuous and $f_0{}^{\rightarrow}c_0{}^{\rightarrow} = c_t{}^{\rightarrow}$, and hence $(L^{c_tX}, c_t\delta) \le (L^{c_0X}, c_0\delta)$, $(L^{c_tX}, c_t\delta)$ is an upper bound of $\mathcal{C}_0$ in $\mathcal{C}(L^X, \delta)$. Now we need only to show that it is the smallest upper bound. Let $(L^{cX}, c\delta) \in \mathcal{C}(L^X, \delta)$ be an upper bound of $\mathcal{C}_0$, then for every $t \in T$, there exists a continuous subspace-mapping $\overline{f_t{}^{\rightarrow}}: \mathcal{P}(c^{\rightarrow}(\underline{1}_X)^-) \rightarrow \mathcal{P}(c_t{}^{\rightarrow}(\underline{1}_X)^-)$ such that $f_t{}^{\rightarrow}c^{\rightarrow} = c_t{}^{\rightarrow}$, where $f_t{}^{\rightarrow}: L^{cX} \rightarrow L^{c_tX}$ is an $L$-fuzzy mapping. Denote $F = \mathop{\triangle}_{t\in T} f_t: cX \rightarrow Y$. Since $p_tF = f_t$ and $\overline{f_t{}^{\rightarrow}} = f_t{}^{\rightarrow}|_{c^{\rightarrow}(\underline{1}_X)^-}$, we have

$$p_t{}^{\rightarrow}F^{\rightarrow}(c^{\rightarrow}(\underline{1}_X)^-) = f_t{}^{\rightarrow}(c^{\rightarrow}(\underline{1}_X)^-) = \overline{f_t{}^{\rightarrow}}(c^{\rightarrow}(\underline{1}_X)^-) \le c_t{}^{\rightarrow}(\underline{1}_X)^- = A_t,$$

$$F^{\rightarrow}(c^{\rightarrow}(\underline{1}_X)^-) \le \bigwedge_{t\in T} p_t{}^{\rightarrow}(A_t) = A.$$

So we obtain a subspace-mapping $\overline{F^{\rightarrow}}: \mathcal{P}(c^{\rightarrow}(\underline{1}_X)^-) \rightarrow \mathcal{P}(A)$. Furthermore, for every $t \in T$ and every $H_t \in (c_t\delta)'$, applying Proposition **11.1.12** twice, we have

$$\begin{aligned}\overline{F^{\rightarrow}}^{\vee}(p_t{}^{\leftarrow}(H_t) \wedge A) &= F^{\leftarrow}(p_t{}^{\leftarrow}(H_t)) \wedge c^{\rightarrow}(\underline{1}_X)^- \\ &= f_t{}^{\leftarrow}(H_t) \wedge c^{\rightarrow}(\underline{1}_X)^- \\ &= \overline{f^{\rightarrow}}^{\vee}(H_t \wedge c^{\rightarrow}(\underline{1}_X)^-) \\ &\in (c\delta)'_{c^{\rightarrow}(\underline{1}_X)^-}.\end{aligned}$$

Hence $\overline{F^{\rightarrow}}$ is continuous. On the other hand, it is easy to find that

$$Fc = \triangle_{t\in T} f_t c = \triangle_{t\in T} c_t = c_0,$$

so by Proposition **11.1.15** and Proposition **11.1.11** (i),

$$\overline{F^{\rightarrow}}(c^{\rightarrow}(\underline{1}_X)^-) \le (\overline{F^{\rightarrow}}(c^{\rightarrow}(\underline{1}_X)))^- = (F^{\rightarrow}c^{\rightarrow}(\underline{1}_X))^- = c_0{}^{\rightarrow}(\underline{1}_X)^-.$$

Denote $C = c^{\rightarrow}(\underline{1}_X)^-$, $D = c_0{}^{\rightarrow}(\underline{1}_X)^-$, then we obtain another subspace-mapping $\overline{F^{\rightarrow}}_{CD}$. By $D \le A$ and the continuity of $\overline{F^{\rightarrow}} = \overline{F^{\rightarrow}}_{CA}$, $\overline{F^{\rightarrow}}_{CD}$ is continuous. Since $F^{\rightarrow}c^{\rightarrow} = c_0{}^{\rightarrow}$, $(L^{c_0X}, c_0\delta) \le (L^{cX}, c\delta)$. The proof is completed.

By Theorem **11.1.5**, $\mathcal{C}(L^X, \delta)$ is always nonempty, so we have

**11.1.18 Corollary** *Every L-fts has a largest compactification.* □

In order to show that the preorder in a family of compactifications of an $L$-fts is not necessarily anti-symmetric, some preparations are needed:

**11.1.19 Lemma** *If a F-lattice L is not Boolean, then every L-fuzzy unit cube* $I(L)^{\kappa}$

*has not any non-trivial crisp closed subset.*

**Proof** Let $F$ be a closed subset in $I(L)^\kappa$ such that $F \neq \underline{0}, \underline{1}$. Note that both $\underline{0}$ and $\underline{1}$ are constant mappings from the carrier domain $I[L]^\kappa$ to $L$ (see Section **9.1**). Suppose

$$F = \bigwedge_{d\in D} \bigvee_{\alpha\in F_d} p_\alpha{}^{\leftarrow}(G_\alpha),$$

where $\{F_d : d \in D\} \subset [\kappa]^{<\omega}$, and for every $d \in D$, every $\alpha \in F_d$, $G_\alpha$ is a closed subset in $I(L)$, i.e. $G_\alpha \in (\mathcal{T}_L^I)'$. By $F \neq \underline{1}$, we can assume that $\bigvee_{\alpha\in F_d} p_\alpha{}^{\leftarrow}(G_\alpha) \neq \underline{1}$ for every $d \in D$. Thus $G_\alpha \neq \underline{1}$ for every $\alpha \in \bigcup_{d\in D} F_d$. Similarly, by $F \neq \underline{0}$ we have $G_\alpha \neq \underline{0}$ for every $\alpha \in \bigcup_{d\in D} F_d$. By Proposition **9.1.13** (i), we can assume $G_\alpha = \bigwedge\{R_{s_i}{}' \vee L_{t_i}{}' : i \in I_\alpha\}$ for every $\alpha \in \bigcup_{d\in D} F_d$, where $I_\alpha$ is an index set, $s_i, t_i \in \mathbf{R}$. Since $R_s{}' = \underline{1}_{I[L]}$ for every $s \geq 1$, $L_t{}' = \underline{1}_{I[L]}$ for every $t \leq 0$, $R_s{}' = \underline{0}_{I[L]}$ for every $s < 0$, $L_t{}' = \underline{0}_{I[L]}$ for every $t > 1$, in general, we can assume that $s_i < 1$ and $t_j > 0$ for arbitrary $i, j$, and for each pair of $s_i$, $t_i$, we have either $s_i \geq 0$ or $t_i \leq 1$.

Since $L$ is not Boolean, there exists an element $c \in L$ such that $c \vee c' \neq 1$. Thus $\{c, c', c \wedge c', c \vee c'\} \subset L\backslash\{0, 1\}$. Consider the element $\lambda$ in $I[L]$, which maps unit interval $[0, 1]$ to the constant value $c$, then for every $G_\alpha$, we have

$$\begin{aligned} G_\alpha(\lambda) &= \bigwedge\{R_{s_i}{}'(\lambda) \vee L_{t_i}{}'(\lambda) : i \in I_\alpha\} = \bigwedge\{\lambda(s_i+)' \vee \lambda(t_i-) : i \in I_\alpha\} \\ &\in \{c, c', c \wedge c', c \vee c'\}. \end{aligned}$$

Take $z \in I[L]^\kappa$ such that every coordinate of $z$ is $\lambda$, then

$$F(z) = (\bigwedge_{d\in D} \bigvee_{\alpha\in F_d} p_\alpha{}^{\leftarrow}(G_\alpha))(z) = \bigwedge_{d\in D} \bigvee_{\alpha\in F_d} G_\alpha(\lambda) \in \{c, c', c \wedge c', c \vee c'\}.$$

Therefore, $F(z) \in L\backslash\{0, 1\}$, $F$ is not a crisp subset. □

**11.1.20 Lemma** *Let $\{(L^{X_t}, \delta_t) : t \in T\}$ be a family of L-fts', $\lambda \in M(L)$, for every $t \in T$, $A_t \in L^{X_t}\backslash\{\underline{0}\}$, and for every $x^t \in X_t$, $A_t(x^t) \in \{0, \lambda\}$. Then in product L-fts $(L^X, \delta) = \prod_{t\in T}(L^{X_t}, \delta_t)$,*

$$(\prod_{t\in T} A_t)^- = \prod_{t\in T} A_t{}^-.$$

**Proof** Clearly we have $(\prod_{t\in T} A_\alpha)^- \leq \prod_{t\in T} A_t{}^-$. For the other inequality, suppose molecule $x_\gamma \in \prod_{t\in T} A_t{}^-$, where $x = (x^t)_{t\in T} \in \prod_{t\in T} X_t$, $\gamma \in M(L)$, then by Theorem **2.3.23** (ii), it is sufficient to prove that $x_\gamma$ is an adherent point of $\prod_{t\in T} A_t$. By the structure of product topology $\prod_{t\in T}\delta_t$, we need only to prove that if $U = \bigwedge\{p_{t_i}{}^{\leftarrow}(U_i) : i \leq n\} \in \mathcal{Q}(x_\gamma)$, where $n \in \mathbf{N}$, $U_i \in \delta_{t_i}$ for every $i \leq n$, then $U \hat{q} \prod_{t\in T} A_t$. Suppose $U = \bigwedge\{p_{t_i}{}^{\leftarrow}(U_i) : i \leq n\} \in \mathcal{Q}(x_\gamma)$ is of this type, then

$$\gamma \not\leq (\bigvee\{p_{t_i}{}^{\leftarrow}(U_i{}') : i \leq n\})(x) = \bigvee\{U_i{}'(x^{t_i}) : i \leq n\}.$$

So $\gamma \not\leq U_i{}'(x^{t_i})$ for every $i \leq n$, $U_i$ is a Q-neighborhood of $(x^{t_i})_\gamma$ in $(L^{X_{t_i}}, \delta_{t_i})$. Since $x_\gamma \leq \prod_{t\in T} A_t{}^- = \bigwedge_{t\in T} p_t{}^{\leftarrow}(A_t{}^-)$, $(x^{t_i})_\gamma \leq A_{t_i}{}^-$. By Theorem **2.3.23** (i), $(x^{t_i})_\gamma$ is an adherent point of $A_{t_i}$, then by $U_i \in \mathcal{Q}((x^{t_i})_\gamma)$, there exists a point $(y^{t_i})_a \leq A_{t_i}$ such that $(y^{t_i})_a \lessdot U_i$, i.e. $a \not\leq U_i{}'(y^{t_i})$. But $A_{t_i}(z^{t_i}) \in \{0, \lambda\}$ for every $z^{t_i} \in X_{t_i}$, so $0 < a \leq \lambda$, $(y^{t_i})_\lambda \leq A_{t_i}$, $U_i$ is also a Q-neighborhood of $(y^{t_i})_\lambda$ in $(L^{X_{t_i}}, \delta_{t_i})$. For every $t \in T\backslash\{t_0, \cdots, t_n\}$, arbitrarily fix a point $y^t \in supp(A_t)$, we have a point $y = (y^t)_{t\in T} \in \prod_{t\in T} X_t$ and, by $A_t(z^t) \in \{0, \lambda\}$ for every $t \in T$ and every $z^t \in X_t$,

$y_\lambda \leq \prod_{t\in T} A_t$. On the other hand, since $\lambda \in M(L)$ and $\lambda \not\leq U_i'(y^{t_i})$ for every $i \leq n$, we have

$$\lambda \not\leq \vee\{U_t'(y^{t_i}) : i \leq n\} = (\vee\{p_{t_i}^{\leftarrow}(U_i{}^{\leftarrow}) : i \leq n\})(y) = U'(y),$$

$y_\lambda \leq \prod_{t\in T} A_t$ quasi-coincides with $U$, $U\hat{q}\prod_{t\in T} A_t$. The proof is completed. □

**11.1.21 Remark** The following two examples show that each of the two conditions in Lemma **11.1.20** cannot be ommitted.

(i) The condition "$\lambda \in M(L)$" in Lemma **11.1.20** cannot be omitted.

Take $L$ as the diamond-type lattice described in Example **1.1.19** and let $a' = b$, then $L$ is a F-lattice. Let $X_1 = X_2 = \{x, y\}$, $\lambda = 1 \in L$, $A_1 = A_2 = x_\lambda$, $U_1 = \underline{a'} \in L^{X_1}$, $U_2 = x_a \vee y_{a'} \in L^{X_2}$. For $i = 1, 2$ let $\delta_i = \{\underline{0}, U_i, \underline{1}\}$, then both $(L^{X_1}, \delta_1)$ and $(L^{X_2}, \delta_2)$ are $L$-fts, $A_i{}^- = \underline{1}_{X_i}$. So $A_1{}^- \times A_2{}^- = \underline{1}_{X_1\times X_2}$. Denote $F = p_1{}^{\leftarrow}(U_1') \vee p_2{}^{\leftarrow}(U_2')$, then $F$ is closed in $(L^{X_1}, \delta_1) \times (L^{X_2}, \delta_2)$ and $F(x, x) = a \vee a' = 1$. So $A_1 \times A_2 = (x, x)_\lambda \leq F$, $(A_1 \times A_2)^- \leq F$. But $F(x, y) = a < 1$, so $(A_1 \times A_2)^- \neq \underline{1}_{X_1\times X_2} = A_1{}^- \times A_2{}^-$.

(ii) The condition "All $A_t$'s take values 0 and a common non-zero join-irreducible element" in Lemma **11.1.20** cannot be omitted. This lemma is not true even if we suppose that each $A_t$ takes values 0 and a non-zero join-irreducible element.

Take $L$, $X_1$, $X_2$ and $A_1$ as defined above. Let $A_2 = \underline{a'}$, $\delta_i = \{\underline{0}, \underline{1}\}$ for $i = 1, 2$, then $A_i{}^- = \underline{1}$, $A_1 \times A_2 = \underline{0}$, $(A_1 \times A_2)^- \neq A_1{}^- \times A_2{}^-$.

**11.1.22 Lemma** *Let $X$ be a singleton $\{x\}$, $\delta \subset L^X$ such that $\delta = \delta'$, then $(L^X, \delta)$ is sub-$T_0$ and completely regular.*

**Proof** $(L^X, \delta)$ is obviously sub-$T_0$. For $U \in \delta$, denote $U(x) = b$, then $x_b$ is both open and closed. Take $W = U$ and $\lambda \in I[L]$ such that $\lambda$ maps $[0, 1]$ to the constant value $b$. Define $f : X \to I[L]$ by $f(x) = \lambda$, then for the open sets $L_s$ and $R_t$, $f^{\leftarrow}(L_s') = f^{\leftarrow}(R_t) = x_b \in \delta$, $f^{\rightarrow} : (L^X, \delta) \to I(L)$ is continuous. Since $W = f^{\leftarrow}(L_1') = f^{\leftarrow}(R_0) = U$, so $(L^X, \delta)$ is completely regular. □

**11.1.23 Example** In an $L$-fts $(L^X, \delta)$, if $X$ is a singleton $\{x\}$, $L \neq \{0, 1\}$, $1 \in M(L)$, $\delta' = \delta$, then $(L^X, \delta)$ has two largest compatifications which are not equivalent to each other.

In fact, by Theorem **10.2.12**, $(L^X, \delta)$ is N-compact. By the definition of the preorder in compactifications, $(L^X, \delta)$ is a largest compactification of itself. By Lemma **11.1.22**, $(L^X, \delta)$ is sub-$T_0$ and completely regular. By Theorem **11.1.6**, $(L^X, \delta)$ can be embedded into an $L$-fuzzy unit cube with the evaluation mapping $F$ to obtain the Stone-Čech compactification $(L^{\beta X}, \beta\delta)$ of it. Since $L \neq \{0, 1\}$, $1 \in M(L)$, $L$ is not Boolean. By Lemma **11.1.19**, $I(L)^{\mathcal{F}(L^X,\delta)}$ has not any non-trivial crisp closed subset. But $\beta^{\rightarrow}(\underline{1}) = F(\underline{1})$ is a crisp point, $\beta^{\rightarrow}(\underline{1}) \neq \underline{1}$, so $\beta^{\rightarrow}(\underline{1})$ is not closed, namely $\beta^{\rightarrow}(\underline{1})^- \neq \beta^{\rightarrow}(\underline{1})$.

Suppose $\mathcal{C}(L^X, \delta) = \{(L^{cX}, c\delta) : c \in \Gamma\}$. According to the proof of Theorem **11.1.17**, we can build a largest compatification $(L^{c_0X}, c_0\delta)$ of $(L^X, \delta)$ with the evaluation mapping $c_0{}^{\rightarrow} = (\Delta\Gamma)^{\rightarrow}$. Since $X$ is a singleton $\{x\}$, $c_0{}^{\rightarrow}(\underline{1})$ is a crisp point in $\prod_{c\in\Gamma}(L^{cX}, c\delta)$, or more exactly, $c_0{}^{\rightarrow}(\underline{1}) = ((c(x))_{c\in\Gamma})_1$. Hence $c_0{}^{\rightarrow}(\underline{1}) = \prod_{c\in\Gamma} c^{\rightarrow}(\underline{1})$.

Since every $c^{\rightarrow}(\underline{1})$ only takes values from $\{0,1\}$ and $1 \in M(L)$, by Lemma **11.1.20**, $c_0{}^{\rightarrow}(\underline{1})^- = \prod_{c\in\Gamma} c^{\rightarrow}(\underline{1})^-$. But it has been shown above that for $\beta \in \Gamma$, $\beta^{\rightarrow}(\underline{1})^- \neq \beta^{\rightarrow}(\underline{1})$, so there exists an $L$-fuzzy point $y_a \leq \beta^{\rightarrow}(\underline{1})^-$ such that $y_a \not\leq \beta^{\rightarrow}(\underline{1})$. Fix an ordinary point $z \in \prod_{c\in\Gamma} cX$ such that $p_c(z) = c(x)$ for every $c \in \Gamma\backslash\{\beta\}$ and $p_\beta(z) = y$, then the $L$-fuzzy point

$$z_a \leq \bigwedge_{c\in\Gamma} p_c{}^{\leftarrow}(c^{\rightarrow}(\underline{1})^-) = \prod_{c\in\Gamma} c^{\rightarrow}(\underline{1})^- = c_0{}^{\rightarrow}(\underline{1})^-.$$

Since $p_\beta{}^{\rightarrow}(z_a) = y_a \not\leq \beta^{\rightarrow}(\underline{1})$, we have

$$z_a \not\leq \bigwedge_{c\in\Gamma} p_c{}^{\leftarrow}(c^{\rightarrow}(\underline{1})) = \prod_{c\in\Gamma} c^{\rightarrow}(\underline{1}) = c_0{}^{\rightarrow}(\underline{1}).$$

Therefore, since $c_0{}^{\rightarrow}(\underline{1}) = ((c(x))_{c\in\Gamma})_1$ is crisp, $z \notin supp(c_0{}^{\rightarrow}(\underline{1}))$. But certainly we have $z \in supp(c_0{}^{\rightarrow}(\underline{1})^-)$, so $(L^{c_0X}, c_0\delta)$ is not equivalent to $(L^X, \delta)$, i.e. $(L^X, \delta)$ has two largest compactifications which are equivalent to each other.

The above example shows that in general, the preorder in compactifications of an $L$-fts is not necessarily a partial order. But if some proper separation property is assumed, the preorder will be a partial order:

**11.1.24 Lemma** *Let $\mathcal{P}(A)$, $\mathcal{P}(B)$ be pseudo-subspaces of L-fts $(L^X, \delta)$ and level-$T_2$ L-fts $(L^Y, \mu)$ respectively, $\overline{f^{\rightarrow}}, \overline{g^{\rightarrow}}: \mathcal{P}(A) \rightarrow \mathcal{P}(B)$ be continuous subspace-mappings, and let*

$$\triangle(\overline{f^{\rightarrow}}, \overline{g^{\rightarrow}}) = \{x \in supp(A):\ f(x) = g(x)\},$$

*then $\chi_{\triangle(\overline{f^{\rightarrow}}, \overline{g^{\rightarrow}})} \wedge A \in \delta'_A$.*

**Proof** Since $(L^Y, \mu)$ is level-$T_2$, for every $x \in supp(A)\backslash \triangle(\overline{f^{\rightarrow}}, \overline{g^{\rightarrow}})$ and every $\alpha \in M(\downarrow A(x))$, $f(x) \neq g(x)$, there exist $U_\alpha \in \mathcal{Q}(f(x)_\alpha)$ and $V_\alpha \in \mathcal{Q}(g(x)_\alpha)$ such that $U_\alpha \wedge V_\alpha = \underline{0}$. Denote $P_x^\alpha = (f^{\leftarrow}(U_\alpha') \vee g^{\leftarrow}(V_\alpha')) \wedge A$. Since $\alpha \in M(L)$, we have $\alpha \not\leq P_x^\alpha(x)$. Since both of $\overline{f^{\rightarrow}}$ and $\overline{g^{\rightarrow}}$ are continuous, by Proposition **11.1.12**,

$$\begin{aligned} P_x^\alpha &= (f^{\leftarrow}(U_\alpha') \wedge A) \vee (g^{\leftarrow}(V_\alpha') \wedge A) \\ &= (f^{\leftarrow}(U_\alpha') \wedge f^{\leftarrow}(B) \wedge A) \vee (g^{\leftarrow}(V_\alpha') \wedge g^{\leftarrow}(B) \wedge A) \\ &= (f^{\leftarrow}(U_\alpha' \wedge B) \wedge A) \vee (g^{\leftarrow}(V_\alpha' \wedge B) \wedge A) \\ &= \overline{f^{\rightarrow}}^{\vee}(U_\alpha' \wedge B) \vee \overline{g^{\rightarrow}}^{\vee}(V_\alpha' \wedge B) \\ &\in \delta'_A. \end{aligned}$$

Meanwhile for every $z \in \triangle(\overline{f^{\rightarrow}}, \overline{g^{\rightarrow}})$, $f(z) = g(z)$,

$$\begin{aligned} P_x^\alpha(z) &= (U_\alpha' f(z) \vee V_\alpha' g(z)) \wedge A(z) \\ &= (U_\alpha' \vee V_\alpha')(f(z)) \wedge A(z) \\ &= (U_\alpha \wedge V_\alpha)'(f(z)) \wedge A(z) \\ &= \underline{0}'(f(z)) \wedge A(z) \\ &= 1_L \wedge A(z) \\ &= A(z). \end{aligned}$$

So $\chi_{\triangle(\overline{f^{\rightarrow}}, \overline{g^{\rightarrow}})} \wedge A \leq P_x^\alpha$. Now we want to prove

$$\chi_{\triangle(\overline{f^{\rightarrow}}, \overline{g^{\rightarrow}})} \wedge A = \bigwedge\{P_x^\alpha :\ x \in supp(A)\backslash \triangle(\overline{f^{\rightarrow}}, \overline{g^{\rightarrow}}),\ \alpha \in M(\downarrow A(x))\}. \tag{11.1}$$

If it is false, then there exists $x \in supp(A) \setminus \triangle(\overline{f^{\to}}, \overline{g^{\to}})$ such that

$$a = (\bigwedge\{P_x^\alpha : \alpha \in M(\downarrow A(x))\})(x) > 0.$$

Since every $P_x^\alpha \leq A$, $a \leq A(x)$. Take $\gamma \in M(\downarrow a) \subset M(\downarrow A(x))$, then $P_x^\gamma(x) \geq a \geq \gamma$. But this contradicts with the property $P_x^\gamma(x) \not\geq \gamma$. So (11.1) is true. Then since every $P_x^\alpha \in \delta'_A$, we have $\chi_{\triangle(\overline{f^{\to}}, \overline{g^{\to}})} \wedge A \in \delta'_A$. □

**11.1.25 Proposition** *Let $(L^X, \delta)$, $(L^Y, \mu)$ be an L-fts and a level-$T_2$ L-fts respectively, $A \in L^X \setminus \{\underline{0}\}$, $B \in L^Y \setminus \{\underline{0}\}$, $\overline{f^{\to}}, \overline{g^{\to}} : \mathcal{P}(A) \to \mathcal{P}(B)$ be continuous, $D \leq A \leq D^-$ and $\overline{f^{\to}}|_{\mathcal{P}(D)} = \overline{g^{\to}}|_{\mathcal{P}(D)}$. Then $\overline{f^{\to}} = \overline{g^{\to}}$.*

**Proof** Since $\overline{f^{\to}}|_{\mathcal{P}(D)} = \overline{g^{\to}}|_{\mathcal{P}(D)}$, $supp(D) \subset \triangle(\overline{f^{\to}}, \overline{g^{\to}})$, so by Lemma **11.1.24**, $D \leq \chi_{\triangle(\overline{f^{\to}}, \overline{g^{\to}})} \wedge A \in \delta'_A$. Then we have

$$A = D^- \wedge A \leq (\chi_{\triangle(\overline{f^{\to}}, \overline{g^{\to}})} \wedge A)^- \wedge A = \chi_{\triangle(\overline{f^{\to}}, \overline{g^{\to}})} \wedge A.$$

So $A \leq \chi_{\triangle(\overline{f^{\to}}, \overline{g^{\to}})}$, and hence $\overline{f^{\to}} = \overline{g^{\to}}$. □

**11.1.26 Theorem** *Let $\mathcal{C}_0$ be the family of all the level-$T_2$ compactifications of an L-fts $(L^X, \delta)$, then the preorder in $\mathcal{C}_0$ is a partial order. More exactly, for all $(L^{c_1X}, c_1\delta)$, $(L^{c_2X}, c_2\delta) \in \mathcal{C}_0$,*

*$(L^{c_1X}, c_1\delta)$ is equivalent to $(L^{c_2X}, c_2\delta)$ if and only if $(L^{c_1X}, c_1\delta) \leq (L^{c_2X}, c_2\delta) \leq (L^{c_1X}, c_1\delta)$.*

**Proof** The necessity is obvious. Let

$$\overline{f^{\to}} : \mathcal{P}(c_1(\underline{1})^-) \to \mathcal{P}(c_2(\underline{1})^-),$$
$$\overline{g^{\to}} : \mathcal{P}(c_2(\underline{1})^-) \to \mathcal{P}(c_1(\underline{1})^-)$$

be continuous and $f^{\to}c_1{}^{\to} = c_2{}^{\to}$, $g^{\to}c_2{}^{\to} = c_1{}^{\to}$. Then $f^{\to}g^{\to}c_2{}^{\to} = c_2{}^{\to}$, $g^{\to}f^{\to}c_1{}^{\to} = c_1{}^{\to}$. So we have

$$\overline{f^{\to}g^{\to}}|_{\mathcal{P}(c_2(\underline{1}))} = (\overline{f^{\to}}\overline{g^{\to}})|_{\mathcal{P}(c_2(\underline{1}))} = id_{\mathcal{P}(c_2(\underline{1}))} : \mathcal{P}(c_2(\underline{1})) \to \mathcal{P}(c_2(\underline{1})),$$
$$\overline{g^{\to}f^{\to}}|_{\mathcal{P}(c_1(\underline{1}))} = (\overline{g^{\to}}\overline{f^{\to}})|_{\mathcal{P}(c_1(\underline{1}))} = id_{\mathcal{P}(c_1(\underline{1}))} : \mathcal{P}(c_1(\underline{1})) \to \mathcal{P}(c_1(\underline{1})).$$

Then by Proposition **11.1.25**, $\overline{f^{\to}}\overline{g^{\to}} = id_{\mathcal{P}(c_2(\underline{1})^-)}$, $\overline{g^{\to}}\overline{f^{\to}} = id_{\mathcal{P}(c_1(\underline{1})^-)}$. Hence by Proposition **11.1.14**, $\overline{f^{\to}}$ is bijective and $\overline{g^{\to}}$ is its inverse mapping. So $\overline{f^{\to}}$ is a homeomorphism. □

## 11.2 Stone-Čech Compactification

Although it has been proved previously that every sub-$T_0$ completely regualr $L$-fts possesses a compactification — Stone-Čech compactifiction — contined in an $L$-fuzzy unit cube, but since the separation property of $L$-fuzzy unit interval is too weak, it is not even quasi-$T_0$ (see Example **9.1.15**), so the property of the Stone-Čech compactification is not ideal. In fact, Liu and Luo proved that the Stone-Čech compactification of the $L$-valued induced space of $[0, 1]$ is not the largest Tychonoff compactification of it.[86] Another problem is also revealed in last section: In general, a compactification, even the Stone-Čech compactification of an $L$-fts, is not a space-compactification, it can only be represented with pseudo-subspaces. Successively, Liu and Luo,[83, 84, 85, 86] Wang and Xu[170] tried to remedy these shortages, their work is what will be introduced in this section.

**11.2.1 Definition** An $(L^X, \delta)$ is called *para-Tychonoff* or *p-Tychonoff* for short, if it is sub-$T_0$ and compaletely p-regular.

Let $(L^X, \delta)$ be a weakly induced and p-Tychonoff $L$-fts, $\mathcal{F}$ be the family of all the $L$-fuzzy continuous mappings from $(L^X, \delta)$ to $I^*(L)$. Then by Theorem **4.3.5**,

$$(\Delta\mathcal{F})^{\rightarrow}(\underline{1}_X)^- = (\chi_{(\Delta\mathcal{F})[X]})^- = \chi_{((\Delta\mathcal{F})[X])^-} \tag{11.2}$$

is still crisp. Let $\mu_0$ and $\mu_1$ be the subspace topologies of $(\Delta\mathcal{F})[X]$ and $((\Delta\mathcal{F})[X])^-$ in $I^*(L)^{\mathcal{F}}$ respectively, then by $(\Delta\mathcal{F})^{\rightarrow}(\underline{1}_X) = \chi_{(\Delta\mathcal{F})[X]}$ and the Embedding Lemma (Theorem **9.2.62**), $(L^X, \delta)$ is homeomorphic to the subspace $(L^{(\Delta\mathcal{F})[X]}, \mu_0)$ of $I^*(L)^{\mathcal{F}}$, and hence by (11.2) and Theorem **9.2.46**, $(L^{((\Delta\mathcal{F})[X])^-}, \mu_1)$ is a space-compactification of $(L^X, \delta)$. So we have the following

**11.2.2 Definition** Let $(L^X, \delta)$ be a weakly induced and p-Tychonoff $L$-fts, $\mathcal{F}$ be the family of all the $L$-fuzzy continuous mappings from $(L^X, \delta)$ to $I^*(L)$.

Denote $\beta^* = \Delta\mathcal{F}: X \rightarrow I^{\mathcal{F}}$, let

$$\beta^* X = supp((\beta^*)^{\rightarrow}(\underline{1}_X)^-) = supp((\Delta\mathcal{F})^{\rightarrow}(\underline{1}_X)^-) = ((\Delta\mathcal{F})[X])^-,$$

and denote the subspace topology of $\beta^* X$ in $I^*(L)^{\mathcal{F}}$ by $\beta^*\delta$. Call the $L$-fts $(L^{\beta^* X}, \beta^*\delta)$ the *$*$-Stone-Čech compactification* of $(L^X, \delta)$.

**11.2.3 Theorem** *The $*$-Stone-Čech compactification of a weakly induced and p-Tychonoff is a space-compactification.* □

Since $(L^{\beta^* X}, \beta^*\delta)$ is a closed subspace of an $L$-fuzzy $*$-unit cube, so by the virtue of Theorem **9.2.52**,

**11.2.4 Theorem** *An L-fts $(L^X, \delta)$ can be embedded into an L-fuzzy $*$-unit cube if and only if $(L^X, \delta)$ is weakly induced and p-$T_{3\frac{1}{2}}$.* □

Since $I^*(L)$ is induced, the $*$-Stone-Čech compactification of an $L$-fts has strong property, it can be induced from the ordinary one:

**11.2.5 Theorem** *Let $(L^X, \delta)$ be an induced L-fts and $(X, [\delta])$ be $T_{3\frac{1}{2}}$, $(\overline{\beta}X, \overline{\beta}[\delta])$ be the Stone-Čech compactification of $(X, [\delta])$, where $\overline{\beta}: (X, [\delta]) \rightarrow I^\kappa$ is the correspondent embedding. Then*

(i) $(\overline{\beta})^{\rightarrow}: (L^X, \delta) \rightarrow I^*(L)^\kappa$ *is an embedding.*

(ii) $\overline{\beta}X = \beta^* X$.

(iii) $\omega_L(\overline{\beta}[\delta]) = \beta^*\delta$.

**Proof** (i) Since $(L^X, \delta)$ is induced, $\delta = \omega_L([\delta])$. By the definition of $I^*(L)$, $I^*(L) = (L^I, \omega_L(\Omega(I)))$. By Theorem **4.4.14**, $I^*(L)^\kappa = (L^{I^\kappa}, \omega_L(\Omega(I)^\kappa))$. Since $\overline{\beta}: (X, [\delta]) \rightarrow (I, \Omega(I))^\kappa = (I^\kappa, \Omega(I)^\kappa)$ is an embedding, so $(X, [\delta])$ is homeomorphic to the subspace $(\overline{\beta}[X], \Omega(I)^\kappa|_{\overline{\beta}[X]})$ of $(I, \Omega(I))^\kappa$. Since $\overline{\beta}$ is continuous, by Lemma **4.3.24**, $(\overline{\beta})^{\rightarrow}: (L^X, \delta) \rightarrow I^*(L)^\kappa$ is continuous. Since $\overline{\beta}$ is injective, by Theorem **2.1.22** (i), $(\overline{\beta})^{\rightarrow}$

is injective. Restricting the value domain of $(\overline{\beta})^{\rightarrow}$, we obtain a bijective $L$-fuzzy mapping

$$(\overline{\beta})^{\rightarrow}: (L^X,\delta) \to (L^{\overline{\beta}[X]}, \omega_L(\Omega(I)^{\kappa})|_{\overline{\beta}[X]}). \quad (11.3)$$

Then what still needs to be proved is that $(\overline{\beta})^{\rightarrow}$ is an $L$-fuzzy open mapping. By Theorem **4.3.17** (iii), $\delta$ has a base

$$\mathcal{B} = \{a\chi_U : a \in L, U \in [\delta]\}.$$

Since $L$-fuzzy mappings are arbitrary join preserving, it is sufficient to prove

$$(\overline{\beta})^{\rightarrow}(a\chi_U) \in \omega_L(\Omega(I)^{\kappa})|_{\overline{\beta}[X]}$$

for every $a \in L$ and every $U \in [\delta]$. But by Theorem **4.4.11**,

$$\omega_L(\Omega(I)^{\kappa})|_{\overline{\beta}[X]} = \omega_L(\Omega(I)^{\kappa}|_{\overline{\beta}[X]}),$$

so by the virtue of that

$$\overline{\beta}: (X,[\delta]) \to (\overline{\beta}[X], \Omega(I)^{\kappa}|_{\overline{\beta}[X]})$$

is a homeomorphism and hence an open mapping, we have

$$(\overline{\beta})^{\rightarrow}(a\chi_U) = (\overline{\beta})^{\rightarrow}(\underline{a}) \wedge (\overline{\beta})^{\rightarrow}(\chi_U) = a\chi_{\overline{\beta}[U]} \in \omega_L(\Omega(I)^{\kappa}|_{\overline{\beta}[X]}) = \omega_L(\Omega(I)^{\kappa})|_{\overline{\beta}[X]},$$

where $(\overline{\beta})^{\rightarrow}(a\chi_U) = (\overline{\beta})^{\rightarrow}(\underline{a}) \wedge (\overline{\beta})^{\rightarrow}(\chi_U)$ is because of that $(\overline{\beta})^{\rightarrow}$ is injective. So the $L$-fuzzy mapping (11.3) is open, and hence $(\overline{\beta})^{\rightarrow}: (L^X,\delta) \to I^*(L)^{\kappa}$ is an embedding.

(ii) Suppose $\mathcal{G}$ is the family of all the continuous mappings from $(X,[\delta])$ to $I$, $\mathcal{F}$ is the family of all the $L$-fuzzy continuous mappings from $(L^X,\delta)$ to $I^*(L)$, then by Proposition **4.2.2** and Lemma **4.3.24**, $\mathcal{F} = \{f^{\rightarrow} : f \in \mathcal{G}\}$. So

$$\overline{\beta}X = (\overline{\beta}[X])^- = ((\Delta\mathcal{G})[X])^- = ((\Delta\mathcal{F})[X])^- = \beta^*X.$$

(iii) The Stone-Čech compactification of $(X,[\delta])$ is $(\overline{\beta}X, \Omega(I)^{\kappa}|_{\overline{\beta}X})$. By (ii) and Theorem **4.4.11**,

$$(L^{\overline{\beta}X}, \omega_L(\Omega(I)^{\kappa}|_{\overline{\beta}X})) = (L^{\overline{\beta}X}, \omega_L(\Omega(I)^{\kappa})|_{\overline{\beta}X}) = (L^{\beta^*X}, \omega_L(\Omega(I)^{\kappa})|_{\beta^*X}).$$

So

$$\omega_L(\overline{\beta}[\delta]) = \omega_L(\Omega(I)^{\kappa}|_{\overline{\beta}X}) = \omega_L(\Omega(I)^{\kappa})|_{\beta^*X} = \beta^*\delta. \qquad \square$$

Opposite to the above theorem, we have

**11.2.6 Theorem** *Let $(L^X,\delta)$ be an induced L-fts, $(L^{\beta^*X},\beta^*\delta)$ be the $*$-Stone-Čech compactification of $(L^X,\delta)$. Then $(L^{\beta^*X},\beta^*\delta)$ is the L-valued induced space of the Stone-Čech compactification $(\overline{\beta}X,\overline{\beta}[\delta])$ of $(X,[\delta])$.* □

The proofs of the following conclusions are left as exercises:

**11.2.7 Theorem** *Let $(L^X,\delta)$ be a weakly induced and p-Tychonoff L-fts, $(L^Y,\mu)$ be a weakly induced, N-compact and $\alpha$-$T_2$ L-fts for some $\alpha \in M(L)$, $f^{\rightarrow}: (L^X,\delta) \to (L^Y,\mu)$ be an L-fuzzy continuous mapping. Then $f^{\rightarrow}$ can be extended to an L-fuzzy continuous mapping $F^{\rightarrow}: (L^{\beta^*X},\beta^*\delta) \to (L^Y,\mu)$, i.e. $F^{\rightarrow}\circ(\beta^*)^{\rightarrow} = f^{\rightarrow}$.* □

**11.2.8 Theorem** *Let $(L^X,\delta)$ be a weakly induced and p-Tychonoff L-fts. Then in the family of all the weakly induced and level-$T_2$ (or $T_2$, or s-$T_2$, or $\alpha$-$T_2$ for some*

*$\alpha \in M(L)$) compactifications of $(L^X, \delta)$, the $*$-Stone-Čech compatification $(L^{\beta^* X}, \beta^* \delta)$ is the unique largest compactification.* □

Let $(L^X, \delta)$ be a sub-$T_0$ and completely e-regular $L$-fts, $\mathcal{F}$ be the family of all the $L$-fuzzy continuous mappings from $(L^X, \delta)$ to $\tilde{I}(L)$. By Theorem **9.2.69**, $(L^X, \delta)$ can be embedded into an $L$-fuzzy enhanced unit cube $\tilde{I}(L)^\kappa$ via the evaluation mapping $(\Delta\mathcal{F})^\rightarrow : (L^X, \delta) \to \tilde{I}(L)^\mathcal{F}$. By Theorem **10.3.6**, $(\Delta\mathcal{F})^\rightarrow(\underline{1})^-$ is N-compact in $\tilde{I}(L)^\mathcal{F}$. By Theorem **9.1.30** and Theorem **4.3.5**, $(\Delta\mathcal{F})^\rightarrow(\underline{1})^-$ is crisp. By Theorem **9.2.69**, $(\Delta\mathcal{F})^\rightarrow(\underline{1})^-$ is sub-$T_0$ and completely e-regular. So the following definition is reasonable:

**11.2.9 Definition** Let $(L^X, \delta)$ be a sub-$T_0$ and completely e-regular $L$-fts, $\mathcal{F}$ the family of all the $L$-fuzzy continuous mappings from $(L^X, \delta)$ to $\tilde{I}(L)$.

Denote $\tilde{\beta} = \Delta\mathcal{F} : X \to I[L]^\mathcal{F}$, let

$$\tilde{\beta}X = supp((\tilde{\beta})^\rightarrow(\underline{1}_X)^-) = supp((\Delta\mathcal{F})^\rightarrow(\underline{1}_X)^-) = ((\Delta\mathcal{F})[X])^-,$$

and denote the subspace topology of $\tilde{\beta}X$ in $\tilde{I}(L)^\mathcal{F}$ by $\tilde{\beta}\delta$. Call the $L$-fts $(L^{\tilde{\beta}X}, \tilde{\beta}\delta)$ the *enhanced Stone-Čech compactification* of $(L^X, \delta)$.

**11.2.10 Theorem** *Let $(L^X, \delta)$ be a sub-$T_0$ and completely e-regular $L$-fts. Then*

(i) *$(L^{\tilde{\beta}X}, \tilde{\beta}\delta)$ is a sub-$T_0$ and completely e-regular space-compactification of $(L^X, \delta)$.*

(ii) *$(\tilde{\beta}X, [\tilde{\beta}\delta])$ is a compactification of $(X, [\delta])$.* □

# Chapter 12

# Paracompactness

Undoubtedly, local finiteness has important applications in topology, and paracompactness is the major theory in topology using this property. In fuzzy topology, Luo introduced reasonable notions of paracompactness and systematically studied paracompactness for the case $L = [0, 1]$ early, found and proved their many nice properties and relations.[110, 111, 112, 113] Then Fan and Xu extended the investigation into $L$-fuzzy topological spaces.[22, 23, 183, 184, 185] Fan also introduced a kind of flinty paracompactness to overcome some limitation of a kind of original fuzzy paracompactness defined by Luo.

## 12.1 Local Finiteness and Flinty Finiteness

**12.1.1 Definition** Let $(L^X, \delta)$ be an $L$-fts, $\mathcal{A} = \{A_t : t \in T\} \subset L^X$, $x_\lambda \in M(L^X)$. $\mathcal{A}$ is called *locally finite* at $x_\lambda$, if there exists $U \in \mathcal{Q}(x_\lambda)$ and a finite subset $T_0$ of $T$ such that

$$t \in T \backslash T_0 \implies A_t \neg \hat{q} U. \tag{12.1}$$

$\mathcal{A}$ is called *locally finite* for short, if $\mathcal{A}$ is locally finite at every molecule $x_\lambda \in M(L^X)$.

$\mathcal{A}$ is called *discrete* at $x_\lambda$, if there exists $U \in \mathcal{Q}(x_\lambda)$ and a singleton $T_0 = \{t_0\} \subset T$ such that the implication (12.1) is fulfilled. $\mathcal{A}$ is called *discrete* for short, if $\mathcal{A}$ is discrete at every molecule $x_\lambda \in M(L^X)$.

**12.1.2 Theorem** *Let $(L^X, \delta)$ be an $L$-fts, $\mathcal{A} = \{A_t : t \in T\} \subset L^X$. Then*

(i) *$\mathcal{A}$ is a locally finite family of subsets $\implies \mathcal{A}^- = \{A_t^- : t \in T\}$ is a locally finite family of subsets.*

(ii) *$\mathcal{A}$ is a discrete family of subsets $\implies \mathcal{A}^- = \{A_t^- : t \in T\}$ is a discrete family of subsets.*

(iii) *$\mathcal{A}$ is a locally finite family of subsets, $\mathcal{A}_0 \subset \mathcal{A} \implies \mathcal{A}_0$ is a locally finite family of subsets.*

(iv) *$\mathcal{A}$ is a discrete family of subsets, $\mathcal{A}_0 \subset \mathcal{A} \implies \mathcal{A}_0$ is a discrete family of subsets.* □

**12.1.3 Definition** Let $(L^X, \delta)$ be an $L$-fts, $\mathcal{A} = \{A_t : t \in T\} \subset L^X$. $\mathcal{A}$ is called *closure preserving*, if for every subfamily $\mathcal{A}_0$ of $\mathcal{A}$,

$$(\bigvee \mathcal{A}_0)^- = \bigvee \mathcal{A}_0^-,$$

i.e.

$$(\bigvee_{t \in T} A_t)^- = \bigvee_{t \in T} A_t^-.$$

**12.1.4 Proposition** *Let $(L^X, \delta)$ be an L-fts, $\mathcal{A} \subset L^X$ is closure preserving. Then for every subfamily $\mathcal{A}_0 = \{A_t : t \in T\} \subset \mathcal{A}$, $\bigvee_{t\in T} A_t^-$ is a closed subset.* □

**12.1.5 Theorem** *Every locally finite family of subsets is closure preserving; particularly, every discrete family of subsets is closure preserving.*

**Proof** Suppose $\mathcal{A} \subset L^X$ is locally finite, $\mathcal{A}_0 = \{A_t : t \in T\} \subset \mathcal{A}$, then $\mathcal{A}_0$ is locally finite. Since $\bigvee \mathcal{A}_0^- \leq (\bigvee \mathcal{A}_0)^-$ is obvious, it is sufficient to prove its reverse. Suppose $x_\lambda \in M(\downarrow(\bigvee \mathcal{A}_0)^-)$, since $\mathcal{A}_0$ is locally finite, there exists $U \in \mathcal{Q}(x_\lambda)$ and a finite subset $T_0$ of $T$ such that $A_t \neg\hat{q} U$ for every $t \in T\backslash T_0$, i.e. $A_t \leq U'$ for every $t \in T\backslash T_0$. If $x_\lambda \not\leq \bigvee \mathcal{A}_0^-$, then $x_\lambda \not\leq A_t^-$ for every $t \in T_0$, and hence there exists $U_t \in \mathcal{Q}(x_\lambda)$ such that $A_t \leq U_t'$. Since $T_0$ is finite, $V = U \wedge (\bigwedge_{t\in T_0} U_t) \in \mathcal{Q}(x_\lambda)$ and $A_t \leq V'$ for every $t \in T$. So $\bigvee_{t\in T} A_t \leq V'$, and hence

$$x_\lambda \leq (\bigvee \mathcal{A}_0)^- = (\bigvee_{t\in T} A_t)^- \leq (V')^- = V'.$$

That is to say, $x_\lambda \not\leq V$, this contradicts with $V \in \mathcal{Q}(x_\lambda)$. Thus $x_\lambda \in \bigvee \mathcal{A}_0^-$, so $(\bigvee \mathcal{A}_0)^- = \bigvee \mathcal{A}_0^-$. □

**12.1.6 Corollary** *If $\mathcal{A}$ is a locally finite family of subsets, then $\bigvee \mathcal{A}^-$ is closed.* □

**12.1.7 Corollary** *If $\mathcal{A}$ is a discrete family of subsets, then $\bigvee \mathcal{A}^-$ is closed.* □

**12.1.8 Theorem** *Let $(L^X, \delta)$, $(L^Y, \mu)$ be L-fts', $f^\to : (L^X, \delta) \to (L^Y, \mu)$ be a continuous and closed L-fuzzy mapping, $\mathcal{A} \subset L^X$ be closure preserving. Then $f^\to[\mathcal{A}] = \{f^\to(A) : A \in \mathcal{A}\}$ is closure preserving in $(L^Y, \mu)$.*

**Proof** Suppose $\mathcal{A}_0 = \{A_t : t \in T\} \subset \mathcal{A}$, we need only prove

$$(\bigvee_{t\in T} f^\to(A_t))^- \leq \bigvee_{t\in T} f^\to(A_t)^-. \tag{12.2}$$

In fact, by Theorem **2.1.15**,

$$(\bigvee_{t\in T} f^\to(A_t))^- = f^\to(\bigvee_{t\in T} A_t)^- \leq f^\to(\bigvee_{t\in T} A_t^-)^-. \tag{12.3}$$

Since $f^\to$ is an $L$-fuzzy closed mapping, and by Corollary **12.1.6**, $\bigvee_{t\in T} A_t^-$ is closed, so by the Theorem **2.4.5** and the inequality (12.3),

$$(\bigvee_{t\in T} f^\to(A_t))^- \leq f^\to(\bigvee_{t\in T} A_t^-)^- = f^\to(\bigvee_{t\in T} A_t^-) = \bigvee_{t\in T} f^\to(A_t^-) \leq \bigvee_{t\in T} f^\to(A_t)^-.$$

Hence the inequality (12.2) is true. □

Clearly, the condition "closure preserving" in the above theorem cannot be substituted with "locally finite", because it is possible that there are infinitely many molecules in $(L^X, \delta)$ are mapped to one molecule in $(L^X, \mu)$ by $f^\to$. But, however, the following reverse result holds:

**12.1.9 Theorem** *Let $(L^X, \delta)$, $(L^Y, \mu)$ be L-fts', $f^\to : (L^X, \delta) \to (L^Y, \mu)$ be an L-fuzzy continuous mapping, $\mathcal{B} \subset L^Y$ be locally finite. Then $f^\leftarrow[\mathcal{B}] = \{f^\leftarrow(B) : B \in \mathcal{B}\}$ is locally finite in $(L^X, \delta)$.*

**Proof** Suppose $x_\lambda \in M(L^X)$, then $f^\to(x_\lambda) \in M(L^Y)$. Since $\mathcal{B}$ is locally finite, there exists $V \in \mathcal{Q}(f^\to(x_\lambda))$ and $\{B_i : i \leq n\} \subset \mathcal{B}$ such that

$$B \in \mathcal{B}\backslash\{B_i : i \leq n\} \Longrightarrow B \leq V'.$$

Then by the continuity of $f^{\rightarrow}$, $f^{\leftarrow}(V) \in \mathcal{Q}(x_\lambda)$, and

$$B \in \mathcal{B}\backslash\{B_i : i \leq n\} \Longrightarrow f^{\leftarrow} \leq (f^{\leftarrow}(V))',$$

i.e. $B \neg\hat{q} f^{\leftarrow}(V)$ for every $\mathcal{B}\backslash\{B_i : i \leq n\}$, $f^{\leftarrow}[\mathcal{B}]$ is locally finite. □

The previous notions "locally finite family" and "discrete family" are defined for $L$-fts'. They can be also defined for $L$-fuzzy subsets:

**12.1.10 Definition** Let $(L^X, \delta)$ be an $L$-fts, $A \in L^X$, $\mathcal{A} = \{A_t : t \in T\} \subset L^X$. $\mathcal{A}$ is called *locally finite in* $A$, if $\mathcal{A}$ is locally finite at every molecule $x_\lambda \in M(\downarrow A)$. $\mathcal{A}$ is called *discrete in* $A$, if $\mathcal{A}$ is discrete at every molecule $x_\lambda \in M(\downarrow A)$.

It is easy to find that the above definition coincides with Definition **12.1.1** provided $A = \underline{1}$.

The proofs of following conclusions are left as exercises:

**12.1.11 Proposition** *If $\mathcal{A}$ is locally finite in $A$ and $B \leq A$, then $\mathcal{A}$ is locally finite in $B$.* □

**12.1.12 Proposition** *If $\mathcal{A}$ is discrete in $A$, $B \leq A$, then $\mathcal{A}$ is discrete in $B$.* □

**12.1.13 Proposition** *If $\mathcal{A}$ is locally finite in $A$, then $\mathcal{A}^-$ is locally finite in $A$.* □

**12.1.14 Proposition** *If $\mathcal{A}$ is discrete in $A$, then $\mathcal{A}^-$ is discrete in $A$.* □

**12.1.15 Theorem** *If $\mathcal{A}$ is locally finite in $A$, then*

$$(\bigvee_{t\in T} A_t)^- \wedge A = \bigvee_{t\in T} (A_t^- \wedge A).$$ □

Luo[113] introduce the following stronger local finiteness:

**12.1.16 Definition** Let $(L^X, \delta)$ be an $L$-fts, $A \in L^X$, $\mathcal{A} = \{A_t : t \in T\} \subset L^X$, $x_\lambda \in M(L^X)$.

$\mathcal{A}$ is called $*$-*locally finite* at $x_\lambda$, if there exists $U \in \mathcal{Q}(x_\lambda)$ and a finite subset $T_0$ of $T$ such that

$$t \in T\backslash T_0 \Longrightarrow \chi_{(A_t)_{(0)}} \neg\hat{q} U. \tag{12.4}$$

$\mathcal{A}$ is called $*$-*locally finite* in $A$, if $\mathcal{A}$ is $*$-locally finite for every molecule $x_\lambda \in M(\downarrow A)$. $\mathcal{A}$ is called $*$-*locally finite* for short, if $\mathcal{A}$ is $*$-locally finite in $\underline{1}$.

$\mathcal{A}$ is called $*$-*discrete* at $x_\lambda$, if there exists $U \in \mathcal{Q}(x_\lambda)$ and a singleton $T_0 = \{t_0\} \subset T$ such that the implication (12.4) is fulfilled. $\mathcal{A}$ is called $*$-*discrete* in $A$, if $\mathcal{A}$ is $*$-discrete for every molecule $x_\lambda \in M(\downarrow A)$. $\mathcal{A}$ is called $*$-*discrete* for short, if $\mathcal{A}$ is $*$-discrete in $\underline{1}$.

**12.1.17 Remark** (1) In Luo's paper [113], $*$-local finiteness was defined for $L = [0, 1]$ as

*$\mathcal{A} = \{A_t : t \in T\}$ is called $*$-locally finite in $A$, if for every $x_\lambda \in M(\downarrow A)$ there exists $U \in \mathcal{Q}(x_\lambda)$ and a finite subset $T_0$ of $T$ such that*

$$t \in T\backslash T_0 \Longrightarrow A_t \wedge U = \underline{0}. \tag{12.5}$$

In the case of $L = [0, 1]$, this definition is equivalent to Definition **12.1.16**. But for a F-lattice $L$, in general, condition (12.4) is strictly stronger than (12.5).

(2) If $\mathcal{A}$ is a fmaily of $L$-fuzzy subsets which is $*$-locally finite in an $L$-fuzzy subset $A$, then $\mathcal{A}^-$ is not necessarily $*$-locally finite in $A$.[113, 168]

The symbol $crs(\delta)$ appearing in the following definition means the family of all the crisp subsets in $\delta$ (see Definition **2.1.1**).

**12.1.18 Definition** Let $(L^X, \delta)$ be an $L$-fts, $A \in L^X$, $\mathcal{A} = \{A_t : t \in T\} \subset L^X$, $x_\lambda \in M(L^X)$.

$\mathcal{A}$ is called *flintily locally finite* at $x_\lambda$, if there exists crisp $U \in \mathcal{Q}(x_\lambda) \cap crs(\delta)$ and a finite subset $T_0$ of $T$ such that the condition (12.1) is fulfilled. $\mathcal{A}$ is called *flintily locally finite* in $A$, if $\mathcal{A}$ is flintily locally finite at every molecule $x_\lambda \in M(\downarrow A)$. $\mathcal{A}$ is called *flintily locally finite* for short, if $\mathcal{A}$ is flintily locally finite in $\underline{1}$.

$\mathcal{A}$ is called *flintily discrete* at $x_\lambda$, if there exists $U \in \mathcal{Q}(x_\lambda) \cap crs(\delta)$ and a singleton $T_0 = \{t_0\} \subset T$ such that the implication (12.1) is fulfilled. $\mathcal{A}$ is called *flintily discrete* in $A$, if $\mathcal{A}$ is flintily discrete at every molecule $x_\lambda \in M(\downarrow A)$. $\mathcal{A}$ is called *flintily discrete* for short, if $\mathcal{A}$ is flintily discrete in $\underline{1}$.

**12.1.19 Remark** In Fan's original papers [22] and [23], flinty local finiteness is called "strong local finiteness", and the "flinty paracompactness" will be defined in the sequel is called "strong paracompactness" there. To avoid possible confusion with ordinary strong paracompactness, we replace "strong" with "flinty".

**12.1.20 Theorem** *In L-fts' the following implications hold:*

*Flinty local finiteness $\Longrightarrow$ $*$-local finiteness $\Longrightarrow$ local finiteness.* □

**12.1.21 Remark** Each of the implications in Theorem **12.1.20** is strict.

**12.1.22 Proposition** *Let $(L^X, \delta)$ be an L-fts, $\{A_t : t \in T\} \subset L^X$, $x_\lambda \in M(L^X)$. Then*

(i) *$\{A_t : t \in T\}$ is $*$-locally finite at $x_\lambda \Longrightarrow \{\chi_{(A_t)_{(0)}} : t \in T\}$ is $*$-locally finite at $x_\lambda$.*

(ii) *$\{A_t : t \in T\}$ is flintily locally finite at $x_\lambda \Longrightarrow \{\chi_{(A_t)_{(0)}} : t \in T\}$ is flintily locally finite at $x_\lambda$.* □

**12.1.23 Theorem** *If $\mathcal{A}$ is flintily locally finite in $A$, then $\mathcal{A}^-$ is flintily locally finite in $A$.* □

**12.1.24 Definition** Let $(L^X, \delta)$ be an $L$-fts, $a \in L$. $(L^X, \delta)$ is called *weakly a-induced,* if $U_{(a)} \in [\delta]$ for every $U \in \delta$.

**12.1.25 Proposition** *Let $(L^X, \delta)$ be an L-fts. Then the following conditions are equivalent:*

(i) *$(L^X, \delta)$ is weakly induced.*

(ii) *$(L^X, \delta)$ is weakly $\gamma$-induced for every $\gamma \in pr(L)$.*

(iii) *$(L^X, \delta)$ is weakly a-induced for every $a \in L$.* □

It is mentioned previously that flinty local finiteness is strictly stronger than $*$-local finiteness. But in weakly 0-induced $L$-fts', they are coincident with each other:

**12.1.26 Theorem** *Let $(L^X, \delta)$ be an weakly 0-induced L-fts, $A \in L^X$, $\mathcal{A} = \{A_t :$*

*$t \in T\} \subset L^X$. Then $\mathcal{A}$ is flintily locally finite in $A$ if and only if $\mathcal{A}$ is $*$-locally finite in $A$.*

**Proof** By Theorem **12.1.20**, it is sufficient to prove that $*$-local finiteness implies flinty local finiteness. Suppose $\mathcal{A}$ is $*$-locally finite in $A$, $x_\lambda) \in M(\downarrow A)$. Then there exists $U \in \mathcal{Q}(x_\lambda)$ and a finite subset $T_0$ of $T$ such that condition (12.4) is fulfilled. Since $(L^X, \delta)$ is weakly 0-induced, $U_{(0)} \in [\delta]$. Let $t \in T \backslash T_0$, $y \in (A_t)_{(0)}$, then by (12.4),

$$U'(y) \geq \chi_{(A_t)_{(0)}}(y) = 1.$$

So $y \in (U')_{[1]} = X \backslash U_{(0)}$ and hence

$$(\chi_{U_{(0)}})'(y) = 1 = \chi_{(A_t)_{(0)}}(y).$$

That is to say, $\chi_{(A_t)_{(0)}} \leq (\chi_{U_{(0)}})'$, $\chi_{(A_t)_{(0)}} \neg \hat{q} \chi_{U_{(0)}}$. Since $\chi_{U_{(0)}} \in \delta$, $\chi_{U_{(0)}} \in \mathcal{Q}(x_\lambda) \cap crs(\delta)$. Hence $\mathcal{A}$ is flintily locally finite. □

## 12.2 Paracompactness and Flinty Paracompactness

Paracompactness is a generalization of compactness. In fuzzy topology, there are various kinds of compactness just as investigated in Chapter 10. Based on local finiteness or flinty local finiteness defined in last section, one can define various kinds of paracompactness for fuzzy compactness, strong compactness, N-compactness and ultra-compactness. In the sequel, only N-compactness will be considered in the investigation of paracompactness.

**12.2.1 Definition** Let $(L^X, \delta)$ be an $L$-fts, $\mathcal{A}, \mathcal{B} \subset L^X$. $\mathcal{A}$ is called a *refinement* of $\mathcal{B}$, or say $A$ *refines* $B$, if for every $A \in \mathcal{A}$, there exists $B \in \mathcal{B}$ such that $A \leq B$.

**12.2.2 Definition** Let $(L^X, \delta)$ be an $L$-fts, $A \in L^X$, $\alpha \in M(L)$.

$A$ is called *$\alpha$-paracompact*, if for every open $\alpha$-Q-cover $\Phi$ of $A$ there exists an open refinement $\Psi$ of $\Phi$ such that $\Psi$ is locally finite in $A$ and $\Psi$ is an $\alpha$-Q-cover of $A$. $A$ is called *paracompact*, if $A$ is $\alpha$-paracompact for every $\alpha \in M(L)$. $(L^X, \delta)$ is called *paracompact*, if $\underline{1}$ is paracompact.

$A$ is called *$\alpha^*$-paracompact*, if for every open $\alpha$-Q-cover $\Phi$ of $A$ there exists an open refinement $\Psi$ of $\Phi$ such that $\Psi$ is $*$-locally finite in $A$ and $\Psi$ is an $\alpha$-Q-cover of $A$. $A$ is called *$*$-paracompact*, if $A$ is $\alpha$-paracompact for every $\alpha \in M(L)$. $(L^X, \delta)$ is called *$*$-paracompact*, if $\underline{1}$ is paracompact.

**12.2.3 Proposition** *Let $(L^X, \delta)$ be an $L$-fts, $A \in L^X$, $\alpha \in M(L)$. Then*

(i) *$A$ is $\alpha^*$-paracompact $\Longrightarrow$ $A$ is $\alpha$-paracompact.*

(ii) *$A$ is $*$-paracompact $\Longrightarrow$ $A$ is paracompact.* □

**12.2.4 Theorem** *Every strongly compact $L$-fuzzy subsets is $*$-paracompact.*

**Proof** By Theorem **10.4.11**. □

But for fuzzy compactness in $L$-fts', even in F-ts', the similar implication does not hold:

**12.2.5 Example** There exists a F-ts which is not 1-paracompact.

Let $X = [0, +\infty)$, and for every $x \in X$, let

$$U_x(y) = \begin{cases} e^{-y}, & 0 \le y < x, \\ 0, & y \ge x, \end{cases}$$

$$U_\infty(y) = e^{-y}, \quad y \in X,$$

$$\delta = \{\underline{0}, U_\infty, \underline{1}\} \cup \{U_x : x \in X\},$$

then it can be easily verified that $(X, \delta)$ is a fuzzy compact F-ts. Take an open 1-Q-cover $\mathcal{U} = \{U_x : x \in X\}$ of $\underline{1}$. Let $\mathcal{V}$ be an open refinement of $\mathcal{U}$ and a 1-Q-cover of $\underline{1}$, then by the virtue of the structure of $\delta$, $\mathcal{V} \subset \mathcal{U}$. Furthermore, we have $\bigvee\{x : U_x \in \mathcal{V}\} = +\infty$, so $|\mathcal{V}| \ge \omega$. Hence $\mathcal{V}$ is not locally finite at fuzzy point $0_1$, $(X, \delta)$ is not 1-paracompact.

**12.2.6 Theorem** *Every $T_2$ and fuzzy compact L-fts is $*$-paracompact.*

**Proof** By Theorem **10.4.28.** □

Paracompactness and $*$-paracompactness are hereditary with respect to closed subsets:

**12.2.7 Theorem** *Let $(L^X, \delta)$ be an L-fts, $\alpha \in M(L)$, $A \in L^X$, $B \in \delta'$. Then*

(i) *$A$ is $\alpha$-paracompact $\Longrightarrow$ $A \wedge B$ is $\alpha$-paracompact.*

(ii) *$A$ is paracompact $\Longrightarrow$ $A \wedge B$ is paracompact.*

**Proof** Clearly, we need only prove (i). Suppose $\mathcal{U}$ is an open $\alpha$-Q-cover of $A \wedge B$, let $\mathcal{V} = \mathcal{U} \cup \{B'\}$, $x_\alpha \le A$. If $x_\alpha \le B$, $x_\alpha \in M(\downarrow(A \wedge B))$, there exists $U \in \mathcal{U} \subset \mathcal{V}$ such that $x_\alpha \lessdot U$. If $x_\alpha \not\le B$, then $x_\alpha \lessdot B' \in \mathcal{V}$. So $\mathcal{V}$ is an open $\alpha$-Q-cover of $A$. By the $\alpha$-paracompactness of $A$, $\mathcal{V}$ has an open refinement $\mathcal{W}$ such that $\mathcal{W}$ is locally finite in $A$ and is an $\alpha$-Q-cover of $A$. Take

$$\mathcal{W}_0 = \{W \in \mathcal{W} : \exists U \in \mathcal{U},\ W \le U\},$$

then $\mathcal{W}_0$ is locally finite in $A \wedge B$. Let $x_\alpha \le A \wedge B \le A$, then there exists $W \in \mathcal{W}$ such that $x_\alpha \lessdot W$. Since $x_\alpha \le B$, $B \not\le W'$, i.e. $W \not\le B'$. Since $\mathcal{W}$ is a refinement of $\mathcal{V} = \mathcal{U} \cup \{B'\}$, there exists $U \in \mathcal{U}$ such that $W \le U$. So $x_\alpha \lessdot W \in \mathcal{W}_0$, $\mathcal{W}_0$ is an $\alpha$-Q-cover of $A \wedge B$. Hence $A \wedge B$ is $\alpha$-paracompact. □

Similarly we have

**12.2.8 Theorem** *Let $(L^X, \delta)$ be an L-fts, $\alpha \in M(L)$, $A \in L^X$, $B \in \delta'$. Then*

(i) *$A$ is $\alpha^*$-paracompact $\Longrightarrow$ $A \wedge B$ is $\alpha^*$-paracompact.*

(ii) *$A$ is $*$-paracompact $\Longrightarrow$ $A \wedge B$ is $*$-paracompact.* □

**12.2.9 Corollary** *$\alpha$-paracompactness, $\alpha^*$-paracompactness, paracompactness and $*$-paracompactness for L-fuzzy subsets are hereditary with respect to closed subsets.* □

For $\alpha^*$-paracompactness and $*$-paracompactness in L-fts', we have

**12.2.10 Theorem** *Let $(L^X, \delta)$ be a weakly induced L-fts. Then the following conditions are equivalent:*

(i) *$(L^X, \delta)$ is $*$-paracompact.*

(ii) *There exists* $\alpha \in M(L)$ *such that* $(L^X, \delta)$ *is* $\alpha^*$*-paracompact.*

(iii) $(X, [\delta])$ *is paracompact.*

**Proof** (i)$\Longrightarrow$(ii): Obvious.

(ii)$\Longrightarrow$(iii): Let $\mathcal{U} \subset [\delta]$ be an open cover of $X$, then $\{\chi_U : U \in \mathcal{U}\}$ is an open $\alpha$-Q-cover of $\underline{1}$, it has an open refinement $\mathcal{V}$ such that $\mathcal{V}$ is locally finite and an $\alpha$-Q-cover of $\underline{1}$. Let $\mathcal{W} = \{V_{(\alpha')} : V \in \mathcal{V}\}$, then $\mathcal{W}$ is both a refinement of $\mathcal{U}$ and a cover of $X$. Since $(L^X, \delta)$ is weakly induced, $\mathcal{W} \subset [\delta]$. Now we need only prove $\mathcal{W}$ is locally finite. Suppose $x \in X$, then by the $\alpha^*$-paracompactness of $(L^X, \delta)$, there exists $O \in \mathcal{Q}(x_\alpha)$ such that $\chi_{O_{(0)}}$ only quasi-coincides with a finite number of members $V_0$, $\cdots$, $V_n$ of $\mathcal{V}$. Then $x \in O_{(0)}$. By the weakly induced property of $(L^X, \delta)$, $O_{(0)} \in [\delta]$, so $O_{(0)} \in \mathcal{N}_{[\delta]}(x)$. For every $V \in \mathcal{V}$, if $O_{(0)} \cap V_{(\alpha')} \neq \emptyset$, then there exists an ordinary point $y \in O_{(0)} \cap V_{(\alpha')}$, $V(y) \not\leq \alpha'$, $V(y) > 0$, $V(y)' < 1$. So $\chi_{O_{(0)}}(y) = 1 \not\leq V(y)'$, $\chi_{O_{(0)}} \hat{q} V$, $V \in \{V_0, \cdots, V_n\}$. Therefore, the neighborhood $O_{(0)}$ of $x$ intersects at most a finite number of members $(V_0)_{(\alpha')}$, $\cdots$, $(V_n)_{(\alpha')}$ of $\mathcal{W}$, $\mathcal{W}$ is locally finite in $X$.

(iii)$\Longrightarrow$(i): Suppose $\alpha \in M(L)$, $\mathcal{U} \subset \delta$ is an open $\alpha$-Q-cover of $\underline{1}$. Since $(L^X, \delta)$ is weakly induced, $\{U_{(\alpha')} : U \in \mathcal{U}\}$ is an open cover of $(X, [\delta])$. Since $(X, [\delta])$ is paracompact, there exists a locally finite and open refinement $\mathcal{V} \subset [\delta]$ of $\{U_{(\alpha')} : U \in \mathcal{U}\}$ which is a cover of $X$. For every $V \in \mathcal{V}$ take $U_V \in \mathcal{U}$ such that $V \subset (U_V)_{(\alpha')}$ and let $\mathcal{W} = \{\chi_V \wedge U_V : V \in \mathcal{V}\}$. Then $\mathcal{W} \subset \delta$ is both a refinement of $\mathcal{U}$ and an open $\alpha$-Q-cover of $\underline{1}$. For every $x_\lambda \in M(L^X)$, take $O \in \mathcal{N}_{[\delta]}(x)$ such that $O$ intersects with only a finite number of members $V_0$, $\cdots$, $V_n$ of $\mathcal{V}$. Then $\chi_O \in \mathcal{Q}_\delta(x_\lambda)$, $\chi_{(\chi_O)_{(0)}} = \chi_O$, we have

$$\begin{aligned}
\chi_{(\chi_O)_{(0)}} \hat{q} (\chi_V \wedge U_V) \in \mathcal{W} &\Longrightarrow \chi_O \hat{q} (\chi_V \wedge U_V) \\
&\Longrightarrow \exists y \in X,\ \chi_O(y) \not\leq (\chi_V \wedge U_V)(y)' = \chi_V(y)' \vee U_V(y)' \\
&\Longrightarrow y \in O,\ y \in V \\
&\Longrightarrow O \cap V \neq \emptyset \\
&\Longrightarrow V \in \{V_0, \cdots, V_n\} \\
&\Longrightarrow \chi_V \wedge U_V \in \{\chi_{V_0} \wedge U_{V_0}, \cdots, \chi_{V_n} \wedge U_{V_n}\}.
\end{aligned}$$

So $\mathcal{W}$ is $*$-locally finite, $(L^X, \delta)$ is $*$-paracompact. □

**12.2.11 Corollary** *$\alpha^*$-paracompactness and $*$-paracompactness for L-fts are good L-extensions of ordinary paracompactness.* □

Similar conclusions for $\alpha$-paracompactness and paracompactness hold in F-ts':

**12.2.12 Theorem** *Let $(X, \delta)$ be a weakly induced F-ts. Then the following conditions are equivalent:*

(i) $(X, \delta)$ *is paracompact.*

(ii) *There exists* $\alpha \in (0, 1)$ *such that* $(X, \delta)$ *is* $\alpha$*-paracompact.*

(iii) $(X, [\delta])$ *is paracompact.*

**Proof** (i)$\Longrightarrow$(ii): Obvious.

(ii)$\Longrightarrow$(iii): Let $\mathcal{U} \subset [\delta]$ be an open cover of $X$, then $\{\chi_U : U \in \mathcal{U}\}$ is an open $\alpha$-Q-cover of $\underline{1}$, it has an open refinement $\mathcal{V}$ such that $\mathcal{V}$ is locally finite and an $\alpha$-Q-cover of $\underline{1}$. Let $\mathcal{W} = \{V_{(\alpha')} : V \in \mathcal{V}\}$, then $\mathcal{W}$ is both a refinement of $\mathcal{U}$ and a cover of $X$. Since $(X,\delta)$ is weakly induced, $\mathcal{W} \subset [\delta]$. Now we need only prove that $\mathcal{W}$ is locally finite. Suppose $x \in X$, then since $\mathcal{V}$ is locally finite and $\alpha \in (0,1)$, $\alpha' > 0$, $x_{\alpha'} \in M(I^X)$, there exists $O_1 \in \mathcal{Q}(x_{\alpha'})$ such that $O_1$ only quasi-coincides with a finite number of members $V_0, \cdots, V_n$ of $\mathcal{V}$. Denote $O = (O_1)_{(\alpha)}$, by the weakly induced property of $(X,\delta)$, $O \in [\delta]$. For every $V \in \mathcal{V}$, if $O \cap V_{(\alpha')} \neq \emptyset$, there exists an ordinary point $y \in O \cap V_{(\alpha')}$, and hence $O_1(y) \not\leq \alpha$, $V(y) \not\leq \alpha'$. Since $\alpha \in (0,1]$, $O_1(y) > \alpha$, $V(y) > \alpha'$, $V(y)' < \alpha < O_1(y)$, $O_1(y) \not\leq V(y)'$, $O_1\hat{q}V$. So $V \in \{V_0, \cdots, V_n\}$, $O \in \mathcal{N}_{[\delta]}(x)$ intersects with only a finite number of members $(V_0)_{(\alpha')}, \cdots, (V_n)_{(\alpha')}$ of $\mathcal{W}$.

(iii)$\Longrightarrow$(i): By Theorem **12.2.10** (iii)$\Longrightarrow$(i) and Proposition **12.2.3** (ii). □

**12.2.13 Corollary** *$\alpha$-paracompactness and paracompactness for F-ts' are good extensions of ordinary paracompactness.* □

**12.2.14 Corollary** *Let $(X,\delta)$ be a weakly induced F-ts. Then the following conditions are equivalent:*

(i) *$(X,\delta)$ is paracompact.*
(ii) *There exists $\alpha \in (0,1)$ such that $(X,\delta)$ is $\alpha$-paracompact.*
(iii) *$(X,\delta)$ is $*$-paracompact.*
(iv) *There exists $\alpha \in (0,1]$ such that $(X,\delta)$ is $\alpha^*$-paracompact.*
(v) *$(X,[\delta])$ is paracompact.* □

The condition "$\alpha \in (0,1)$" in the above theorem and corollary cannot be substituted by "$\alpha \in (0,1]$", even if the F-ts is induced:

**12.2.15 Example** Let $X = [0,1)$ and

$$\mathcal{P} = \{\underline{\alpha} : \alpha \in (0,1]\} \cup \{\chi_{[0,x)} : x \in X\},$$

suppose $\delta$ is the fuzzy topology on $X$ generated by $\mathcal{P}$, then $(X,\delta)$ is an induced F-ts.

Let $\mathcal{U} \subset \delta$ be a 1-Q-cover of $\underline{1}$. Not loss any of generality, we can suppose that $U \neq \underline{0}$ and $\{U\}$ is not a 1-Q-cover of $\underline{1}$ for every $U \in \mathcal{U}$. Then by the structure of $\mathcal{P}$, for every $U \in \mathcal{U}$ there exists $x_U \in X$ such that

$$U|_{[0,x_U)} > 0, \quad U[[x_U,1)] = \{0\}.$$

Since $\mathcal{U}$ is a 1-Q-cover of $\underline{1}$, $\bigvee\{x_U : U \in \mathcal{U}\} = 1$. Take a countable family $\{U_n : n < \omega\} \subset \mathcal{U}$ such that $\bigvee\{x_{U_n} : n < \omega\} = 1$ and let $V_n = (\frac{1}{2}X) \wedge U_n$ for every $n < \omega$, then $\mathcal{V} = \{V_n : n < \omega\}$ is both an open refinement of $\mathcal{U}$ and a 1-Q-cover of $\underline{1}$. For every $x \in X$ and every $\lambda \in (0,1]$, take $k < \omega$ such that $\frac{1}{k} < \lambda$, then $(1 - \frac{1}{k})X \in \mathcal{Q}(x_\lambda)$ and is not quasi-coincident with $V_n$ for every $n \geq k$. Hence $\mathcal{V}$ is locally finite, $(X,\delta)$ is 1-paracompact.

But it is easy to show that

$$[\delta] = \{\emptyset, X\} \cup \{[0,x) : x \in X\},$$

so $\mathcal{W} = \{[0,x) : x \in X\}$ is an open cover of $(X,[\delta])$. Clearly, $\mathcal{W}$ has not any locally finite open refinement which is a cover of $X$, so $(X,[\delta])$ is not paracompact.

**12.2.16 Remark** By Theorem **12.2.12**, the F-ts defined in Example **12.2.15** is 1-paracompact but for every $\alpha \in (0,1)$, it is not $\alpha$-paracompact.

**12.2.17 Lemma** *For every weakly induced F-ts $(X,\delta)$, if there exists $\alpha \in (0,1)$ such that every open $\alpha$-Q-cover of $\underline{1}$ has a locally finite and closed refinement which is an $\alpha$-Q-cover of $\underline{1}$, then $(X,\delta)$ is $\alpha$-paracompact.*

**Proof** Let $\mathcal{U}$ be an open $\alpha$-Q-cover of $\underline{1}$. Take a locally finite refinement $\mathcal{A} = \{A_t : t \in T\}$ of $\mathcal{U}$ such that $\mathcal{A}$ is a closed $\alpha$-Q-cover of $\underline{1}$. Denote $\gamma = min\{\alpha, 1-\alpha\}$, then $\gamma \in (0,1)$. For every $x \in X$, take $U_x \in \mathcal{Q}(x_\gamma)$ such that $U_x$ is quasi-coincident with only a finite number of members of $\mathcal{A}$ and let $\tilde{U}_x = \chi_{(U_x)_{(1-\gamma)}} \wedge U_x$, then $\mathcal{U}_1 = \{\tilde{U}_x : x \in X\}$ is an open $\alpha$-Q-cover of $\underline{1}$ and has a locally finite and closed refinement $\mathcal{F}$ which is also an $\alpha$-Q-cover of $\underline{1}$. For every $t \in T$ let

$$W_t = \bigwedge\{F' : F \in \mathcal{F},\ F' \geq \chi_{(A_t)_{(1-\alpha)}}\},$$

then by Corollary **12.1.6**, $W_t \in \delta'$, $W_t \in \delta$ and for every $F \in \mathcal{F}$ we have

$$W_t \hat{q} F \iff \chi_{(A_t)_{(1-\alpha)}} \hat{q} F. \tag{12.6}$$

Take $U_t \in \mathcal{U}$ for every $t \in T$ such that $A_t \leq U_t$ and let

$$V_t = \chi_{(W_t)_{(1-\gamma)}} \wedge U_t.$$

Then by the weakly induce property of $(X,\delta)$, $\mathcal{V} = \{V_t : t \in T\}$ is an open refinement of $\mathcal{U}$. For every $x \in X$, take $t \in T$ such that $A_t(x) > 1-\alpha$. Since $W_t \geq \chi_{(A_t)_{(1-\alpha)}}$, $W_t(x) = 1 > 1-\gamma$. So

$$V_t(x) = U_t(x) \geq A_t(x) > 1-\alpha.$$

Hence $\mathcal{V}$ is also an $\alpha$-Q-cover of $\underline{1}$.

Finally, we prove that $\mathcal{V}$ is locally finite. For every $x \in X$ and every $\lambda \in (0,1]$, let $\rho = min\{\lambda, \gamma\}$, take $U \in \mathcal{Q}(x_\rho)$ such that $U$ is quasi-coincident with only a finite number of members $F_0, \cdots, F_n$ of $\mathcal{F}$. Since $\mathcal{F}$ is a refinement of $\mathcal{U}_1$, for every $i \leq n$, we can take $x^i \in X$ such that $F_i \leq \tilde{U}_{x^i}$. Then by the definition of $\mathcal{U}_1$, $T_0 = \{t \in T : A_t \hat{q}(\bigvee_{i \leq n} \tilde{U}_{x^i})\}$ is finite. We say for every $t \in T$ and every $i \leq n$,

$$\chi_{(A_t)_{(1-\alpha)}} \hat{q} F_i \implies t \in T_0. \tag{12.7}$$

In fact, if $\chi_{(A_t)_{(1-\alpha)}} \hat{q} F_i$, then there exists $y \in X$ such that $A_t(y) > 1-\alpha$, $\tilde{U}_{x^i}(y) \geq F_i(y) > 0$, $\tilde{U}_{x^i}(y) > 1-\gamma$. So

$$A_t(y) > 1-\alpha \geq \gamma > \tilde{U}_{x^i}(y)',$$

we have $A_t \not\leq (\tilde{U}_{x^i})'$, $A_t \hat{q} \tilde{U}_{x^i}$, $A_t \hat{q}(\bigvee_{i \leq n} \tilde{U}_{x^i})$, $t \in T_0$, the implication (12.7) is true. Take $V = \chi_{U_{(1-\rho)}} \wedge U$, then $V \in \mathcal{Q}(x_\lambda)$. If $V \hat{q} V_t$, then there exists $y \in X$ such that $V(y) > 1-\rho \geq 1-\gamma$, $V_t(y) > 0$. Since $\mathcal{F}$ is an $\alpha$-Q-cover of $\underline{1}$, there exists $F \in \mathcal{F}$ such that

$$F(y) > 1-\alpha \geq \gamma \geq V(y)'.$$

So $F \not\leq V'$, $V \hat{q} F$, $F \in \{F_0, \cdots, Fn\}$. On the other hand, since $V_t(y) >$, $W_t(y) > 1-\gamma \geq \alpha > F(y)'$, $W_t \hat{q} F$. By relation (12.6), $\chi_{(A_t)_{(1-\alpha)}} \hat{q} F$; by relation (12.7), $t \in T_0$. So $V$ quasi-coincides with only the members of the finite subfamily $\{V_t : t \in T_0\}$ of

$\mathcal{V}$, $\mathcal{V}$ is locally finite. □

**12.2.18 Definition** Let $(L^X, \delta)$ be an $L$-fts, $\mathcal{A} \subset L^X$, $B \in L^X$.

$\mathcal{A}$ is called *$\sigma$-locally finite* in $B$, if $\mathcal{A}$ is a countable union of subfamilies which are locally finite in $B$. $\mathcal{A}$ is called *$\sigma$-locally finite* for short if $\mathcal{A}$ is $\sigma$-locally finite in $\underline{1}$.

$\mathcal{A}$ is called *$\sigma^*$-locally finite* in $B$, if $\mathcal{A}$ is a countable union of subfamilies which are $*$-locally finite in $B$. $\mathcal{A}$ is called *$\sigma^*$-locally finite* for short if $\mathcal{A}$ is $\sigma^*$-locally finite in $\underline{1}$.

$\mathcal{A}$ is called *$\sigma$-discrete* in $B$, if $\mathcal{A}$ is a countable union of subfamilies which are discrete in $B$. $\mathcal{A}$ is called *$\sigma$-discrete* for short if $\mathcal{A}$ is $\sigma$-discrete in $\underline{1}$.

$\mathcal{A}$ is called *$\sigma^*$-discrete* in $B$, if $\mathcal{A}$ is a countable union of subfamilies which are $*$-discrete in $B$. $\mathcal{A}$ is called *$\sigma^*$-discrete* for short if $\mathcal{A}$ is $\sigma^*$-discrete in $\underline{1}$.

**12.2.19 Lemma** *Let $(L^X, \delta)$ be a weakly $\alpha'$-induced L-fts, where $\alpha \in M(L)$. Then every $\sigma$-locally finite and open $\alpha$-Q-cover of $\underline{1}$ has a locally finite refinement which is an $\alpha$-Q-cover of $\underline{1}$.*

**Proof** Let $\mathcal{V} = \bigcup_{i<\omega} \mathcal{V}_i$ be an open $\alpha$-Q-cover of $\underline{1}$, where every $\mathcal{V}_i = \{V_t : t \in T_i\}$ is locally finite in $\underline{1}$ and $T_i \cap T_j = \emptyset$ whenever $i \neq j$. Let $T = \bigcup_{i<\omega} T_i$. For every $i < \omega$ and every $t \in T_i$, let

$$A_t = V_t \wedge \Big(\bigvee_{k<i} \bigvee_{s\in T_k} \chi_{(V_s)_{(\alpha')}}\Big)', \quad \mathcal{A} = \{A_t : t \in T\},$$

then $\mathcal{A}$ is a refinement of $\mathcal{V}$.

For every $x \in X$, let $i_0 = min\{i < \omega : \bigvee \mathcal{V}_i \in \mathcal{Q}(x_\alpha)\}$ and take $t_0 \in T_{i_0}$ such that $V_{t_0} \in \mathcal{Q}(x_\alpha)$. Then for every $k < i_0$ and every $t \in T_k$, $x_\alpha \not\prec V_t$, $V_t(x) \leq \alpha'$, and $A_{t_0}(x) = V_{t_0}(x) \not\leq \alpha'$. So $\mathcal{A}$ is an $\alpha$-Q-cover of $\underline{1}$. For every $\lambda \in M(L)$ and every $i < i_0$, take $U_i \in \mathcal{Q}(x_\lambda)$ such that $U_i$ quasi-coincides with only a finite number of members of $\mathcal{V}_i$. Let

$$U = U_1 \wedge \cdots \wedge U_{i_0} \wedge \chi_{(V_{i_0})_{(\alpha')}},$$

then $U \in \mathcal{Q}(x_\lambda)$ and for every $i < \omega$, $i > i_0$, and every $t \in T_i$, we have

$$A_t{}' = V_t{}' \vee \bigvee_{k<i} \bigvee_{s\in T_k} \chi_{(V_t)_{(\alpha')}} \geq \chi_{(V_{t_0})_{(\alpha')}} \geq U,$$

$U$ and $A_t$ are not quasi-coincident with each other for every $i > i_0$. Since for every $i < i_0$, $U_i \leq U$ and $A_t \leq V_t$ for every $t \in T_i$, so $U$ quasi-coincides with only a finite number of members of $\{A_t : t \in T_i\}$. Hence $U$ quasi-coincides with only a finite number of members of $\mathcal{A}$, $\mathcal{A}$ is locally finite. □

**12.2.20 Remark** There is an example showing that the condition "$(X, \delta)$ is weakly $\alpha'$-induced" in the above lemma cannot be removed.[113]

**12.2.21 Lemma** *Let $(L^X, \delta)$ be a regular L-fts, $A, B \in L^X$. If every open Q-cover of $A$ has a refinement which is locally finite in $B$ and is a Q-cover of $A$, then for every open Q-cover $\{U_t : t \in T\}$ of $A$ such that $(\bigvee_{t\in T} U_t)^- \subset B$, there exists a closed Q-cover $\{F_t : t \in T\}$ of $A$ which is locally finite in $B$ such that $F_t \leq U_t$ for every $t \in T$.*

**Proof** By Theorem **8.3.6**, $A$ has an open Q-cover $\mathcal{W}$ such that $\mathcal{W}^- = \{W^- : W \in$

$\mathcal{W}\}$ is a refinement of $\{U_t : t \in T\}$. Take a Q-cover $\{A_s : s \in S\}$ of $A$ such that $\{A_s : s \in S\}$ is both a refinement of $\mathcal{W}$ and locally finite in $B$, then there exists a mapping $f : S \to T$ such that $A_s^- \leq U_{f(s)}$. Let $F_t = \bigvee\{A_s^- : f(s) = t\}$, then by Theorem **12.1.15**, $F_t$ is a closed subset. Suppose $x_\lambda \in M(\downarrow B)$, then as a result of $\{A_t : t \in T\}$ being locally finite in $B$ and Proposition **12.1.13**, there exists $U \in \mathcal{Q}(x_\lambda)$ and a finite subset $S_0 \subset S$ such that $A_s^- \leq U'$ for every $s \in S \backslash S_0$. Suppose $f[S_0] = T_0$, then

$$f(s) \in T \backslash T_0 \implies s \in S \backslash S_0.$$

So for every $t \in T \backslash T_0$ and every $s \in S$ such that $f(s) = t$, we have $s \in S \backslash S_0$, and hence $A_s^- \leq U'$. This implies $F_t = \bigvee\{A_s^- : f(s) = t\} \leq U'$, $F_t \neg\hat{q} U$. Therefore, the closed Q-cover $\{F_t : t \in T\}$ of $A$ is also locally finite in $B$. Moreover, for every $t \in T$ we have $F_t \leq U_t$. □

**12.2.22 Lemma** *Let $(X, \delta)$ be a weakly induced, regular and paracompact F-ts, $\alpha \in (0,1]$. Then every open $\alpha$-Q-cover of $\underline{1}$ has a $\sigma$-discrete refinement which is an open $\alpha$-Q-cover of $\underline{1}$.*

**Proof** Suppose $\Phi$ is an open $\alpha$-Q-cover of $\underline{1}$, then by the virtue of Theorem **4.2.17**, $\mathcal{U} = \{U_{(\alpha')} : U \in \Phi\}$ is an open cover of $(X, [\delta])$. As a result of Theorem **12.2.12**, $(X, [\delta])$ is paracompact. By Theorem **8.3.10**, $(X, [\delta])$ is regular. Therefore, $(L^X, \delta)$ is paracompact in the sense of [50]. By Theorem 28 in Chapter 5 of [50], $\mathcal{U}$ has an $\sigma$-discrete refinement $\mathcal{V} = \bigcup_{n<\omega} \mathcal{V}_n$ such that $\mathcal{V}$ is an open cover of $(X, [\delta])$, where every $\mathcal{V}_n$ is a discrete family of open subsets in $(X, [\delta])$. For every $V \in \mathcal{V}$, take $U_V \in \Phi$ such that $V \subset (U_V)_{(\alpha')}$, let $W_V = U_V \wedge \chi_V$, $\Psi_n = \{W_V : V \in \mathcal{V}_n\}$, $\Psi = \bigcup_{n<\omega} \Psi_n$.

For every $x \in X$, since $\mathcal{V}$ is a cover of $X$, there exists $V \in \mathcal{V}$ such that $x \in V$. Then $x \in (U_V)_{(\alpha')}$, $U_V(x) \not\leq \alpha'$, $U_V \in \mathcal{Q}(x_\alpha)$, $W_V = U_V \wedge \chi_V \in \mathcal{Q}(x_\alpha)$, $\Psi$ is an open $\alpha$-Q-cover of $\underline{1}$. For every $W_V \in \Psi$, $W_V \leq U_V \in \Phi$, so $\Psi$ is a refinement of $\Phi$. Arbitrarily fix a number $n < \omega$ and a F-point $x_\lambda$ in $(X, \delta)$, since $\mathcal{V}_n$ is discrete in $(X, [\delta])$, there exists a neighborhood $O$ of $x$ in $(X, [\delta])$ such that $O$ intersects with at most one member $V_0$ of $\mathcal{V}_n$. Take $U = \chi_O$, then for every $V \in \mathcal{V}_n \backslash \{V_0\}$, $O \subset V'$. So by Theorem **4.3.5**,

$$U = \chi_O \leq \chi_{V'} = (\chi_V)' \leq (W_V)'.$$

It means $U \neg\hat{q} W_V$ for every $V \in \mathcal{V}_n \backslash \{V_0\}$, $\Psi_n$ is discrete in $(X, \delta)$. Therefore, $\Psi$ is $\sigma$-discrete in $(X, \delta)$. □

**12.2.23 Theorem** *Let $(X, \delta)$ be a weakly induced and regular F-ts. Then the following conditions are equivalent:*

(i) *$(X, \delta)$ is paracompact.*

(ii) *For every $\alpha \in (0,1]$, every open $\alpha$-Q-cover of $\underline{1}$ has a $\sigma$-locally finite and open refinement which is an $\alpha$-Q-cover of $\underline{1}$.*

(iii) *There exists $\alpha \in (0,1)$ such that every open $\alpha$-Q-cover of $\underline{1}$ has a $\sigma$-locally finite and open refinement which is an $\alpha$-Q-cover of $\underline{1}$.*

(iv) *For every $\alpha \in (0,1]$, every open $\alpha$-Q-cover of $\underline{1}$ has a $\sigma$-discrete and open* *refinement which is an $\alpha$-$Q$-cover of $\underline{1}$.*

(v) *There exists $\alpha \in (0,1)$ such that every open $\alpha$-$Q$-cover of $\underline{1}$ has a $\sigma$-discrete and open refinement which is an $\alpha$-$Q$-cover of $\underline{1}$.*

(vi) *For every $\alpha \in (0,1]$, every open $\alpha$-$Q$-cover of $\underline{1}$ has a locally finite refinement which is an $\alpha$-$Q$-cover of $\underline{1}$.*

(vii) *There exists $\alpha \in (0,1)$ such that every open $\alpha$-$Q$-cover of $\underline{1}$ has a locally finite refinement which is an $\alpha$-$Q$-cover of $\underline{1}$.*

(viii) *For every $\alpha \in (0,1]$, every open $\alpha$-$Q$-cover of $\underline{1}$ has a locally finite and closed refinement which is an $\alpha$-$Q$-cover of $\underline{1}$.*

(ix) *There exists $\alpha \in (0,1)$ such that every open $\alpha$-$Q$-cover of $\underline{1}$ has a locally finite and closed refinement which is an $\alpha$-$Q$-cover of $\underline{1}$.*

**Proof** (i)$\Longrightarrow$(ii)$\Longrightarrow$(iii): Obvious.

(iii)$\Longrightarrow$(vii): By Lemma **12.2.19**.

(vii)$\Longrightarrow$(ix): By Lemma **12.2.21**.

(ix)$\Longrightarrow$(i): By Lemma **12.2.17** and Theorem **12.2.12**.

(i)$\Longrightarrow$(vi): Obvious.

(vi)$\Longrightarrow$(viii): By Lemma **12.2.21**.

(viii)$\Longrightarrow$(i): Clearly, we have (viii)$\Longrightarrow$(ix). On the other hand, it has been proved that (ix)$\Longrightarrow$(i), so (viii)$\Longrightarrow$(i).

(i)$\Longrightarrow$(iv): By Lemma **12.2.22**.

(iv)$\Longrightarrow$(v): Obvious.

(v)$\Longrightarrow$(i): Since (v)$\Longrightarrow$(iii), so by (iii)$\Longrightarrow$(i) proved above, we have (v)$\Longrightarrow$(i).□

**12.2.24 Remark** In [168], Wang investigated some results related to paracompactness in F-ts' and $L$-fts' in Luo's sense[113] and Fan's sense[23] respectively (including many results parallel to the conclusions in the sequel related to regularity and paracompactness or flinty paracompactness). But the notion "regularity" used in [168] (called *inclusive regularity* there) is different from the previous one used in this book, defined as

*An L-fts $(L^X, \delta)$ is called regular, if for every L-fuzzy open subset $U \in \delta$ and every L-fuzzy point $x_a \leq U$, there exists L-fuzzy open subset $V \in \delta$ such that $x_a \leq V \leq V^- \leq U$.*

Regularity in this sense is not equivalent to the one defined in this book.

Similar to the above theorem, for $\alpha^*$-paracompactness and $*$-paracompactness the following conclusions hold (proofs are left as exercises):

**12.2.25 Theorem** *Let $(X, \delta)$ be a weakly induced and regular F-ts. Then the following conditions are equivalent:*

(i) *$(X, \delta)$ is paracompact.*

(ii) *For every $\alpha \in (0,1]$, every open $\alpha$-$Q$-cover of $\underline{1}$ has a $\sigma^*$-locally finite and open refinement which is an $\alpha$-$Q$-cover of $\underline{1}$.*

(iii) *There exists $\alpha \in (0,1)$ such that every open $\alpha$-$Q$-cover of $\underline{1}$ has a $\sigma^*$-locally finite and open refinement which is an $\alpha$-$Q$-cover of $\underline{1}$.*

(iv) *For every $\alpha \in (0,1]$, every open $\alpha^*$-Q-cover of $\underline{1}$ has a $\sigma^*$-discrete and open refinement which is an $\alpha$-Q-cover of $\underline{1}$.*

(v) *There exists $\alpha \in (0,1)$ such that every open $\alpha$-Q-cover of $\underline{1}$ has a $\sigma^*$-discrete and open refinement which is an $\alpha$-Q-cover of $\underline{1}$.*

(vi) *For every $\alpha \in (0,1]$, every open $\alpha$-Q-cover of $\underline{1}$ has a $*$-locally finite refinement which is an $\alpha$-Q-cover of $\underline{1}$.*

(vii) *There exists $\alpha \in (0,1)$ such that every open $\alpha$-Q-cover of $\underline{1}$ has a $*$-locally finite refinement which is an $\alpha$-Q-cover of $\underline{1}$.*

(viii) *For every $\alpha \in (0,1]$, every open $\alpha$-Q-cover of $\underline{1}$ has a $*$-locally finite and closed refinement which is an $\alpha$-Q-cover of $\underline{1}$.*

(ix) *There exists $\alpha \in (0,1)$ such that every open $\alpha$-Q-cover of $\underline{1}$ has a $*$-locally finite and closed refinement which is an $\alpha$-Q-cover of $\underline{1}$.* □

**12.2.26 Remark** In [113], a F-ts $(X,\delta)$ is constructed such that $(X,\delta)$ is regular and satisfies condition (vi) of Theorem **12.2.25**; but for every $\alpha \in (0,1]$, it is not $\alpha$-paracompact.

**12.2.27 Theorem** *Let $(L^X,\delta)$ and $(L^Y,\mu)$ be L-fts', $A \in L^X$ is strongly compact in $(L^X,\delta)$, $B \in L^Y$ is paracompact in $(L^Y,\mu)$. Then $A \times B$ is paracompact in product L-fts $(L^X,\delta) \times (L^Y,\mu)$.*

**Proof** Let $\alpha \in M(L)$, $\Phi$ be an open $\alpha$-Q-cover of $A \times B$.

(i) If $A(x) \not\geq \alpha$ for every $x \in X$, then for every $y \in Y$,

$$(A \times B)((x,y)) = A(x) \wedge B(y) \not\geq \alpha,$$

$A \times B$ does not contain any molecule with height $\alpha$. Then arbitrarily take a finite subfamily $\Psi$ of $\Phi$, $\Psi$ is an open $\alpha$-Q-cover of $A \times B$ and a refinement of $\Phi$. Since $\Psi$ is finite, it is obviously locally finite. Similarly, if $B(y) \not\geq \alpha$ for every $y \in Y$, the same conclusion will be obtained.

(ii) Suppose there exists $x \in X$ such that $A(x) \geq \alpha$ and there exists $y \in Y$ such that $B(y) \geq \alpha$. Let $x_\alpha \leq A$ and $y_\alpha \in B$, then $(A \times B)((x,y)) \geq \alpha$, $(x,y)_\alpha \in M(\downarrow(A \times B))$. Since $\Phi$ is an open $\alpha$-Q-cover of $A \times B$, there exists $H(x,y) \in \Phi$ such that $H(x,y) \in \mathcal{Q}_{\delta\times\mu}((x,y)_\alpha)$. By the definition of product topology of $\delta$ and $\mu$, there exist $U_{x,y} \in \delta$ and $V_{x,y} \in \mu$ such that

$$p_1^{\leftarrow}(U_{x,y}) \wedge p_2^{\leftarrow}(V_{x,y}) \leq H(x,y), \quad p_1^{\leftarrow}(U_{x,y}) \wedge p_2^{\leftarrow}(V_{x,y}) \in \mathcal{Q}_{\delta\times\mu}((x,y)_\alpha).$$

Then $U_{x,y} \in \mathcal{Q}_\delta(x_\alpha)$, $V_{x,y} \in \mathcal{Q}_\mu(y_\alpha)$. For every $y \in X$ such that $y_\alpha \leq B$ let $W_y = \{U_{x,y} : x_\alpha \leq A\}$, then $W_y$ is an open $\alpha$-Q-cover of $A$. Since $A$ is strongly compact in $(L^X,\delta)$, by Theorem **10.4.11**, $W_y$ has a finite sub-$\alpha$-Q-cover $W_y^0 = \{U_{x(y)_i,y} : i \leq n(y)\}$. Let $V_y = \bigwedge_{i\leq n(y)} V_{x(y)_i,y}$, then $V_y \in \mathcal{Q}_\mu(y_\alpha)$. Let $\Delta = \{V_y : y_\alpha \leq B\}$, then $\Delta$ is an open $\alpha$-Q-cover of $B$. Since $B$ is paracompact, there exists a family $\Sigma \subset \mu$ such that

(a) $\Sigma$ is a refinement of $\Delta$.

(b) $\Sigma$ is an open $\alpha$-Q-cover of $B$.

(c) $\Sigma$ is locally finite in $B$.

For every $\sigma \in \Sigma$, since $\Sigma$ is a refinement of $\Delta$, there exists $y(\sigma) \in Y$ such that $\sigma \leq V_{y(\sigma)}$. Let

$$\Gamma = \{p_1{}^{\leftarrow}(U_{x(y(\sigma))_i,y(\sigma)}) \wedge p_2{}^{\leftarrow}(\sigma) : \sigma \in \Sigma,\ i \leq n(y(\sigma))\}.$$

then

(a′) $\Gamma$ is an open $\alpha$-Q-cover of $A \times B$.

In fact, suppose $(A \times B)((x,y)) \geq \alpha$, then $x_\alpha \leq A$, $y_\alpha \leq B$. Since $\Sigma$ is an open $\alpha$-Q-cover of $B$, there $\sigma \in \Sigma$ such that $\sigma \in \mathcal{Q}_\mu(y_\alpha)$. Since $W^0_{y(\sigma)}$ is an open $\alpha$-Q-cover of $A$, there exists $i \leq n(y(\sigma))$ such that $U_{x(y(\sigma))_i,y(\sigma)} \in \mathcal{Q}_\delta(x_\alpha)$. Let $O = p_1{}^{\leftarrow}(U_{x(y(\sigma))_i,y(\sigma)}) \wedge p_2{}^{\leftarrow}(\sigma)$, then $O \in \Gamma$, $O \in \mathcal{Q}_{\delta\times\mu}((x,y)_\alpha)$, $\Gamma$ is an open $\alpha$-Q-cover of $A \times B$.

(b′) $\Gamma$ is a refinement of $\Phi$.

This is because of that every element of $\Gamma$ fulfils

$$\begin{aligned} p_1{}^{\leftarrow}(U_{x(y(\sigma))_i,y(\sigma)}) \wedge p_2{}^{\leftarrow}(\sigma) &\leq p_1{}^{\leftarrow}(U_{x(y(\sigma))_i,y(\sigma)}) \wedge p_2{}^{\leftarrow}(V_{y(\sigma)}) \\ &\leq p_1{}^{\leftarrow}(U_{x(y(\sigma))_i,y(\sigma)}) \wedge p_2{}^{\leftarrow}(V_{x(y(\sigma))_i,y(\sigma)}) \\ &\leq H(x(y(\sigma)_i, y(\sigma)) \in \Phi. \end{aligned}$$

(c′) $\Gamma$ is locally finite in $A \times B$.

In fact, suppose $(x,y)_\lambda \in M(\Downarrow(A \times B))$, then $y_\lambda$ is a molecule in $B$. Since $\Sigma$ is locally finite in $B$, there exists $V \in \mathcal{Q}_\mu(y_\lambda)$ such that there exists $\{\sigma_0, \cdots, \sigma_k\} \subset \Sigma$ fulfilling

$$\sigma \in \Sigma \backslash \{\sigma_0, \cdots, \sigma_k\} \implies \sigma \neg\hat{q} V \implies \sigma \leq V'.$$

Let $O = p_2{}^{\leftarrow}(V)$, then $O \in \mathcal{Q}_{\delta\times\mu}((x,y)_\lambda)$, and

$$\begin{aligned} &\sigma \in \Sigma \backslash \{\sigma_0, \cdots, \sigma_k\} \implies \\ &\quad p_1{}^{\leftarrow}(U_{x(y(\sigma))_i,y(\sigma)}) \wedge p_2{}^{\leftarrow}(\sigma) \leq p_2{}^{\leftarrow}(\sigma) \leq p_2{}^{\leftarrow}(V') = (p_2{}^{\leftarrow}(V))' = O' \\ &\quad \implies p_1{}^{\leftarrow}(U_{x(y(\sigma))_i,y(\sigma)}) \wedge p_2{}^{\leftarrow}(\sigma) \neg\hat{q} O. \end{aligned}$$

So $\Gamma$ is locally finite in $A \times B$.

Synthesizing the above conclusions, we have completed the proof of the paracompactness of $A \times B$. □

**12.2.28 Corollary** *Every product of N-compact L-fuzzy subset and paracompact L-fuzzy subset is paracompact.* □

**12.2.29 Theorem** *Let $(X, \delta)$ and $(Y, \mu)$ be weakly induced F-ts', $f^{\rightarrow} : (X, \delta) \rightarrow (Y, \mu)$ be a fuzzy continuous, closed and surjective mapping. If $(X, \delta)$ is paracompact and $T_2$, then $(Y, \mu)$ is paracompact.*

**Proof** Since $(X, \delta)$ is weakly induced, by Theorem **12.2.12**, $(X, [\delta])$ is paracompact; by Proposition **8.2.5**, $(X, [\delta])$ is $T_2$. So $(X, [\delta])$ is paracompact in the sense of [20]. By Theorem **2.1.22** and Proposition **4.2.2**, $f : (X, [\delta]) \rightarrow (Y, [\mu])$ is surjective and continuous. Similarly we can find that $f$ is also closed. So by the Michael Theorem (see Theorem 5.1.33 in [20]), $(Y, [\mu])$ is paracompact. Applying Theorem **12.2.12** once more, we know that $(Y, \mu)$ is paracompact. □

In original papers, paracompactness and $*$-compactness were considered in fuzzy topological spaces, i.e. for the case of value domain $I$. The previous results show that paracompactness is a reasonable and nice notion in F-ts'. But it encountered some problems when the related investigations were extended to the case of general F-lattices. So another kind of paracompactness were introduced into $L$-fuzzy topological spaces by Fan[22, 23] to overcome these limitations.

**12.2.30 Definition** Let $(L^X, \delta)$ be an $L$-fts, $A \in L^X$, $\alpha \in M(L)$.

$A$ is called *flintily $\alpha$-paracompact,* if for every open $\alpha$-Q-cover $\Phi$ of $A$ there exists an open refinement $\Psi$ of $\Phi$ such that $\Psi$ is flintily locally finite in $A$ and $\Psi$ is an $\alpha$-Q-cover of $A$. $A$ is called *flintily paracompact,* if $A$ is flintily $\alpha$-paracompact for every $\alpha \in M(L)$. $(L^X, \delta)$ is called *flintily paracompact* for short, if $\underline{1}$ is flintily paracompact.

By Theorem **12.1.20**, the following implications hold:

**12.2.31 Proposition** *Let $(L^X, \delta)$ be an L-fts, $A \in L^X$, $\alpha \in M(L)$. Then*

(i) *$A$ is flintily $\alpha$-paracompact $\Longrightarrow$ $A$ is $\alpha^*$-paracompact $\Longrightarrow$ $A$ is $\alpha$-paracompact.*

(ii) *$A$ is flintily paracompact $\Longrightarrow$ $A$ is $*$-paracompact $\Longrightarrow$ $A$ is paracompact.* □

Finite family is certainly flintily locally finite, so we have

**12.2.32 Theorem** *Every stronly compact L-fuzzy subset is flintily paracompact.* □

Fan[23] constructed an example to show the difference between flinty paracompactness and paracompactness:

**12.2.33 Example** Let $L = [0, 1]$, $X = (-\infty, +\infty)$, $\mathcal{A} = \{A_k : k \in \mathbf{Z}\}$, where $A_k \in L^X$ is defined for every $k \in \mathbf{Z}$ as follows:

$$A_k(x) = \begin{cases} 0, & x \le k-1, \\ x-(k-1), & k-1 < x \le k, \\ 1, & k < x \le k+1, \\ 0, & x > k+1. \end{cases}$$

Let $\delta$ denote the F-topology on $X$ generated by the subbase $\mathcal{A}$, then $(X, \delta)$ is a F-ts and

(i) $(X, \delta)$ is paracompact.

Suppose $\Phi$ is an open $\alpha$-Q-cover of $\underline{1}$. If $\alpha < 1$, let $\Psi = \mathcal{A}$, then $\Psi$ is a refinement of $\Phi$. In fact, for every $A_k \in \Psi$, suppose $x = k+1-\alpha$, then since $\Phi$ is an open $\alpha$-Q-cover of $\underline{1}$, there exists $U \in \Phi$ such that $U \in \mathcal{Q}(x_\alpha)$. Since $\mathcal{A}$ is a subbase of $\delta$, there exist $\{k_0, \cdots, k_m\} \subset \mathbf{Z}$ such that

$$U \ge A_{k_0} \wedge \cdots \wedge A_{k_m} \in \mathcal{Q}(x_\alpha).$$

Then $A_{k_i}(x) > \alpha' = 1 - \alpha$ for every $i \in \{0, \cdots, m\}$. Since for every $i \in \{0, \cdots, m\}$ we have

$$A_{k_i}(x) = A_{k_i}(k+1-\alpha) = \begin{cases} 0, & k_i < k, \\ 1, & k_i = k, \\ 1-\alpha, & k_i = k+1, \\ 0, & k_i > k+1, \end{cases}$$

so $k_i = k$ for every $i \in \{0, \cdots, m\}$, and hence $U \geq A_k \in \Psi$. Moreover, by the structure of $\Psi = \mathcal{A}$, it is easy to find that $\Psi$ is a locally finite open $\alpha$-Q-cover of $\underline{1}$.

If $\alpha = 1$. Suppose $\Phi$ is an open 1-Q-cover of $\underline{1}$. Take $\Psi = \{A_{k-1} \wedge A_k : k \in \mathbf{Z}\}$, then $\Psi$ is a refinement of $\Phi$. In fact, suppose $A_{k-1} \wedge A_k \in \Psi$, take $U \in \Phi$ such that $U \in \mathcal{Q}(x_1)$, where $x = k$. Since $\mathcal{A}$ is a subbase of $\delta$, there exists $\{k_0, \cdots, k_m\} \subset \mathbf{Z}$ such that

$$U \geq A_{k_0} \wedge \cdots \wedge A_{k_m} \in \mathcal{Q}(x_1).$$

Then $A_{k_i}(k) = A_{k_i}(x) > 0$ for every $i \in \{0, \cdots, m\}$. But this is possible only if $k_i = k - 1$ or $k_i = k$, so $U \geq A_{k-1} \wedge A_k \in \Psi$. Easy to find that $\Psi$ is a locally finite open $\alpha$-Q-cover of $\underline{1}$. So $(X, \delta)$ is paracompact.

(ii) $(X, \delta)$ is not flintily 1-paracompact.

In fact, let $\Phi = \{A_{k-1} \wedge A_k : k \in \mathbf{Z}\}$, then $\Phi$ is an open 1-Q-cover of $\underline{1}$. Suppose $\Psi$ is a refinement of $\Phi$ and an open 1-Q-cover of $\underline{1}$, then easy to verify that $\Psi = \Phi$ or $\Psi = \Phi \cup \{\underline{0}\}$. For every molecule $x_\lambda$ in $(X, \delta)$, there exists only one crisp Q-neighborhood $\underline{1}$ of $x_\lambda$. But for this crisp Q-neighborhood, $\Psi$ has infinitely many elements quasi-coincide with it, so $\Psi$ is not flintily locally finite in $\underline{1}$, $(X, \delta)$ is not flintily 1-paracompact.

**12.2.34 Definition** An $L$-fts $(L^X, \delta)$ is called *extremally disconnected*, if $U^- \in \delta$ for every $U \in \delta$.

Wang[168] offered another example as follows:

**12.2.35 Example** Let $L = [0, 1]$, $X = \mathbf{N}$ be the set of natural numbers. Let

$$\delta = \{A \in L^X : \forall n \in \mathbf{N},\ A(n) = 0 \implies A(n+1) = 0\}.$$

Then $\delta$ is a F-topology on $X$. $(X, \delta)$ possesses the following properties:

(i) $(X, \delta)$ is stratified. This is obvious.

(ii) $(X, \delta)$ is paracompact. In fact, suppose $\Phi$ is an open $\alpha$-Q-cover of $\underline{1}$, $\alpha \in (0, 1]$. For every $k \in \mathbf{N}$, take $U_k \in \Phi$ such that $U_k(k) > 1 - \alpha$. Let $\Phi_0 = \{U_k : k \in X\}$ and

$$W_k = \tfrac{1}{k+1} X \vee \chi_{\uparrow k}, \quad V_k = U_k \wedge W_k \wedge \tfrac{(1-\alpha)k+1}{k+1} X$$

for every $k \in \mathbf{N}$, and let $\Psi = \{V_k : k \in \mathbf{N}\}$, then $\Psi$ is a refinement of $\Phi_0$ and hence a refinement of $\Phi$. Easy to verify that $\Psi$ is an open $\alpha$-Q-cover of $\underline{1}$, we need only prove that $\Psi$ is locally finite. Suppose $x_\lambda$ is a molecule in $(X, \delta)$, take $\gamma \in (1 - \alpha, 1)$, take $U \in L^X$ as follows:

$$U(y) = \begin{cases} 1 - \frac{\lambda}{2}, & y = x, \\ 1 - \gamma, & y \neq x, \end{cases}$$

then $U \in \mathcal{Q}(x_\lambda)$. Take $m \in \mathbf{N}$ such that $m > x$ and $\frac{1}{m} < min\{\gamma - (1 - \alpha), \frac{\lambda}{2}\}$, then for every $k \in \mathbf{N}$, $k \geq m$,

$$V_k \leq \tfrac{(1-\alpha)k+1}{k+1} X \leq (1 - \alpha + \tfrac{1}{k}) X \leq \gamma X,$$

and by $k \geq m > x$,

$$V_k(x) \leq W_k(x) = \tfrac{1}{k+1} < \tfrac{1}{m} < \tfrac{\lambda}{2}.$$

So $V_k \leq U'$ for every $k \geq m$, $\Psi$ is locally finite.

(iii) $(X, \delta)$ is not flintily 1-paracompact. In fact, let $\Phi = \{A_k : k \in X\}$, where $A_k \in L^X$ for every $k \in X$ is defined as

$$A_k(x) = \begin{cases} 1, & x < k, \\ 0, & x \geq k, \end{cases}$$

then $\Phi$ is an open 1-Q-cover of $\underline{1}$. Suppose $\Psi$ is an open 1-Q-cover of $\underline{1}$ and a refinement of $\Phi$, it is sufficient to show that $\Psi$ is not flintily locally finite. For every $k \in X$, since $\Psi$ is an open 1-Q-cover of $\underline{1}$, there exists $V_k \in \Psi$ such that $V_k(k) > 0$. By the structure of $\delta \supset \Psi$, $V_k(x) > 0$ for every $x \leq k$. On the other hand, since $\Psi$ is a refinement of $\Phi$, there exists $A_m \in \Phi$ such that $V_k \leq A_m$. So $V_k(x) > 0$ for only a finite number of points $x$'s in $X$. Then there exist infinitely many $V_k$'s in $\Psi$ such that each one is different from others and $V_k(k) > 0$. Suppose these $V_k$'s are $V_{k_0}$, $V_{k_1}$, $\cdots$, then for every $i < \omega$ and every $x \leq k_i$, $V_{k_i}(x) > 0$. If $\Psi$ is flintily locally finite, then every F-point $x_\lambda$ has a crisp Q-neighborhood $U$ such that $U'$ contains infinitely many $V_{k_i}$'s. So $U' = \underline{1}$, $U = \underline{0}$. But this contradicts with $U \in \mathcal{Q}(x_\lambda)$, hence $(X, \delta)$ is not flintily locally finite.

(iv) $(X, \delta)$ is normal. In fact, suppose $A \in \delta'$, $U \in \delta$, $A \leq U$. Let

$$V(x) = \begin{cases} A(x), & A(x) > 0, \\ U(x), & A(x) = 0,\ U(x) < 1, \\ \frac{1}{2}, & A(x) = 0,\ U(x) = 1, \end{cases}$$

then $A \leq V \leq U$. To complete the verification of normality we need only prove that $V$ is both open and closed. Suppose $V(n) = 0$, then by the definition of $V$, $U(n) = 0$. Since $U$ is open, $U(n+1) = 0$. Then $A(n+1) = 0$, hence $V(n+1) = U(n+1) = 0$. So $V \in \delta$. Moreover, suppose $V(n) = 1$, then $A(n) = 1$. Since $A \in \delta'$, we have $A(n+1) = 1$, so $V(n+1) = A(n+1) = 1$. Hence $V \in \delta'$, $V$ is both open and closed.

(v) $(X, \delta)$ is not regular. In fact, arbitrarily fix $x \in X$, then $x_1$ is a molecule in $(X, \delta)$. Take $U \in L^X$ as follows:

$$U(y) = \begin{cases} 1, & y \leq x, \\ 0, & y > x, \end{cases}$$

then $U \in \delta$ and $U \geq x_1$. If $A$ is a closed subset such that $A \geq x_1$, then $A(y) = 1$ for every $y \geq x$. So $A \not\leq U$. By Theorem **8.3.7**, $(X, \delta)$ is not regular.

(vi) $(X, \delta)$ is both connected and extremally disconnected. Easy to show that $(X, \delta)$ is connected. Suppose $U \in \delta$. If $U(x) < 1$ for every $x \in X$, then $U^- = U \in \delta$. If there exists $x \in X$ such that $U(x) = 1$, denote $k = min\{x \in X : U(x) = 1\}$, then $U(y) \neq 0$ for every $y < k$. By the structure of $\delta$,

$$U^-(y) = \begin{cases} 1, & y \geq k, \\ U(y), & y < k, \end{cases}$$

so $U^-(y) \neq 0$ for every $y \in X$, $U^- \in \delta$.

Flinty paracompactness is hereditary with respect to closed subsets:

**12.2.36 Theorem** *Let $(L^X, \delta)$ be an L-fts, $\alpha \in M(L)$, $A \in L^X$, $B \in \delta'$. Then*

(i) *$A$ is flintily $\alpha$-paracompact $\Longrightarrow$ $A \wedge B$ is flintily $\alpha$-paracompact.*

(ii) *$A$ is flintily paracompact $\Longrightarrow$ $A \wedge B$ is flintily paracompact.* □

**12.2.37 Theorem** *Let $(L^X, \delta)$ be a weakly induced L-fts. Then the following conditions are equivalent:*

(i) *$(L^X, \delta)$ is flintily paracompact.*

(ii) *There exists $\alpha \in M(L)$ such that $(L^X, \delta)$ is flintily $\alpha$-paracompact.*

(iii) *$(X, [\delta])$ is paracompact.*

**Proof** (i)$\Longrightarrow$(ii): Obvious.

(ii)$\Longrightarrow$(iii): Suppose $\mathcal{U}$ is an open cover of $(X, [\delta])$, let $\Phi = \{\chi_U : U \in \mathcal{U}\}$. By (ii), $\Phi$ has an open and flintily locally finite refinement $\Psi = \{W_s : s \in S\}$ such that $\Psi$ is an $\alpha$-Q-cover of $\underline{1}$. For every $s \in S$, let $V_s = (W_s)_{(\alpha')}$, $\mathcal{V} = \{V_s : s \in S\}$, then by the weakly induced property of $(L^X, \delta)$, $\mathcal{V}$ is an open cover of $(X, [\delta])$.

Now prove $\mathcal{V}$ is locally finite refinement of $\mathcal{U}$. Suppose $V_s \in \mathcal{V}$. Since $\Psi$ is a refinement of $\Phi$, there exists $U \in \mathcal{U}$ such that $W_s \leq \chi_U$. Suppose $x \in V_s$, then $W_s(x) \not\leq \alpha'$, so $\chi_U(x) \neq 0$, $x \in U$, $V_s \subset U$. So $\mathcal{V}$ is a refinement of $\mathcal{U}$.

Let $x \in X$. Since $\Psi$ is flintily locally finite, $x_\alpha$ has a crisp Q-neighborhood $O$ such that $O\hat{q}W_s$ for only a finite number of $W_s$'s in $\Psi$. Since $O \in \mathcal{Q}_\delta(x_\alpha)$ is crisp, $O_{(0)} \in \mathcal{N}_{[\delta]}(x)$. For every $s \in S$, if $W_s \neg\hat{q} O$, then $V_s \cap O_{(0)}$. So $U_{(0)}$ intersects with only a finite number of members of $\mathcal{V}$, $\mathcal{V}$ is locally finite. Hence $(X, [\delta])$ is paracompact.

(iii)$\Longrightarrow$(i): Suppose $\alpha \in M(L)$, $\Phi = \{A_t : t \in T\} \subset \delta$ is an open $\alpha$-Q-cover of $\underline{1}$. For every $t \in T$, let $U_t = (A_t)_{(\alpha')}$, $\mathcal{U} = \{U_t : t \in T\}$, then as a result of $\Phi$ being an open $\alpha$-Q-cover of $\underline{1}$ and $(L^X, \delta)$ being weakly induced, $\mathcal{U}$ is an open cover of $(X, [\delta])$. By the paracompactness of $(X, [\delta])$, there exists an open and locally finite refinement $\mathcal{V} = \{V_s : s \in S\}$ of $\mathcal{U}$ which is still a cover of $(X, [\delta])$. For every $s \in S$, take $t(s) \in T$ such that $V_s \subset U_{t(s)}$, let $B_s = A_{t(s)} \wedge \chi_{V_s}$, then $B_s$ is an open subset in $(L^X, \delta)$ and $B_s \leq A_{t(s)}$ for every $s \in S$. So $\Psi = \{B_s : s \in S\}$ is an open refinement of $\Phi$.

Suppose $x_\alpha \in M(L^X)$, take $s \in S$ such that $x \in V_s$, and hence $x \in U_{t(s)}$. So $A_{t(s)}(x) \not\leq \alpha'$, $\alpha \not\leq A_{t(x)}(x)'$, $A_{t(s)} \in \mathcal{Q}(x_\alpha)$. Since $x \in V_s$, $\chi_{V_s} \in \mathcal{Q}(x_\alpha)$, we have $B_s = A_{t(s)} \wedge \chi_{V_s} \in \mathcal{Q}(x_\alpha)$. Therefore, $\Psi$ is an open $\alpha$-Q-cover of $\underline{1}$.

Suppose $x_\lambda \in M(L^X)$, then by the virtue of $\mathcal{V}$ being locally finite in $(X, [\delta])$, there exists a neighborhood $W$ of $x$ in $(X, [\delta])$ such that $W$ intersects with only finitely many members of $\mathcal{V}$. Suppose these members are $V_{s_0}, \cdots, V_{s_k}$, then for every $s \in S \backslash \{s_0, \cdots, s_k\}$, $V_s \cap W = \emptyset$, $W \subset V_s'$, and hence

$$\chi_W \leq \chi_{V_s'} \leq A_{t(s)}' \vee (\chi_{V_s})' = B_s'.$$

So the crisp Q-neighborhood $\chi_W$ of $x_\lambda$ is not quasi-coincident with $B_s$, $\Psi$ is flintily locally finite.

As a result of the above proved conclusions, $(L^X, \delta)$ is flintily paracompact. □

**12.2.38 Corollary** *Flinty paracompactness for L-fts is a good L-extension of ordi-*

*nary paracompactness.* □

Since flinty paracompactness implies paracompactness, we have

**12.2.39 Corollary** *Let $(L^X, \delta)$ be a weakly induced L-fts, $(X, [\delta])$ be paracompact, then $(L^X, \delta)$ is paracompact.* □

Corollary **12.2.14** and Theorem **12.2.37** imply the following

**12.2.40 Theorem** *Let $(X, \delta)$ be a weakly induced F-ts. Then the following conditions are equivalent:*

(i) *$(X, \delta)$ is flintily paracompact.*

(ii) *There exists $\alpha \in (0, 1]$ such that $(X, \delta)$ is flintily $\alpha$-paracompact.*

(iii) *$(X, \delta)$ is $*$-paracompact.*

(iv) *There exists $\alpha \in (0, 1]$ such that $(X, \delta)$ is $\alpha^*$-paracompact.*

(v) *$(X, \delta)$ is paracompact.*

(vi) *There exists $\alpha \in (0, 1)$ such that $(X, \delta)$ is $\alpha$-paracompact.*

(vii) *$(X, [\delta])$ is paracompact.* □

Parallel to the proof of Lemma **12.2.21**, one can easily prove the following

**12.2.41 Lemma** *Let $(L^X, \delta)$ be a regular L-fts, $A, B \in L^X$. If every open Q-cover of A has a refinement which is flintily locally finite in B and is a Q-cover of A, then for every open Q-cover $\{U_t : t \in T\}$ of A such that $(\bigvee_{t \in T} U_t)^- \subset B$, there exists a closed Q-cover $\{F_t : t \in T\}$ of A which is flintily locally finite in B such that $F_t \leq U_t$ for every $t \in T$.* □

**12.2.42 Definition** Let $(L^X, \delta)$ be an L-fts, $\mathcal{A} \subset L^X$, $B \in L^X$.

$\mathcal{A}$ is called *flintily $\sigma$-locally finite* in $B$, if $\mathcal{A}$ is a countable union of subfamilies which are flintily locally finite in $B$. $\mathcal{A}$ is called *flintily $\sigma$-locally finite* for short, if $\mathcal{A}$ is flintily $\sigma$-locally finite in $\underline{1}$.

$\mathcal{A}$ is called *flintily $\sigma$-discrete* in $B$, if $\mathcal{A}$ is a countable union of subfamilies which are flintily discrete in $B$. $\mathcal{A}$ is called *flintily $\sigma$-discrete* for short, if $\mathcal{A}$ is flintily $\sigma$-discrete in $\underline{1}$.

Similar to Lemma **12.2.22** and Lemma **12.2.19**, one can easily verify the following two lemmas:

**12.2.43 Lemma** *Let $(L^X, \delta)$ be a weakly induced, regular and flintily paracompact L-fts, $\alpha \in M(L)$. Then every open $\alpha$-Q-cover of $\underline{1}$ has a flintily $\sigma$-discrete refinement which is an open $\alpha$-Q-cover of $\underline{1}$.* □

**12.2.44 Lemma** *Let $(L^X, \delta)$ be a weakly $\alpha'$-induced L-fts, where $\alpha \in M(L)$. Then every flintily $\sigma$-locally finite and open $\alpha$-Q-cover of $\underline{1}$ has a flintily locally finite refinement which is an $\alpha$-Q-cover of $\underline{1}$.* □

**12.2.45 Lemma** *Let $(L^X, \delta)$ be a weakly induced and regular L-fts. If there exists $\alpha \in M(L)$ such that every open $\alpha$-Q-cover of $\underline{1}$ has a flintily $\sigma$-locally finite refinement which is an open $\alpha$-Q-cover of $\underline{1}$, then $(L^X, \delta)$ is flintily paracompact.*

**Proof** By Theorem **12.2.37**, it is sufficient to verify the paracompactness of $(X, [\delta])$. Suppose $\mathcal{U}$ is an open cover of $(X, [\delta])$. For every $U \in \mathcal{U}$, let $\Phi = \{\chi_U : U \in \mathcal{U}\}$, then $\Phi$ is an $\alpha$-Q-cover of $\underline{1}$, there exists an open $\alpha$-Q-cover $\Psi = \bigcup_{n<\omega} \Psi_n$ of $\underline{1}$ such that $\Psi$ is a refinement of $\Phi$ and $\Psi_n$ is flintily locally finite for every $n < \omega$. Hence for every $x \in X$, there exists $V \in \Psi$ such that $V \in \mathcal{Q}(x_\alpha)$, $\alpha \not\leq V(x)'$, $x \in V_{(\alpha')}$. So by Theorem **4.2.17**, $\mathcal{V} = \{V_{(\alpha')} : V \in \Psi\}$ is an open cover of $(X, [\delta])$. By the definition of $\Phi$, it is clear that $\mathcal{V}$ is a refinement of $\mathcal{U}$. Since $(L^X, \delta)$ is weakly induced and regular, as a consequence of Theorem **8.3.10**, $(X, [\delta])$ is regular. So in order to prove the paracompactness of $(X, [\delta])$, by Theorem 28 in Chapter 5 of [50], it is sufficient to prove that $\mathcal{V}$ is $\sigma$-locally finite.

Let $n < \omega$, $\mathcal{V}_n = \{V_{(\alpha')} : V \in \Psi_n\}$. Since $\Psi_n$ flintily locally finite, for every $x \in X$, there exists a crisp Q-neighborhood $W$ of $x_\alpha$ and a finite subset $\{V_0, \cdots, V_k\} \subset \Psi_n$ such that $V \leq W'$ for every $V \in \Psi_n \backslash \{V_0, \cdots, V_k\}$. Since $(L^X, \delta)$ is weakly induced, $W_{(0)} \in \mathcal{N}_{[\delta]}(x)$. For every $V \in \Psi_n \backslash \{V_0, \cdots, V_k\}$, if $y \in W_{(0)}$, then by $V \leq W'$ and $W$ being crisp, $y \notin V_{(\alpha')}$, $V_{(\alpha')} \cap W_{(0)} = \emptyset$. That is to say, $\mathcal{V}_n$ is locally finite in $(X, [\delta])$, and hence $\mathcal{V} = \bigcup_{n<\omega} \mathcal{V}_n$ is $\sigma$-locally finite. □

**12.2.46 Lemma** *Let $(L^X, \delta)$ be a weakly induced and regular L-fts, $\alpha \in M(L)$. If every open $\alpha$-Q-cover of $\underline{1}$ has a flintily locally finite refinement which is an $\alpha$-Q-cover of $\underline{1}$, then $(L^X, \delta)$ is flintily paracompact.*

**Proof** By Theorem **12.2.37**, it is sufficient to verify the paracompactness of $(X, [\delta])$. Suppose $\mathcal{U}$ is an open cover of $(X, [\delta])$, then $\mathcal{U}^* = \{\chi_U : U \in \mathcal{U}\}$ is an open $\alpha$-Q-cover of $\underline{1}$, $\mathcal{U}^*$ has a flintily locally finite refinement $\mathcal{A}$ which is an $\alpha$-Q-cover of $\underline{1}$. Let $\mathcal{B} = \{A_{(\alpha')} : A \in \mathcal{A}\}$, then $\mathcal{B}$ is a cover of $(X, [\delta])$. Let $A \in \mathcal{A}$, since $\mathcal{A}$ is a refinement of $\mathcal{U}^*$, there exists $U \in \mathcal{U}$ such that $A \leq \chi_U$. So $A_{(\alpha')} \subset U$, $\mathcal{B}$ is a refinement of $\mathcal{U}$. Suppose $x \in X$, since $\mathcal{A}$ is flintily locally finite, there exists a crisp neighborhood $O$ of $x_\alpha$ such that $O$ is quasi-coincident with only finitely many members $A_0, \cdots, A_k$ of $\mathcal{A}$. Then $x \in O_{(0)} \in [\delta]$ and for every $A \in \mathcal{A} \backslash \{A_0, \cdots, A_k\}$, by $A \leq O'$ and $O$ being crisp, we have $A_{(\alpha')} \subset O_{(0)}{}'$, $A_{(\alpha')} \cap O_{(0)} = \emptyset$. So the neighborhood $O_{(0)}$ of $x$ in $X$ intersects with at most finitely many members $(A_0)_{(\alpha')}, \cdots, (A_k)_{(\alpha')}$ of $\mathcal{B}$, $\mathcal{B}$ is locally finite in $(X, [\delta])$. Hence for every open cover $\mathcal{U}$ of $(X, [\delta])$, there exists a locally finite refinement $\mathcal{B}$ of $\mathcal{U}$ such that $\mathcal{B}$ is a cover of $(X, [\delta])$. By Theorem **8.3.10**, $(X, [\delta])$ is regular. Therefore, as a result of correspondent result in ordinary topology (e.g., Theorem 28 in Chapter 5 of [50]), $(X, [\delta])$ is paracompact. □

**12.2.47 Theorem** *Let $(L^X, \delta)$ be a weakly induced and regular L-fts. Then the following conditions are equivalent:*

(i) *$(L^X, \delta)$ is flintily paracompact.*

(ii) *$(X, [\delta])$ is paracompact.*

(iii) *There exists $\alpha \in M(L)$ such that every open $\alpha$-Q-cover of $\underline{1}$ has a flintily locally finite refinement which is an open $\alpha$-Q-cover (or closed $\alpha$-Q-cover, or $\alpha$-Q-cover) of $\underline{1}$.*

(iv) *For every $\alpha \in M(L)$, every open $\alpha$-Q-cover of $\underline{1}$ has a flintily locally finite* *refinement which is an open $\alpha$-Q-cover (or closed $\alpha$-Q-cover, or $\alpha$-Q-cover) of $\underline{1}$.*

(v) *There exists $\alpha \in M(L)$ such that every open $\alpha$-Q-cover of $\underline{1}$ has a flintily $\sigma$-discrete refinement which is an open $\alpha$-Q-cover of $\underline{1}$.*

(vi) *For every $\alpha \in M(L)$, every open $\alpha$-Q-cover of $\underline{1}$ has a flintily $\sigma$-discrete refinement which is an open $\alpha$-Q-cover of $\underline{1}$.*

(vii) *There exists $\alpha \in M(L)$ such that every open $\alpha$-Q-cover of $\underline{1}$ has a flintily $\sigma$-locally finite refinement which is open $\alpha$-Q-cover of $\underline{1}$.*

(viii) *For every $\alpha \in M(L)$, every open $\alpha$-Q-cover of $\underline{1}$ has a flintily $\sigma$-locally finite refinement which is an open $\alpha$-Q-cover of $\underline{1}$.*

**Proof** (i)$\Longrightarrow$(iv): By Lemma **12.2.41**.

(iv)$\Longrightarrow$(iii): Obvious.

(iii)$\Longrightarrow$(ii): By Lemma **12.2.46** and Theorem **12.2.37**.

(ii)$\Longrightarrow$(i): By Theorem **12.2.37**.

(i)$\Longrightarrow$(vi): By Lemma **12.2.43**.

(vi)$\Longrightarrow$(v): Obvious.

(v)$\Longrightarrow$(vii): Obvious.

(vii)$\Longrightarrow$(i): By Lemma **12.2.45**. □

Parallel to Theorem **12.2.29**, one can verify the following

**12.2.48 Theorem** *Let $(L^X, \delta)$ and $(L^Y, \mu)$ be weakly induced L-fts', $f^{\rightarrow} : (L^X, \delta) \rightarrow (L^Y, \mu)$ be an L-fuzzy continuous, closed and surjective mapping. If $(L^X, \delta)$ is flintily paracompact and $T_2$, then $(L^Y, \mu)$ is flintily paracompact.* □

## 12.3 Separations, Lindelöf Property and Paracompactness

**12.3.1 Theorem** *Let $(X, \delta)$ be a weakly induced, $T_2$ and paracompact F-ts, then $(X, \delta)$ is s-p-regular and s-p-normal.*

**Proof** By Theorem **12.2.12**, $(X, [\delta])$ is paracompact. By Corollary **8.2.21**, $(X, [\delta])$ is $T_2$. Then as known results in ordinary topology (e.g., see [50]), $(X, [\delta])$ is regular and normal. By Theorem **8.3.12** (ii), $(X, \delta)$ is s-p-regular. By Theorem **9.2.57** (ii), $(X, \delta)$ is s-p-normal. □

Similarly one can prove the following

**12.3.2 Theorem** *Let $(L^X, \delta)$ be a weakly induced, $T_2$ and $*$-paracompact L-fts, then $(L^X, \delta)$ is s-p-regular and s-p-normal.* □

Since flinty paracompactness implies $*$-paracompactness, so we have

**12.3.3 Corollary** *Let $(L^X, \delta)$ be a weakly induced, $T_2$ and flintily paracompact L-fts, then $(L^X, \delta)$ is s-p-regular and s-p-normal.* □

As flinty paracompactness has stronger properties than paracompactness and $*$-paracompactness, stronger conclusions should be expected for it:

**12.3.4 Theorem** *Let $(L^X, \delta)$ be a $T_2$ and flintily paracompact L-fts, then $(L^X, \delta)$ is p-regular.*

**Proof** Suppose $A$ is a pseudo-crisp closed subset in $(L^X, \delta)$, $x_\lambda \in M(L^X)$, $x \notin supp(A)$. Take $\alpha \in M(L)$ such that for every $y \in supp(A)$, $A(y) \geq \alpha$, then $x \neq y$. By the $T_2$ property of $(L^X, \delta)$, there exist $U_y \in \mathcal{Q}(x_\lambda)$ and $V_y \in \mathcal{Q}(y_\alpha)$ such that

$$U_y \wedge V_y = \underline{1}. \tag{12.8}$$

Take $\Phi = \{V_y : y \in supp(A)\}$, then $\Phi$ is an open $\alpha$-Q-cover of $A$. Since $A$ is closed, by Theorem **12.2.36**, $A$ is flintily paracompact, there exists a flintily locally finite refinement $\Psi = \{W_t : t \in T\}$ of $\Phi$ which is an open $\alpha$-Q-cvoer of $A$. Then there exists a crisp Q-neighborhood $U$ of $x_\lambda$ and a finite subset $T_0$ of $T$ such that $W_t \leq U'$ for every $t \in T \backslash T_0$, $\bigvee_{t \in T \backslash T_0} W_t \leq U'$. Since $U$ is crisp, we have

$$\bigvee_{t \in T \backslash T_0} (W_t \wedge U) = \underline{0}. \tag{12.9}$$

Since $\Psi$ is a refinement of $\Phi$, for every $t \in T_0$, there exists $y_t \in supp(A)$ such that $W_t \leq V_{y_t}$. Take $U^* = U \wedge \bigwedge_{t \in T_0} U_{y_t}$, $W^* = \bigvee_{t \in T} W_t$, then $U^*$ is still a Q-neighborhood of $x_\lambda$, $W^*$ is a Q-neighborhood of $A$. Then by (12.8),

$$(\bigwedge_{s \in T_0} U_{y_s}) \wedge (\bigvee_{t \in T_0} V_{y_t}) = \bigvee_{t \in T_0} ((\bigwedge_{s \in T_0} U_{y_s}) \wedge V_{y_t}) \leq \bigvee_{t \in T_0} (U_{y_t} \wedge V_{y_t}) = \underline{0}. \tag{12.10}$$

By (12.9) and $W_t \leq V_{y_t}$ for every $t \in T_0$,

$$\begin{aligned} U^* \wedge V^* &= (U^* \wedge (\bigvee_{t \in T \backslash T_0} W_t)) \vee (U^* \wedge (\bigvee_{t \in T_0} W_t)) \\ &\leq (U \wedge (\bigvee_{t \in T \backslash T_0} W_t)) \vee ((\bigwedge_{t \in T_0} U_{y_t}) \wedge (\bigvee_{t \in T_0} V_{y_t})) \\ &= \underline{0}. \end{aligned}$$

So $(L^X, \delta)$ is p-regular. □

Similarly one can prove the following

**12.3.5 Theorem** *Let $(L^X, \delta)$ be a s-$T_2$ and flintily paracompact L-fts, then $(L^X, \delta)$ is s-p-regular.* □

**12.3.6 Theorem** *Let $(L^X, \delta)$ be a s-$T_2$ and flintily paracompact L-fts, then $(L^X, \delta)$ is s-p-normal.* □

**Proof** Suppose $A$, $B$ are pseudo-crisp closed subsets in $(L^X, \delta)$, and $supp(A) \cap supp(B) = \emptyset$. Take $\lambda, \alpha \in M(L)$ such that $supp(A) = A_{[\lambda]}$, $supp(B) = B_{[\alpha]}$. For every $x \in supp(A)$, by Theorem **12.3.5**, there exist $U_x \in \mathcal{Q}(x_\lambda)$ and $V_x \in \mathcal{Q}(B)$ such that $(U_x)_{(0)} \cap (V_x)_{(0)} = \emptyset$. Let $\Phi = \{U_x : x \in supp(A)\}$, then $\Phi$ is an open $\lambda$-Q-cover of $A$. Since $A$ is closed in flintily paracompact $L$-fts $(L^X, \delta)$, by Theorem **12.2.36**, $A$ is flintily paracompact, $\Phi$ has a flintily locally finite refinement $\Psi = \{W_t : t \in T\}$ which is an open $\lambda$-Q-cover of $A$. Let $\Gamma = \{\chi_{(W_t)_{(0)}} : t \in T\}$, then by Proposition **12.1.22**, $\Gamma$ is still flintily locally finite. Take $V = (\bigvee_{t \in T} (\chi_{(W_t)_{(0)}})^-)'$, by the flintily locally finiteness of $\Gamma$, $V$ is open. Let $U = \bigvee_{t \in T} W_t$, since $\Psi$ is an open $\lambda$-Q-cover of $A$, $U$ is a Q-neigbborhood of $A$. Then it is sufficient to verify that $V$ quasi-coincides with $B$ at every $y \in supp(B)$ and $U_{(0)} \cap V_{(0)} = \emptyset$.

For every $y \in supp(B)$, since $\Gamma$ is flintily locally finite, there exists a crisp Q-neighborhood $V^*$ of $y_\alpha$ and a finite subset $T_0$ of $T$ such that $\chi_{(W_t)_{(0)}} \leq (V^*)'$ and hence $(\chi_{(W_t)_{(0)}})^- \leq (V^*)'$ for every $t \in T \backslash T_0$. Then $\bigvee_{t \in T\backslash T_0} (\chi_{(W_t)_{(0)}})^- \leq (V^*)'$. Since $\Psi$ is a refinement of $\Phi$, for every $t \in T_0$, there exists $x_t \in supp(A)$ such that $W_t \leq U_{x_t}$. By $(U_x)_{(0)} \cap (V_x)_{(0)} = \emptyset$ for every $x \in supp(A)$, $(W_t)_{(0)} \subset (U_{x_t})_{(0)} \subset (V_{x_t})_{(0)}{}'$. So for every $t \in T_0$,

$$\chi_{(W_t)_{(0)}} \leq \chi_{(V_{x_t})_{(0)}{}'} = (\chi_{(V_{x_t})_{(0)}})' \leq V_{x_t}{}',$$

and hence $(\chi_{(W_t)_{(0)}})^- \leq V_{x_t}{}'$. Denote $O = V^* \wedge (\bigwedge_{t \in T_0} V_{x_t})$. By $supp(B) = B_{[\alpha]}$, $B(y) \geq \alpha$, $V^* \in \mathcal{Q}(y_\alpha) \subset \mathcal{Q}(y_{B(y)})$. On the other hand, for every $t \in T_0$, $V_{x_t} \in \mathcal{Q}(B)$, so $V_{x_t} \in \mathcal{Q}(y_{B(y)})$, $\bigwedge_{t \in T_0} V_{x_t} \in \mathcal{Q}(y_{B(y)})$. Hence we have $O \in \mathcal{Q}(y_{B(y)})$ and

$$V' = \bigvee_{t \in T} (\chi_{(W_t)_{(0)}})^- = (\bigvee_{t \in T\backslash T_0} (\chi_{(W_t)_{(0)}})^-) \vee (\bigvee_{t \in T_0} (\chi_{(W_t)_{(0)}})^-) \leq (V^*)' \vee (\bigvee_{t \in T_0} V_{x_t}{}') = O'.$$

Therefore, by $y_{B(y)} \not\leq O'$ we have $y_{B(y)} \not\leq V'$, $V$ quasi-coincides with $B$ at $y \in supp(B)$.

If $x \in U_{(0)}$, then by Proposition **4.2.4** (i), there exists $r \in T$ such that $x \in (W_r)_{(0)}$. Then

$$(\chi_{(W_r)_{(0)}})^-(x) \geq \chi_{(W_r)_{(0)}}(x) = 1,$$
$$((\chi_{(W_r)_{(0)}})^-)'(x) = 0,$$
$$V(x) = \bigwedge_{t \in T} ((\chi_{(W_r)_{(0)}})^-)'(x) = 0.$$

So $x \notin V_{(0)}$, $U_{(0)} \cap V_{(0)} = \emptyset$. □

In [113], Luo introduced the notion of fuzzy Lindelöf subset, investigated its properties and relations between fuzzy Lindelöf property and various fuzzy paracompact properties. But the value domain used there was $[0,1]$. In the sequel, the value domain will be generalized to be a F-lattice $L$, then the correspondent generalized conclusions will be investigated.[168]

**12.3.7 Definition** Let $(L^X, \delta)$ be an $L$-fts, $A \in L^X$, $\alpha \in M(L)$. $A$ is called *$\alpha$-Lindelöf*, if every open $\alpha$-Q-cover of $A$ has a countable subfamily which is still an $\alpha$-Q-cover of $A$. $A$ is called *Lindelöf*, if $A$ is $\alpha$-Lindelöf for every $\alpha \in M(L)$. $(L^X, \delta)$ is called *Lindelöf*, if $\underline{1}$ is Lindelöf.

**12.3.8 Proposition** *Strongly compact L-fts is Lindelöf.* □

Lindelöf property is hereditary to closed subsets:

**12.3.9 Theorem** *Let $(L^X, \delta)$ be an L-fts, $\alpha \in M(L)$, $A \in L^X$, $B \in \delta'$. Then*

(i) *$A$ is $\alpha$-Lindelöf $\Longrightarrow$ $A \wedge B$ is $\alpha$-Lindelöf.*

(ii) *$A$ is Lindelöf $\Longrightarrow$ $A \wedge B$ is Lindelöf.* □

**12.3.10 Corollary** *Lindelöf property for L-fts' is hereditary to closed subsets.* □

**12.3.11 Theorem** *Let $(L^X, \delta)$ be a weakly induced F-ts. Then the following conditions are equivalent:*

(i) *$(L^X, \delta)$ is Lindelöf.*

(ii) *There exists $\alpha \in M(L)$ such that $(L^X, \delta)$ is $\alpha$-Lindelöf.*

(iii) *$(X, [\delta])$ is Lindelöf.*

**Proof** (i)$\Longrightarrow$(ii): Obvious.

(ii)$\Longrightarrow$(iii): Suppose $\mathcal{U}$ is an open cover of $(X, [\delta])$, then $\mathcal{U}^* = \{\chi_U : U \in \mathcal{U}\}$ is an open $\alpha$-Q-cover of $\underline{1}$, it has a countable subfamily $\mathcal{U}^*{}_0$ which is still an $\alpha$-Q-cover of $\underline{1}$. Hence $\mathcal{U}_0 = \{supp(V) : V \in \mathcal{U}^*{}_0\}$ is a countable subfamily of $\mathcal{U}$ and a cover of $(X, [\delta])$, $(X, [\delta])$ is Lindelöf.

(iii)$\Longrightarrow$(i): Suppose $\alpha \in M(L)$, $\mathcal{U}$ is an open $\alpha$-Q-cover of $\underline{1}$. Since $(L^X, \delta)$ is weakly induced, by Theorem **4.2.17**, $\mathcal{U}^* = \{U_{(\alpha')} : U \in \mathcal{U}\}$ is an open cover of $(X, [\delta])$. Then $\mathcal{U}$ has a countable subfamily $\mathcal{U}_0$ such that $\mathcal{U}^*{}_0 = \{U_{(\alpha')} : U \in \mathcal{U}_0\}$ is still a cover of $(X, [\delta])$. Hence $\mathcal{U}_0$ is a countable subfamily of $\mathcal{U}$ which is still an $\alpha$-Q-cover of $\underline{1}$, $(L^X, \delta)$ is $\alpha$-Lindelöf. Since $\alpha$ is arbitrarily chosen from $M(L)$, $(L^X, \delta)$ is Lindelöf. □

**12.3.12 Theorem** *Every weakly induced, regular and Lindelöf L-fts is flintily paracompact.*

**Proof** Suppose $(L^X, \delta)$ is a weakly induced, regular and Lindelöf $L$-fts, $\alpha \in M(L)$, $\Phi$ is an open $\alpha$-Q-cover of $\underline{1}$, then $\Phi$ has a countable subfamily $\Psi$ which is still an $\alpha$-Q-cover of $\underline{1}$. Clearly, $\Psi$ is a refinement of $\Phi$ and is flintily locally finite. So by Theorem **12.2.47**, $(L^X, \delta)$ is flintily paracompact. □

**12.3.13 Corollary** *Every weakly induced, regular and Lindelöf L-fts is paracompact.* □

**12.3.14 Proposition** *Every secondly countable L-fts is Lindelöf.* □

The condition "weakly induced" in the above theorem and corollary cannot be omitted:

**12.3.15 Example** There exists a regular and $T_2$ F-ts $(X, \delta)$ which has a countable base (and hence is Lindelöf), but it is not 1-paracompact.

Take $X = [0, +\infty)$, let $\mathcal{T}_0$ denote the ordinary subspace topology of $X$ in $\mathbf{R}$, and

$$U_0 = \tfrac{1}{2}\chi_{\{0\}}, \quad U_1 = (\tfrac{1}{2}\chi_{\{0\}}) \vee \chi_{X \setminus (\{0\} \cup \{\frac{1}{n}:\, n \in \mathbf{N}\})},$$
$$\mathcal{P} = \{U_0, U_0', U_1\} \cup \{\chi_U : U \in \mathcal{T}_0\},$$

then let $\delta$ denote the F-topology on $X$ generated by $\mathcal{P}$, we have

(i) $(X, \delta)$ is a $T_2$ F-ts. This is obvious.

(ii) $(X, \delta)$ is regular. In fact, for every F-point $e$ and every $U \in \mathcal{P} \cap \mathcal{Q}(e)$, there exists $V \in \mathcal{Q}(e)$ such that $V^- \leq U$. By Theorem **8.3.7**, $(X, \delta)$ is regular.

(iii) $(X, \delta)$ has a countable base. Since $\mathcal{T}_0$ has a countable base, one can find a countable base $\mathcal{P}_1 \subset \mathcal{P}$ of $\delta$.

(iv) The open Q-cover $\mathcal{U} = \{U_1\} \cup \{\chi_{(\frac{1}{n} - \frac{1}{2n(n+1)}, \frac{1}{n} + \frac{1}{2n(n+1)})} : n \in \mathbf{N}\}$ of $(X, \delta)$ has not any refinement $\mathcal{V}$ which is an open 1-Q-cover of $\underline{1}$ and is locally finite at $(0)_{\frac{1}{2}}$. In fact, if $\mathcal{V}$ is a refinement of $\mathcal{U}$ which is an open 1-Q-cover of $\underline{1}$, then for every $n \in \mathbf{N}$, let $V_n \in \mathcal{V}$ such that $V_n \leq \chi_{(\frac{1}{n} - \frac{1}{2n(n+1)}, \frac{1}{n} + \frac{1}{2n(n+1)})}$, we have $V_m \wedge V_n = \underline{0}$

whenever $m, n \in \mathbf{N}$, $m \neq n$. Hence the mapping $\frac{1}{n} \mapsto V_n$ is one to one. On the other hand, we have $U(0) > \frac{1}{2}$ for every $U \in \mathcal{Q}((0)_{\frac{1}{2}})$, so by the structure of $\mathcal{P}$ there exists $W \in \mathcal{T}_0$ such that $U \geq \chi_W$. Therefore, $U$ is quasi-coincident with an infinite number of $V_n$'s.

**12.3.16 Remark** In F-ts', Luo proved the following conclusions:[113]

*Every weakly* $(1-\alpha)$*-induced and Lindelöf F-ts is* $\alpha$*-paracompact.*

An example was also constructed there to show a weakly 0-induced, regular and 1-Lindelöf F-ts which is not 1-paracompact.

# Chapter 13

# Uniformity and Proximity

Fuzzy versions of uniformity theory were established by B. Hutton,[38], R. Lowen,[96], U. Höhle,[34], A. K. Katsaras[48], Ji-Hua Liang[58, 59, 60], etc. Fuzzy uniformity in Hutton's sense has been accepted by many authors and has attracted wide attention in the literature. In an analogous way to that in general topology, A.K. Katsaras[47] introduced proximity structures in fuzzy spaces and investigated the F-topology generated by these proximities. However, as was pointed in Ref.[57] that those F-topologies are always crisp. Thus a new and more reasonable definition was given in Refs.[57] and [65] independently. An bijective correspondence between the fuzzy uniformities in the sense of Hutton and the proximities on a set $X$ was obtained in Ref.[57], and it was also proved that an $L$-fts is completely regular if and only if it can be generated by a proximity. The embedding theorem for fuzzy proximity spaces was established in Ref.[57].

## 13.1 Uniformity

**13.1.1 Definition** Let $L$, $L_0$, $L_1$ be complete lattices. Denote by $\mathcal{J}(L_0, L_1)$ the family of all the arbitrary join preserving mappings from $L_0$ to $L_1$. Equip $\mathcal{J}(L_0, L_1)$ with the partial order $\leq$ as follows:

$$\forall f, g \in \mathcal{J}(L_0, L_1), \quad f \leq g \Longrightarrow \forall a \in L, \ f(a) \leq g(a).$$

A self mapping $f: L \to L$ on $L$ is called *value increasing*, if $f(a) \geq a$ for every $a \in L$; called *value decreasing*, if $f(a) \leq a$ for every $a \in L$. Denote by $\mathcal{E}(L)$ the family of all the value increasing and arbitrary join preserving self mappings on $L$. Equip $\mathcal{E}(L)$ with the partial order $\leq$ defined in $\mathcal{J}(L, L) \supset \mathcal{E}(L)$, i.e.

$$\forall f, g \in \mathcal{E}(L), \quad f \leq g \Longrightarrow \forall a \in L, \ f(a) \leq g(a).$$

**13.1.2 Remark** In completely distributive lattices, Hutton[38] and Liu[67, 69] established a series of notions and results related to meet (intersection) operation and inverse operation. With He, Liu also introduced the notion of induced mapping into $L$-fuzzy topology.[81] In the sequel, some of these notions and results related to our topic will be introduced.

**13.1.3 Theorem** *Let $L_0$, $L_1$ be complete lattices. Then*

(i) *$\mathcal{J}(L_0, L_1)$ is a complete lattice.*

(ii) *In $\mathcal{J}(L_0, L_1)$, for every $\mathcal{A} \subset \mathcal{J}(L_0, L_1)$ and every $a \in L_0$,*

$$(\bigvee \mathcal{A})(a) = \bigvee_{f \in \mathcal{A}} f(a). \qquad \square$$

**13.1.4 Theorem** *Let $L$ be a complete lattice. Then*

(i) *$\mathcal{E}(L)$ is a complete lattice.*

(ii) *In $\mathcal{E}(L)$, for every $\mathcal{A} \subset \mathcal{E}(L)$ and every $a \in L$,*

$$(\bigvee \mathcal{A})(a) = \bigvee_{f \in \mathcal{A}} f(a).$$ □

**13.1.5 Theorem** *Let $L_0$, $L_1$ be completely distributive lattices. Then*

(i) $\forall f, g \in \mathcal{J}(L_0, L_1)$, $\forall a \in L_0$,

$$(f \wedge g)(a) = \bigvee_{\lambda \in \beta^*(a)} (f(\lambda) \wedge g(\lambda)). \tag{13.1}$$

(ii) $\forall \{f_0, \cdots, f_n\} \in [\mathcal{J}(L_0, L_1)]^{<\omega}$, $\forall a \in L$,

$$(f_0 \wedge \cdots \wedge f_n)(a) = \bigvee_{\lambda \in \beta^*(a)} (f_0(\lambda) \wedge \cdots \wedge f_n(\lambda)). \tag{13.2}$$

(iii) $\forall f, g \in \mathcal{J}(L_0, L_1)$, $\forall a \in L_0$,

$$(f \wedge g)(a) = \bigwedge_{b \vee c = a} (f(b) \vee g(c)). \tag{13.3}$$

(iv) $\forall \{f_0, \cdots, f_n\} \in [\mathcal{J}(L_0, L_1)]^{<\omega}$, $\forall a \in L$,

$$(f_0 \wedge \cdots \wedge f_n)(a) = \bigwedge_{b_0 \vee \cdots \vee b_n = a} (f_0(b_0) \vee \cdots \vee f_n(b_n)). \tag{13.4}$$

**Proof** (i) This is a consequence of (ii) which will be proved in following.

(ii) Take $h: L_1 \to L_0$ as

$$\forall a \in L_0, \quad h(a) = \bigvee_{\lambda \in \beta^*(a)} (f_0(\lambda) \wedge \cdots \wedge f_n(\lambda)),$$

then $h$ is value increasing. By Theorem **1.3.24** (ii), $h$ is arbitrary join preserving. So $h \in \mathcal{J}(L_0, L_1)$. For every $\lambda \in \beta^*(a)$,

$$(f_0 \wedge \cdots \wedge f_n)(\lambda) \leq f_0(\lambda) \wedge \cdots \wedge f_n(\lambda) \leq \bigvee_{\gamma \in \beta^*(a)} (f_0(\gamma) \wedge \cdots \wedge f_n(\gamma)),$$

so as a result of $f_0 \wedge \cdots \wedge f_n \in \mathcal{F}(L_0, L_1)$ and $\bigvee \beta^*(a) = a$, $(f_0 \wedge \cdots \wedge f_n)(a) \leq h(a)$, $f_0 \wedge \cdots \wedge f_n \leq h$. On the other hand, for every $i \leq n$,

$$h(a) \leq \bigvee_{\lambda \in \beta^*(a)} f_i(\lambda) = f_i(\bigvee_{\lambda \in \beta^*(a)} \lambda) = f_i(a);$$

so $h \leq f_0 \wedge \cdots \wedge f_n$. Therefore, $h = f \wedge \cdots \wedge f_n$.

(iii) By (i) proved above and the complete distributive law,

$$(f \wedge g)(a) = \bigvee_{\lambda \in \beta^*(a)} (f(\lambda) \wedge g(\lambda)) = \bigwedge_{D \subset \beta^*(a)} ((\bigvee_{\lambda \in D} f(\lambda)) \vee (\bigvee_{\gamma \in \beta^*(a) \setminus D} g(\gamma))).$$

Since for a fixed $D \subset \beta^*(a)$, $(\bigvee D) \vee (\bigvee(\beta^*(a) \setminus D)) = \bigvee \beta^*(a) = a$, and $f$, $g$ are arbitrary join preserving, so by the equation proved above, we have actually proved $(f \wedge g)(a) \geq \bigwedge_{b \vee c = a} (f(b) \vee g(c))$. For its reverse, suppose $b, c \in L$ such that $b \vee c = a$, then

$$f(b) \vee g(c) \geq (f \wedge g)(b) \vee (f \wedge g)(c) = (f \wedge g)(b \vee c) = (f \wedge g)(a),$$

so (13.3) holds.

(iv) It has been proved that formula (13.4) holds for $n = 2$. Suppose (13.4) holds for some $n \geq 2$, it is sufficient to prove that

$$(f_0 \wedge \cdots \wedge f_{n+1})(a) = \bigwedge_{b_0 \vee \cdots \vee b_{n+1} = a} (f_0(b_0) \vee \cdots \vee f_{n+1}(b_{n+1})). \tag{13.5}$$

By (ii) and the infinite distributivity of $L_0$, $L_1$,

$$\begin{aligned}(f_0\wedge\cdots\wedge f_{n+1})(a) &= \bigwedge_{b\vee b_{n+1}=a}((f_0\wedge\cdots\wedge f_n)(b)\vee f_{n+1}(b_{n+1}))\\ &= \bigwedge_{b\vee b_{n+1}=a}((\bigwedge_{b_0\vee\cdots\vee b_n=b}(f_0(b_0)\vee\cdots\vee f_n(b_n)))\vee f_{n+1}(b_{n+1}))\\ &= \bigwedge_{b\vee b_{n+1}=a}\ \bigwedge_{b_0\vee\cdots\vee b_n=b}(f_0(b_0)\vee\cdots\vee f_{n+1}(b_{n+1}))\\ &= \bigwedge_{b_0\vee\cdots\vee b_{n+1}=a}(f_0(b_0)\vee\cdots\vee f_{n+1}(b_{n+1})).\end{aligned}$$

So (13.5) holds, and hence (13.4) is true. □

**13.1.6 Definition** Let $L_0$, $L_1$ be complete lattices, $f:\ L_0\to L_1$ be a mapping. Define

$$f^{\vee}:\ L_1\to L_0,\quad b\mapsto\bigvee\{a\in L_0:\ f(a)\le b\},$$
$$f^{\wedge}:\ L_1\to L_0,\quad b\mapsto\bigvee\{a\in L_0:\ f(a)\ge b\}.$$

Call $f^{\vee}$ and $f^{\wedge}$ the *join induced mapping* and the *meet induced mapping* of $f$ respectively.

**13.1.7 Definition** Let $P_0$, $P_1$ be posets, $f:\ P_0\to P_1$ and $g:\ P_1\to P_0$ be order preserving mappings. The ordered pair $(g,f)$ is called a *Galois connection* between $P_0$ and $P_1$, if

$$\forall(a,b)\in P_0\times P_1,\quad f(a)\le b\iff a\le g(b).$$

Denote the relation "$(g,f)$ is a Galois connection" by $f\dashv g$, call $f$ the *left adjoint* of $g$ and call $g$ the *right adjoint* of $f$. Also call $f\dashv g$ an *adjunction.*

**13.1.8 Proposition** *Let $f:\ P_0\to P_1$ and $g:\ P_1\to P_0$ be order preserving mappings between posets. Then the following conditions are equivalent:*

(i) $f\dashv g$.

(ii) $gf\ge id_{P_0}$, $fg\le id_{P_1}$.

**Proof** (i)$\Longrightarrow$(ii): For every $a\in P_0$, $f(a)\le f(a)$ implies $a\le gf(a)$, so $id_{P_0}\le gf$. Similarly prove $id_{P_1}\ge fg$.

(ii)$\Longrightarrow$(i): If $(a,b)\in P_0\times P_1$, $f(a)\le b$, then by $gf\ge id_{P_0}$ and $f$ being order preserving, $a\le gf(a)\le g(b)$, i.e. $a\le g(b)$. Similarly prove $a\le g(b)\Longrightarrow f(a)\le b$. □

**13.1.9 Proposition** *Let $f:\ L_0\to L_1$ be a mapping between complete lattices. Then*

(i) *$f^{\vee}$ and $f^{\wedge}$ are order preserving.*

(ii) *$f^{\vee}$ is arbitrary meet preserving and $f^{\wedge}$ is arbitrary join preserving provided $f$ is order preserving.* □

**13.1.10 Theorem** *Let $f:\ L_0\to L_1$ be a mapping between complete lattices. Then*

(i) $f^{\vee}\circ f\ge id_{L_0}$.

(ii) *If $f$ is arbitrary join preserving, then $f\circ f^{\vee}\le id_{L_1}$.*

(iii) $f(a)\le b\Longrightarrow a\le f^{\vee}(b)$.

(iv) *If $f$ is arbitrary join preserving, then $a\le f^{\vee}(b)\Longrightarrow f(a)\le b$.*

(v) *If $f$ is arbitrary join preserving, then $(f^{\vee},f)$ is a Galois connection.*

(vi) *$f$ is arbitrary join preserving if and only if $f=f^{\vee\wedge}$.*

(vii) *$f$ is arbitrary meet preserving if and only if $f=f^{\wedge\vee}$.*

(viii) *$f$ is value increasing $\Longrightarrow$ $f^{\vee}$ and $f^{\wedge}$ are value decreasing.*

(ix) *$f$ is value decreasing $\Longrightarrow$ $f^{\vee}$ and $f^{\wedge}$ are value increasing.* □

**13.1.11 Theorem** *Let $f,g:\ L_0\to L_1$ be mappings between complete lattices. Then*

(i) $f \leq g \Longrightarrow f^{\vee} \geq g^{\vee},\ f^{\wedge} \geq g^{\wedge}.$

(ii) *If $f$ and $g$ are arbitrary join preserving, then* $f^{\vee} \geq g^{\vee} \Longrightarrow f \leq g.$

(iii) *If $f$ and $g$ are arbitrary meet preserving, then* $f^{\wedge} \geq g^{\wedge} \Longrightarrow f \leq g.$ □

**13.1.12 Theorem** *Let $f: L_0 \to L_1$, $g: L_1 \to L_2$ be mappings between complete lattices. Then*

(i) $(g\circ f)^{\vee} \leq f^{\vee}\circ g^{\vee} \Longrightarrow (g\circ f)^{\wedge} \geq f^{\wedge}\circ g^{\wedge}.$

(ii) *If $g$ is arbitrary join preserving (arbitrary meet preserving, respectively), or $g$ is order preserving and both $f$ and $f^{\vee}$ ($f^{\wedge}$, respectively) are arbitrary join preserving (arbitrary meet preserving), then*

$$(g\circ f)^{\vee} = f^{\vee}\circ g^{\vee} \quad ((g\circ f)^{\wedge} = f^{\wedge}\circ g^{\wedge}, \text{ respectively}).$$ □

**13.1.13 Definition** Let $L_0$ be a complete lattice, $L_1$ be a complete lattice with an order-reversing involution $'\colon L_1 \to L_1$. For every mapping $f: L_0 \to L_1$, define the *reverse-meet induced mapping* $f^{\triangleleft}: L_1 \to L_0$ of $f$ as follows:

$$\forall b \in L_1,\quad f^{\triangleleft}(b) = \bigwedge\{a \in L_0 : f(a') \leq b'\}.$$

**13.1.14 Proposition** *Let $L_0$, $L_1$ be complete lattices, $r: L_1 \to L_1$ be an order-reversing involution, $f: L_0 \to L_1$ be a mapping. Then*

$$r\circ f^{\triangleleft} = f^{\vee}\circ r.$$ □

**13.1.15 Theorem** *Let $L$ be a F-lattice. Then*

(i) $\forall a, b \in L,\ \forall f \in \mathcal{E}(L),\quad f(a') \leq b' \iff f^{\triangleleft}(b) \leq a.$

(ii) $f \in \mathcal{E}(L) \implies f^{\triangleleft} \in \mathcal{E}(L).$

(iii) $f \in \mathcal{E}(L) \implies (f^{\triangleleft})^{\triangleleft} = f.$

(iv) $f, g \in \mathcal{E}(L),\ f \leq g \implies f^{\triangleleft} \leq g^{\triangleleft}.$

(v) $\{f_t : t \in T\} \subset \mathcal{E}(L) \implies (\bigvee_{t\in T} f_t)^{\triangleleft} = \bigvee_{t\in T} f_t^{\triangleleft}.$

(vi) $f, g \in \mathcal{E}(L) \implies (f \wedge g)^{\triangleleft} = f^{\triangleleft} \wedge g^{\triangleleft}.$

(vii) $f, g \in \mathcal{E}(L) \implies (f\circ g)^{\triangleleft} = g^{\triangleleft}\circ f^{\triangleleft}.$

**Proof** (i) Suppose $f(a') \leq b'$, then by the definition of $f^{\triangleleft}$, $f^{\triangleleft}(b) \leq a$. Conversely, if $f^{\triangleleft}(b) \leq a$, then $a' \leq (f^{\triangleleft}(b))'$. By the definition of $f^{\triangleleft}$ and $f$ being arbitrary join preserving,

$$\begin{aligned} f(a') &\leq f((f^{\triangleleft}(b))') \\ &= f((\bigwedge\{c \in L : f(c') \leq b'\})') \\ &= f(\bigvee\{c' : c \in L,\ f(c') \leq b'\}) \\ &= \bigvee\{f(c') : c \in L,\ f(c') \leq b'\} \\ &\leq b'. \end{aligned}$$

(ii) Suppose $b \in L$. If $a \in L$ such that $f(a') \leq b'$, then by $f(a') \geq a'$ we have $a' \leq b'$, $b \leq a$. So $f(a') \leq b'$ implies $b \leq a$, and hence by the definition, $f^{\triangleleft}$ is value increasing. By the definition, $f^{\triangleleft}$ is order preserving. Suppose $B \subset L$, then by (i) proved above, for every $c \in L$,

$$\begin{aligned} \bigvee_{b\in B} f^{\triangleleft}(b) \leq c &\iff \forall b \in B,\ f^{\triangleleft}(b) \leq c \\ &\iff \forall b \in B,\ f(c') \leq b' \\ &\iff f(c') \leq \bigwedge_{b\in B} b' = (\bigvee B)' \\ &\iff f^{\triangleleft}(\bigvee B) \leq c. \end{aligned}$$

Since $c \in L$ is arbitrarily fixed, we have $f^{\triangleleft}(\bigvee B) = \bigvee_{b\in B} f^{\triangleleft}(b)$, $f^{\triangleleft}$ is arbitrary join preserving. Hence $f^{\triangleleft} \in \mathcal{E}(L)$.

(iii) For every $a, c \in L$, by (i) proved above,

$$(f^{\triangleleft})^{\triangleleft}(a) \le c \iff f^{\triangleleft}(c') \le a' \iff f(a) \le c.$$

Hence $(f^{\triangleleft})^{\triangleleft} = f$.

(iv) Suppoe $f, g \in \mathcal{E}(L)$, $a, b \in L$, then it follows from $g(a') \le b' \Longrightarrow f(a') \le b'$, we have

$$\{a \in L:\ g(a') \le b'\} \subset \{a \in L:\ f(a' \le b'\}.$$

So

$$f^{\triangleleft}(b) = \bigwedge\{a \in L:\ f(a') \le b'\} \le \bigwedge\{a \in L:\ g(a') \le b'\} = g^{\triangleleft}(b),$$

i.e. $f^{\triangleleft} \le g^{\triangleleft}$.

(v) First of all, we prove that for every $\{A_t:\ t \in T\} \subset \mathcal{P}(L)$,

$$\forall t \in T,\ \uparrow\! A_t = A_t,\ \bigwedge A_t \in A_t \Longrightarrow \bigwedge(\bigcap_{t\in T} A_t) = \bigvee_{t\in T}(\bigwedge A_t). \tag{13.6}$$

For every $s \in T$ and every $a \in \bigcap_{t\in T} A_t$, we have $a \ge \bigwedge A_s$, so $\bigwedge(\bigcap_{t\in T} A_t) \ge \bigvee_{t\in T}(\bigwedge A_t)$. On the other hand, since $\uparrow\! A_t = A_t$ and $\bigwedge A_t \in A_t$ for every $t \in T$, we have $\bigvee_{t\in T}(\bigwedge A_t) \in \bigcap_{t\in T} A_t$, so $\bigwedge(\bigcap_{t\in T} A_t) \le \bigvee_{t\in T}(\bigwedge A_t)$. Therefore, we have proved $\bigwedge(\bigcap_{t\in T} A_t) = \bigvee_{t\in T}(\bigwedge A_t)$, (13.6) holds.

Note that for every $f \in \mathcal{E}(L)$ and every $b \in L$, denote $A(f, b) = \{a:\ f(a) \le b\}$, then $A(f, b)$ satisfies

$$\downarrow\! A(f,b) = A(f,b), \quad \bigvee A(f,b) \in A(f,b),$$

we have

$$\uparrow\!(A(f,b)') = A(f,b)', \quad \bigwedge(A(f,b)') \in A(f,b)'.$$

So for every $b \in L$, by Proposition **13.1.14**,

$$\begin{aligned}(\bigvee_{t\in T} f_t)^{\triangleleft}(b) &= r\circ(\bigvee_{t\in T} f_t)^{\vee}\circ r(b)\\ &= (\bigvee\{a:\ (\bigvee_{t\in T} f_t)(a) \le b'\})'\\ &= (\bigvee(\bigcap_{t\in T}\{a:\ f_t(a) \le b'\}))'\\ &= \bigwedge(\bigcap_{t\in T}\{a':\ f_t(a) \le b'\})\\ &= \bigwedge(\bigcap_{t\in T} A(f_t, b')')\\ &= \bigvee_{t\in T}(\bigwedge A(f_t, b')')\\ &= \bigvee_{t\in T}\bigwedge\{a':\ f_t(a) \le b'\}\\ &= \bigvee_{t\in T}(\bigvee\{a:\ f_t(a) \le b'\})'\\ &= (\bigvee_{t\in T} r\circ f_t^{\vee}\circ r)(b)\\ &= (\bigvee_{t\in T} f_t^{\triangleleft})(b),\end{aligned}$$

(v) is proved.

(vi) Suppose $b \in L$, $\gamma \in \beta^*(b)$, then

$$\begin{aligned}(f\wedge g)^{\triangleleft}(b) &= \bigwedge\{a:\ (f\wedge g)(a')\le b'\}\\ &= \bigwedge\{a:\ \bigwedge_{c'\vee d'=a'}(f(c')\vee g(d'))\le b'\}\\ &= \bigwedge\{a:\ \bigvee_{c\wedge d=a}(f(c')\vee g(d'))'\ge b\}\\ &= \bigwedge\{a:\ \forall\lambda\in\beta^*(b),\ \exists c,d,\ c\wedge d=a,\ (f(c')\vee g(d'))'\ge\lambda\}\\ &= \bigwedge\{a:\ \forall\lambda\in\beta^*(b),\ \exists c,d,\ c\wedge d=a,\ f(c')\vee g(d')\le\lambda'\}\\ &\ge \bigwedge\{a:\ \exists c,d,\ c\wedge d=a,\ f(c')\vee g(d')\le\gamma'\}\\ &= \bigwedge\{c\wedge d:\ f(c')\le\gamma',\ g(d')\le\gamma'\}\\ &= (\bigwedge\{c:\ f(c')\le\gamma'\})\wedge(\bigwedge\{d:\ g(d')\le\gamma'\})\\ &= f^{\triangleleft}(\gamma)\wedge g^{\triangleleft}(\gamma).\end{aligned}$$

So

$$(f\wedge g)^{\triangleleft}(b)\ge\bigvee_{\gamma\in\beta^*(b)}(f^{\triangleleft}(\gamma)\vee g^{\triangleleft}(\gamma))=(f^{\triangleleft}\wedge g^{\triangleleft})(b).$$

By (iv), the reverse of the above inequality is obvious. So $(f\wedge g)^{\triangleleft}=f^{\triangleleft}\wedge g^{\triangleleft}$.

(vii) By Proposition **13.1.14** and Theorem **13.1.12** (ii),

$$(f\circ g)^{\triangleleft}=r\circ(f\circ g)^{\vee}\circ r=r\circ(g^{\vee}\circ f^{\vee})\circ r=(r\circ g^{\vee}\circ r)\circ(r\circ f^{\vee}\circ r)=g^{\triangleleft}\circ f^{\triangleleft}.\qquad\square$$

Now we can turn to define $L$-fuzzy uniformites. In the sequel, unless particularly declare, $L^X$ always means an $L$-fuzzy space such that $L$ is a F-lattice.

**13.1.16 Definition** Let $\mathcal{D}\subset\mathcal{E}(L^X)$. $\mathcal{D}$ is called an *$L$-fuzzy quasi-unformity* on $X$, if $\mathcal{D}$ fulfils the following conditions (UF1) – (UF3):

(UF1) $f\in\mathcal{E}(L^X),\ g\in\mathcal{D},\ f\ge g\implies f\in\mathcal{D}$.

(UF2) $f,g\in\mathcal{D}\implies f\wedge g\in\mathcal{D}$.

(UF3) $f\in\mathcal{D}\implies\exists g\in\mathcal{D},\ g\circ g\le f$.

$\mathcal{D}$ is called an *$L$-fuzzy uniformity* on $X$, if $\mathcal{D}$ fulfils the above conditions (UF1) – (UF3) and the following condition:

(UF4) $f\in\mathcal{D}\implies f^{\triangleleft}\in\mathcal{D}$.

Call $(L^X,\mathcal{D})$ an *$L$-fuzzy quasi-uniform space* (or *$L$-fuzzy uniform space*, respectively), if $\mathcal{D}$ is an $L$-fuzzy quasi-uniformity (or an $L$-fuzzy uniformity, respectively) on $X$.

**13.1.17 Definition** Let $\mathcal{D}\subset\mathcal{E}(L^X)$ be an $L$-fuzzy quasi-uniformity (uniformity, respectively) on $X$. A subfamily $\mathcal{B}\subset\mathcal{D}$ is called a *base* of $\mathcal{D}$, if

$$\mathcal{D}=\{f\in\mathcal{E}(L^X):\ \exists g\in\mathcal{B},\ f\ge g\}.$$

$\mathcal{B}$ is called a *subbase* of $\mathcal{D}$, if

$$\{f_0\wedge\cdots\wedge f_n:\ n<\omega,\ \{f_0,\cdots,f_n\}\subset\mathcal{B}\}$$

is a base of $\mathcal{D}$.

**13.1.18 Remark** Recall the uniformity theory in ordinary topology, it is described in the form of subsets of the product set $X\times X$ for some space $X$. Since $L$-fuzzy subsets are lattice-valued mappings, it is natural to represent uniformity structure in $L$-fuzzy topology in the form of mappings. Wang[168] analyzed the relation between these two forms in detail. On the other hand, Liang[63] established a geometrical characterization for $L$-fuzzy uniformity structure defined in the form of mappings.

**13.1.19 Proposition** *Let $\mathcal{B}\subset\mathcal{E}(L^X)$. Then*

(i) *$\mathcal{B}$ is a base of an L-fuzzy quasi-uniformity $\mathcal{D}$ on X if and only if $\mathcal{B}$ fulfils the following conditions* (BUF1) *and* (BUF2).

(ii) *$\mathcal{B}$ is a base of an L-fuzzy uniformity $\mathcal{D}$ on X if and only if $\mathcal{B}$ fulfils the following conditions* (BUF1) – (BUF3).

(BUF1) $f, g \in \mathcal{B} \implies \exists h \in \mathcal{B},\ h \leq f \wedge g$.

(BUF2) $f \in \mathcal{B} \implies \exists g \in \mathcal{B},\ g \circ g \leq f$.

(BUF3) $f \in \mathcal{B} \implies \exists g \in \mathcal{B},\ g \leq f^{\triangleleft}$.

**Proof** (i) (Necessity) Suppose $\mathcal{B}$ is a base of an $L$-fuzzy quasi-uniformity $\mathcal{D}$ on $X$. If $f, g \in \mathcal{B}$, then by (UF2), $f \wedge g \in \mathcal{D}$. So there exists $h \in \mathcal{B}$ such that $h \leq f \wedge g$, (BUF1) is fulfilled. If $f \in \mathcal{B} \subset \mathcal{D}$, then by (UF3), there exists $h \in \mathcal{D}$ such that $h \circ h \leq f$. Since $\mathcal{B}$ is a base of $\mathcal{D}$, there exists $g \in \mathcal{B}$ such that $g \leq h$. Hence $g \circ g \leq h \circ h \leq f$, (BUF2) is fulfilled.

(Sufficiency) Suppose $\mathcal{B}$ fulfils conditions (BUF1) and (BUF2). Let

$$\mathcal{D} = \{f \in \mathcal{E}(L^X) :\ \exists g \in \mathcal{B},\ f \geq g\},$$

then (UF1) is fulfilled obviously. If $f, g \in \mathcal{D}$, then there exists $h_1, h_2 \in \mathcal{B}$ such that $f \geq h_1$, $g \geq h_2$. By (BUF1), there exists $h_3 \in \mathcal{B}$ such that $h_3 \leq h_1 \wedge h_2 \leq f \wedge g$. So $f \wedge g \in \mathcal{D}$, (UF2) is fulfilled. If $f \in \mathcal{D}$, then there exists $g \in \mathcal{B}$ such that $f \geq g$. By (BUF2), there exists $h \in \mathcal{B} \subset \mathcal{D}$ such that $h \circ h \leq g \leq f$. So (UF3) is also fulfilled, $\mathcal{D}$ is an $L$-fuzzy quasi-uniformity on $X$ and $\mathcal{B}$ is just a base of $\mathcal{D}$.

(ii) (Necessity) Suppose $\mathcal{B}$ is a base of an $L$-fuzzy uniformity on $X$, it is sufficient to verify (BUF3). If $f \in \mathcal{B} \subset \mathcal{D}$, then by (UF4), $f^{\triangleleft} \in \mathcal{D}$. So there exists $g \in \mathcal{B}$ such that $g \leq f^{\triangleleft}$, (BUF3) is fulfilled.

(Sufficiency) Suppose $\mathcal{B}$ fulfils conditions (BUF1) – (BUF3). Let

$$\mathcal{D} = \{f \in \mathcal{E}(L^X) :\ \exists g \in \mathcal{B},\ f \geq g\},$$

then by (i) proved above, $\mathcal{D}$ is an $L$-fuzzy quasi-uniformity on $X$. If $f \in \mathcal{D}$, then there exists $g \in \mathcal{B}$ such that $f \geq g$. By (BUF3), there exists $h \in \mathcal{B}$ such that $h \leq g^{\triangleleft}$. By Theorem **13.1.15** (iv), $f^{\triangleleft} \geq g^{\triangleleft} \geq h$. So $f^{\triangleleft} \in \mathcal{D}$, (UF4) is fulfilled, $\mathcal{D}$ is an $L$-fuzzy uniformity on $X$. □

**13.1.20 Proposition** *Let $\mathcal{P} \subset \mathcal{E}(L^X)$.*

(i) *$\mathcal{P}$ is a subbase of an L-fuzzy quasi-uniformity $\mathcal{D}$ on X if and only if $\mathcal{P}$ fulfils the following condition* (SBUF1).

(ii) *$\mathcal{P}$ is a subbase of an L-fuzzy uniformity $\mathcal{D}$ on X if and only if $\mathcal{P}$ fulfils the following conditions* (SBUF1) – (SBUF2).

(SBUF1) $f \in \mathcal{P} \implies \exists \{f_0, \cdots, f_n\} \in [\mathcal{P}]^{<\omega},\ (f_0 \wedge \cdots \wedge f_n) \circ (f_0 \wedge \cdots \wedge f_n) \leq f$.

(SBUF2) $f \in \mathcal{P} \implies \exists \{f_0, \cdots, f_n\} \in [\mathcal{P}]^{<\omega},\ f_0 \wedge \cdots \wedge f_n \leq f^{\triangleleft}$.

**Proof** Let

$$\mathcal{B} = \{f_0 \wedge \cdots \wedge f_n :\ n < \omega,\ \{f_0, \cdots, f_n\} \subset \mathcal{P}\},$$

(i) (Necessity) If $\mathcal{P}$ is a subbase of an $L$-fuzzy quasi-uniformity $\mathcal{D}$ on $X$, then $\mathcal{B}$ defined above is a base of $\mathcal{D}$. By Proposition **13.1.19**, $\mathcal{B}$ fulfils (BUF2), for every $f \in \mathcal{P} \subset \mathcal{B}$, there exists $g \in \mathcal{B}$ such that $g \circ g \leq f$. Then there exists $\{g_0, \cdots, g_n\} \in [\mathcal{P}]^{<\omega}$ such that $g = f_0 \wedge \cdots \wedge f_n$. So

$$(f_0 \wedge \cdots \wedge f_n) \circ (f_0 \wedge \cdots \wedge f_n) = g \circ g \leq f,$$

$\mathcal{P}$ fulfils (SBUF1).

(Sufficiency) Suppose $\mathcal{P}$ fulfils the condition (SBUF1), then $\mathcal{B}$ is clearly fulfils the condition (BUF1). Let $f \in \mathcal{B}$, then there exists $\{f_0, \cdots, f_n\} \in [\mathcal{P}]^{<\omega}$ such that $f = f_0 \wedge \cdots \wedge f_n$. For every $i \leq n$, by (SBUF1), there exists $\{g_{i,0}, \cdots, g_{i,j_i}\} \in [\mathcal{P}]^{<\omega}$ such that $(g_{i,0} \wedge \cdots \wedge g_{i,j_i}) \circ (g_{i,0} \wedge \cdots \wedge g_{i,j_i}) \leq f_i$. Then

$$(\bigwedge_{i \leq n} \bigwedge_{j \leq j_i} g_{i,j}) \circ (\bigwedge_{i \leq n} \bigwedge_{j \leq j_i} g_{i,j}) \leq \bigwedge_{i \leq n} ((g_{i,0} \wedge \cdots \wedge g_{i,j_i}) \circ (g_{i,0} \wedge \cdots \wedge g_{i,j_i})) \leq \bigwedge_{i \leq n} f_i = f.$$

Since $\bigwedge_{i \leq n} \bigwedge_{j \leq j_i} g_{i,j} \in \mathcal{B}$, (BUF2) is true, so $\mathcal{B}$ is a base of an $L$-fuzzy quasi-uniformity on $X$.

(ii) (Necessity) If $\mathcal{P}$ is a subbase of an $L$-fuzzy uniformity $\mathcal{D}$ on $X$, then by (i) proved above, $\mathcal{P}$ fulfils (SBUF1). Since $\mathcal{B}$ defined above is a base of $\mathcal{D}$, by Proposition **13.1.15**, $\mathcal{B}$ fulfils (BUF3). So for every $f \in \mathcal{P} \subset \mathcal{B}$, there exists $g \in \mathcal{B}$ such that $g \leq f^{\triangleleft}$. By the definition of $\mathcal{B}$, this just implies (SBUF2).

(Sufficiency) Suppose $\mathcal{P}$ fulfils (SBUF1) and (SBUF2), then by (i) proved above, $\mathcal{B}$ fulfils (BUF1) and (BUF2), we need only prove that $\mathcal{B}$ also fulfils (BUF3). For every $f \in \mathcal{B}$, then there exists $\{f_0, \cdots, f_n\} \in [\mathcal{P}]^{<\omega}$ such that $f = f_0 \wedge \cdots \wedge f_n$. By (SBUF2), for every $i \leq n$, there exists $\{g_{i,0}, \cdots, g_{i,j_i}\} \in [\mathcal{P}]^{<\omega}$ such that $g_{i,0} \wedge \cdots \wedge g_{i,j_i} \leq f_i^{\triangleleft}$. So by Theorem **13.1.15** (vi),

$$\bigwedge_{i \leq n} \bigwedge_{j \leq j_i} g_{i,j} \leq \bigwedge_{i \leq n} f_i^{\triangleleft} = (\bigwedge_{i \leq n} f_i)^{\triangleleft} = f^{\triangleleft}.$$

Since $\bigwedge_{i \leq n} \bigwedge_{j \leq j_i} g_{i,j} \in \mathcal{B}$, this implies that $\mathcal{B}$ fulfils (BUF3). □

The following proposition is a direct consequence of Proposition **13.1.20**:

**13.1.21 Proposition** *Let $\{\mathcal{P}_t : t \in T\} \subset \mathcal{E}(L^X)$ be a family of subbases of L-fuzzy quasi-uniformities (uniformities, respectively) on $X$, then $\mathcal{P} = \bigcup_{t \in T} \mathcal{P}_t$ is a subbase of an L-fuzzy quasi-uniformity (uniformity, respectively) on $X$.* □

Then the following definitions are reasonable:

**13.1.22 Definition** Let $\mathcal{B} \subset \mathcal{E}(L^X)$ fulfil conditions (BUF1) – (BUF2) ((BUF1) – (BUF3), respectively). Then

$$\mathcal{D} = \{f \in \mathcal{E}(L^X) : \exists g \in \mathcal{B}, \ f \geq g\}$$

is called the $L$-fuzzy quasi-uniformity ($L$-fuzzy uniformity, respectively) *generated by the base $\mathcal{B}$*.

**13.1.23 Definition** Let $\mathcal{P} \subset \mathcal{E}(L^X)$ fulfil conditions (SBUF1) ((SBUF1) – (SBUF2), respectively). Then

$$\mathcal{D} = \{f \in \mathcal{E}(L^X) : \exists \{g_0, \cdots, g_n\} \subset \mathcal{P}, \ f \geq g_0 \wedge \cdots \wedge g_n\}$$

is called the $L$-fuzzy quasi-uniformity ($L$-fuzzy uniformity, respectively) *generated by the subbase $\mathcal{P}$*.

**13.1.24 Theorem** *Let $\mathcal{D} \subset \mathcal{E}(L^X)$ fulfil conditions* (UF1) *and* (UF2)*. Then $\mathcal{D}$ is an L-fuzzy uniformity on $X$ if and only if for every $f \in \mathcal{D}$, there exists $g \in \mathcal{D}$ such that $g = g^{\triangleleft}$ and $g \circ g \leq f$.*

**Proof** (Necessity) By (UF3), there exists $h \in \mathcal{D}$ such that $h \circ h \leq f$. By (UF4), $h^{\triangleleft} \in \mathcal{D}$, and hence $g = h \wedge h^{\triangleleft} \in \mathcal{D}$. By Theorem **13.1.15** (iii) and (vi),

$$g^{\triangleleft} = (h \wedge h^{\triangleleft})^{\triangleleft} = h^{\triangleleft} \wedge (h^{\triangleleft})^{\triangleleft} = h^{\triangleleft} \wedge h = g.$$

Finally, since every element of $\mathcal{E}(L^X) \supset \mathcal{D}$ is order preserving, so by $h \circ h^{\triangleleft} \leq h$ we have

$$g \circ g = (h \wedge h^{\triangleleft}) \circ (h \wedge h^{\triangleleft}) \leq h \circ h \leq f.$$

(Sufficiency) Suppose $f \in \mathcal{D}$ and there exists $g \in \mathcal{D}$ such that $g = g^{\triangleleft}$ and $g \circ g \leq f$, then (UF3) is fulfilled. Since $g$ is value increasing, $g \circ g \geq g$. By Theorem **13.1.15** (iv),

$$f^{\triangleleft} \geq (g \circ g)^{\triangleleft} \geq g^{\triangleleft} = g \in \mathcal{D}.$$

So by (UF1), $f^{\triangleleft} \in \mathcal{D}$, (UF4) is fulfilled. □

**13.1.25 Theorem** *Let $(L^X, \mathcal{D})$ be an L-fuzzy quasi-uniform space, mapping $i : L^X \to L^X$ be defined as follows:*

$$\forall A \in L^X, \quad i(A) = \bigvee\{C \in L^X : \exists f \in \mathcal{D}, \ f(C) \leq A\}, \tag{13.7}$$

*then $i$ is an interior operator on $L^X$.*

**Proof** Since $f(C) \leq \underline{1}$ for every $f \in \mathcal{D} \subset \mathcal{E}(L^X)$ and every $C \in L^X$, so $i(\underline{1}) = \underline{1}$, condition (IO1) of interior opertor is fulfilled. Every $f \in \mathcal{D}$ is value increasing, so for arbitrary $A, C \in L^X$ such that there exists $f \in \mathcal{D}$ such that $f(C) \leq A$, we have $C \leq f(A) \leq A$. So $i(A) \geq A$, condition (IO2) is fulfilled. Since every $f \in \mathcal{D}$ is order preserving, so is $i$. Hence to prove condition (IO3) of interior operator we need only prove $i(A) \wedge i(B) \leq i(A \wedge B)$ for arbitrary $A, B \in L^X$. In fact, since for arbitrary $f, g \in \mathcal{D}$ and arbitrary $A, B, C, D \in L^X$ such that $f(C) \leq A$ and $g(D) \leq B$, we have

$$(f \wedge g)(C \wedge D) \leq f(C) \wedge g(D) \leq A \wedge B,$$

so

$$\begin{aligned} i(A) \wedge i(B) &= \bigvee\{C \wedge D : \ C, D \in L^X, \ \exists f, g \in \mathcal{D}, \ f(C) \leq A, \ g(D) \leq B\} \\ &\leq \bigvee\{C \wedge D : \ C, D \in L^X, \ \exists f, g \in \mathcal{D}, \ (f \wedge g)(C \wedge D) \leq A \wedge B\} \\ &\leq \bigvee\{C \in L^X : \ \exists f \in \mathcal{D}, \ f(C) \leq A \wedge B\} \\ &= i(A \wedge B). \end{aligned}$$

Hence (IO3) is satisfied. For every $A \in L^X$, since every $g \in \mathcal{D}$ is value increasing, it follows from (UF3),

$$\begin{aligned} C \in L^X, \ f \in \mathcal{D}, \ f(C) \leq A &\Longrightarrow \ \exists g \in \mathcal{D}, \ g \circ g(C) \leq f(C) \leq A \\ &\Longrightarrow \ g(C) \leq i(A) \\ &\Longrightarrow \ C \leq i(i(A)). \end{aligned}$$

So

$$i(A) = \bigvee\{C \in L^X : \ \exists f \in \mathcal{D}, \ f(C) \leq A\} \leq ii(A).$$

By the virtue of (IO2), $ii(A) = i(A)$, (IO4) is fulfilled. □

**13.1.26 Definition** Let $\mathcal{D}$ be an $L$-fuzzy quasi-uniformity on $X$.

The interior operator defined in (13.7) is called the interior operator on $L^X$ *generated by the L-fuzzy quasi-uniformity* $\mathcal{D}$. The $L$-fuzzy topology on $X$ generated by the interior opertor defined by (13.7) is called the $L$-fuzzy topology *generated by the L-fuzzy quasi-uniformity* $\mathcal{D}$, denoted by $\delta(\mathcal{D})$. $(L^X, \delta(\mathcal{D}))$ is called the $L$-fts *corresponding to* $(L^X, \mathcal{D})$.

**13.1.27 Lemma** *Let $(L^X, \mathcal{D})$ be an L-fuzzy quasi-uniform space, $\mathcal{B}$ a base of $\mathcal{D}$, $A \in L^X$. Then*

(i) $i(A) = \bigvee\{C \in L^X : \exists f \in \mathcal{B},\ f(C) \leq A\}$.
(ii) $i(A) = \bigvee\{C \in L^X : \exists f \in \mathcal{B},\ f \circ f(C) \leq A\}$.
(iii) $i(A) = \bigvee\{f(C) : C \in L^X,\ f \in \mathcal{B},\ f \circ f(C) \leq A\}$.
(iv) $i(A) = \bigvee\{(f^{\triangleleft}(A'))' : f \in \mathcal{B}\}$.

**Proof** (i) Denote $\mathcal{A} = \{C \in L^X : \exists f \in \mathcal{B},\ f(C) \leq A\}$, then by $\mathcal{B} \subset \mathcal{D}$, $\bigvee \mathcal{A} \leq i(A)$. If $C \in L^X$ and $f \in \mathcal{D}$ such that $f(C) \leq A$, then there exists $g \in \mathcal{B}$ such that $g \leq f$, $g(C) \leq f(C) \leq A$, $C \in \mathcal{A}$. So $i(A) \leq \bigvee \mathcal{A}$, and hence $i(A) = \bigvee \mathcal{A}$.

(ii) If $C \in L^X$ and $g \in \mathcal{D}$ such that $g(C) \leq A$, then by Proposition **13.1.19** (ii), there exists $f \in \mathcal{B}$ such that $f \circ f \leq g$, $(f \circ f)(C) \leq g(C) \leq A$. So

$$i(A) = \bigvee\{C \in L^X : \exists g \in \mathcal{D},\ g(C) \leq A\} \leq \bigvee\{C \in L^X : \exists f \in \mathcal{B},\ f \circ f(C) \leq A\}.$$

Conversely, if $C \in L^X$ and $g \in \mathcal{B}$ such that $g \circ g(C) \leq A$, then by $g \circ g \in \mathcal{D}$,

$$\bigvee\{C \in L^X : \exists g \in \mathcal{B},\ g \circ g(C) \leq A\} \leq \bigvee\{C \in L^X : \exists f \in \mathcal{D},\ f(C) \leq A\} = i(A).$$

(iii) Since every member of $\mathcal{B} \subset \mathcal{D}$ is value increasing, by (ii) proved above,

$$\bigvee\{f(C) : C \in L^X,\ f \in \mathcal{B},\ f \circ f(C) \leq A\} \geq i(A).$$

On the other hand, if $C \in L^X$, $f \in \mathcal{B}$ such that $f \circ f(C) \leq A$, then $f(C) \in \{D \in L^X : \exists g \in \mathcal{D},\ g(D) \leq A\}$, and hence

$$\bigvee\{f(C) : C \in L^X, f \in \mathcal{B}, f \circ f(C) \leq A\} \leq \bigvee\{D \in L^X : \exists g \in \mathcal{D}, g(D) \leq A\} = i(A).$$

(iv) By (i) proved above,

$$\begin{aligned} i(A) &= \bigvee\{C \in L^X : \exists f \in \mathcal{B},\ f(C) \leq A\} \\ &= \bigvee\{C' : C \in L^X, \exists f \in \mathcal{B},\ f(C') \leq (A')'\} \\ &= \bigvee\{\bigvee\{C' : C \in L^X,\ f(C') \leq (A')'\} : f \in \mathcal{B}\} \\ &= \bigvee\{(f^{\triangleleft}(A'))' : f \in \mathcal{B}\}. \end{aligned}$$

□

Dual to the $L$-fuzzy topologies generated by interior operators on $L^X$, we can also consider the $L$-fuzzy topologies generated by closure operators on $L^X$. In fact, some authors just generated $L$-fuzzy topologies in $L$-fuzzy uniform spaces with closure operators.[168]

**13.1.28 Theorem** *Let $(L^X, \mathcal{D})$ be an $L$-fuzzy quasi-uniform space, mapping $c : L^X \to L^X$ be defined as follows:*

$$\forall A \in L^X, \quad c(A) = \bigwedge\{f(A) : f \in \mathcal{D}\}, \tag{13.8}$$

*then $c$ is a closure operator on $L^X$.*

**Proof** Since every $f \in \mathcal{D}$ is arbitrary join preserving, so $f(\underline{0}) = \underline{0}$, $c(\underline{0}) = \underline{0}$, condition (CO1) of closure opertor is fulfilled. Every $f \in \mathcal{D}$ is value increasing, so $c(A) \geq A$, condition (CO2) is fulfilled. Since every $f \in \mathcal{D}$ is order preserving, so is $c$. Hence in order to prove condition (CO3) of closure operator we need only prove $c(A \vee B) \leq c(A) \vee c(B)$ for arbitrary $A, B \in L^X$. Suppose $e \in M(L^X)$ such that $e \not\leq c(A) \vee c(B)$, then $e \not\leq c(A)$, $e \not\leq c(B)$, and hence there exist $f, g \in \mathcal{D}$ such that $e \not\leq f(A)$, $e \not\leq g(B)$. Then $e \not\leq (f \wedge g)(A \vee B)$, $e \not\leq c(A \vee B)$. So $c(A \vee B) \leq c(A) \vee c(B)$, (CO3) is fulfilled. For every $A \in L^X$, since

$$cc(A) = \bigwedge\{f(c(A)) : f \in \mathcal{D}\} \leq \bigwedge\{f(f(A)) : f \in \mathcal{D}\}.$$

By condition (UF4) of $L$-fuzzy uniformity,

$$cc(A) \leq \bigwedge\{f \circ f(A) : f \in \mathcal{D}\} \leq \bigwedge\{g(A) : g \in \mathcal{D}\} = c(A).$$

So $cc(A) = c(A)$, condition (CO4) is fulfilled. □

**13.1.29 Definition** Let $\mathcal{D}$ be an $L$-fuzzy quasi-uniformity on $X$. The closure operator defined in (13.8) is called the closure operator on $L^X$ *generated by the L-fuzzy quasi-uniformity* $\mathcal{D}$.

Now in an $L$-fuzzy quasi-uniform space $(L^X, \mathcal{D})$, two $L$-fuzzy topologies $\delta_i$ and $\delta_c$ can be generated respectively by the interior operator $i$ and the closure operator $c$ which are generated by $\mathcal{D}$. A natural question is: "Is $\delta_i$ coincident with $\delta_c$"? For $L$-fuzzy quasi-uniform spaces, unfortunately, the answer is negative, just as the following example[168] shows. However, for $L$-fuzzy uniform spaces, the answer is positive, just as the following theorem **13.1.31** shows.

**13.1.30 Example** Let

$$X = [0,1],\ L = \{0,1\},$$
$$D = \{(x,y):\ 0 \le x \le y \le 1\},$$
$$g_D:\ L^X \to L^X,\quad g_D(A) = \bigvee\{y_1:\ \exists x \in supp(A),\ (x,y) \in D\},$$

then $g_D \in \mathcal{E}(L^X)$. Since $D \circ D = D$ if consider $D$ as a relation on $X$, so $g_D \circ g_D = g_D$. By Proposition **13.1.19** (i), $\{g_D\}$ is a base of the $L$-fuzzy quasi-uniformity $\mathcal{D} = \{f \in \mathcal{E}(L^X):\ f \ge g_D\}$. Then one can verify that every element of $\delta_c$ is a lower set in $[0,1]$ and every element of $\delta_i$ is an upper set in $[0,1]$. Especially, $[0, \frac{1}{2}] \in \delta_c \backslash \delta_i$. So $\delta_c \ne \delta_i$.

**13.1.31 Theorem** *Let $(L^X, \mathcal{D})$ be an $L$-fuzzy uniform space, $i$ and $c$ be the interior operator and the closure operator on $L^X$ generated by $\mathcal{D}$ respectively, $\delta_i$ and $\delta_c$ be the $L$-fuzzy topologies generated by $i$ and $c$ respectively. Then*

(i) *For every $A \in L^X$, $i(A)' = c(A')$.*

(ii) $\delta_i = \delta_c$.

**Proof** It is sufficient to prove (i). Suppose $A \in L^X$. For every $f \in \mathcal{D}$, by Theorem **13.1.24**, there exists $g_f \in \mathcal{D}$ such that $(g_f)^{\triangleleft} = g_f$ and $g_f \circ g_f \le f$. Since $g_f$ is value increasing, $g_f \le g_f \circ g_f \le f$. By Theorem **13.1.15** (iv), $(g_f)^{\triangleleft} \le f^{\triangleleft}$. So by Lemma **13.1.27** (iv) and $\{g_f:\ f \in \mathcal{D}\} \subset \{f:\ f \in \mathcal{D}\}$,

$$\begin{aligned} i(A)' &= \bigwedge\{f^{\triangleleft}(A'):\ f \in \mathcal{D}\} \\ &\ge \bigwedge\{(g_f)^{\triangleleft}(A'):\ f \in \mathcal{D}\} \\ &= \bigwedge\{g_f(A'):\ f \in \mathcal{D}\} \\ &\ge \bigwedge\{f(A'):\ f \in \mathcal{D}\} \\ &= c(A'). \end{aligned}$$

On the other hand, by $\{(g_f)^{\triangleleft}:\ f \in \mathcal{D}\} \subset \{f^{\triangleleft}:\ f \in \mathcal{D}\}$ and $g_f \le f$ for every $f \in \mathcal{D}$,

$$\begin{aligned} i(A)' &= \bigwedge\{f^{\triangleleft}(A'):\ f \in \mathcal{D}\} \\ &\le \bigwedge\{(g_f)^{\triangleleft}(A'):\ f \in \mathcal{D}\} \\ &\le \bigwedge\{g_f(A'):\ f \in \mathcal{D}\} \\ &\le \bigwedge\{f(A'):\ f \in \mathcal{D}\} \\ &= c(A'). \end{aligned}$$

So $i(A)' = c(A')$. □

Theorem **13.1.25** shows that every $L$-fuzzy quasi-uniformity can generates an $L$-fuzzy topology; but the unexpected result is that its reverse is also true:

**13.1.32 Theorem** *Let $(L^X, \delta)$ be an L-fts, then there exists an L-fuzzy quasi-uniformity $\mathcal{D}$ on $X$ such that $\delta = \delta(\mathcal{D})$.*

**Proof** For every $U \in \delta$, define a self mapping $f_U$ on $L^X$ as follows:

$$\forall A \in L^X, \quad f_U(A) = \begin{cases} \underline{1}, & A \not\leq U, \\ U, & \underline{0} \neq A \leq U, \\ \underline{0}, & A = \underline{0}, \end{cases} \tag{13.9}$$

then it is easy to find that $f_U$ is value increasing. Suppose $\mathcal{A} \subset L^X$,

$$f_U(\bigvee \mathcal{A}) = \begin{cases} \underline{1}, & \bigvee \mathcal{A} \not\leq U, \\ U, & \underline{0} \neq \bigvee \mathcal{A} \leq U, \\ \underline{0}, & \bigvee \mathcal{A} = \underline{0}. \end{cases} \tag{13.10}$$

$$\bigvee_{A \in \mathcal{A}} f_U(A) = \begin{cases} \underline{1}, & \exists A \in \mathcal{A},\ A \not\leq U, \\ U, & \forall A \in \mathcal{A},\ A \leq U,\ \exists A_0 \in \mathcal{A},\ A_0 \neq \underline{0}, \\ \underline{0}, & \forall A \in \mathcal{A},\ A = \underline{0}. \end{cases} \tag{13.11}$$

Since the conditions in the right sides of (13.10) and (13.11) are equivalent respectively, so $f_U(\bigvee \mathcal{A}) = \bigvee_{A \in \mathcal{A}} f_U(A)$, $f_U$ is arbitrary join preserving. So $f_U \in \mathcal{E}(L^X)$.

For every $A \in L^X$,

$$(f_U \circ f_U)(A) = \begin{cases} \underline{1}, & f_U(A) \not\leq U, \\ U, & \underline{0} \neq f_U(A) \leq U, \\ \underline{0}, & f_U(A) = \underline{0}. \end{cases}$$

But $f_U(A) \not\leq U$ if and only if $A \not\leq U$, so $f_U(A) \leq U$ if and only if $A \leq U$. Moreover, $f_U(A) = \underline{0}$ if and only if $A = \underline{0}$, so $f_U \circ f_U = f_U$.

Let

$$\mathcal{D} = \{f \in \mathcal{E}(L^X) : \ \exists \mathcal{A} \in [\delta]^{<\omega},\ f \geq \bigwedge_{U \in \mathcal{A}} f_U\},$$

then conditions (UF1) and (UF2) are obviously satified. Suppose $f \in \mathcal{D}$, take $\mathcal{A} \in [\delta]^{<\omega}$ such that

$$f \geq \bigwedge_{U \in \mathcal{A}} f_U = \bigwedge_{U \in \mathcal{A}} (f_U \circ f_U).$$

Since for every $V \in \mathcal{A}$,

$$f_V \circ f_V \geq (\bigwedge_{U \in \mathcal{A}} f_U) \circ (\bigwedge_{U \in \mathcal{A}} f_U),$$

so take $g = \bigwedge_{U \in \mathcal{A}} f_U$ we have $g \in \mathcal{D}$ and $g \circ g \leq f$, condition (UF3) is satisfied, $\mathcal{D}$ is an $L$-fuzzy quasi-uniformity on $X$. Then by the viertue of Theorem **13.1.25**, $\delta(\mathcal{D})$ generates a interior operator $i$ on $L^X$ defined as (13.7). In order to prove $\delta(\mathcal{D}) = \delta$, we need only prove that $i(A) = A$ if and only if $A \in \delta$.

Let $A \in \delta$, then $f_A(A) = A$. Since $C \leq f(C)$ for every $C \in L^X$ and every $f \in \mathcal{D}$, so by the virtue of $f_A \in \mathcal{D}$,

$$i(A) = \bigvee\{C \in L^X : \ \exists f \in \mathcal{D},\ f(C) \leq A\} = A.$$

Conversely, suppose $i(A) = A$, then by Lemma **13.1.27** (ii) and (13.7),

$$\begin{aligned} i(A) &= \bigvee\{C \in L^X : \exists g \in \mathcal{D},\ g\circ g(C) \le A\} \\ &= \bigvee\{C \in L^X : \exists g \in \mathcal{E}(L^X),\ \exists \mathcal{A} \in [\delta]^{<\omega},\ g \ge \bigwedge_{U\in\mathcal{A}} f_U,\ g\circ g(C) \le A\} \\ &\le \bigvee\{(\bigwedge_{U\in\mathcal{A}} f_U)(C) : C \in L^X,\ \mathcal{A} \in [\delta]^{<\omega},\ \exists g \in \mathcal{E}(L^X),\ g \ge \bigwedge_{U\in\mathcal{A}} f_U, \\ &\qquad g((\bigwedge_{U\in\mathcal{A}} f_U)(C)) \le A\} \\ &\le \bigvee\{D \in L^X : \exists g \in \mathcal{D},\ g(D) \le A\} \\ &= i(A). \end{aligned}$$

Let $P$ denote the condition

$$C \in L^X,\ \mathcal{A} \in [\delta]^{<\omega},\ \exists g \in \mathcal{E}(L^X),\ g \ge \bigwedge_{U\in\mathcal{A}} f_U,\ g((\bigwedge_{U\in\mathcal{A}} f_U)(C)) \le A,$$

then $i(A) = \bigvee\{(\bigwedge_{U\in\mathcal{A}} f_U)(C) : P\}$. Since $f_U(C) \in \delta$ for every $U \in \delta$, so by (13.2), every $(\bigwedge_{U\in\mathcal{A}} f_U)(C)$ is open $(L^X, \delta)$, and hence $A = i(A) \in \delta$. □

For convenience, it is necessary to simplify the statement of Theorem **13.1.32**:

**13.1.33 Definition** An $L$-fts $(L^X, \delta)$ is called *quasi-uniformizable* (or *uniformizable*, respectively), if there exists an $L$-fuzzy quasi-uniformity (or an $L$-fuzzy uniformity, respectively) $\mathcal{D}$ on $X$ such that $\delta$ can be generated by $\mathcal{D}$.

Then Theorem **13.1.32** can be restated as: "Every $L$-fts is quasi-uniformizable". But how about the uniformizability of an $L$-fts? The following well known example shown by Hutton[38] is the first one in this respect:

**13.1.34 Example** $L$-fuzzy unit interval $I(L)$ is uniformizable.

For convenience, define

$$\begin{aligned} L_{-\infty} &= \underline{0}_{I[L]},\ L_{+\infty} = \underline{1}_{I[L]}, \\ R_{-\infty} &= \underline{1}_{I[L]},\ R_{+\infty} = \underline{0}_{I[L]}, \\ \mathcal{S} &= \mathcal{S}_L^I \cup \{L_{-\infty}, L_{+\infty}, R_{-\infty}, R_{+\infty}\}, \end{aligned}$$

and denote $X = I[L]$, $\delta = \mathcal{T}_L^I$, then $\mathcal{S}$ is a subbase of $\delta$. For every $A \in L^X$, let

$$\begin{aligned} S(A) &= \{s \in \mathbf{R} : A \le L_s{}'\},\quad u(A) = \bigvee S(A), \\ T(A) &= \{t \in \mathbf{R} : A \le R_t{}'\},\quad l(A) = \bigwedge T(A), \end{aligned} \tag{13.12}$$

then we always have $(-\infty, 0) \subset S(A)$, $(1, +\infty) \subset T(A)$. So both $S(A)$ and $T(A)$ are always nonempty. For every $\varepsilon > 0$ and every $A \in L^X$, let

$$f_\varepsilon(A) = R_{u(A)-\varepsilon}, \tag{13.13}$$

then the family $\{f_\varepsilon : \varepsilon > 0\}$ possesses the following properties:

(i) $f_\varepsilon \in \mathcal{E}(L^X)$, and $f_\varepsilon \ge f_\rho$ whenever $\varepsilon \ge \rho > 0$.

In fact, for every $A \in L^X$, every $x = [\lambda] \in X$ and every $s \in S(A)$, we have $A(x) \le \lambda(s-)$. So

$$\begin{aligned} A(x) &\le \bigwedge\{\lambda(s-) : s \in S(A)\} \\ &\le \lambda((u(A) - \varepsilon)+) \\ &= R_{u(A)-\varepsilon}(x) \\ &= f_\varepsilon(A)(x), \end{aligned}$$

$f_\varepsilon$ is value increasing.

Now prove that $f_\varepsilon$ is arbitrary join preserving. Form (13.12) we know

$$S(\bigvee_{\alpha\in\Gamma} A_\alpha) = \bigcap_{\alpha\in\Gamma} S(A_\alpha).$$

Note that $S(A)$ is a lower set, possesses the form of $(-\infty, b)$ or $(-\infty, b]$, we have

$$u(\bigvee_{\alpha\in\Gamma} A_\alpha) = \bigwedge_{\alpha\in\Gamma} u(A_\alpha). \tag{13.14}$$

By (13.13), (13.14) and Proposition **9.1.2** (iv),

$$\bigvee_{\alpha\in\Gamma} f_\varepsilon(A_\alpha) = \bigvee_{\alpha\in\Gamma} R_{u(A_\alpha)-\varepsilon} = R_{(\bigwedge_{\alpha\in\Gamma} u(A_\alpha))-\varepsilon} = R_{u(\bigvee_{\alpha\in\Gamma} A_\alpha)-\varepsilon} = f_\varepsilon(\bigvee_{\alpha\in\Gamma} A_\alpha),$$

$f_\varepsilon$ is arbitrary join preserving. So $f_\varepsilon \in \mathcal{E}(L^X)$.

As for the other conclusion, clearly we have $\varepsilon \geq \rho > 0 \implies f_\varepsilon \geq f_\rho$.

(ii) For every $\varepsilon > 0$ and every $A \in L^X$,

$$f_\varepsilon{}^\triangleleft(A) = L_{l(A)+\varepsilon}. \tag{13.15}$$

And hence $f_\varepsilon{}^\triangleleft \geq f_\rho{}^\triangleleft$ whenever $\varepsilon \geq \rho$.

In fact, supppose $A \neq \underline{0}$, we have

$$f_\varepsilon{}^\triangleleft(A) = \bigwedge\{B:\ f_\varepsilon(B') \leq A'\} = \bigwedge\{B:\ R_{u(B')-\varepsilon} \leq A'\}. \tag{13.16}$$

Suppose $R_{u(B')-\varepsilon} \leq A'$, then $A \leq (R_{u(B')-\varepsilon})'$. It follows from the definition (13.12), $u(B') - \varepsilon \in T(A)$. So $u(B') - \varepsilon \geq l(A)$, and hence

$$u(B') \geq l(A) + \varepsilon. \tag{13.17}$$

This implies

$$R_{u(B')-\varepsilon} \leq A' \implies u(B') \geq l(A) + \varepsilon. \tag{13.18}$$

Conversely, suppose (13.17) holds, then for every $\rho > 0$, $B \geq L_{l(A)+\varepsilon-\rho}$. This is true because otherwise we should have $B' \not\leq (L_{l(A)+\varepsilon-\rho})'$, then by the property of $L_s$, the following implication holds:

$$B' \leq L_s{}' \implies s < l(A) + \varepsilon - \rho,$$

and hence $M(B') + \varepsilon - \rho$, contradicts with (13.17). So

$$R_{u(B')-\varepsilon} \leq A' \implies \forall \rho > 0,\ B \geq L_{l(A)+\varepsilon-\rho}.$$

Then by Proposition **9.1.2** (iii),

$$R_{u(B')-\varepsilon} \leq A' \implies B \geq \bigvee_{\rho>0} L_{l(A)+\varepsilon-\rho} = L_{\bigvee_{\rho>0}(l(A)+\varepsilon-\rho)} = L_{l(A)+\varepsilon}.$$

So by the virtue of (13.16), $f_\varepsilon{}^\triangleleft(A) \geq L_{l(A)+\varepsilon}$.

On the other hand, suppose $B = L_{l(A)+\varepsilon}$ we have

$$u(B') - \varepsilon = l(A) = \bigwedge\{t \in \mathbf{R}:\ A \leq R_t{}'\}.$$

Then by Proposition **9.1.2** (iv),

$$\begin{aligned} R_{u(B')-\varepsilon} &= R_{\bigwedge\{t\in\mathbf{R}:\ A\leq R_t{}'\}} \\ &= \bigvee\{R_t:\ t\in\mathbf{R},\ A \leq R_t{}'\} \\ &= \bigvee\{R_t:\ t\in\mathbf{R},\ R_t \leq A'\} \\ &\leq A'. \end{aligned}$$

So $R_{u(B')-\varepsilon} \leq A'$, as a result of (13.16), $f_\varepsilon{}^\triangleleft(A) \leq L_{l(A)+\varepsilon}$. Therefore, (13.15) is proved.

(iii) For every $\varepsilon > 0$ and every $\rho > 0$,

$$f_\varepsilon \circ f_\rho = f_{\varepsilon+\rho}. \tag{13.19}$$

In fact, for every $A \in L^X$,

$$(f_\varepsilon \circ f_\rho)(A) = f_\varepsilon(f_\rho(A)) = R_{u(f_\rho(A))-\varepsilon}. \tag{13.20}$$

Since

$$\begin{aligned} u(f_\rho(A)) &= \bigvee\{s \in \mathbf{R} :\ f_\rho(A) \leq L_s{}'\} \\ &= \bigvee\{s \in \mathbf{R} :\ R_{u(A)-\varepsilon} \leq L_s{}'\}, \end{aligned}$$

and we have

$$\begin{aligned} R_{u(A)-\rho} \leq L_s{}' &\Longleftrightarrow \forall x = [\lambda] \in X,\ R_{u(A)-\rho}(x) \leq L_s{}'(x) \\ &\Longleftrightarrow \lambda((u(A) - \rho)+) \leq \lambda(s-) \\ &\Longleftrightarrow u(A) - \rho \geq s, \end{aligned}$$

so

$$u(f_\rho(A)) = \bigvee\{s \in \mathbf{R} :\ u(A) - \rho \geq s\} = u(A) - \rho.$$

Then by (13.20),

$$(f_\varepsilon \circ f_\rho)(A) = R_{u(A)-\varepsilon-\rho} = f_{\varepsilon+\rho}(A). \tag{13.21}$$

Hence (13.19) is true.

(iv) $\mathcal{E} = \{f \in \mathcal{E}(L^X) :\ \exists \varepsilon > 0,\ f \geq f_\varepsilon \wedge f_\varepsilon{}^\triangleleft\}$ is an $L$-fuzzy uniformity on $X$.

In fact, condition (UF1) clearly holds. Suppose $f, g \in \mathcal{E}$, then there exsit $\varepsilon > 0$, $\rho > 0$ such that $f \geq f_\varepsilon \wedge f_\varepsilon{}^\triangleleft$, $g \geq f_\rho \wedge f_\rho{}^\triangleleft$. No loss any generality, suppose $\varepsilon \geq \rho$, then $f_\varepsilon \geq f_\rho$, $f_\varepsilon{}^\triangleleft \geq f_\rho{}^\triangleleft$, hence

$$f \wedge g \geq (f_\varepsilon \wedge f_\varepsilon{}^\triangleleft) \wedge (f_\rho \wedge f_\rho{}^\triangleleft) = f_\rho \wedge f_\rho{}^\triangleleft.$$

According to the definition of $\mathcal{E}$, $f \wedge g \in \mathcal{E}$, condition (UF2) is satisfied. As a consequence of (13.19), $f_\varepsilon = f_{\frac{\varepsilon}{2}} \circ f_{\frac{\varepsilon}{2}}$, so $\mathcal{E}$ also fulfils condition (UF3). At last, suppose $f \in \mathcal{E}$, $f \geq f_\varepsilon \wedge f_\varepsilon{}^\triangleleft$, then by conclusions (iii), (iv) and (vi) in Theorem **13.1.15**,

$$f^\triangleleft \geq (f_\varepsilon \wedge f_\varepsilon{}^\triangleleft)^\triangleleft = f_\varepsilon{}^\triangleleft \wedge (f_\varepsilon{}^\triangleleft)^\triangleleft = f_\varepsilon{}^\triangleleft \wedge f_\varepsilon,$$

we have $f^\triangleleft \in \mathcal{E}$, condition (UF4) is also fulfilled. Therefore, $\mathcal{E}$ is an $L$-fuzzy uniformity on $X$.

(v) The $L$-fuzzy topology on $X$ generated by $\mathcal{E}$ is just $\mathcal{T}_L^I$.

In fact, suppose the interior operator generated by $\mathcal{E}$ is $i$ and the $L$-fuzzy topology on $X$ generated by $\mathcal{E}$ is $\delta$, then if $A \in \delta$, we have $A = i(A)$. By Lemma **13.1.27** (ii) and the definition (13.7) of $i$,

$$\begin{aligned} A &= \bigvee\{C \in L^X :\ \exists g \in \mathcal{D},\ g \circ g(C) \leq A\} \\ &= \bigvee\{C \in L^X :\ \exists g \in \mathcal{E}(L^X),\ \exists \varepsilon > 0,\ g \geq f_\varepsilon \wedge f_\varepsilon{}^\triangleleft,\ g \circ g(C) \leq A\} \\ &\leq \bigvee\{(f_\varepsilon \wedge f_\varepsilon{}^\triangleleft)(C) :\ C \in L^X,\ \varepsilon > 0,\ \exists g \in \mathcal{E}(L^X),\ g \geq f_\varepsilon \wedge f_\varepsilon{}^\triangleleft, \\ &\qquad g \circ (f_\varepsilon \wedge f_\varepsilon{}^\triangleleft)(C) \leq A\} \\ &\leq \bigvee\{D \in L^X :\ \exists g \in \mathcal{D},\ g(D) \leq A\} \\ &= i(A) \\ &= A. \end{aligned}$$

Let $P$ denote the condition

$$C \in L^X,\ \varepsilon > 0,\ \exists g \in \mathcal{E}(L^X),\ g \geq f_\varepsilon \wedge f_\varepsilon^\triangleleft,\ g\circ(f_\varepsilon \wedge f_\varepsilon^\triangleleft)(C) \leq A,$$

then $A = \bigvee\{(f_\varepsilon \wedge f_\varepsilon^\triangleleft)(C) : P\}$. By (13.1),

$$A = \bigvee\{\bigvee_{K \in \beta^*(C)} (f_\varepsilon(K) \wedge f_\varepsilon^\triangleleft(K)) :\ P\}.$$

By (13.13) and (13.15), every $f_\varepsilon(K)$ and every $f_\varepsilon^\triangleleft(K)$ in the above equation are open in $I(L)$, so $A \in \mathcal{T}_L^I$.

Conversely, suppose $A \in \mathcal{T}_L^I$, then $A$ can be represented as a join of some finite meets of members in $\{R_s, L_t :\ s, t \in \mathbf{R}\}$. So in order to prove $A \in \delta$ it is sufficient to prove $R_s, L_t \in \delta$ for all $s, t \in \mathbf{R}$.

First of all, since $s < t \implies R_s > R_t$, so

$$l(R_s') = \bigwedge\{t \in \mathbf{R} :\ R_s' \leq R_t'\} = s,$$

by Proposition **9.1.2** (iv),

$$\begin{aligned} c(R_s') &\leq \bigwedge\{(f_\varepsilon)^\triangleleft(R_\varepsilon)' :\ \varepsilon > 0\} = \bigwedge\{L_{l(R_s')+\varepsilon} :\ \varepsilon > 0\} \\ &= \bigwedge_{\varepsilon>0} L_{s+\varepsilon} \leq \bigwedge_{\varepsilon>0} R_{s+2\varepsilon}' = (\bigvee_{\varepsilon>0} R_{s+2\varepsilon})' = R_s'. \end{aligned}$$

On the other hand, by (CO2), $c(R_s') \geq R_s'$. So $R_s' = c(R_s')$. By Theorem **13.1.31** (i), $R_s = c(R_s')' = i(R_s) \in \delta$, $\mathcal{T}_L^I \subset \delta$, $\mathcal{T}_L^I = \delta$. □

**13.1.35 Definition** The $L$-fuzzy uniformity $\mathcal{E}$ on $I[L]$ defined in Example **13.1.34** is called the *canonical uniformity* of $I(L)$.

**13.1.36 Remark** Example **13.1.34** was originally provided by Hutton.[38] But in his proof two incorrect formulas

$$\text{“}\bigwedge_{\varepsilon>0} R_{s-\varepsilon} = R_s\text{” and “}\bigvee_{\varepsilon>0} L_{s+\varepsilon} = L_s\text{”}$$

were used (see Proposition **9.1.2** (iii), (iv)). Wang[168] provided a neat proof for this example with $L$-fuzzy topology generated by closure operator. The previous proof of Example **13.1.34** is also based on Wang's proof.

Moreover, it has been proved that a necessary and sufficient condition for an $L$-fts being uniformizable is that it is complete regular:

**13.1.37 Definition** Let $(L^X, \mathcal{D}_0)$, $(L^Y, \mathcal{D}_1)$ be $L$-fuzzy quasi-uniform spaces (uniform spaces, respectively), $F^\to :\ L^X \to L^Y$ an $L$-fuzzy mapping. $F^\to$ is called *quasi-uniformly continuous* (*uniformly continuous*, respectively), if for every $g \in \mathcal{D}_1$, there exists $f \in \mathcal{D}_0$ such that $F^\to\circ f \leq g\circ F^\to$. Particularly, an $L$-fuzzy uniformly continuous mapping $F^\to :\ (L^X, \mathcal{D}) \to I(L)$ defined on an $L$-fuzzy uniform space $(L^X, \mathcal{D})$ is called an *$L$-fuzzy uniformly continuous function.*

**13.1.38 Theorem** *Let $F^\to :\ (L^X, \mathcal{D}_0) \to (L^Y, \mathcal{D}_1)$ be an $L$-fuzzy quasi-uniformly continuous mapping, then $F^\to :\ (L^X, \delta(\mathcal{D}_0)) \to (L^Y, \delta(\mathcal{D}_1))$ is continuous.*

**Proof** Suppose $B \in (\delta(\mathcal{D}_1))'$, then by the definition (13.8) of the closure operator generated by $\mathcal{D}_1$,

$$B = \bigwedge\{g(B) :\ g \in \mathcal{D}_1\}.$$

For every $g \in \mathcal{D}_1$, take $f \in \mathcal{D}_0$ such that $F^\to\circ f \leq g\circ F^\to$, then

$$f(F^\leftarrow(B)) \leq F^\leftarrow F^\to f F^\leftarrow(B) \leq F^\leftarrow g F^\to F^\leftarrow(B) \leq F^\leftarrow(g(B)).$$

Since $F^\leftarrow$ is arbitrary meet preserving,

$F^{\leftarrow}(B) = \bigwedge\{F^{\leftarrow}(g(B)) : g \in \mathcal{D}_1\} \geq \bigwedge\{f(F^{\leftarrow}(B)) : f \in \mathcal{D}_0\} = c(F^{\leftarrow}(B)) \in (\delta(\mathcal{D}_0))'$. So $F^{\rightarrow} : (L^X, \delta(\mathcal{D}_0) \to (L^Y, \delta(\mathcal{D}_1))$ is continuous. □

**13.1.39 Theorem** *Let $(L^X, \mathcal{D})$ be an L-fuzzy uniform space, $f \in \mathcal{D}$, $A, B \in L^X$ such that $f(A) \leq B$. Then there exists an L-fuzzy uniformly continuous mapping $F^{\rightarrow} : (L^X, \mathcal{D}) \to I(L)$ such that*

$$A \leq F^{\leftarrow}(L_1{}') \leq F^{\leftarrow}(R_0) \leq B. \tag{13.22}$$

**Proof** For every $r \in \mathbf{R}$, take $A_r \in L^X$ as follows: As the first step, let

$$A_r = \begin{cases} \underline{1}, & r < 0, \\ B, & r = 0, \\ A, & r = 1, \\ \underline{0}, & r > 1. \end{cases}$$

Denote

$$\begin{aligned} &\mathbf{N}_0 = \{0\} \cup \mathbf{N}, \\ &\mathbf{B}(n) = \{\tfrac{2i-1}{2^k} : \ k \leq n, \ 0 < i \leq 2^{k-1}\}, \ \forall n \in \mathbf{N}_0, \\ &\mathbf{B}_0 = \{\tfrac{2i-1}{2^k} : \ k \in \mathbf{N}_0, \ 0 < i \leq 2^{k-1}\}. \end{aligned}$$

Then

$$\forall n \in \mathbf{N}_0, \quad \mathbf{B}(n) \subset \mathbf{B}(n+1) \subset \mathbf{B}_0, \quad \bigcup_{n \in \mathbf{N}_0} \mathbf{B}(n) = \mathbf{B}_0. \tag{13.23}$$

We shall construct $\{h_{\frac{1}{2^n}} : \ n \in \mathbf{N}_0\} \subset \mathcal{D}$ and $\mathcal{A} = \{A_r : \ r \in \mathbf{B}_0\} \subset L^X$ such that for every $n \in \mathbf{N}_0$ and $1 \leq i \leq 2^{n-1}$,

(I) $k \in \mathbf{N} \implies h_{\frac{1}{2^k}} \circ h_{\frac{1}{2^k}} \leq h_{\frac{1}{2^{k-1}}}$,

(II) $m \leq n, \ \frac{1}{2^m} < \frac{2i-1}{2^n} \implies h_{\frac{1}{2^m}}(A_{\frac{2i-1}{2^n}}) \leq A_{\frac{2i-1}{2^n} - \frac{1}{2^m}}$,

(III) $r, s \in \mathbf{B}(n), \ r \leq s \implies A_r \geq A_s$.

Let $h_1 = f \in \mathcal{D}$. Suppose for some $n \in \mathbf{N}$ and every $k \leq n$ we have constructed $h_{\frac{1}{2^k}}$ fulfilling condition (I), then by condition (UF3) of uniformity, there exists $h_{\frac{1}{2^{n+1}}} \in \mathcal{D}$ such that $h_{\frac{1}{2^{n+1}}} \circ h_{\frac{1}{2^{n+1}}} \leq h_{\frac{1}{2^n}}$. By inductive method, we have constructed $\{h_{\frac{1}{2^n}} : n \in \mathbf{N}_0\} \subset \mathcal{D}$ satisfying condition (I). Suppose $n \in \mathbf{N}_0, \ \ i \leq 2^n$, denote

$$\xi(n,i) = \{h_{\frac{1}{2^{m_0}}} \circ \cdots \circ h_{\frac{1}{2^{m_k}}} : k \in \mathbf{N}_0, m_0, \cdots, m_k \leq n, \sum_{j=0}^{k} \tfrac{1}{2^{m_j}} \leq 1 - \tfrac{i}{2^n}\}, \tag{13.24}$$

then

$$m, n \in \mathbf{N}_0, \ i \leq 2^m, \ j \leq 2^n, \ \tfrac{i}{2^m} \leq \tfrac{j}{2^n} \implies \xi(m,i) \supset \xi(n,j). \tag{13.25}$$

We have had $A_0 \geq A_1$, which satisfy condition (II) and (III). Suppose for some $n \in \mathbf{N}_0$ we have constructed $\{A_r : \ r \in \mathbf{B}(n)\} \subset L^X$ satisfying conditions (II) and (III). Define $\{A_{\frac{2i-1}{2^{n+1}}} : \ 1 \leq i \leq 2^n\}$ as follows:

$$A_{\frac{2i-1}{2^{n+1}}} = \bigvee\{g(A_1) : \ g \in \xi(n+1, 2i-1)\},$$

then by (13.23), we have constructed the family $\{A_r : \ r \in \mathbf{B}(n+1)\} \subset L^X$. Let $1 \leq i \leq 2^n, \ \ m \leq n+1$, we have

$$\begin{aligned}
h_{\frac{1}{2^m}}(A_{\frac{2i-1}{2^{n+1}}}) &= h_{\frac{1}{2^m}}(\bigvee\{g(A_1):\ g\in\xi(n+1,2i-1)\}) \\
&= \bigvee\{h_{\frac{1}{2^m}}\circ h_{\frac{1}{2^{m_0}}}\circ\cdots\circ h_{\frac{1}{2^{m_k}}}(A_1):\ k\in\mathbf{N}_0,\ m_0,\cdots,m_k\le n+1, \\
&\qquad \sum_{j=0}^{k}\tfrac{1}{2^{m_j}}\le 1-\tfrac{2i-1}{2^{n+1}}\} \\
&= \bigvee\{h_{\frac{1}{2^m}}\circ h_{\frac{1}{2^{m_0}}}\circ\cdots\circ h_{\frac{1}{2^{m_k}}}(A_1):\ k\in\mathbf{N}_0,\ m_0,\cdots,m_k\le n+1, \\
&\qquad \tfrac{1}{2^m}+\sum_{j=0}^{k}\tfrac{1}{2^{m_j}}\le 1-(\tfrac{2i-1}{2^{n+1}}-\tfrac{1}{2^m})\} \\
&\le \bigvee\{h_{\frac{1}{2^{m_0}}}\circ\cdots\circ h_{\frac{1}{2^{m_k}}}(A_1):\ k\in\mathbf{N}_0,\ m_0,\cdots,m_k\le n+1, \\
&\qquad \sum_{j=0}^{k}\tfrac{1}{2^{m_j}}\le 1-(\tfrac{2i-1}{2^{n+1}}-\tfrac{1}{2^m})\} \\
&= \bigvee\{g(A_1):\ g\in\xi(n+1,(2i-1)-2^{n+1-m})\} \\
&= A_{\frac{(2i-1)-2^{n+1-m}}{2^{n+1}}} \\
&= A_{\frac{2i-1}{2^{n+1}}-\frac{1}{2^m}},
\end{aligned}$$

condition (II) holds for $n+1$. To prove (III), note that every number in the form of $\frac{2i}{2^n}$ can be represented as a member in $\mathbf{B}_0$, we need only prove the following inequalities for $i\in\{1,\cdots,2^n\}$:

$$A_{\frac{2i-2}{2^{n+1}}}\ge A_{\frac{2i-1}{2^{n+1}}}\ge A_{\frac{2i}{2^{n+1}}}. \tag{13.26}$$

Suppose the irreducible form of $\frac{2i-2}{2^{n+1}}$ is $\frac{2j-1}{2^k}$, i.e. $\frac{2i-2}{2^{n+1}}=\frac{2j-1}{2^k}$, then $\frac{2j-1}{2^k}<\frac{2i-1}{2^{n+1}}$, by (13.25), the first inequality of (13.26) is true. Similarly, the second inequality of (13.26) holds also. Therefore, for $n+1$, we have proved conditions (I) – (III). By inductive method and (13.23), we have constructed $\{h_{\frac{1}{2^n}}:\ n\in\mathbf{N}_0\}\subset\mathcal{D}$ and $\mathcal{A}=\{A_r:\ r\in\mathbf{B}_0\}\subset L^X$ such that conditions (I) – (III) hold.

Since every $h_{\frac{1}{2^m}}$ is arbitrary join preserving, $h_{\frac{1}{2^m}}(\underline{0})=\underline{0}$, and hence by the previous definition of $A_r$, for every $n<\omega$ and $i>2^n$, we have

$$h_{\frac{1}{2^m}}(A_{\frac{i}{2^n}})=h_{\frac{1}{2^m}}(\underline{0})=\underline{0}\le A_{\frac{i}{2^n}-\frac{1}{2^m}};$$

since $h_{\frac{1}{2^m}}$ is value increasing, for $i\le 0$,

$$h_{\frac{1}{2^m}}(A_{\frac{i}{2^n}})=h_{\frac{1}{2^n}}(\underline{1})\ge\underline{1},$$
$$h_{\frac{1}{2^m}}(A_{\frac{i}{2^n}})=\underline{1}=A_{\frac{i}{2^n}-\frac{1}{2^m}}.$$

Therefore, condition (II) holds for every $i\in\mathbf{Z}$.

Denote

$$\mathbf{B}=\{\tfrac{2i-1}{2^n}:\ n<\omega,\ i\in\mathbf{Z}\}. \tag{13.27}$$

For every $t\in\mathbf{R}$ and every $m\in\mathbf{N}_0$, let

$$\mathbf{B}_m(<t)=\{\tfrac{2i-1}{2^n}\in\mathbf{B}:\ n\ge m,\ \tfrac{2i-1}{2^n}<t\},$$
$$\mathbf{B}_m(>t)=\{\tfrac{2i-1}{2^n}\in\mathbf{B}:\ n\ge m,\ \tfrac{2i-1}{2^n}>t\},$$
$$A_t=\bigvee\{A_s:\ s\in\mathbf{B},\ s>t\},\quad \forall t\in[0,1],$$

then $\mathbf{B}$ is dense in $\mathbf{R}$, for every $m\in\mathbf{N}_0$, $\mathbf{B}_m(<t)$ is dense in $(-\infty,t)$, $\mathbf{B}_m(>t)$ is dense in $(t, +\infty)$. By the previous definition of $A_r$, we obtain a family $\mathcal{A}^* = \{A_t : t \in \mathbf{R}\} \subset L^X$. For every $x \in X$ and every $t \in \mathbf{R}$, if let $\lambda(t) = A_t(x)$, then by condition (III), $\lambda : \mathbf{R} \to L$ is monotonically decreasing and $\lambda(t) = 1$ for every $t < 0$, $\lambda(t) = 0$ for every $t > 1$. So $\lambda \in md_I(L)$ (see Section **9.1**), $[\lambda] \in I[L]$. Therefore, following the stipulation about the meaning of $F(x)(t)$ in the statement of this theorem, we can reasonablly define an ordinary mapping $F : X \to I[L]$ as follows:

$$\forall x \in X,\ \forall t \in \mathbf{R}, \quad F(x)(t) = A_t(x).$$

Then

$$A(x) = A_1(x) \le \bigwedge_{\varepsilon>0} A_{1-\varepsilon}(x) = F(x)(1-) = F^{\leftarrow}(L_1{}')(x),$$
$$B(x) = A_0(x) \ge \bigvee_{\varepsilon>0} A_\varepsilon(x) = F(x)(0+) = F^{\leftarrow}(R_0)(x),$$

(13.22) holds.

Now we turn to the proof of the uniform continuity of $F^{\rightarrow}$. First of all, we prove the following two equalities: $\forall r \in \mathbf{R}$,

$$m \in \mathbf{N}_0,\ r \in \mathbf{R} \implies F^{\leftarrow}(R_r) = \bigvee_{s\in\mathbf{B}_m(>r)} A_s, \quad F^{\leftarrow}(L_r) = \bigvee_{s\in\mathbf{B}_m(<r)} A_s{}'. \quad (13.28)$$

To prove (13.28), suppose $x \in X$, $F(x) = [\lambda]$, then for every $t \in \mathbf{R}$, $A_t(x) = \lambda(t)$. Since $\mathbf{B}_m(> r)$ is dense in $(r, +\infty)$,

$$\begin{aligned} F^{\leftarrow}(R_r)(x) &= R_r(F(x)) = R_r([\lambda]) = \lambda(r+) \\ &= \bigvee_{s\in\mathbf{B}_m(>r)} \lambda(s) = \bigvee_{s\in\mathbf{B}_m(>r)} A_s(x) \\ &= (\bigvee_{s\in\mathbf{B}_m(>r)} A_s)(x). \end{aligned}$$

So $F^{\leftarrow}(R_r) = \bigvee_{s\in\mathbf{B}_m(>r)} A_s$. Similarly prove the second equality in (13.28).

To consider the members of the canonical uniformity of $I(L)$, suppose $f_\varepsilon$ is defined by (13.13) for $\varepsilon > 0$, then $f_\varepsilon{}^{\triangleleft}$ is determined by (13.15), and there exist $g_1, g_2 \in \mathcal{D}$ such that

$$F^{\rightarrow}\circ g_1 \le f_\varepsilon \circ F^{\rightarrow}, \qquad F^{\rightarrow}\circ g_2 \le f_\varepsilon{}^{\triangleleft}\circ F^{\rightarrow}. \quad (13.29)$$

In fact, fix a $\frac{1}{2^m} \in (0, \varepsilon)$, then for every $r \in \mathbf{R}$, by condition (II) (It has been proved above, (II) holds for every $i \in \mathbf{Z}$) and (13.28),

$$\begin{aligned} F^{\rightarrow} h_{\frac{1}{2^m}} F^{\leftarrow}(R_r) &= F^{\rightarrow} h_{\frac{1}{2^m}}(\bigvee_{s\in\mathbf{B}_m(>r)} A_s) \\ &= F^{\rightarrow}(\bigvee_{s\in\mathbf{B}_m(>r)} h_{\frac{1}{2^m}}(A_s)) \\ &\le F^{\rightarrow}(\bigvee_{s\in\mathbf{B}_m(>r)} A_{s-\frac{1}{2^m}}) \\ &= F^{\rightarrow} F^{\leftarrow}(R_{r-\frac{1}{2^m}}) \\ &\le R_{r-\frac{1}{2^m}} \\ &\le R_{r-\varepsilon}. \end{aligned}$$

On the other hand, since $R_r \le L_s{}'$ if and only if $R_r \le R_s$. By (13.13),

$$f_\varepsilon(R_r) = R_{u(R_r)-\varepsilon} = R_{\bigvee\{s:\ R_r \le L_s{}'\}-\varepsilon} = R_{\bigvee\{s:\ R_r\le R_s\}-\varepsilon} = R_{r-\varepsilon}.$$

So

$$F^{\rightarrow}h_{\frac{1}{2^m}}F^{\leftarrow}(R_r) \leq f_\varepsilon(R_r).$$

Suppose $C \in L^X \backslash \{\underline{0}\}$. If $u(C) > 0$ for the function $u$ defined in (13.12), take $\sigma \in (0, \varepsilon) \cap (0, u(C))$, then by (13.19),

$$f_\varepsilon(C) = f_{\varepsilon-\sigma}(f_\sigma(C)) = f_{\varepsilon-\sigma}(R_{u(C)-\sigma}).$$

Take $m \in \mathbf{N}$ such that $\frac{1}{2^m} < \varepsilon - \sigma$, then by $C \leq f_\sigma(C) = R_{u(C)-\sigma}$,

$$\begin{aligned} F^{\rightarrow}h_{\frac{1}{2^m}}F^{\leftarrow}(C) &\leq F^{\rightarrow}h_{\frac{1}{2^m}}F^{\leftarrow}(R_{u(C)-\sigma}) \\ &\leq f_{\varepsilon-\sigma}(R_{u(C)-\sigma}) \\ &= f_{\varepsilon-\sigma}(f_\sigma(C)) \\ &= f_\varepsilon(C). \end{aligned}$$

So the following implication holds for arbitrary $C \in L^X \backslash \{\underline{0}\}$ such that $u(C) > 0$:

$$\varepsilon > 0,\ m \in \mathbf{N},\ \tfrac{1}{2^m} < \varepsilon \implies F^{\rightarrow}h_{\frac{1}{2^m}}F^{\leftarrow}(C) \leq f_\varepsilon(C). \tag{13.30}$$

If $u(C) \not> 0$, then by $u(C) \geq 0$ we have $u(C) = \bigvee\{s \in \mathbf{R} : C \leq L_s{}'\} = 0$. Hence

$$f_\varepsilon(C) = R_{u(C)-\varepsilon} = R_{-\varepsilon} = \underline{1},$$

the implication (13.30) still holds. Since (13.30) is true for $C = \underline{0}$, so (13.30) holds for every $C \in L^X$. Therefore, $F^{\rightarrow}h_{\frac{1}{2^m}}F^{\leftarrow} \leq f_\varepsilon$. Take $g_1 = h_{\frac{1}{2^m}}$ we obtain the first inequality in (13.29). Similarly prove the second inequality in (13.29).

Suppose $h$ is a member of the canonical uniformity $\mathcal{E}$ of $I(L)$, then there exists $\varepsilon > 0$ such that $h \geq f_\varepsilon \wedge f_\varepsilon{}^{\triangleleft}$. Take $g_1, g_2 \in \mathcal{D}$ such that (13.29) holds, let $g = g_1 \wedge g_2$, then

$$F^{\rightarrow}gF^{\leftarrow} \leq (F^{\rightarrow}g_1F^{\leftarrow}) \wedge (F^{\rightarrow}g_2F^{\leftarrow}) \leq f_\varepsilon \wedge f_\varepsilon{}^{\triangleleft} \leq h,$$

hence $F^{\rightarrow} \circ g \leq h \circ F^{\rightarrow}$. This completes the proof of the uniform continuity of $F^{\rightarrow}$. □

**13.1.40 Remark** In [38], Hutton introduced Theorem **13.1.39** and the outline of its proof. Different from $\{h_{\frac{1}{2^n}} : n \in \mathbf{N}_0\} \subset \mathcal{D}$, $\{A_r : r \in \mathbf{B}_0\} \subset L^X$ and condition (II) in the proof of Theorem **13.1.39**, without proof, Hutton affirmed that one could find $\{h_r : r > 0\} \subset \mathcal{D}$ and $\{A_s : s \in \mathbf{R}\} \subset L^X$ such that

$$h_r(A_s) \leq A_{s-r}. \tag{13.31}$$

This skeleton of a proof was widely adopted later. To complete the proof, some detailed proof of the above inequality for all $r$'s in $\mathbf{B}$ which is defined by (13.27) was introduced.[168] The family $\{h_r : r \in \mathbf{B}\} \subset \mathcal{D}$ defined there satisfies condition (I) in the proof of Theorem **13.1.39**. But the proof has some gaps: After inductively constructed

$$A_{\frac{2i-1}{2^{n+1}}} = h_{\frac{1}{2^{n+1}}}(A_{\frac{k}{2^n}}), \tag{13.32}$$

it was affirmed that "Let $h_{\frac{i}{2^n}} = \underbrace{h_{\frac{1}{2^n}} \circ \cdots \circ h_{\frac{1}{2^n}}}_{i}$ and let $h_{\frac{i}{2^n}+\frac{j}{2^m}} = h_{\frac{i}{2^n}} \circ h_{\frac{j}{2^m}}$, then it is not hard to verify that (13.31) holds for arbitrary $r, s \in \mathbf{B} \cap [0,1]$". But, in general, because we can only ensure $h_{\frac{2}{2^n}} \geq h_{\frac{1}{2^n}} \circ h_{\frac{1}{2^n}}$ but not $h_{\frac{2}{2^n}} = h_{\frac{1}{2^n}} \circ h_{\frac{1}{2^n}}$, and $h_{\frac{i}{2^n}}$ had always been defined previously as (13.32) in the inductive proof for every even number $i = 2k \leq 2^n$, so $h_{\frac{i}{2^n}} = h_{\frac{k}{2^{n-1}}}$ cannot be redefined as $h_{\frac{i}{2^n}} = \underbrace{h_{\frac{1}{2^n}} \circ \cdots \circ h_{\frac{1}{2^n}}}_{i}$; otherwise every defined $h_{\frac{2k}{2^n}}$ will be continuously changed in follow-up inductive steps, and consequently no family $\{h_r : r \in \mathbf{B} \cap [0,1]\} \subset \mathcal{D}$ can be inductively determined. Similarly, the constructive step "let $h_{\frac{i}{2^n}+\frac{j}{2^m}} = h_{\frac{i}{2^n}} \circ h_{\frac{j}{2^m}}$" is not reasonable either.

Since arbitrary join preserving property is preserved by composition of mappings, the following definition is reasonable:[69]

**13.1.41 Definition** Let $L_0$, $L_1$, $L$ be complete lattices. For every $g \in \mathcal{J}(L_0, L_1)$ and every $h \in \mathcal{J}(L_1, L_0)$, define

$$\Theta_{g,h} : \mathcal{E}(L_1) \to \mathcal{E}(L_0), \quad \Theta_{g,h}(f) = h \circ f \circ g.$$

Call $\Theta_{g,h}$ the *parallel transference* with respect to $g$ and $h$.

**13.1.42 Proposition** *$\Theta_{g,h}$ is order preserving.* □

**13.1.43 Proposition** *Let $h \in \mathcal{J}(L_1, L_0)$ be finite meet preserving. Then for arbitrary $g \in \mathcal{J}(L_0, L_1)$, $f_0, f_1 \in \mathcal{E}(L_1)$,*

$$\Theta_{g,h}(f_0 \wedge f_1) = \Theta_{g,h}(f_0) \wedge \Theta_{g,h}(f_1).$$

**Proof** By Proposition **13.1.42**, it is sufficient to prove $\Theta_{g,h}(f_0 \wedge f_1) \geq \Theta_{g,h}(f_0) \wedge \Theta_{g,h}(f_1)$. By Theorem **13.1.12** (i), for every $a \in L_0$,

$$\begin{aligned}
\Theta_{g,h}(f_0 \wedge f_1)(a) &= h \circ (f_0 \wedge f_1) \circ g(a) \\
&= h\Big(\bigvee_{\lambda \in \beta^*(a)} (f_0(g(\lambda)) \wedge f_1(g(\lambda)))\Big) \\
&= \bigvee_{\lambda \in \beta^*(a)} (h \circ f_0 \circ g(\lambda) \wedge h \circ f_1 \circ g(\lambda)) \\
&= \bigvee_{\lambda \in \beta^*(a)} (\Theta_{g,h}(f_0)(\lambda) \wedge \Theta_{g,h}(f_1)(\lambda)) \\
&= (\Theta_{g,h}(f_0) \wedge \Theta_{g,h}(f_1))(a),
\end{aligned}$$

this completes the proof. □

**13.1.44 Proposition** *Let $g \in \mathcal{J}(L_0, L_1)$, $h \in \mathcal{J}_1(L_1)$, $g \circ h \leq id_{L_1}$. Then for arbitrary $f_0, f_1 \in \mathcal{E}(L_1)$,*

$$\Theta_{g,h}(f_0 \circ f_1) \geq \Theta_{g,h}(f_0) \circ \Theta_{g,h}(f_1).$$ □

**13.1.45 Proposition** *Let $L_0$, $L_1$ be complete lattices with order-reversing involutions $'$; $g \in \mathcal{J}(L_0, L_1)$, $h \in \mathcal{J}_1(L_1)$ satisfy the following conditions:*

(i) $g \circ h \leq id_{L_1}$,

(ii) $h \circ g \geq id_{L_0}$,

(iii) $\forall b \in L_1, \quad h(b') = h(b)'$,

(iv) *$h$ is arbitrary meet preserving.*

*Then for every $f \in \mathcal{E}(L_1)$,*

$$\Theta_{g,h}(f^{\triangleleft}) = (\Theta_{g,h}(f))^{\triangleleft}.$$

**Proof** Suppose $a \in L_0$,

$$\begin{aligned}\Theta_{g,h}(f^{\triangleleft})(a) &= h\circ f^{\triangleleft}\circ g(a)\\ &= h(\bigwedge\{b\in L_1:\ f(b')\leq g(a)'\})\\ &= \bigwedge\{h(b):\ b\in L_1,\ f(b')\leq g(a)'\},\\ (\Theta_{g,h}(f))^{\triangleleft}(a) &= \bigwedge\{c\in L_0:\ \Theta_{g,h}(f)(c')\leq a'\}.\end{aligned}$$

If $b \in L_1$ such that $f(b') \leq g(a)'$, then

$$\Theta_{g,h}(f)(h(b)') = \Theta_{g,h}(f)(h(b')) = hfgh(b') \leq hf(b') \leq h(g(a)') = (hg(a))' \leq a'.$$

So $h(b) \in \{c \in L_0 : \Theta_{g,h}(f)(c') \leq a'\}$, and hence

$$\{h(b):\ b\in L_1,\ f(b')\leq g(a')\} \subset \{c\in L_0:\ \Theta_{g,h}(f)(c')\leq a'\},$$

it implies

$$\begin{aligned}\Theta_{g,h}(f^{\triangleleft})(a) &= \bigwedge\{h(b):\ b\in L_1,\ f(b')\leq g(a')\}\\ &\geq \bigwedge\{c\in L_0:\ \Theta_{g,h}(f)(c')\leq a'\}\\ &= (\Theta_{g,h}(f))^{\triangleleft}(a).\end{aligned}$$

So we have proved

$$\Theta_{g,h}(f^{\triangleleft}) \geq (\Theta_{g,h}(f))^{\triangleleft}. \tag{13.33}$$

Replace $f^{\triangleleft}$ in (13.33) by $(f^{\triangleleft})^{\triangleleft}$, then it follows from Theorem **13.1.15** (iii), (iv) and inequality (13.33),

$$\Theta_{g,h}(f^{\triangleleft}) = ((\Theta_{g,h}(f^{\triangleleft}))^{\triangleleft})^{\triangleleft} \leq (\Theta_{g,h}((f^{\triangleleft})^{\triangleleft}))^{\triangleleft} = (\Theta_{g,h}(f))^{\triangleleft}.$$

This completes the proof. □

**13.1.46 Corollary** *Let $L$ be a complete lattice with an order-reversing involution $'$, $F^{\rightarrow}: L^X \to L^Y$ be an $L$-fuzzy mapping. Then*

$$\Theta_{F^{\rightarrow},F^{\leftarrow}}(f^{\triangleleft}) = (\Theta_{F^{\rightarrow},F^{\leftarrow}}(f))^{\triangleleft}.$$ □

**13.1.47 Theorem** *Let $F^{\rightarrow}: L^X \to L^Y$ be an $L$-fuzzy mapping, and*

$$\mathcal{G} \subset \mathcal{E}(L^X),\ \ \mathcal{C} \subset \mathcal{G},\ \ \mathcal{B} = \{\Theta_{F^{\rightarrow},F^{\leftarrow}}(f):\ f \in \mathcal{C}\}.$$

(i) *If $\mathcal{G}$ is an $L$-fuzzy quasi-uniformity (uniformity, respectively) on $Y$ and $\mathcal{C}$ is a base of $\mathcal{G}$, then $\mathcal{B}$ is a base of an $L$-fuzzy quasi-uniformity (uniformity, respectively) on $X$.*

(ii) *If $\mathcal{G}$ is an $L$-fuzzy quasi-uniformity (uniformity, respectively) on $Y$ and $\mathcal{C}$ is a subbase of $\mathcal{G}$, then $\mathcal{B}$ is a subbase of an $L$-fuzzy quasi-uniformity (uniformity, respectively) on $X$.*

**Proof** (i) If $\mathcal{G}$ is an $L$-fuzzy quasi-uniformity on $Y$, $\mathcal{C}$ is a base of $\mathcal{G}$, then by Proposition **13.1.19**, $\mathcal{C}$ fulfils conditions (BUF1) and (BUF2). Suppose $f, g \in \mathcal{C}$, by (BUF1), there exists $h \in \mathcal{C}$ such that $h \leq f \wedge g$. Then by Proposition **13.1.42** and Proposition **13.1.43**,

$$\Theta_{F^{\rightarrow},F^{\leftarrow}}(h) \leq \Theta_{F^{\rightarrow},F^{\leftarrow}}(f\wedge g) = \Theta_{F^{\rightarrow},F^{\leftarrow}}(f) \wedge \Theta_{F^{\rightarrow},F^{\leftarrow}}(g),$$

$\mathcal{B}$ fulfils (BUF1). For every $f \in \mathcal{C}$, since $\mathcal{C}$ fulfils (BUF2), there exists $g \in \mathcal{C}$ such that $g \circ g \leq f$. Then by Proposition **13.1.44** and Proposition **13.1.42**,

$$\Theta_{F^{\rightarrow},F^{\leftarrow}}(g) \circ \Theta_{F^{\rightarrow},F^{\leftarrow}}(g) \leq \Theta_{F^{\rightarrow},F^{\leftarrow}}(g \circ g) \leq \Theta_{F^{\rightarrow},F^{\leftarrow}}(f),$$

$\mathcal{B}$ fulfils (BUF2). By Proposition **13.1.19**, $\mathcal{B}$ is a base of an $L$-fuzzy quasi-uniformity on $X$.

Moreover, if $\mathcal{G}$ is an $L$-fuzzy uniformity on $Y$, then besides (BUF1) and (BUF2), $\mathcal{C}$ also fulfils (BUF3). Suppose $f \in \mathcal{C}$, then there exists $g \in \mathcal{C}$ such that $g \leq f^{\triangleleft}$. By Proposition **13.1.45**,

$$\Theta_{F^{\rightarrow},F^{\leftarrow}}(g) \leq \Theta_{F^{\rightarrow},F^{\leftarrow}}(f^{\triangleleft}) = (\Theta_{F^{\rightarrow},F^{\leftarrow}}(f))^{\triangleleft},$$

$\mathcal{B}$ fulfils (BUF3). By Proposition **13.1.19**, $\mathcal{B}$ is a base of an $L$-fuzzy uniformity on $X$.

(ii) By Proposition **13.1.20**, the proof is similar to (i). □

Now we can introduce the following importan theorem of Hutton,[38] the proof introduced below is based on Liu's "parallel transferences":[69]

**13.1.48 Theorem** *Let $(L^X, \delta)$ be an L-fts be an L-fts, then $(L^X, \delta)$ is uniformizable if and only if $(L^X, \delta)$ is completely regular.*

**Proof** (Necessity) Suppose $\mathcal{D}$ is an $L$-fuzzy uniformity on $X$ such that $\delta(\mathcal{D}) = \delta$, $U \in \delta$. By Theorem **13.1.31** (i), $U = \bigvee\{(f(U'))' : f \in \mathcal{D}\}$. For every $f \in \mathcal{D}$, let $W_f = (f(U'))'$, then $U = \bigvee\{W_f : f \in \mathcal{D}\}$. By Theorem **13.1.15** (i) and $f(U') = W_f{}'$, $f^{\triangleleft}(W_f) \leq U$. Then by Theorem **13.1.39**, there exists an $L$-fuzzy continuous mapping $F_f{}^{\rightarrow} : (L^X, \mathcal{D}) \rightarrow I(L)$ such that for every $x \in X$,

$$W_f(x) \leq F_f{}^{\rightarrow}(x)(1-) \leq F_f{}^{\rightarrow}(x)(0+) \leq U(x).$$

Then condition (CR3) in Remark **9.2.3** is satisfied, $(L^X, \delta)$ is completely regular.

(Sufficiency) Suppose $(L^X, \delta)$ is completely regular, $\mathcal{E} = \{f \in \mathcal{E}(L^X) : \exists \varepsilon > 0,\ f \geq f_\varepsilon \wedge f_\varepsilon{}^{\triangleleft}\}$ is the canonical uniformity of $I(L)$, $\mathcal{F}$ is the family of all the $L$-fuzzy continuous mappings from $(L^X, \delta)$ to $I(L)$. Let

$$\mathcal{C} = \{f_\varepsilon \wedge f_\varepsilon{}^{\triangleleft} : \varepsilon > 0\}, \tag{13.34}$$

$$\forall F^{\rightarrow} \in \mathcal{F},\ \forall \varepsilon > 0, \quad g(F, \varepsilon) = \Theta_{F^{\rightarrow},F^{\leftarrow}}(f_\varepsilon \wedge f_\varepsilon{}^{\triangleleft}) = F^{\leftarrow}(f_\varepsilon \wedge f_\varepsilon{}^{\triangleleft})F^{\rightarrow},$$

$$\forall F^{\rightarrow} \in \mathcal{F}, \quad \mathcal{B}_F = \{g(F, \varepsilon) : \varepsilon > 0\},$$

$$\mathcal{P} = \bigcup_{F^{\rightarrow} \in \mathcal{F}} \mathcal{B}_F,$$

$$\mathcal{D}^* = \{g(F_1, \varepsilon_1) \wedge \cdots \wedge g(F_k, \varepsilon_k) : F_i{}^{\rightarrow} \in \mathcal{F},\ \varepsilon_i > 0,\ 1 \leq i \leq k,\ k \in \mathrm{N}\}, \tag{13.35}$$

$$\mathcal{D} = \{g \in \mathcal{E}(L^X) : \exists h \in \mathcal{D}^*,\ g \geq h\}, \tag{13.36}$$

then by the proof of Example **13.1.34**, $\mathcal{C}$ is a base of the canonical uniformity $\mathcal{E}$ of $I(L)$. By Theorem **13.1.47** (i), $\mathcal{B}_F$ is a base of an $L$-fuzzy uniformity on $X$ for every $F^{\rightarrow} \in \mathcal{F}$. By Proposition **13.1.21**, $\mathcal{P}$ is a subbase and hence $\mathcal{D}^*$ is a base of an $L$-fuzzy uniformity on $X$, $\mathcal{D}$ is the $L$-fuzzy uniformity on $X$ generated by $\mathcal{D}^*$.

Now we need to prove $\delta = \delta(\mathcal{D})$.

(i) $\delta(\mathcal{D}) \subset \delta$.

Suppose $i$ is the interior operator on $L^X$ generated by $\mathcal{D}$, then by Lemma **13.1.27** (iii), for every $A \in \delta(\mathcal{D})$,

$$A = \bigvee\{h(B) : B \in L^X,\ h \in \mathcal{D}^*,\ h \circ h(B) \le A\}.$$

To prove $\delta(\mathcal{D}) \subset \delta$, it is sufficient to prove $A \in \delta$, or prove $h(B) \in \delta$ for every $h \in \mathcal{D}^*$. Suppose $h = g(F_1, \varepsilon_1) \wedge \cdots \wedge g(F_k, \varepsilon_k) \in \mathcal{D}^*$, then by Proposition **13.1.43**,

$$\begin{aligned} h &= g(F_1, \varepsilon_1) \wedge \cdots \wedge g(F_k, \varepsilon_k) \\ &= \Theta_{F_1{}^{\rightarrow}, F_1{}^{\leftarrow}}(f_{\varepsilon_1} \wedge f_{\varepsilon_1}{}^{\triangleleft}) \wedge \cdots \wedge \Theta_{F_k{}^{\rightarrow}, F_k{}^{\leftarrow}}(f_{\varepsilon_k} \wedge f_{\varepsilon_k}{}^{\triangleleft}) \\ &= \Theta_{F_1{}^{\rightarrow}, F_1{}^{\leftarrow}}(f_{\varepsilon_1}) \wedge \Theta_{F_1{}^{\rightarrow}, F_1{}^{\leftarrow}}(f_{\varepsilon_1}{}^{\triangleleft}) \wedge \cdots \wedge \Theta_{F_k{}^{\rightarrow}, F_k{}^{\leftarrow}}(f_{\varepsilon_k}) \wedge \Theta_{F_k{}^{\rightarrow}, F_k{}^{\leftarrow}}(f_{\varepsilon_k}{}^{\triangleleft}). \end{aligned}$$

So by (13.2), we need only prove

$$(F_i{}^{\leftarrow} f_{\varepsilon_i} F_i{}^{\rightarrow})(B) \in \delta, \quad (F_i{}^{\leftarrow} f_{\varepsilon_i}{}^{\triangleleft} F_i{}^{\rightarrow})(B) \in \delta \tag{13.37}$$

for every $i \in \{1, \cdots, k\}$. By (13.13) and (13.15), both $f_{\varepsilon_i}(F_i{}^{\rightarrow})(B))$ and $f_{\varepsilon_i}{}^{\triangleleft}(F_i{}^{\rightarrow})(B))$ are open subsets in $I(L)$, so by the continuity of $F_i{}^{\rightarrow}$ we have (13.37).

Now we have proved $\delta(\mathcal{D}) \subset \delta$ already. Note that the complete regularity of $(L^X, \delta)$ has not been used yet up to the present.

(ii) $\delta \subset \delta(\mathcal{D})$.

First of all, by (13.13) and Proposition **9.1.2** (iv),

$$f_\varepsilon = \bigvee_{t \in (0,\varepsilon)} f_{\varepsilon - t}. \tag{13.38}$$

Next we have

$$\forall F^{\rightarrow} \in \mathcal{F},\ \forall \varepsilon > 0,\ \forall A \in L^Y, \quad F^{\leftarrow} f_\varepsilon F^{\rightarrow}(A) \in \delta(\mathcal{D}). \tag{13.39}$$

In fact, in

$$i(F^{\leftarrow} f_\varepsilon F^{\rightarrow}(A)) = \bigvee\{C \in L^X : \exists g \in \mathcal{D},\ g(B) \le F^{\leftarrow} f_\varepsilon F^{\rightarrow}(A)\},$$

let $g = F^{\leftarrow} f_{\frac{\varepsilon}{n}} F^{\rightarrow}$, $C = F^{\leftarrow} f_{(1-\frac{1}{n})\varepsilon} F^{\rightarrow}(A)$ for every $n > 1$, then by (13.19),

$$g(C) \le F^{\leftarrow} f_\varepsilon F^{\rightarrow}(A),$$

so

$$i(F^{\leftarrow} f_\varepsilon F^{\rightarrow}(A)) \ge F^{\leftarrow} f_{(1-\frac{1}{n})\varepsilon} F^{\rightarrow}(A), \quad n = 2, 3, \cdots.$$

Then by (13.38),

$$F^{\leftarrow} f_\varepsilon F^{\rightarrow}(A) = i(F^{\leftarrow} f_\varepsilon F^{\rightarrow}(A)) \in \delta(\mathcal{D}),$$

(13.39) is true.

Since $(L^X, \delta)$ is completely regular, suppose $U \in \delta$, there exist $\{U_t : t \in T\} \subset L^X$ and $\{F_t{}^{\rightarrow} : t \in T\} \subset \mathcal{F}$ such that

$$\bigvee_{t \in T} U_t = U, \quad U_t \le F_t{}^{\leftarrow}(L_1{}') \le F_t{}^{\leftarrow}(R_0) \le U, \quad \forall t \in T. \tag{13.40}$$

Particulaly, $F_t{}^{\rightarrow}(U_t) \le L_1{}' \le R_{\frac{1}{2}}$. Take $\varepsilon = \frac{1}{2}$, then

$$U_t \le F^{\leftarrow} f_\varepsilon F^{\rightarrow}(U_t) \le F^{\leftarrow}(R_{u(R_{\frac{1}{2}}) - \frac{1}{2}}) = F_t{}^{\leftarrow}(R_0),$$

so by (13.40),

$$U_t \le F^{\leftarrow} f_\varepsilon F^{\rightarrow}(U_t) \le U, \quad \forall t \in T. \tag{13.41}$$

By the virtue of (13.39), (13.40) and (13.41), $U \in \delta(\mathcal{D})$. So $\delta \subset \delta(\mathcal{D})$.

By (i) and (ii) proved above, $\delta = \delta(\mathcal{D})$. □

In Example **13.1.34**, it is proved that $I(L)$ is uniformizable. Rodabaugh[144] ever posed a question: Is the stratification of $I(L)$ uniformizable? By the virtue of Theorem **9.2.35** and Theorem **13.1.48**, we obtain the following:[195]

**13.1.49 Proposition** *Let L be a F-lattice, then the stratification of $I(L)$ is uniformizable.* □

Just as it is investigated above, different from the definition of ordinary uniformity, $L$-fuzzy uniformity ($L$-fuzzy quasi-uniformity) is defined in the form of mappings. But it can be also described in the form of subsets just as ordinary uniformity. Ji-Hua Liang[63] completed this work as follows:

**13.1.50 Definition** Let $L$ be an F-lattice, $a \in L$, denote

$$\hat{Q}(a) = \{b \in L : \ b \not\leq a'\}, \qquad \hat{Q}^*(a) = \hat{Q}(a) \cap M(L).$$

**13.1.51 Lemma** *Let L be a F-lattice, $a \in L$, $\alpha \in \beta^*(a)$. Then*

(i) $\beta^*(\bigwedge \hat{Q}(\alpha)) \cap \hat{Q}^*(a) \neq \emptyset$.

(ii) $\lambda \in \beta^*(\bigwedge \hat{Q}(\alpha)) \implies \bigwedge \hat{Q}(\lambda) \geq \alpha$.

**Proof** (i) Suppose $\forall \lambda \in \beta^*(\bigwedge \hat{Q}(\alpha))$, $\lambda \notin \hat{Q}^*(a)$, then $\lambda \leq a'$ for every $\lambda \in \beta^*(\bigwedge \hat{Q}(\alpha))$, and hence

$$\bigwedge \hat{Q}(\alpha) = \bigvee \beta^*(\bigwedge \hat{Q}(\alpha)) \leq a'.$$

So $a \leq \bigvee\{b' : \ b \in \hat{Q}(\alpha)\}$. Since $\alpha \in \beta^*(a)$, there exists $b \in \hat{Q}(\alpha)$ such that $\alpha \leq b'$. But this contradicts with the definition of $\hat{Q}(\alpha)$.

(ii) Suppose $b \in \hat{Q}(\lambda)$. If $b \not\geq \alpha$, then $b' \in \hat{Q}(\alpha)$. So $\lambda \leq \bigwedge \hat{Q}(\alpha) \leq b'$. But this contradicts with $b \in \hat{Q}(\lambda)$. So $b \geq \alpha$, hence $\bigwedge \hat{Q}(\lambda) \geq \alpha$. □

**13.1.52 Lemma** *Let L be a F-lattice, $a, b \in L$, $A \subset L$. Then*

(i) $\hat{Q}^*(a) \subset \hat{Q}^*(b) \iff a \leq b$.

(ii) $\hat{Q}^*(\bigvee A) = \bigcup_{a \in A} \hat{Q}^*(a)$. □

**13.1.53 Definition** Let $L$ be a F-lattice.

Denote

$$\Delta^* = \{(\alpha, \beta) \in M(L) \times M(L) : \ \alpha \in \hat{Q}(\beta)\}.$$

call $\Delta^*$ the *quasi-diagonal* of $L \times L$.

A subset $U \in M(L) \times M(L)$ is called a *basic element* in $L \times L$, if $U$ fulfils the following conditions (BE1) – (BE3):

(BE1) $\Delta^* \subset U$.

(BE2) $\uparrow U = U$.

(BE3) If $\{\alpha_s : \ s \in S\} \subset M(L)$, $\{\beta_t : \ t \in T\} \subset M(L)$ and $(\bigvee_{s \in S} \alpha_s, \bigvee_{t \in T} \beta_t) \in U$, then there exists $(s, t) \in S \times T$ such that $(\alpha_s, \beta_t) \in U$.

Denote

$$\mathcal{U}(L) = \{U \subset M(L) \times M(L) : \ U \text{ is a basic element in } L \times L\},$$

equip $\mathcal{U}(L)$ with the inclusion order.

**13.1.54 Definition** Let $L$ be a F-lattice, $U \subset M(L) \times M(L)$. Denote

$$\forall a \in L\backslash\{0\}, \quad H_U(a) = \{b \in L : \ \exists \alpha \in \beta^*(a), \ \forall \beta \in \hat{Q}^*(b), \ (\alpha, \beta) \in U\},$$

$$D_U(a) = \begin{cases} \bigvee H_U(a), & a \neq 0, \\ 0, & a = 0. \end{cases}$$

**13.1.55 Lemma** *Let $L$ be a F-lattice, $U, V \subset M(L) \times M(L)$. Then*

(i) $U \subset V \implies D_U \leq D_V$.

(ii) $U \in \mathcal{U}(L) \implies D_U \in \mathcal{E}(L)$.

**Proof** (i) It follows from the definition of $D_U$.

(ii) Suppose $U \in \mathcal{U}(L)$, $a \in L\backslash\{0\}$. For every $\alpha \in \beta^*(a)$, by (BE1), $(\alpha, \beta) \in \Delta^* \subset U$ for every $\beta \in \hat{Q}^*(\alpha)$, so $\alpha \in H_U(a)$, $\alpha \leq D_U(a)$. Hence

$$D_U(a) = \bigvee H_U(a) \geq \bigvee \beta^*(a) = a,$$

$D_U$ is value increasing. Since $D_U(0) = 0$, to prove $D_U \in \mathcal{E}(L)$, it is sufficient to prove the following equality holds for every nonempty subset $A \subset L$:

$$D_U(\bigvee A) = \bigvee_{a \in A} D_U(a). \tag{13.42}$$

Since $U$ is a basic element in $L \times L$, it is easy to verify that $D_U : L \to L$ is order preserving, so $D_U(\bigvee A) \geq \bigvee_{a\in A} D_U(a)$. On the other hand, for every $b \in H_U(\bigvee A)$, then there exists $\alpha \in \beta^*(\bigvee A)$ such that $(\alpha, \beta) \in U$ for every $\beta \in \hat{Q}^*(b)$. By Theorem **1.3.24** (ii), there exists $a \in A$ such that $\alpha \in \beta^*(a)$, so $b \in H_U(a)$. Since $b$ is arbitrarily take from $H_U(\bigvee A)$, we have $H_U(\bigvee A) \subset \bigcup_{a\in A} H_U(a)$, and hence $D_U(\bigvee A) \leq \bigvee_{a\in A} D_U(a)$. This completes the proof. □

**13.1.56 Definition** Let $L$ be a F-lattice. For every $f \in \mathcal{E}(L)$, denote

$$K_L(f) = \{(\alpha, \beta) \in M(L) \times M(L) : \beta \in \hat{Q}^*(f(\alpha))\}. \tag{13.43}$$

**13.1.57 Theorem** *Let $L$ be a F-lattice. Then*

(i) *For every $f \in \mathcal{E}(L)$, $K_L(f) \in \mathcal{U}(L)$.*

(ii) *$K_L : \mathcal{E}(L) \to \mathcal{U}(L)$ is an ismorphism.*

**Proof** (i) Suppose $f \in \mathcal{E}(L)$. By Lemma **13.1.52** and the definitions of $K_L(f)$ and $\hat{Q}^*(f(\alpha))$, $K_L(f)$ fulfils (BE2) and (BE3). Since $f$ is value increasing, by Lemma **13.1.52**, $K_L(f)$ also fulfils (BE1). So $K_L(f) \in \mathcal{U}(L)$.

(ii) We shall prove that $K_L$ is bijective, $K_L$ and $K_L{}^{-1}$ are order preserving.

(1) $K_L$ is injective.

Suppose $f, g \in \mathcal{E}(L)$ such that $f \neq g$, then there exists $\alpha \in M(L)$ such that $f(\alpha) \neq g(\alpha)$. By Lemma **13.1.52** (i), $\hat{Q}^*(f(\alpha)) \neq \hat{Q}^*(g(\alpha))$. Without loss of any generality, suppose $\hat{Q}^*(f(\alpha))\backslash\hat{Q}^*(g(\alpha)) \neq \emptyset$, then there exists $\beta \in \hat{Q}^*(f(\alpha))\backslash\hat{Q}^*(g(\alpha))$. So $(\alpha, \beta) \in K_L(f)\backslash K_L(g)$, $K_L(f) \neq K_L(g)$.

(2) $K_L$ is surjective.

Suppose $U \in \mathcal{U}(L)$. By Lemma **13.1.55** (ii), $D_U \in \mathcal{E}(L)$. We need only prove $K_L(D_U) = U$.

Suppose $(\alpha, \beta) \in K_L(D_U)$, then $\beta \in \hat{Q}^*(D_U(\alpha))$. By $D_U(\alpha) = \bigvee H_U(\alpha)$, there exists $\gamma \in H_U(\alpha)$ such that $\beta \in \hat{Q}^*(\gamma)$. By $\gamma \in H_U(\alpha)$, there exists $\alpha_0 \in \beta^*(\alpha)$ such that $(\alpha_0, \lambda) \in U$ for every $\lambda \in \hat{Q}^*(\gamma)$, and hence $(\alpha_0, \beta) \in U$. By (BE2), $(\alpha, \beta) \in U$. So $K_L(D_U) \subset U$. Conversely, if $(\alpha, \beta) \in U$, by (BE3) and Lemma **13.1.52**, there exist $\alpha_0 \in \beta^*(\alpha)$, $\beta_0 \in \beta^*(\beta)$ such that $(\alpha_0, \beta_0) \in U$. By Lemma **13.1.51**, there exists $\lambda \in \beta^*(\bigwedge \hat{Q}(\beta_0)) \cap \hat{Q}^*(\beta)$, so $\gamma \geq \beta_0$ for every $\gamma \in \hat{Q}^*(\lambda)$. By (BE2), $(\alpha_0, \gamma) \in U$, it means $\lambda \in H_U(\alpha)$, and hence $\beta \in \hat{Q}^*(D_U(\alpha))$. So $(\alpha, \beta) \in K_L(D_U)$, $U \subset K_L(D_U)$. Therefore, $K_L(D_U) = U$.

(3) Both $K_L$ and $K_L{}^{-1}$ are order preserving.

Easy to verify the order preserving property of $K_L$. Suppose $U, V \in \mathcal{U}(L)$ such that $U \subset V$, then by the equality $K_L(D_U) = U$ proved above, $K_L{}^{-1}(U) = D_U$, $K_L{}^{-1}(V) = D_V$. For every $a \in L\backslash\{0\}$ and every $b \in H_U(a)$, then there exists $\alpha \in \beta^*(a)$ such that $(\alpha, \beta) \in U \subset V$ for every $\beta \in \hat{Q}^*(b)$, so $b \in H_U(a)$. Hence

$$D_U(a) = \bigvee H_U(a) \leq \bigvee H_V(a) = D_V(a),$$

i.e. $D_U \leq D_V$, $K_L{}^{-1}$ is order preserving. □

**13.1.58 Definition** Let $L$ be a F-lattice, $U, V \subset M(L) \times M(L)$. Define

$$\begin{aligned} U \circ V = \{(\alpha, \beta) \in M(L) \times M(L) :\ & \exists \gamma \in M(L),\ (\gamma, \beta) \in U, \\ & \forall \lambda \in \hat{Q}^*(\gamma),\ (\alpha, \lambda) \in V\}, \end{aligned}$$

$$U^{-1} = \{(\alpha, \beta) :\ (\beta, \alpha) \in U\}.$$

**13.1.59 Lemma** *Let $L$ be a F-lattice, then for arbitrary $f, g \in \mathcal{E}(L)$,*

$$K_L(f) \circ K_L(g) = K_L(f \circ g).$$

**Proof** Suppose $(\alpha, \beta) \in K_L(f) \circ K_L(g)$, then there exists $\gamma \in M(L)$ such that $(\gamma, \beta) \in K_L(f)$ and $(\alpha, \lambda) \in K_L(g)$ for every $\lambda \in \hat{Q}^*(\gamma)$. This is equivalent to $\beta \in \hat{Q}^*(f(\gamma))$ and $\hat{Q}^*(\gamma) \subset \hat{Q}^*(g(\alpha))$. By Lemma **13.1.52**, $\gamma \leq g(\alpha)$. Since $f$ is order preserving, $f(\gamma) \leq f(g(\alpha))$, $\beta \in \hat{Q}^*(f(\gamma)) \subset \hat{Q}^*(f(g(\alpha)))$, $(\alpha, \beta) \in K_L(f \circ g)$, $K_L(f) \circ K_L(g) \subset K_L(f \circ g)$. Conversely, let $(\alpha, \beta) \in K_L(f \circ g)$, then

$$\beta \in \hat{Q}^*(f(g(\alpha)) = \hat{Q}^*(f(\bigvee \beta^*(g(\alpha))) = \bigcup_{\gamma \in \beta^*(g(\alpha))} \hat{Q}^*(f(\gamma)),$$

so there exists $\gamma \in \beta^*(g(\alpha))$ such that $\beta \in \hat{Q}^*(f(\gamma))$. Hence $\gamma \leq g(\alpha)$ and $(\gamma, \beta) \in K_L(f)$. By Lemma **13.1.52**, $\hat{Q}^*(\gamma) \subset \hat{Q}^*(g(\alpha))$, so $(\alpha, \lambda) \in K_L(g)$ for every $\lambda \in \hat{Q}^*(\gamma)$. Then we have $(\alpha, \beta) \in K_L(f) \circ K_L(g)$, $K_L(f \circ g) \subset K_L(f) \circ K_L(g)$. This completes the proof. □

**13.1.60 Lemma** *Let $L$ be a F-lattice, $\alpha, \beta \in M(L)$, $f \in \mathcal{E}(L)$. Then*

$$\beta \in \hat{Q}^*(f(\alpha)) \iff \alpha \in \hat{Q}^*(f^{\triangleleft}(\beta)).$$

**Proof** By Theorem **13.1.15** (i),

$$\beta \in \hat{Q}^*(f(\alpha)) \iff f(\alpha) \not\leq \beta' \iff f^{\triangleleft}(\beta) \not\leq \alpha' \iff \alpha \in \hat{Q}^*(f^{\triangleleft}(\beta)).$$ □

**13.1.61 Lemma** *Let $L$ be a F-lattice, $f \in \mathcal{E}(L)$, then $K_L(f^{\triangleleft}) = K_L(f)^{-1}$.*

**Proof** By Lemma **13.1.60**,

$$\begin{aligned} (\alpha, \beta) \in K_L(f^{\triangleleft}) &\iff \beta \in \hat{Q}^*(f^{\triangleleft}(\alpha)) \\ &\iff \alpha \in \hat{Q}^*(f(\beta)) \\ &\iff (\beta, \alpha) \in K_L(f) \\ &\iff (\alpha, \beta) \in K_L(f)^{-1}. \end{aligned}$$ □

The definition of $L$-fuzzy quasi-uniformity ($L$-fuzzy uniformity) on a nonempty ordinary set $X$ can be changed to a more general one for general F-lattices:[38, 40]

**13.1.62 Definition** Let $L$ be a F-lattice. $\mathcal{D} \subset \mathcal{E}(L)$. $\mathcal{D}$ is called a *quasi-unformity* on $L$, if $\mathcal{D}$ fulfils the following conditions (UFL1) – (UFL3):

(UFL1) $f \in \mathcal{E}(L),\ g \in \mathcal{D},\ f \geq g \implies f \in \mathcal{D}$.

(UFL2) $f, g \in \mathcal{D} \implies f \wedge g \in \mathcal{D}$.

(UFL3) $f \in \mathcal{D} \implies \exists g \in \mathcal{D},\ g \circ g \leq f$.

$\mathcal{D}$ is called a *uniformity* on $L$, if $\mathcal{D}$ fulfils the above conditions (UFL1) – (UFL3) and the following condition:

(UFL4) $f \in \mathcal{D} \implies f^{\triangleleft} \in \mathcal{D}$.

By Theorem **13.1.57**, Lemma **13.1.59** and Lemma **13.1.61**, we obtain a geometrical characterization of quasi-uniformity (uniformity) on lattices:

**13.1.63 Theorem** *Let $L$ be a F-lattice, $\mathcal{D} \subset \mathcal{E}(L)$. Then $\mathcal{D}$ is a quasi-uniformity (uniformity, respectively) on $L$ if and only if $K_L[\mathcal{D}]$ fulfils the following conditions* (GUF1) – (GUF3) ((GUF1) – (GUF4), *respectively*)*:*

(GUF1) $U \in \mathcal{E}(L),\ V \in K_L[\mathcal{D}],\ U \supset V \implies U \in K_L[\mathcal{D}]$.

(GUF2) $U, V \in K_L[\mathcal{D}] \implies U \cap V \in K_L[\mathcal{D}]$.

(GUF3) $U \in K_L[\mathcal{D}] \implies \exists V \in K_L[\mathcal{D}],\ V \circ V \subset U$.

(GUF4) $U \in K_L[\mathcal{D}] \implies U^{-1} \in K_L[\mathcal{D}]$. □

To consider the products of $L$-fuzzy quasi-uniform spaces or $L$-fuzzy uniform spaces, we need a lemma as following:

**13.1.64 Lemma** *Let $\{(L^{X_t}, \mathcal{D}_t) : t \in T\}$ be a family of $L$-fuzzy quasi-uniform spaces ($L$-fuzzy uniform spaces, respectively), $X = \prod_{t \in T} X_t$. Then*

$$\mathcal{B} = \{ \bigwedge_{i \leq k} p_{t_i}{}^{\leftarrow} f_{t_i} p_{t_i}{}^{\rightarrow} : \ \{t_0, \cdots, t_k\} \in [T]^{<\omega},\ \forall i \leq k,\ f_{t_i} \in \mathcal{D}_{t_i} \}$$

*is a base of an $L$-fuzzy quasi-uniformity ($L$-fuzzy uniformity, respectively) on $X$.*

**Proof** For every $f \in \bigcup_{t \in T} \mathcal{D}_t$ and every $t \in T$, denote $g(f, t) = p_t{}^{\leftarrow} f p_t{}^{\rightarrow}$. We need to verify (BUF1) – (BUF2) ((BUF1) – (BUF3), respectively). (BUF1) is obviously satisfied. Suppose $\bigwedge_{i \leq k} g(f_{t_i}, t_i) \in \mathcal{B}$, then $f_{t_i} \in \mathcal{D}_{t_i}$ for every $i \leq k$. By (UF3), there exists $g_{t_i} \in \mathcal{D}_{t_i}$ such that $g_{t_i} \circ g_{t_i} \leq f_{t_i}$. Then

$$\begin{aligned} \bigwedge_{i \leq k} g(f_{t_i}, t_i) &\geq \bigwedge_{i \leq k} p_{t_i}{}^{\leftarrow} (g_{t_i} \circ g_{t_i}) p_{t_i}{}^{\rightarrow} \\ &\geq \bigwedge_{i \leq k} (p_{t_i}{}^{\leftarrow} g_{t_i} p_{t_i}{}^{\rightarrow}) \circ (p_{t_i}{}^{\leftarrow} g_{t_i} p_{t_i}{}^{\rightarrow}) \\ &\geq (\bigwedge_{i \leq k} p_{t_i}{}^{\leftarrow} g_{t_i} p_{t_i}{}^{\rightarrow}) \circ (\bigwedge_{i \leq k} p_{t_i}{}^{\leftarrow} g_{t_i} p_{t_i}{}^{\rightarrow}) \\ &= (\bigwedge_{i \leq k} g(g_{t_i}, t_i)) \circ (\bigwedge_{i \leq k} g(g_{t_i}, t_i)). \end{aligned}$$

Since $g(g_{t_i}, t_i) \in \mathcal{D}_{t_i}$ for every $i \leq k$, so $\bigwedge_{i \leq k} g(g_{t_i}, t_i) \in \mathcal{B}$, (BUF2) is fulfilled. So the conclusion for $L$-fuzzy quasi-uniformities has been proved.

As for the case of $L$-fuzzy uniformity, we need only prove (BUF3) under the assumption that every $(L^{X_t}, \mathcal{D}_t)$ is an $L$-fuzzy uniform space. For every $t \in T$, denote the order-reserving involution $' : L^{X_t} \to L^{X_t}$ by $r_t$, then by Theorem **13.1.15** (iii), Proposition **13.1.14** and Theorem **13.1.10** (ii), for arbitrary $A, B \in L^X$ and every $f \in \mathcal{D}_t$ such that $f^{\triangleleft} = f$,

$$\begin{aligned}
g(f,t)(A) \le B &\iff p_t^{\leftarrow} f p_t^{\rightarrow}(A) \le B \\
&\iff p_t^{\leftarrow}((f p_t^{\rightarrow}(A))') \ge B' \\
&\iff ((f p_t^{\rightarrow}(A))' \ge p_t^{\rightarrow}(B') \\
&\iff r_t f^{\vee} r_t p_t^{\rightarrow}(A) = f^{\triangleleft} p_t^{\rightarrow}(A) = f p_t^{\rightarrow}(A) \le r_t p_t^{\rightarrow} r_t(B) \\
&\iff f^{\vee} r_t p_t^{\rightarrow}(A) \ge p_t^{\rightarrow} r_t(B) \\
&\iff r_t p_t^{\rightarrow}(A) \ge f \circ f^{\vee} r_t p_t^{\rightarrow}(A) \ge f p_t^{\rightarrow} r_t(B) \\
&\iff p_t^{\leftarrow} f p_t^{\rightarrow}(B') \le p_t^{\leftarrow}((p_t^{\rightarrow}(A))') = (p_t^{\leftarrow} p_t^{\rightarrow}(A))' \le A' \\
&\iff g(f,t)(B') \le A'.
\end{aligned}$$

On the other hand, by Proposition **13.1.14** and Theorem **13.1.10** (ii),

$$\begin{aligned}
g(f,t)^{\triangleleft}(A) \le B &\iff r_t g(f,t)^{\vee} r_t(A) \le B \\
&\iff g(f,t)^{\vee} r_t(A) \ge r_t(B) \\
&\iff g(f,t)(r_t(B)) \le g(f,t) g(f,t)^{\vee} r_t(A) \le r_t(A) \\
&\iff g(f,t)(B') \le A'.
\end{aligned}$$

So for arbitrary $A, B \in L^X$,

$$g(f,t)(A) \le B \iff g(f,t)(B') \le A' \iff g(f,t)^{\triangleleft}(A) \le B,$$

and hence $g(f,t)^{\triangleleft} = g(f,t)$. Since by Theorem **13.1.24** we can always find $g_{t_i} \in \mathcal{D}_t$ such that $g_{t_i}{}^{\triangleleft} = g_{t_i}$ and $g_{t_i} \le g_{t_i} \circ g_{t_i} \le f_{t_i}$ for every $f_{t_i}$ in an arbitrary element $\bigwedge_{i \le k} p_{t_i}{}^{\leftarrow} f_{t_i} p_{t_i}{}^{\rightarrow} \in \mathcal{B}$, so by Theorem **13.1.15** (vi), we have

$$\bigwedge_{i \le k} g(f_{t_i}, t_i) \ge \bigwedge_{i \le k} p_{t_i}{}^{\leftarrow} g_{t_i} p_{t_i}{}^{\rightarrow} = \bigwedge_{i \le k} g(g_{t_i}, t_i) = \bigwedge_{i \le k} g(g_{t_i}, t_i)^{\triangleleft} = (\bigwedge_{i \le k} g(g_{t_i}, t_i))^{\triangleleft}.$$

Since $\bigwedge_{i \le k} g(g_{t_i}, t_i) \in \mathcal{B}$, and by Theorem **13.1.15** (iii), (iv) we have

$$\bigwedge_{i \le k} g(g_{t_i}, t_i) = ((\bigwedge_{i \le k} g(g_{t_i}, t_i))^{\triangleleft})^{\triangleleft} \le (\bigwedge_{i \le k} g(f_{t_i}, t_i))^{\triangleleft},$$

so (BUF3) is fulfilled by $\mathcal{B}$, $\mathcal{B}$ is a base of an $L$-fuzzy uniformity on $X$. □

Then we can reasonably define $L$-fuzzy product quasi-uniformity ($L$-fuzzy product uniformity) as follows:

**13.1.65 Definition** Let $\{(L^{X_t}, \mathcal{D}_t) : t \in T\}$ be a family of $L$-fuzzy quasi-uniform spaces ($L$-fuzzy uniform spaces, respectively), $X = \prod_{t \in T} X_t$. Define the *$L$-fuzzy product quasi-uniformity* (*$L$-fuzzy product uniformity*, respectively) of $\{\mathcal{D}_t : t \in T\}$ on $X$ as the $L$-fuzzy quasi-uniformity ($L$-fuzzy uniformity, respectively) generated by the following base:

$$\mathcal{B} = \{\bigwedge_{i \le k} p_{t_i}{}^{\leftarrow} f_{t_i} p_{t_i}{}^{\rightarrow} : \{t_0, \cdots, t_k\} \in [T]^{<\omega},\ \forall i \le k,\ f_{t_i} \in \mathcal{D}_{t_i}\},$$

and denote it by $\prod_{t \in T} \mathcal{D}_t$.

To end this section, we leave the following conclusion[58] as an exercise:

**13.1.66 Theorem** *Let $\{(L^{X_t}, \mathcal{D}_t) : t \in T\}$ be a family of $L$-fuzzy quasi-uniform spaces, $\mathcal{D} = \prod_{t \in T} \mathcal{D}_t$. Then $\delta(\prod_{t \in T} \mathcal{D}) = \prod_{t \in T} \delta(\mathcal{D}_t)$.* □

**13.1.67 Remark** On fuzzy uniformity, there are other approaches by R. Lowen[96] and Höhle:[34]

(1) Approach of Lowen

For every $\mu \in I^X$, $\gamma \in I^{X\times X}$, define $\gamma\langle\mu\rangle \in I^X$ as follows:

$$\forall x \in X, \quad \gamma\langle\mu\rangle(x) = \bigvee\{\mu(y) \wedge \gamma(y,x) : y \in X\},$$

define the composition of $\gamma, \zeta \in I^{X\times X}$ by

$$\forall (x,y) \in X \times X, \quad \gamma\circ\zeta(x,y) = \bigvee\{\zeta(x,z) \wedge \gamma(z,y) : z \in X\}.$$

A fuzzy uniformity on $X$ is a subset $\mathcal{U} \subset I^{X\times X}$ such that

(LUF1) $\mathcal{U}$ is a filter.

(LUF2) $\forall\{\gamma_\varepsilon : \varepsilon \in I\backslash\{0\}\} \subset \mathcal{U}, \ \bigvee\{\gamma_\varepsilon - \varepsilon : \varepsilon \in I\backslash\{0\}\} \in \mathcal{U}$.

(LUF3) $\forall \gamma \in \mathcal{U}, \ \forall x \in X, \ \gamma(x,x) = 1$.

(LUF4) $\forall \gamma \in \mathcal{U}, \ {}_s\gamma \in \mathcal{U}$, where ${}_s\gamma(x,y) = \gamma(y,x)$ for every $(x,y) \in X \times X$.

(LUF5) $\forall \gamma \in \mathcal{U}, \ \forall \varepsilon \in I\backslash\{0\}, \ \exists \gamma_\varepsilon \in \mathcal{U}, \ \gamma_\varepsilon\circ\gamma_\varepsilon - \varepsilon \le \gamma$.

If $\mathcal{U}$ is a fuzzy uniformity on $X$, then $(I^X, \mathcal{U})$ is called a uniform space. The fuzzy uniformity determines a closure operator $^- : I^X \to I^X$ as follos:

$$\forall \mu \in I^X, \quad \mu^- = \bigwedge\{\gamma\langle\mu\rangle : \gamma \in \mathcal{U}\}.$$

The fuzzy topology (in Lowen's sense, which is stratified) induced by this closure operator is denoted by $t(\mathcal{U})$.

A mapping $f^\to : (I^X, \mathcal{U}_1) \to (I^Y, \mathcal{U}_2)$ is called uniformly continuous if for every $\gamma_2 \in \mathcal{U}_2$ such that $(f \times f)^\leftarrow(\gamma_2) \in \mathcal{U}_1$. It was proved that the uniform continuity of $f^\to$ implies the continuity of $f^\to : (I^X, t(\mathcal{U}_1)) \to (I^Y, t(\mathcal{U}_2))$.

For a systematical study of the properties of Lowen's fuzzy uniform spaces and their relation to ordinary uniform spaces we refer the reader to Refs. [96, 107].

(2) Approach of Höhle

For every triangle norm $T$ on $[0,1]$, Höhle[35] introduced the notion of $T$-fuzzy uniformity. A triangle norm $T$ is a mapping $T : [0,1] \times [0,1] \to [0,1]$ which fulfils:

(i) $T(0,0) = 0$, $T(x,1) = x$.

(ii) $x \le x'$, $y \le y' \implies T(x,y) \le T(x',y')$.

(iii) $T(T(x,y),z) = T(x,T(y,z))$.

(iv) $T(x,y) = T(y,x)$.

The definition of $T$-fuzzy uniformity on a set $X$ is obtained by replacing (LUF5) in Lowen's definition of fuzzy uniformity by

(U5)* $\forall \gamma \in \mathcal{U}, \ \forall \varepsilon > 0, \ \exists \gamma_\varepsilon \in \mathcal{U}, \ \gamma_\varepsilon\circ_T\gamma_\varepsilon - \varepsilon \le \gamma$, where the composition $\gamma_\varepsilon\circ_T\gamma_\varepsilon - \varepsilon$ is defined as follows:

$$\forall (x,y) \in X \times X, \quad \gamma_\varepsilon\circ_T\gamma_\varepsilon - \varepsilon(x,y) = \bigvee\{T(\gamma_\varepsilon(x,z), \gamma_\varepsilon(z,y)) : z \in X\}.$$

Trivially $T = \wedge$ implies that the $T$-fuzzy uniformity is just the fuzzy uniformity in Lowen's sense.

In the case $T = T_m$, Höhle proved that the $T_m$-fuzzy uniformities on a set $X$ correspond bijectively to the fuzzy unifomities of Hutton on $X$ which fulfil some conditions, where

$$\forall x, y \in I, \quad T_m(x,y) = max\{x + y - 1, 0\}.$$

He also investigated the relation between fuzzy $T$-uniformities on $X$ and the probabilistic metrics on $X$.

## 13.2 Proximity

In this section, unless special declaration, $L$ in an $L$-fuzzy space $L^X$ always means a F-lattice.

**13.2.1 Definition** Let $L^X$ be an $L$-fuzzy space, $\varrho$ be a relation on $L^X$, i.e. $\varrho \subset L^X \times L^X$. Denote $(A, B) \in \varrho$ by $A\varrho B$, $(A, B) \notin \varrho$ by $A\bar{\varrho}B$ for short. $\varrho$ is called an *L-fuzzy proximity* on $X$, if $\varrho$ fulfils the following conditions (PX1) – (PX5):

(PX1) $A\varrho B \implies B\varrho A$.

(PX2) $(A \vee B)\varrho C \implies A\varrho C$ or $B\varrho C$.

(PX3) $A\varrho B \implies A \neq \underline{0},\ B \neq \underline{0}$.

(PX4) $A\bar{\varrho}B \implies \exists C \in L^X,\ A\bar{\varrho}C,\ C'\bar{\varrho}B$.

(PX5) $A\hat{\varrho}B \implies A\varrho B$.

If $\varrho$ is an $L$-fuzzy proximity on $X$, call $(L^X, \varrho)$ an *L-fuzzy proximity space*.

The following conclusions are obvious:

**13.2.2 Proposition** *Let $(L^X, \varrho)$ be an L-fuzzy proximity space. Then*

(i) $A\varrho B,\ A \leq A_1,\ B \leq B_1 \implies A_1\varrho B_1$.

(ii) $\forall A \in L^X,\ \underline{0}\bar{\varrho}A$.

(iii) $(A \vee B)\varrho C \iff A\varrho C$ *or* $B\varrho C$. □

**13.2.3 Theorem** *Let $(L^X, \varrho)$ be an L-fuzzy proximity space, mapping $i: L^X \to L^X$ be defined as follows:*

$$\forall A \in L^X, \quad i(A) = \bigvee\{C \in L^X : C\bar{\varrho}A'\}, \tag{13.44}$$

*then $i$ is an interior operator on $L^X$.*

**Proof** By (PX3), $\underline{1}\bar{\varrho}\underline{0} = (\underline{1})'$, so $i(\underline{1}) = \underline{1}$, (IO1) is fulfilled by $i$. If $C\bar{\varrho}A'$, by (PX5), $C \leq A$. So $i(A) \leq A$, (IO2) is fulfilled by $i$. To prove (IO3), verify the order preserving property of $i$ in the first. Suppose $A, B \in L^X$, $A \leq B$, $C\varrho B'$, then by $B' \leq A'$ and Proposition **13.2.2** (i), $C\varrho A'$. So $C\bar{\varrho}A' \implies C\bar{\varrho}B'$, $i(A) \leq i(B)$, $i$ is order preserving. So $i(A \wedge A) \leq i(A) \wedge i(B)$, we need only prove its reverse. Suppose $C, D \in L^X$, $C\bar{\varrho}A'$, $D\bar{\varrho}B'$, then by (PX1) and Proposition **13.2.2** (i), $A'\bar{\varrho}(C \wedge D)$, $B'\bar{\varrho}(C \wedge D)$. By (PX2), $(A' \vee B')\bar{\varrho}(C \wedge D)$, so $(C \wedge D)\bar{\varrho}(A \wedge B)'$. That is to say,

$$\begin{aligned} i(A) \wedge i(B) &= \bigvee\{C \wedge D : C, D \in L^X,\ C\bar{\varrho}A',\ D\bar{\varrho}B'\} \\ &\leq \bigvee\{C \in L^X : C\bar{\varrho}(A \wedge B)'\} \\ &= i(A \wedge B), \end{aligned}$$

so $i$ fulfils (IO3). By (IO2) proved above, we need only prove $ii(A) \geq i(A)$. Suppose $C\bar{\varrho}A'$, then by (PX4), there exists $D \in L^X$ such that $C\bar{\varrho}D$, $D'\bar{\varrho}A'$. By (13.44), $D' \leq i(A)$, so $D \geq i(A)'$. By $C\bar{\varrho}D$ and Proposition **13.2.2** (i), $C\bar{\varrho}i(A)'$. So

$$i(A) = \bigvee\{C \in L^X : C\bar{\varrho}A'\} \leq \bigvee\{C \in L^X : C\bar{\varrho}i(A)'\} = i(i(A)),$$

$i$ fulfils (IO4). Hence $i$ is an interior operator on $L^X$. □

**13.2.4 Definition** Let $\varrho$ be an $L$-fuzzy proximity on $X$. Then interior operator defined in (13.44) is called the interior operator on $L^X$ *generated by the L-fuzzy proximity $\varrho$*. The $L$-fuzzy topology on $X$ generated by the interior operator defined by (13.44) is called the $L$-fuzzy topology *generated by the L-fuzzy proximity* $\varrho$, denoted by $\delta(\varrho)$.

By $A^- = A'^{\circ\prime}$ and (13.44), we have

**13.2.5 Corollary** *Let $\varrho$ be an L-fuzzy proximity on X, then in L-fts $(L^X, \delta(\varrho))$, for every $A \in L^X$,*

$$A^- = \bigwedge\{C \in L^X : C'\bar{\varrho}A\}. \qquad \square$$

To consider the relation between $L$-fuzzy proximity and $L$-fuzzy uniformity, define a kind of relation on $L^X$ as follows:[57]

**13.2.6 Definition** Let $\prec$ be a relation on $L^X$. $\prec$ is called a *topological order* on $L^X$, if $\prec$ fulfils the following conditions (TO1) – (TO6):

(TO1) $\underline{0} \prec \underline{0}$, $\underline{1} \prec \underline{1}$.

(TO2) $A \prec B \implies A \leq B$.

(TO3) $A \leq B \prec C \leq D \implies A \prec D$.

(TO4) $A \prec C$, $B \prec D$, $\implies (A \vee B) \prec (C \vee D)$, $(A \wedge B) \prec (C \wedge D)$.

(TO5) $A \prec B \implies B' \prec A'$.

(TO6) $A \prec B \implies \exists C \in L^X$, $A \prec C \prec B$.

For every pair $A, B \in L^X$, denote $A \not\prec B$ if $A \prec B$ is not true.

**13.2.7 Proposition** *Let $\prec$ be a topological order on $L^X$, then $\underline{0} \prec A \prec \underline{1}$ for every $A \in L^X$.* $\square$

**13.2.8 Theorem** *Let $\prec$ be a topological order on $L^X$, mapping $i : L^X \to L^X$ be defined as follows:*

$$\forall A \in L^X, \quad i(A) = \bigvee\{C \in L^X : C \prec A\}, \tag{13.45}$$

*then i is an interior operator on $L^X$.*

**Proof** It follows from (TO1), $\underline{1} \prec \underline{1}$, so $i(\underline{1}) = \underline{1}$, (IO1) is fulfilled. By (TO2), $i$ fulfils (IO2). It also follows from (TO2) that $i$ is order preserving, so $i(A \wedge B) \leq i(A) \wedge i(B)$, to prove (IO3) we need only prove $i(A \wedge B) \geq i(A) \wedge i(B)$. Suppose $C \prec A$, $D \prec B$, then by (TO4), $(C \wedge D) \prec (A \wedge B)$. So

$$\begin{aligned} i(A) \wedge i(B) &= (\bigvee\{C \in L^X : C \prec A\}) \wedge (\bigvee\{D \in L^X : D \prec B\}) \\ &= \bigvee\{C \wedge D : C, D \in L^X, C \prec A, D \prec B\} \\ &\leq \bigvee\{C \wedge D : C, D \in L^X, (C \wedge D) \prec (A \wedge B)\} \\ &= i(A \wedge B), \end{aligned}$$

(IO3) is fulfilled. Since $i(A) \leq A$ for every $A \in L^X$, we have $ii(A) \leq i(A)$. Suppose $C \prec A$, by (TO6), there exists $D \in L^X$ such that $C \prec D \prec A$. Then by (13.45), $C \leq i(D)$, $D \leq i(A)$. Since $i$ is order preserving, we have $C \leq i(D) \leq i(i(A))$, so $i(A) \leq i(i(A))$, (IO4) has been verified, $i$ is an interior operator on $L^X$. $\square$

**13.2.9 Definition** Let $\prec$ be a topological order on $L^X$. Then the interior operator defined in (13.45) is called the interior operator on $L^X$ *generated by the topological order* $\prec$. The $L$-fuzzy topology on $X$ generated by the interior operator defined by (13.45) is called the $L$-fuzzy topology *generated by the topological order* $\prec$, denoted by $\delta(\prec)$.

By $A^- = A'^{\circ\prime}$, (13.45) and (TO5), we have

**13.2.10 Corollary** *Let $\prec$ be a topological order on $X$, then in L-fts $(L^X, \delta(\prec))$, for every $A \in L^X$,*

$$A^- = \bigwedge\{C \in L^X : A \prec C\}. \qquad \square$$

By the above Corollary **13.2.10**, the following definition is reasonable and coincides with Definition **13.2.9**:

**13.2.11 Definition** Let $\prec$ be a topological order on $L^X$. Then the closure operator $c: L^X \to L^X$ defined by the following equation is called the closure operator on $L^X$ *generated by the topological order* $\prec$:

$$\forall A \in L^X, \quad c(A) = \bigwedge\{C \in L^X : A \prec C\}. \tag{13.46}$$

**13.2.12 Definition** If $\varrho$ is an $L$-fuzzy proximity on $X$, then define a relation $\prec(\varrho)$ on $L^X$ as follows:

$$\forall A, B \in L^X, \quad A \prec(\varrho) B \iff A \bar{\varrho} B'. \tag{13.47}$$

If $\prec$ is a topological order on $L^X$, define a relation $\varrho(\prec)$ on $L^X$ as follows:

$$\forall A, B \in L^X, \quad A \varrho(\prec) B \iff A \not\prec B'. \tag{13.48}$$

**13.2.13 Theorem** (i) *If $\varrho$ is an L-fuzzy proximity on $X$, then $\prec(\varrho)$ is a topological order on $L^X$, and*

$$\varrho(\prec(\varrho)) = \varrho, \quad \delta(\prec(\varrho)) = \delta(\varrho).$$

(ii) *If $\prec$ is a topological order on $L^X$, then $\varrho(\prec)$ is an L-fuzzy proximity on $X$, and*

$$\prec(\varrho(\prec)) = \prec, \quad \delta(\varrho(\prec)) = \delta(\prec).$$

**Proof** (i) Suppose $\varrho$ is an $L$-fuzzy proximity on $L^X$, $\prec = \prec(\varrho)$, then (TO1) and (TO2) can be deduced from (PX3) and (PX5) respectively. (TO3) can be deduced from Proposition **13.2.2** (i). (TO4) and (TO5) are deduced from (PX2) and (PX1) respectively. Finally, suppose $A \prec B$, by (13.47), $A\bar{\varrho}B'$, so by (PX4) there exists $C \in L^X$ such that $A\bar{\varrho}C'$ and $C\bar{\varrho}B'$. So $A \prec C \prec B$, (TO6) also holds. By (13.47) we can verify $\varrho(\prec(\varrho)) = \varrho$ and $\delta(\prec(\varrho)) = \delta(\varrho)$ straightforward.

(ii) Conversely, suppose $\prec$ is a topological order on $L^X$, it can be similarly proved that $\varrho = \varrho(\prec)$ is an $L$-fuzzy proximity on $X$, and by (13.48) we have $\prec(\varrho(\prec)) = \prec$ and $\delta(\varrho(\prec)) = \delta(\prec)$. $\square$

**13.2.14 Definition** For every $L$-fuzzy proximity $\varrho$ on $X$, call $\prec(\varrho)$ the *topological order generated by the L-fuzzy proximity $\varrho$*. For every topological order $\prec$ on $L^X$, call $\varrho(\prec)$ the *L-fuzzy proximity generated by the tological order* $\prec$.

**13.2.15 Definition** Let $\mathcal{D}$ be an $L$-fuzzy uniformity on $X$, define a relation $\varrho(\mathcal{D})$ on $L^X$ as follows:

$$\forall A, B \in L^X, \quad A\varrho(\mathcal{D})B \iff \forall f \in \mathcal{D},\ f(A)\hat{q}B \text{ or } f(B)\hat{q}A. \tag{13.49}$$

**13.2.16 Theorem** *Let $\mathcal{D}$ be an L-fuzzy uniformity on $X$. Then $\varrho = \varrho(\mathcal{D})$ is an L-fuzzy proximity on $X$, and $\delta(\varrho) = \delta(\mathcal{D})$.*

**Proof** By the definition of $\varrho$ in (13.49), $A\varrho B \Longrightarrow B\varrho A$ obviously, (PX1) is fulfilled. Suppose $A\bar{\varrho}C$ and $B\bar{\varrho}C$, then there exist $f, g \in \mathcal{D}$ such that

$$f(A) \leq C',\ f(C) \leq A',\ g(B) \leq C',\ g(C) \leq B'.$$

Let $h = f \wedge g$, then $g \in \mathcal{D}$ and

$$h(A \vee B) \le f(A) \vee g(B) \le C',$$
$$h(C) \le f(C) \wedge g(C) \le A' \wedge B' = (A \vee B)'.$$

So by (13.49), $(A \vee B)\overline{\varrho}C$, (PX2) holds. If $A\varrho B$, by (13.49), for every $f \in \mathcal{D}$, $f(A) \not\le B'$ or $f(B) \not\le A'$. So $A \ne \underline{0}$, $B \ne \underline{0}$, (PX3) holds. Suppose $A\overline{\varrho}B$, then there exists $f \in \mathcal{D}$ such that $f(A) \le B'$ and $f(B) \le A'$. By Theorem **13.1.24**, there exists $g \in \mathcal{D}$ such that $g = g^{\triangleleft}$ and $g \circ g \le f$. Let $C = (g(A))'$, then $g(A) \le C'$, and hence

$$g(C) = g^{\triangleleft}(C) = \bigwedge\{D \in L^X : \ g(D') \le C'\} \le A'.$$

So $A\overline{\varrho}C$. Since

$$g(C') = g \circ g(A) \le f(A) \le B',$$

so

$$g(B) = g^{\triangleleft}(B) = \bigwedge\{D \in L^X : \ g(D') \le B'\} \le C = (C')'.$$

Hence $C'\overline{\varrho}B$. Therefore, (PX4) is fulfilled. Suppose $A\hat{q}B$, then $A \not\le B'$. For every $f \in \mathcal{D}$, since $f$ is value increasing, $f(A) \not\le B'$, so $A\varrho B$, (PX5) is true. So $\varrho$ is an $L$-fuzzy proximity on $X$.

Since $\mathcal{D}$ is an $L$-fuzzy uniformity on $X$, for every $A \in L^X$, by Theorem **13.1.24** and Theorem **13.1.15** (i), we have

$$\begin{aligned}
\exists f \in \mathcal{D},\ f(C) \le A,\ f(A') \le C' &\Longrightarrow \exists f \in \mathcal{D},\ f(C) \le A,\ f^{\triangleleft}(C) \le A \\
&\Longrightarrow \exists g \in \mathcal{D},\ g = g^{\triangleleft},\ g(C) \le A \\
&\Longrightarrow \exists f \in \mathcal{D},\ f(C) \le A,\ f^{\triangleleft}(C) \le A \\
&\Longrightarrow \exists f \in \mathcal{D},\ f(C) \le A,\ f(A') \le C'.
\end{aligned}$$

So denote the interior operators generated respectively by $\mathcal{D}$ and $\varrho$ by $i_{\mathcal{D}}$ and $i_{\varrho}$, by (13.44) and (13.7),

$$\begin{aligned}
i_{\varrho}(A) &= \bigvee\{C \in L^X : \ C\overline{\varrho}A'\} \\
&= \bigvee\{C \in L^X : \ \exists f \in \mathcal{D},\ f(C) \le A,\ f(A') \le C'\} \\
&= \bigvee\{C \in L^X : \ \exists g \in \mathcal{D},\ g = g^{\triangleleft},\ g(C) \le A\} \\
&= \bigvee\{C \in L^X : \ \exists f \in \mathcal{D},\ f(C) \le A\} \\
&= i_{\mathcal{D}}(A).
\end{aligned}$$

Hence $\delta(\varrho) = \delta(\mathcal{D})$. □

**13.2.17 Definition** Let $\mathcal{D}$ be an $L$-fuzzy uniformity on $X$, then the $L$-fuzzy proximity $\varrho$ on $X$ defined in (13.49) is called the *$L$-fuzzy proximity generated by $L$-fuzzy uniformity* $\mathcal{D}$, is denoted by $\varrho(\mathcal{D})$.

It follows from Theorem **13.2.16** that for every $L$-fuzzy uniformity $\mathcal{D}$ on $X$ there exists an $L$-fuzzy proximity $\varrho(\mathcal{D})$ on $X$ such that $\delta(\varrho(\mathcal{D})) = \delta(\mathcal{D})$. But what about its reverse? The answer is also positive, but the procedure is not still straight:

**13.2.18 Lemma** *Let $\varrho$ be an $L$-fuzzy proximity on $X$, $\delta = \delta(\varrho)$ be the $L$-fuzzy topology generated by $\varrho$, $A, B \in L^X$ such that $A\overline{\varrho}B'$. Then there exists an $L$-fuzzy continuous mapping $F^{\to} : \ (L^X, \delta) \to I(L)$ such that*

$$A \le F^{\leftarrow}(L_1{}') \le F^{\leftarrow}(R_0) \le B. \tag{13.50}$$

**Proof** Let $\prec = \prec(\varrho)$ as is defined in (13.47), then by Theorem **13.2.13**, $\prec$ is a topological order on $L^X$. Denote $\mathbf{B} = \{\frac{i}{2^n} : \ n \in \mathbf{N},\ i \le 2^n\}$, we shall inductively construct a family $\mathcal{A} = \{A_r : \ r \in \mathbf{B}\}$ such that

$$A_0 = B,\ A_1 = A,\ A_{\frac{i+1}{2^n}} \prec A_{\frac{i}{2^n}}, \quad \forall n \in \mathbf{N},\ \forall i \le 2^n. \tag{13.51}$$

Let $A_0 = B$, $A_1 = A$, then $A_{\frac{1}{2^0}} \prec A_{\frac{0}{2^0}}$ holds. Suppose (13.51) holds for every $k \le n \in \mathbf{N}$, then since $\prec$ is a topological order on $L^X$, by (TO6), for every $i \le 2^n$, there exists $C_{\frac{i}{2^n}} \in L^X$ such that

$$A_{\frac{i+1}{2^n}} \prec C_{\frac{i}{2^n}} \prec A_{\frac{i}{2^n}}.$$

Let

$$A_{\frac{2i+1}{2^{n+1}}} = C_{\frac{i}{2^n}}, \quad \forall i \le 2^{n-1},$$

then $A_{\frac{i+1}{2^{n+1}}} \prec A_{\frac{i}{2^{n+1}}}$ for every $i \le 2^{n+1}$. By inductive method, $\mathcal{A} = \{A_r : r \in \mathbf{B}\}$ has been constructed to satisfy (13.51).

For every $t \in \mathbf{R}$, denote $\mathbf{B}(<t) = \{r \in \mathbf{B} : r < t\}$, let

$$A_t = \begin{cases} \underline{1}, & t \le 0, \\ \bigwedge_{r \in \mathbf{B}(<t)} A_r, & 0 < t < 1, \\ \underline{0}, & t \ge 1, \end{cases}$$

then by (TO2),

$$r, s \in \mathbf{R},\ r < s \implies A_r \ge A_s. \tag{13.52}$$

So the relation

$$\forall x \in X,\ \forall t \in \mathbf{R}, \qquad F(x)(t) = A_t(x)$$

defines an ordinary mapping $F: X \to md_I(L)$ (see Definition **9.1.1**). Identify every $F(x) \in md_I(L)$ with $[F(x)] \in I[L]$, we obtain an ordinary mapping $F: X \to I[L]$ and hence an $L$-fuzzy mapping $F^{\to}: (L^X, \delta) \to I(L)$ which satisfies (13.50) as follows: for every $x \in X$,

$$F^{\leftarrow}(L_1')(x) = L_1'(F(x)) = F(x)(1-) = \bigwedge_{t<1} A_t(x) \ge A_1(x) = A(x),$$
$$F^{\leftarrow}(R_0)(x) = R_0(F(x)) = F(x)(0+) = \bigvee_{t>0} A_t(x) \le A_0(x) = B(x).$$

Moreover, it follows from (13.45) and (13.51) that $r < s \implies A_s \le i(A_r)$, and similar with $F^{\leftarrow}(R_0) = \bigvee_{t>0} A_t$ proved above we can show $F^{\leftarrow}(R_s) = \bigvee_{t>s} A_t$ for every $s \in \mathbf{R}$, so

$$F^{\leftarrow}(R_s) = \bigvee_{t>s} A_t = \bigvee_{t>s} (\bigvee_{r>t} A_r) \le \bigvee_{t>s} i(A_t).$$

On the other hand, clearly we have

$$F^{\leftarrow}(R_s) = \bigvee_{t>s} A_t \ge \bigvee_{t>s} i(A_t),$$

so for every $s \in \mathbf{R}$ we have

$$F^{\leftarrow}(R_s) = \bigvee_{t>s} i(A_t) \in \delta(\varrho) = \delta.$$

Similarly prove

$$F^{\leftarrow}(L_s') = \bigwedge_{t<s} A_t^- \in \delta(\varrho)', \quad \forall s \in \mathbf{R},$$

hence $F^{\to}$ is continuous.

**13.2.19 Corollary** *Let $\varrho$ be an L-fuzzy proximity on X, $\delta = \delta(\varrho)$ be the L-fuzzy topology generated by $\varrho$, $A, B \in L^X$, $A \prec(\varrho) B$. Then there exists an L-fuzzy continuous mapping $(L^X, \delta) \to I(L)$ and $\varepsilon > 0$ such that*

$$F^{\leftarrow}(f_\varepsilon \wedge f_\varepsilon{}^{\triangleleft})F^{\rightarrow}(A) \leq B, \tag{13.53}$$

*where $f_\varepsilon$ is defined by* (13.13).

**Proof** By Lemma **13.2.18**, take an L-fuzzy continuous mapping $F^{\rightarrow} : (L^X, \delta) \to I(L)$ fulfilling (13.50), then $F^{\rightarrow}(A) \leq L_{\frac{1}{2}}{}' \leq R_0$. Let $\varepsilon = \frac{1}{2}$, then by (13.13),

$$(f_\varepsilon \wedge f_\varepsilon{}^{\triangleleft})F^{\rightarrow}(A) \leq f_\varepsilon F^{\rightarrow}(A) \leq f_{\frac{1}{2}}(L_{\frac{1}{2}}{}') = R_{u(L_{\frac{1}{2}}{}')-\frac{1}{2}} = R_0.$$

So by (13.50),

$$F^{\leftarrow}(f_\varepsilon \wedge f_\varepsilon{}^{\triangleleft})F^{\rightarrow}(A) \leq F^{\leftarrow}(R_0) \leq B. \qquad \square$$

Now we can prove the reverse of Theorem **13.2.16**:

**13.2.20 Theorem** *Let $\varrho$ be an L-fuzzy proximity on X. Then there exists an L-fuzzy uniformity $\mathcal{D}(\varrho)$ on X such that*

(i) $\delta(\mathcal{D}(\varrho)) = \delta(\varrho)$.

(ii) $\varrho(\mathcal{D}(\varrho)) \subset \varrho$.

**Proof** (i) Denote the family of all L-fuzzy continuous mappings from $(L^X, \delta(\varrho))$ to $I(L)$ by $\mathcal{F}$. Suppose $\delta(\varrho)$ is the L-fuzzy topology on $L^X$ generated by $\varrho$, define $\mathcal{D}(\varrho)$ as (13.36), then it is proved in Theorem **13.1.48** that $\mathcal{D}(\varrho)$ is an L-fuzzy uniformity on X and $\delta(\mathcal{D}(\varrho)) \subset \delta(\varrho)$ (note that in the proof of Theorem **13.1.48** we prove this inclusion relation without the assumption of complete regularity). Now we prove $\delta(\varrho) \subset \delta(\mathcal{D}(\varrho))$. In fact, suppose $U \in \delta(\varrho)$, then $U = \bigvee\{C \in L^X : C \prec(\varrho) U\}$. For every $C \in L^X$ such that $C \prec(\varrho) U$, by the virtue of Corollary **13.2.19**, there exists an L-fuzzy continuous mapping $F^{\rightarrow} : (L^X, \delta(\varrho)) \to I(L)$ and $\varepsilon > 0$ such that (13.53) holds. Then

$$\begin{aligned} int_{\delta(\mathcal{D}(\varrho))}(U) &= \bigvee\{C \in L^X : \exists g \in \mathcal{D}(\varrho),\ g(C) \leq U\} \\ &\geq \bigvee\{C \in L^X : F^{\leftarrow}(f_\varepsilon \wedge f_\varepsilon{}^{\triangleleft})F^{\rightarrow}(C) \leq U,\ F' \in \mathcal{F},\ \varepsilon > 0\} \\ &\geq \bigvee\{C \in L^X : C \prec(\varrho) U\} = int_{\delta(\varrho)}(U) = U, \end{aligned}$$

so $U \in \delta(\mathcal{D}(\varrho))$.

(ii) Denote $\rho = \varrho(\mathcal{D}(\varrho))$, i.e.

$$A\rho B \iff \forall g \in \mathcal{D}(\varrho),\ g(A)\hat{q}B \text{ or } g(B)\hat{q}A,$$

or equivalently,

$$A\bar{\rho}B' \iff \exists g \in \mathcal{D}(\varrho),\ g(A) \leq B,\ g(B') \leq A'. \tag{13.54}$$

Suppose $A\bar{\varrho}B'$, by Corollary **13.2.19**, there exists $F^{\rightarrow} \in \mathcal{F}$ and $\varepsilon > 0$ such that (13.53) holds. Then by Theorem **13.1.15** (vi), (i),

$$\begin{gathered} F^{\leftarrow}((f_\varepsilon \wedge f_\varepsilon{}^{\triangleleft})F^{\rightarrow}(A))' \geq B', \\ ((f_\varepsilon \wedge f_\varepsilon{}^{\triangleleft})F^{\rightarrow}(A))' \geq F^{\rightarrow}(B'), \\ (f_\varepsilon \wedge f_\varepsilon{}^{\triangleleft})^{\triangleleft}F^{\rightarrow}(A) \leq (F^{\rightarrow}(B'))', \\ (f_\varepsilon \wedge f_\varepsilon{}^{\triangleleft})F^{\rightarrow}(B') \leq (F^{\rightarrow}(A))', \\ F^{\leftarrow}(f_\varepsilon \wedge f_\varepsilon{}^{\triangleleft})F^{\rightarrow}(B') \leq F^{\leftarrow}((F^{\rightarrow}(A))') = (F^{\leftarrow}F^{\rightarrow}(A))' \leq A'. \end{gathered}$$

Let $g = F^{\leftarrow}(f_\varepsilon \wedge f_\varepsilon{}^{\triangleleft})F^{\rightarrow}$, then $g \in \mathcal{D}(\varrho)$ and $g(A) \le B$, $g(B') \le A'$. So by (13.54) we have $A\bar{\rho}B'$. This completes the proof of $\rho \subset \varrho$. □

By the virtue of Theorem **13.1.48**, Theorem **13.2.16** and Theorem **13.2.20**, we have the following

**13.2.21 Theorem** *Let $(L^X, \delta)$ be an L-fts. Then $\delta$ can generated by an L-fuzzy proximity on $X$ if and only if $(L^X, \delta)$ is completely regular.* □

Suppose $\prec$ is a topological order on $L^X$, then it follows from Theorem **13.2.13** (ii), $\varrho(\prec)$ is an $L$-fuzzy proximity on $X$ and $\delta(\varrho(\prec)) = \delta(\prec)$. By the virtue of Theorem **13.2.20**, $\varrho(\prec)$ generates an $L$-fuzzy uniformity $\mathcal{D}(\varrho(\prec))$ and $\delta(\mathcal{D}(\varrho(\prec))) = \delta(\varrho(\prec))$. So the topological order $\prec$ on $L^X$ generates an $L$-fuzzy uniformity $\mathcal{D}(\varrho(\prec))$ on $X$ and $\delta(\mathcal{D}(\varrho(\prec))) = \delta(\prec)$. This is just the following

**13.2.22 Theorem** *Let $\prec$ be a topological order on $L^X$. Then $\mathcal{D}(\varrho(\prec))$ is an L-fuzzy uniformity on $X$ and*

$$\delta(\mathcal{D}(\varrho(\prec))) = \delta(\prec).$$ □

From Theorem **13.2.13**, one can find that between topological orders and $L$-fuzzy proximities there are direct and neat relations. But for $L$-fuzzy uniformities, the situation is not so clear, although Theorem **13.2.22** provides a indirect relation between topological orders and $L$-fuzzy uniformities. However, if the property of a topological order becomes stronger, we can still show a direct relation between it and an $L$-fuzzy uniformity:

**13.2.23 Definition** Let $\prec$ be a topological order on $L^X$. $\prec$ is called *complete*, if the following implication holds for arbitrary $\{A_t : t \in T\}, \{B_t : t \in T\} \subset L^X$:

$$\forall t \in T,\ A_t \prec B_t \Longrightarrow (\bigvee_{t\in T} A_t) \prec (\bigvee_{t\in T} B_t),\ (\bigwedge_{t\in T} A_t) \prec (\bigwedge_{t\in T} B_t).$$

**13.2.24 Theorem** *Let $\prec$ be a complete topological order on $L^X$, $c_\prec : L^X \to L^X$ be the closure operator generated by $\prec$ as* (13.46). *Then*

(i) $A \prec c_\prec(A)$.

(ii) $c_\prec(A) \le B \iff A \prec B$.

(iii) $c_\prec \in \mathcal{E}(L^X)$.

(iv) $c_\prec{}^{\triangleleft} = c_\prec$.

**Proof** (i) By the completeness of $\prec$.

(ii) By (i) and (TO3), $c_\prec(A) \le B \Longrightarrow A \prec B$. Conversely, if $A \prec B$, then $c_\prec(A) \le B$ follows from (13.46).

(iii) By the virtue of (ii), (TO3) and the completeness of $\prec$,

$$\begin{aligned} c_\prec(\bigvee_{t\in T} A_t) \le B &\iff \bigvee_{t\in T} A_t \prec B \iff \forall t \in T,\ A_t \prec B \\ &\iff \forall t \in T,\ c_\prec(A_t) \le B \iff \bigvee_{t\in T} c_\prec(A_t) \le B. \end{aligned}$$

So $c_\prec(\bigvee_{t\in T} A_t) = \bigvee_{t\in T} c_\prec(A_t)$, $c_\prec$ is arbitrary join preserving. Since $c_\prec$ is value increasing, (iii) is true.

(iv) By (ii) prove above and (TO5),

$$c_\prec{}^{\triangleleft}(A) = \bigwedge\{C : c_\prec(C') \le A'\} = \bigwedge\{C : C' \prec A'\} = \bigwedge\{C : A \prec C\} = c_\prec(A).$$ □

**13.2.25 Theorem** *Let $\prec$ be a complete topological order on $L^X$, $c_\prec$ be the closure operator generated by $\prec$, $\mathcal{D} \subset (L^X)^{L^X}$ be defined as follows:*

$$\mathcal{D} = \{f \in \mathcal{E}(L^X) : \ f \geq c_\prec\}. \tag{13.55}$$

*Then $\mathcal{D}$ is an $L$-fuzzy uniformity on $X$ and $\delta(\mathcal{D}) = \delta(\prec)$.*

**Proof** (UF1) and (UF2) are fulfilled clearly. Since $c_\prec$ is a closure operator on $L^X$, $c_\prec \circ c_\prec = c_\prec$, so (UF3) is also fulfilled. Finally, (UF4) follows from Theorem **13.2.24** (iv). So $\mathcal{D}$ is an $L$-fuzzy uniformity on $X$. By Theorem **13.2.24** (iii), $c_\prec$ is the smallest element of $\mathcal{D}$. So for every $A \in L^X$, we have

$$c_\prec(A) = \bigwedge\{C : \ c_\prec(A) \leq C\} = \bigwedge\{f(A) : \ f \in \mathcal{D}\} = cl_{\delta(\mathcal{D})}(A),$$

this just means $\delta(\mathcal{D}) = \delta(\prec)$. □

# Chapter 14

# Metric Spaces

As a natural generalization of the ordinary situatiom, $L$-fuzzy quasi (pseudo-quasi) metric was considered by B. Hutton[38] in the first. Parallel to an equivalent form of pseudo metric in orindary topology, $L$-fuzzy pseudo-quasi metric space is defined as an $L$-fuzzy quasi-uniform space possessing a countable base. In 1979, M.A. Erceg[21] introduced a notion of $L$-fuzzy pseudo (pseudo-quasi) metric by generalizing the Hausdorff distance between usual sets. From the point of view of topology, the notion of $L$-fuzzy pseudo (pseudo-quasi) metric in Hutton's sense is equivalent to that in Erceg's sense; but different from the point of view of metric.

## 14.1 Metrics in Hutton's Sense and Erceg's Sense

**14.1.1 Definition** Let $(L^X, \mathcal{D})$ be an $L$-fuzzy uniform space ($L$-fuzzy quasi-uniform space, respectively). $(L^X, \mathcal{D})$ is called a *L-fuzzy pseudo metric space* (*L-fuzzy psedu-quasi metric space*, respectively), if $\mathcal{D}$ has a countable base. Also call an $L$-fuzzy pseudo metric space ($L$-fuzzy pseudo-quasi metric space, respectively) an *L-fuzzy p. metric space* (*L-fuzzy p.q. metric space*, respectively) for short.

**14.1.2 Theorem** *Let $(L^X, \mathcal{D})$ be an L-fuzzy p.q. metric space, then $\mathcal{D}$ has a base $\mathcal{D}^* = \{f_r : r > 0\}$ satisfying*

(i) $f_r \circ f_s \leq f_{r+s}$. (14.1)

(ii) $f_r = \bigvee_{s<r} f_s$. (14.2)

*If $(L^X, \mathcal{D})$ is a p. metric space, then $\mathcal{D}^*$ can be also required to satisfy*

(iii) $f_r = f_r^{\triangleleft}$. (14.3)

**Proof** Since $(L^X, \mathcal{D})$ is a p.q. metric space, $\mathcal{D}$ has a countable base $\mathcal{D}_0 = \{g_n : n \in \mathbf{N}\}$. Take $g_{n_1} = g_1 \in \mathcal{D}_0$, then we can inductively take $\{n_2, n_3, \cdots\} \subset \mathcal{N}$ such that

$$g_{n_2} \circ g_{n_2} \circ g_{n_2} \leq g_{n_1} \wedge g_2,$$

$$\vdots$$

$$g_{n_{i+1}} \circ g_{n_{i+1}} \circ g_{n_{i+1}} \leq g_{n_i} \wedge g_{i+1},$$

$$\vdots$$

then for arbitrary $f, g \in \mathcal{D}$, there exists $i, j \in \mathbf{N}$ such that $g_i \leq f$, $g_j \leq g$. Suppose $i \leq j$, then

$$f \wedge g \geq g_i \wedge g_j \geq g_{n_i} \wedge g_{n_j} = g_{n_j}.$$

So $\{g_{n_i} : i \in \mathbf{N}\}$ is still a countable base of $\mathcal{D}$. For convenience, we can assume $\mathcal{D}_0 = \{g_n : n \in \mathbf{N}\}$ fulfils the following condition by itself:

$$g_{n+1} \circ g_{n+1} \circ g_{n+1} \leq g_n, \quad n \in \mathbf{N}.$$

For every $r > 0$, define a mapping $\varphi_r : L^X \to L^X$ as follows:

$$\forall A \in L^X, \quad \varphi_r(A) = \begin{cases} g_n(A), & \frac{1}{2^n} < r \le \frac{1}{2^{n-1}}, \\ 1, & r > 1,\ A \ne \underline{0}, \\ 0, & r > 1,\ A = \underline{0}, \end{cases}$$

then $\varphi_r \in \mathcal{E}(L^X)$ obviously, and easy to verify the following two relations:

$$\varphi_r \circ \varphi_r \circ \varphi_r \le \varphi_{2r}, \tag{14.4}$$

$$\varphi_r = \bigvee_{s<r} \varphi_s,\ r > 0. \tag{14.5}$$

For every $r > 0$, define $f_r : L^X \to L^X$ as follows:

$$f_r = \bigvee\{\varphi_{r_0} \circ \cdots \circ \varphi_{r_k} : \sum_{i=0}^{k} r_k = r,\ \forall i \le k,\ r_i > 0,\ k < \omega\}. \tag{14.6}$$

Let $\mathcal{D}^* = \{f_r : r > 0\}$, then $\mathcal{D}^*$ is the base of $\mathcal{D}$ fulfilling the requirements.

In fact, since $\mathcal{E}(L^X)$ is closed under compositions and arbitrary joins, $f_r \in \mathcal{E}(L^X)$ for every $r > 0$. Suppose $g \in \mathcal{D}$, then there exists $n \in \mathbf{N}$ such that $g_n \le g$. Take $r > 0$ such that $\frac{1}{2^n} < 2r \le \frac{1}{2^{n-1}}$, then $\varphi_{2r} = g_n$. We affirm $f_r \le \varphi_{2r}$, hence $f_r \le g$. To verify this inequality, by (14.6), it is sufficient to verify $\varphi_{r_0} \circ \cdots \circ \varphi_{r_k} \le \varphi_{2r}$ for every $k < \omega$, where $r_i > 0$ for every $i \le k$, $\sum_{i \le k} r_i = r$. This inequality holds for every $r > 0$ and $k = 0, 1$. Suppose $n \ge 2$ and the following relations hold for every $k < n$:

$$r > 0,\ \forall i \le k,\ r_i > 0,\ \sum_{i=0}^{k} r_i = r \implies \varphi_{r_0} \circ \cdots \circ \varphi_{r_k} \le \varphi_{2r}. \tag{14.7}$$

Now suppose $s_j > 0$ for every $j \le n$ and $\sum_{j \le n} s_j = r$. If $s_1 > \frac{r}{2}$, then $\sum_{j=1}^{n} s_j < \frac{r}{2}$, by (14.7),

$$\varphi_{s_0} \circ (\varphi_{s_1} \circ \cdots \circ \varphi_{s_n}) \le \varphi_r \circ \varphi_r \le \varphi_{2r}.$$

If $s_0 \le \frac{r}{2}$, let $m = max\{p < \omega : \sum_{j \le p} s_j \le \frac{r}{2}\}$, then by (14.4) we have

$$\varphi_{s_0} \circ \cdots \circ \varphi_{s_n} = (\varphi_{s_1} \circ \cdots \circ \varphi_{s_m}) \circ \varphi_{s_{m+1}} \circ (\varphi_{s_{m+2}} \circ \cdots \circ \varphi_{s_n}) \le \varphi_r \circ \varphi_r \circ \varphi_r \le \varphi_{2r}.$$

So $\varphi_{s_0} \circ \cdots \circ \varphi_{s_n} \le \varphi_{2r}$ always holds for $\sum_{j \le n} s_j = r$. By inductive method, (14.7) holds for every $k < \omega$. Then the following relations follows from (14.7):

$$f_r \le \varphi_{2r} = g_n \le g.$$

That is to say, $\mathcal{D}^*$ is a base of $\mathcal{D}$. $\mathcal{D}^*$ obviously fulfils (14.1); especially it follows from (14.1) that $s < r \implies f_s \le f_r$. So to verify (14.2) it is sufficient to prove $\bigvee_{s<r} f_s \ge f_r$. For $\varphi_{r_0} \circ \cdots \circ \varphi_{r_k}$ in the right side of (14.6), where $r_i > 0$ for every $i \le k$, $\sum_{i \le k} r_i = r$, by (14.5), $\varphi_{r_0} = \bigvee_{t < r_0} \varphi_t$. For every $t < r_0$, $s = t + \sum_{i=1}^{k} r_i < r$, so by $f_s \ge \varphi_t \circ \varphi_{r_1} \circ \cdots \circ \varphi_{r_k}$,

$$\bigvee_{s<r} f_s \ge \bigvee_{t<r_0} (\varphi_t \circ \varphi_{r_1} \circ \cdots \circ \varphi_{r_k}) = (\bigvee_{t<r_0} \varphi_t) \circ \varphi_{r_1} \circ \cdots \circ \varphi_{r_k} = \varphi_{r_0} \circ \cdots \circ \varphi_{r_k}.$$

By (14.6), $\bigvee_{s<r} f_s \ge f_r$, (14.2) holds.

If $(L^X, \mathcal{D})$ is a p. metric space, then by Theorem **13.1.24** and the constructing procedure of $\{g_{n_i} : i \in \mathbf{N}\}$ in the beginning of this proof, we can assume $g_n{}^{\triangleleft} = g_n$ for every memeber $g_n \in \mathcal{D}_0$. Then it can be easily verified that $\varphi_r{}^{\triangleleft} = \varphi_r$ and hence by Theorem **13.1.15** (vii),

$$(\varphi_{r_0}\circ\cdots\circ\varphi_{r_k})^{\triangleleft} = \varphi_{r_k}{}^{\triangleleft}\circ\cdots\circ\varphi_{r_0}{}^{\triangleleft} = \varphi_{r_k}\circ\cdots\circ\varphi_{r_0}.$$

So by (14.6) and Theorem **13.1.15** (v),

$$\begin{aligned} f_r{}^{\triangleleft} &= \bigvee\{(\varphi_{r_0}\circ\cdots\circ\varphi_{r_k})^{\triangleleft} : \sum_{i=0}^{k} r_i = r,\ \forall i \le k,\ r_i > 0,\ k \in \mathbf{N}\} \\ &= \bigvee\{\varphi_{r_0}\circ\cdots\circ\varphi_{r_k}) : \sum_{i=0}^{k} r_i = r,\ \forall i \le k,\ r_i > 0,\ k \in \mathbf{N}\} \\ &= f_r. \end{aligned}$$

This completes the proof of (14.3). □

**14.1.3 Definition** Let $\mathcal{D} \subset \mathcal{E}(L^X)$, $\mathbf{R}_0 = (0, +\infty)$ or $\mathbf{R}_0 = \{r : r \in \mathbf{Q},\ r > 0\}$, $\mathcal{D}^* = \{f_r : r \in \mathbf{R}_0\} \subset \mathcal{E}(L^X)$. $\mathcal{D}^*$ is called a *standard p.q. metrc base* (*standard p. metric base*, respectively) on $L^X$, if $\mathcal{D}^*$ fulfils conditions (14.1) – (14.2) ((14.1) – (14.3), respectively). Also call a standard p.q. metric base or a p. metric base a *standard base* for short sometimes. Call $\mathcal{D}^*$ a *standard p.q. metric base* (standard p. metric base, respcetively) of $\mathcal{D}$ or of $(L^X, \mathcal{D})$ or say *$\mathcal{D}$ is generated by $\mathcal{D}^*$*, if $\mathcal{D}^*$ is a standard p.q. metric base (standard p. metric base, respectively) on $L^X$ and

$$\mathcal{D} = \{f \in \mathcal{E}(L^X) : \exists g \in \mathcal{D}^*,\ f \ge g\}.$$

Then the following three corollaries are easy to be verified:

**14.1.4 Corollary** *Let $(L^X, \mathcal{D})$ be a p.q. metric space, then $\mathcal{D}$ has a countable base $\mathcal{D}^* = \{f_r : r \in \mathbf{Q},\ r > 0\}$ such that* (14.1) *and* (14.2) *hold. If $(L^X, \mathcal{D})$ is a p. metric space, then $\mathcal{D}^*$ can be also required to satisfy* (14.3). □

**14.1.5 Corollary** *Let $\mathcal{D} \subset \mathcal{E}(L^X)$, then the following conditions are equivalent:*

(i) *$(L^X, \mathcal{D})$ is a p.q. metric space.*

(ii) *$\mathcal{D}$ is generated by a countable standard p.q. metric base $\mathcal{D}^* = \{f_r : r \in \mathbf{Q},\ r > 0\}$.*

(iii) *$\mathcal{D}$ is generated by a standard p.q. metric base $\mathcal{D}^* = \{f_r : r > 0\}$.* □

**14.1.6 Corollary** *Let $\mathcal{D} \subset \mathcal{E}(L^X)$, then the following conditions are equivalent:*

(i) *$(L^X, \mathcal{D})$ is a p. metric space.*

(ii) *$\mathcal{D}$ is generated by a countable standard p.q. metric base $\mathcal{D}^* = \{f_r : r \in \mathbf{Q},\ r > 0\}$.*

(iii) *$\mathcal{D}$ is generated by a standard p.q. metric base $\mathcal{D}^* = \{f_r : r > 0\}$.* □

**14.1.7 Example** $L$-fuzzy unit interval $I(L)$ is a p. metric space.

In fact, $I(L)$ has an $L$-fuzzy uniformity $\mathcal{E}$ which has a base $\mathcal{D}^* = \{f_\varepsilon \wedge f_\varepsilon{}^{\triangleleft} : \varepsilon > 0\}$, where $f_\varepsilon$ is defined by (13.13). Then $\mathcal{D}^*{}_0 = \{f_\varepsilon \wedge f_\varepsilon{}^{\triangleleft} : \varepsilon \in \mathbf{Q},\ \varepsilon > 0\}$ is a countable base of $\mathcal{E}$, so $I(L)$ is a p. metric space.

To describe p.q. metric space from angle of "distance", M.A. Erceg[21] introduced a "distance" function between every pair of $L$-fuzzy subsets to define $L$-fuzzy p.q. metric:

**14.1.8 Definition** A mapping $P : L^X \times L^X \to [0, +\infty]$ is called an *$L$-fuzzy p.q. metric* on $X$, if $P$ fulfils the following conditions (EM1) – (EM4):

(EM1) $A \ne \underline{0} \implies P(\underline{0}, A) = +\infty$,
$P(A, \underline{0}) = P(A, A) = 0.$

(EM2) $P(A, B) \le P(A, C) + P(C, B)$.

(EM3) (i) $A \le B \implies P(A,C) \ge P(B,C)$.
(ii) $P(A, \bigvee_{t\in T} B_t) = \bigvee_{t\in T} P(A, B_t)$.

(EM4) If $P(A_t, C) < r \Longrightarrow C \le B$ for every $C \in L^X$ and every $t \in T$, then the following implication holds for every $D \in L^X$:

$$P(\bigvee_{t\in T} A_t, D) < r \implies D \le B.$$

**14.1.9 Definition** Let $P$ be an $L$-fuzzy p.q. metric on $X$. Define mapping $f_r$ : $L^X \to L^X$ for every $r > 0$ as follows:

$$\forall A \in L^X, \quad f_r(A) = \bigvee\{C \in L^X : P(A,C) < r\}. \tag{14.8}$$

Call $\mathcal{D}_P{}^* = \{f_r : r > 0\}$ the *associated neighborhood mappings* of $P$.

**14.1.10 Definition** Let $P$ be an $L$-fuzzy p.q. metric on $X$, $\mathcal{D}_P{}^* = \{f_r : r > 0\}$ be the family of the associated neighborhood mappings of $P$. $P$ is called an $L$-fuzzy p. metric on $X$, if every $f_r \in \mathcal{D}_P{}^*$ satisfies the following condition (EM5):

(EM5) $f_r = f_r{}^{\triangleleft}$.

Now we have two different ways to describe $L$-fuzzy p.q. metric spaces: From Hutton's angle and from Erceg's angle. But what is the relation between these two ways? The following theorem tells us that every Erceg's p.q. metric determines a Hutton's p.q. metric space:

**14.1.11 Theorem** *Let $P: L^X \times L^X \to [0,+\infty]$ be an $L$-fuzzy p.q. metric on $X$, $\mathcal{D}_P{}^* = \{f_r : r > 0\}$ be the associated neighborhood mappings of $P$. Then $\mathcal{D}_P{}^* \subset \mathcal{E}(L^X)$ and fulfils* (14.1) *and* (14.2).

**Proof** Suppose $f_r \in \mathcal{D}_P{}^*$, then by $P(A,A) = 0$ in (EM1) we know that $f_r(A) \ge A$, $f_r$ is value increasing. Suppose $\{A_t : t \in T\} \subset L^X$, then $f_r$ being order preserving is deduced from (EM3) (i), so $f_r(\bigvee_{t\in T} A_t) \ge \bigvee_{t\in T} f_r(A_t)$. Conversely, suppose $B \ge \bigvee_{t\in T} f_r(A_t)$, then $B \ge f_r(A_t)$ for every $t \in T$, and hence by (14.8) we have

$$C \in L^X,\ t \in T,\ P(A_t, C) < r \implies C \le B.$$

Therefore, by (EM4) we have

$$D \in L^X,\ P(\bigvee_{t\in T} A_t, D) < r \implies D \le B.$$

So it follows from (14.8) that $f_r(\bigvee_{t\in T} A_t) \le B$. That is to say, $f_r(\bigvee_{t\in T} A_t) \le \bigvee_{t\in T} f_r(A_t)$, and hence $f_r(\bigvee_{t\in T} A_t) = \bigvee_{t\in T} f_r(A_t)$. So $f_r \in \mathcal{E}(L^X)$, $\mathcal{D}_P{}^* \subset \mathcal{E}(L^X)$.

By (14.8), to verify (14.1), it is sufficient to verify that for arbitrary $r, s > 0$ and arbitrary $A, B \in L^X$,

$$P(\bigvee\{C : P(A,C) < s\}, B) < r \implies P(A,B) < r + s. \tag{14.9}$$

In fact, by (EM3) (ii),

$$P(A, \bigvee\{C : P(A,C) < s\}) = \bigvee\{P(A,C) : P(A,C) < s\} \le s,$$

so by the triangle inequality (EM2), (14.9) holds.

To verify (14.2), we need only verify $f_r \le \bigvee_{s<r} f_s$. For every $A \in L^X$, if $P(A,B) < r$, then there exists $s < r$ such that $P(A,B) < s$, so $B \le f_s(A)$. Hence $f_r(A) \le \bigvee_{s<r} f_s(A)$, $f_r \le \bigvee_{s<r} f_s$. □

By Theorem **14.1.11** and Corollary **14.1.5**, we have

**14.1.12 Corollary** *Let* $P: L^X \times L^X \to [0,+\infty]$ *be an L-fuzzy p.q. metric on* $X$, $\mathcal{D} \subset \mathcal{E}(L^X)$ *be defined by*

$$\mathcal{D} = \{f \in \mathcal{E}(L^X) : \exists f_r \in \mathcal{D}_P{}^*,\ f \geq f_r\}.$$

*Then* $(L^X, \mathcal{D})$ *is an L-fuzzy p.q. metric space with a standard base* $\mathcal{D}_P{}^*$. □

As the reverse of Theorem **14.1.11**, the following theorem tells us that every Hutton's p.q. metric space also determines a Erceg's p.q. metric:

**14.1.13 Theorem** *Let* $(L^X, \mathcal{D})$ *be an L-fuzzy p.q. metric space, then there exists an L-fuzzy p.q. metric* $P: L^X \times L^X \to [0,+\infty]$ *such that the family* $\mathcal{D}_P{}^*$ *of the associated neighborhood mappings of* $P$ *is a standard p.q. metric base of* $\mathcal{D}$.

**Proof** By Theorem **14.1.2**, $\mathcal{D}$ has a standard p.q. metric base $\mathcal{D}^*$. For every pair $A, B \in L^X$, define

$$P(A,B) = \bigwedge\{r : B \leq f_r(A)\}, \tag{14.10}$$

where we let $\bigwedge\emptyset = +\infty$, then $P$ fulfils (EM1). To verify (EM2), without loss any generality, suppose

$$P(A,C) = \bigwedge\{s : C \leq f_s(A)\} < +\infty,$$
$$P(C,B) = \bigwedge\{t : B \leq f_t(C)\} < +\infty.$$

For every $\varepsilon > 0$, take $s > 0$ such that $C \leq f_s(A)$ and $s < P(A,C) + \frac{\varepsilon}{2}$. Take $t > 0$ such that $B \leq f_s(C)$ and $t < P(C,B) + \frac{\varepsilon}{2}$. Then

$$B \leq f_t(C) \leq f_t \circ f_s(A) \leq f_{t+s}(A),$$

so

$$P(A,B) \leq t + s < P(A,C) + P(C,B) + \varepsilon.$$

Since $\varepsilon > 0$ is arbitrary, so (EM2) holds.

(i) in (EM3) is obviously holds. To verify (ii) in (EM3), by (EM1), it is sufficient to verify the case $A \neq \underline{0}$. Suppose

$$s = P(A, \bigvee_{t\in T} B_t) = \bigwedge\{r : \bigvee_{t\in T} B_t \leq f_r(A)\} < +\infty.$$

Arbitrarily fix an $\varepsilon > 0$, then by (14.10) there exists $r > 0$ such that $r < s + \varepsilon$ such that $\bigvee_{t\in T} B_t \leq f_r(A)$, so $B_t \leq f_r(A)$ for every $t \in T$, hence

$$\bigvee_{t\in T} P(A, B_t) \leq r < s + \varepsilon.$$

By the arbitrariness of $\varepsilon > 0$, the following inequality holds for every $s < +\infty$:

$$P(A, \bigvee_{t\in T} B_t) \geq \bigvee_{t\in T} P(A, B_t).$$

Since the above inequality also holds for $s = +\infty$, so it always holds in any case. Conversely, suppose $s = \bigvee_{t\in T} P(A, B_t) < +\infty$. Arbitrarily fix an $\varepsilon > 0$. For every $t \in T$, by $P(A, B_t) \leq s$ we can take a $r_t \in [s, s+\varepsilon)$ such that

$$B_t \leq f_{r_t}(A) \leq f_{s+\varepsilon}(A).$$

Then $\bigvee_{t\in T} B_t \leq f_{s+\varepsilon}(A)$, and hence $P(A, \bigvee_{t\in T} B_t) \leq s+\varepsilon$, i.e. the following inequality holds for $s < +\infty$:

$$P(A, \bigvee_{t\in T} B_t) \leq \bigvee_{t\in T} P(A, B_t).$$

Since the above inequality also holds for $s = +\infty$, so it always holds in any case, (ii) in (EM3) is verified.

Suppose

$$C \in L^X,\ t \in T,\quad P(A_t, C) < r \implies C \le B \tag{14.11}$$

and $P(\bigvee_{t\in T} A_t, D) < r$. Then there exists $s < r$ such that

$$D \le f_s(\bigvee_{t\in T} A_t) = \bigvee_{t\in T} f_s(A_t). \tag{14.12}$$

It follows from (14.2) that $f_r$ is monotonically increasing with respect to $r$, so for every $t \in T$,

$$P(A_t, f_s(A_t)) = \bigwedge\{u:\ f_s(A_t) \le f_u(A_t)\} = s < r.$$

By (14.11), $f_s(A_t) \le B$ for every $t \in T$. By (14.12),

$$D \le \bigvee_{t\in T} f_s(A_t) \le B.$$

Hence (EM4) is proved, $P$ is an $L$-fuzzy p.q. metric on $X$.

Suppose $\mathcal{D}_P{}^* = \{g_r:\ r > 0\}$ is the family of the associated neighborhood mappings of $P$, then by Definition **14.1.9**, (14.10) and $\mathcal{D}^*$ fulfilling (14.2), for every $A \in L^X$ and every $r > 0$,

$$\begin{aligned} g_r(A) &= \bigvee\{B:\ P(A,B) < r\} \\ &= \bigvee\{B:\ \bigwedge\{s:\ B \le f_s(A)\} < r\} \\ &= \bigvee\{B:\ \exists s < r,\ B \le f_s(A)\} \\ &= \bigvee\{f_s(A):\ s < r\} \\ &= f_r(A). \end{aligned}$$

So $\mathcal{D}_P{}^* = \mathcal{D}^*$ is a standard p.q. metric base of $\mathcal{D}$. □

By Theorem **14.1.13**, the following definition is reasonable:

**14.1.14 Definition** Let $(L^X, \mathcal{D})$ be an $L$-fuzzy p.q. metric space with a standard base $\mathcal{D}^* = \{f_r:\ r \in \mathbf{R}_0\}$. Call the $L$-fuzzy p.q. metric $P$ on $X$ defined by (14.10) the *$L$-fuzzy p.q. metric generated by* $\mathcal{D}$, denote it by $P_{\mathcal{D}}$.

By Theorem **14.1.11**, the following definition is reasonable:

**14.1.15 Definition** Let $P:\ L^X \times L^X \to [0, +\infty]$ be an $L$-fuzzy p.q. metric on $X$, $\mathcal{D}_P{}^* = \{f_r:\ r > 0\}$ be the family of associated neighborhood mappings of $P$. Call the $L$-fuzzy quasi-uniformity $\mathcal{D}$ generated by the base $\mathcal{D}_P{}^*$ the *$L$-fuzzy quasi-uniformity generated by* $P$, denote it by $\mathcal{D}_P$.

Then we can exactly describe the relation between Hutton's p.q. metric space and Erceg's p.q. metric:

**14.1.16 Theorem** (i) *If $(L^X, \mathcal{D})$ is an $L$-fuzzy p.q. metric space, then $\mathcal{D}_{P_{\mathcal{D}}} = \mathcal{D}$.*

(ii) *If $P$ is an $L$-fuzzy p.q. metric on $X$, then $P_{\mathcal{D}_P} = P$.*

**Proof** (i) This is deduced from Theorem **14.1.13**.

(ii) Suppose $P$ is an $L$-fuzzy p.q. metric on $X$, $\mathcal{D}_P{}^* = \{f_r:\ r > 0\}$ is the family of associated neighborhood mappings of $P$, then by (14.8), for every $s > 0$,

$$P(A,B) < s \Longrightarrow B \le f_s(A) \Longrightarrow P_{\mathcal{D}_P}(A,B) \le s.$$

On the other hand, if $P_{\mathcal{D}_P}(A,B) \le s$, then by

$$P_{\mathcal{D}_P}(A,B) = \bigwedge\{r:\ B \le f_r(A)\} = \bigwedge\{r:\ B \le \bigvee\{C:\ P(A,C) < r\}\},$$

we have $B \le f_{s+\varepsilon}(A)$ for every $\varepsilon > 0$, and hence $B \le \bigvee\{C:\ P(A,C) < s+\varepsilon\}$. So by condition (ii) in (EM3),

$$P(A,B) \le P(A, \bigvee\{C: P(A,C) < s+\varepsilon\}) = \bigvee\{P(A,C): P(A,C) < s+\varepsilon\} \le s+\varepsilon.$$

Therefore, we have proved that for every $s > 0$ and every $\varepsilon > 0$,

$$P(A,B) < s \Longrightarrow P_{\mathcal{D}_P}(A,B) \le s \Longrightarrow P(A,B) \le s+\varepsilon.$$

This implies $P_{\mathcal{D}_P}(A,B) = P(A,B)$, even if $P_{\mathcal{D}_P}(A,B) = +\infty$ or $P(A,B) = +\infty$. □

## 14.2 Pointwise Characterizations of Metrics

Both $L$-fuzzy p.q. metric space in Hutton's sense and $L$-fuzzy p.q. metric in Erceg's sense are based on $L$-fuzzy subsets but not $L$-fuzzy points as orindary metrics. Especially, this structure makes the geometric intuition of the symmetry of a metric dubious. To offset this point, Liang[61] introduced a pointwise description of $L$-fuzzy p.q. metric. Liu[74] also proved the Fuzzy Urysohn Metrization Theorem with the imbedding theory. The $L$-fuzzy topology generated by this kind of p.q. metric is coincident with the one generated by Erceg's p.q. metric, but they are not coincident with each other by themself; more exactly, Liang's distance function is not a restriction of the one of Erceg on the set of molecules. This problem was solved by Peng.[131] He equivalently simplified Erceg's definition of $L$-fuzzy p.q. metric and then restricted the distance function on the set of molecules to obtain a pointwise characterization.[133] These results will be introduced in this section.

**14.2.1 Lemma** *Let $P:\ L^X \times L^X \to [0,+\infty]$ be an $L$-fuzzy p.q. metric on $X$, $\{f_r:\ r>0\}$ be the family of the associate neighborhood mappings of $P$. Then*

(i) $P(A,B) < r \implies B \le f_r(A) \implies P(A,B) \le r.$

(ii) $e \in \beta^*(f_r(A)) \implies P(A,e) < r.$

**Proof** (i) If $P(A,B) < r$, then by (14.8), $B \le \bigvee\{C:\ P(A,C) < r\} = f_r(A)$. If $B \le f_r(B)$, then by (14.8) and (EM3) (ii),

$$\begin{aligned} P(A,B) &\le P(A, f_r(A)) \\ &= P(A, \bigvee\{C:\ P(A,C) < r\}) \\ &= \bigvee\{P(A,C):\ P(A,C) < r\} \le r. \end{aligned}$$

(ii) If $e \in \beta^*(f_r(A))$, then by $f_r(A) = \bigvee\{C:\ P(A,C) < r\}$ there exists $C \in L^X$ such that $P(A,C) < r$ and $e \le C$. Then by (EM3) (ii), $P(A,e) \le P(A,C) < r$. □

**14.2.2 Theorem** *$P:\ L^X \times L^X \to [0,+\infty]$ is an $L$-fuzzy p.q. metric (p. metric, respectively) on $X$ if and only if $P$ satisfies the following conditions* (SEM1) – (SEM3) ((SEM1) – (SEM4), *respectively*):

(SEM1) $A \ge B \implies P(A,B) = 0,$
$\qquad B \ne \underline{0} \implies P(\underline{0},B) = +\infty.$

(SEM2) $P(A,B) \le P(A,C) + P(C,B).$

(SEM3) $A, B \ne \underline{0} \implies P(A,B) = \bigvee_{d\in\beta^*(B)} \bigwedge_{e\in\beta^*(A)} P(e,d).$

(SEM4) "$P(A,C) < r \Longrightarrow C \le B$" $\Longleftrightarrow$ "$P(B',D) < r \Longrightarrow D \le A'$".

**Proof** (Sufficiency) Easy to find that (EM1), (EM2) and (EM3) (i) hold. Now verify (EM3) (ii). By Theorem **1.3.24** (ii) and (SEM3),

$$\begin{aligned}P(A, \bigvee_{t\in T} B_t) &= \bigvee\{\bigwedge\{P(e,d):\ e\in\beta^*(A)\}:\ d\in\beta^*(\bigvee_{t\in T} B_t)\}\\ &= \bigvee\{\bigwedge\{P(e,d):\ e\in\beta^*(A)\}:\ \exists t\in T,\ d\in\beta^*(B_t)\}\\ &= \bigvee\{\bigvee\{\bigwedge\{P(e,d):\ e\in\beta^*(A)\}:\ d\in\beta^*(B_t)\}:\ t\in T\}\\ &= \bigvee_{t\in T} P(A,B_t).\end{aligned}$$

For (EM4), suppose $P(A_t, C) < r \Longrightarrow C \leq B$ for every $C \in L^X$ and every $t \in T$. If $P(\bigvee_{t\in T} A_t, D) < r$, then by (SEM3),

$$\bigvee\{\bigwedge\{P(e,d):\ e\in\beta^*(\bigvee_{t\in T} A_t)\}:\ d\in\beta^*(D)\} = P(\bigvee_{t\in T} A_t, D) < r.$$

So for every $d \in \beta^*(D)$, by Theorem **1.3.24** (ii), there exists $e \in \beta^*(\bigvee_{t\in T} A_t) = \bigcup_{t\in T}\beta^*(A_t)$ such that $P(e,d) < r$. Hence by (EM3) (i), there exists $t \in T$ and $e \in \beta^*(A_t)$ such that $P(A_t, d) \leq P(e,d) < r$. By the previous assumption, $d \leq B$, and hence $D = \bigvee\beta^*(D) \leq B$, (EM4) holds. As for (EM5), by Theorem **14.1.16** (ii), Theorem **13.1.15** (i) and (SEM4),

$$\begin{aligned}f_r(A)\leq B &\Longleftrightarrow \bigvee\{C:\ P(A,C)<r\}\leq B\\ &\Longleftrightarrow \text{“}P(A,C)<r\Longrightarrow C\leq B\text{”}\\ &\Longleftrightarrow \text{“}P(B',D)<r\Longrightarrow D\leq A'\text{”}\\ &\Longleftrightarrow \bigvee\{D:\ P(B',D)<r\}\leq A'\\ &\Longleftrightarrow f_r(B')\leq A'\\ &\Longleftrightarrow f_r^{\triangleleft}(A)\leq B,\end{aligned}$$

so $f_r(A) = f_r^{\triangleleft}(A)$, $f_r = f_r^{\triangleleft}$, (EM5) is proved.

(Necessity) (SEM2) is just (EM2), (SEM4) is equivalent to (EM5). Now verify (SEM1): If $A \geq B$, then for the family $\mathcal{D}_P{}^* = \{f_r :\ r > 0\}$ of associated neighborhood mappings, $f_r(A) \geq A \geq B$ for every $r > 0$. So by Theorem **14.1.16** (ii),

$$P(A,B) = \bigwedge\{r:\ B\leq f_r(A)\} = 0,$$

(SEM1) holds. For (SEM3), by Theorem **1.3.24** (ii), for arbitrary $A, B \neq \underline{0}$,

$$\beta^*(f_r(A)) = \beta^*(f_r(\bigvee\beta^*(A))) = \beta^*(\bigvee_{e\in\beta^*(A)} f_r(e)) = \bigvee_{e\in\beta^*(A)}\beta^*(f_r(e)),$$

so by Lemma **14.2.1**,

$$\begin{aligned}P(A,B)<r &\Longrightarrow B\leq f_r(A)\\ &\Longrightarrow \forall d\in\beta^*(B),\ d\in\beta^*(f_r(A))\\ &\Longrightarrow \forall d\in\beta^*(B),\ \exists e\in\beta^*(A),\ d\in\beta^*(f_r(e))\\ &\Longrightarrow \forall d\in\beta^*(B),\ \exists e\in\beta^*(A),\ P(e,d)<r\\ &\Longrightarrow \bigvee_{d\in\beta^*(B)}\bigwedge_{e\in\beta^*(A)} P(e,d)<r\\ &\Longrightarrow \forall d\in\beta^*(B),\ \exists e_d\in\beta^*(A),\ P(e_d,d)<r\\ &\Longrightarrow \forall d\in\beta^*(B),\ \exists e_d\in\beta^*(A),\ d\leq f_r(e_d)\\ &\Longrightarrow B=\bigvee\beta^*(B)\leq \bigvee_{d\in\beta^*(B)} f_r(e_d)\\ &\qquad = f_r(\bigvee_{d\in\beta^*(B)} e_d)\leq f_r(\bigvee\beta^*(A)) = f_r(A)\\ &\Longrightarrow P(A,B)<r,\end{aligned}$$

$$P(A,B)<r \iff \bigvee_{d\in\beta^*(B)}\bigwedge_{e\in\beta^*(A)} P(e,d)<r,$$

$$P(A,B)=\bigvee_{d\in\beta^*(B)}\bigwedge_{e\in\beta^*(A)} P(e,d). \qquad \square$$

**14.2.3 Remark** In Peng's original papers [131, 133] for pointwise characterization of $L$-fuzzy p.q. metric, investigations were carried on for a F-lattice $L$ but not an $L$-fuzzy space $L^X$ with a F-lattice $L$. But there is no essential difference between these two kinds of approaches for our topic in this chapter. Besides this, Peng's original proofs have some gaps, they are filled in the proofs introduced in the sequel.

**14.2.4 Lemma** *Let $A,B\in L^X$, $e,d\in M(L^X)$, $f\in\mathcal{E}(L^X)$. Then*

(i) $A\hat{q}f(B)\iff B\hat{q}f^{\triangleleft}(A)$.

(ii) $e\lessdot f(d)\iff d\lessdot f^{\triangleleft}(e)$.

**Proof** (i) By Theorem **13.1.15** (i), we have the following equivalent relations:

$$\begin{aligned} f(B)\le A' &\iff f^{\triangleleft}(A)\le B',\\ A\not\le f(B)' &\iff B\not\le f^{\triangleleft}(A)',\\ A\hat{q}f(B) &\iff B\hat{q}f^{\triangleleft}(A). \end{aligned}$$

(ii) Deduced from (i). $\square$

**14.2.5 Lemma** *Let $P:\ L^X\times L^X\to[0,1]$ be an $L$-fuzzy p.q. metric on $X$, $\mathcal{D}_P{}^*=\{f_r:\ r>0\}$ be the family of associated neighborhood mappings of $P$, $A\in L^X$, $r>0$. Then*

$$f_r(A)=\bigvee\{e\in M(L^X):\ P(A,e)<r\}.$$

**Proof** Denote

$$\mathcal{A}=\{e\in M(L^X):\ P(A,e)<r\},$$

then $\bigvee\mathcal{A}\le\bigvee\{C:\ P(A,C)<r\}=f_r(A)$. Conversely, if $C\in L^X$ such that $P(A,C)<r$, then by (EM3) (ii), for every $e\in M(\Downarrow C)$ we have $e\le C$, $P(A,e)\le P(A,C)<r$, $e\in\mathcal{A}$, $M(\Downarrow C)\subset\mathcal{A}$, $C=\bigvee M(\Downarrow C)\le\bigvee\mathcal{A}$, $f_r(A)=\bigvee\{C:\ P(A,C)<r\}\le\bigvee\mathcal{A}$. So $f_r(A)=\bigvee\mathcal{A}$. $\square$

**14.2.6 Lemma** *Let $P:\ L^X\times L^X\to[0,1]$ be an $L$-fuzzy p.q. metric on $X$, $\mathcal{D}_P{}^*=\{f_r:\ r>0\}$ be the family of associated neighborhood mappings of $P$. Then the following conditions are equivalent:*

(i) $\forall r>0,\quad f_r=f_r{}^{\triangleleft}$.

(ii) $\forall r>0,\ \forall e,d\in M(L^X)$,

$$e\lessdot f_r(d)\iff d\lessdot f_r(e).$$

(iii) $\forall r>0,\ \forall e,d\in M(L^X)$,

$$\exists u\in M(L^X),\ e\lessdot u,\ P(d,u)<r\iff\exists v\in M(L^X),\ d\lessdot v,\ P(e,v)<r.$$

**Proof** (i)$\iff$(ii): Since both $f_r$ and $f_r{}^{\triangleleft}$ are arbitrary join preserving for every $r>0$ and $M(L^X)$ is a join-generating set of $L^X$, by Lemma **14.2.4** (ii),

$$\begin{aligned} f_r=f_r{}^{\triangleleft} &\iff \forall e\in M(L^X),\quad f_r{}^{\triangleleft}(e)=f_r(e)\\ &\iff \forall e,d\in M(L^X),\quad \text{“}d\lessdot f_r{}^{\triangleleft}(e)\iff d\lessdot f_r(e)\text{”}\\ &\iff \forall e,d\in M(L^X),\quad \text{“}e\lessdot f_r(d)\iff d\lessdot f_r(e)\text{”}. \end{aligned}$$

(ii)$\Longrightarrow$(iii): By Lemma **14.2.5** and (ii),

$$\begin{aligned}\exists u \in M(L^X),\ e \lessdot u,\ P(d,u) < r &\iff e \lessdot \bigvee\{u \in M(L^X):\ P(d,u) < r\}\\ &\iff e \lessdot f_r(d)\\ &\iff d \lessdot f_r(e)\\ &\iff d \lessdot \bigvee\{v \in M(L^X):\ P(e,v) < r\}\\ &\iff \exists v \in M(L^X),\ d \lessdot v,\ P(e,v) < r.\end{aligned}$$

(iii)$\Longrightarrow$(ii): By Lemma **14.2.5** and (ii),

$$\begin{aligned}e \lessdot f_r(d) &\iff e \lessdot \bigvee\{u \in M(L^X):\ P(d,u) < r\}\\ &\iff \exists u \in M(L^X),\ e \lessdot u,\ P(d,u) < r\\ &\iff \exists v \in M(L^X),\ d \lessdot v,\ P(e,v) < r\\ &\iff d \lessdot \bigvee\{v \in M(L^X):\ P(e,v) < r\}\\ &\iff d \lessdot f_r(e).\end{aligned}$$

This completes the proof. □

For an $L$-fuzzy p.q. metric $P$ on $X$, restrict $P$ on $M(L^X)$, we obtain an $L$-fuzzy pointwise p.q. metric on $X$ as follows:

**14.2.7 Definition** Mapping $p:\ M(L^X) \times M(L^X) \to [0, +\infty)$ is called an $L$-fuzzy pointwise p.q. metric (p. metric, respectively) on $X$, if $p$ satisfies the following conditions (PM1) – (PM3) ((PM1) – (PM4), respectively):

(PM1) $e \geq d \Longrightarrow p(e,d) = 0$.

(PM2) $p(e,d) \leq p(e,u) + p(u,d)$.

(PM3) $p(e,d) = \bigvee\limits_{u \in \beta^*(d)} \bigwedge\limits_{v \in \beta^*(e)} p(u,v)$.

(PM4) $\exists u \in M(L^X),\ e \lessdot u,\ P(d,u) < r \iff \exists v \in M(L^X),\ d \lessdot v,\ P(e,v) < r$.

**14.2.8 Definition** Let $p$ be an $L$-fuzzy pointwise p.q. metric on $X$. Define mapping $\hat{f}_r:\ L^X \to L^X$ for every $r > 0$ as follows:

$$\forall A \in L^X, \quad \hat{f}_r(A) = \bigvee\{d \in M(L^X):\ \exists e \in \beta^*(A),\ p(e,d) < r\}. \tag{14.13}$$

Call $\mathcal{D}_p{}^* = \{\hat{f}_r:\ r > 0\}$ the *associated neighborhood mappings* of $p$.

Then we obtain a pointwise characterization of $L$-fuzzy p.q. metric (p. metric, respectively) on $X$:

**14.2.9 Theorem** *Let $L^X$ be an $L$-fuzzy space with F-lattice $L$.*

(i) *If $P:\ L^X \times L^X \to [0, +\infty]$ is an $L$-fuzzy p.q. metric (p. metric, respcetively) on $X$, then $p = P|_{M(L^X)\times M(L^X)}:\ M(L^X) \times M(L^X) \to [0, +\infty]$ is an $L$-fuzzy pointwise p.q. metric (p. metric, respcetively) on $X$ and for the family $\mathcal{D}_p{}^* = \{\hat{f}_r:\ r > 0\}$ of the associate neighborhood mappings of $p$,*

$$p(e,d) = \bigwedge\{r:\ d \leq \hat{f}_r(e)\}.$$

(ii) *If $p:\ M(L^X) \times M(L^X) \to [0, +\infty]$ is an $L$-fuzzy pointwise p.q. metric (p. metric, respcetively) on $X$, then the mapping $P:\ L^X \times L^X \to [0, +\infty]$ defined as*

$$P(A,B) = \begin{cases} 0, & B = \underline{0},\\ +\infty, & A = \underline{0},\ B \neq \underline{0},\\ \bigvee\limits_{d \in \beta^*(B)} \bigwedge\limits_{e \in \beta^*(A)} p(e,d), & A, B \neq \underline{0}\end{cases}$$

*is an L-fuzzy p.q. metric (p. metric, respectively) on $X$ and is precisely an extension of $p$ on $L^X \times L^X$, and for the family $\mathcal{D}_p{}^* = \{\hat{f}_r : r > 0\}$ of the associate neighborhood mappings of $p$ and the family $\mathcal{D}_P{}^* = \{f_r : r > 0\}$ of the associate neighborhood mappings of $P$, $f_r = \hat{f}_r$ for every $r > 0$.*

**Proof** (i) Deduced from Theorem **14.2.2**, Lemma **14.2.6** (i)$\Longleftrightarrow$(iii) and Theorem **14.1.16** (ii).

(ii) By the virtue of Theorem **14.2.2**, we need only prove that $P$ satifies conditions (SEM1) – (SEM3) ((SEM1) – (SEM4)).

To verify (SEM1), by definition of $P$, it is sufficient to verify "$A \geq B > 0 \Longrightarrow P(A, B) = 0$". In fact, suppose $A \geq B > 0$, if $d \in \beta^*(B)$, then $d \preceq B \leq A = \bigvee \beta^*(A)$. So there exists $e \in \beta^*(A)$ such that $d \leq e$. By (PM1), $p(e, d) = 0$, so $\bigwedge\{d(e, d) : e \in \beta^*(A)\} = 0$ for every $d \in \beta^*(B)$, and hence

$$P(A, B) = \bigvee_{d\in\beta^*(B)} \bigwedge_{e\in\beta^*(A)} d(e, d) = 0.$$

As for (SEM2), if one of $A$, $B$ and $C$ is $\underline{0}$, (SEM2) can be easily verified. Suppose $A, B, C \neq \underline{0}$, then for arbitrary $e \in \beta^*(A)$, $d \in \beta^*(B)$ and $u \in \beta^*(C)$, by (PM2), $p(e, d) \leq p(e, u) + p(u, d)$, and hence

$$\bigwedge_{e\in\beta^*(A)} p(e, d) \leq \bigwedge_{e\in\beta^*(A)} p(e, u) + p(u, d).$$

Since the above inequality holds for every $u \in \beta^*(C)$, so

$$\bigwedge_{e\in\beta^*(A)} p(e, d) \leq \bigwedge_{e\in\beta^*(A)} p(e, u) + \bigwedge_{u\in\beta^*(C)} p(u, d),$$

$$\begin{aligned} P(A, B) &= \bigvee_{d\in\beta^*(B)} \bigwedge_{e\in\beta^*(A)} p(e, d) \\ &\leq \bigwedge_{e\in\beta^*(A)} p(e, u) + \bigvee_{d\in\beta^*(B)} \bigwedge_{u\in\beta^*(C)} p(u, d) \\ &\leq \bigvee_{u\in\beta^*(C)} \bigwedge_{e\in\beta^*(A)} p(e, u) + \bigvee_{d\in\beta^*(B)} \bigwedge_{u\in\beta^*(C)} p(u, d) \\ &= P(A, C) + P(C, B), \end{aligned}$$

To verify (SEM3), it is sufficient to verify that $P$ is an extension of $p$ on $M(L^X) \times M(L^X)$, i.e. $P(e, d) = p(e, d)$ for arbitriary $e, d \in M(L^X)$. But by (PM3) we precisely have

$$P(e, d) = \bigvee_{u\in\beta^*(d)} \bigwedge_{v\in\beta^*(e)} p(u, v) = p(e, d),$$

so (SEM3) holds.

By the facts proved above, if $p$ is an $L$-fuzzy pointwise p.q. metric on $X$, then $P$ is an $L$-fuzzy p.q. metric on $X$ and an extension of $p$ on $M(L^X) \times M(L^X)$. Moreover, by Lemma **14.2.6**, $P$ is an $L$-fuzzy p. metric if and only if $p$ is an $L$-fuzzy pointwise p. metric.

At last, we turn to prove $f_r = \hat{f}_r$ for every $r > 0$. Without loss of any generality, suppose $A \neq \underline{0}$. If for $d \in M(L^X)$ there exists $e \in \beta^*(A)$ such that $p(e, d) < r$, then $P(e, d) < r$ by $p = P|_{M(L^X)\times M(L^X)}$ proved above. Hence $d \leq f_r(e) \leq f_r(A)$. By (14.13), $\hat{f}_r(A) \leq f_r(A)$. Conversely, if $u \in \beta^*(f_r(A))$, then by Lemma **14.2.1** (ii),

$$\bigvee_{d\in\beta^*(u)}\bigwedge_{e\in\beta^*(A)} p(e,d) = P(A,u) < r.$$

So for every $d \in \beta^*(u)$ there exists $e \in \beta^*(A)$ such that $p(e,d) < r$. By (14.13), $u = \bigvee \beta^*(u) \leq \hat{f}_r(A)$. Hence $f_r(A) = \bigvee \beta^*(f_r(A)) \leq \hat{f}_r(A)$, we have proved $f_r(A) = \hat{f}_r(A)$ for an arbitrary $A$, so $f_r = \hat{f}_r$. □

## 14.3 Metrization

In this section, some properties of $L$-fuzzy p.q. metric (p. metric, respectively), especially, the metriability, will be investigated.

**14.3.1 Theorem** *Let $(L^X, \mathcal{D})$ be an L-fuzzy p.q. metric space. If L has a countable join-generating set, then $(L^X, \delta(\mathcal{D}))$ is firstly countable.*

**Proof** By Theorem **1.1.28** (i), there exists a countable join-generating set $G \subset M(L)$. By Corollary **14.1.5**, $\mathcal{D}$ has a countable standard p.q. metric base $\mathcal{D}^* = \{f_r : r \in \mathbf{Q},\ r > 0\}$. For $x_\lambda \in M(L^X)$, let

$$\mathcal{B} = \{(f_r(x_\mu))^\circ :\ \mu \in G,\ \mu \not\leq \lambda',\ f_r \in \mathcal{D}^*\},$$

then $\mathcal{B} \subset \delta(\mathcal{D})$ is countable, we need only prove that it is a Q-neighborhood base of $x_\lambda$. First of all, we prove $\mathcal{B} \subset \mathcal{Q}(x_\lambda)$. Suppose $\mu \in G$, $\mu \not\leq \lambda'$, $f_r \in \mathcal{D}^*$, then

$$(f_r(x_\mu))^\circ = \bigvee\{C :\ \exists g \in \mathcal{D},\ g(C) \leq f_r(x_\mu)\} \geq x_\mu.$$

So by $\mu \not\leq \lambda'$, $x_\lambda \lessdot x_\mu$, $x_\lambda \lessdot (f_r(x_\mu))^\circ$, $(f_r(x_\mu))^\circ \in \mathcal{Q}(x_\lambda)$, $\mathcal{B} \subset \mathcal{Q}(x_\lambda)$. Secondly, suppose $U \in \mathcal{Q}(x_\lambda)$, by Lemma **13.1.27** (i),

$$x_\lambda \lessdot U = \bigvee\{C :\ \exists f_r \in \mathcal{D}^*,\ f_r(C) \leq U\},$$

so there exists $C \in L^X$ and $f_r \in \mathcal{D}^*$ such that $f_r(C) \leq U$ and $x_\lambda \lessdot C$. Take $c = f_r(C)(x)$, then by $C \leq f_r(C)$ and $x_\lambda \not\leq C'$, $\lambda \not\leq c'$, $c \not\leq \lambda'$. Since $c = \bigvee(G \cap \downarrow c)$, so there exists $\mu \in G \cap \downarrow c$ such that $\mu \not\leq \lambda'$. Hence $(f_r(x_\mu))^\circ \in \mathcal{B}$ and

$$(f_r(x_\mu))^\circ \leq f_r(x_\mu) \leq f_r(x_c) \leq f_r(C) \leq U,$$

this just completes the proof. □

Recall the symbol $\hat{Q}^*(a)$ defined in Definition **13.1.50**, we have the following

**14.3.2 Lemma** *Let $\mathcal{D}^* = \{f_r :\ r \in \mathbf{R}_0\}$ be a standard p.q. metric base on $L^X$, $\mathcal{D} \subset \mathcal{E}(L^X)$ be generated by $\mathcal{D}^*$, $A \in L^X$, $x_\lambda \in M(L^X)$, $r, s \in \mathbf{R}_0$. Then for the L-fuzzy p.q. metric space $(L^X, \mathcal{D})$,*

(i) $\{f_r(C) :\ C \in L^X,\ r \in \mathbf{R}_0\}$ *is a base of* $\delta(\mathcal{D})$.

(ii) $\{f_r(x_\mu) :\ \mu \in \hat{Q}^*(\lambda),\ r \in \mathbf{R}_0\}$ *is a Q-neighborhood base of* $x_\lambda$ *in* $(L^X, \delta(\mathcal{D}))$.

(iii) $f_{\frac{r+s}{2}}(A) \leq f_r(A) \vee f_s(A)$.

**Proof** (i) Note that

$$int_{\delta(\mathcal{D})}(f_r(A)) = \bigvee\{C :\ \exists f \in \mathcal{D},\ f(C) \leq f_r(A)\}.$$

For every $t \in (0,r) \cap \mathbf{R}_0$, by (14.1), $f_{r-t} \circ f_t(A) \leq f_r(A)$. So $f_t(A) \leq int_{\delta(\mathcal{D})}(f_r(A))$. By (14.2),

$$f_r(A) = \bigvee_{s\in(0,r)\cap\mathbf{R}_0} f_s(A) \leq int_{\delta(\mathcal{D})}(f_r(A)),$$

so $f_r(A) = int_{\delta(\mathcal{D})}(f_r(A)) \in \delta(\mathcal{D})$. Then by the virtue of Lemma **13.1.27** (iii) we know that $\{f_r : r \in \mathbf{R}_0\}$ is a base of $\delta(\mathcal{D})$.

(ii) Suppose $U \in \mathcal{Q}_{\delta(\mathcal{D})}(x_\lambda)$, then there exists $\mu \in \beta^*(U(x)) \cap \hat{Q}^*(\lambda)$. By Theorem **1.3.14** (i), there exists $\xi \in M(L)$ such that $\mu \preceq \xi \preceq U(x)$. By (i) proved above, we can suppose $U = \bigvee_{s \in S} f_{r_s}(C_s)$, then $U(x) = \bigvee_{s \in S} f_{r_s}(C_s)(x)$. Since $\xi \preceq U(x)$, so there exists $t \in S$ such that $\xi \leq f_{r_t}(C_t)(x)$. By (14.2), $f_{r_t}(C_t)(x) = \bigvee_{u \in (0,r_t) \cap \mathbf{R}_0} f_u(C_t)(x)$. Then since by $\mu \preceq \xi \leq f_{r_t}(C_t)(x)$ and Theorem **1.3.6** (ii) we have $\mu \preceq f_{r_t}(C_t)(x)$, so there exists $u \in (0, r_t) \cap \mathbf{R}_0$ such that $\mu \leq f_u(C_t)(x)$, $x_\mu \leq f_u(C_t)$. Take $r = r_t - u$, then by (14.1),

$$f_r(x_\mu) \leq (f_r \circ f_u)(C_t) \leq f_{r+u}(C_t) = f_{r_t}(C_t) \leq U.$$

Since $\mu \in \hat{Q}^*(\lambda)$, $\mu \not\leq \lambda'$, so by $x_\mu \leq f_r(x_\mu)$ we have $f_r(x_\mu)(x) \not\leq \lambda'$, by (i) proved above we have $f_r(x_\mu) \in \mathcal{Q}_{\delta(\mathcal{D})}(x_\lambda)$. Since we have proved $f_r(x_\mu) \leq U$, (ii) has been proved.

(iii) If $r \leq s$, then $\frac{r+s}{2} \leq \frac{s+s}{2} = s$; if $r > s$, then $\frac{r+s}{2} \leq \frac{r+r}{2} = r$. So

$$f_{\frac{r+s}{2}}(A) \leq f_r(A) \vee f_s(A). \qquad \square$$

Erceg[21] proved the following important results on $L$-fuzzy p. metric space related separation:

**14.3.3 Theorem** *Let $(L^X, \mathcal{D})$ be an $L$-fuzzy p. metric space. Then $(L^X, \delta(\mathcal{D}))$ is normal.*

**Proof** Suppose $A \in (\delta(\mathcal{D}))'$, $U \in \delta(\mathcal{D})$, $A \leq U$. Denote

$$\mathcal{U} = \{C \in L^X : \exists r(C) > 0,\ f_{r(C)}(C) \leq U\},$$

then $U = \bigvee \mathcal{U}$ and for every $C \in \mathcal{U}$, there exists $r(C) > 0$ such that $f_{r(C)}(C) \leq U$. Let

$$V = \bigvee\{f_{\frac{r(C)}{2}}(A \wedge C) : C \in \mathcal{U}\},$$

then by Lemma **14.3.2**, $V \in \delta(\mathcal{D})$, and since $A \leq U = \bigvee \mathcal{U}$ we have

$$A = A \wedge U = \bigvee\{A \wedge C : C \in \mathcal{U}\} \leq \bigvee\{f_{\frac{r(C)}{2}}(A \wedge C) : C \in \mathcal{U}\} = V.$$

On the other hand, by Theorem **13.1.31** (ii) and Lemma **14.3.2** (iii),

$$\begin{aligned}
cl_{\delta(\mathcal{D})}(V) &= \bigwedge\{f_s(V) : s > 0\} \\
&= \bigwedge\{f_{\frac{s}{2}}(V) : s > 0\} \\
&= \bigwedge_{s>0} \bigvee\{f_{\frac{s}{2}} f_{\frac{r(C)}{2}}(A \wedge C) : C \in \mathcal{U}\} \\
&\leq \bigwedge_{s>0} \bigvee\{f_{\frac{s+r(C)}{2}}(A \wedge C) : C \in \mathcal{U}\} \\
&\leq \bigwedge_{s>0} \bigvee\{f_s(A \wedge C) \vee f_{r(C)}(A \wedge C) : C \in \mathcal{U}\} \\
&\leq \bigwedge_{s>0} \bigvee\{f_s(A) \vee U : C \in \mathcal{U}\} \\
&= \bigwedge_{s>0} (f_s(A) \vee U) \\
&= (\bigwedge_{s>0} f_s(A)) \vee U \\
&= cl_{\delta(\mathcal{D})}(A) \vee U \\
&= A \vee U \\
&= U.
\end{aligned}$$

So $A \leq V \leq V^- \leq U$, $(L^X, \delta(\mathcal{D}))$ is normal. □

Hutton and Reilly [41] introduced the following notion:

**14.3.4 Definition** An $L$-fts $(L^X, \delta)$ is called $R_0$, if every open subset in $(L^X, \delta)$ is a join of closed subsets.

An obvious example of $R_0$ $L$-fts is $T_1$ $L$-fts. Regularity also implies $R_0$ property.

**14.3.5 Definition** Let $(L^X, \delta)$ be an $L$-fts. $(L^X, \delta)$ is called p.q. metrizable (p. metrizable, respectively), if there exists an $L$-fuzzy quasi-uniformity (uniformity, respectively) $\mathcal{D}$ on $X$ such that $\delta = \delta(\mathcal{D})$.

Liang[58] proved the following theorem. But the proof introduced in the sequel is based on the simplified one belonging to Wang:[168]

**14.3.6 Theorem** *Let $(L^X, \delta)$ be a secondly countable $L$-fts. Then $(L^X, \delta)$ is p. metrizable if and only if $(L^X, \delta)$ is normal and $R_0$.*

**Proof** (Sufficiency) Suppose $L$-fts $(L^X, \delta)$ is normal and $R_0$, and has a countable base $\{U_n : n \in \mathbf{N}\}$, then it can be easily verified that for every $U \in \delta$ there exists $\mathcal{C}_U \subset \{U_n : n \in \mathbf{N}\}$ such that

$$U = \bigvee_{V \in \mathcal{C}_U} V = \bigvee_{V \in \mathcal{C}_U} V^-. \tag{14.14}$$

So for every $n \in \mathbf{N}$, there exists $m \in \mathbf{N}$ such that $U_m{}^- \leq U_n$. Let $\mathcal{B} = \{(U_m, U_n) : U_m{}^- \leq U_n,\ m, n \in \mathbf{N}\}$, then $\mathcal{B}$ is nonempty and countable. Suppose $\mathcal{B} = \{W_n : n \in \mathbf{N}\}$, then by the normality of $(L^X, \delta)$ and the Urysohn's Lemma **9.2.13**, for every $W_n = (U_i, U_j) \in \mathcal{B}$, there exists an $L$-fuzzy continuous mapping $(g_n)^\rightarrow : (L^X, \delta) \to I(L)$ such that

$$U_i{}^- \leq (g_n)^\leftarrow(L_1{}') \leq (g_n)^\leftarrow(R_0) \leq U_j. \tag{14.15}$$

Let $\mathcal{D}^* = \{\Theta_{(g_n)^\rightarrow,(g_n)^\leftarrow}(f_\varepsilon \wedge f_\varepsilon{}^\triangleleft) : n \in \mathbf{N},\ \varepsilon \in \mathbf{Q},\ \varepsilon > 0\}$, then since it is easy to find in the proof of Example **13.1.34** that $\{f_\varepsilon \wedge f_\varepsilon{}^\triangleleft) : \varepsilon \in \mathbf{Q},\ \varepsilon > 0\}$ is a base of the canonical uniformity of $I(L)$, by Theorem **13.1.47** (i) and Proposition **13.1.21**, $\mathcal{D}^*$ is a countable base of an $L$-fuzzy uniformity $\mathcal{D}$ on $X$, where

$$\mathcal{D} = \{f \in \mathcal{E}(L^X) : \exists \Theta_{(g_n)^\rightarrow,(g_n)^\leftarrow}(f_\varepsilon \wedge f_\varepsilon{}^\triangleleft) \in \mathcal{D}^*,\ f \geq \Theta_{(g_n)^\rightarrow,(g_n)^\leftarrow}(f_\varepsilon \wedge f_\varepsilon{}^\triangleleft)\}.$$

So $(L^X, \mathcal{D})$ is an $L$-fuzzy p. metric space, we need only prove $\delta = \delta(\mathcal{D})$.

Suppose $U \in \delta(\mathcal{D})$. For every $\varepsilon > 0$, by (13.19) and Theorem **13.1.15** (vii),

$$(f_\varepsilon \wedge f_\varepsilon{}^\triangleleft) \circ (f_\varepsilon \wedge f_\varepsilon{}^\triangleleft) \leq (f_\varepsilon \circ f_\varepsilon) \wedge (f_\varepsilon{}^\triangleleft \circ f_\varepsilon{}^\triangleleft) = (f_\varepsilon \circ f_\varepsilon) \wedge (f_\varepsilon \circ f_\varepsilon)^\triangleleft = f_{2\varepsilon} \wedge f_{2\varepsilon}{}^\triangleleft.$$

So by Lemma **13.1.27** (iii) and Proposition **13.1.44** we have

$$U = \bigvee\{\Theta_{(g_n)^\rightarrow,(g_n)^\leftarrow}(f_\varepsilon \wedge f_\varepsilon{}^\triangleleft)(C) : \Theta_{(g_n)^\rightarrow,(g_n)^\leftarrow}(f_{2\varepsilon} \wedge f_{2\varepsilon}{}^\triangleleft)(C) \leq U\}.$$

But by (13.13) and (13.15), both $f_\varepsilon((g_n)^\rightarrow(D))$ and $f_\varepsilon{}^\triangleleft((g_n)^\rightarrow(D))$ are open in $I(L)$ for every $D \in L^X$, so by Proposition **13.1.43**, (13.2) and the continuity of $(g_n)^\rightarrow$,

$$\begin{aligned}\Theta_{(g_n)^\rightarrow,(g_n)^\leftarrow}(f_\varepsilon \wedge f_\varepsilon{}^\triangleleft)(C) &= (\Theta_{(g_n)^\rightarrow,(g_n)^\leftarrow}(f_\varepsilon) \wedge \Theta_{(g_n)^\rightarrow,(g_n)^\leftarrow}(f_\varepsilon{}^\triangleleft))(C) \\ &= \bigvee_{D \in \beta^*(C)} (\Theta_{(g_n)^\rightarrow,(g_n)^\leftarrow}(f_\varepsilon)(D) \wedge \Theta_{(g_n)^\rightarrow,(g_n)^\leftarrow}(f_\varepsilon{}^\triangleleft)(D) \\ &= \bigvee_{D \in \beta^*(C)} ((g_n)^\leftarrow(f_\varepsilon((g_n)^\rightarrow(D))) \wedge ((g_n)^\leftarrow(f_\varepsilon{}^\triangleleft((g_n)^\rightarrow(D))))\end{aligned}$$

is an open subset in $(L^X, \delta)$. Hence $U \in \delta$, $\delta(\mathcal{D}) \subset \delta$.

Conversely, since $\{U_n : n \in \mathbf{N}\}$ is a counable base of $\delta$, so to prove $\delta \subset \delta(\mathcal{D})$, it is sufficient to prove $U_n \in \delta(\mathcal{D})$ for every $n \in \mathbf{N}$. By (14.14), there exists $A_n \subset \mathbf{N}$ such that

$$U_n = \bigvee_{m \in A_n} U_m = \bigvee_{m \in A_n} U_m^{-}. \tag{14.16}$$

By (14.15), for every $m \in A_n$, there exists $k_m \in \mathbf{N}$ such that

$$U_m^{-} \leq (g_{k_m})^{\leftarrow}(R_0) \leq U_n. \tag{14.17}$$

Arbitrarily fix an $A \in L^X \backslash \{\underline{0}\}$, let $\varepsilon_m = u((g_{k_m})^{\rightarrow}(A))$ for every $m \in A_n$, where $u$ is defined by (13.12), then $f_{\varepsilon_m} \circ (g_{k_m})^{\rightarrow}(A) = R_0$ by (13.13), and hence by (14.16) and (14.17) we have

$$U_n = \bigvee_{m \in A_n} (g_{k_m})^{\leftarrow}(R_0) = \bigvee_{m \in A_n} (g_{k_m})^{\leftarrow} \circ f_{\varepsilon_m} \circ (g_{k_m})^{\rightarrow}(A).$$

So to prove $U_n \in \delta(\mathcal{D})$, we need only prove $(g_{k_m})^{\leftarrow} \circ f_{\varepsilon_m} \circ (g_{k_m})^{\rightarrow}(A) \in \delta(\mathcal{D})$ for every $m \in A_n$, or more generally, prove $(g_k)^{\leftarrow} \circ f_r \circ (g_k)^{\rightarrow}(A) \in \delta(\mathcal{D})$ for every $k \in \mathbf{N}$ and every $r > 0$. Suppose $k \in \mathbf{N}$, $r > 0$, let $h_r = (g_k)^{\leftarrow} \circ f_r \circ (g_k)^{\rightarrow}$, fix an $\varepsilon \in \mathbf{Q}$ such that $0 < \varepsilon < r$, then

$$h_r \geq (g_k)^{\leftarrow}(f_\varepsilon \wedge f_\varepsilon{}^{\triangleleft})(g_k)^{\rightarrow} = \Theta_{(g_n)^{\rightarrow},(g_n)^{\leftarrow}}(f_\varepsilon \wedge f_\varepsilon{}^{\triangleleft}) \in \mathcal{D}^*.$$

So $h_r \in \mathcal{D}$, we obtain a subfamily $\{h_r : r > 0\} \subset \mathcal{D}$. By (13.13) and Proposition **9.1.2**, $f_r = \bigvee_{s<r} f_s$, so

$$h_r = \bigvee_{s<r} h_s,$$

$\{h_r : r > 0\}$ fulfils the condition (14.1). By (13.19) and Proposition **13.1.44**,

$$\begin{aligned} h_r \circ h_s &= \Theta_{(g_k)^{\rightarrow},(g_k)^{\leftarrow}}(f_r) \circ \Theta_{(g_k)^{\rightarrow},(g_k)^{\leftarrow}}(f_s) \\ &\leq \Theta_{(g_k)^{\rightarrow},(g_k)^{\leftarrow}}(f_r \circ f_s) \\ &\leq \Theta_{(g_k)^{\rightarrow},(g_k)^{\leftarrow}}(f_{r+s}) \\ &= h_{r+s}, \end{aligned}$$

$\{h_r : r > 0\}$ also fulfils the condition (14.2). By Lemma **14.3.2** (i),

$$(g_{k_m})^{\leftarrow} \circ f_{\varepsilon_m} \circ (g_{k_m})^{\rightarrow}(A) = h_r(A) \in \delta(\mathcal{D}).$$

Hence we have proved $\delta = \delta(\mathcal{D})$, $(L^X, \delta)$ is p. metrizable.

(Necessity) Suppose $(L^X, \delta)$ is p. metrizable, then by Theorem **14.3.3**, $(L^X, \delta)$ is normal, we need only prove that $(L^X, \delta)$ is $R_0$. Suppose $\mathcal{D}$ is an $L$-fuzzy uniformity on $X$ with a standard p. metric base $\mathcal{D}^* = \{f_r : r > 0\}$. Let $U \in \delta = \delta(\mathcal{D})$, by Lemma **14.3.2** (i), $f_r(U')$ is open and hence $f_r(U')'$ is closed in $(L^X, \delta(\mathcal{D})$ for every $r > 0$. Hence it follows from Theorem **13.1.31** (i) that $U = \bigvee\{f_r(U')' : r > 0\}$ is a join of closed subsets, $(L^X, \delta)$ is $R_0$. □

The following theorems[58] are deduced from Lemma **13.1.64**, Definition **13.1.65** and Theorem **13.1.66**:

**14.3.7 Theorem** *Let $\{(L^{X_n}, \mathcal{D}_n) : n < \omega\}$ be a countable family of L-fuzzy p.q. metric (p. metric, respectively) spaces, $X = \prod_{n<\omega} X_n$, $\mathcal{D} = \prod_{n<\omega} \mathcal{D}_n$. Then $(L^X, \mathcal{D})$ is an L-fuzzuy p.q. metric (p. metric, respectively) space.* □

**14.3.8 Theorem** *Let $\{(L^{X_n}, \delta_n) : n < \omega\}$ be a countable family of p.q. metrizable (p. metrizable, respectively) L-fts', $X = \prod_{n<\omega} X_n$, $\delta = \prod_{n<\omega} \delta_n$. Then $(L^X, \delta)$ is p.q.*

*metrizable (p. metrizable, respectively).* □

**14.3.9 Remark** Since there exists counterexamples to show that first countability is not $\kappa$-multiplicative for $\kappa > \aleph_0$, so in general, p.q. metrizability (p. metrizability) is not $\kappa$-multiplicative for $\kappa > \aleph_0$.

**14.3.10 Remark** Using her pointwise description of $L$-fuzzy p.q. metric (p. metric, respectively),[61] Liang[62] ever proved a sufficient condition of the Smirnov-Nagata Metrization Theorem in $L$-fts'.

The following two theorems were skillfully proved by Liang[64] via her pointwise description of $L$-fuzzy p.q. metric (p. metric, respectively):

**14.3.11 Theorem** *An induced $L$-fts is p. metrizable if and only if its background space is p. metrizable.* □

**14.3.12 Theorem** *Let $(L^X, \delta)$ be an induced $L$-fts, $L$ have a countable join-generating set. Then the following conditions are equivalent:*

(i) *$(L^X, \delta)$ is p. metrizable.*

(ii) *$(L^X, \delta)$ is regular and has a $\sigma$-discrete base.*

(iii) *$(L^X, \delta)$ is regular and has a $\sigma$-locally finite base.* □

We can also consider metrizability problem in $L$-fts':

**14.3.13 Definition** Let $\mathcal{D}^*, \mathcal{D} \subset \mathcal{E}(L^X)$. $\mathcal{D}^*$ is called a *standard metric base* on $L^X$, if $\mathcal{D}^*$ is a standard p. metric base on $L^X$ and fulfils the following condition:

$$\forall e \in M(L^X), \quad e = \bigwedge_{r \in \mathbf{R}_0} f_r(e). \tag{14.18}$$

$\mathcal{D}^*$ is called a *standard metric base* of $\mathcal{D}$ or of $(L^X, \mathcal{D})$ or say *$\mathcal{D}$ is generated by $\mathcal{D}^*$*, if $\mathcal{D}^*$ is a standard metric base on $L^X$ and for every $f \in \mathcal{D}$ there exists $g \in \mathcal{D}^*$ such that $f \geq g$. An $L$-fts $(L^X, \delta)$ is called *metrizable*, if there exists an $L$-fuzzy uniformity $\mathcal{D}$ on $X$ generated by a standard metric base $\mathcal{D}^*$ on $L^X$ such that $\delta = \delta(\mathcal{D})$.

$\mathcal{D}^* \subset \mathcal{E}(L^X)$ is called a *standard metric base* of $\mathcal{D}$ or $(L^X, \mathcal{D})$

**14.3.14 Theorem** *Let $(L^X, \mathcal{D})$ be an $L$-fuzzy p. metric space. Then $(L^X, \mathcal{D})$ is an $L$-fuzzy metric space if and only if $(L^X, \delta(\mathcal{D}))$ is $T_1$.*

**Proof** If $(L^X, \delta(\mathcal{D}))$ is $T_1$, then by Theorem **8.1.23**, (14.18) follows from Theorem **13.1.31** and (13.8). Conversely, if (14.18) holds, then by Theorem **13.1.31** and (13.8), $e$ is a closed subset in $(L^X, \delta(\mathcal{D}))$. By Theorem **8.1.23**, $(L^X, \delta(\mathcal{D}))$ is $T_1$. □

Since every $L$-fuzzy subset can be represented as a join of molecules, so by theorems **14.3.6** and **14.3.14** we have

**14.3.15 Theorem** *Let $(L^X, \delta)$ be a secondly countable $L$-fts. Then $(L^X, \delta)$ is metrizable if and only if $(L^X, \delta)$ is normal and $T_1$.* □

In general topology, there exists close connections between paracompactness and metrizability. Just as mentioned above, Liang[62] also proved that a regular $L$-fts possessing a $\sigma$-locally finite base is p. metrizable. On the other hand, from Chapter 12 we find that paracompactness has not been strong enough to describe some local properties in $L$-fts', although it is very nice and reasonable in F-ts'. But, however, even to closure preserving property which is strictly weaker than locally finite property, an $L$-fuzzy metric space's property may be weaker than we hope. Luo's counterexample[110] shows this point:

**14.3.16 Lemma** *Let $(L^X, \mathcal{D})$ be an $L$-fuzzy p.q. metric space with a standard base $\mathcal{D}^* = \{f_r : r > 0\}$. Then the following conditions are equivalent:*

(i) $\forall r > 0, \quad f_r = f_r^{\triangleleft}$.

(ii) $\forall r > 0, \ \forall A, B \in L^X, \quad f_r(A) \leq B' \iff f_r(B) \leq A'$.

(iii) $\forall r > 0, \ \forall e \in M(L^X), \ \forall A \in L^X,$

$$f_r(f_r(e)') \leq e', \tag{14.19}$$

$$f_r(A) \leq e' \implies A \leq f_r(e)'. \tag{14.20}$$

**Proof** (i)$\Longrightarrow$(ii): By Theorem **13.1.15**.

(ii)$\Longrightarrow$(iii): Since $f_r(e) = (f_r(e)')'$, so by (i),

$$f_r(f_r(e)') \leq e',$$

$$f_r(A) \leq e' \implies f_r(e) \leq A' \implies A \leq f_r(e)'.$$

(iii)$\Longrightarrow$(i): Since $f_r^{\triangleleft}$ is arbitrary join preserving, so to verify (i) we need only verify (i) for every molecule $e \in M(L^X)$. For every $e \in M(L^X)$, by (14.20),

$$f_r^{\triangleleft}(e) = \bigwedge\{A : f_r(A') \leq e'\} \leq f_r(e).$$

By (14.20),

$$f_r(e) = \bigwedge\{A' : f_r(e) \leq A'\} \leq \bigwedge\{A' : f_r(A) \leq e'\} = f_r^{\triangleleft}(e).$$

So $f_r = f_r^{\triangleleft}$. □

**14.3.17 Definition** Let $(L^X, \delta)$ be an $L$-fts, $A \in L^X$, $\mathcal{A} \subset L^X$. $\mathcal{A}$ is called *point-countable* (point-finite, respectively) in $A$, if for every $e \in M(\downarrow A)$, there exists a countable (finite, respectively) subfamily $\mathcal{A}_0$ of $\mathcal{A}$ such that

$$C \in \mathcal{A} \backslash \mathcal{A}_0 \implies e \not\leq C.$$

$\mathcal{A}$ is called *$\sigma$-point-finite* in $A$, if $\mathcal{A}$ can be represented as a union of countably many subfamilies which are point-finite in $A$.

**14.3.18 Example** There exists a fuzzy metric space $(I^X, \mathcal{D})$ such that for every $\alpha \in (0,1)$,

(i) $(I^X, \delta(\mathcal{D}))$ has an open $\alpha$-Q-cover $\mathcal{U}$ such that $\mathcal{U}$ has no closure preserving refinement which is an $\alpha$-Q-cover of $(I^X, \delta(\mathcal{D}))$.

(ii) $(I^X, \delta(\mathcal{D}))$ has an open $\alpha$-Q-cover $\mathcal{U}$ such that $\mathcal{U}$ has no point-countable refinement which is an open $\alpha$-Q-cover of $(I^X, \delta(\mathcal{D}))$.

(iii) $(I^X, \delta(\mathcal{D}))$ has no point-finite base.

Take $X = [0, +\infty)$, let

$$f : \{(x, \lambda, r) \in (0, +\infty) \times [0,1] \times (0, +\infty) : \lambda \geq 1 - \tfrac{r}{x}\} \to [0,1],$$

$$f(x, \lambda, r) = \lambda[1 - \tfrac{(1-\lambda)x}{r}],$$

$$g : (0, +\infty) \times [0,1] \times (0, +\infty) \to [0,1],$$

$$g(x, \lambda, r) = \tfrac{1}{2x}[r + x - \sqrt{r^2 + 2rx(1 - 2\lambda) + x^2}].$$

Since in $dom(g)$ we have

$$r^2 + 2rx(1 - 2\lambda) + x^2 \geq r^2 - 2rx + x^2 = (r - x)^2 \geq 0,$$

$$0 = g(x,0,r) \leq g(x,\lambda,r) \leq g(x,1,r) = \tfrac{1}{2x}(r + x - |r - x|) = 1 \wedge \tfrac{r}{x}, \tag{14.21}$$

so $g$ is well-defined.

It is not hard to verify

$$f(x, 1 - g(x,\lambda,r), r) = 1 - \lambda. \tag{14.22}$$

Arbitrarily fix $x$, $r$, it can be easily found from the definition of $g$ that $g$ goes through every point in $[0, 1 \wedge \frac{r}{x}]$ when $\lambda$ goes through every point in $[0,1]$, so for every $\mu \in [0 \vee (1 - \frac{r}{x}), 1]$, there exists $\lambda \in [0,1]$ such that $g(x,\lambda,r) = 1 - \mu$, and hence by (14.22) we have

$$f(x,\mu,r) = f(x, 1 - g(x,\lambda,r), r) = 1 - \lambda,$$
$$g(x, 1 - f(x,\mu,r), r) = g(x,\lambda,r) = 1 - \mu,$$

thus

$$\forall \lambda \in [0 \vee (1 - \tfrac{r}{x}), 1], \quad g(x, 1 - f(x,\lambda,r), r) = 1 - \lambda. \tag{14.23}$$

Obviously we have

$$f(x,\lambda,r) \leq \lambda. \tag{14.24}$$

Arbitrarily fix $x$ and $r$ in $f(x,\lambda,r)$, then $f$ goes through every point in $[0,1]$ when $\lambda$ goes through every point in $[0 \vee (1 - \frac{r}{x}), 1]$, so for every $\lambda \in [0,1]$, there exists $\mu \in [0 \vee (1 - \frac{r}{x}), 1]$ such that $f(x,\mu,r) = 1 - \lambda$. Then by (14.24) we have

$$g(x,\lambda,r) = g(x, 1 - f(x,\mu,r), r) = 1 - \mu \leq 1 - f(x,\mu,r) = \lambda,$$
$$g(x,\lambda,r) \leq \lambda. \tag{14.25}$$

Easy to find:

*$f$ is continuous, is strictly increasing with respect to $\lambda$, is strictly increasing with respect to $r$ when $\lambda \in (0,1)$.* (14.26)

Clearly, $g$ is continuous and strictly increasing with respect to $\lambda$. Since $|1-2\lambda| \leq 1$ and the equation holds if and only if $\lambda\{0,1\}$, so by the virtue of

$$[r + (1-2\lambda)x]^2 = r^2 + 2rx(1-2\lambda) + (1-2\lambda)^2x^2 \leq r^2 + 2rx(1-2\lambda) + x^2$$

the following relations hold, and the equation holds if and only if ] $\lambda\{0,1\}$:

$$\frac{\partial g}{\partial r} = \frac{1}{2x}\Big[1 - \frac{r + (1-2\lambda)x}{\sqrt{r^2 + 2rx(1-2\lambda) + x^2}}\Big] \geq 0.$$

Then we have

*$g$ is continuous, is strictly increasing with respect to $\lambda$, is strictly increasing with respect to $r$ when $\lambda \in (0,1)$.* (14.27)

For every $x \in (0,+\infty)$ and every $\lambda \in (0,1]$, let

$$f_r(0_\lambda)(y) = \begin{cases} \lambda, & y = 0, \\ f(y, \frac{2\alpha\lambda}{1+\alpha}, r), & y \in (0, \frac{r}{1 - \frac{2\alpha\lambda}{1+\alpha}}), \\ 0, & y \in [\frac{r}{1 - \frac{2\alpha\lambda}{1+\alpha}}, +\infty), \end{cases} \tag{14.28}$$

$$f_r(x_\lambda)(y) = \begin{cases} (1 - \frac{1+\alpha}{2\alpha}(1 - g(x,\lambda,r))) \vee 0, & y = 0, \\ f(y, g(x,\lambda,r), r), & y \in (0, \frac{r}{1 - g(x,\lambda,r)})\backslash\{x\}, \\ \lambda, & y = x, \\ 0, & y \in [\frac{r}{1-g(x,\lambda,r)}, +\infty)\backslash\{x\}, \end{cases} \tag{14.29}$$

in the above and the sequel $\frac{r}{1-g(x,\lambda,r)}$ means $+\infty$ if $g(x,\lambda,r)=1$. For every $A \in I^X$, let

$$f_r(A) = \bigvee\{f_r(x_{A(x)}) : x \in supp(A)\}. \tag{14.30}$$

$$\mathcal{D}^* = \{f_r : r > 0\}.$$

(1) Now $\mathcal{D}^*$ fufils $f_r(\underline{0}) = \underline{0}$ and $f_r(A) \geq A$. By the virtue of (14.26) and (14.27), $\mathcal{D}^*$ also fulfils (14.2).

(2) Prove $f_r(\bigvee_{t\in T} A_t) = \bigvee_{t\in T} f_r(A_t)$. By the virtue of (14.26) and (14.27),

$$\forall x \in X,\ \forall C \subset (0,1],\ C \neq \emptyset,\ \bigvee C = \lambda \implies f_r(x_\lambda) = \bigvee\{f_r(x_\mu) : \mu \in C\},$$

so if denote $A = \bigvee_{t\in T} A_t$, by (14.30) we have

$$\begin{aligned} f_r(A) &= \bigvee\{f_r(x_{A(x)}) : x \in supp(A)\} \\ &= \bigvee\{\bigvee\{f_r(x_{A_t(x)}) : t \in T\} : x \in supp(A)\} \\ &= \bigvee\{f_r(x_{A_t(x)}) : \exists t \in T,\ x \in supp(A_t)\} \\ &= \bigvee_{t\in T} f_r(A_t). \end{aligned}$$

(3) Prove (14.1). Since for every $y > 0$ by (14.24) and (14.25) we have

$$\begin{aligned} f_r(y_{f(y,\frac{2\alpha\lambda}{1+\alpha},r)}) &\leq (1-(1-g(y,f(y,\tfrac{2\alpha\lambda}{1+\alpha},r),r)))\chi_{\{0\}} \vee y_{f(y,\frac{2\alpha\lambda}{1+\alpha},r)} \vee \\ &\qquad \bigvee\{z_{f(z,g(y,f(y,\frac{2\alpha\lambda}{1+\alpha},r),r),r)} : z \in (0, \tfrac{r}{1-g(y,f(y,\frac{2\alpha\lambda}{1+\alpha},r),r)})\backslash\{y\}\} \\ &\leq 0_\lambda \vee y_{f(y,\frac{2\alpha\lambda}{1+\alpha},r)} \vee \bigvee\{z_{f(z,\frac{2\alpha\lambda}{1+\alpha},r)} : z \in (0, \tfrac{r}{1-\frac{2\alpha\lambda}{1+\alpha}})\backslash\{y\}\} \\ &= f_r(0_\lambda), \end{aligned}$$

so by $f_r(A) \geq A$ and (14.2) which have been fulfilled by $\mathcal{D}^* = \{f_r : r > 0\}$, we have

$$f_r(0_\lambda) \leq (f_r \circ f_r)(0_\lambda) = f_r(0_\lambda) \vee \bigvee\{f_r(y_{f(y,\frac{2\alpha\lambda}{1+\alpha},r)}) : y \in (0, \tfrac{r}{1-\frac{2\alpha\lambda}{1+\alpha}})\} \leq f_r(0_\lambda),$$

$$(f_r \circ f_s)(0_\lambda) \leq (f_{r+s} \circ f_{r+s})(0_\lambda) = f_{r+s}(0_\lambda).$$

On the other hand, since

$$(1 - \tfrac{1+\alpha}{2\alpha}(1-g(x,\lambda,r))) \vee 0 \leq 1-(1-g(x,\lambda,r)) = g(x,\lambda,r),$$

so by (14.24) and (14.25) for every $x > 0$ we have

$$\begin{aligned} f_r(((1-\tfrac{1+\alpha}{2\alpha}(1-g(x,\lambda,r)))\vee 0)\chi_{\{0\}}) &\leq ((1-\tfrac{1+\alpha}{2\alpha}(1-g(x,\lambda,r)))\vee 0)\chi_{\{0\}} \vee \\ &\qquad \bigvee\{z_{f(z,g(x,\lambda,r),r)} : z \in (0, \tfrac{r}{1-g(x,\lambda,r)})\} \\ &\leq f_r(x_\lambda), \end{aligned}$$

$$\begin{aligned} f_r(y_{f(y,g(x,\lambda,r),r)}) &\leq ((1-\tfrac{1+\alpha}{2\alpha}(1-g(x,\lambda,r)))\vee 0)\chi_{\{0\}} \vee \\ &\qquad \bigvee\{z_{f(z,g(x,\lambda,r),r)} : z \in (0, \tfrac{r}{1-g(x,\lambda,r)})\} \\ &\leq f_r(x_\lambda), \end{aligned}$$

$$\begin{aligned} f_r(x_\lambda) &\leq (f_r \circ f_r)(x_\lambda) \\ &= f_r(x_\lambda) \vee f_r(((1-\tfrac{1+\alpha}{2\alpha}(1-g(x,\lambda,r)))\vee 0)\chi_{\{0\}} \vee \\ &\qquad \bigvee\{f_r(y_{f(y,g(x,\lambda,r),r)}) : y \in (0, \tfrac{r}{1-g(x,\lambda,r)})\backslash\{x\}\} \\ &\leq f_r(x_\lambda), \end{aligned}$$

$$(f_r \circ f_s)(x_\lambda) \leq (f_{r+s} \circ f_{r+s})(x_\lambda) = f_{r+s}(x_\lambda).$$

So by (14.30) we have proved (14.1).

(4) Prove (14.3). By Lemma **14.3.16**, we need only prove the condition (iii) in Lemma **14.3.16**. By the definition,

$$f_r((1-\lambda)\chi_{\{0\}})(0) = 1-\lambda.$$

For every $y \in (0, \frac{r}{1-\frac{2\alpha\lambda}{1+\alpha}})$, we have $\frac{2\alpha\lambda}{1+\alpha} > 1 - \frac{r}{y}$, so by (14.23),

$$\begin{aligned} f_r(y_{1-f(y,\frac{2\alpha\lambda}{1+\alpha},r)})(0) &= (1 - \tfrac{1+\alpha}{2\alpha}(1 - g(y, 1 - f(y, \tfrac{2\alpha\lambda}{1+\alpha}, r), r))) \vee 0 \\ &= (1 - \tfrac{2\alpha\lambda}{1+\alpha}(1 - (1 - \tfrac{2\alpha\lambda}{1+\alpha}))) \vee 0 \\ &= 1 - \lambda. \end{aligned}$$

For every $y \in [\frac{r}{1-\frac{2\alpha\lambda}{1+\alpha}}, +\infty)$, we have $y \geq \frac{r}{1-\frac{2\alpha\lambda}{1+\alpha}} > r$, so

$$g(y,1,r) = \tfrac{r}{y} \leq 1 - \tfrac{2\alpha\lambda}{1+\alpha},$$

by the definition we have

$$f_r(y_1)(0) = (1 - \tfrac{1+\alpha}{2\alpha}(1 - (1 - \tfrac{2\alpha\lambda}{1+\alpha}))) \vee 0 = 1 - \lambda,$$

$$\begin{aligned} f_r(f_r(0_\lambda)') = {} & f_r((1-\lambda)\chi_{\{0\}}) \vee \bigvee\{f_r(y_{1-f(y,\frac{2\alpha\lambda}{1+\alpha},r)}) : y \in (0, \tfrac{r}{1-\frac{2\alpha\lambda}{1+\alpha}})\} \vee \\ & \bigvee\{f_r(y_1) : y \in [\tfrac{r}{1-\frac{2\alpha\lambda}{1+\alpha}}, +\infty)\} \\ \leq {} & (0_\lambda)'. \end{aligned}$$

If $A \not\leq f_r(0_\lambda)'$, then there exists $y \in [0, \frac{r}{1-\frac{2\alpha\lambda}{1+\alpha}})$ such that $\mu = A(y) > 1 - f_r(0_\lambda)(y)$. If $y = 0$, then $\mu > 1-\lambda$, we have had $f_r(A) \not\leq (0_\lambda)'$; if $y \in (0, \frac{r}{1-\frac{2\alpha\lambda}{1+\alpha}})$, then $\frac{2\alpha\lambda}{1+\alpha} \in (1-\frac{r}{y}, 1)$, by $\mu > 1 - f(y, \frac{2\alpha\lambda}{1+\alpha}, r)$, (14.27) and (14.23) we have

$$\begin{aligned} 1 - \tfrac{1+\alpha}{2\alpha}(1 - g(y,\mu,r)) &> 1 - \tfrac{1+\alpha}{2\alpha}(1 - g(y, 1 - f(y, \tfrac{2\alpha\lambda}{1+\alpha}, r), r)) \\ &= 1 - \tfrac{1+\alpha}{2\alpha}(1 - (1 - \tfrac{2\alpha\lambda}{1+\alpha})) \\ &= 1 - \lambda \\ &\geq 0, \end{aligned}$$

$$f_r(A)(0) \geq f_r(y_\mu)(0) = 1 - \tfrac{1+\alpha}{2\alpha}(1 - g(y,\mu,r)) > 1 - \lambda,$$

we still have $f_r(A) \not\leq (0_\lambda)'$. So $0_\lambda$ fulfils Lemma **14.3.16** (iii).

Suppose $x > 0$. If $\frac{1+\alpha}{2\alpha}(1 - g(x,\lambda,r)) < 1$, then

$$\mu = f_r(0_{(\frac{1}{2\alpha}(1+\alpha)(1-g(x,\lambda,r)))\wedge 1})(x) = 0 \leq 1 - \lambda,$$

or (by (14.22))

$$\mu = f(x, \tfrac{2\alpha\lambda}{1+\alpha}\tfrac{1}{2\alpha}(1+\alpha)(1 - g(x,\lambda,r)), r) = 1 - \lambda.$$

If $\frac{1+\alpha}{2\alpha}(1 - g(x,\lambda,r)) \geq 1$, then $\mu = 0 \leq 1 - \lambda$ or (by (14.22))

$$\mu = f(x, \tfrac{2\alpha\lambda}{1+\alpha}, r) \leq f(x, 1 - g(x,\lambda,r), r) = 1 - \lambda.$$

So we always have

$$f_r(f_r(x_\lambda)'(0)\chi_{\{0\}})(x) \leq 1 - \lambda.$$

If $y \in (0, \frac{r}{1-g(x,\lambda,r)})\backslash\{x\}$, we have $g(x,\lambda,r) \in (1 - \frac{r}{y}, 1]$, by (14.23) and (14.22) we have

$$\begin{aligned} f_r(f_r(x_\lambda)'(y)\chi_{\{y\}})(x) &= f(x, g(y, f_r(x_\lambda)'(y), r), r) \\ &= f(x, g(y, 1 - f(y, g(x, \lambda, r), r), r), r) \\ &= f(x, 1 - g(x, \lambda, r), r) \\ &= 1 - \lambda; \end{aligned}$$

if $y \in [\frac{r}{1-g(x,\lambda,r)}, +\infty)\backslash\{x\}$,

$$\mu = f_r(f_r(x_\lambda)'(y)\chi_{\{y\}})(x) = 0,$$

or (by $y \geq \frac{r}{1-g(x,\lambda,r)} > r$, (14.21) and (14.22))

$$\mu = f(x, g(y, 1, r), r) = f(x, \tfrac{r}{y}, r) \leq f(x, 1 - g(x, \lambda, r), r) = 1 - \lambda.$$

So for $x > 0$ we have

$$f_r(f_r(x_\lambda)') = \bigvee\{f_r(f_r(x_\lambda)'(y)\chi_{\{y\}}) : \ y \in X\} \subset (x_\lambda)'.$$

Suppose $A \not\leq f_r(x_\lambda)'$, then there exists $y \in [0, \frac{r}{1-g(x,\lambda,r)}) \cup \{x\}$ such that

$$\mu = A(y) > 1 - f_r(x_\lambda)(y).$$

If $y = x$,

$$f_r(A)(x) \geq f_r(y_\mu)(x) = \mu > 1 - \lambda, \quad f_r(A) \not\leq (x_\lambda)'.$$

If $y = 0$, then

$$1 \geq \mu > \tfrac{1+\alpha}{2\alpha}(1 - g(x, \lambda, r)) \wedge 1 = \tfrac{1+\alpha}{2\alpha}(1 - g(x, \lambda, r)),$$

by (14.26) and (14.22) we have

$$f_r(A)(x) \geq f_r(y_\mu)(x) = f(x, \tfrac{2\alpha\mu}{1+\alpha}, r) > f(x, 1 - g(x, \lambda, r), r) = 1 - \lambda,$$
$$f_r(A) \not\leq (x_\lambda)'.$$

If $y \in (0, \frac{r}{1-g(x,\lambda,r)})\backslash\{x\}$, then $\mu > 1 - f(y, g(x, \lambda, r), r)$, $g(x, \lambda, r) \in (1 - \frac{r}{y}, 1]$, by (14.27) and (14.23) we have

$$\frac{r}{1-g(y,\mu,r)} > \frac{r}{1-g(y,1-f(y,g(x,\lambda,r),r),r)} = \frac{r}{g(x,\lambda,r)} \geq \frac{r}{g(x,1,r)} = \frac{r}{1 \wedge \frac{r}{x}} \geq x,$$

$$\begin{aligned} f_r(A)(x) &\geq f_r(y_\mu)(x) \\ &= f(x, g(y, \mu, r), r) \\ &> f(x, g(y, 1 - f(y, g(x, \lambda, r), r), r), r) \\ &= f(x, 1 - g(x, \lambda, r), r) \\ &= 1 - \lambda, \end{aligned}$$
$$f_r(A) \not\leq (x_\lambda)'.$$

So $x_\lambda$ also fulfils Lemma **14.3.16** (iii) for every $x > 0$, hence (14.3) is fulfilled.

(5) Prove (14.18). This point can be verified by $\frac{2\alpha}{1+\alpha} \in (0, 1)$, $g(x, 1, r) = 1 \wedge \frac{r}{x}$ and the related definitions.

So $\mathcal{D}^* = \{f_r : \ r > 0\}$ is a standard metric base on $L^X$, $(I^X, \mathcal{D})$ is a fuzzy metric space for

$$\mathcal{D} = \{f \in \mathcal{E}(L^X) : \ \exists r > 0, \ f \geq f_r\}.$$

Now we prove (i) – (iii) respectively.

(i) Since $1 - \alpha \in (0, 1)$, so

$$\tfrac{1-\alpha^2}{2\alpha} = \tfrac{1+\alpha}{2\alpha}(1 - \alpha) > 1 - \alpha.$$

Let

$$\lambda = \tfrac{1-\alpha^2}{2\alpha} \wedge 1, \qquad U_0 = f_1(0_\lambda),$$

then

$$\tfrac{2\alpha\lambda}{1+\alpha} \le 1 - \alpha < \lambda,$$
$$U_0(y) > 1 - \alpha \iff y = 0. \tag{14.31}$$

Fix $\mu \in (0,1)$, then $r \to 0$ implies $g \to 0$, so for every $x > 0$, there exists $r_x > 0$ such that

$$g(x, 1 - \tfrac{\alpha}{2}, r_x) \le \tfrac{1}{3}(1 - \alpha). \tag{14.32}$$

Then let

$$U_x = f_{r_x}(x_{1-\frac{\alpha}{2}}), \qquad \mathcal{U} = \{U_x : x \in X\},$$

by (14.32), for every $x > 0$ we have

$$U_x(y) > 1 - \alpha \iff y = x, \tag{14.33}$$

so $\mathcal{U}$ is an $\alpha$-Q-cover of $(L^X, \delta(\mathcal{D}))$. Since by Lemma **14.3.2** (i) we have $f_r(A) \in \delta(\mathcal{D})$ for every $r > 0$ and every $A \in L^X$, so $\mathcal{U}$ is an open $\alpha$-Q-cover of $(L^X, \delta(\mathcal{D}))$.

Suppose $\mathcal{A}$ is an $\alpha$-Q-cover of $\underline{1}$ and a refinement of $\mathcal{U}$, then by (14.31) and (14.33) we know that for every $x \in X$ there exists $A_x \in \mathcal{A}$ such that

$$A_x \subset U_x, \qquad A_x(x) > 1 - \alpha. \tag{14.34}$$

If $x > 0$, it follows from (14.32)

$$\begin{aligned}
U_x(0) &= f_{r_x}(x_{1-\frac{\alpha}{2}})(0)\\
&= (1 - \tfrac{1+\alpha}{2\alpha}(1 - g(x, 1 - \tfrac{\alpha}{2}, r_x))) \vee 0\\
&\le (1 - (1 - g(x, 1 - \tfrac{\alpha}{2}, r_x))) \vee 0\\
&= g(x, 1 - \tfrac{\alpha}{2}, r_x)\\
&\le \tfrac{1}{3}(1 - \alpha)\\
&= f_{r_x}(0_{1-\frac{1}{3}(1-\alpha)})'(0),\\
U_x(x) &= 1 - \tfrac{\alpha}{2}\\
&= 1 - f(x, 1 - g(x, 1 - \tfrac{\alpha}{2}, r_x), r_x)\\
&\le 1 - f(x, 1 - \tfrac{1-\alpha}{3}, r_x)\\
&\le 1 - f(x, \tfrac{2\alpha}{1+\alpha}(1 - \tfrac{1-\alpha}{3}), r_x)\\
&= f_{r_x}(0_{1-\frac{1}{3}(1-\alpha)})'(x);
\end{aligned}$$

if $y \in X \backslash \{0, x\}$,

$$\begin{aligned}
U_x(y) &= f_{r_x}(x_{1-\frac{\alpha}{2}}(y)\\
&\le f(y, g(x, 1 - \tfrac{\alpha}{2}, r_x), r_x)\\
&\le g(x, 1 - \tfrac{\alpha}{2}, r_x)\\
&\le \tfrac{1-\alpha}{3}\\
&= f_{r_x}(0_{1-\frac{1}{3}(1-\alpha)})'(0)\\
&\le f_{r_x}(0_{1-\frac{1}{3}(1-\alpha)})'(y),
\end{aligned}$$

so for every $x > 0$, since every $f_r(A)$ is open in $(L^X, \delta(\mathcal{D}))$, we have

$$\begin{aligned} U_x &\leq f_{r_x}(0_{1-\frac{1}{3}(1-\alpha)})', \\ A_x^- \leq U_x^- &\leq f_{r_x}(0_{1-\frac{1}{3}(1-\alpha)})', \\ A_x^-(0) \leq f_{r_x}&(0_{1-\frac{1}{3}(1-\alpha)})'(0) = \tfrac{1}{3}(1-\alpha), \\ (\bigvee_{x>0} A_x^-)(0) &\leq \tfrac{1}{3}(1-\alpha). \end{aligned} \tag{14.35}$$

Let $\mu = \frac{1}{2}(1-\alpha)$, then $0_\mu \not\leq \bigvee_{x>0} A_x^-$ by (14.35). But for every $U \in \mathcal{Q}(0_\mu)$, by Lemma **14.3.2** (ii) and (14.28), there exists $\xi \in (1-\mu, 1]$ and $r > 0$ such that $f_r(0_\xi) \leq U$. Then by $\frac{2\alpha\xi}{1+\alpha} > \frac{2\alpha}{1+\alpha}(1-\mu) = \alpha$ there exists $\varepsilon > 0$ such that

$$\begin{aligned} (0,\varepsilon) &\subset \{x \in X: \ f(x, \tfrac{2\alpha\xi}{1+\alpha}, r) \geq \alpha\} \\ &\subset \{x \in X: \ f_r(0_\xi)(x) \geq \alpha\} \\ &\subset \{x \in X: \ U(x) \geq \alpha\}. \end{aligned}$$

So it follows from (14.34) that $U\hat{q}A_x$ for every $x \in (0,\varepsilon)$, so $U\hat{q}(\bigvee_{x>0} A_x)$. By the virtue of Theorem **2.3.23** (i), $0_\mu \leq (\bigvee_{x>0} A_x)^-$. Since $0_\mu \not\leq \bigvee_{x>0} A_x^-$, we have $(\bigvee_{x>0} A_x)^- \neq \bigvee_{x>0} A_x^-$, $\mathcal{A}$ is not closure preserving.

(ii) For the open $\alpha$-Q-cover $\mathcal{U}$ of $(L^X, \delta(\mathcal{D}))$ defined above, suppose $\mathcal{V}$ is an open $\alpha$-Q-cover of $(L^X, \delta(\mathcal{D}))$ and a refinement of $\mathcal{U}$, then for every $x > 0$, there exists $V_x \in \mathcal{V}$ such that $V_x \leq U_x$ and $V_x(x) > 1-\alpha$. By Lemma **14.3.2** (ii), for every $x > 0$, there exists $s_x > 0$ and $\lambda_x > 1-\alpha$ such that $f_{s_x}(x_{\lambda_x}) \leq V_x$. Denote

$$\beta_x = \frac{s_x}{1-g(x,\lambda_x,s_x)} > 0,$$

then by $V_x \geq f_{s_x}(x_{\lambda_x})$ and the definition of $f_{s_x}(x_{\lambda_x})$ we have

$$x > 0, \ y \in (0, \beta_x) \implies V_x(y) > 0. \tag{14.36}$$

Easy to know that there exists $\beta^* > 0$ such that

$$|\{x \in X: \ x > 0, \ \beta_x \geq \beta^*\}| \geq \aleph_1,$$

so for $C = \{x \in X: \ x > 0, \ \beta_x \geq \beta^*\}$, by (14.36) we can easily find that there exists $\xi \in (0,1)$ such that

$$|\{x \in C: \ V_x(\tfrac{1}{2}\beta^*) + \xi > 1\}| \geq \aleph_1,$$

$\{V_x : \ x \in C\} \subset \mathcal{V}$ is not point-countable at the molecule $(\frac{1}{2}\beta^*)_\xi$, so $\mathcal{V}$ is not point-countable at this molecule.

(iii) If $(I^X, \delta(\mathcal{D}))$ has a point-countable base, then every family of open subsets in $(I^X, \delta(\mathcal{D}))$ has a point-countable open refinement covering the same subset. But it follows from (ii) that this is impossible, so $(I^X, \delta(\mathcal{D}))$ has no point-countable base.

# Chapter 15

# Relations Between Fuzzy Topological Spaces and Locales

Since all the open subsets in an $L$-fts naturally forms a locale, it is a natural requirement to study the relation between $L$-fuzzy topological spaces and locales. S.E. Rodabaugh's paper [147] is undoubtedly the first systematic and important work on this respect. Applying the notion of "point" in locales to $L$-fuzzy topological spaces, M.W. Warner's research [172] investigated the membership relation between points and $L$-fuzzy subsets in $L$-fuzzy topology. In Luo's research [115] on locales, the relation between $L$-fts' and locales was also investigated. The Stone Representation Theorem for Boolean algebras and its generalized form – the Stone Representation Theorem for distributive lattices generated important influence in modern mathematics. Rodabaugh's work [147] and the work [197] of Zhang and Liu established the $L$-fuzzy topological representation of distributive lattices from diferent ways.

## 15.1 Related Results in Locales

**15.1.1 Definition**[44] Define the category **Loc** of locales as the category $\mathbf{Frm}^{op}$. Every object in **Loc** is called a *locale* and every morphism in **Loc** is called a *continuous map*.

**15.1.2 Definition**[44] Let $\mathcal{L} = \{L_t : t \in T\}$ be a family of locales. Denote the categorical product of $\mathcal{L}$ in **Loc** by $\prod_\ell \mathcal{L}$ or $\prod_{t\in T} {}_\ell L_t$. Especially, denote $\prod_\ell \mathcal{L}$ by $L_0 \times_\ell L_1$ for short if $\mathcal{L} = \{L_0, L_1\}$.

**15.1.3 Definition**[44] Let $L$ be a locale. A *point* in $L$ is a frame homomorphism $\delta: L \to 2$, where $2 = \{0, 1\}$ equipped with the natural partial order. Denote the set of all the points of $L$ by $pt(L)$.

**15.1.4 Definition**[44] Let $L$ be a locale, $L_0 \subset L$. A mapping $j: L \to L$ is called a *nucleus* on $L$, if

(i) $a \leq j(a)$.

(ii) $j(a \wedge b) = j(a) \wedge j(b)$.

(iii) $jj(a) \leq j(a)$.

Denote $j[L]$ by $L_j$. $L_0$ equipped with the relative partial order in $L$ is called a *sublocale* of $L$, if there exists a nucleus $j: L \to L$ such that $L_0 = L_j$.

A sublocale $L_j$ of $L$ is called *closed*, if there exists $a \in L$ such that $j(b) = a \vee b$ for every $b \in L$. Denote this nucleus $j$ by $c(a)$.

A sublocale $L_j$ of $L$ is called *open*, if there exists $a \in L$ such that $j(b) = (a \to b)$ (see Definition **9.1.9**) for every $b \in L$. Denote this nucleus $j$ by $u(a)$.

A sublocale $L_j$ of $L$ is called *dense*, if $0_L \in L_j$, i.e. $j(0_L) = 0_L$.

**15.1.5 Proposition**[44] *Let $L$ be a locale, $a \in L$. Then both $L_{c(a)}$ and $L_{u(a)}$ are sublocales of $L$.* □

**15.1.6 Proposition**[44] *Let $L$ be a locale, then*

$$\gamma \mapsto \chi_{L \backslash \downarrow \gamma} : pr(L) \to pt(L)$$

*is a bijection.* □

**15.1.7 Proposition**[44] *Let $L$ be a locale, $j : L \to L$ be a nucleus, then*

(i) *$L_j$ is a locale.*

(ii) *$j : L \to L_j$ is a frame homomorphism.* □

**15.1.8 Theorem** *Let $L$ be a locale, $j, k : L \to L$ be nuclei. Then $j \leq k$ if and only if $L_j \supset L_k$.*

**Proof** (Necessity) For every $a \in L$,

$$k(a) \leq jk(a) \leq kk(a) = k(a),$$
$$k(a) = jk(a) \in L_j,$$
$$L_k \subset L_j.$$

(Sufficiency) Suppose $L_k \supset L_j$, then for every $a \in L$, $k(a) \in L_j$, $k(a) = jk(a)$. So

$$a \wedge j(a) \leq k(a) \wedge j(a) = jk(a) \wedge j(a) = j(k(a) \wedge a),$$
$$j(a) = j(a) \wedge jj(a) = j(a \wedge j(a)) \leq jj(k(a) \wedge a) = jk(a) \wedge j(a) = k(a) \wedge j(a),$$
$$j(a) \leq k(a).$$

Hence $j \leq k$. □

**15.1.9 Definition** Let $L$ be a locale.

$\gamma \in pr(L)$ is called a *sub-prime element* of $L$, if

$$\bigwedge\{\sigma \in pr(L) : \sigma > \gamma\} > \gamma.$$

Denote the set of all the sub-prime elements of $L$ by $supr(L)$.

$\gamma \in pr(L)$ is called a *quasi-prime element* of $L$, if $\gamma$ is a maximal element in $pr(L)$. Denote the set of all the quasi-prime elements of $L$ by $qupr(L)$.

$\gamma \in L \backslash \{1\}$ is called a *really prime element*, if $\gamma$ is a maximal element in $L \backslash \{1\}$. Denote the set of all the really prime elements of $L$ by $repr(L)$. $L$ is called *point-real*, if $pr(L) \subset repr(L)$.

**15.1.10 Proposition** *Let $L$ be a locale, $\gamma \in L$. Then $\gamma$ is really prime if and only if $a \vee \gamma = 1$ for every $a \not\leq \gamma$.* □

**15.1.11 Proposition** *Let $L$ be a locale, $\gamma \in repr(L)$. Then there exists at most one non-zero element $a$ in $L$ such that $a \wedge \gamma = 0$.* □

**15.1.12 Proposition** *Let $L$ be a locale, then*

$$repr(L) \subset qupr(L) \subset supr(L) \subset pr(L).$$ □

**15.1.13 Theorem** *Let $L$ be a locale, $L_0$ be a sublocale of $L$. Then*

(i) $pr(L_0) = pr(L) \cap L_0$.

(ii) $supr(L_0) \supset supr(L) \cap L_0$.

(iii) $qupr(L_0) \supset qupr(L) \cap L_0$.

(iv) $repr(L_0) \supset repr(L) \cap L_0$.

**Proof** Suppose $j$ is a nucleus on $L$ such that $L_0 = L_j$, denote respectively the join operation and meet operation in $L_j$ by $sup_{L_j}$ and $inf_{L_j}$.

(i) Let $\gamma \in pr(L_j)$. If $a, b \in L$ such that $a \wedge b \leq \gamma$, then

$$inf_{L_j}\{j(a), j(b)\} = j(a) \wedge j(b) = j(a \wedge b) \leq j(\gamma) = \gamma.$$

Since $\gamma \in pr(L_j)$, $\gamma \geq j(a)$ or $\gamma \geq j(b)$. But $j(a) \geq a$, $j(b) \geq b$, so $\gamma \geq a$ or $\gamma \geq b$. Since $\gamma < 1_{L_j} = 1_L$, $\gamma \in pr(L)$.

Let $\gamma \in pr(L) \cap L_j$. If $a, b \in L_j$ such that $inf_{L_j}\{a, b\} \leq \gamma$, then

$$a \wedge b \leq j(a \wedge b) = j(a) \wedge j(b) = inf_{L_j}\{j(a), j(b)\} = inf_{L_j}\{a, b\} \leq \gamma.$$

Since $\gamma \in pr(L)$, $\gamma \geq a$ or $\gamma \geq b$. By $\gamma < 1_L = 1_{L_j}$, $\gamma \in pr(L_j)$.

(ii) Let $\gamma \in supr(L) \cap L_j$. Since $j: L \to L_j$ is order preserving, by (i) we have

$$\begin{aligned}\gamma &< \bigwedge\{\sigma \in pr(L) : \sigma > \gamma\} \\ &\leq j(\bigwedge\{\sigma \in pr(L) : \sigma > \gamma\}) \\ &\leq inf_{L_j}\{j(\sigma) : \sigma \in pr(L), \sigma > \gamma\} \\ &\leq inf_{L_j}\{j(\sigma) : \sigma \in pr(L) \cap L_j, \sigma > \gamma\} \\ &= inf_{L_j}\{j(\sigma) : \sigma \in pr(L_j), \sigma > \gamma\} \\ &= inf_{L_j}\{\sigma \in pr(L_j) : \sigma > \gamma\}.\end{aligned}$$

By $\gamma \in supr(L) \cap L_j \subset pr(L) \cap L_j = pr(L_j)$, we have $\gamma \in supr(L_j)$.

(iii) Let $\gamma \in qupr(L) \cap L_j \subset pr(L) \cap L_j = pr(L_j)$, then $\gamma$ is a maximal element in $pr(L)$, and hence a maximal element in $pr(L_j) \subset pr(L)$, $\gamma \in qupr(L_j)$.

(iv) Suppose $\gamma \in repr(L) \cap L_j$, then $\gamma$ is a maximal element in $L \backslash \{1_L\}$ and an element of $L_j$. Since $1_{L_j} = 1_L$, so $\gamma$ is also a maximal element in $L_j \backslash \{1_{L_j}\}$, $\gamma \in repr(L_j)$. □

**15.1.14 Proposition** *Every sublocale of a point-real locale is point-real.* □

**15.1.15 Definition** For a topological space $(X, \mathcal{T})$, denote its topology $\mathcal{T}$ by $\Omega(X)$, and equip $\Omega(X)$ with the inclusion order, call $\Omega(X)$ the *open subset lattice* of $X$.

**15.1.16 Definition** Let $L$ be a locale. For every $a \in L$, define

$$\mu(a) = \{\gamma \in pr(L) : a \not\leq \gamma\}.$$

Define a topology $\Omega(pr(L))$ on $pr(L)$ as follows:

$$\Omega(pr(L)) = \{\mu(a) : a \in L\}.$$

Call $\mu : L \to \Omega(pr(L))$ the *canonical homomorphism* on L, call $\Omega(pr(L))$ the *canonical topology* of $L$. $L$ is called *spatial*, if the canonical homomorphism $\mu : L \to \Omega(pr(L))$ is an isomorphism.

**15.1.17 Proposition**[44] *Let $L$ be a locale. Then*

(i) *$\Omega(pr(L))$ is a topology on $pr(L)$.*

(ii) *$\mu : L \to \Omega(pr(L))$ is a frame homomorphism.* □

**15.1.18 Proposition**[44] *Let $L$ be a locale, then the following conditions are equivalent:*

(i) *$L$ is spatial.*

(ii) *$L$ is frame isomorphic to the open subset lattice of a topological space.*

(iii) *$\forall a, b \in L$, $a \not\leq b \Longrightarrow \exists \gamma \in pr(L)$ such that $a \not\leq \gamma$, $b \leq \gamma$.* □

**15.1.19 Definition**[44] Let $L$ be a locale. $a \in L$ is called *compact*, if for every $A \subset L$ such that $\bigvee A \geq a$ there exists a finite subset $A_0 \subset A$ such that $\bigvee A_0 \geq a$.

Denote the set of all the compact elements of $L$ by $K(L)$. $L$ is called *coherent*, if $a = \bigvee((\downarrow a) \cap K(L))$ for every $a \in L$ and $K(L)$ is a sublattice of $L$. $L$ is called *compact*, if $1_L$ is compact.

**15.1.20 Proposition**[44] *Let $L$ be a distributive lattice, then $Idl(L)$ equipped with the inclusion order is a locale.* □

**15.1.21 Proposition**[44] *A locale $L$ is coherent if and only if there exists a distributive $D$ such that $L$ is isomorphic to $Idl(D)$.* □

**15.1.22 Proposition**[44] *Let $L$ be a locale, then $pr(Idl(L)) = PrIdl(L)$.* □

**15.1.23 Proposition**[44] *Every coherent locale is spatial.* □

**15.1.24 Definition**[44] Let $L$ be a distributive lattice. Call $pr(Idl(L))$ equipped with the canonical topology the *spectrum* (sometimes *prime spectrum*) of $L$, denoted by $spec(L)$.

**15.1.25 Theorem**[44] *Every locale $L$ is isomorphic to a sublocale of $spec(L)$.* □

**15.1.26 Definition** A topological space $X$ is called $T_D$, if for every $x \in X$, $\{x\}$ is an open subset of $\{x\}^-$; is called *sober*, if for every $P \in pr(\Omega(X))$, there exists a unique $x \in X$ such that $P = X \backslash \{x\}^-$; is called *strongly sober*, if $X$ is $T_D$ and sober; is called $T_{1\frac{1}{2}}$, if for every $P \in pr(\Omega(X))$, there exists a unique $x \in X$ such that $P = X \backslash \{x\}$.

**15.1.27 Theorem** *In category* **Top**, *the following implying relations hold:*

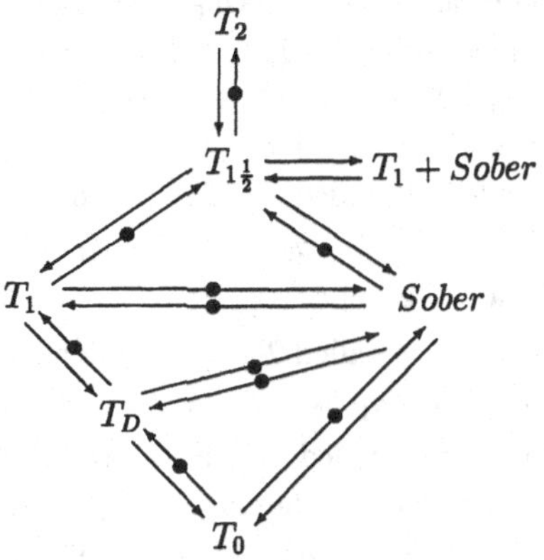


*where "⟶" and "–•→" mean "implies" and "does not imply" respectively.* □

**15.1.28 Theorem** *In category* **Top**,

(i) $T_{1\frac{1}{2}} \Longrightarrow$ *Strongly sober* $\Longrightarrow$ *sober*,

(ii) *Strongly sober* $\Longrightarrow T_D$,

*and every above implication can not be reversed.* □

**15.1.29 Definition** A locale $L$ is called *strongly sober*, if $pr(L) \subset supr(L)$.

**15.1.30 Definition**[45] Let $\mathcal{A} = \{A_t : t \in T\}$ be a family of posets, $S \subset \prod_{t \in T} A_t$. Denote

$$S^+ = \{a \in \prod \mathcal{A} : \forall x \in S, \exists t \in T, p_t(x) \leq p_t(a)\},$$
$$S^- = \{a \in \prod \mathcal{A} : \forall x \in S, \exists t \in T, p_t(x) \geq p_t(a)\}.$$

**15.1.31 Definition**[45] Let $\mathcal{L} = \{L_t : t \in T\}$ be a family of locales. Define

$$\Psi:\ \prod_\ell \mathcal{L} \to \prod_\ell \mathcal{L}, \quad \Psi(I) = I^{+-} \cap [\mathcal{L}].$$

**15.1.32 Definition**[45] Let $\mathcal{L} = \{L_t:\ t \in T\}$ be a family of locales. Define the *weak product* of $\mathcal{L}$ as follows:

$$\bigotimes \mathcal{L} = \bigotimes_{t\in T} L_t = \{I \in \textstyle\prod_\ell \mathcal{L}:\ \Psi(I) = I\}.$$

**15.1.33 Definition**[45] A locale $L$ is called *Hausdorff*, if the following implication holds:

$$I \in L \otimes L,\ I \supset \Delta_*(0) \implies \exists a \in L,\ I = \Delta_*(a).$$

**15.1.34 Definition**[44] Let $L$ be a locale. Define

$$\Delta^*:\ L \times_\ell L \to L, \quad \Delta^*(I) = \vee\{a \in I:\ (a,a) \in I\},$$
$$\Delta_*:\ L \to L \times_\ell L, \quad \Delta_*(a) = \{(b,c) \in L \times L:\ b \wedge c \le a\}.$$

**15.1.35 Definition**[44] A locale $L$ is called *strongly Hausdorff*, if for every $I \in L \times_\ell L$ such that $I \supset \Delta_*(0)$, $I = \Delta_*\, \Delta^*\,(I)$.

**15.1.36 Definition**[44] Let $L$ be a locale, $a, b \in L$. Say *a is well inside b*, denoted by $a \lessdot b$, if there exists $c \in L$ such that $c \wedge a = 0$, $c \vee b = 1$. Denote

$$\Downarrow a = \{b \in L:\ b \lessdot a\}, \quad \Uparrow a = \{b \in L:\ a \lessdot b\}.$$

A locale $L$ is called *regular*, if $a = \bigvee \Downarrow a$ for every $a \in L$.

A locale $L$ is called *normal*, if the following implication holds:

$$a, b \in L,\ a \vee b = 1 \implies \exists c, d \in L,\ c \wedge d = 0,\ c \vee a = 1,\ d \vee b = 1.$$

**15.1.37 Definition** A locale $L$ is called $T_1$, if $supr(L) \subset qupr(L)$.

**15.1.38 Definition** A locale $L$ is called $T_{1\frac{1}{2}}$, if $pr(L) \subset qupr(L)$.

**15.1.39 Definition** A locale $L$ is called $T_2$, if the following implication holds:

$$\gamma_1, \gamma_2 \in pr(L),\ \gamma_1 \ne \gamma_2 \implies \exists a_1, a_2 \in L,\ a_1 \not\le \gamma_1,\ a_2 \not\le \gamma_2,\ a_1 \wedge a_2 = 0.$$

**15.1.40 Definition** A locale $L$ is called $T_{2\frac{1}{2}}$, if $L$ is $T_2$ and point-real.

**15.1.41 Definition** A locale $L$ is called $T_3$, if the following implication holds:

$$a \in L,\ \gamma \in pr(L),\ a \not\le \gamma \implies \exists a_1, a_2 \in L,\ a_1 \vee \gamma = 1,\ a_2 \vee a = 1,\ a_1 \wedge a_2 = 0.$$

**15.1.42 Definition** A locale $L$ is called $T_{3\frac{1}{2}}$, if for every $\gamma \in pr(L)$ and every $a \in L$ such that $a \not\le \gamma$, there exists a frame homomorphism $f:\ \Omega([0,1]) \to L$ such that for every neighborhood $U$ of 0 and every neighborhood $V$ of 1 in $[0,1]$,

$$\gamma \vee f(U) = 1_L, \quad a \vee f(V) = 1_L.$$

**15.1.43 Definition** A locale $L$ is called $T_4$, if $L$ is normal and point-real.

**15.1.44 Definition** Let $P$ be a condition for topological spaces, $\tilde{P}$ be a condition for locales. $\tilde{P}$ is called a *good extension* of $P$, if $\tilde{P}$ fulfils one of the following two conditions:

(i) $P \Longrightarrow T_0$, $\forall X \in Ob(\mathbf{Top})$,

$$X \text{ fulfils } P \Longleftrightarrow X \text{ is } T_0 \text{ and } \Omega(X) \text{ fulfils } \tilde{P}.$$

(ii) $\forall X \in Ob(\mathbf{Top})$,

$$X \text{ fulfils } P \Longleftrightarrow \Omega(X) \text{ fulfils } \tilde{P}. \qquad \square$$

The separations in locales defined above have many nice properties, such as the diagonal characterization, the net characterization and the filter characterization of $T_2$ property, and so on. Some of these properties will be stated in the sequel. But since their proofs do not relate our topics in this book directly, we omit all of them.

**15.1.45 Theorem** *Let $X$ be a topological space, then*

$$X \text{ is } T_1 \Longleftrightarrow X \text{ is } T_D \text{ and } \Omega(X) \text{ is } T_1. \qquad \square$$

**15.1.46 Theorem**

(i) *$T_{1\frac{1}{2}}$ property in locales is a good extension of $T_{1\frac{1}{2}}$ property in topological spaces.*

(ii) *$T_2$ property in locales is a good extension of $T_2$ property in topological spaces.*

(iii) *$T_{2\frac{1}{2}}$ property in locales is a good extension of $T_2$ property in topological spaces.*

(iv) *$T_3$ property in locales is a good extension of $T_3$ property in topological spaces.*

(v) *$T_{3\frac{1}{2}}$ property in locales is a good extension of $T_{3\frac{1}{2}}$ property in topological spaces.*

(vi) *$T_4$ property in locales is a good extension of $T_4$ property in topological spaces.* □

**15.1.47 Theorem**

(i) *In* **Loc**, *there does not exist a good extension of $T_D$ property in* **Top**.

(ii) *In* **Loc**, *there does not exist a good extension of $T_1$ property in* **Top**.

(iii) *In* **Loc**, *there does not exist a good extension of soberiety in* **Top**. □

**15.1.48 Definition** A property $P$ for locales is called *multiplicative*, if for every family $\mathcal{L}$ of locales,

$$\textit{Every element of } \mathcal{L} \textit{ possesses } P \Longrightarrow \textstyle\prod_{\ell} \mathcal{L} \textit{ possesses } P;$$

called *strongly point-multiplicative*, if for every family $\mathcal{L}$ of locales,

$\prod_{\ell} \mathcal{L}$ possesses $P \Longleftrightarrow$ *There exists $L \in \mathcal{L}$, $pr(L) = \emptyset$*; or *every $L \in \mathcal{L}$ possesses $P$.*

**15.1.49 Theorem**

(i) *$T_1$ property in locales is multiplicative.*

(ii) *$T_{1\frac{1}{2}}$ property, $T_2$ property, $T_{2\frac{1}{2}}$ property, $T_3$ property and $T_{3\frac{1}{2}}$ property are strongly point-multiplicative.* □

**15.1.50 Theorem**

(i) *A sublocale $L_0$ of a $T_1$ locale $L$ satisfying $supr(L_0) = supr(L) \cap L_0$ is $T_1$.*

(ii) *$T_{1\frac{1}{2}}$ property, $T_2$ property, $T_{2\frac{1}{2}}$ property, $T_3$ property and $T_{3\frac{1}{2}}$ property are hereditary.*

(iii) *$T_4$ property is hereditary with respect to closed sublocales.* □

**15.1.51 Theorem** *In* **Loc**, *denote the relation "strictly weaker than" by $<$, then*

(i) $T_1 < T_{1\frac{1}{2}} < T_2 < T_{2\frac{1}{2}} < T_3 < T_{3\frac{1}{2}} < T_4$.

(ii) *$T_2$ <Hausdorff<Strongly Hausdorff<Regular.*

(iii) *$T_3$ <Regular.*

(iv) *Strongly sober< $T_{1\frac{1}{2}}$ <Point-real < $T_{2\frac{1}{2}}$.* □

**15.1.52 Theorem** *In* **Loc**, *denote the relation "strictly weaker than" by $<$, then*

(i) *$T_2$+Compact<Regular+Compact.*

(ii) *Normal< $T_4$.*

(iii) $T_2$+*Compact* $\Longrightarrow T_4$. □

**15.1.53 Theorem** *Every $T_{1\frac{1}{2}}$ and compact locale is point-real.* □.

## 15.2 Separations in Fuzzy Topological Spaces and Locales

Clearly, every F-lattice and hence every $L$-fuzzy topology equipped with inclusion order is a locale. So in the sequel, we always consider an $L$-fuzzy topology as a locale as well.

In this section, we consider the relations between separations in $L$-fts' and in locales. Since their proofs are not hard, we leave them as exercises.

**15.2.1 Definition** An $L$-fts $(L^X, \delta)$ is called $T_D$, if $(L^X, \delta)$ is $T_0$ and

$$supr(\delta) = \{e^{-\prime} : e \in M(L^X)\};$$

is called *sober*, if for every $U \in pr(\delta)$, there exists a unique $e \in M(L^X)$ such that $U = e^{-\prime}$; is called $T_{1\frac{1}{2}}$, if $(L^X, \delta)$ is $T_1$ and sober.

**15.2.2 Definition** An $L$-fts $(L^X, \delta)$ is called *strongly* $T_1$, or called $s$-$T_1$ for short, if for every two distinguished molecules $e$ and $d$ in $(L^X, \delta)$, there exists $U \in \mathcal{Q}(e)$ such that $d \not\leq U$ and there exists $V \in \mathcal{Q}(d)$ such that $e \not\leq V$.

**15.2.3 Theorem** *Let $(L^X, \delta)$ be an L-fts, then the following conditions are equivalent:*

(i) $(L^X, \delta)$ *is s-$T_1$.*

(ii) $(L^X, \delta)$ *is $T_1$ and $L$ is $T_{1\frac{1}{2}}$ as a locale.*

(iii) $(L^X, \delta)$ *is $T_D$ and $\delta$ is $T_1$ as a locale.* □

**15.2.4 Theorem** *Let $(L^X, \delta)$ be an L-fts, then the following conditions are equivalent:*

(i) $(L^X, \delta)$ *is $T_{1\frac{1}{2}}$ and $L$ is $T_{1\frac{1}{2}}$ as a locale.*

(ii) $(L^X, \delta)$ *is $T_0$ and $\delta$ is $T_{1\frac{1}{2}}$ as a locale.* □

**15.2.5 Theorem** *Let $(L^X, \delta)$ be an L-fts. Then $\delta$ is $T_{1\frac{1}{2}}$ as a locale if and only if $\delta$ is point-real as a locale.* □

**15.2.6 Definition** An $L$-fts $(L^X, \delta)$ is called *primarily purely* $T_2$, or called $pp$-$T_2$ for short, if the following implication holds:

$$e, d \in M(L^X),\ e \perp d \Longrightarrow \exists U \in \mathcal{Q}(e),\ \exists V \in \mathcal{Q}(d),\ U \wedge V = \underline{0};$$

is called *strongly purely* $T_2$, or called $sp$-$T_2$ for short, if the following implication holds:

$$e, d \in M(L^X),\ e \neq d \Longrightarrow \exists U \in \mathcal{Q}(e),\ \exists V \in \mathcal{Q}(d),\ U \wedge V = \underline{0}.$$

**15.2.7 Proposition** *In $L$-**FTS**, denote the relation "strictly weaker than" by $<$, then*

(i) $T_2 < p$-$T_2 < pp$-$T_2 < sp$-$T_2$.

(ii) $T_2 < s$-$T_2$. □

**15.2.8 Theorem** *Let $(L^X, \delta)$ be a sp-$T_2$ L-fts. Then*

(i) $(L^X, \delta)$ *is $T_{1\frac{1}{2}}$.*

(ii) $(L^X, \delta)$ *is s-$T_1$.*

(iii) *$L$ is $T_2$ as a locale.* □

**15.2.9 Theorem** *An L-fts $(L^X, \delta)$ is sp-$T_2$ if and only if $(L^X, \delta)$ is $T_0$ and $\delta$ is $T_2$ as a locale.* □

**15.2.10 Theorem** *Let $(L^X, \delta)$ be an L-fts. Then $\delta$ is $T_{2\frac{1}{2}}$ as a locale if and only if $\delta$ is $T_2$ as a locale.* □

**15.2.11 Definition** An $L$-fts $(L^X, \delta)$ is called *typically regular*, or called *t-regular* for short, if the following implications holds:

$$e \in M(L^X),\ A \in \delta',\ e \not\leq A \implies \exists U, V \in \delta,\ U \vee e' = \underline{1},\ V \vee A' = \underline{1}, U \wedge V = \underline{0}.$$

Call a $T_1$ and t-regular $L$-fts *t-$T_3$*.

**15.2.12 Definition** An $L$-fts $(L^X, \delta)$ is called *typically normal*, or called *t-normal* for short, if the following implications holds:

$$A, B \in \delta',\ A \wedge B = \underline{0} \implies \exists U, V \in \delta,\ U \vee A' = \underline{1},\ V \vee B' = \underline{1}, U \wedge V = \underline{0}.$$

Call a $T_1$ and t-normal $L$-fts *t-$T_4$*.

**15.2.13 Theorem** *An L-fts $(L^X, \delta)$ is t-$T_3$ if and only if $(L^X, \delta)$ is $T_0$ and $\delta$ is $T_3$ as a locale.* □

**15.2.14 Theorem** *Let $(L^X, \delta)$ be an L-fts. Then*

(i) *$(L^X, \delta)$ is t-normal if and only if $\delta$ is t-normal as a locale.*

(ii) *$(L^X, \delta)$ is t-$T_4$ if and only if $(L^X, \delta)$ is $T_0$ and $\delta$ is t-$T_4$ as a locale.* □

## 15.3 Relations Between Fuzzy Topological Spaces and Locales

For every locale $L$ and a nonempty set $X$, $L$ is always isomorphic to the open subset lattice consisting of all the layers of $X$. But we are not interested by this special kind of situation. We shall investigate the possibility of the existence of a frame isomorphism between a locale $M$ and an $L$-fuzzy topology for a complete lattice $L$. So in sequel, unless the special declaration, $L$ will always mean an infinitely distributive lattice, i.e. a complete lattice fulfilling the both the 1st and the 2nd infinitely distributive laws (IFD1), (IFD2). Under this assumption, the relative notions such as $L$-fuzzy topology and so on can be naturally copied in this section. So we assume that this procedure has been completed in the sequel.

**15.3.1 Definition** Let $(L^X, \delta)$ be an $L$-fts, $x \in X$, $\gamma \in pr(L)$. Define a *co-molecule* $\breve{x}_\gamma \in L^X$ as follows:

$$\breve{x}_\gamma(y) = \begin{cases} \gamma, & y = x, \\ 1, & y \in X\backslash\{x\}. \end{cases}$$

Denote the set of all the co-molecules in $L^X$ by $pr(L^X)$.

Easy to find $pr(L^X)$ is precisely the set of all the prime elements of $L^X$, so this symbol is reasonable.

**15.3.2 Proposition** *Let $(L^X, \delta)$ be an L-fts, $\breve{x}_\gamma \in pr(L^X)$. Then $(\breve{x}_\gamma)^\circ \in pr(\delta)$, and hence*

$$X \neq \emptyset,\ pr(L) \neq \emptyset \implies pr(\delta) \neq \emptyset.$$

**Proof** Suppose $U, V \in \delta$, note that $\gamma \in pr(L)$, we have

$$\begin{aligned}(\check{x}_\gamma)^\circ \geq U \wedge V &\Longrightarrow \check{x}_\gamma \geq U \wedge V \\ &\Longrightarrow \gamma \geq U(x) \wedge V(x) \\ &\Longrightarrow \gamma \geq U(x) \text{ or } \gamma \geq V(x) \\ &\Longrightarrow \check{x}_\gamma \geq U \text{ or } \check{x}_\gamma \geq V \\ &\Longrightarrow (\check{x}_\gamma)^\circ \geq U \text{ or } (\check{x}_\gamma)^\circ \geq V.\end{aligned}$$

Since clearly we have $(\check{x}_\gamma)^\circ < \underline{1}$, so $(\check{x}_\gamma)^\circ \in pr(\delta)$. □

**15.3.3 Definition** A complete lattice $L$ is called *prime element generating*, if every element of $L$ is a meet of elements in $pr(L)$.

**15.3.4 Theorem** *Let $(L^X, \delta)$ be an L-fts, L be prime element generating, then $\delta$ is a spatial locale.*

**Proof** Suppose $U, V \in \delta$, $U \not\leq V$, then there exists $x \in X$ such that $U(x) \not\leq V(x)$. Since $L$ is prime element generating, there exists $\gamma \in pr(L)$ such that $U(x) \not\leq \gamma$ and $V(x) \leq \gamma$. Then by Proposition **15.3.2**, $(\check{x}_\gamma)^\circ \in pr(\delta)$ and $U \not\leq (\check{x}_\gamma)^\circ$, $V \leq (\check{x}_\gamma)^\circ$. By Proposition **15.1.18**, $\delta$ is spatial. □

**15.3.5 Corollary** *Let L be prime element generating. Then every locale which is not spatial can not be frame isomorphic to an L-fuzzy topology.* □

To consider the possibility of the existence of a frame isomorphism between a locale $M$ and an $L$-fuzzy topology, the following definition is natural:

**15.3.6 Definition**[147] A locale $M$ is called *L-spatial*, if there exists an $L$-fts $(L^X, \delta)$ such that $M$ is frame isomorphic to $\delta$.

Clearly, a locale $M$ is spatial if and only if $M$ is 2-spatial.

**15.3.7 Definition** Let $L$, $M$ be complete lattices. Denote the family of all the mappings from $M$ to $L$ preserving arbitrary joins and finite meets by $Fr(M, L)$.

Then we obtain a characterization of $L$-spatial locales as follows, where the condition (iii) belongs to Rodabaugh:[147]

**15.3.8 Theorem** *Let M be a locale, then the following conditions are equivalent:*

(i) *M is L-spatial.*

(ii) *M can be embedded into a product of copies of L via a frame isomorphism.*

(iii) *For arbitrary $a, b \in M$ such that $a \not\leq b$, there exists $f \in Fr(M, L)$ such that $f(a) \not\leq f(b)$.*

**Proof** (i)$\Longrightarrow$(ii): Obvious.

(ii)$\Longrightarrow$(iii): Suppose $\tilde{f}: M \to L^X$ is a frame isomorphism, $a, b \in M$, $a \not\leq b$, then $\tilde{f}(a) \not\leq \tilde{f}(b)$, there exists $x \in X$ such that $\tilde{f}(a)(x) \not\leq \tilde{f}(b)(x)$. Suppose $p_x : L^X \to L$ is the coordinate projection such that $p_x(A) = A(x)$, the $p_x \in Fr(L^X, L)$. Let $f = p_x \circ \tilde{f} : M \to L$, then $f \in Fr(M, L)$ and

$$f(a) = \tilde{f}(a)(x) \not\leq \tilde{f}(b)(x) = f(b).$$

(iii)$\Longrightarrow$(i): Let $X = Fr(M, L)$, then by the virtue of (iii) we know that $M$ is a singleton if $X = \emptyset$, $M$ is clearly frame isomorphic to the unique $L$-fuzzy topology $\delta = \{\emptyset\}$ on $X$. Suppose $X \neq \emptyset$, let

$$\tilde{f}: M \to L^X, \quad \tilde{f}(a)(f) = f(a), \quad a \in M, \ f \in X,$$

then $\tilde{f} \in Fr(M, L^X)$. If $a, b \in M$ such that $a \not\leq b$, then there exists $f \in X$ such that

$$\tilde{f}(a)(f) = f(a) \not\leq f(b) = \tilde{f}(b)(f).$$

So $\tilde{f}(a) \not\leq \tilde{f}(b)$, $\tilde{f}$ is injective. Then $\delta = \tilde{f}[M] \subset L^X$ is an $L$-fuzzy topology on $X$, $\tilde{f}|^{\delta} : M \to \delta$ is a frame isomorphism. □

Theorem **15.3.8** offers some conditions to determine whether or not a locale is $L$-spatial. But in general the family $Fr(M, L)$ of mappings where is hard to determine, even to judge whether it is empty or not. So we offer some other conditions to do this thing from a different way – the way based on subsets but not mappings:

**15.3.9 Proposition**[44] *A subset $S$ of a locale $L$ is a sublocale of $L$ if and only if $S$ fulfils the following conditions:*

(i) *For every $\mathcal{A} \subset A \subset S$, the meet $\bigwedge A$ of $A$ in $L$ belongs to $S$.*

(ii) *$(a \to_L s) \in S$ for every $a \in L$ and every $s \in S$.* □

**15.3.10 Proposition**[44] *Let $L$ be a locale, then*

(i) *$\neg\neg : L \to L$ is a nucleus.*

(ii) *$L_{\neg\neg}$ is the smallest dense sublocale of $L$.* □

**15.3.11 Definition**[42] Let $L$ be a locale. Denote the family of all the nuclei on $L$ by $N(L)$, equipped with a partial order $\leq$ defined as follows:

$$j \leq k \iff \forall a \in L,\ j(a) \leq j(b).$$

**15.3.12 Theorem**[42] *For every locale $L$, $N(L)$ is a complete Heyting algebra.* □

Correspondent to Definition **15.3.11**, we have

**15.3.13 Definition** Let $L$ be a locale. Denote the family of all the sublocales of $L$ by $N_*(L)$, equipped with the inclusion order:

$$L_j \leq L_k \iff L_j \subset L_k.$$

By Theorem **15.1.8** and Theorem **15.3.12** we have

**15.3.14 Theorem** *Let $L$ be a locale. Then*

$$f : N(L) \to N_*(L), \quad f(j) = L_j$$

*is an order-reversing isomorphism, and hence $N_*(L)$ is a complete lattice, $N_*(L)^{op}$ is a complete Heyting algebra.* □

In the proof of Proposition 2.5 in [44], it has been proved in fact that the following conclusions hold:

**15.3.15 Proposition**[44] *Let $L$ be a locale. Then*

(i) $\forall \mathcal{A} \subset N(L)$, $\forall a \in L$, $\quad (\bigwedge \mathcal{A})(a) = \bigwedge\{j(a) : j \in \mathcal{A}\}$.

(ii) $\forall \mathcal{B} \subset N_*(L)$, $\quad \bigwedge \mathcal{B} = \bigcap \mathcal{B}$.

**Proof** (i) has been proved in [44], (ii) can be directly verified by Proposition **15.3.9**. □

Proposition **15.3.9** provides a method to determine whether or not a subset of a locale is a sublocale, but we have not known its correspondent nucleus yet. The following results solves this problem:

**15.3.16 Definition** Let $L$ be a locale, $S \subset L$. Define a mapping $m(S) : L \to L$ as follows:

$$m(S)(a) = \bigwedge(S \cap \downarrow a),$$

where $\bigwedge$ is the meet operation in $L$.

**15.3.17 Proposition** *Let $L$ be a locale, $S \subset L$. Then*

(i) $m(S): L \to L$ *is order preserving.*

(ii) $m(S)(s) = s$.

(iii) $m(S)(a) \geq a$.

(iv) $m(S) \circ m(S)(a) = m(S)(a)$.

**Proof** (i), (ii) and (iii) are clear. Prove (iv): $m(S)(a)$ is certainly a lower bound of $S \cap \uparrow m(S)(a)$ in $L$. If $x \in L$ is a lower bound of $S \cap \uparrow m(S)(a)$, then for every $y \in S \cap \uparrow a$, we have $y \geq m(S)(a)$, so $y \in S \cap \uparrow m(S)(a)$. Then $x \leq y$, $x$ is also a lower bound of $S \cap \uparrow a$ in $L$. So $x \leq m(S)(a)$,

$$m(S) \circ m(S)(a) = \bigwedge(S \cap \uparrow m(S)(a)) = m(S)(a). \qquad \square$$

Then we not only obtain a method to determine sublocales but also obtain the correspondent nucleus:

**15.3.18 Theorem** *Let $L$ be a locale, $S \subset L$. Then*

(i) *$S$ is a sublocale of $L$ if and only if $m(S)[L] \subset S$ and $m(S)$ preserves nonempty finite meets.*

(ii) *If $S$ is a sublocale of $L$, then $m(S)$ is a nucleus and $S = L_{m(S)}$.*

**Proof** (i) (Necessity) Suppose $S$ is a sublocale of $L$, the correspondent nucleus is $j: L \to L$. For every $a \in L$, by Proposition **15.3.9** (i) we have

$$a \leq m(S)(a) \in S = L_j,$$

so

$$a \leq j(a) \leq j(m(S)(a)) = m(S)(a) \in S = L_j.$$

By the definition of $m(S)$ and $m(S)(a) \in S$, $j(a) = m(S)(a)$. Hence $m(S) = j: L \to L$, the conclusion is proved.

(Sufficiency) Suppose $m(S)[L] \subset S$ and $m(S)$ preserves nonempty finite meets, then by Proposition **15.3.17** (ii) we have $S = m(S)[L]$. So $S$ is a sublocale of $L$.

(ii) In the part of necessity of (i), it has been proved that $m(S)$ is the nucleus correspondent to $S$, so the conclusion holds. $\square$

With the mapping $m(S)$ established above, we can obtain the following proposition as a complement of Proposition **15.3.15**:

**15.3.19 Proposition** *Let $L$ be a locale. Then*

(i) $\forall \mathcal{A} \subset N(L), \quad (\bigvee \mathcal{A}) = m(\bigcap\{L_j : j \in \mathcal{A}\})$,

(ii) $\forall \mathcal{B} \subset N_*(L), \quad \bigvee \mathcal{B} = L_{\bigwedge\{m(S):\ S \in \mathcal{B}\}}$.

**Proof** Follows from Theorem **15.3.14**, Theorem **15.3.18** and Proposition **15.3.15**. $\square$

**15.3.20 Proposition** *Let $L_1$ be a sublocale of a locale $L$, $L_0 \subset L_1$. Then $L_0$ is a sublocale of $L$ if and only if $L_0$ is a sublocale of $L_1$.* $\square$

**15.3.21 Definition** Let $L$ be a locale, $a \in L$. Define a mapping $w(a): L \to L$ as follows:

$$w(a)(x) = a \vee ((x \to a) \to a).$$

**15.3.22 Theorem** *Let $L$ be a locale, $a \in L$. Then*

(i) *$w(a)$ is a nucleus.*

(ii) *$L_{w(a)}$ is the smallest sublocale containing $a$.*

**Proof** Consider the nucleus $c(a): L \to L$, $c(a)(x) = a \vee x$. Denote the operations such as $\wedge$, $\vee$, $\neg$ and so on in the sublocale $L_{c(a)}$ by $\wedge^*$, $\vee^*$, $\neg^*$, to distinguish them from the correspoindent original one in $L$.

(i) First of all, we prove $w(a)(x) = \neg^*\neg^*(c(a)(x))$:

$$\begin{aligned}\neg^*(c(a)(x)) &= \bigvee{}^*\{b \in L_{c(a)}: \ b \wedge^* c(a)(x) \le 0^*\}\\ &= \bigvee\{a \vee y: \ y \in L, \ (a \vee y) \wedge (a \vee x) \le a\}\\ &= a \vee \bigvee\{y \in L: \ a \vee (x \wedge y) \le a\}\\ &= a \vee \bigvee\{y \in L: \ x \wedge y \le a\}\\ &= a \vee (x \to a),\\ \neg^*\neg^*(c(a)(x)) &= \neg^*(a \vee (x \to a))\\ &= \neg^*(c(a)(x \to a))\\ &= a \vee ((x \to a) \to a)\\ &= w(a)(x).\end{aligned}$$

Then denote the inclusion $L_{c(a)} \hookrightarrow L$ by $i$, we have

$$w(a) = i \circ (\neg^*\neg^*) \circ c(a).$$

By Proposition **15.3.10** (i) we know that $w(a)$ is a nucleus.

(ii) By Proposition **15.3.15**, the smallest sublocale of $L$ containing $a$ exists, denote it by $L_0$. Since $a \in L_{c(a)}$, so $L_0 \subset L_{c(a)}$. By Proposition **15.3.20**, $L_0$ is the smallest sublocale of $L_{c(a)}$ containing $a = 0_{L_{c(a)}}$. By the virtue of Proposition **15.3.10** (ii),

$$L_0 = (L_{c(a)})_{\neg^*\neg^*} = L_{\neg^*\neg^* c(a)} = L_{w(a)}. \qquad \square$$

**15.3.23 Corollary** *Let $L$ be a locale, $S \subset L$. Denote $j = \bigwedge_{s \in S} w(s)$, then $L_j$ is the smallest sublocale of $L$ containing $S$.*

**Proof** Suppose $L_0$ is a sublocale of $L$ containing $S$, then for every $s \in S$, $L_{w(s)} \subset L_0$. So by Proposition **15.3.19** (ii) and Theorem **15.3.18** (ii),

$$L_0 \supset \bigvee\{L_{w(s)}: \ s \in S\} = L_j.$$

Since clearly we have $S \subset \bigvee\{L_{w(s)}: \ s \in S\} = L_j$, so the conclusion holds. $\square$

**15.3.24 Theorem** *Let $L$, $L_0$ be locales. Then $L_0$ is frame isomorphic to a sublocale of $L$ if and only if there exists a surjective frame homomorphism from $L$ to $L_0$.*

**Proof** The necessity is obvious. Suppose $f: L \to L_0$ is a surjective frame homomorphism, then by Theorem **13.1.10** (v), $f^{\vee}$ is the right adjoint of $f$, preserves arbitrary meet in $L_0$. Then $j = f^{\vee} \circ f: L \to L$ is a nucleus and $L_j$ is frame isomorphic to $L_0$. $\square$

Then we obtain a characterization for the $L$-spatiality of a locale:

**15.3.25 Theore** *Let $M$ be a locale. Then the following conditions are equivalent:*

(i) *$M$ is $L$-spatial.*

(ii) *For arbitrary $a, b \in M$ such that $a \not\le b$, $M$ has a subset $S$ such that $S$ is closed under arbitrary meet, $m(S)$ preserves finite meets, $m(S)(a) \not\le m(S)(b)$ and $L$ has a sublattice $L_0$ such that $L_0$ is closed under arbitrary* *join, finite meet, $L_0$ is isomorphic to $S$.*

(iii) *For arbitrary $a, b \in M$ such that $a \not\leq b$, $M$ has a subset $S$ such that for $j = \bigwedge\{w(s) : s \in S\}$, $j(a) \not\leq j(b)$ and $L$ has a sublattice $L_0$ such that $L_0$ is closed under arbitrary join, finite meet, $L_0$ is isomorphic to $M_j$.*

**Proof** (i)$\Longrightarrow$(ii): By Theorem **15.3.8** (i)$\Longrightarrow$(iii), for every pair $a, b \in M$ such that $a \not\leq b$, there exists $f \in Fr(M, L)$ such that $f(a) \not\leq f(b)$. Then $f[M]$ is a sublattice of $L$ which is closed under arbitrary join and finite meet, and it is a locale. By Theorem **15.3.24**, $f[M]$ is isomorphic to a sublocale $S$ of $M$. By Proposition **9.1.10**, $S$ is closed under arbitrary meet, by Theorem **15.3.18** (i), Proposition **15.3.17** (iii) we know that $m(S)$ preserves finite meet. Clearly we have $m(S)(a) \not\leq m(S)(b)$.

(ii)$\Longrightarrow$(iii): By Proposition **15.3.18**, $S$ is a sublocale of $M$. By Corollary **15.3.23**, $S = M_j$ for $j = \bigwedge\{w(s) : s \in S\}$. By Theorem **15.3.14**, $j = m(S)$. So (iii) is true.

(iii)$\Longrightarrow$(i): Suppose the isomorphism mentioned in the conclusion is $g : M_j \to L_0$, denote the inclusion $L_0 \hookrightarrow L$ by $i$, then $i$ preserves arbitrary joins and finite meets. So $f = i \circ g \circ j : M \to L$ preserves arbitrary joins and finite meets, and since $g$ is an isomorphism, $f(a) \not\leq f(b)$. By Theorem **15.3.8** (iii)$\Longrightarrow$(i), $M$ is $L$-spatial. □

## 15.4 Fuzzy Stone Repesentation Theorem

The notions and the results in this section are mainly based on Zhang and Liu.[197]

Still suppose $L$ be an infinitely distributive lattice.

**15.4.1 Definition** Let $M$ be a distributive lattice. A $L$-fuzzy subset $A : M \to L$ is called an *$L$-fuzzy lower subset* in $M$, if $A$ is order preserving; is called an *$L$-fuzzy ideal* in $M$, if $A$ is finite join preserving; is called an *$L$-fuzzy prime ideal* in $M$, if $A$ is finite join preserving and finite meet preserving.

Then we have directly

**15.4.2 Proposition** *Let $M$ be a distributive lattice.*

(i) *The pointwise join of a family of $L$-fuzzy ideals in $M$ is still an $L$-fuzzy ideal of $M$.*

(ii) *If a family $\{I_t : t \in T\}$ of $L$-fuzzy ideals in $M$ is a down-directed set under the pointwise order, then the pointwise meet of $\{I_t : t \in T\}$ is an $L$-fuzzy ideal in $M$.*

(iii) *The family of all the $L$-fuzzy ideals in $M$ forms a complete lattice under the pointwise order with the smallest element $\underline{0} : M \to L$ and the largest element $\chi_{M\setminus\{0\}} : M \to L$.* □

Then for a distributive lattice $M$, we have

**15.4.3 Definition** Let $M$ be a distributive lattice. Denote the complete lattice of all the $L$-fuzzy ideals in $M$ with the pointwise order by $LDI(M)$, denote respectively the smallest element $\underline{0}$ and the largest element $\chi_{M\setminus\{0\}}$ of $LDI(M)$ by $0^*_M$ and $1^*_M$, and denote respectively the joins and meets in $LDI(M)$ by $\vee^*$ and $\wedge^*$. Call the opposite lattice $LDI(M)^{op}$ of $LDI(M)$ the *lattice of $L$-fuzzy ideals of $M$*, denote it by $LId(M)$.

Easy to find that the joins in $LDI(M)$ are just the pointwise joins. But in general, the meets are not. To describe the meets in $LDI(M)$, we introduce the following symbol and lemma:

**15.4.4 Definition** Let $M$ be a lattice, $a \in M$. Denote

$$M[a] = \{F \in [M]^{<\omega} : \ \bigvee F = a\}.$$

**15.4.5 Lemma** *Let $M$ be a distributive lattice, $f: \ M \to L$ be order preserving. Then the mapping $f^*: \ M \to L$ defined as*

$$\forall a \in M, \qquad f^*(a) = \bigwedge_{F \in M[a]} \bigvee_{c \in F} f(c) \tag{15.1}$$

*is the largest finite join preserving mapping smaller than $f$.*

**Proof** (1) $f^*$ is order preserving.

For every pair $a, b \in M$ such that $a \le b$ and every $F \in M[b]$, by the distributivity of $M$, for $aF = \{a \wedge c : \ c \in F\}$, $\bigvee aF = a$. So

$$f^*(a) \le \bigvee_{c \in aF} f(c) \le \bigvee_{d \in F} f(d).$$

Hence we have

$$f^*(a) \le \bigwedge_{F \in M[b]} \bigvee_{d \in F} f(d) = f^*(b).$$

(2) If $M$ has the smallest element 0, then $f^*(0) = 0$.

In fact, in this situation, $\{F \in [M]^{<\omega} : \ \bigvee F = 0\} = \{\emptyset, \{0\}\}$, so easy to find $f^*(0) = 0$.

(3) $f^*$ preserves binary joins.

For every $a, b \in M$, $f^*(a \vee b) \ge f^*(a) \vee f^*(b)$. On the other hand, since $L$ fulfils the 2nd infinitely distributive law (IFD2),

$$\begin{aligned} f^*(a) \vee f^*(b) &= (\bigwedge_{F \in M[a]} \bigvee_{c \in F} f(c)) \vee (\bigwedge_{G \in M[b]} \bigvee_{d \in G} f(d)) \\ &= \bigwedge_{F \in M[a]} \bigwedge_{G \in M[b]} \bigvee_{e \in F \cup G} f(e) \\ &\ge \bigwedge_{F \in M[a \vee b]} \bigvee_{c \in F} f(c) \\ &= f^*(a \vee b). \end{aligned}$$

So $f^*$ preserves binary joins.

(4) $f^* \le f$.

(5) If $g: \ M \to L$ preserves order, then $f^* \le g^*$.

(6) It follows from the definition of $f^*$ and (3) that

$$f^* = f \iff f \text{ is finite join preserving.}$$

Therefore, we have proved the lemma. □

So we can describe the meets in $LDI(M)$ and hence the joins in $LId(M)$ as follows:

**15.4.6 Proposition** *Let $M$ be a distributive lattice, $\{I_t : \ t \in T\} \subset LDI(M)$. Then the meet $\bigwedge^*_{t \in T} I_t$ of $\{I_t : \ t \in T\}$ in $LDI(M)$ is determined by*

$$\forall a \in M, \qquad (\bigwedge^*_{t \in T} I_t)(a) = \bigwedge_{F \in M[a]} \bigvee_{c \in F} \bigwedge_{t \in T} I_t(c).$$ □

**15.4.7 Proposition** *For every distributive lattice $M$, $LId(M)$ is a locale.*

**Proof** It is sufficient to prove that $LDI(M)$ fulfils the 2nd infinitely distributive law (IFD2), i.e. for every $I \in LDI(M)$ and every $\{I_t : t \in T\} \subset LDI(M)$, prove

$$I \vee^* \bigwedge_{t\in T}^* I_t = \bigwedge_{t\in T}^* (I \vee^* I_t). \tag{15.2}$$

For every $a \in M$,

$$\begin{aligned}(I \vee^* (\bigwedge_{t\in T}^* I_t)(a) &= I(a) \vee \bigwedge_{F\in M[a]} \bigvee_{c\in F} \bigwedge_{t\in T} I_t(c) \\ &= \bigwedge_{F\in M[a]} \bigvee_{c\in F} (I(a) \vee \bigwedge_{t\in T} I_t(c)) \\ &\geq \bigwedge_{F\in M[a]} \bigvee_{c\in F} (I(c) \vee \bigwedge_{t\in T} I_t(c)) \\ &= \bigwedge_{F\in M[a]} \bigvee_{c\in F} \bigwedge_{t\in T} (I(c) \vee I_t(c)) \\ &= (\bigwedge_{t\in T}^* (I \vee I_t))(a).\end{aligned}$$

Conversely, $I \vee^* \bigwedge_{t\in T}^* I_t \leq \bigwedge_{t\in T}^* (I \vee^* I_t)$ clearly holds, so (15.2) is true. □

**15.4.8 Lemma** *Let $f : M \to N$ be a lattice homomorphism between distributive lattices, $J : N \to L$ an L-fuzzy ideal in $N$. Then*

(i) *$J \circ f$ is an L-fuzzy ideal in $M$.*

(ii) *$J \circ f$ is an L-fuzzy prime ideal in $M$ provided $J$ is an L-fuzzy prime ideal in $N$.* □

So we have the following definition:

**15.4.9 Definition** Let $f : M \to N$ be a lattice homomorphism between distributive lattices. Define a mapping $\underline{f}^{\prec} : LId(N) \to LId(M)$ as follows:

$$\forall J \in LId(N), \quad \underline{f}^{\prec}(J) = J \circ f. \tag{15.3}$$

**15.4.10 Lemma** *Let $f : M \to N$ be a lattice homomorphism between distributive lattices. Then*

(i) *For every $I \in LId(M)$, $\underline{f}(I)$ defined as follows is an L-fuzzy ideal in $N$:*

$$\forall I \in LId(M),\ \forall b \in N, \quad \underline{f}(I)(b) = \bigwedge_{f(a)\geq b} I(a). \tag{15.4}$$

(ii) *For every $I \in LId(M)$, $\underline{f}^{\prec} \circ \underline{f}(I) \leq I$.*

(iii) *For every $J \in LId(N)$, $\underline{f}^{\prec}(J) \leq I \implies J \leq \underline{f}(I)$.*

**Proof** (i) For every pair $u, v \in N$,

$$\begin{aligned}\underline{f}(I)(u) \vee \underline{f}(I)(v) &= (\bigwedge_{f(c)\geq u} I(c)) \vee (\bigwedge_{f(d)\geq v} I(d)) \\ &= \bigwedge_{f(c)\geq u} \bigwedge_{f(d)\geq v} (I(c) \vee I(d)) \\ &\geq \bigwedge_{f(c)\geq u\vee v} I(c) \\ &= \underline{f}(u \vee v).\end{aligned}$$

Since the converse inequality is obvious, $\underline{f}(I)$ preserves finite joins.

(ii) For every $a \in M$,

$$\underline{f}^{\prec} \circ \underline{f}(I)(a) = \underline{f}(I)(f(a)) = \bigwedge_{f(c)\geq f(a)} I(c) \leq I(a).$$

(iii) If $\underline{f}^{\prec}(J) \leq I$, then $J \circ f \leq I$. So for every $b \in N$,

$$J(b) \leq \bigwedge_{f(c) \geq b} J(f(c)) \leq \bigwedge_{f(c) \geq b} I(c) = \underline{f}(I)(b).$$

That just means $J \leq \underline{f}(I)$. □

Then we have the following well-defined notion:

**15.4.11 Definition** Let $f: M \to N$ be a lattice homomorphism between distributive lattices. Define a mapping $\underline{f}: LId(M) \to LId(N)$ by (15.4) and define its inverse mapping $\underline{f}^{\prec}: LId(N) \to LId(M)$ by (15.3).

**15.4.12 Proposition** *Let $f: M \to N$ be a lattice homomorphism between distributive lattices. Then*

(i) $\underline{f}^{\prec} \circ \underline{f} \leq id$.

(ii) $\underline{f} \circ \underline{f}^{\prec} \geq id$.

(iii) $\underline{f}^{\prec} \circ \underline{f} \circ \underline{f}^{\prec} = \underline{f}^{\prec}$.

(iv) $\underline{f} \circ \underline{f}^{\prec} \circ \underline{f} = \underline{f}$. □

**15.4.13 Definition** Let $L$ be a complete Heyting algebra with an order-reversing involution $': L \to L$, $M$ a distributive lattice. Call the set of all the $L$-fuzzy prime ideals in $M$ the *$L$-fuzzy spectrum* of $M$, denote it by $spec_L(M)$, equip it with the pointwise order.

**15.4.14 Definition** Let $L$ be a complete Heyting algebra with an order-reversing involution $': L \to L$, $M$ a distributive lattice. For every $I \in LDI(M)$, define an $L$-fuzzy subset $D_I: spec_L(M) \to L$ as follows:

$$\forall J \in spec_L(M), \quad D_I(J) = \bigwedge_{a \in M} (J'(a) \vee I(a)). \tag{15.5}$$

**15.4.15 Lemma** *Let $M$ be a distributive lattice. Then*

(i) $D_{0^*_M} = 0$.

(ii) $D_{1^*_M} = 1_{spec_L(M)}$.

(iii) $D_{I_1 \vee^* I_2} = D_{I_1} \vee D_{I_2}$.

(iv) $D_{\bigwedge^*_{t \in T} I_t} = \bigwedge_{t \in T} D_{I_t}$.

**Proof** (i) and (ii) are straightforward.

(iii) By the definition, $D_{I_1 \vee^* I_2} \geq D_{I_1} \vee D_{I_2}$. For every $J \in spec_L(M)$,

$$\begin{aligned} D_{I_1}(J) \vee D_{I_2}(J) &= (\bigwedge_{a \in M} (J'(a) \vee I_1(a))) \vee (\bigwedge_{b \in M} (J'(b) \vee I_2(b))) \\ &= \bigwedge_{a,b \in M} (J'(a) \vee J'(b) \vee I_1(a) \vee I_2(b)) \\ &= \bigwedge_{a,b \in M} (J'(a \wedge b) \vee I_1(a) \vee I_2(b)) \\ &\geq \bigwedge_{a,b \in M} (J'(a \wedge b) \vee I_1(a \wedge b) \vee I_2(a \wedge b))) \\ &= \bigwedge_{a \in M} (J'(a) \vee (I_1(a) \vee I_2(b)) \\ &= D_{I_1 \vee^* I_2}(J). \end{aligned}$$

(iv) Since

$$\bigwedge_{t \in T} (D_{I_t}(J)) = \bigwedge_{t \in T} \bigwedge_{a \in M} (J'(a) \vee I_t(a)) = \bigwedge_{a \in M} (J'(a) \vee \bigwedge_{t \in T} I_t(a)),$$

so by the definition of $\bigwedge_{t\in T}^{*} I_t$ it is sufficient to prove the following equation for every order preserving mapping $f: M \to L$ and every $J \in spec_L(M)$:

$$\bigwedge_{a\in M}(J'(a) \vee f(a)) = \bigwedge_{a\in M}(J'(a) \vee f^*(a)). \tag{15.6}$$

First of all, by $f^* \leq f$ we have

$$\bigwedge_{a\in M}(J'(a) \vee f^*(a)) \leq \bigwedge_{a\in M}(J'(a) \vee f(a)).$$

Secondly,

$$\begin{aligned}\bigwedge_{a\in M}(J'(a) \vee f^*(a)) &= \bigwedge_{a\in M}(J'(a) \vee (\bigwedge_{F\in M[a]} \bigvee_{c\in F} f(a)))\\ &= \bigwedge_{a\in M} \bigwedge_{F\in M[a]} ((\bigwedge_{c\in F} J'(c)) \vee (\bigwedge_{c\in F} f(c)))\\ &\geq \bigwedge_{a\in M} \bigwedge_{F\in M[a]} \bigwedge_{c\in F} (J'(c) \vee f(c))\\ &= \bigwedge_{a\in M} (J'(c) \vee f(c)).\end{aligned}$$

Hence (15.6) holds.

By Lemma **15.4.15**, $\{D_I' : I \in LDI(M)\}$ is an $L$-fuzzy topology on $spec_L(M)$. So the following definition is reasonable:

**15.4.16 Definition** Let $M$ be a distributive lattice, $L$ a complete Heyting algebra with an order-inversing involution $' : L \to L$. Call $\{D_I' : I \in LDI(M)\}$ the *canonical L-fuzzy topolgy* on $spec_L(M)$, denote it by $\delta_L(M)$, call $(L^{spec_L(M)}, \delta_L(M))$ the *L-fuzzy spectrum space* of $M$.

**15.4.17 Definition** Let $f : M \to N$ be a lattice homomorphism between distributive lattices, $L$ a complete Heyting algebra with an order-inversing involution $' : L \to L$. Define a mapping $spec_L(f) : spec_L(N) \to spec_L(M)$ as follows:

$$\forall J \in spec_L(N), \quad spec_L(f)(J) = \underline{f}^{\prec}(J) = J \circ f.$$

**15.4.18 Proposition** *Let $f : M \to N$ be a lattice homomorphism between distributive lattices, $L$ a complete Heyting algebra with an order-inversing involution $' : L \to L$. Then the L-fuzzy mapping $(spec_L(f))^{\to} : (L^{spec_L(N)}, \delta_L(N)) \to (L^{spec_L(M)}, \delta_L(M))$ is continuous.*

**Proof** It is sufficient to prove for every $I \in LDI(M)$, $(spec_L(f))^{\leftarrow}(D_I) = D_{\underline{f}(I)}$.

For every $J \in spec_L(N)$,

$$\begin{aligned}(spec_L(f))^{\leftarrow}(D_I)(J) &= D_I(J \circ f)\\ &= \bigwedge_{a\in M}((J \circ f(a))' \vee I(a))\\ &= \bigwedge_{b\in N} \bigwedge_{f(a)\geq b} (J'(b) \vee I(a))\\ &= \bigwedge_{b\in N} (J'(b) \vee \bigwedge_{f(a)\geq b} I(a))\\ &= \bigwedge_{b\in N} (J'(b) \vee \underline{f}(I)(b))\\ &= D_{\underline{f}(I)}(J).\end{aligned}$$

This completes the proof. □

By the virtue of Lemma **15.4.15** and Proposition **15.4.18** we have the following

$L$-fuzzy topological representations of distributive lattices:

**15.4.19 Theorem** *For every complete Heyting algebra $L$ with an order-inversing involution $'$ : $L \to L$, $spec_L$ is a contravariant functor from the category of distributive lattices to the category of $L$-fuzzy topological spaces.* □

# Bibliography

[1] Adámek, J., H. Herrlich and G.E. Strecker, *Abstract and Concrete Categories*, John Wiley & Sons, Inc., 1990, New York.

[2] Adnadjević, D., Separation properties of F-spaces, *Matem, Vesnik*, **6**(1982), 1-8.

[3] Ahsanullah, T.M.G. and N.N. Morsi, Invariant probabilistic metrizability of fuzzy neighbourhood groups, *Fuzzy Sets and Systems*, **47**(1992), 223-245.

[4] Amer, M.A. and N.N. Morsi, Characterization of some fuzzy topological notions in probabilistic metric spaces, *Fuzzy Sets and Systems*, **51**(1992), 351-359.

[5] Artico, G. and R. Moresco, Fuzzy proximities and totally bounded fuzzy uniformities, *J. Math. Anal. Appl.*, **99**(1984), 320-337.

[6] Artico, G. and R. Moresco, On fuzzy metrizability, *J. Math. Anal. Appl.*, **107**(1985), 144-147.

[7] Artico, G. and R. Moresco, Fuzzy proximities according with Lowen fuzzy uniformities, *Fuzzy Sets and Systems*, **21**(1987), 85-98.

[8] Artico, G. and R. Moresco, Uniform properties of normal and compact fuzzy topological spaces, *J. Math. Anal. Appl.*, **132**(1988), 514-519.

[9] Artico, G. and R. Moresco, Fuzzy uniformities induced by fuzzy proximities, *Fuzzy Sets and Systems*, **31**(1988), 111-121.

[10] Banaschewskii, B., The duality of distributive continuous lattices, *Canadian J. Math.*, **32**(1980), 385-394.

[11] Birkhoff, G., *Lattice Theory*, Amer. Math. Soc. Colloquium Publ. vol.**35**, 3rd ed., 1967.

[12] Bourbaki, N., *Elements of Mathematics: General Topology, Part I.*, Addison-Wesley Read. Mass., 1966.

[13] Burton, M.H., The relationship getween a fuzzy uniformity and its family of $\alpha$-level uniformities, *Fuzzy Sets and Systems*, **54**(1993), 311-316.

[14] Cerutti, U., The Stone-Čech compactification in the category of fuzzy topological spaces, **Fuzzy Sets and Systems**, **6**(1981), 197-204.

[15] Chang, C.L., Fuzzy topological spaces, *J. Math. Anal. Appl.*, **24**(1968), 182-190.

[16] Dieudonné, J., Une generalization des espaces compacts, *J. Math. Pures et Appl.*, **23**(1944), 65-76.

[17] Ehresmann, C., Gattungen von lokalen structuren, *Jber. Deutch Math. Verein.*, **60**(1975), 59-77.

[18] Eklund, P. and W. G'ahler, Basic notions for fuzzy topology, I, *Fuzzy Sets and Systems*, **26**(1988), 333-356.

[19] Eklund, P. and W. G'ahler, Basic notions for fuzzy topology, II, *Fuzzy Sets and Systems*, **27**(1988), 171-195.

[20] Engelking, R., *General Topology*, 1977, Warzawa.

[21] Erceg, M.A., Metric spaces in fuzzy set theory, *J. Math. Anal. Appl.*, *69*(1979), 205-230.

[22] Fan, Jiu-Lun, Characterizations of two kinds of strong paracompactness in *L*-fuzzy topological spaces, *Journal of Shaanxi Normal University* (Natural Science Edition), **17**(1989), 3:1-4 (in Chinese).

[23] Fan, Jiu-Lun, Paracompactness and strong paracompactness in *L*-fuzzy topological spaces, *Fuzzy System and Mathematics*, **4**(1990), 88-94 (in Chinese).

[24] Gantner, T.E., R.C. Steinlage and R.H. Warren, Compactness in fuzzy topological spacess, *J. Math. Anal. Appl.*, **62**(1978), 547-562.

[25] Gierz, G., K.H. Hofmann, K. Keimel, J.D. Lawson, M. Mislove and D.S. Scott, *A Compendium of Continuous Lattices*, Springer-Verlag, 1980, Berlin.

[26] Goguen, J.A., *L*-fuzzy sets, *J. Math. Anal. Appl.*, **18**(1967), 145-174.

[27] Goguen, J.A., The fuzzy Tychonoff theorem, *J. Math. Anal. Appl.*, **43**(1973), 734-742.

[28] Hahn, H., Uber halbstetige und unstetige Functionen, *Sitzungsber. Akad. Wiss. Wenn Abt*, IIa, **126**(1917), 91-110.

[29] Halpern, J.D., The independence of the axiom of choice from the Boolean prime ideal theorem, *Fund. Math.* **55**, 57-66.

[30] Hausdorff, F., *Grundzöge der Mengenlehre*, Leipzig 1914.

[31] He, Bo-Yong, Upper-class spaces of topological molecular lattices, *Chinese Annals of Math.*, *12*(1991), 4:514-520

[32] He, Bo-Yong, Some theorems for hereditary property and produce property in *L*-fuzzy topological spaces, *ibid.*, **13**(1992), 2:243-247.

[33] Herrlich, H. and G.E. Strecker, *Category Theory*, Sigma Ser. Pure Math. 1, Heldermann Verlag, Berlin, 1979.

[34] Höhle, U., Probabilistic topologies induced by *L*-fuzzy uniformities, *Manuscripta Math.*, **38**(1982), 289-323.

[35] Höhle, U., Fuzzy topologies and the topological space objects in a topos, *Fuzzy Sets and Systems*, *19*(1986), 299-304.

[36] Höhle, U. and L.N. Stout, Foundations of fuzzy sets, *Fuzzy Sets and Systems*, **40**(1991), 257-296.

[37] Hutton, B., Normality in fuzzy topological spaces, *J. Math. Anal. Appl.*, **50**(1975), 74-79.

[38] Hutton, B., Uniformities on fuzzy topological spaces, *ibid.*, **58**(1977), 559-571.

[39] Hutton, B., Products of fuzzy topological spaces, *Top. Appl.*, **11**(1980), 59-67.

[40] Hutton, B., Uniformities on fuzzy topological spaces, Part II, *Fuzzy Mathematics*, **1**(1983).

[41] Hutton, B. and I. Reilly, Separation axioms in fuzzy topological spaces, *Fuzzy Sets and Systems*, **3**(1980), 93-104.

[42] Isbell, J.R., Atomless parts of spaces, *Math. Scan.*, **31**(1972), 5-32.

[43] Jiang, Ji-Guang, Character, density and cellularity of *L*-fuzzy topological spaces, *Acta Math. Phys. Sinica*, **2**(1982), 343-352.

[44] Johnstone, P.T., *Stone Spaces*, Cambridge University Press, Cambridge, 1982.

[45] Johnstone, P.T. and Shu-Hao Sun, Weak products and Hausdorff locales, Springer LNM 1348, 173-193.

[46] Kaleva, O., The completion of fuzzy metric spaces, *J. Math. Anal. Appl.*, **109**(1985), 194-198.

[47] Katsaras, A.K., On fuzzy proximity spaces, *J. Math. Anal. Appl.*, **75**(1980), 571-583.

[48] Katsaras, A.K., C.G. Petalas, A unified theory of fuzzy topologies, fuzzy proximities and fuzzy uniformities, *Rev. Roum. Math. Pures et Appl.*, **28**(1983), 9:845-856.

[49] Kelley, J.L., Convergence in topology, *Duke Math.Journ.*, **17**(1950), 277-283.

[50] Kelley, J.L., *General Topology*, Van Nostrand Reinhold, New York, 1955.

[51] Klein, A.J., Generalizing the *L*-fuzzy unit interval, *Fuzzy Sets and Systems*, **12**(1984), 271-279.

[52] Klimovsky, G., El teorema de Zorn y la existencia de filtros y ideales maximales en los reticulados distributivos, *Rev. Un. Mat. Argentina* **18**, 160-164.

[53] Kubiak, T., Extending continuous *L*-real functions, *Math. Japonica*, **31**(1986), 875-887.

[54] Kubiak, T., *L*-fuzzy normal spaces and Tietze Extension Theorem, *J. Math. Anal. Appl.*, **125**(1987), 141-153.

[55] Kubiak, T., On *L*-Tychonoff spaces, *Fuzzy Sets and Systems*, **73**(1995), 25-54.

[56] Li, Zhong-Fu, Compactness in fuzzy topological spaces, KEXUE TONGBAO , **29**(1984), 321-323 (in Chinese).

[57] Liang, Ji-Hua, Fuzzy proximity spaces, *Fuzzy Math.*, 1983, 4:23-34 (in Chinese).

[58] Liang, Ji-Hua, Several problems in fuzzy metric spaces, *Chinese Annals of Math.*, **6A**(1984), 1:59-67 (in Chinese).

[59] Liang, Ji-Hua, Completeness of fuzzy uniform spaces, *J. Sichuan Univ.* (Natural Sci. Ed.), 1985, 2:10-15 (in Chinese).

[60] Liang, Ji-Hua, On embedding theorem of F.P. spaces and metrization problem, *Fuzzy Math.*, 1986, 2:1-6 (in Chinese).

[61] Liang, Ji-Hua, Pointed characterization of fuzzy metric and its applications, *Acta Math. Sinica*, **30**(1987), 6:733-741 (in Chinese).

[62] Liang, Ji-Hua, Fuzzy Smirnov-Nagata Metrization Theorem, *Northeastern Math. J.*, **3**(1987), 4:419-428 (in Chinese).

[63] Liang, Ji-Hua, Geometrical characterization of uniformity on lattices, *J. Sichuan Univ.* (Natural Sci. Ed.), 1988, 2:140-144 (in Chinese).

[64] Liang, Ji-Hua, Metrization problem in induced spaces, *Advances in Math.*, **19**(1990), 4:473-477 (in Chinese).

[65] Liu, Wang-Jin, Fuzzy proximity spaces redefined, *Fuzzy Sets and Systems*, **15**(1985), 241-248.

[66] Liu, Xiao-Shi, Cardinal functions of pointwise topology on lattices, *J. Math. Research and Exposition*, **3**(1986), 5-11.

[67] Liu, Ying-Ming, Intersection operation on union-preserving mappings in completely distributive lattices, *J. Math. Anal. Appl.*, **84**(1981), 249-255.

[68] Liu, Ying-Ming, Compactness in fuzzy topological spaces and Tychonoff product theorem, *Acta Math. Sinica*, **24**(1981), 260-268 (in Chinese).

[69] Liu, Ying-Ming, Inverse operation on union-preserving mappings in lattices and its applications to fuzzy uniform spaces, *Proc. 12th International Symposium of Multiple-valued Logic, 1982*, Paris, IEEE, 280-288.

[70] Liu, Ying-Ming, Progress in general topology, 1982, *Science Yearbook*, Shanghai.

[71] Liu, Ying-Ming, Pointwise characterization of complete regularity and imbedding theorem in fuzzy topological spaces, *Scientia Sinica*, Ser.A, *26*(1983), 138-147.

[72] Liu, Ying-Ming, On fuzzy Stone-Čech compactification, *Acta Math. Sinica*, **26**(1983), 507-512.

[73] Liu, Ying-Ming, An analysis on fuzzy membership relation in fuzzy set theory, Chinese Annals of Math., **5A**(1984), 431-486; also see E. Sanchez, Ed., *Fuzzy Information, Knowledge Representation and Decision Analysis*, Pergamon, 1984, 115-122.

[74] Liu, Ying-Ming, Fuzzy imbedding theory and its applications, *Busefal*, **19**(1984), 4-12.

[75] Liu, Ying-Ming, Some properties of convex fuzzy sets, *J. Math. Anal. Appl.*, **111**(1985), 119-129.

[76] Liu, Ying-Ming, Topology on lattice and fuzzy topology, *J. Sichuan Univ.*, **3**(1986), 43-52.

[77] Liu, Ying-Ming, Structure of fuzzy order homomorphisms, *Fuzzy Sets and Systems*, **21**(1987), 43-51.

[78] Liu, Ying-Ming, Fuzzy metrization – an application of imbedding theory, (Ed. J. Bezdek) *Analysis of Fuzzy Information*, CRC, 1987, 203-209.

[79] Liu, Ying-Ming, On fuzzy convergence classes, *Fuzzy Sets and Systems*, **30**(1989), 47-51.

[80] Liu, Ying-Ming, Fuzzy topology, stratifications and category theory, in *Between Mind and Computer*, Eds. Wang Loe, 1993, Singapore.

[81] Liu, Ying-Ming and Ming He, Induced mappings in completely distributive lattices, *Proc. of International Symposium on Multiple-valued Logic*, Ontario, IEEE, 1985, 346-353.

[82] Liu, Ying-Ming, Ji-Hua Liang, Fuzzy topology — stratification and pointwise disposition, *Advances in Math.*, **23**(1994), 4:304-321 (in Chinese).

[83] Liu, Ying-Ming and Mao-Kang Luo, Induced spaces and fuzzy Stone-Čech compactification, *Sientia Sinica*, Ser. A, **30**(1987), 1034-1044.

[84] Liu, Ying-Ming and Mao-Kang Luo, Preorder relations in fuzzy compactifications, *Chinese Science Bulletin*, **34**(1989), 9: 705-709.

[85] Liu, Ying-Ming and Mao-Kang Luo, Fuzzy Stone-Čech-type compactifications, *Fuzzy Sets and Systems*, **33**(1989), 355-372.

[86] Liu, Ying-Ming and Mao-Kang Luo, Fuzzy Stone-Čech compactifications and the largest Tychonoff compactifications, *Chin. Ann. of Math.*, **10B**(1989), 1:74-84.

[87] Liu, Ying-Ming and Mao-Kang Luo, Separations in lattice-valued induced spaces, *Fuzzy Sets and Systems*, **36**(1990), 55-66.

[88] Liu, Ying-Ming and Mao-Kang Luo, Lattice-valued mappings, completely distributive law and induced spaces, *Fuzzy Sets and Systems*, **42**(1991), 43-56.

[89] Liu, Ying-Ming and Mao-Kang Luo, Lattice-valued Hahn-Dieudonné-Tong insertion theorem and stratification structure, *Top. and its Appl.*, **45**(1992), 173-188.

[90] Liu, Ying-Ming and Mao-Kang Luo, $T_D$ property and spatial sublocales, *Acta Math. Sinica*, New Ser., *11*(1995), 3:324-336.

[91] Liu, Ying-Ming, Mao-Kang Luo and Qian Peng, Analytic and toplogical characterizations of completely distributive law, *Chinese Science Bulletin*, **35**(1990), 15: 1237-1240.

[92] Lowen, R., Fuzzy topological spaces and fuzzy compactness, *J. Math. Anal. Appl.*, **56**(1976), 621-633.

[93] Lowen, R., Initial and final fuzzy topologies and the fuzzy Tychonoff theorem, *J. Math. Anal. Appl.*, **58**(1977), 11-21.

[94] Lowen, R., A comparison of different compactness notions in fuzzy topological spaces, *J. Math. Anal. Appl.*, **64**(1978), 446-454.

[95] Lowen, R., Convergence in fuzzy topological spaces, *General Top. Appl.*, **10**(1979), 147-160.

[96] Lowen, R., Fuzzy uniform spaces, *J. Math. Anal. Appl.*, **82**(1981), 370-385.

[97] Lowen, R., Compact Hausdorff fuzzy topological spaces are topological, *Topology and its Appl.*, **12**(1981), 65-74.

[98] Lowen, R., Fuzzy neighborhood spaces, *Fuzzy Sets and Systems*, **7**(1982), 165-189.

[99] Lowen, R., The relation between filter and net-convergence in fuzzy topological spaces, *Fuzzy Math.*, **4**(1983), 41-52.

[100] Lowen, R., $I^X$, the hyperspace of fuzzy sets, a natural non-topological fuzzy topological space, *Trans. Amer. Math. Soc.*, **278**(1983), 547-564.

[101] Lowen, R., On The Existence of Natural Nontopological, Fuzzy Topological Spaces, Heldermann Verlag, Berlin, 1985.

[102] Lowen, R., The order aspect of the fuzzy real line, *Manuscripta Math.*, **39**(1985), 293-309.

[103] Lowen, R., A topological universe extension of FTS, in *Applications of Category Theory to Fuzzy Subsets*, Eds. S.E. Rodabaugh, E.P. Klement and U. Höhle, Kluwer Academic Publishers, 1992, Netherlands.

[104] Lowen, E. and R. Lowen, A characterization of convergence in fuzzy topological spaces, *Int. J. Math. and Math. Sci.*, **8**(1985), 497-511.

[105] Lowen, E., R. Lowen and P. Wuyts, The categorical topology approach to fuzzy topology and fuzzy convergence, *Fuzzy Sets and Systems*, **40**(1991, 347-373.

[106] Lowen, R. and P. Wuyts, Completeness, compactnessand precompactness in fuzzy uniform spaces, I, *J. Math. Anal. Appl.*, **90**(1982), 563-583.

[107] Lowen, R. and P. Wuyts, Completeness, compactness and precompactness in fuzzy uniform spaces, II, *J. Math. Anal. Appl.*, *92*(1983), 342-371.

[108] Lowen, R. and P. Wuyts, Concerning the constants in fuzzy topology, *J. Math. Anal. Appl.*, **129**(1988), 256-268.

[109] Lowen, R., P. Wuyts and E. Lowen, On the reflectiveness and coreflectiveness of subcategories of FTS, *Math. Nachr.*, **141**(1989), 55-65.

[110] Luo, Mao-Kang, A note on fuzzy paracompactness and fuzzy metric property, *J. Sichuan Univ.*, Natural Sci. Edition, 1985, No.4: 141-150 (in Chinese).

[111] Luo, Mao-Kang, Weak paracompactness and compactness in fuzzy topological spaces, *Chinese Ann. Math.*, **7A**(1986), 5:590-596 (in Chinese).

[112] Luo, Mao-Kang, Paracompactness and compactness in fuzzy topological spaces, *Acta Math. Sinica*, **30**(1987), 4:548-552 (in Chinese).

[113] Luo, Mao-Kang, Paracompactness in fuzzy topological spaces, *J. Math. Anal. Appl.*, **130**(1988), 1: 55-77.

[114] Luo, Mao-Kang, Lattice-valued semicontinuous mappings and induced topologies, *Acta Math. Sinica*, New Ser., **6**(1990), 3: 193-205.

[115] Luo, Mao-Kang, Pointed disposition of topology on lattices, PhD Thesis, Sichuan Univ., 1992 (in Chinese).

[116] Martin, H.W., Weakly induced fuzzy topological spaces, *J. Math. Anal. Appl.*, **78**(1980), 634-639.

[117] Martin, H.W., A Stone-Čech ultra fuzzy compactification, *J. Math. Anal. Appl.*, **73**(1980), 453-456.

[118] Moore, E.H. and H.L. Smith, A general theory of limits, *Amer. Journ. of Math.*, **44**(1922), 102-121.

[119] Morsi, M.N., The Urysohn lemma for fuzzy neighborhood spaces, **Fuzzy Sets and Systems**, **39**(1991), 347-360.

[120] Mrowka, S., On the ideals extension theorem and its equivalence to the axiom of choice, *Fund. Math.* **43**, 46-49.

[121] Mrowka, S., Two remarks to my paper 'On the ideals extension theorem and its equivalence to the axiom of choice', *Fund. Math.* **46**, 165-166.

[122] Nobeling, G., *Gundlagen der Analytischen Topoligie*, Springer, 1954.

[123] Peng, Qian, Local N-compactness in fuzzy topological spaces, *Chinese Annals of Math.*, **13A**: (supplement) (1992), 99-103 (in Chinese).

[124] Peng, Yu-Wei, Topological structure of a fuzzy function space – the pointwise convergent topology and compact open topology, *KEXUE TONGBAO*, **29**(1984), 3:289-292.

[125] Peng, Yu-Wei, Algebraic property of join-irreducible elements in completely distributive lattices and structure of molecular lattices, *Journal of Engineering Math.*, **1985**, 2:114-117 (in Chinese).

[126] Peng, Yu-Wei, N-compactness in *L*-fuzzy topological spaces, *Acta Math. Sinica*, **29**(1986), 555-558 (in Chinese).

[127] Peng, Yu-Wei, Geometrical characterization of N-compactness in *L*-fuzzy topological spaces, *Advances in Math.*, **16**(1987), 87-90 (in Chinese).

[128] Peng, Yu-Wei, Latticed-valued semicontinuous mappings and Hausdorff N-compactness spaces, *Chinese Annals of Math.*, **11A**(1990), 753-760 (in Chinese).

[129] Peng, Yu-Wei, Some advances o fpointwise p.q.(p.) metric theory on a lattice, *J. Sichuan Univ.*, **27**(1990), 3:275-281.

[130] Peng, Yu-Wei, Sufficient and necessary conditions of a Hausdorff N-compact space being ultra-compact, *J. Math. Research and Exposition*, **12**(1992), 4:509-512 (in Chinese).

[131] Peng, Yu-Wei, Pointwise p.q. metric and induced mappings on completely distributive lattices, *Chinese Annals of Math.*, **13A**(1992), 3:353-359 (in Chinese).

[132] Peng, Yu-Wei, On two open problems on lattice-valued induced spaces, *Acta Math. Sinica*, **35**(1992), 6:751-757.

[133] Peng, Yu-Wei, Simplification of Erceg's metric function and its applications, *Fuzzy Sets and Systems*, **54**(1993), 2:181-189.

[134] Peng, Yu-Wei, Some characterization theorems on lattice-valued induced spaces, *Fuzzy Sets and Systems*, **55**(1993), 209-217.

[135] Peng, Yu-Wei and Xiao-Zhan Xu, Product of p.q. metric topological molecular lattices, *Northeastern Math. J.*, **9**(1993), 3: 331-337.

[136] Peng, Yu-Wei, Notes on p.q. (p.) metric in Erceg's sense on lattices, *J. Math. Research and Exposition*, **16**(1996), 1:135-138 (in Chinese).

[137] Pu, Bao-Ming and Ying-Ming Liu, Fuzzy topology I, neighborhood structure of a fuzzy point, *J. Sichuan Univ.*, Natural Sci. Edition, 1977, No.1: 31-50 (in Chinese), and English edition see *J. Math. Anal. Appl.*, **76**(1980), 541-599.

[138] Pu, Bao-Ming and Ying-Ming Liu, Fuzzy topology II, product and quotient spaces, *J. Math. Anal. Appl.*, **77**(1980), 20-37.

[139] Pu, Bao-Ming and Ying-Ming Liu, A survey of some aspects on the research work of fuzzy topology in China, *Advances in Fuzzy Sets, Possibility THeory, and Applications*, Plenum, New York-London, 1983, 31-36.

[140] Rodabaugh, S.E., The Hausdorff separation axiom for fuzzy topological spaces, *Gen. Top. Appl.*, **11**(1980), 319-334.

[141] Rodabaugh, S.E., Connectivity and the *L*-fuzzy unit interval, *Rocky Mountain J. Math.*, **12**(1982), 113-121.

[142] Rodabaugh, S.E., The categorical accommodation of various notions to fuzzy topology, *Fuzzy Sets and Systems*, **9**(1983), 241-265.

[143] Rodabaugh, S.E., A point-set lattice-theoretic framework **T** for toplogy which contins **LOC** as a subcategory of singleton spaces and in which there are general classes of Stone Representations and Compactification theorems, Youngstown State University Cental Printing Office (first edition Febrary 1986, second edition April 1987), Youngstown (Ohio).

[144] Rodabaugh, S.E., A theory of fuzzy uniformities with applications to the fuzzy real lines, *J. Math. Anal. Appl.*, **128**(1988), 37-70.

[145] Rodabaugh, S.E., Dynamic topologies and their applications to crisp topologies, fuzzifications of crisp topologies, and fuzzy topologies on the crisp real line, *J. Math. Anal. Appl.*, **131**(1988), 25-66.

[146] Rodabaugh, S.E., Lowen, para-Lowen, and $\alpha$-level functors and fuzzy topologies on the crisp real line, *ibid.*, **131**(1988), 157-169.

[147] Rodabaugh, S.E., Point-Set lattice-theoretic topology, *Fuzzy Sets and Systems*, **40**(1991), 297-345.

[148] Rodabaugh, S.E., Separation axioms and the fuzzy real lines, *Fuzzy Sets and Systems*, **11**(1983), 163-183.

[149] Rodabaugh, S.E., Categorical Frameworkds for Stone Representation Theories, in "*Applications of Category Theory to Fuzzy Subsets*", 177-231, Ed. by S.E. Rodabaugh *et al.*, Kluwer Academic Publishers, 1992, Netherlands.

[150] Scott, D.S., The theorem on maximal ideals in lattices and the axiom of choice, *Bull. Amer. Math. Soc.* **60**, 83.

[151] Shostak, A.P., On a fuzzy topological structure, *Suppl. Rend. Circ. Mat. Palermo*, Ser. II, *11*(1985), 89-105.

[152] Shostak, A.P., A fuzzy modification of the category of linearly ordered spaces, *Comm. Math. Univ. Carol.*, **26**(1985), 421-442.

[153] Shostak, A.P., Two decades of fuzzy topology: basic ideas, notions, and results, *Russian Math. Surveys*, **44**(1989), 6: 125-186.

[154] Sierpinski, W., *General Topology*, Toronto, 1952.

[155] Stone, M.H., The theory of representations for Boolean algebras, *Trans. Amer. Math. Soc.* **40**, 37-111.

[156] Sun, Shu-Hao, Topological structure of locales and Grothendieck generalized sheaf theory, PhD Thesis, Sichuan Univ., 1987 (in Chinese).

[157] Tong, H., Some characterizations of normal and perfecty normal spaces, **Bull. Amer. Math. Soc.**, **54**(1948), 65; also see: it Duke Math. J., **19**(1952), 289-292.

[158] Wang, Ge-Ping and Lan-Fang Hu, On induced fuzzy topological spaces, *J. Math. Anal. Appl.*, *108*(1985), 495-506.

[159] Wang, Ge-Ping, Induced $I(L)$-fuzzy topological spaces and N-compactness, *Kexue Tongbao*, **5**(1989), 333-335.

[160] Wang, Guo-Jun, Pointwise topology on completely distributive lattices, *Fuzzy Sets and Systems*, **30**(1980), 1:53-62.

[161] Wang, Guo-Jun, Topological molecular lattices (I), *J. Shaanxi Normal Univ.*, Natural Sci. Edition, 1979, 1-15 (in Chinese); also see *KEXUE TONGBAO*, **28**(1983), 1089-1091.

[162] Wang, Guo-Jun, A new fuzzy compactness defined by fuzzy nets, *J. Math. Anal. Appl.*, 94(1983), 1-23.

[163] Wang, Guo-Jun, Generalized topological molecular lattices, *Scientia Sinica*, **8**(1984), 785-798.

[164] Wang, Guo-Jun, Order-homomorphisms on fuzzies, *Fuzzy Sets and Systems*, *12*(1984), 280-288.

[165] Wang, Guo-Jun, On the structure of fuzzy lattices, *Acta Math. Sinica*, **4**(1986), 539-543.

[166] Wang, Guo-Jun, On the structure of fuzzy lattices, *Acta Math. Sinica*, **4**(1986), 539-543.

[167] Wang, Guo-Jun, Some research aspects of fuzzy topolgoy, *J. Math. Res. Expo.*, **2**(1987), 357-366.

[168] Wang, Guo-Jun, *Theory of L-fuzzy Topological Spaces*, Shaanxi Normal University Press, 1988 (in Chinese).

[169] Wang, Guo-Jun, Pointwise topology on completely distributive lattices, *Fuzzy Sets and Systems*, **30**(1989), 1:53-62.

[170] Wang, Guo-Jun and Luo-Shan Xu, Internal topology and refining of Hutton's unit interval, *Scientia Sinica*, Ser. A, **35**(1992), 705-712 (in Chinese).

[171] Warner, M.W., Aspects of topology, (Ed. I.M. James) *London Math. Society Lecture Note*, 93, 1985, 127-140.

[172] Warner, M.W., Frame-fuzzy points and membership, *Fuzzy Sets and Systems*, **42**(1991), 453-456.

[173] Warren, R.H., Converegence in fuzzy topology, *Rocky Mountain J. Math.*, **13**(1983), 31-36.

[174] Weiss, M.D., Fixed points, separation, and induced topologies for fuzzy sets, *J. Math. Anal. Appl.*, **50**(1975), 142-150.

[175] Wong, C.K., covering properties in fuzzy topological spaces, *J. Math. Anal. Appl.*, **43**(1973), 697-703.

[176] Wong, C.K., Fuzzy topology: product and quotient theorems, *J. Math. Anal. Appl.*, **45**(1974), 512-521.

[177] Wong, C.K., Fuzzy points and local properties of fuzzy topology, *J. Math. Anal. Appl.*, **46**(1974), 316-328.

[178] Wuyts, P. and R. Lowen, On separation axioms in fuzzy topological spaces, *J. Math. Anal. Appl.*, **93**(1983), 24-41.

[179] Wuyts, P. and R. Lowen, On local and global measures of separation in fuzzy topological spaces, *Fuzzy Sets and Systems*, *19*(1986), 51-80.

[180] Wuyts, P., R. Lowen and E. Lowen, Reflectors and coreflectors in the category of fuzzy topological spaces, *Comput. Math. Appl.*, **16**(1988), 823-836.

[181] Xu, Jian-Jun, On the I-F paracompactness concerning the F-compactness in L-fuzzy topological spaces, *Proc. of Fifth International Fuzzy Systems Association World Congress'93*, Vol.I, Seoul, 344-345.

[182] Xu, Jian-Jun, On the II-F paracompactness concerning the F-compactness in L-Fuzzy topological spaces, *Proc. of Fifth International Fuzzy Systems Association World Congress'93*, Vol.I, Seoul, 346-347.

[183] Xu, Luo-Shan, Properties of strongly R-neighborhood families and local R-neighborhood families in *L*-fts, *Journal of Yangzhou Normal College*, **9**(1989), 1:15-21 (in Chinese).

[184] Xu, Luo-Shan, Paracompactness in *L*-fts, *Journal of Yangzhou Normal College*, **10**(1990), 2:20-28 (in Chinese).

[185] Xu, Luo-Shan, LN-paracompactness, *Proc. of "Fuzzy System and Mathematics", 5th Congress of System Engineering Academy of China*, Chengdu, 1990, 242-244 (in Chinese).

[186] Xu, Xiao-Quan, Level-structure of N-compact sets and Wallace theorem, *Kexue Tongbao*, **30**(1989), 1052-1054 (in Chinese).

[187] Xu, Xiao-Quan, Product and coproduct in the category of completely distributive lattices, *Chinese Science Bull.*, **31**(1990), 643-646.

[188] Yi, Yun, Characterize interior operator in induced spaces by stratifications, *Chinese Science Bulletin*, **34**(1989), 15: 1124-1126.

[189] Yi, Yun, Stratifications of fuzzy topologies, *Acta Math. Sinica*, **34**(1991), 6: 730-736 (in Chinese).

[190] Ying, Ming-Sheng, A new approach for fuzzy topology (I), *Fuzzy Sets and Systems*, **39**(1991), 303-321.

[191] Ying, Ming-Sheng, A new approach for fuzzy topology (II), *ibid.*, **47**(1992), 221-232.

[192] Ying, Ming-Sheng, A new approach for fuzzy topology (III), *ibid.*, **55**(1993), 193-207.

[193] Zadeh, L.A., Fuzzy sets, *Inform. Control.*, **8**(1965), 338-353.

[194] Zhang, De-Xue, Stratification structure of lattice-valued semicontinuous functions, *J. Sichuan Univ.* (Natural Sci. Ed.), **28**(1991), 4: 459-463 (in Chinese).

[195] Zhang, De-Xue, The stratified canonical fuzzy topology on $I(L)$ is uniformizable, *Fuzzy Sets and Systems*, **54**(1993), 225-227.

[196] Zhang, De-Xue and Ying-Ming Liu, Weakly induced modifications of *L*-fuzzy topological spaces, *Acta Math. Sinica*, **36**(1993), 1: 68-73 (in Chinese).

[197] Zhang, De-Xue and Ying-Ming Liu, *L*-fuzzy topological representation of distributive lattices and *L*-fuzzy ideals, *Chinese Annals of Math.*, **15A**(1994), 1:59-68 (in Chinese).

[198] Zhao, Dong-Sheng, The N-compactness in *L*-fuzzy topological spaces, *J. Math. Anal. Appl.* **128**(1987), 64-79.

[199] Zhao, Xiao-Dong, Connectedness on fuzzy topological spaces, *Fuzzy Sets and Systems*, **20**(1986), 223-240.

[200] Zheng, Chong-You, *L*-fuzzy unit interval and fuzzy connectedness, *Fuzzy Sets and Systems*, **27**(1988), 73-76.

[201] Zheng, Chong-You, Lei Fan, Hong-Bin Cui, *Frame and Continuous Lattices*, Capital Normal University Press, 1994 (in Chinese).

# Index